T0213568

Lecture Notes in Computer Science   10626

Commenced Publication in 1973
Founding and Former Series Editors:
Gerhard Goos, Juris Hartmanis, and Jan van Leeuwen

Editorial Board

More information about this series at http://www.springer.com/series/7410

Tsuyoshi Takagi · Thomas Peyrin (Eds.)

Advances in Cryptology – ASIACRYPT 2017

23rd International Conference on the Theory
and Applications of Cryptology and Information Security
Hong Kong, China, December 3–7, 2017
Proceedings, Part III

 Springer

Editors
Tsuyoshi Takagi
The University of Tokyo
Tokyo
Japan

Thomas Peyrin
Nanyang Technological University
Singapore
Singapore

ISSN 0302-9743   ISSN 1611-3349   (electronic)
Lecture Notes in Computer Science
ISBN 978-3-319-70699-3   ISBN 978-3-319-70700-6   (eBook)
https://doi.org/10.1007/978-3-319-70700-6

Library of Congress Control Number: 2017957984

LNCS Sublibrary: SL4 – Security and Cryptology

Printed on acid-free paper

This Springer imprint is published by Springer Nature
The registered company is Springer International Publishing AG
The registered company address is: Gewerbestrasse 11, 6330 Cham, Switzerland

Preface

ASIACRYPT 2017, the 23rd Annual International Conference on Theory and Application of Cryptology and Information Security, was held in Hong Kong, SAR China, during December 3–7, 2017.

The conference focused on all technical aspects of cryptology, and was sponsored by the International Association for Cryptologic Research (IACR).

ASIACRYPT 2017 received 243 submissions from all over the world. The Program Committee selected 67 papers (from which two were merged) for publication in the proceedings of this conference. The review process was made by the usual double-blind peer review by the Program Committee consisting of 48 leading experts of the field. Each submission was reviewed by at least three reviewers, and five reviewers were assigned to submissions co-authored by Program Committee members. This year, the conference operated a two-round review system with rebuttal phase. In the first-round review the Program Committee selected the 146 submissions that were considered of value for proceeding to the second round. In the second-round review the Program Committee further reviewed the submissions by taking into account their rebuttal letter from the authors. All the selection process was assisted by 334 external reviewers. These three-volume proceedings contain the revised versions of the papers that were selected. The revised versions were not reviewed again and the authors are responsible for their contents.

The program of ASIACRYPT 2017 featured three excellent invited talks. Dustin Moody gave a talk entitled "The Ship Has Sailed: The NIST Post-Quantum Cryptography 'Competition'," Wang Huaxiong spoke on "Combinatorics in Information-Theoretic Cryptography," and Pascal Paillier gave a third talk. The conference also featured a traditional rump session that contained short presentations on the latest research results of the field. The Program Committee selected the work "Identification Protocols and Signature Schemes Based on Supersingular Isogeny Problems" by Steven D. Galbraith, Christophe Petit, and Javier Silva for the Best Paper Award of ASIACRYPT 2017. Two more papers, "Kummer for Genus One over Prime Order Fields" by Sabyasachi Karati and Palash Sarkar, and "A Subversion-Resistant SNARK" by Behzad Abdolmaleki, Karim Baghery, Helger Lipmaa, and Michał Zając were solicited to submit the full versions to the *Journal of Cryptology*. The program chairs selected Takahiro Matsuda and Bart Mennink for the Best PC Member Award.

Many people have contributed to the success of ASIACRYPT 2017. We would like to thank the authors for submitting their research results to the conference. We are very grateful to all of the Program Committee members as well as the external reviewers for their fruitful comments and discussions on their areas of expertise. We are greatly indebted to Duncan Wong and Siu Ming Yiu, the general co-chairs, for their efforts and overall organization. We would also like to thank Allen Au, Catherine Chan, Sherman S.M. Chow, Lucas Hui, Zoe Jiang, Xuan Wang, and Jun Zhang, the local

Organizing Committee, for their continuous supports. We thank Duncan Wong and Siu Ming Yiu for expertly organizing and chairing the rump session.

Finally, we thank Shai Halevi for letting us use his nice software for supporting all the paper submission and review process. We also thank Alfred Hofmann, Anna Kramer, and their colleagues for handling the editorial process of the proceedings published at Springer LNCS.

December 2017   Tsuyoshi Takagi
Thomas Peyrin

ASIACRYPT 2017

The 23rd Annual International Conference on Theory and Application of Cryptology and Information Security

Sponsored by the International Association for Cryptologic Research (IACR)

December 3–7, 2017, Hong Kong, SAR China

General Co-chairs

Duncan Wong   CryptoBLK Limited
Siu Ming Yiu   The University of Hong Kong, SAR China

Program Co-chairs

Tsuyoshi Takagi   University of Tokyo, Japan
Thomas Peyrin   Nanyang Technological University, Singapore

Program Committee

Shweta Agrawal   IIT Madras, India
Céline Blondeau   Aalto University, Finland
Joppe W. Bos   NXP Semiconductors, Belgium
Chris Brzuska   TU Hamburg, Germany
Jie Chen   East China Normal University, China
Sherman S.M. Chow   The Chinese University of Hong Kong, SAR China
Kai-Min Chung   Academia Sinica, Taiwan
Nico Döttling   University of California, Berkeley, USA
Thomas Eisenbarth   Worcester Polytechnic Institute, USA
Dario Fiore   IMDEA Software Institute, Madrid, Spain
Georg Fuchsbauer   Inria and ENS, France
Steven Galbraith   Auckland University, New Zealand
Jian Guo   Nanyang Technological University, Singapore
Viet Tung Hoang   Florida State University, USA
Jérémy Jean   ANSSI, France
Jooyoung Lee   KAIST, South Korea
Dongdai Lin   Chinese Academy of Sciences, China
Feng-Hao Liu   Florida Atlantic University, USA
Stefan Mangard   Graz University of Technology, Austria
Takahiro Matsuda   AIST, Japan
Alexander May   Ruhr University Bochum, Germany
Bart Mennink   Radboud University, The Netherlands

Amir Moradi — Ruhr University Bochum, Germany
Pratyay Mukherjee — Visa Research, USA
Mridul Nandi — Indian Statistical Institute, India
Khoa Nguyen — Nanyang Technological University, Singapore
Miyako Ohkubo — NICT, Japan
Tatsuaki Okamoto — NTT Secure Platform Laboratories, Japan
Arpita Patra — Indian Institute of Science, India
Bart Preneel — KU Leuven, Belgium
Matthieu Rivain — CryptoExperts, France
Reihaneh Safavi-Naini — University of Calgary, Canada
Yu Sasaki — NTT Secure Platform Laboratories, Japan
Peter Schwabe — Radboud University, The Netherlands
Fang Song — Portland State University, USA
Francois-Xavier Standaert — UCL, Belgium
Damien Stehlé — ENS Lyon, France
Ron Steinfeld — Monash University, Australia
Rainer Steinwandt — Florida Atlantic University, USA
Mehdi Tibouchi — NTT Secure Platform Laboratories, Japan
Dominique Unruh — University of Tartu, Estonia
Gilles Van Assche — STMicroelectronics, Belgium
Serge Vaudenay — EPFL, Switzerland
Ingrid Verbauwhede — KU Leuven, Belgium
Ivan Visconti — University of Salerno, Italy
Lei Wang — Shanghai Jiaotong University, China
Meiqin Wang — Shandong University, China
Jiang Zhang — State Key Laboratory of Cryptology, China

Additional Reviewers

Masayuki Abe
Arash Afshar
Divesh Aggarwal
Shashank Agrawal
Ahmad Ahmadi
Mamun Akand
Gorjan Alagic
Joel Alwen
Abdelrahaman Aly
Miguel Ambrona
Elena Andreeva
Diego Aranha
Nuttapong Attrapadung
Sepideh Avizheh
Saikrishna
   Badrinarayanan

Shi Bai
Fatih Balli
Subhadeep Banik
Zhenzhen Bao
Hridam Basu
Alberto Batistello
Balthazar Bauer
Carsten Baum
Georg T. Becker
Christof Beierle
Sonia Beläd
Fabrice Benhamouda
Francesco Berti
Guido Bertoni
Sanjay Bhattacherjee
Jean-Francois Biasse

Begül Bilgin
Olivier Blazy
Johannes Bloemer
Sonia Mihaela Bogos
Sasha Boldyreva
Charlotte Bonte
Raphael Bost
Leif Both
Florian Bourse
Sébastien Canard
Brent Carmer
Wouter Castryck
Dario Catalano
Gizem Çetin
Avik Chakraborti
Nishanth Chandran

Melissa Chase
Binyi Chen
Cong Chen
Long Chen
Yi-Hsiu Chen
Yu Chen
Yu-Chi Chen
Nai-Hui Chia
Gwangbae Choi
Wutichai Chongchitmate
Chi-Ning Chou
Ashish Choudhury
Chitchanok
Chuengsatiansup
Hao Chung
Michele Ciampi
Thomas De Cnudde
Katriel Cohn-Gordon
Henry Corrigan-Gibbs
Craig Costello
Geoffroy Couteau
Eric Crockett
Tingting Cui
Edouard Cuvelier
Joan Daemen
Wei Dai
Pratish Datta
Bernardo David
Marguerite Delcourt
Jeroen Delvaux
Yi Deng
David Derler
Julien Devigne
Claus Diem
Christoph Dobraunig
Yarkin Doroz
Léo Ducas
Dung H. Duong
Ratna Dutta
Stefan Dziembowski
Maria Eichlseder
Muhammed Esgin
Thomas Espitau
Xiong Fan
Antonio Faonio

Sebastian Faust
Björn Fay
Serge Fehr
Luca De Feo
Nils Fleischhacker
Jean-Pierre Flori
Tore Kasper Frederiksen
Thomas Fuhr
Marc Fyrbiak
Tommaso Gagliardoni
Chaya Ganesh
Flavio Garcia
Pierrick Gaudry
Rémi Géraud
Satrajit Ghosh
Irene Giacomelli
Benedikt Gierlichs
Junqing Gong
Louis Goubin
Alex Grilo
Hannes Gross
Vincent Grosso
Chun Guo
Hui Guo
Helene Haagh
Patrick Haddad
Harry Halpin
Shuai Han
Yoshikazu Hanatani
Jens Hermans
Gottfried Herold
Julia Hesse
Felix Heuer
Minki Hhan
Fumitaka Hoshino
Yin-Hsun Huang
Zhenyu Huang
Andreas Hülsing
Jung Yeon Hwang
Ilia Iliashenko
Mehmet Inci
Vincenzo Iovino
Ai Ishida
Takanori Isobe
Tetsu Iwata

Malika Izabachène
Michael Jacobson
Abhishek Jain
David Jao
Zhengfeng Ji
Dingding Jia
Shaoquan Jiang
Anthony Journault
Jean-Gabriel Kammerer
Sabyasachi Karati
Handan Kilinç
Dongwoo Kim
Jihye Kim
Jon-Lark Kim
Sam Kim
Taechan Kim
Elena Kirshanova
Ágnes Kiss
Fuyuki Kitagawa
Susumu Kiyoshima
Thorsten Kleinjung
Miroslav Knezevic
Alexander Koch
François Koeune
Konrad Kohbrok
Lisa Kohl
Ilan Komargodski
Yashvanth Kondi
Robert Kuebler
Frédéric Lafitte
Ching-Yi Lai
Russell W.F. Lai
Adeline Langlois
Gregor Leander
Changmin Lee
Hyung Tae Lee
Iraklis Leontiadis
Tancrède Lepoint
Debbie Leung
Yongqiang Li
Jyun-Jie Liao
Benoit Libert
Fuchun Lin
Wei-Kai Lin
Patrick Longa

Julian Loss
Steve Lu
Xianhui Lu
Atul Luykx
Chang Lv
Vadim Lyubashevsky
Monosij Maitra
Mary Maller
Giorgia Azzurra Marson
Marco Martinoli
Daniel Masny
Sarah Meiklejohn
Peihan Miao
Michele Minelli
Takaaki Mizuki
Ahmad Moghimi
Payman Mohassel
Maria Chiara Molteni
Seyyed Amir Mortazavi
Fabrice Mouhartem
Köksal Mus
Michael Naehrig
Ryo Nishimaki
Anca Nitulescu
Luca Nizzardo
Koji Nuida
Kaisa Nyberg
Adam O'Neill
Tobias Oder
Olya Ohrimenko
Emmanuela Orsini
Elisabeth Oswald
Elena Pagnin
Pascal Paillier
Jiaxin Pan
Alain Passelègue
Sikhar Patranabis
Roel Peeters
Chris Peikert
Alice Pellet-Mary
Ludovic Perret
Peter Pessl
Thomas Peters
Christophe Petit
Duong Hieu Phan
Antigoni Polychroniadou

Romain Poussier
Ali Poustindouz
Emmanuel Prouff
Kexin Qiao
Baodong Qin
Sebastian Ramacher
Somindu C. Ramanna
Shahram Rasoolzadeh
Divya Ravi
Francesco Regazzoni
Jean-René Reinhard
Ling Ren
Joost Renes
Oscar Reparaz
Joost Rijneveld
Damien Robert
Jérémie Roland
Arnab Roy
Sujoy Sinha Roy
Vladimir Rozic
Joeri de Ruiter
Yusuke Sakai
Amin Sakzad
Simona Samardjiska
Olivier Sanders
Pascal Sasdrich
Alessandra Scafuro
John Schanck
Tobias Schneider
Jacob Schuldt
Gil Segev
Okan Seker
Binanda Sengupta
Sourav Sengupta
Jae Hong Seo
Masoumeh Shafienejad
Setareh Sharifian
Sina Shiehian
Kazumasa Shinagawa
Dave Singelée
Shashank Singh
Javier Silva
Luisa Siniscalchi
Daniel Slamanig
Benjamin Smith
Ling Song

Pratik Soni
Koutarou Suzuki
Alan Szepieniec
Björn Tackmann
Mostafa Taha
Raymond K.H. Tai
Katsuyuki Takashima
Atsushi Takayasu
Benjamin Hong
    Meng Tan
Qiang Tang
Yan Bo Ti
Yosuke Todo
Ni Trieu
Roberto Trifiletti
Thomas Unterluggauer
John van de Wetering
Muthuramakrishnan
    Venkitasubramaniam
Daniele Venturi
Dhinakaran
    Vinayagamurthy
Vanessa Vitse
Damian Vizár
Satyanarayana Vusirikala
Sebastian Wallat
Alexandre Wallet
Haoyang Wang
Minqian Wang
Wenhao Wang
Xiuhua Wang
Yuyu Wang
Felix Wegener
Puwen Wei
Weiqiang Wen
Mario Werner
Benjamin Wesolowski
Baofeng Wu
David Wu
Keita Xagawa
Zejun Xiang
Chengbo Xu
Shota Yamada
Kan Yang
Kang Yang
Kan Yasuda

Local Organizing Committee

Co-chairs

Duncan Wong   CryptoBLK Limited
Siu Ming Yiu   The University of Hong Kong, SAR China

Members

Lucas Hui (Chair)   The University of Hong Kong, SAR China
Catherine Chan (Manager)   The University of Hong Kong, SAR China
Jun Zhang   The University of Hong Kong, SAR China
Xuan Wang   Harbin Institute of Technology, Shenzhen, China
Zoe Jiang   Harbin Institute of Technology, Shenzhen, China
Allen Au   The Hong Kong Polytechnic University, SAR China
Sherman S.M. Chow   The Chinese University of Hong Kong, SAR China

Invited Speakers

The Ship Has Sailed: the NIST Post-quantum Cryptography "Competition"

Dustin Moody

Computer Security Division, National Institute of Standards and Technology

Abstract. In recent years, there has been a substantial amount of research on quantum computers – machines that exploit quantum mechanical phenomena to solve mathematical problems that are difficult or intractable for conventional computers. If large-scale quantum computers are ever built, they will compromise the security of many commonly used cryptographic algorithms. In particular, quantum computers would completely break many public-key cryptosystems, including those standardized by NIST and other standards organizations.

Due to this concern, many researchers have begun to investigate post-quantum cryptography (also called quantum-resistant cryptography). The goal of this research is to develop cryptographic algorithms that would be secure against both quantum and classical computers, and can interoperate with existing communications protocols and networks. A significant effort will be required to develop, standardize, and deploy new post-quantum algorithms. In addition, this transition needs to take place well before any large-scale quantum computers are built, so that any information that is later compromised by quantum cryptanalysis is no longer sensitive when that compromise occurs.

NIST has taken several steps in response to this potential threat. In 2015, NIST held a public workshop and later published NISTIR 8105, Report on Post-Quantum Cryptography, which shares NIST's understanding of the status of quantum computing and post-quantum cryptography. NIST also decided to develop additional public-key cryptographic algorithms through a public standardization process, similar to the development processes for the hash function SHA-3 and the Advanced Encryption Standard (AES). To begin the process, NIST issued a detailed set of minimum acceptability requirements, submission requirements, and evaluation criteria for candidate algorithms, available at http://www.nist.gov/pqcrypto. The deadline for algorithms to be submitted was November 30, 2017.

In this talk, I will share the rationale on the major decisions NIST has made, such as excluding hybrid and (stateful) hash-based signature schemes. I will also talk about some open research questions and their potential impact on the standardization effort, in addition to some of the practical issues that arose while creating the API. Finally, I will give some preliminary information about the submitted algorithms, and discuss what we've learned during the first part of the standardization process.

Combinatorics in Information-Theoretic Cryptography

Huaxiong Wang

School of Physical and Mathematical Sciences,
Nanyang Technological University, Singapore
hxwang@ntu.edu.sg

Abstract. Information-theoretic cryptography is an area that studies cryptographic functionalities whose security does not rely on hardness assumptions from computational intractability of mathematical problems. It covers a wide range of cryptographic research topics such as one-time pad, authentication code, secret sharing schemes, secure multiparty computation, private information retrieval and post-quantum security etc., just to mention a few. Moreover, many areas in complexity-based cryptography are well known to benefit or stem from information-theoretic methods. On the other hand, combinatorics has been playing an active role in cryptography, for example, the hardness of Hamiltonian cycle existence in graph theory is used to design zero-knowledge proofs. In this talk, I will focus on the connections between combinatorics and information-theoretic cryptography. After a brief (incomplete) overview on their various connections, I will present a few concrete examples to illustrate how combinatorial objects and techniques are applied to the constructions and characterizations of information-theoretic schemes. Specifically, I will show

1. how perfect hash families and cover-free families lead to better performance in certain secret sharing schemes;
2. how graph colouring from planar graphs is used in constructing secure multiparty computation protocols over non-abelian groups;
3. how regular intersecting families are applied to the constructions of private information retrieval schemes.

Part of this research was funded by Singapore Ministry of Education under Research Grant MOE2016-T2-2-014(S).

Contents – Part III

Symmetric Key Designs

Asiacrypt 2017 Award Paper II

A Subversion-Resistant SNARK

Behzad Abdolmaleki, Karim Baghery, Helger Lipmaa$^{(\boxtimes)}$, and Michał Zając

Institute of Computer Science, University of Tartu, Tartu, Estonia
abdolmaleki.behzad.ir@gmail.com, baghery.karim@gmail.com,
helger.lipmaa@gmail.com, m.p.zajac@gmail.com

Abstract. While zk-SNARKs are widely studied, the question of what happens when the CRS has been subverted has received little attention. In ASIACRYPT 2016, Bellare, Fuchsbauer and Scafuro showed the first negative and positive results in this direction, proving also that it is impossible to achieve subversion soundness and (even non-subversion) zero knowledge at the same time. On the positive side, they constructed an involved sound and Sub-ZK argument system for NP. We make Groth's zk-SNARK for CIRCUIT-SAT from EUROCRYPT 2016 computationally knowledge-sound and perfectly composable Sub-ZK with minimal changes. We just require the CRS trapdoor to be extractable and the CRS to be publicly verifiable. To achieve the latter, we add some new elements to the CRS and construct an efficient CRS verification algorithm. We also provide a definitional framework for sound and Sub-ZK SNARKs and describe implementation results of the new Sub-ZK SNARK.

Keywords: Common reference string · Generic group model · Non-interactive zero knowledge · SNARK · Subversion zero knowledge

1 Introduction

Combined effort of a large number of recent research papers (to only mention a few, [17,22,26,27,30,31,35]) has made it possible to construct very efficient succinct non-interactive zero-knowledge arguments of knowledge (zk-SNARKs) for both the Boolean and the Arithmetic CIRCUIT-SAT and thus for NP. The most efficient known approach for constructing zk-SNARKs for the Arithmetic CIRCUIT-SAT is based on Quadratic Arithmetic Programs (QAP, [22]).

In a QAP, the prover builds a set of polynomial equations that are then checked by the verifier by using a small number of pairings. QAP-based zk-SNARKs have additional nice properties that make them applicable in verifiable computation [21,22,35] where the client outsources some computation to the server, who returns the computation result together with a succinct efficiently-verifiable correctness argument. Especially due to this application, zk-SNARKs have several heavily optimized implementations [7,8,15,35]. Other applications of zk-SNARK include cryptocurrencies [4]. See, e.g., [27] for more references.

One drawback of zk-SNARKs is that they are all based on non-falsifiable assumptions (like the knowledge assumptions [16], or the generic bilinear group

© International Association for Cryptologic Research 2017
T. Takagi and T. Peyrin (Eds.): ASIACRYPT 2017, Part III, LNCS 10626, pp. 3–33, 2017.
https://doi.org/10.1007/978-3-319-70700-6_1

model, GBGM [12,33,34,36]). In fact, Gentry and Wichs [23] showed that non-falsifiable assumptions are needed to construct zk-SNARKs for non-trivial languages. The currently most efficient zk-SNARK for Arithmetic CIRCUIT-SAT was proposed by Groth (EUROCRYPT 2016, [27]) who proved it to be knowledge-sound in the GBGM. In Groth's zk-SNARK, the argument consists of only 3 bilinear group elements and the verifier has to check a single pairing equation, dominated by the computation of only 3 bilinear (type III [20]) pairings and m_0 exponentiations, where m_0 is the statement size.

After the Snowden revelations, there has been a recent surge of interest in constructing cryptographic primitives and protocols secure against active subversion. In the context of zk-SNARKs, while the common reference string (CRS) model [11] is widely accepted to be the proper model, one has to be very careful to guarantee that the CRS has been created correctly. In [2], Bellare, Fuchsbauer and Scafuro tackled this problem by studying how much security one can still achieve when the CRS generator cannot be trusted. They proved several negative and positive results. In particular, they showed that it is impossible to achieve subversion soundness and (even non-subversion) zero knowledge simultaneously, the essential reason being that the zero knowledge simulator can be used to break subversion soundness.

In one of their positive solutions, Bellare *et al.* show that it is possible to get (non-subversion) soundness and computational subversion zero knowledge (Sub-ZK, ZK even if the the CRS is not trusted). Their main new idea is to use a knowledge assumption in the Sub-ZK proof, so that the simulator can extract a "trapdoor" from the untrusted CRS and then use this trapdoor to simulate the argument. While neat, the resulting argument system is quite complicated. Moreover, the non-interactive Sub-ZK argument system of [2] has linear communication; in the case of zk-SNARKs one presumably has to employ different techniques. We also need to take care to define and implement statistical Sub-ZK.

Our Contributions. We will take Groth's zk-SNARK from EUROCRYPT 2016 [27] as a starting point since, as mentioned before, it is currently the most efficient and thus the most attractive zk-SNARK.[1] We propose a minimal modification to Groth's zk-SNARK that makes it computationally knowledge-sound in what we call the "subversion generic bilinear group model" (Sub-GBGM) and perfect composable Sub-ZK. In fact, we consider two different versions of perfect Sub-ZK: (i) the version with an efficient subverter, where we assume the existence of an efficient extractor and prove Sub-ZK under a knowledge assumption, and (ii) the version with a computationally unbounded subverter, where we assume the existence of a computationally unbounded extractor and prove Sub-ZK unconditionally.

---

[1] While less efficient zk-SNARKs like Pinocchio [35] are used more widely, this situation might change. Moreover, Groth's zk-SNARK is also more efficient from the theoretical perspective: it has both a more complex CRS and more sophisticated soundness proof; hence, we expect that it is easier to achieve Sub-ZK for Pinocchio.

We change Groth's zk-SNARK by adding extra elements to the CRS so that the CRS will become publicly verifiable; this minimal step (clearly, some public verifiability of the CRS is needed in the case the CRS generator cannot be trusted) will be sufficient to obtain Sub-ZK. However, choosing which elements to add to the CRS is not straightforward since the zk-SNARK must remain knowledge-sound even given enlarged CRS; adding too many or just "wrong" elements to the CRS can break the knowledge-soundness. (We will give a simple example about that in Sect. 6). On the other hand, importantly, the prover and the verifier of the new zk-SNARK are unchanged compared to Groth's SNARK [27]. In the rest of the introduction, we will only outline the novel properties of the new SNARK as compared to [27].

We start by defining perfect subversion-complete (this includes the requirement that an honestly generated CRS is accepted by the CRS verification), computationally adaptively knowledge-sound, and statistically unbounded (or composable) Sub-ZK SNARKs. These definitions are similar to but subtly different from the non-subversion security definitions as given in, say, [25]. First, since one cannot check whether the subverter uses perfectly uniform coins (or, the *CRS trapdoor*) to generate the CRS, we divide the CRS generation into three different algorithms:

- generation of the CRS trapdoor tc (a probabilistic algorithm K_{tc}),
- creation of the CRS from tc (a deterministic algorithm K_{crs}), and
- creation of the simulation trapdoor from tc (a deterministic algorithm K_{ts}).

While we cannot check that K_{tc} works correctly, we will guarantee that given a fixed tc, K_{crs} has been executed on this tc. More precisely, we require that a Sub-ZK SNARK satisfies a CRS trapdoor extractability property that allows one to extract tc used by the subverter, s.t. if the subverted CRS crs is accepted by the CRS verification algorithm (see below) then K_{tc} maps tc to crs. The extractability requirement forces our ZK proof to use either a computationally unbounded extractor or a knowledge assumption. While we use the Sub-ZK definition with an efficient subverter and extractor throughout this introduction and the paper (mainly, since it is actually more difficult to achieve[2]), we will discuss the case of computationally unbounded subverter and extractor in Sect. 9.

In the proof of knowledge-soundness, we use (a version of) the GBGM. Using GBGM seems to be the best we can do since Groth's non-Sub zk-SNARK is proven knowledge-sound in GBGM and as mentioned above, the use of a knowledge assumption or the generic model in the knowledge-soundness proof is necessary due to the impossibility result of Gentry and Wichs [23]. However, following Bellare *et al.* [2], we weaken the usual definition of GBGM by allowing the generic adversary to create (under realistic restrictions) random elements in the source groups without knowing their discrete logarithms. We call the resulting somewhat weaker model the *subversion generic bilinear group model* (Sub-GBGM). Following Groth [27] (the main difference being that modeling a more powerful

---

[2] Moreover, security against unbounded subverter does not guarantee security against efficient subverter, since in the last case also the extractor has to be efficient.

generic adversary and taking into account new CRS elements will complicate the proof somewhat), we prove that the new SNARK is (adaptively) knowledge-sound in the Sub-GBGM even in the case of type-I pairings. (We emphasize once more that Groth's zk-SNARK is proven knowledge-sound in the GBGM, and that Sub-GBGM is actually a weaker model than the GBGM.) This provides a hedge against possible future cryptanalysis that finds an efficient isomorphism between the two source groups.

Consider the case of efficient subverter. In the proof of perfect composable Sub-ZK, we use a well-known knowledge assumption (see, e.g., [17]) that we call BDH-KE. We argue that BDH-KE makes sense in the Sub-GBGM. The Sub-ZK proof of the only previously known non-interactive Sub-ZK argument system by Bellare *et al.* [2] also relies on knowledge assumptions. We follow the main idea of [2] by first using BDH-KE to extract the CRS trapdoor tc from the CRS and then construct a non-subversion simulator (that gets a part of the tc as an input) to simulate the argument. However, since we construct a zk-SNARK, our concrete approach is different from [2].

Also here, we rely on the existence of the efficient CRS verification algorithm CV. We show that if CV accepts a crs, then crs has been computed correctly by K_{crs} from a tc bijectively fixed by crs. From this, it follows under the BDH-KE assumption that for any subverter that produces a crs accepted by CV, there exists an extractor that produces tc such that K_{crs} given tc outputs crs.

We emphasize that our security proofs of knowledge-soundness and of Sub-ZK are in incomparable models. The knowledge-soundness proof uses the full power of Sub-GBGM in the case of any pairings (including type-I). The Sub-ZK proof, on the other hand, uses a concrete standard-looking knowledge assumption BDH-KE that holds in the the GBGM but does not hold in the Sub-GBGM in the case of type-I pairings. (In the case of computationally unbounded subverter, we even do not need BDH-KE.) This enables us to construct an efficient Sub-ZK SNARK that uses type-III pairings, while guaranteeing its knowledge-soundness even in the case of type-I pairings.

General Design Recommendations. We do not expect that constructing Sub-ZK SNARKs can be done automatically, in particular since our framework points to the necessity of making CRS publicly verifiable which potentially means adding new elements to the CRS. Since knowledge-soundness proofs of many SNARKs are very subtle, it seems to be difficult to give a general "theorem" about which SNARKs can be modified to be Sub-ZK or even whether their CRS can be made verifiable without a major reconstruction. Whether a SNARK remains sound after that must be proven separately in each case.

However, we can still give a few recommendations for designing a Sub-ZK SNARK from a non subversion-secure SNARK (or from scratch) when using the same approach as the current paper:

1. <u>Division of duties:</u> make sure that K can be divided into randomized K_{tc}, deterministic K_{ts}, and deterministic K_{crs}.
2. <u>CRS trapdoor extractability:</u> for each element of tc, make sure that it can be extracted from the CRS. For this, one can use a generic proof of knowledge,

a specific knowledge-assumption, or a computationally unbounded extractor. A few additional properties described in Sect. 7 can also help.

3. CRS verifiability: the CRS must be publicly verifiable.
4. Sound approach: make sure that the previous steps do not hurt the knowledge-soundness. To achieve it, one should aim at designing a SNARK with a very simple CRS or where CRS verifiability comes naturally. Depending on the SNARK in question, this step may be the most difficult one.

On Efficiency. Since the new zk-SNARK is closely based on the most efficient known non-subversion zk-SNARK of Groth [27], it has comparable efficiency. Importantly, the new CRS verification algorithm CV has to be executed only by the prover (this is since we achieve Sub-ZK and non-subversion knowledge-soundness). This means that it suffices for CV to have the same computational complexity as the prover's algorithm. The initial CV we describe in Fig. 1 is quite inefficient. However, as we show in Sect. 8 (see the full version [1] for more information), by using batching techniques the CV algorithm can be sped up to be faster than the prover's algorithm (at the information-theoretical security level 2^{-80}) and even faster at the information-theoretical security level 2^{-40}. We implemented the new SNARK by using the `libsnark` [8] library and in the full version [1], we back up the last claim by concrete numbers.

An interesting open question is to minimize the computational complexity of the CRS verification. In particular, in most of the known zero-knowledge argument systems, either the argument length is at least linear and hence the verification algorithm takes at least linear time (e.g., the Groth-Sahai proofs [28]) or the CRS length is at least linear and hence the CRS verification algorithm takes at least linear time. A SNARK where both CRS and the argument are succinct is called *fully succinct*. See [7,9,15,26,30] for work on zk-SNARKs with sublinear CRS. However, existing fully succinct zk-SNARK are not really practical, see [7] for discussions. Moreover, it is not clear a priori how to make it Sub-ZK. An important open research topic hence is to construct an efficient fully succinct non-interactive Sub-ZK argument system.

Due to the lack of space, comparison to the MPC approach and implementation results are only given in the full version [1].

2 Preliminaries

For a matrix $M \in \mathbb{Z}_p^{n \times m}$, we denote by $\boldsymbol{M}_i$ the ith row of M and by $\boldsymbol{M}^{(j)}$ the jth column of M. PPT stands for probabilistic polynomial-time, and NUPPT stands for non-uniform PPT. We write $f(\kappa) \approx_\kappa g(\kappa)$, if $f(\kappa) - g(\kappa)$ is negligible as a function of κ. For an algorithm A, let range(A) be the range of A, i.e., the set of of valid outputs of A. For an algorithm A, let RND(A) denote the random tape of A, and let $r \leftarrow_r$ RND(A) denote the random choice of the randomizer r from RND(A). By $y \leftarrow \mathsf{A}(\mathsf{x}; r)$ we denote the fact that A, given an input x and a randomizer r, outputs y. When we use this notation then r represents the full random tape of A. For algorithms A and $\mathsf{X_A}$, we write $(y \, \| \, y') \leftarrow (\mathsf{A} \, \| \, \mathsf{X_A})(\chi; r)$

Algorithm 1. Computing $(\ell_i(\chi))_{i=1}^n$

---

1 $\zeta \leftarrow (\chi^n - 1)/n; \; \omega' \leftarrow 1;$
2 **if** $\chi = \omega'$ **then** $\ell_1(\chi) \leftarrow 1;$ **else** $\ell_1(\chi) \leftarrow \zeta/(\chi - \omega');$
3 **for** $i = 2$ **to** n **do**
4 $\quad\mid \quad \zeta \leftarrow \omega\zeta; \; \omega' \leftarrow \omega\omega';$
5 $\quad\mid \quad$ **if** $\chi = \omega'$ **then** $\ell_i(\chi) \leftarrow 1;$ **else** $\ell_i(\chi) \leftarrow \zeta/(\chi - \omega');$

---

as a shorthand for $y \leftarrow \mathsf{A}(\chi; r)$, $y' \leftarrow \mathsf{X_A}(\chi; r)$. In both cases, $\mathsf{X_A}$ and A use internally the same randomness r.

In the new argument system, we will use a small number of indeterminates, and we assume that each indeterminate has a canonical concrete value (a uniformly random element of $\mathbb{Z}_p$ or $\mathbb{Z}_p^*$). The canonical value of $X, X_\alpha, X_\beta, X_\gamma, X_\delta$ (resp., Y_i) will be $\chi, \alpha, \beta, \gamma, \delta$ (resp., v_i, for $i \in \{1, 2, 3\}$). The canonical value of the vector $\boldsymbol{X} = (X, X_\alpha, X_\beta, X_\gamma, X_\delta)$ (resp., $\boldsymbol{Y} = (Y_1, Y_2, Y_3)$) will be $\boldsymbol{x} = (\chi, \alpha, \beta, \gamma, \delta)$ (resp., $\boldsymbol{y} = (v_1, v_2, v_3)$).

Interpolation. Assume n is a power of two, and let ω be the n-th primitive root of unity modulo p. Such ω exists, given that $n \mid (p-1)$. Then,

- $\ell(X) := \prod_{i=1}^n (X - \omega^{i-1}) = X^n - 1$ is the unique degree n monic polynomial such that $\ell(\omega^{i-1}) = 0$ for all $i \in [1 \mathinner{.\,.} n]$.
- For $i \in [1 \mathinner{.\,.} n]$, let $\ell_i(X)$ be the *ith Lagrange basis polynomial*, i.e., the unique degree $n - 1$ polynomial s.t. $\ell_i(\omega^{i-1}) = 1$ and $\ell_i(\omega^{j-1}) = 0$ for $i \neq j$. Clearly,

$$\ell_i(X) := \frac{\ell(X)}{\ell'(\omega^{i-1})(X - \omega^{i-1})} = \frac{(X^n - 1)\omega^{i-1}}{n(X - \omega^{i-1})}. \tag{1}$$

Thus, $\ell_i(\omega^{i-1}) = 1$ while $\ell_i(\chi) = (\chi^n - 1)\omega^{i-1}/(n(\chi - \omega^{i-1}))$ for $\chi \neq \omega^{i-1}$. Given any $\chi \in \mathbb{Z}_p$, Algorithm 1 (see, e.g., [5]) computes $\ell_i(\chi)$ for $i \in [1 \mathinner{.\,.} n]$. It can be implemented by using $4n - 2$ multiplications and divisions in $\mathbb{Z}_p$.

Clearly, $L_{\boldsymbol{a}}(X) := \sum_{i=1}^n a_i \ell_i(X)$ is the interpolating polynomial of $\boldsymbol{a}$ at points ω^{i-1}, with $L_{\boldsymbol{a}}(\omega^{i-1}) = a_i$, and its coefficients can thus be computed by executing an inverse Fast Fourier Transform in time $\Theta(n \log n)$. Moreover, $(\ell_j(\omega^{i-1}))_{i=1}^n = \boldsymbol{e}_j$ (the jth unit vector) and $(\ell(\omega^{i-1}))_{i=1}^n = \boldsymbol{0}_n$.

Elliptic Curves and Bilinear Maps. On input 1^κ, a *bilinear map generator* genbp returns $\mathsf{gk} = (p, \mathbb{G}_1, \mathbb{G}_2, \mathbb{G}_T, \hat{e}, \mathfrak{g}_1, \mathfrak{g}_2)$, where $\mathbb{G}_1$, $\mathbb{G}_2$ and $\mathbb{G}_T$ are three additive cyclic groups of prime order p (with $\log p = \Omega(\kappa)$) and $\mathfrak{g}_z$ is a random generator of $\mathbb{G}_z$ for $z \in \{1, 2, T\}$. We denote the elements of $\mathbb{G}_1$, $\mathbb{G}_2$, and $\mathbb{G}_T$ by using Fraktur as in $\mathfrak{g}_1$. Additionally, $\hat{e}$ is an efficient bilinear map $\hat{e} \colon \mathbb{G}_1 \times \mathbb{G}_2 \to \mathbb{G}_T$ that satisfies in particular the following two properties: (i) $\hat{e}(\mathfrak{g}_1, \mathfrak{g}_2) \neq 1$, and (ii) $\hat{e}(a\mathfrak{g}_1, b\mathfrak{g}_2) = (ab)\hat{e}(\mathfrak{g}_1, \mathfrak{g}_2)$.

We give genbp another input, n (intuitively, the input length), and allow p to depend on n. We assume that all algorithms that handle group elements verify by default that their inputs belong to corresponding groups and reject if not.

Usually, arithmetic in (say) $\mathbb{G}_1$ is considerably cheaper than in $\mathbb{G}_2$; hence, we count separately exponentiations in both groups.

We use the bracket notation. That is, for an integer x, we denote $[x]_z := x\mathfrak{g}_z$ even when x is unknown. We denote $[x]_1 \bullet [y]_2 := \hat{e}([x]_1, [y]_2)$, hence $[xy]_T = [x]_1 \bullet [y]_2$. We denote $[a_1, \ldots, a_s]_z = ([a_1]_z, \ldots, [a_s]_z)$.

To speed up interpolation and other related computation, we will need the existence of the n-th, where n is a power of 2, primitive root of unity modulo p. For this, it suffices that $(n+1) \mid (p-1)$ (recall that p is the elliptic curve group order). Fortunately, given κ and a practically relevant value of n, one can easily find a Barreto-Naehrig curve such that $(n+1) \mid (p-1)$ holds [8].

Quadratic Arithmetic Programs. Quadratic Arithmetic Program (QAP) was introduced by Gennaro *et al.* [22] as a language where for an input x and witness w, (x, w) $\in \mathbf{R}$ can be verified by using a parallel quadratic check, and that has an efficient reduction from the well-known language (either Boolean or Arithmetic) CIRCUIT-SAT. Hence, an efficient zk-SNARK for QAP results in an efficient zk-SNARK for CIRCUIT-SAT.

For an m-dimensional vector $\boldsymbol{A}$, let $\mathsf{aug}(\boldsymbol{A}) = \left(\begin{smallmatrix} 1 \\ \boldsymbol{A} \end{smallmatrix}\right)$. For an n-dimensional vector $\boldsymbol{M}^{(0)}$ and an $n \times m$ matrix M over finite field $\mathbb{F}$, let $\mathsf{aug}(M) := (\boldsymbol{M}^{(0)}, M)$. Let $m_0 < m$ be a non-negative integer. An instance $\mathcal{Q}$ of the QAP language is specified by $(\mathbb{F}, m_0, \mathsf{aug}(U), \mathsf{aug}(V), \mathsf{aug}(W))$ where $U, V, W \in \mathbb{F}^{n \times m}$.

In the case of Arithmetic CIRCUIT-SAT, n is the number of multiplication gates and m to the number of wires in the circuit. Here, we consider arithmetic circuits that consist only of fan-in-2 multiplication gates, but either input of each multiplication gate can be a weighted sum of some wire values, [22].

For a fixed instance $\mathcal{Q}$ of QAP, define the relation $\mathbf{R}$ as follows:

$$\mathbf{R}_{\mathcal{Q}} = \left\{ \begin{array}{l} (\mathsf{x}, \mathsf{w}) \colon \mathsf{x} = (A_1, \ldots, A_{m_0})^\top \wedge \mathsf{w} = (A_{m_0+1}, \ldots, A_m)^\top \wedge \\ (\mathsf{aug}(U) \cdot \mathsf{aug}(\boldsymbol{A})) \circ (\mathsf{aug}(V) \cdot \mathsf{aug}(\boldsymbol{A})) = \mathsf{aug}(W) \cdot \mathsf{aug}(\boldsymbol{A}) \end{array} \right\}$$

where $\boldsymbol{a} \circ \boldsymbol{b} = (a_i b_i)_{i=1}^n$ denotes the entrywise product of vectors $\boldsymbol{a}$ and $\boldsymbol{b}$.

In a cryptographic setting, it is more convenient to work with the following alternative definition of QAP and of the relation $\mathbf{R}_{\mathcal{Q}}$. (This corresponds to the original definition of QAP in [22]). Let $\mathbb{F} = \mathbb{Z}_p$, such that ω is the n-th primitive root of unity modulo p. (This requirement is needed for the sake of efficiency, and we will make it implicitly throughout the current paper.) Let $\mathcal{Q} = (\mathbb{Z}_p, m_0, \mathsf{aug}(U), \mathsf{aug}(V), \mathsf{aug}(W))$ be a QAP instance. For $j \in [0 .. m]$, define $u_j(X) := L_{\boldsymbol{U}^{(j)}}(X)$, $v_j(X) := L_{\boldsymbol{V}^{(j)}}(X)$, and $w_j(X) := L_{\boldsymbol{W}^{(j)}}(X)$. Thus, $u_j, v_j, w_j \in \mathbb{Z}_p^{(\leq n-1)}[X]$.

An QAP instance $\mathcal{Q}_p$ is specified by the so defined $(\mathbb{Z}_p, m_0, \{u_j, v_j, w_j\}_{j=0}^m)$. This instance defines the following relation, where we assume that $A_0 = 1$:

$$\mathbf{R}_{\mathcal{Q}_p} = \left\{ \begin{array}{l} (\mathsf{x}, \mathsf{w}) \colon \mathsf{x} = (A_1, \ldots, A_{m_0})^\top \wedge \mathsf{w} = (A_{m_0+1}, \ldots, A_m)^\top \wedge \\ \left(\sum_{j=0}^m A_j u_j(X)\right)\left(\sum_{j=0}^m A_j v_j(X)\right) \equiv \sum_{j=0}^m A_j w_j(X) \pmod{\ell(X)} \end{array} \right\}$$

Alternatively, $(x, w) \in \mathbf{R}$ if there exists a (degree $\leq n - 2$) polynomial $h(X)$, s.t.

$$\left(\sum_{j=0}^{m} A_j u_j(X) \right) \left(\sum_{j=0}^{m} A_j v_j(X) \right) - \sum_{j=0}^{m} A_j w_j(X) = h(X)\ell(X).$$

Clearly, $\mathbf{R}_{\mathcal{Q}} = \mathbf{R}_{\mathcal{Q}_p}$, given that $\mathcal{Q}_p$ is constructed from $\mathcal{Q}$ as above.

3   Definitions: SNARKs and Subversion Zero Knowledge

Next, we define subversion zk-SNARKs and all their properties. To achieve subversion zero knowledge (Sub-ZK), we augment a zk-SNARK by requiring the existence of an *efficient* CRS verification algorithm. As outlined in Sect. 1, we also subdivide the CRS generation algorithm into three efficient algorithms.

Our definition of (statistical unbounded) Sub-ZK for SNARKs is motivated by the definition of [2]. We also define statistical composable Sub-ZK. As in [25], the definition of unbounded zero knowledge guarantees security for any (polynomial) number of queries to the prover or simulator, while the definition of composable zero knowledge guarantees security only in the case of a single query. Following [25] we prove that a statistical composable Sub-ZK argument system is also statistical unbounded Sub-ZK.

3.1   Syntax

Let $\mathcal{R}$ be a relation generator, such that $\mathcal{R}(1^\kappa)$ returns a polynomial-time decidable binary relation $\mathbf{R} = \{(x, w)\}$. Here, x is the statement and w is the witness. We assume that κ is explicitly deductible from the description of $\mathbf{R}$. The relation generator also outputs auxiliary information $z_{\mathbf{R}}$ that will be given to the honest parties and the adversary. As in [27, Sect. 2.3], $z_{\mathbf{R}}$ will be equal to $gk \leftarrow \mathsf{genbp}(1^\kappa, n)$ for a well-defined n. Because of this, we will also give $z_{\mathbf{R}}$ as an input to the honest parties; if needed, one can include an additional auxiliary input as an input to the adversary. We recall that the choice of p and thus of the groups $\mathbb{G}_z$ depends on n. Let $\mathcal{L}_{\mathbf{R}} = \{x : \exists w, (x, w) \in \mathbf{R}\}$ be an **NP**-language.

A (subversion-resistant) *non-interactive zero-knowledge argument system* Ψ for $\mathcal{R}$ consists of seven PPT algorithms:

CRS trapdoor generator: $\mathsf{K_{tc}}$ is a probabilistic algorithm that, given $(\mathbf{R}, z_{\mathbf{R}}) \in \mathrm{range}(\mathcal{R}(1^\kappa))$, outputs a *CRS trapdoor* tc. Otherwise, it outputs a special symbol $\perp$.

Simulation trapdoor generator: $\mathsf{K_{ts}}$ is a *deterministic* algorithm that, given $(\mathbf{R}, z_{\mathbf{R}}, \mathsf{tc})$ where $(\mathbf{R}, z_{\mathbf{R}}) \in \mathrm{range}(\mathcal{R}(1^\kappa))$ and $\mathsf{tc} \in \mathrm{range}(\mathsf{K_{tc}}(\mathbf{R}, z_{\mathbf{R}})) \setminus \{\perp\}$, outputs the *simulation trapdoor* ts. Otherwise, it outputs $\perp$.

CRS generator: $\mathsf{K_{crs}}$ is a *deterministic* algorithm that, given $(\mathbf{R}, z_{\mathbf{R}}, \mathsf{tc})$, where $(\mathbf{R}, z_{\mathbf{R}}) \in \mathrm{range}(\mathcal{R}(1^\kappa))$ and $\mathsf{tc} \in \mathrm{range}(\mathsf{K_{tc}}(\mathbf{R}, z_{\mathbf{R}})) \setminus \{\perp\}$, outputs crs. Otherwise, it outputs $\perp$. For the sake of efficiency and readability, we divide crs into $\mathsf{crs_P}$ (the part needed by the prover), $\mathsf{crs_V}$ (the part needed by the verifier), and $\mathsf{crs_{CV}}$ (the part needed only by CV and not by P or V).

CRS verifier: CV is a probabilistic algorithm that, given $(\mathbf{R}, z_{\mathbf{R}}, crs)$, returns either 0 (the CRS is incorrectly formed) or 1 (the CRS is correctly formed),

Prover: P is a probabilistic algorithm that, given $(\mathbf{R}, z_{\mathbf{R}}, crs_P, x, w)$ for $CV(\mathbf{R}, z_{\mathbf{R}}, crs) = 1$ and $(x, w) \in \mathbf{R}$, outputs an argument π. Otherwise, it outputs $\perp$.

Verifier: V is a probabilistic algorithm that, given $(\mathbf{R}, z_{\mathbf{R}}, crs_V, x, \pi)$, returns either 0 (reject) or 1 (accept).

Simulator: S is a probabilistic algorithm that, given $(\mathbf{R}, z_{\mathbf{R}}, crs, ts, x)$ where $CV(\mathbf{R}, z_{\mathbf{R}}, crs) = 1$, outputs an argument π.

We also define the (non-subverted) CRS generation algorithm $K(\mathbf{R}, z_{\mathbf{R}})$ that first sets $tc \leftarrow K_{tc}(\mathbf{R}, z_{\mathbf{R}})$ and then outputs $(crs \| ts) \leftarrow (K_{crs} \| K_{ts})(\mathbf{R}, z_{\mathbf{R}}, tc)$.

One can remove S from the definition of Ψ, and instead require that for each PPT verifier V^* there exists a corresponding PPT simulator S. We follow [27] and other SNARK literature by letting Ψ to fix S; this guarantees that there exists a single simulator that simulates the CRS for all subverters. In the case of non subversion-resistant QANIZKs, existence of a single simulator is required [29]. See Definition 5 and Remark 1 for a longer discussion.

SNARKs. A non-interactive argument system is *succinct* if the proof size is polynomial in κ and the verifier runs in time polynomial in $\kappa + |x|$. A succinct non-interactive argument of knowledge is usually called *SNARK*. A zero knowledge SNARK is abbreviated to *zk-SNARK*.

Discussions. A (non subversion-resistant) non-interactive zero-knowledge argument system is defined as a tuple (K, P, V, S), see, e.g., [27]. We will now briefly motivate the differences compared to the established syntax of non-interactive zero-knowledge argument systems. Section 3.2 will give formal security definitions where the above syntactic definition will become an important part.

The division of K into 3 subalgorithms K_{tc}, K_{ts}, and K_{crs} is usually not needed, but the K algorithm of many known non-interactive zero-knowledge argument systems (and all SNARKs that we know) satisfies such a division. K_{tc} just generates all randomness (tc), needed to compute crs and ts, and then K_{crs} and K_{ts} compute from tc deterministically crs and ts. We note that such division can be formalized by requiring the crs to be witness-sampleable [29] that also seems to be a reasonable requirement in the case one is interested in subversion-resistance. Really, an important subgoal of Sub-ZK is to guarantee that the subverted CRS is consistent with at least some choice of tc. This means that for each tc there must exist corresponding crs (accepted by CV) and ts (that can be used by the simulator to simulate subverted crs corresponding to tc).

The existence of *efficient* CV will be crucial to obtain Sub-ZK. To guarantee Sub-ZK, it is intuitively clear that the honest prover should *at least* check the correctness of the CRS. Efficiency-wise, since the prover in known SNARK constructions (including say [22,27] and the current paper) takes superlinear time, this is fine unless the CRS verification will be even slower. For the sake of clarity, however, we do not assume that CV is a part of the P *algorithm*; instead we assume that an honest prover first runs CV and given that it accepts, runs P. This is since, in practice one could execute many zero-knowledge arguments by using the same CRS; it is natural to require that the prover executes CV only once. On the other hand, since we are not aiming to get subversion (knowledge-)soundness, the honest verifier does not have to execute CRS verification.

Finally, $\mathsf{crs_P}$ (resp., $\mathsf{crs_V}$) is the part of the CRS given to an honest prover (resp., an honest verifier), and $\mathsf{crs_{CV}}$ is the part of CRS not needed by the prover or the verifier except to run CV. The distinction between $\mathsf{crs_P}$, $\mathsf{crs_V}$, and $\mathsf{crs_{CV}}$ is not important from the security point of view, but in many cases $\mathsf{crs_V}$ is significantly shorter than $\mathsf{crs_P}$. Keeping $\mathsf{crs_{CV}}$ separate helps one to evaluate better the additional efficiency penalty introduced by subversion security.

3.2 Security Definitions

A Sub-ZK SNARK has to satisfy various security definitions. The most important ones are *subversion-completeness* (an honest prover convinces an honest verifier, and an honestly generated CRS passes the CRS verification test), *computational knowledge-soundness* (if a prover convinces an honest verifier, then he knows the corresponding witness), and *statistical Sub-ZK* (given a possibly subverted CRS, an argument created by the honest prover reveals no side information). In the case of Sub-ZK, we will consider the case of an efficient subverter and a computationally unbounded distinguisher; in Sect. 9 we will discuss the case when also the subverter is unbounded. Next, we will give definitions of those properties that guarantee both composability and subversion resistance.

To keep the new security definitions as close to the accepted security definitions of zk-SNARKs as possible, we will start with non subversion-resistant security definitions from [27] (that will be stated below for the sake of completeness), albeit by using our notation, and modify them by adding elements of subversion as in Bellare *et al.* [2]. To ease the reading, we will *emphasize* the differences between non-subversion and subversion definitions. We use the division of CRS generation into three different algorithms also in the non subversion-resistant case. As motivated in Sect. 3.1, we also give $\mathsf{z_R}$ as an input to all honest parties. Finally, the notions of unbounded (ZK is guaranteed against an adversary who can arbitrarily query an oracle that outputs either proofs or simulations) and composable (ZK is guaranteed against an adversary who has to distinguish a single argument from a simulation) zero knowledge follow the definitions given in [25] and are in fact equal in our case.

As [27], we define all security notions against a non-uniform adversary. However, since our security reductions are uniform, it is a simple matter to consider only uniform adversaries, as it was done by Bellare *et al.* [2] (see also [24]).

Definition 1 (Perfect Completeness [27]). *A non-interactive argument Ψ is* perfectly complete *for $\mathcal{R}$, if for all κ, all $(\mathbf{R}, \mathsf{z_R}) \in \mathrm{range}(\mathcal{R}(1^\kappa))$, $\mathsf{tc} \in \mathrm{range}(\mathsf{K_{tc}}(\mathbf{R}, \mathsf{z_R})) \setminus \{\bot\}$, and $(\mathsf{x}, \mathsf{w}) \in \mathbf{R}$,*

$$\Pr\left[\mathsf{crs} \leftarrow \mathsf{K_{crs}}(\mathbf{R}, \mathsf{z_R}, \mathsf{tc}) : \mathsf{V}(\mathbf{R}, \mathsf{z_R}, \mathsf{crs_V}, \mathsf{x}, \mathsf{P}(\mathbf{R}, \mathsf{z_R}, \mathsf{crs_P}, \mathsf{x}, \mathsf{w})) = 1\right] = 1.$$

Definition 2 (Perfect Subversion-Completeness). *A non-interactive argument Ψ is* perfectly subversion-complete *for $\mathcal{R}$, if for all κ, all $(\mathbf{R}, \mathsf{z_R}) \in$* range$(\mathcal{R}(1^{\kappa}))$, $\mathsf{tc} \in$ range$(\mathsf{K_{tc}}(\mathbf{R}, \mathsf{z_R})) \setminus \{\bot\}$, *and* $(\mathsf{x}, \mathsf{w}) \in \mathbf{R}$,

$$\Pr\left[\begin{array}{l} \mathsf{crs} \leftarrow \mathsf{K_{crs}}(\mathbf{R}, \mathsf{z_R}, \mathsf{tc}) : \boxed{\mathsf{CV}(\mathbf{R}, \mathsf{z_R}, \mathsf{crs}) = 1 \wedge} \\ \mathsf{V}(\mathbf{R}, \mathsf{z_R}, \mathsf{crs_V}, \mathsf{x}, \mathsf{P}(\mathbf{R}, \mathsf{z_R}, \mathsf{crs_P}, \mathsf{x}, \mathsf{w})) = 1 \end{array}\right] = 1.$$

Definition 3 (Computational Knowledge-Soundness [27]). *Ψ is computationally (adaptively)* knowledge-sound *for $\mathcal{R}$, if for every NUPPT A, there exists a NUPPT extractor $\mathsf{X_A}$, s.t. for all κ,*

$$\Pr\left[\begin{array}{l} (\mathbf{R}, \mathsf{z_R}) \leftarrow \mathcal{R}(1^{\kappa}), (\mathsf{crs} \,\|\, \mathsf{ts}) \leftarrow \mathsf{K}(\mathbf{R}, \mathsf{z_R}), \\ r \leftarrow_r \mathsf{RND}(\mathsf{A}), ((\mathsf{x}, \pi) \,\|\, \mathsf{w}) \leftarrow (\mathsf{A} \,\|\, \mathsf{X_A})(\mathbf{R}, \mathsf{z_R}, \mathsf{crs}; r) : \\ (\mathsf{x}, \mathsf{w}) \notin \mathbf{R} \wedge \mathsf{V}(\mathbf{R}, \mathsf{z_R}, \mathsf{crs_V}, \mathsf{x}, \pi) = 1 \end{array}\right] \approx_{\kappa} 0.$$

Here, $\mathsf{z_R}$ can be seen as a common auxiliary input to A and $\mathsf{X_A}$ that is generated by using a benign [10] relation generator; we recall that we just think of $\mathsf{z_R}$ as being the description of a secure bilinear group. A knowledge-sound argument system is called an *argument of knowledge*.

Next, we define statistically unbounded ZK. Unbounded (non-Sub) ZK was not defined in [27], presumably because it is a corollary of composable non-Sub ZK as shown in [25]. Hence, we will first give a modified version of the definition of non-Sub ZK from [25]; in [25], K will only output a crs and a CRS simulator $\mathsf{S_{crs}}$ will return $(\mathsf{crs} \,\|\, \mathsf{ts})$. As in [27], we find it more convenient to let $\mathsf{S_{crs}}$ (whom we call K) to generate also the honest crs.

Definition 4 (Statistically Unbounded ZK [25]). *Ψ is* statistically unbounded Sub-ZK *for $\mathcal{R}$, if for all κ, all $(\mathbf{R}, \mathsf{z_R}) \in$* range$(\mathcal{R}(1^{\kappa}))$, *and all computationally unbounded A, $\varepsilon_0^{unb} \approx_{\kappa} \varepsilon_1^{unb}$, where*

$$\varepsilon_b^{unb} = \Pr[(\mathsf{crs} \,\|\, \mathsf{ts}) \leftarrow \mathsf{K}(\mathbf{R}, \mathsf{z_R}) : \mathsf{A}^{\mathsf{O}_b(\cdot, \cdot)}(\mathbf{R}, \mathsf{z_R}, \mathsf{crs}) = 1].$$

Here, the oracle $\mathsf{O}_0(\mathsf{x}, \mathsf{w})$ returns $\bot$ (reject) if $(\mathsf{x}, \mathsf{w}) \notin \mathbf{R}$, and otherwise it returns $\mathsf{P}(\mathbf{R}, \mathsf{z_R}, \mathsf{crs_P}, \mathsf{x}, \mathsf{w})$. Similarly, $\mathsf{O}_1(\mathsf{x}, \mathsf{w})$ returns $\bot$ (reject) if $(\mathsf{x}, \mathsf{w}) \notin \mathbf{R}$, and otherwise it returns $\mathsf{S}(\mathbf{R}, \mathsf{z_R}, \mathsf{crs}, \mathsf{ts}, \mathsf{x})$. Ψ is perfectly unbounded Sub-ZK *for $\mathcal{R}$ if one requires that $\varepsilon_0^{unb} = \varepsilon_1^{unb}$.*

The following definition of unbounded Sub-ZK differs from this as follows. Since we allow the CRS to be subverted, the CRS is generated by a subverter who also returns $\mathsf{z_{\Sigma}}$. The adversary's access to $\mathsf{z_{\Sigma}}$ models the possibility that the subverter and the adversary collaborate. The extractor $\mathsf{X_{\Sigma}}$ extracts tc from Σ, and then tc is used to generate the simulation trapdoor ts that is then given as an auxiliary input to the adversary and to the oracle O_1. In [25], tc was not given to the adversary because K was not required to return tc and thus was not guaranteed to exist; adding this input to the adversary only increases the power of the adversary. In our construction, it does not matter since a computationally unbounded A could compute ts herself (see Algorithm 5 in Fig. 3). If we allow both the subverter and the extractor to be computationally unbounded, we would not need to rely on a knowledge assumption; the Sub-ZK proof in Sect. 7 would also simplify.

A weaker version of Sub-ZK definition would require that the extractor only outputs ts that is used then for simulation. We prefer the current definition since it is stronger and allows us to prove CRS trapdoor extractability (see Definition 8). Since we consider statistical ZK, achieving our stronger definition will concur almost no extra cost if we also allow X_Σ to be computationally unbounded.

Definition 5 (Statistically Unbounded Sub-ZK). Ψ *is* statistically unbounded Sub-ZK for $\mathcal{R}$, *if for any NUPPT subverter Σ there exists a NUPPT X_Σ, such that for all κ, all $(\mathbf{R}, \mathbf{z_R}) \in \mathrm{range}(\mathcal{R}(1^\kappa))$, and all computationally unbounded* A, $\varepsilon_0^{unb} \approx_\kappa \varepsilon_1^{unb}$, *where ε_b^{unb} is defined as*

$$\mathrm{Pr}\left[\begin{array}{l} r \leftarrow_r \mathsf{RND}(\Sigma), (\mathsf{crs}, \mathsf{z}_\Sigma \,\|\, \mathsf{tc}) \leftarrow (\Sigma \,\|\, X_\Sigma)(\mathbf{R}, \mathbf{z_R}; r), \\ \mathsf{ts} \leftarrow \mathsf{K_{ts}}(\mathbf{R}, \mathbf{z_R}, \mathsf{tc}) \; : \; \mathsf{CV}(\mathbf{R}, \mathbf{z_R}, \mathsf{crs}) = 1 \wedge \mathsf{A}^{\mathsf{O}_b(\cdot,\cdot)}(\mathbf{R}, \mathbf{z_R}, \mathsf{crs}, \mathsf{ts}, \mathsf{z}_\Sigma) = 1 \end{array}\right].$$

Here, the oracle $\mathsf{O}_0(\mathsf{x}, \mathsf{w})$ returns $\perp$ (reject) if $(\mathsf{x}, \mathsf{w}) \notin \mathbf{R}$, and otherwise it returns $\mathsf{P}(\mathbf{R}, \mathbf{z_R}, \mathsf{crs_P}, \mathsf{x}, \mathsf{w})$. Similarly, $\mathsf{O}_1(\mathsf{x}, \mathsf{w})$ returns $\perp$ (reject) if $(\mathsf{x}, \mathsf{w}) \notin \mathbf{R}$, and otherwise it returns $\mathsf{S}(\mathbf{R}, \mathbf{z_R}, \mathsf{crs}, \mathsf{ts}, \mathsf{x})$. Ψ is perfectly unbounded Sub-ZK for $\mathcal{R}$ if one requires that $\varepsilon_0^{unb} = \varepsilon_1^{unb}$.

Remark 1 (Comparison to Sub-ZK Definition of [2]). Let O_b be defined as before. In [2], it is required that for any (non-uniform) PPT subverter Σ there exists a (non-uniform) PPT simulator (X_Σ, S), such that for all κ, $(\mathbf{R}, \mathbf{z_R}) \in \mathrm{range}(\mathcal{R}(1^\kappa))$, all $(\mathsf{x}, \mathsf{w}) \in \mathbf{R}$, and all (non-uniform) PPT A, $\varepsilon_0^{bfs} \approx_\kappa \varepsilon_1^{bfs}$, where

$$\varepsilon_b^{bfs} = \mathrm{Pr}\left[\begin{array}{l} \textbf{if } b = 0 \textbf{ then } r \leftarrow_r \mathsf{RND}(\Sigma), \mathsf{crs} \leftarrow \Sigma(\mathbf{R}, \mathbf{z_R}; r) \\ \textbf{else } (\mathsf{crs}, r) \leftarrow X_\Sigma(\mathbf{R}, \mathbf{z_R}) \textbf{ fi} : \mathsf{A}^{\mathsf{O}_b(\cdot,\cdot)}(\mathbf{R}, \mathbf{z_R}, \mathsf{crs}, r) \end{array}\right].$$

First, [2] defines computational Sub-ZK while we define statistical Sub-ZK. This by itself changes several aspects of the definition. E.g., we could just let A to compute ts and z_Σ from crs instead of giving them as extra inputs to A.

Second, compared to [2], we give to A extra information, ts and z_Σ. Having access to ts means that one can implement SNARKs for different relations using the same CRS [25]. Really, given ts, A will be able to both form and simulate arguments for any of the considered relations. Having access to z_Σ models, as already mentioned, the possibility that Σ and A are collaborating. (Bellare *et al.* only allow the subverter to communicate r, the secret coins.) In this sense, our definition is stronger compared to [2].

Third, Bellare *et al.* required that for each Σ there exists a simulator (X_Σ, S). We consider it to be more natural to think of X_Σ as an extractor and allow only X_Σ to depend on Σ. In the SNARK literature, extractors usually depend on the adversary (here, Σ) while there is a single simulator that works for all adversaries. In particular, this is a formal requirement in the case of QANIZKs, [29]. Also

Bellare *et al.* used X_Σ as an extractor in their construction. In this sense, our definition is stronger compared to [2].

Fourth, we give to X_Σ the same coins r as to Σ, while Bellare *et al.* allow X_Σ to generate its own coins, only requiring that the distribution of (crs, r) is computationally indistinguishable from the output of Σ. Also in this sense, our definition seems to be stronger.

Finally, we use explicitly the syntax of subversion-resistant SNARKs from Sect. 3.1, assuming the existence of algorithms K_{ts} and CV. While it might seem to be restrictive, as argued in Sect. 3.1, existence of both K_{ts} and CV seems to be necessary for subversion-resistance. □

In the case of composable ZK, the adversary can only see one purported (i.e., real or simulated) argument π, instead of being able to make many queries to a purported prover oracle as in the case of unbounded ZK. If composable ZK is defined carefully, it will be at least as strong as unbounded ZK while potentially allowing for simpler ZK proofs, [25]. We will show that the same holds in the case of Sub-ZK.

Definition 6 (Statistically Composable ZK [27]). Ψ *is statistically composable Sub-ZK for* $\mathcal{R}$*, if for any NUPPT subverter* Σ *there exists a NUPPT* X_Σ*, such that for all* κ*,* $(\mathbf{R}, z_{\mathbf{R}}) \in \mathrm{range}(\mathcal{R}(1^\kappa))$*, all* $(x, w) \in \mathbf{R}$*, and all computationally unbounded* A*,* $\varepsilon_0^{comp} \approx_\kappa \varepsilon_1^{comp}$*, where*

$$\varepsilon_b^{comp} = \Pr \left[\begin{array}{l} (\mathsf{crs} \,\|\, \mathsf{ts}) \leftarrow \mathsf{K}(\mathbf{R}, z_{\mathbf{R}}), \\ \text{if } b = 0 \text{ then } \pi \leftarrow \mathsf{P}(\mathbf{R}, z_{\mathbf{R}}, \mathsf{crs_P}, x, w) \\ \text{else } \pi \leftarrow \mathsf{S}(\mathbf{R}, z_{\mathbf{R}}, \mathsf{crs}, \mathsf{ts}, x) \text{ fi} : \mathsf{A}(\mathbf{R}, z_{\mathbf{R}}, \mathsf{crs}, \mathsf{ts}, \pi) = 1 \end{array} \right].$$

Ψ *is perfectly composable Sub-ZK for* $\mathcal{R}$ *if one requires that* $\varepsilon_0^{comp} = \varepsilon_1^{comp}$*.*

Next, we define statistical composable subversion zero knowledge (Sub-ZK). This definition is related to but crucially different from the definition of (computational) composable ZK from [25]. Most importantly, [25] defines two properties, the first being *reference string indistinguishability*, meaning that the CRS generated by *honest* K and the CRS simulated by a simulator S_{crs} should be indistinguishable. We will use the CRS generated by the subverter in both the real and the simulated case. The second property in [25] is simulation indistinguishability. Our definition of composable Sub-ZK is similar to the definition of simulation indistinguishability in [25]. However, instead of simulating the CRS, we use crs generated by the subverter, assume that an extractor extracts tc from crs, compute ts from tc by using K_{ts}, and finally verify that CV accepts crs. We also quantify over all valid $(x, w) \in \mathbf{R}$ instead of letting the adversary to choose it; since we deal with statistical Sub-ZK this results in an equivalent definition.

Moreover, we differ from the definition of composable non-Sub ZK in [27] as follows. The CRS is generated by Σ who also returns z_Σ. A's access to z_Σ models the possibility that Σ and A collaborate; although of course A could just recompute it. X_Σ extracts tc from Σ, and then tc is used to generate the simulation trapdoor ts that is then given as an auxiliary input to A.

Definition 7 (Statistically Composable Sub-ZK). Ψ *is statistically composable Sub-ZK for* $\mathcal{R}$, *if for any NUPPT subverter* Σ *there exists a NUPPT* X_Σ, *such that for all* κ, *all* $(\mathbf{R}, z_\mathbf{R}) \in \mathrm{range}(\mathcal{R}(1^\kappa))$, *all* $(x, w) \in \mathbf{R}$, *and all computationally unbounded* A, $\varepsilon_0^{comp} \approx_\kappa \varepsilon_1^{comp}$, *where* ε_b^{comp} *is defined as*

$$\Pr \left[\begin{array}{l} r \leftarrow_r \mathsf{RND}(\Sigma), (\mathsf{crs}, z_\Sigma \,\|\, \mathsf{tc}) \leftarrow (\Sigma \,\|\, X_\Sigma)(\mathbf{R}, z_\mathbf{R}; r),\ \mathsf{ts} \leftarrow \mathsf{K_{ts}}(\mathbf{R}, z_\mathbf{R}, \mathsf{tc}), \\ \text{if } b = 0 \text{ then } \pi \leftarrow \mathsf{P}(\mathbf{R}, z_\mathbf{R}, \mathsf{crs_P}, x, w) \text{ else } \pi \leftarrow \mathsf{S}(\mathbf{R}, z_\mathbf{R}, \mathsf{crs}, \mathsf{ts}, x) \text{ fi} : \\ \mathsf{CV}(\mathbf{R}, z_\mathbf{R}, \mathsf{crs}) = 1 \wedge \mathsf{A}(\mathbf{R}, z_\mathbf{R}, \mathsf{crs}, \mathsf{ts}, \pi, z_\Sigma) = 1 \end{array} \right] .$$

Ψ *is perfectly composable Sub-ZK for* $\mathcal{R}$ *if one requires that* $\varepsilon_0^{comp} = \varepsilon_1^{comp}$.

Now, we will prove the following result that makes it possible to operate in the rest of the paper with the simpler composable Sub-ZK definition. It is motivated by a similar result of Groth (ASIACRYPT 2006, [25]) that considers computational non subversion-resistant zero knowledge. As seen from below, we can establish the same result for statistical zero knowledge, but then we have to restrict the number of oracle calls to a polynomial number.

Theorem 1. *(i) Statistical composable Sub-ZK implies statistical unbounded Sub-ZK, assuming that* A *is given access to polynomially many oracle calls. (ii) Perfect composable Sub-ZK implies perfect unbounded Sub-ZK, even if given access to an unbounded number of oracle calls.*

Proof. (I) STATISTICAL SUB-ZK. Assume that the adversary can make up to $q(\kappa)$ oracle queries for some fixed polynomial q. We define a sequence of $q(\kappa) + 1$ oracles $\mathsf{O}_0'(x, w), \ldots, \mathsf{O}_{q(\kappa)}'(x, w)$. Given the jth adversarial query (x_j, w_j), the oracle $\mathsf{O}_k'(\cdot, \cdot)$ responds with $\mathsf{O}_1(x_j, w_j)$ for $j \in [1 .. k]$ and $\mathsf{O}_0(x_j, w_j)$ for $j \in [k + 1 .. q(\kappa)]$. Hence, $\mathsf{O}_0' = \mathsf{O}_0$ and $\mathsf{O}_{q(\kappa)}' = \mathsf{O}_1$.

Due to the statistical composable Sub-ZK property, we get that for $i \in [0 .. q(\kappa) - 1]$, $\varepsilon_i \approx_\kappa \varepsilon_{i+1}$, where ε_i is defined as

$$\Pr \left[\begin{array}{l} r \leftarrow_r \mathsf{RND}(\Sigma), (\mathsf{crs}, z_\Sigma \,\|\, \mathsf{tc}) \leftarrow (\Sigma \,\|\, X_\Sigma)(\mathbf{R}, z_\mathbf{R}; r), \\ \mathsf{ts} \leftarrow \mathsf{K_{ts}}(\mathbf{R}, z_\mathbf{R}, \mathsf{tc}) : \mathsf{CV}(\mathbf{R}, z_\mathbf{R}, \mathsf{crs}) = 1 \wedge \mathsf{A}^{\mathsf{O}_i'(\cdot, \cdot)}(\mathbf{R}, z_\mathbf{R}, \mathsf{crs}, \mathsf{ts}, z_\Sigma) = 1 \end{array} \right] ,$$

since the oracles can be efficiently implemented given crs and ts, and inserting π as the answer to the ith query. Since $\varepsilon_0 = \varepsilon_0^{unb}$, q is a polynomial, and $\varepsilon_0^{unb} = \varepsilon_0 \approx_\kappa \cdots \approx_\kappa \varepsilon_{q(\kappa)} = \varepsilon_1^{unb}$, we get that $\varepsilon_0^{unb} \approx_\kappa \varepsilon_1^{unb}$ and hence the claim holds.

(II) PERFECT SUB-ZK. As above, but assume q is the actual (possibly unbounded) number of queries. We get $\varepsilon_0^{unb} = \varepsilon_0 = \cdots = \varepsilon_{q(\kappa)} = \varepsilon_1^{unb}$, and hence $\varepsilon_0^{unb} = \varepsilon_1^{unb}$, and the claim holds. $\square$

In [25], composable ZK was a stronger requirement than unbounded ZK since in the case of composable ZK, (i) the simulated CRS was required to be indistinguishable from the real CRS, and (ii) the adversary got access to ts. In the case of our Sub-ZK definitions, there is no such difference and it is easy to see that composable Sub-ZK and unbounded Sub-ZK are in fact equal notions.

In the proof that the new SNARK is Sub-ZK (see Theorem 5), we require that if Σ generates a crs accepted by CV then there exists a NUPPT extractor X_Σ that generates a CRS trapdoor tc that corresponds to crs; that is, K_{crs} on input tc outputs crs. We formalize this property—that intuitively gives us a stronger version of witness-sampleability—as follows.

Definition 8 (Statistical CRS Trapdoor Extractability). Ψ *has statistical CRS trapdoor extractability for* $\mathcal{R}$, *if for any NUPPT subverter* Σ *there exists a NUPPT* X_Σ, *s.t. for all* κ *and all* $(\mathbf{R}, \mathbf{z_R}) \in \mathrm{range}(\mathcal{R}(1^\kappa))$,

$$
\Pr\left[
\begin{array}{l}
r \leftarrow_r \mathsf{RND}(\Sigma), (\mathsf{crs}, z_\Sigma \,\|\, \mathsf{tc}) \leftarrow (\Sigma \,\|\, X_\Sigma)(\mathbf{R}, \mathbf{z_R}; r): \\
\mathsf{CV}(\mathbf{R}, \mathbf{z_R}, \mathsf{crs}) = 1 \wedge \mathsf{K_{crs}}(\mathbf{R}, \mathbf{z_R}, \mathsf{tc}) \neq \mathsf{crs}
\end{array}
\right] \approx_\kappa 0.
$$

Ψ *has perfect CRS trapdoor extractability for* $\mathcal{R}$ *if the same property holds but with* $\approx_\kappa$ *changed to* $=$.

4 GBGM and Sub-GBGM

Preliminaries: Generic Bilinear Group Model. In Sect. 6, we will prove that the new zk-SNARK is knowledge-sound in the subversion generic bilinear group model (Sub-GBGM). In the current subsection, we will introduce the GBGM [12,33,34,36], by following the exposition in [33]. After that, we will introduce the Sub-GBGM.

We start by picking a random asymmetric bilinear group gk := $(p, \mathbb{G}_1, \mathbb{G}_2, \mathbb{G}_T, \hat{e}) \leftarrow \mathsf{genbp}(1^\kappa, n)$. Consider a black box $\mathbf{B}$ that can store values from additive groups $\mathbb{G}_1, \mathbb{G}_2, \mathbb{G}_T$ in internal state variables $\mathsf{cell}_1, \mathsf{cell}_2, \ldots$, where for simplicity we allow the storage space to be infinite (this only increases the power of a generic adversary). The initial state consists of some values $(\mathsf{cell}_1, \mathsf{cell}_2, \ldots, \mathsf{cell}_{|inp|})$, which are set according to some probability distribution. Each state variable cell_i has an accompanying type $\mathsf{type}_i \in \{1, 2, T, \bot\}$. We assume initially $\mathsf{type}_i = \bot$ for $i > |inp|$. The black box allows computation operations on internal state variables and queries about the internal state. No other interaction with $\mathbf{B}$ is possible.

Let Π be an allowed set of computation operations. A computation operation consists of selecting a (say, t-ary) operation $f \in \Pi$ together with $t + 1$ indices $i_1, i_2, \ldots, i_{t+1}$. Assuming inputs have the correct type, $\mathbf{B}$ computes $f(\mathsf{cell}_{i_1}, \ldots, \mathsf{cell}_{i_t})$ and stores the result in $\mathsf{cell}_{i_{t+1}}$. For a set Σ of relations, a query consists of selecting a (say, t-ary) relation $\varrho \in \Sigma$ together with t indices $i_1, i_2, \ldots, i_t$. Assuming inputs have the correct type, $\mathbf{B}$ replies to the query with $\varrho(\mathsf{cell}_{i_1}, \ldots, \mathsf{cell}_{i_t})$. In the GBGM, we define $\Pi = \{+, \hat{e}\}$ and $\Sigma = \{=\}$, where

1. On input $(+, i_1, i_2, i_3)$: if $\mathsf{type}_{i_1} = \mathsf{type}_{i_2} \neq \bot$ then set $\mathsf{cell}_{i_3} \leftarrow \mathsf{cell}_{i_1} + \mathsf{cell}_{i_2}$ and $\mathsf{type}_{i_3} \leftarrow \mathsf{type}_{i_1}$.
2. On input $(\hat{e}, i_1, i_2, i_3)$: if $\mathsf{type}_{i_1} = 1$ and $\mathsf{type}_{i_2} = 2$ then set $\mathsf{cell}_{i_3} \leftarrow \hat{e}(\mathsf{cell}_{i_1}, \mathsf{cell}_{i_2})$ and $\mathsf{type}_{i_3} \leftarrow T$.

3. On input $(=, i_1, i_2)$: if $\mathsf{type}_{i_1} = \mathsf{type}_{i_2} \neq \perp$ and $\mathsf{cell}_{i_1} = \mathsf{cell}_{i_2}$ then return 1. Otherwise return 0.

Since we are proving lower bounds, we will give a generic adversary A additional power. We assume that all relation queries are for free. We also assume that A is successful if after τ operation queries, he makes an equality query $(=, i_1, i_2)$, $i_1 \neq i_2$, that returns 1; at this point A quits. Thus, if $\mathsf{type}_i \neq \perp$, then $\mathsf{cell}_i = F_i(\mathsf{cell}_1, \ldots, \mathsf{cell}_{|inp|})$ for a polynomial F_i known to A.

Sub-GBGM. By following Bellare *et al.* [2], we enhance the power of generic bilinear group model [12,33,34,36]. Since the power of the generic adversary will increase, security proofs in the resulting *Sub-GBGM* will be somewhat more realistic than in the GBGM model.

As noted by Bellare *et al.* [2], it is known how to hash into elliptic curves and thus create group elements without knowing their discrete logarithms. However, it is not known how to create four elements $[1]_z$, $[a]_z$, $[b]_z$, and $[ab]_z$ without knowing either a or b. The corresponding assumption—that may also be true in the case of symmetric pairings—was named *DH-KE(A)* in [2].

However, asymmetric pairings are much more efficient than symmetric pairings. If we work in the type III pairing setting where there is no efficient isomorphism either from $\mathbb{G}_1$ to $\mathbb{G}_2$ or from $\mathbb{G}_2$ to $\mathbb{G}_1$, then clearly an adversary cannot, given $[a]_z$ for $z \in \{1, 2\}$ and an unknown a, compute $[a]_{3-z}$. In the same vein, it seems reasonable to make a stronger assumption (that we call *BDH-KE*, a simplification of the asymmetric PKE assumption of [17]) that if an adversary creates $[a]_1$ and $[a]_2$ then she knows a. Really, since there is no polynomial-time isomorphism from $\mathbb{G}_1$ to $\mathbb{G}_2$ (or back), it seems to be natural to assume that one does not have to worry about an adversary knowing some trapdoor that would break the BDH-KE assumption. Since BDH-KE is not a falsifiable assumption, this does not obviously mean that it must hold for each type III pairing. Instead, the BDH-KE assumption can be interpreted as a *stronger* definition of the type III pairing setting. We formalize the added adversarial power as follows.

We give the generic model adversary an additional power to effectively create new indeterminates Y_i in groups $\mathbb{G}_1$ and $\mathbb{G}_2$ (e.g., by hashing into elliptic curves), without knowing their values. We note since $[Y]_1 \bullet [1]_2 = [Y]_T$ and $[1]_1 \bullet [Y]_2 = [Y]_T$, the adversary that has generated an indeterminate Y in $\mathbb{G}_z$ can also operate with Y in $\mathbb{G}_T$. Formally, this means that Π will contain one more operation create, with the following semantics:

4. On input (create, i, t): if $\mathsf{type}_i = \perp$ and $t \in \{1, 2, T\}$ then set $\mathsf{cell}_i \leftarrow \mathbb{Z}_p$ and $\mathsf{type}_i \leftarrow t$.

The semantics of create dictates that the actual value of the indeterminate Y_i is uniformly random in $\mathbb{Z}_p$, that is, the adversary cannot create indeterminates for which she does not know the discrete logarithm and that yet are not random. This assumption is needed for the lower bound on the generic adversary's time to be provable in Theorem 3.

In the type III setting, this semantics does *not* allow the adversary to create the same indeterminate Y_i in both groups $\mathbb{G}_1$ and $\mathbb{G}_2$; she can only create a representation of a known to her integer in both groups. We formalize this by making the following Bilinear Diffie-Hellman Knowledge of Exponents (*BDH-KE*) assumption: if the adversary, given random generators $\mathfrak{g}_1 = [1]_1 \in \mathbb{G}_1$ and $\mathfrak{g}_2 = [1]_2 \in \mathbb{G}_2$, can generate elements $[\alpha_1]_1 \in \mathbb{G}_1$ and $[\alpha_2]_2 \in \mathbb{G}_2$, such that $[1]_1 \bullet [\alpha_2]_2 = [\alpha_1]_1 \bullet [1]_2$, then the adversary knows the value $\alpha_1 = \alpha_2$. To simplify the further use of the BDH-KE assumption in security reductions, we give the adversary access to $(\mathbf{R}, \mathsf{z}_\mathbf{R}) \in \mathrm{range}(\mathcal{R}(1^\kappa))$. As before, $\mathsf{z}_\mathbf{R}$ just contains gk, that is, the description of the bilinear group together with $[1]_1$ and $[1]_2$.

Definition 9 (BDH-KE). *We say that* genbp *is BDH-KE secure for* $\mathcal{R}$ *if for any* κ, $(\mathbf{R}, \mathsf{z}_\mathbf{R}) \in \mathrm{range}(\mathcal{R}(1^\kappa))$, *and NUPPT adversary* A *there exists a NUPPT extractor* X_A, *such that*

$$\Pr\left[\begin{array}{l} r \leftarrow_r \mathsf{RND}(\mathsf{A}), ([\alpha_1]_1, [\alpha_2]_2 \parallel a) \leftarrow (\mathsf{A} \parallel \mathsf{X}_\mathsf{A})(\mathbf{R}, \mathsf{z}_\mathbf{R}; r): \\ [\alpha_1]_1 \bullet [1]_2 = [1]_1 \bullet [\alpha_2]_2 \wedge a \neq \alpha_1 \end{array} \right] \approx_\kappa 0.$$

The BDH-KE assumption is a simple special case of the PKE assumption as used in the case of asymmetric pairings say in [17]. In the PKE assumption of [17], adversary is given as an input the tuple $\{([\chi^i]_1, [\chi^i]_2)\}_{i=0}^n$ for some $n \geq 0$, and it is assumed that if an adversary outputs $([\alpha]_1, [\alpha]_2)$ then she knows $(a_0, a_1, \ldots, a_n)$, such that $\alpha = \sum_{i=0}^n a_i \chi^i$. In our case, $n = 0$. BDH-KE can also be seen as an asymmetric-pairing version of the original KE assumption [16].

We think that for the following reasons, the BDH-KE assumption is more natural than the DH-KE assumption by Bellare *et al.* [2] which states that if the adversary can create elements $[\alpha_1]_z$, $[\alpha_2]_z$ and $[\alpha_1\alpha_2]_z$ of the group $\mathbb{G}_z$ then she knows either α_1 or α_2.

First, the BDH-KE assumption is well suited to type-III pairings that are by far the most efficient pairings. The DH-KE assumption is tailored to type-I pairings. In the case of type-III pairings, DH-KE assumption can still be used, but it results in inefficient protocols. For example in [2], in security proofs the authors employs an adversary that extracts either α_1 or α_2. Since it is not known a priori which value will be extracted, several elements in the argument system have to be doubled, for the case α_1 is extracted and for the case α_2 is extracted.

Second, most of the efficient SNARKs are constructed to be sound and zero-knowledge in the (most efficient) type-III setting. While the SNARK of Groth [27] is known to be sound in the case of both symmetric and asymmetric pairings, in the case of symmetric pairings it will be much less efficient. To take the advantage of already known efficient SNARKs, it is only natural to keep the type-III setting. In the current paper, we are able to have the best of both worlds. As in the case of [27], we construct a SNARK that uses type-III pairings. On the one hand, we prove it to be Sub-ZK solely under the BDH-KE assumption. On the other hand, we prove that it is (adaptively) knowledge-sound in the Sub-GBGM, independently of whether one uses type-I, type-II, or type-III pairings. This provides a partial hedge against cryptanalysis: even if one were to later find an efficient isomorphism between $\mathbb{G}_1$ and $\mathbb{G}_2$, this would only break the Sub-ZK of the new SNARK but leave the soundness property intact.

5   New Sub-ZK Secure SNARK

Consider a QAP instance $\mathcal{Q}_p = (\mathbb{Z}_p, m_0, \{u_j, v_j, w_j\}_{j=0}^m)$. The goal of the prover of a QAP argument of knowledge [22,27] is to show that for public $(A_1, \ldots, A_{m_0})$ and $A_0 = 1$, the prover knows $(A_{m_0+1}, \ldots, A_m)$ and a degree $\leq n-2$ polynomial $h(X)$, such that

$$h(X) = \frac{a(X)b(X) - c(X)}{\ell(X)}, \tag{2}$$

where $a(X) = \sum_{j=0}^m A_j u_j(X)$, $b(X) = \sum_{j=0}^m A_j v_j(X)$, $c(X) = \sum_{j=0}^m A_j w_j(X)$.

5.1   Construction

Next, we describe a Sub-ZK SNARK for $\mathcal{R}$ that is closely based on the (non subversion-resistant) SNARK by Groth from EUROCRYPT 2016 [27]. See Figs. 1 and 2. As always, we assume implicitly that each algorithm checks that their inputs belong to correct groups and that $(\mathbf{R}, \mathsf{z_R}) \in \mathrm{range}(\mathcal{R}(1^\kappa))$.

This SNARK uses crucially several random variables, $\chi, \alpha, \beta, \gamma, \delta$. As in [27], α and β (and the inclusion of $\alpha\beta$ in the verification equation) will guarantee that $\mathfrak{a}$, $\mathfrak{b}$, and $\mathfrak{c}$ are computed by using the same coefficients A_i. The role of γ and δ is to make the three products in the verification equation "independent" of each other. Due to the lack of space, we omit a more precise intuition behind Groth's SNARK and refer an interested reader to [27].

As emphasized before, the new Sub-ZK SNARK is closely based on Groth's zk-SNARK. In fact, the differences between the construction of the two SNARKs can be summarized very briefly:

(i) We add to the CRS $2n + 3$ new elements (see the variable $\mathsf{crs_{CV}}$ in Fig. 1) that are needed for CV to work efficiently.
(ii) We divide the CRS generation algorithm into three algorithms, $\mathsf{K_{tc}}$, $\mathsf{K_{ts}}$, and $\mathsf{K_{crs}}$. Groth's CRS generation algorithm returns $\mathsf{K_{crs}}(\mathbf{R}, \mathsf{z_R}, \mathsf{K_{tc}}(\mathbf{R}, \mathsf{z_R}))$ (minus the mentioned $\mathsf{crs_{CV}}$ part) as the CRS and $\mathsf{K_{ts}}(\mathbf{R}, \mathsf{z_R}, \mathsf{K_{tc}}(\mathbf{R}, \mathsf{z_R}))$ as the simulation trapdoor.
(iii) We describe an efficient CRS verification algorithm CV (see Fig. 1).

It is straightforward to see that Groth's original zk-SNARK does not achieve Sub-ZK. Really, since neither $[\ell_i(\chi)]_1$ nor $[\chi^i]_1$ are given to the prover, he cannot check the correctness of $[u_j(\chi)]_1$ and $[v_j(\chi)]_1$. This means that a subverter can change those values to some bogus values and due to that, the proof computed by an honest prover and a simulated proof (that relies on the knowledge of trapdoor elements α and β and does not use the CRS elements $[u_j(\chi)]_1$ and $[v_j(\chi)]_1$; see Fig. 3) will have different distributions.

We prove the completeness of the new SNARK in the rest of this section. We postpone the full proof of knowledge-soundness to Sect. 6 and of zero knowledge to Sect. 7. We analyze the efficiency of this SNARK in Sect. 8.

$K_{tc}(\mathbf{R}, z_{\mathbf{R}})$: Generate $tc = (\chi, \alpha, \beta, \gamma, \delta) \leftarrow_r \mathbb{Z}_p^3 \times (\mathbb{Z}_p^*)^2$.

$K_{ts}(\mathbf{R}, z_{\mathbf{R}}, tc)$: Set $ts \leftarrow (\chi, \alpha, \beta, \delta)$.

$K_{crs}(\mathbf{R}, z_{\mathbf{R}}, tc)$: Compute $(\ell_i(\chi))_{i=1}^n$ by using Alg. 1. Set $u_j(\chi) \leftarrow \sum_{i=1}^n U_{ij}\ell_i(\chi)$, $v_j(\chi) \leftarrow \sum_{i=1}^n V_{ij}\ell_i(\chi)$, $w_j(\chi) \leftarrow \sum_{i=1}^n W_{ij}\ell_i(\chi)$ for all $j \in \{0, \dots, m\}$. Let

$$crs_P \leftarrow \left(\begin{bmatrix} \alpha, \beta, \delta, \left(\frac{u_j(\chi)\beta + v_j(\chi)\alpha + w_j(\chi)}{\delta} \right)^m_{j=m_0+1} \end{bmatrix}_1, \\ \left[(\chi^i\ell(\chi)/\delta)_{i=0}^{n-2}, (u_j(\chi), v_j(\chi))_{j=0}^m \right]_1, \left[\beta, \delta, (v_j(\chi))_{j=0}^m \right]_2 \right),$$

$$crs_V \leftarrow \left(\left[\left(\frac{u_j(\chi)\beta + v_j(\chi)\alpha + w_j(\chi)}{\gamma} \right)^{m_0}_{j=0} \right]_1, [\gamma, \delta]_2, [\alpha\beta]_T \right) ,$$

$$crs_{CV} \leftarrow \left([\gamma, (\chi^i)_{i=1}^{n-1}, (\ell_i(\chi))_{i=1}^n]_1, [\alpha, \chi, \chi^{n-1}]_2 \right) .$$

Return $crs \leftarrow (crs_{CV}, crs_P, crs_V)$.

$K(\mathbf{R}, z_{\mathbf{R}})$: Let $tc \leftarrow K_{tc}(\mathbf{R}, z_{\mathbf{R}})$. Return $(crs \| ts) \leftarrow (K_{crs} \| K_{ts})(\mathbf{R}, z_{\mathbf{R}}, tc)$.

$CV(\mathbf{R}, z_{\mathbf{R}}, crs)$:

1. For $\iota \in \{\gamma, \delta\}$: check that $[\iota]_1 \neq [0]_1$
2. For $\iota \in \{\alpha, \beta, \gamma, \delta\}$: check that $[\iota]_1 \bullet [1]_2 = [1]_1 \bullet [\iota]_2$,
3. For $i = 1$ to $n - 1$: check that $[\chi^i]_1 \bullet [1]_2 = [\chi^{i-1}]_1 \bullet [\chi]_2$,
4. Check that $([\ell_i(\chi)]_1)_{i=1}^n$ is correctly computed by using Alg. 2,
5. For $j = 0$ to m:
   (a) Check that $[u_j(\chi)]_1 = \sum_{i=1}^n U_{ij} [\ell_i(\chi)]_1$,
   (b) Check that $[v_j(\chi)]_1 = \sum_{i=1}^n V_{ij} [\ell_i(\chi)]_1$,
   (c) Set $[w_j(\chi)]_1 \leftarrow \sum_{i=1}^n W_{ij} [\ell_i(\chi)]_1$,
   (d) Check that $[v_j(\chi)]_1 \bullet [1]_2 = [1]_1 \bullet [v_j(\chi)]_2$,
6. For $j = m_0 + 1$ to m: check that $[(u_j(\chi)\beta + v_j(\chi)\alpha + w_j(\chi))/\delta]_1 \bullet [\delta]_2 = [u_j(\chi)]_1 \bullet [\beta]_2 + [v_j(\chi)]_1 \bullet [\alpha]_2 + [w_j(\chi)]_1 \bullet [1]_2$,
7. Check that $[\chi^{n-1}]_1 \bullet [1]_2 = [1]_1 \bullet [\chi^{n-1}]_2$,
8. For $i = 0$ to $n-2$: check that $[\chi^i\ell(\chi)/\delta]_1 \bullet [\delta]_2 = [\chi^{i+1}]_1 \bullet [\chi^{n-1}]_2 - [\chi^i]_1 \bullet [1]_2$,
9. Check that $[\alpha]_1 \bullet [\beta]_2 = [\alpha\beta]_T$.

Fig. 1. The CRS generation and verification of the Sub-ZK SNARK for $\mathcal{R}$

Algorithm 2. Checking that $[(\ell_i(\chi))_{i=1}^n]_1$ is correctly computed

---

Input: $([\chi^{n-1}, (\ell_i(\chi))_{i=1}^n]_1, [1, \chi]_2, [1]_T)$

// $i = 1$

1 $[\zeta]_T \leftarrow ([\chi^{n-1}]_1 \bullet [\chi]_2 - [1]_T)/n; \ [\omega']_2 \leftarrow [1]_2;$

2 Check that $[\ell_1(\chi)]_1 \bullet ([\chi]_2 - [\omega']_2) = [\zeta]_T;$

3 **for** $i = 2$ **to** n **do**

4 $\quad [\zeta]_T \leftarrow \omega [\zeta]_T; \ [\omega']_2 \leftarrow \omega [\omega']_2;$

5 $\quad$ Check that $[\ell_j(\chi)]_1 \bullet ([\chi]_2 - [\omega']_2) = [\zeta]_T;$

5.2 Subversion Completeness

In the proof of subversion-completeness, we need the following result that hopefully is of independent interest.

$P(\mathbf{R}, z_{\mathbf{R}}, crs_P, x = (A_1, \ldots, A_{m_0}), w = (A_{m_0+1}, \ldots, A_m))$:

 /* After executing CV & assuming $A_0 = 1$, the prover does: */

 1. Let $a^\dagger(X) \leftarrow \sum_{j=0}^m A_j u_j(X)$, $b^\dagger(X) \leftarrow \sum_{j=0}^m A_j v_j(X)$,

 2. Let $c^\dagger(X) \leftarrow \sum_{j=0}^m A_j w_j(X)$,

 3. Set $h(X) = \sum_{i=0}^{n-2} h_i X^i \leftarrow (a^\dagger(X) b^\dagger(X) - c^\dagger(X))/\ell(X)$,

 4. Set $[h(\chi)\ell(\chi)/\delta]_1 \leftarrow \sum_{i=0}^{n-2} h_i [\chi^i \ell(\chi)/\delta]_1$,

 5. Set $r_a \leftarrow_r \mathbb{Z}_p$; Set $\mathfrak{a} \leftarrow \sum_{j=0}^m A_j [u_j(\chi)]_1 + [\alpha]_1 + r_a [\delta]_1$,

 6. Set $r_b \leftarrow_r \mathbb{Z}_p$; Set $\mathfrak{b} \leftarrow \sum_{j=0}^m A_j [v_j(\chi)]_2 + [\beta]_2 + r_b [\delta]_2$,

 7. Set $\mathfrak{c} \leftarrow r_b \mathfrak{a} + r_a \left(\sum_{j=0}^m A_j [v_j(\chi)]_1 + [\beta]_1 \right) +$
    $\sum_{j=m_0+1}^m A_j [(u_j(\chi)\beta + v_j(\chi)\alpha + w_j(\chi))/\delta]_1 + [h(\chi)\ell(\chi)/\delta]_1$,

 8. Return $\pi \leftarrow (\mathfrak{a}, \mathfrak{b}, \mathfrak{c})$.

$V(\mathbf{R}, z_{\mathbf{R}}, crs_V, x = (A_1, \ldots, A_{m_0}), \pi = (\mathfrak{a}, \mathfrak{b}, \mathfrak{c}))$: assuming $A_0 = 1$, check that

$$\mathfrak{a} \bullet \mathfrak{b} = \mathfrak{c} \bullet [\delta]_2 + \left(\sum_{j=0}^{m_0} A_j \left[\frac{u_j(\chi)\beta + v_j(\chi)\alpha + w_j(\chi)}{\gamma} \right]_1 \right) \bullet [\gamma]_2 + [\alpha\beta]_T \ .$$

Fig. 2. The prover and the verifier of the Sub-ZK SNARK for $\mathcal{R}$ (unchanged from Groth's zk-SNARK)

Lemma 1. *Given* $([\chi^{n-1}, (\ell_i(\chi))_{i=1}^n]_1, [1, \chi]_2, [1]_T)$ *as an input, Algorithm 2 checks that* $[\ell_i(\chi)]_1$ *has been correctly computed for all* $i \in [1 .. n]$. *It can be implemented by using* $n + 1$ *pairings,* $n - 1$ *exponentiations in* $\mathbb{G}_2$, *and* $n - 1$ *exponentiations in* $\mathbb{G}_T$.

Proof. Recalling that $\ell_i(X) = (X^n - 1)\omega^{i-1}/(n(X - \omega^{i-1}))$ are as given by Eq. (1), the proof is straightforward. Really, by induction on i, at any concrete value of i, $\zeta = (\chi^n - 1)\omega^{i-1}/n$ and $\omega' = \omega^{i-1}$. Thus $[\ell_i(\chi)]_1 \bullet ([\chi]_2 - [\omega']_2) = [(\chi - \omega^{i-1}) \cdot \ell_i(\chi)]_T = [\zeta]_T$ iff $[\ell_i(\chi)]_1$ was correctly computed if $\chi \neq \omega^{i-1}$. If $\chi = \omega^{i-1}$, then $(\chi - \omega^{i-1}) \cdot \ell_i(\chi) = (\chi^n - 1)\omega^{i-1}/n = 0$, since ω is the n-th primitive root of unity. $\square$

For $z \in \{1, 2\}$, let crs_z be the subset of the (honest or subverted) crs that consists of all elements of $\mathbb{G}_z$.

Theorem 2. *The new SNARK of Sect. 5.1 is perfectly subversion-complete.*

Proof. CV ACCEPTS AN HONESTLY GENERATED CRS: this can be established by a simple but tedious calculation. Basically, CV accepts due to the properties of the bilinear map, and due to the definitions of $\ell(X)$ and $\ell_i(X)$. We only prove the correctness of two of the least obvious steps.

Step 6 of CV holds since $[(u_j(\chi)\beta + v_j(\chi)\alpha + w_j(\chi))/\delta]_1 \bullet [\delta]_2 = [u_j(\chi)]_1 \bullet [\beta]_2 + [v_j(\chi)]_1 \bullet [\alpha]_2 + [w_j(\chi)]_1 \bullet [1]_2$ iff $[(u_j(\chi)\beta + v_j(\chi)\alpha + w_j(\chi))/\delta \cdot \delta]_T = [u_j(\chi)\beta + v_j(\chi)\alpha + w_j(\chi)]_T$ which is a tautology.

Similarly, Lemma 1 has established that Algorithm 2 works correctly. Thus, Step 8 of CV holds since $[\chi^i \ell(\chi)/\delta]_1 \bullet [\delta]_2 = [\chi^{i+1}]_1 \bullet [\chi^{n-1}]_2 - [\chi^i]_1 \bullet [1]_2$ iff $[\chi^i \ell(\chi)/\delta \cdot \delta]_T = [\chi^{i+1} \cdot \chi^{n-1} - \chi^i]_T$ iff $[\chi^i \ell(\chi)]_T = [\chi^i(\chi^n - 1)]_T$, which is a tautology since $\ell(\chi) = \chi^n - 1$.

V ACCEPTS AN HONESTLY GENERATED ARGUMENT: In the honest case, see Fig. 2, $\mathfrak{a} = [A(\boldsymbol{x})]_1$, $\mathfrak{b} = [B(\boldsymbol{x})]_2$, and $\mathfrak{c} = [C(\boldsymbol{x})]_1$, where $A(\boldsymbol{x}) = \sum_{j=0}^{m} A_j u_j(\chi) + \alpha + r_a \delta$, $B(\boldsymbol{x}) = \sum_{j=0}^{m} A_j v_j(\chi) + \beta + r_b \delta$, and $C(\boldsymbol{x}) = r_b A(\boldsymbol{x}) + r_a(B(\boldsymbol{x}) - r_b \delta) + \sum_{j=m_0+1}^{m}(u_j(\chi)\beta + v_j(\chi)\alpha + w_j(\chi))/\delta + h(\chi)\ell(\chi)/\delta$. Clearly,

$$A(\boldsymbol{x})B(\boldsymbol{x}) - r_b A(\boldsymbol{x})\delta - r_a(B(\boldsymbol{x}) - r_b \delta)\delta$$
$$= (\textstyle\sum_{j=0}^{m} A_j u_j(\chi) + \alpha) \cdot (\sum_{j=0}^{m} A_j v_j(\chi) + \beta)$$
$$= (\textstyle\sum_{j=0}^{m} A_j u_j(\chi)) \cdot (\sum_{j=0}^{m} A_j v_j(\chi)) + \alpha \sum_{j=0}^{m} A_j v_j(\chi) +$$
$$\beta \textstyle\sum_{j=0}^{m} A_j u_j(\chi) + \alpha\beta.$$

Let $V(\boldsymbol{x}) = A(\boldsymbol{x})B(\boldsymbol{x}) - C(\boldsymbol{x})\delta - \sum_{j=0}^{m_0} A_j(u_j(\chi)\beta + v_j(\chi)\alpha + w_j(\chi)) - \alpha\beta$. Hence, $V(\boldsymbol{x}) = a^\dagger(\chi)b^\dagger(\chi) - c^\dagger(\chi) - h(\chi)\ell(\chi)$. Due to the definition of $h(\boldsymbol{X})$, $V(\boldsymbol{x}) = 0$. Thus, $[V(\boldsymbol{x})]_T = [0]_T$, which is what the verification equation ascertains. □

6   Proof of Knowledge-Soundness

Since we are proving knowledge-soundness (and not subversion knowledge-soundness), the following security proof is similar to the corresponding proof in [27]. The main difference is in the need to take into account added elements to the CRS and to incorporate new indeterminates Y_i created by the adversary. This means that the proof will be slightly (but not much) more complicated than the proof in [27].

Remark 2. It is easy to see that the knowledge-soundness of the new SNARK (and hence also Groth's SNARK) can be trivially broken by adding a single well-chosen element to the CRS. Namely, after adding $[\alpha\beta/\delta]_1$ to the CRS in Fig. 1, a malicious prover P* can set $\mathfrak{a} \leftarrow (\sum_{j=0}^{m_0} A_j [(u_j(\chi)\beta + v_j(\chi)\alpha + w_j(\chi))/\gamma]_1)$, $\mathfrak{b} \leftarrow [\gamma]_2$, and $\mathfrak{c} \leftarrow [-\alpha\beta/\delta]_1$. Clearly, V accepts this argument. □

Next, we will prove adaptive knowledge-soundness even in the case when one uses symmetric pairings, as it was done in [27]. The use of symmetric pairings means that the generic adversary gets additional power compared to Sub-GBGM: in particular, we do not assume that BDH-KE holds, and moreover, in the asymmetric variant of the following theorem, the generic adversary will be allowed to create new indeterminates Y_i in both $\mathbb{G}_1$ and $\mathbb{G}_2$, without knowing their discrete logarithms. Since this only increases the power of the generic adversary, it provides some hedge against future cryptanalytic attacks that say make it possible to compute an efficient isomorphism between $\mathbb{G}_1$ and $\mathbb{G}_2$, or at least show that BDH-KE does not hold in the concrete groups.

Theorem 3 (Knowledge-soundness). *Consider the new argument system of Sect. 5.1. It is adaptively knowledge-sound in the Sub-GBGM even in the case of symmetric pairings. More precisely, any generic adversary attacking the knowledge-soundness of the new argument system in the symmetric setting requires $\Omega(\sqrt{p/n})$ computation.*

Proof. Assume the symmetric setting, $\mathbb{G}_1 = \mathbb{G}_2$. Let $\boldsymbol{X}$ be the vector of indeterminates created by K_{tc} and $\boldsymbol{Y} = (Y_1, \ldots, Y_q)^\top$ (for a non-negative integer q) be the vector of indeterminates created by the generic adversary.

The three elements output by a generic adversary are equal to $\mathfrak{a} = [A(\boldsymbol{x}, \boldsymbol{y})]_1$, $\mathfrak{b} = [B(\boldsymbol{x}, \boldsymbol{y})]_1$, and $\mathfrak{c} = [C(\boldsymbol{x}, \boldsymbol{y})]_1$, where, for $T \in \{A, B, C\}$,

$$T(\boldsymbol{X}, \boldsymbol{Y}) = T_\alpha X_\alpha + T_\beta X_\beta + T_\gamma X_\gamma + T_\delta X_\delta + T_c(X) +$$
$$\sum_{j=0}^{m_0} T_j \cdot \frac{u_j(X)X_\beta + v_j(X)X_\alpha + w_j(X)}{X_\gamma} +$$
$$\sum_{j=m_0+1}^{m} T_j \cdot \frac{u_j(X)X_\beta + v_j(X)X_\alpha + w_j(X)}{X_\delta} + \frac{T_h(X)\ell(X)}{X_\delta} + \sum_{i=1}^{q} T_{yi} Y_i.$$

Here, $T_c(X)$ is a degree $\leq n - 1$ and $T_h(X)$ is a degree $\leq n - 2$ polynomial. Thus, $T(\boldsymbol{X}, \boldsymbol{Y}) \cdot X_\gamma X_\delta$ is a degree $(n - 2) + n - 1 + 2 = 2n - 1$ polynomial.

Since the only difference compared to the knowledge-soundness proof in [27] is in the addition of the terms $\sum_{i=1}^{q} T_{yi} Y_i$ to those three polynomials, it suffices that we show that $T_{yi} = 0$ for $T \in \{A, B, C\}$ and $i \in [1..q]$. After that, the knowledge-soundness of the new SNARK follows from the knowledge-soundness of Groth's SNARK.

Motivated by the verification equation in Fig. 2, define

$$V(\boldsymbol{X}, \boldsymbol{Y}) := A(\boldsymbol{X}, \boldsymbol{Y})B(\boldsymbol{X}, \boldsymbol{Y}) - C(\boldsymbol{X}, \boldsymbol{Y})X_\delta -$$
$$\sum_{j=0}^{m_0} A_j \left(u_j(X)X_\beta + v_j(X)X_\alpha + w_j(X) \right) - X_\alpha X_\beta,$$

where the Laurent polynomials $A(\boldsymbol{X}, \boldsymbol{Y})$, $B(\boldsymbol{X}, \boldsymbol{Y})$, and $C(\boldsymbol{X}, \boldsymbol{Y})$ are as given before. The verification equation states that $[V(\boldsymbol{x}, \boldsymbol{y})]_T = [0]_T$ and hence in the case of a generic adversary, $V(\boldsymbol{X}, \boldsymbol{Y}) \cdot X_\gamma^2 X_\delta^2 = 0$ as a polynomial or equivalently, each of the coefficients of $V(\boldsymbol{X}, \boldsymbol{Y})$ is 0.

First, the coefficient of X_α^2 in $V(\boldsymbol{X}, \boldsymbol{Y})$ is $A_\alpha B_\alpha$. Thus, from $V(\boldsymbol{X}, \boldsymbol{Y}) = 0$ it follows that $A_\alpha B_\alpha = 0$. Since $A(\boldsymbol{X}, \boldsymbol{Y})$ and $B(\boldsymbol{X}, \boldsymbol{Y})$ play dual roles in the symmetric case, we can assume, w.l.o.g., that $B_\alpha = 0$ (the same assumption was made in [27]).

Because $B_\alpha = 0$, the following claims hold:

- from the coefficient of $X_\alpha X_\beta$, $A_\alpha B_\beta + A_\beta B_\alpha - 1 = 0$. Thus, $A_\alpha B_\beta = 1$ (and in particularly, neither of them is equal to 0).
- from the coefficient of $X_\alpha Y_j$, $j \in [1..q]$, $A_\alpha B_{yj} + A_{yj} B_\alpha = 0$. Thus, $B_{yj} = 0$,
- from the coefficient of $X_\beta Y_j$, $j \in [1..q]$, $A_\beta B_{yj} + A_{yj} B_\beta = 0$. Thus, $A_{yj} = 0$,
- from the coefficient of $X_\delta Y_j$, $j \in [1..q]$, $C_{yj} = B_{yj} r_a + A_{yj} r_b$. Thus, $C_{yj} = 0$,

Since the coefficients A_{yj}, B_{yj}, and C_{yj} were the only coefficients that are new compared to the knowledge-soundness proof of [27], the rest of the current proof follows from the knowledge-soundness proof of [27].

Let us now compute a lower bound to the efficiency of a generic adversary (this was not done in [27], our bound clearly also holds in the case of Groth's SNARK). Assume that after some τ steps, the adversary has made a successful equality query $(=, i_1, i_2)$, i.e., $\mathrm{cell}_{i_1} = \mathrm{cell}_{i_2}$ for $i_1 \neq i_2$. Hence, she has found a collision $D_1(\boldsymbol{x}, \boldsymbol{y}) = D_2(\boldsymbol{x}, \boldsymbol{y})$ such that $D_1(\boldsymbol{X}, \boldsymbol{Y}) \neq D_2(\boldsymbol{X}, \boldsymbol{Y})$.

Redefine $D_j(\boldsymbol{X}, \boldsymbol{Y}) := D_j(\boldsymbol{X}, \boldsymbol{Y}) \cdot X_\gamma X_\delta$ (if $\mathsf{type}_{i_1} \in \{1, 2\}$) and $D_j(\boldsymbol{X}, \boldsymbol{Y}) := D_j(\boldsymbol{X}, \boldsymbol{Y}) \cdot X_\gamma^2 X_\delta^2$ (if $\mathsf{type}_{i_1} = T$) for $j \in \{1, 2\}$, this guarantees that $D_j(\boldsymbol{X}, \boldsymbol{Y})$ is a polynomial. Thus,

$$D_1(\boldsymbol{x}, \boldsymbol{y}) - D_2(\boldsymbol{x}, \boldsymbol{y}) \equiv 0 \pmod{p}. \tag{3}$$

Note that

- If $\mathsf{type}_{i_1} = 1$, then $\deg D_j(\boldsymbol{X}, \boldsymbol{Y}) \leq 2n - 1 =: d_1$,
- If $\mathsf{type}_{i_1} = 2$, then $\deg D_j(\boldsymbol{X}, \boldsymbol{Y}) \leq 2n - 1 =: d_2$, and thus
- If $\mathsf{type}_{i_1} = T$, then $\deg D_j(\boldsymbol{X}, \boldsymbol{Y}) \leq 2 \cdot (2n - 1) = 4n - 2 =: d_T$.

Clearly, $\boldsymbol{x} = (\chi, \alpha, \beta, \gamma, \delta)$ is chosen uniformly random from $\mathbb{Z}_p^3 \times (\mathbb{Z}_p^*)^2$. Due to the assumption that the canonical values of Y_i are uniformly random in $\mathbb{Z}_p$, $\boldsymbol{y} = (v_1, v_2, v_3)$ is a uniformly random value in $\mathbb{Z}_p^3$. Hence, due to the Schwartz-Zippel lemma and since $D_1(\boldsymbol{X}, \boldsymbol{Y}) \neq D_2(\boldsymbol{X}, \boldsymbol{Y})$ as a polynomial, Eq. (3) holds with probability at most $\deg D_j(\boldsymbol{X}, \boldsymbol{Y})/(p-1) \leq d_{\mathsf{type}_{i_1}}/(p-1)$. Clearly, an adversary working in time τ can generate up to τ new group elements. Then the probability that there exists a collision between any two of those group elements is upper bounded by $\binom{\tau}{2} \cdot \deg D_j(\boldsymbol{X}, \boldsymbol{Y})/(p-1) \leq \binom{\tau}{2} \cdot d_{\mathsf{type}_{i_1}}/(p-1) \leq \tau^2/2 \cdot d_{\mathsf{type}_{i_1}}/(p-1)$. Thus, a successful adversary on average requires time at least τ, where $\tau^2 \geq 2(p-1)/d_{\mathsf{type}_{i_1}} \geq 2(p-1)/d_T = 2(p-1)/(4n-2)$, to produce a collision. Simplifying, we get $\tau = \Omega(\sqrt{p/n})$.   □

7   Proof of Perfect Composable Subversion ZK

Before proving Theorem 4, we first prove the following helpful lemma that has a simple but tedious proof.

Lemma 2. *Let* $(\mathbf{R}, \mathsf{z_R}) \in \mathrm{range}(\mathcal{R}(1^\kappa))$, *and let* crs *be any CRS such that* $\mathsf{CV}(\mathbf{R}, \mathsf{z_R}, \mathsf{crs}) = 1$. *Then, with probability* 1, $\mathsf{crs} = \mathsf{K_{crs}}(\mathbf{R}, \mathsf{z_R}, \mathsf{tc})$ *for a tc bijectively corresponding to* $[\chi, \alpha, \beta, \gamma, \delta]_1$.

Proof. In what follows, we will consider each line in the construction of CV in Fig. 1 separately, and write down the corollary from that line. We note that the CRS verification equations in Fig. 1 were written down as if the CRS were already correctly formed; e.g., there we have a check that $[\beta]_1 \bullet [1]_2 = [1]_1 \bullet [\beta]_2$ which may fail (and then obviously there exists no such β). However, before these equations are checked, it is of course not known that β in the left hand side (LHS) and in the right hand side (RHS) are equal. Therefore, in this proof only, in each step below, we use D_1 as the temporary name of the yet-unestablished LHS variable and D_2 as the temporary name of the yet-unestablished RHS variable.

We assume that ι in $[\iota]_1$ is already established for $\iota \in \{\chi, \alpha, \beta, \gamma, \delta\}$. (In particular, χ is established on Step 3 when $i = 1$.) We can do it since $[\iota]_1$ information-theoretically fixes ι.

Algorithm 3: $\Sigma^\iota(\mathbf{R}, \mathsf{z_R}; r)$	**Algorithm 4:** $\mathsf{X}_\Sigma(\mathbf{R}, \mathsf{z_R}; r)$
1 $(\mathsf{crs}, \mathsf{z}_\Sigma) \leftarrow \Sigma(\mathbf{R}, \mathsf{z_R}; r);$	**1 for** $\lambda \in \{\chi, \alpha, \beta, \gamma, \delta\}$ **do**
2 return $([\iota]_1, [\iota]_2);$	**2** $\quad \lfloor \; \lambda \leftarrow \mathsf{X}_\Sigma^\lambda(\mathbf{R}, \mathsf{z_R}; r);$
	3 $\mathsf{tc} \leftarrow (\chi, \alpha, \beta, \gamma, \delta);$
	4 return $\mathsf{tc};$

Algorithm 5: $\mathsf{S}(\mathbf{R}, \mathsf{z_R}, \mathsf{crs}, \mathsf{ts}, x)$
1 $\sigma, \tau \leftarrow_r \mathbb{Z}_p;$
2 $(\mathfrak{a}, \mathfrak{b}) \leftarrow ([\sigma]_1, [\tau]_2);$
3 $\mathfrak{c} \leftarrow [(\sigma\tau - \alpha\beta - \sum_{j=m_0+1}^m (u_j(\chi)\beta + v_j(\chi)\alpha + w_j(\chi)))/\delta]_1;$
4 return $\pi \leftarrow (\mathfrak{a}, \mathfrak{b}, \mathfrak{c});$

Fig. 3. Algorithms used in extraction and simulation, where $\iota \in \{\chi, \alpha, \beta, \gamma, \delta\}$

1. For $\iota \in \{\gamma, \delta\}$: after that we know that $\iota \neq 0$, and hence there is a bijection between a valid $\mathsf{tc} \in \mathbb{Z}_p^3 \times (\mathbb{Z}_p^*)^2$ and the value hidden in $[\chi, \alpha, \beta, \gamma, \delta]_1$.
2. For $\iota \in \{\alpha, \beta, \gamma, \delta\}$: from $[\iota]_1 \bullet [1]_2 = [1]_1 \bullet [D_2]_2$ we get $[\iota]_T = [D_2]_T$. This implies $[\iota - D_2]_T = [0]_T$. Since $[1]_T$ is not the unity element, $[D_2]_2 = [\iota]_2$.
3. For $i = 1$ to $n - 1$:k we can assume by induction that we have already established $[\chi^{i-1}]_1$. Since $[D_1]_T = [\chi^{i-1}]_1 \bullet [\chi]_2$, we get $[D_1]_T = [\chi^i]_T$.
4. For $i = 1$ to n: $\omega' = \omega^{i-1}$ and $\zeta = (\chi^n - 1)\omega^{i-1}/n$ in the algorithm Algorithm 2 are computed by the CRS verifier. For $\chi \neq \omega^{i-1}$, the equation $[D_1]_1 \bullet ([\chi]_2 - [\omega']_2) = [\zeta]_T$ implies $[D_1]_1 = [\zeta/(\chi - \omega')]_1 = [(\chi^n - 1)\omega^{i-1}/(n(\chi - \omega^{i-1}))]_1 = [\ell_i(\chi)]_1$. For $\chi = \omega^{i-1}$, we get $D_1 \cdot (\chi - \omega^{i-1}) = \zeta$, since ω^{i-1} is a root of unity and $\zeta = (\chi^n - 1)\omega^{i-1}/n$, thus both sides of the equation are 0.
5. For $j = 0$ to m:
   (a) $[D_1]_1 = \sum_{i=1}^n U_{ij} [\ell_i(\chi)]_1$ implies $[D_1]_1 = [\sum_{i=1}^n U_{ij}\ell_i(\chi)]_1 = [u_j(\chi)]_1$.
   (b) $[D_1]_1 = \sum_{i=1}^n V_{ij} [\ell_i(\chi)]_1$ implies $[D_1]_1 = [\sum_{i=1}^n V_{ij}\ell_i(\chi)]_1 = [v_j(\chi)]_1$.
   (c) Clearly, $[w_k(\chi)]_1$ is computed correctly.
   (d) $[v_j(\chi)]_1 \bullet [1]_2 = [1]_1 \bullet [D_2]_2$ implies $[D_2]_2 = [v_j(\chi)]_2$.
6. For $j = m_0+1$ to m: $[D_1]_1 \bullet [\delta]_2 = [u_j(\chi)]_1 \bullet [\beta]_2 + [v_j(\chi)]_1 \bullet [\alpha]_2 + [w_j(\chi)]_1 \bullet [1]_2$ implies $[D_1]_1 = [(u_j(\chi)\beta + v_j(\chi)\alpha + w_j(\chi))/\delta]_1$.
7. $[\chi^{n-1}]_1 \bullet [1]_2 = [1]_1 \bullet [D_2]_2$ implies $[D_2]_2 = [\chi^{n-1}]_2$,
8. For $i = 0$ to $n - 2$: $[D_1]_1 \bullet [\delta]_2 = [\chi^{i+1}]_1 \bullet [\chi^{n-1}]_2 - [\chi^i]_1 \bullet [1]_2$ implies $[D_1]_1 = [\chi^i \ell(\chi)/\delta]_1$.
9. $[\alpha]_1 \bullet [\beta]_2 = [D_2]_T$ implies $[D_2]_T = [\alpha\beta]_T$.

By direct observation, it is clear that we have now established all elements of the crs_P in Fig. 1. Hence, $\mathsf{crs} = \mathsf{K}_{\mathsf{crs}}(\mathbf{R}, \mathsf{z_R}, \mathsf{tc})$. $\qquad\square$

Theorem 4. *The SNARK from Sect. 5.1 has perfect CRS trapdoor extractability under the BDH-KE assumption.*

Proof. Let Σ be a subverter, and let $r \leftarrow_r \mathsf{RND}(\Sigma)$. Let $\iota \in \{\chi, \alpha, \beta, \gamma, \delta\}$. Let $\Sigma^\iota(\mathbf{R}, \mathbf{z_R}; r)$ be as in Algorithm 3 in Fig. 3. Since $\mathsf{CV}(\mathbf{R}, \mathbf{z_R}, \mathsf{crs})$ accepts, crs must contain $([\iota]_1, [\iota]_2)$. Hence, by the BDH-KE assumption, there exists a NUPPT extractor X_Σ^ι, such that if $\mathsf{CV}(\mathbf{R}, \mathbf{z_R}, \mathsf{crs}) = 1$ then $\iota \leftarrow \mathsf{X}_\Sigma^\iota(\mathbf{R}, \mathbf{z_R}; r)$.

Finally, we construct the (NUPPT) extractor $\mathsf{X}_\Sigma(\mathbf{R}, \mathbf{z_R}; r)$ in Algorithm 4 in Fig. 3. By the BDH-KE assumption, if $\mathsf{CV}(\mathbf{R}, \mathbf{z_R}, \mathsf{crs}) = 1$ then $\mathsf{tc} \leftarrow \mathsf{X}_\Sigma(\mathbf{R}, \mathbf{z_R}; r)$ satisfies $\mathsf{tc} \in \mathrm{range}(\mathsf{K_{tc}}(\mathbf{R}, \mathbf{z_R}))$.

To prove CRS trapdoor extractability, assume that crs is returned by Σ and tc is returned by X_Σ. In addition, assume that $\mathsf{CV}(\mathbf{R}, \mathbf{z_R}, \mathsf{crs}) = 1$. By $\mathsf{CV}(\mathbf{R}, \mathbf{z_R}, \mathsf{crs}) = 1$ and Lemma 2, $\mathsf{K_{crs}}(\mathbf{R}, \mathbf{z_R}, \mathsf{tc}) = \mathsf{crs}$. The claim follows.   □

In Theorem 5, we also need the next simple lemma.

Lemma 3. *Let* $(\mathbf{R}, \mathbf{z_R}) \in \mathrm{range}(\mathcal{R}(1^\kappa))$, *and let* crs *be any CRS such that* $\mathsf{CV}(\mathbf{R}, \mathbf{z_R}, \mathsf{crs}) = 1$. *Consider any values of* $\mathfrak{a}$, $\mathfrak{b}$, *and* $(A_j)_{j=0}^{m_0}$. *Then there exists at most one value* $\mathfrak{c}$, *such that* $\mathsf{V}(\mathbf{R}, \mathbf{z_R}, \mathsf{crs_V}, \mathsf{x}, (\mathfrak{a}, \mathfrak{b}, \mathfrak{c})) = 1$.

Proof. Assume that $\mathfrak{c}_0$ and $\mathfrak{c}_1$ are both accepted by the verifier. That means (see Fig. 2) that $\mathfrak{c}_k \bullet [\delta]_2 = [s]_T$ for $k \in \{0, 1\}$, where s is some k-independent value. By bilinearity, $(\mathfrak{c}_1 - \mathfrak{c}_0) \bullet [\delta]_2 = [0]_T$. Since $\delta \neq 0$ (this is guaranteed by CV accepting crs) and the pairing is non-degenerate, we have $\mathfrak{c}_0 = \mathfrak{c}_1$.   □

Theorem 5. *The SNARK from Sect. 5.1 is perfectly composable Sub-ZK under the BDH-KE assumption.*

Proof. We use the same simulator $\mathsf{S}(\mathbf{R}, \mathbf{z_R}, \mathsf{crs}, \mathsf{ts}, \mathsf{x})$ as defined in [27], see Algorithm 5 in Fig. 3. Fix a subverter Σ, and let X_Σ be as defined in Theorem 4, see Algorithm 4 in Fig. 3. Fix κ, $(\mathbf{R}, \mathbf{z_R}) \leftarrow \mathcal{R}(1^\kappa)$, $(\mathsf{x}, \mathsf{w}) \in \mathbf{R}$, and an adversary A. As in Definition 7, assume $r \leftarrow_r \mathsf{RND}(\Sigma)$, $(\mathsf{crs}, \mathbf{z}_\Sigma) \leftarrow \Sigma(\mathbf{R}, \mathbf{z_R}; r)$ such that $\mathsf{CV}(\mathbf{R}, \mathbf{z_R}, \mathsf{crs}) = 1$, $\mathsf{tc} \leftarrow \mathsf{X}_\Sigma(\mathbf{R}, \mathbf{z_R}; r)$, $\mathsf{ts} \leftarrow \mathsf{K_{ts}}(\mathbf{R}, \mathbf{z_R}, \mathsf{tc})$. Assume that $\pi_0 \leftarrow \mathsf{P}(\mathbf{R}, \mathbf{z_R}, \mathsf{crs_P}, \mathsf{x}, \mathsf{w})$ ($b = 0$) and $\pi_1 \leftarrow \mathsf{S}(\mathbf{R}, \mathbf{z_R}, \mathsf{crs}, \mathsf{ts}, \mathsf{x})$ ($b = 1$). It is sufficient to show that π_0 and π_1 have the same distribution.

Case $b = 0$. The honest prover creates $\mathfrak{a} \leftarrow \ldots + r_a [\delta]_1$ and $\mathfrak{b} \leftarrow \ldots + r_b [\delta]_2$ for uniformly random r_a and r_b. Since $\delta \neq 0$ (this is guaranteed by CV accepting crs) and the pairing is non-degenerate, $\mathfrak{a}$ and $\mathfrak{b}$ are uniformly random. We know by the perfect CRS trapdoor extractability (Theorem 4) that $\mathsf{K_{crs}}(\mathbf{R}, \mathbf{z_R}, \mathsf{tc}) = \mathsf{crs}$. Thus, $(\mathsf{crs}, \mathsf{ts})$ are created as in the first step of the definition of perfect subversion-completeness (Definition 2). Since the new SNARK satisfies perfect subversion-completeness (Theorem 2), $\mathsf{V}(\mathbf{R}, \mathbf{z_R}, \mathsf{crs_V}, \mathsf{x}, \pi_0) = 1$. By Lemma 3, the verification equation, crs, $\mathfrak{a}$, and $\mathfrak{b}$ uniquely determine the acceptable $\mathfrak{c}$.

Case $b = 1$. Here, by the definition of the simulator, $\mathfrak{a}$ and $\mathfrak{b}$ are uniformly random. Moreover, π_1 is explicitly created so that $\mathsf{V}(\mathbf{R}, \mathbf{z_R}, \mathsf{crs_V}, \mathsf{x}, \pi_1) = 1$. Since CV accepts then by Lemma 3, the verification equation, crs, $\mathfrak{a}$, and $\mathfrak{b}$ uniquely determine the acceptable $\mathfrak{c}$.

Since CV accepts crs, in both cases $\mathfrak{a}$ and $\mathfrak{b}$ are uniformly random and $\mathfrak{c}$ is uniquely determined by them, crs, and the verification equation, the real and the simulated arguments have identical distributions.   □

The following result follows directly from Theorems 1 and 5.

Theorem 6. *The SNARK from Sect. 5.1 is perfectly unbounded Sub-ZK under the BDH-KE assumption.*

8 Efficiency

CRS Length. The CRS contains gk (that includes $([1]_1, [1]_2)$) and a number of additional elements. Not counting gk, the number of CRS elements in different groups is given by the following table. Hence, the total size of the CRS is $4m + 3n + 13$ group elements.

	$\mathbb{G}_1$	$\mathbb{G}_2$	$\mathbb{G}_T$	Total
crs$_P$	$3m + n - m_0 + 4$	$m + 3$	0	$4m + n - m_0 + 7$
crs$_V$	$m_0 + 1$	2	1	$m_0 + 4$
crs$_{CV}$	$2n$	3	0	$2n + 3$
Total	$3m + 3n + 5$	$m + 7$	1	$4m + 3n + 13$

One element (namely, $[\delta]_2$) belongs both to crs$_P$ and crs$_V$ and thus the numbers in the "total" row are not equal to the sum of the numbers in previous rows.

In Groth's zk-SNARK [27] the CRS consists of $m + 2n$ elements of $\mathbb{G}_1$ and n elements of $\mathbb{G}_2$. On top of it, we added $2n + 3$ group elements to make the CRS verification possible and also some elements to speed up the prover's computation and the verifier's computation; the latter elements can alternatively be computed from the rest of the CRS.

CRS Generation: Computational Complexity. Assume that gk has already been computed. One can compute crs by first computing all CRS elements within brackets, and then compute their bracketed versions. One can evaluate $u_j(\chi)$, $v_j(\chi)$, and $w_j(\chi)$ for each $j \in [0..m]$ in time $\Theta(n)$ by using precomputed values $\ell_i(\chi)$ for $i \in [1..n]$ and the fact that the matrices U, V, W contain $\Theta(n)$ non-zero elements. The rest of the CRS can be computed efficiently by using straightforward algorithms.

By using Algorithm 1, the whole CRS generation dominated by $3m + 3n + 5$ exponentiations in $\mathbb{G}_1$, $m + 7$ exponentiations in $\mathbb{G}_2$, and 1 exponentiation in $\mathbb{G}_T$ (one per CRS element) and $\Theta(n)$ multiplications/divisions in $\mathbb{Z}_p$.

CV's Computational Complexity. We assume that it is difficult to subvert gk; this makes sense assuming that the SNARK uses well-known bilinear groups (say, the Barreto-Naehrig curves). Consider the CRS verification algorithm in Fig. 1. It is clear that all other steps but Step 4 are efficient (computable in $\Theta(n)$ cryptographic operations); this follows from the fact that U, V, and W are sparse. Computation in those steps is dominated by $6m + 5n - 4m_0 + 8$ pairings. On top of it, one has to execute $s(U) + s(V) + s(W)$ exponentiations in $\mathbb{G}_1$, where $s(M)$ is the number of "large" (i.e., large enough so that exponentiating with them is expensive) entries in the matrix M. Often, $s(M)$ are very small.

By using Algorithm 2, one can check that $[\ell_i(\chi)]_1$ has been correctly computed for all $i \in [1 .. n]$ in $n + 1$ pairings, $n - 1$ exponentiations in $\mathbb{G}_2$ and n exponentiations in $\mathbb{G}_T$. Hence, the whole CRS verification algorithm is dominated by $6m + 6n - 4m_0 + 9$ pairings, $s(U) + s(V) + s(W)$ exponentiations in $\mathbb{G}_1$, $n - 1$ exponentiations in $\mathbb{G}_2$, and n exponentiations in $\mathbb{G}_T$.

In addition, one can speed up CV by using batching [3]. Namely, clearly if $\sum_{i=1}^{s} t_i([a_i]_1 \bullet [b_i]_2) = [c]_T$ for uniformly random t_i, then w.h.p., $[a_i]_1 \bullet [b_i]_2 = [c]_T$ for each individual $i \in [1 .. s]$. The speed up follows from the use of bilinear properties and from the fact that exponentiation is faster than pairing. Moreover, one can further slightly optimize this by assuming $t_s = 1$ [18, 32].

Full batched version of CV is described in the full version [1]. As we show there, a batched CV will be dominated by $5(m + n) - 4m_0 + s(U) + s(V)$ (mostly, short-exponent) exponentiations in $\mathbb{G}_1$ and $m + s(W)$ (mostly, short-exponent) exponentiations in $\mathbb{G}_2$. Since an exponentiation with short exponent is significantly less costly than a pairing, this will decrease the execution time of CV significantly. (See the full version [1] for concrete numbers).

We note that after taking batching into account, CV will become a probabilistic algorithm, and will accept incorrect CRSs with negligible probability. This means that one has to modify some of the previous security results. For example, Theorems 4 and 5 will be modified as follows.

Theorem 7. *After batching* CV, *the SNARK from Sect. 5.1 has statistical CRS trapdoor extractability under the BDH-KE assumption.*

Theorem 8. *After batching* CV, *the SNARK from Sect. 5.1 is statistically composable Sub-ZK under the BDH-KE assumption.*

As in [27], the prover's computational complexity is dominated by the need to compute $h(X)$ (3 interpolations, 1 polynomial multiplication, and 1 polynomial division; in total $\Theta(n \log n)$ non-cryptographic operations in $\mathbb{Z}_p$), followed by $(n - 1) + (s(A) + 1) + 1 + (s(A) + 1) + s(A_1, \ldots, A_{m_0}) \leq n + 3s(A) + 2$ exponentiations in $\mathbb{G}_1$ and $s(A) + 1$ exponentiations in $\mathbb{G}_2$, where $s(A)$ is the number of large elements in A (i.e., large enough so that exponentiating with them would be expensive). This means that the prover's computation is dominated by $\Theta(n \log n)$ non-cryptographic operations and $\Theta(n)$ cryptographic operations.

The verifier executes a single pairing equation that is dominated by 3 pairings and m_0 exponentiations in $\mathbb{G}_1$. The exponentiations can be done offline since they do not depend on the argument π but only on the common input $(A_1, \ldots, A_{m_0})$. Hence, the verifier's computation is dominated by $\Theta(m_0)$ cryptographic operations but her online computation is only dominated by 3 pairings.

The argument consists of 2 elements from $\mathbb{G}_1$ and 1 element from $\mathbb{G}_2$.

9   The Case of Unbounded Subverter

Since we consider statistical Sub-ZK, it is natural to ask what will happen if also the subverter is computationally unbounded. It comes out that in this case, several definitions and proofs would actually simplify. For this reason, we decided to first present the case of efficient subverter.

Assume now that the subverter Σ is computationally unbounded. Then we need a computationally unbounded extractor X_Σ (otherwise, it will not be able to even execute Σ). corresponding version of Definition 5 with changes being *emphasized*.

Definition 10 (Statistically Unbounded USub-ZK). *Ψ is statistically unbounded Sub-ZK with unbounded subverter (USub-ZK) for $\mathcal{R}$, if for any computationally unbounded subverter Σ there exists a computationally unbounded X_Σ, such that for all κ, all $(\mathbf{R}, z_{\mathbf{R}}) \in \text{range}(\mathcal{R}(1^\kappa))$, and all computationally unbounded A, $\varepsilon_0^{unb} \approx_\kappa \varepsilon_1^{unb}$, where ε_b^{unb} is defined as*

$$\Pr \left[\begin{array}{l} r \leftarrow_r \text{RND}(\Sigma), (\text{crs}, z_\Sigma \| \text{tc}) \leftarrow (\Sigma \| X_\Sigma)(\mathbf{R}, z_{\mathbf{R}}; r), \\ \text{ts} \leftarrow K_{\text{ts}}(\mathbf{R}, z_{\mathbf{R}}, \text{tc}) : \text{CV}(\mathbf{R}, z_{\mathbf{R}}, \text{crs}) = 1 \wedge \mathsf{A}^{O_b(\cdot,\cdot)}(\mathbf{R}, z_{\mathbf{R}}, \text{crs}, \text{ts}, z_\Sigma) = 1 \end{array} \right].$$

Here, the oracle $O_0(x, w)$ returns $\bot$ (reject) if $(x, w) \notin \mathbf{R}$, and otherwise it returns $P(\mathbf{R}, z_{\mathbf{R}}, \text{crs}_P, x, w)$. Similarly, $O_1(x, w)$ returns $\bot$ (reject) if $(x, w) \notin \mathbf{R}$, and otherwise it returns $S(\mathbf{R}, z_{\mathbf{R}}, \text{crs}, \text{ts}, x)$. Ψ is perfectly unbounded USub-ZK for $\mathcal{R}$ if one requires that $\varepsilon_0^{unb} = \varepsilon_1^{unb}$.

An unbounded Σ can obviously break the BDH-KE assumption by creating elements of $\mathbb{G}_1$ and $\mathbb{G}_2$ as he wants; however, because she is unbounded, we can of course still argue that Σ will know their discrete logarithms, or more formally, that X_Σ will be able to extract them. Since the extraction is now unconditional (i.e., it does not depend on any assumptions), it means that X_Σ can extract the discrete logarithm of any element of $\mathbb{G}_z$.

In particular, there is no need anymore to include $[\gamma]_1$ to the CRS (this makes the CRS shorter by 1 element), or handle it in CV (that is, one could remove the check that $[\gamma]_1 \bullet [1]_2 = [1]_1 \bullet [\gamma]_2$) or in security proofs (e.g., in Lemma 2, one would not have to use the equation $[\iota]_1 \bullet [1]_2 = [1]_1 \bullet [\iota]_2$ to establish $[\gamma]_1$, or define $\Sigma^\gamma(\mathbf{R}, z_{\mathbf{R}}; r)$ inside Theorem 4).

Moreover, if the CRS has the (easily satisfied) bijectivity property required in Sect. 7 (see Lemma 2), it means that the requirement that X_Σ returns tc (instead of just returning ts) is not restrictive anymore.

10   Related Work

Bellare *et al.* [2] also showed that one can achieve simultaneously subversion soundness and non-interactive witness-indistinguishability (NIWI) by using a

(CRS-less) NIWI argument system. However, the known (CRS-less) NIWI argument systems are not succinct. Construction of more efficient NIWI argument systems is a very interesting open question.

Ben-Sasson *et al.* [6] proposed an efficient MPC approach to achieve security in the case one can trust at least one CRS creator. We emphasize that [2] and the current paper study the scenario where you can trust none. In such a case, the approach of [6] still works but it is not efficient. For example, the computational cost of CV (see Sect. 8) in the new SNARK is very small compared to the cost of the joint CRS creation and verification protocol in [6]. Nevertheless, while the starting point of their approach is different, it actually resulted in a somewhat similar solution. See [1] for a longer comparison.

Independently, Fuchsbauer [19] also constructed subversion-resistant SNARKs. The importance of subversion-resistant SNARKs was also recognized in [13,14] although no construction was given. In particular, [14] described a practical attack against SNARKs as used in known contingent payment protocols and proposed to use a Sub-ZK SNARK.

Acknowledgment. We thank Janno Siim for his help in the optimization of the batched CV algorithm, Rosario Gennaro for pointing out [14], and Georg Fuchsbauer for helpful discussion after [19] was made public.This work was supported by the European Union's Horizon 2020 research and innovation programme under grant agreement No 653497 (project PANORAMIX), and by institutional research funding IUT2-1 of the Estonian Ministry of Education and Research.

References

1. Abdolmaleki, B., Baghery, K., Lipmaa, H., Zajac, M.: A Subversion-Resistant SNARK. TR 2017/599, IACR (2017). http://eprint.iacr.org/2017/599
2. Bellare, M., Fuchsbauer, G., Scafuro, A.: NIZKs with an untrusted CRS: security in the face of parameter subversion. In: Cheon, J.H., Takagi, T. (eds.) ASIACRYPT 2016. LNCS, vol. 10032, pp. 777–804. Springer, Heidelberg (2016). doi:10.1007/978-3-662-53890-6_26
3. Bellare, M., Garay, J.A., Rabin, T.: Batch verification with applications to cryptography and checking. In: Lucchesi, C.L., Moura, A.V. (eds.) LATIN 1998. LNCS, vol. 1380, pp. 170–191. Springer, Heidelberg (1998). doi:10.1007/BFb0054320
4. Ben-Sasson, E., Chiesa, A., Garman, C., Green, M., Miers, I., Tromer, E., Virza, M.: Zerocash: decentralized anonymous payments from bitcoin. In: IEEE SP 2014, pp. 459–474 (2014)
5. Ben-Sasson, E., Chiesa, A., Genkin, D., Tromer, E., Virza, M.: SNARKs for C: verifying program executions succinctly and in zero knowledge. In: Canetti, R., Garay, J.A. (eds.) CRYPTO 2013. LNCS, vol. 8043, pp. 90–108. Springer, Heidelberg (2013). doi:10.1007/978-3-642-40084-1_6
6. Ben-Sasson, E., Chiesa, A., Green, M., Tromer, E., Virza, M.: Secure sampling of public parameters for succinct zero knowledge proofs. In: IEEE SP 2015, pp. 287–304 (2015)

7. Ben-Sasson, E., Chiesa, A., Tromer, E., Virza, M.: Scalable zero knowledge via cycles of elliptic curves. In: Garay, J.A., Gennaro, R. (eds.) CRYPTO 2014. LNCS, vol. 8617, pp. 276–294. Springer, Heidelberg (2014). doi:10.1007/978-3-662-44381-1_16

8. Ben-Sasson, E., Chiesa, A., Tromer, E., Virza, M.: Succinct non-interactive zero knowledge for a Von Neumann architecture. In: USENIX 2014, pp. 781–796 (2014)

9. Bitansky, N., Canetti, R., Chiesa, A., Tromer, E.: Recursive composition and bootstrapping for SNARKs and proof-carrying data. In: STOC 2013, pp. 241–250 (2013)

10. Bitansky, N., Canetti, R., Paneth, O., Rosen, A.: On the existence of extractable one-way functions. In: STOC 2014, pp. 505–514 (2014)

11. Blum, M., Feldman, P., Micali, S.: Non-interactive zero-knowledge and its applications. In: STOC 1988, pp. 103–112 (1988)

12. Boneh, D., Boyen, X., Goh, E.-J.: Hierarchical identity based encryption with constant size ciphertext. In: Cramer, R. (ed.) EUROCRYPT 2005. LNCS, vol. 3494, pp. 440–456. Springer, Heidelberg (2005). doi:10.1007/11426639_26

13. Bowe, S., Gabizon, A., Green, M.D.: A multi-party protocol for constructing the public parameters of the Pinocchio zk-SNARK. TR 2017/602, IACR (2017). http://eprint.iacr.org/2017/602. Accessed 25 June 2017

14. Campanelli, M., Gennaro, R., Goldfeder, S., Nizzardo, L.: Zero-knowledge contingent payments revisited: attacks and payments for services. TR 2017/566, IACR (2017). http://eprint.iacr.org/2017/566

15. Costello, C., Fournet, C., Howell, J., Kohlweiss, M., Kreuter, B., Naehrig, M., Parno, B., Zahur, S.: Geppetto: versatile verifiable computation. In: IEEE SP 2015, pp. 253–270 (2015)

16. Damgård, I.: Towards practical public key systems secure against chosen ciphertext attacks. In: Feigenbaum, J. (ed.) CRYPTO 1991. LNCS, vol. 576, pp. 445–456. Springer, Heidelberg (1992). doi:10.1007/3-540-46766-1_36

17. Danezis, G., Fournet, C., Groth, J., Kohlweiss, M.: Square span programs with applications to succinct NIZK arguments. In: Sarkar, P., Iwata, T. (eds.) ASIACRYPT 2014. LNCS, vol. 8873, pp. 532–550. Springer, Heidelberg (2014). doi:10.1007/978-3-662-45611-8_28

18. Fauzi, P., Lipmaa, H., Zając, M.: A shuffle argument secure in the generic model. In: Cheon, J.H., Takagi, T. (eds.) ASIACRYPT 2016. LNCS, vol. 10032, pp. 841–872. Springer, Heidelberg (2016). doi:10.1007/978-3-662-53890-6_28

19. Fuchsbauer, G.: Subversion-zero-knowledge SNARKs. TR 2017/587, IACR (2017). http://eprint.iacr.org/2017/587

20. Galbraith, S.D., Paterson, K.G., Smart, N.P.: Pairings for cryptographers. Discrete Appl. Math. 156(16), 3113–3121 (2008)

21. Gennaro, R., Gentry, C., Parno, B.: Non-interactive verifiable computing: outsourcing computation to untrusted workers. In: Rabin, T. (ed.) CRYPTO 2010. LNCS, vol. 6223, pp. 465–482. Springer, Heidelberg (2010). doi:10.1007/978-3-642-14623-7_25

22. Gennaro, R., Gentry, C., Parno, B., Raykova, M.: Quadratic span programs and succinct NIZKs without PCPs. In: Johansson, T., Nguyen, P.Q. (eds.) EUROCRYPT 2013. LNCS, vol. 7881, pp. 626–645. Springer, Heidelberg (2013). doi:10.1007/978-3-642-38348-9_37

23. Gentry, C., Wichs, D.: Separating succinct non-interactive arguments from all falsifiable assumptions. In: STOC 2011, pp. 99–108 (2011)

24. Goldreich, O.: A uniform-complexity treatment of encryption and zero-knowledge. J. Cryptol. 6(1), 21–53 (1993)

25. Groth, J.: Simulation-sound NIZK proofs for a practical language and constant size group signatures. In: Lai, X., Chen, K. (eds.) ASIACRYPT 2006. LNCS, vol. 4284, pp. 444–459. Springer, Heidelberg (2006). doi:10.1007/11935230_29

26. Groth, J.: Short pairing-based non-interactive zero-knowledge arguments. In: Abe, M. (ed.) ASIACRYPT 2010. LNCS, vol. 6477, pp. 321–340. Springer, Heidelberg (2010). doi:10.1007/978-3-642-17373-8_19

27. Groth, J.: On the size of pairing-based non-interactive arguments. In: Fischlin, M., Coron, J.-S. (eds.) EUROCRYPT 2016. LNCS, vol. 9666, pp. 305–326. Springer, Heidelberg (2016). doi:10.1007/978-3-662-49896-5_11

28. Groth, J., Sahai, A.: Efficient non-interactive proof systems for bilinear groups. In: Smart, N. (ed.) EUROCRYPT 2008. LNCS, vol. 4965, pp. 415–432. Springer, Heidelberg (2008). doi:10.1007/978-3-540-78967-3_24

29. Jutla, C.S., Roy, A.: Shorter quasi-adaptive NIZK proofs for linear subspaces. In: Sako, K., Sarkar, P. (eds.) ASIACRYPT 2013. LNCS, vol. 8269, pp. 1–20. Springer, Heidelberg (2013). doi:10.1007/978-3-642-42033-7_1

30. Lipmaa, H.: Progression-free sets and sublinear pairing-based non-interactive zero-knowledge arguments. In: Cramer, R. (ed.) TCC 2012. LNCS, vol. 7194, pp. 169–189. Springer, Heidelberg (2012). doi:10.1007/978-3-642-28914-9_10

31. Lipmaa, H.: Succinct non-interactive zero knowledge arguments from span programs and linear error-correcting codes. In: Sako, K., Sarkar, P. (eds.) ASIACRYPT 2013. LNCS, vol. 8269, pp. 41–60. Springer, Heidelberg (2013). doi:10.1007/978-3-642-42033-7_3

32. Lipmaa, H.: Prover-efficient commit-and-prove zero-knowledge SNARKs. In: Pointcheval, D., Nitaj, A., Rachidi, T. (eds.) AFRICACRYPT 2016. LNCS, vol. 9646, pp. 185–206. Springer, Cham (2016). doi:10.1007/978-3-319-31517-1_10

33. Maurer, U.: Abstract models of computation in cryptography. In: Smart, N.P. (ed.) Cryptography and Coding 2005. LNCS, vol. 3796, pp. 1–12. Springer, Heidelberg (2005). doi:10.1007/11586821_1

34. Nechaev, V.I.: Complexity of a determinate algorithm for the discrete logarithm. Mathematical Notes 55(2), 165–172 (1994). Matematicheskie Zapiski 55(2), 91–101 (1994)

35. Parno, B., Gentry, C., Howell, J., Raykova, M.: Pinocchio: nearly practical verifiable computation. In: IEEE SP 2013, pp. 238–252 (2013)

36. Shoup, V.: Lower bounds for discrete logarithms and related problems. In: Fumy, W. (ed.) EUROCRYPT 1997. LNCS, vol. 1233, pp. 256–266. Springer, Heidelberg (1997). doi:10.1007/3-540-69053-0_18

Cryptographic Protocols

Two-Round PAKE from Approximate SPH and Instantiations from Lattices

Jiang Zhang[1](✉) and Yu Yu[1,2,3]

[1] State Key Laboratory of Cryptology, P.O. Box 5159, Beijing 100878, China
jiangzhang09@gmail.com, yuyuathk@gmail.com
[2] Department of Computer Science and Engineering, Shanghai Jiao Tong University, Shanghai, China
[3] Westone Cryptologic Research Center, Beijing 100070, China

Abstract. Password-based authenticated key exchange (PAKE) enables two users with shared low-entropy passwords to establish cryptographically strong session keys over insecure networks. At Asiacrypt 2009, Katz and Vaikuntanathan showed a generic *three-round* PAKE based on any CCA-secure PKE with associated approximate smooth projective hashing (ASPH), which helps to obtain the first PAKE from lattices. In this paper, we give a framework for constructing PAKE from CCA-secure PKE with associated ASPH, which uses only *two-round* messages by carefully exploiting a *splittable* property of the underlying PKE and its associated *non-adaptive* ASPH. We also give a *splittable* PKE with associated *non-adaptive* ASPH based on the LWE assumption, which finally allows to instantiate our two-round PAKE framework from lattices.

1 Introduction

As one of the most fundamental and widely used cryptographic primitives, key exchange (KE) dates back to the seminal work of Diffie and Hellman (DH) [23], and it enables users to establish a session key via public exchanges of messages. Due to lack of authentication, the original DH protocol, and key exchange in general, only ensures two users to share a secure session key in presence of passive eavesdroppers, and it is insecure against an active adversary who has full control of all communication. To overcome this issue, authenticated key exchange (AKE) enables each user to authenticate the identities of others with the help of some pre-shared information, and thus provides the additional guarantee that only the intended users can access the session key. Typically, the shared information can be either a high-entropy cryptographic key (such as a secret key for symmetric-key encryption, or a public key for digital signature) or a low-entropy password. After decades of development, the community has witnessed great success in designing AKE based on high-entropy cryptographic keys, even in the setting of lattices [6,50,59]. However, people rarely make full use of the large character set in forming passwords and many tend to pick easily memorizable ones from a relatively small dictionary. AKEs based on high-entropy cryptographic keys

ⓒ International Association for Cryptologic Research 2017
T. Takagi and T. Peyrin (Eds.): ASIACRYPT 2017, Part III, LNCS 10626, pp. 37–67, 2017.
https://doi.org/10.1007/978-3-319-70700-6_2

usually do not apply to the case where only low-entropy passwords are available. Indeed, as shown in [32,37], it can be trivially insecure to use a low-entropy password as a cryptographic key.

Informally, a secure password-based AKE (PAKE) should resist *off-line dictionary attacks* in which the adversary tries to determine the correct password using only information obtained during previous protocol executions, and limit the adversary to the trivial *on-line attacks* where the adversaries simply run the protocol with honest users using (a bounded number of) password trials. Formal security models for PAKE were developed in [9,16]. Later, many provably secure PAKE protocols based on various hardness assumptions were proposed, where the research mainly falls into two lines:[1] the first line starts from the work of Bellovin and Merritt [11], followed by plenty of excellent work focusing on PAKE in the random oracle/ideal cipher models and aiming at achieving the highest possible levels of performance [9,16,17,45]; the second line dates back to the work of Katz et al. [37], from which Gennaro and Lindell [29] abstracted out a generic PAKE framework (in the CRS model) based on smooth projective hash (SPH) functions [22]. This line of research devoted to seeking more efficient PAKE in the standard model [2,3,12,21,35,39].

As noted in [38], none of the above PAKEs can be instantiated from lattices. In particular, it is an open problem [52] to instantiate SPH functions [22] on lattice assumptions. Despite the great success in lattice-based cryptography in the past decade, little progress was made on lattice-based PAKE until the work of Katz and Vaikuntanathan [38]. Concretely, they [38] introduced the notion of approximate smooth projective hashing (ASPH) so as to be instantiatable from lattices, and plugged it into an adapted version of the GL-framework [29] to yield the first lattice-based PAKE by using only three-round messages in the standard model (just like the counterparts in [29,37]). Up until now (seven years after the publication of [38]), the Katz-Vaikuntanathan PAKE remained the most efficient lattice-based PAKE to the best of our knowledge. This raises the following questions: *is it possible to construct more efficient PAKE from lattices (e.g., a PAKE with less message rounds/communication overheads), and does there exist other generic PAKE framework that fits better with lattices?*

1.1   Our Contribution

In this paper, we first give a new PAKE framework (also in the CRS model) from PKE with associated ASPH, which uses only *two-round* messages. We mainly benefit from two useful features of the underlying primitives: (1) the PKE is *splittable*, which informally requires that each ciphertext of the PKE scheme consists of two relatively independent parts, where the first part is designed for realizing the "functionality" of encryption, while the second part helps to achieve CCA-security; and (2) the ASPH is *non-adaptive* [39], i.e., the projection function only depends on the hash key, and the smoothness property holds even when the ciphertext depends on the projection key. By carefully exploiting the above

---

[1] Please refer to Sect. 1.3 for other related works.

features, we overcome several obstacles (e.g., the "approximate correctness" of ASPH) to obtain a generic two-round PAKE in the standard model.

We also propose a concrete construction of splittable PKE with associated non-adaptive ASPH from learning with errors (LWE). Note that the PKEs with associated SPH (based on either DDH or decisional linear assumptions) in [39] can be used to instantiate our framework, but the only known lattice-based PKE with associated ASPH in [38] does not satisfy our requirements. We achieve our goal by first developing an adaptive smoothing lemma for q-ary lattices, and then combining it with several recent techniques. Technically, the lemma is needed for achieving the strong smoothness of our *non-adaptive* ASPH, and may be of independent interest. As in [39], our PKE construction relies on simulation-sound non-interactive zero-knowledge (SS-NIZK) proofs, and thus, in general, is computationally inefficient. Fortunately, we can construct an efficient SS-NIZK from lattices in the random oracle model,[2] and finally obtain an efficient two-round lattice-based PAKE (also in the random oracle model), which is at least $O(\log n)$ times more efficient in the communication overhead than the three-round lattice-based PAKE (in the standard model) [38].

1.2 Our Techniques

We begin with the GL-framework [29] from CCA-secure public-key encryption (PKE) with associated smooth projective hash (SPH) functions. Informally, the SPH for a PKE scheme is a keyed hash function which maps a ciphertext-plaintext pair into a hash value, and can be computed in two ways: either using the hash key hk or using a projection key hp (which can be efficiently determined from hk and a targeted ciphertext c). The GL-framework for PAKE roughly relies on the following two properties of SPH:

Correctness: if c is an encryption of the password pw using randomness r, then the hash value $H_{hk}(c, pw) = Hash(hp, (c, pw), r)$, where both functions H and Hash can be efficiently computed from the respective inputs.

Smoothness: if c is not an encryption of pw, the value $H_{hk}(c, pw)$ is statistically close to uniform given hp, c and pw (over the random choice of hk).

Specifically, the GL-framework for PAKE has three-round messages: (1) the client computes an encryption c_1 of the password pw using randomness r_1, and sends c_1 to the server; (2) the server randomly chooses a hash key hk_2, computes a projection key hp_2 (from hk_2 and c_1) together with an encryption c_2 of the password pw using randomness r_2, and sends (hp_2, c_2) to the client; (3) the client sends a projection key hp_1 corresponding to a randomly chosen hash key hk_1 and c_2. After exchanging the above three messages, both users can compute the same session key $sk = H_{hk_1}(c_2, pw) \oplus Hash(hp_2, (c_1, pw), r_1) = Hash(hp_1, (c_2, pw), r_2) \oplus H_{hk_2}(c_1, pw)$ by the correctness of the SPH. Note that if the PKE scheme is CCA-secure, no user can obtain useful information about the password held by the

---

[2] We leave it as an open problem to directly construct an SS-NIZK from lattice problems in the standard model.

other user from the received ciphertext. Thus, if the client (resp., the server) does not hold the correct password pw, his view is independent from the "session key" computed by the server (resp., the client) by the smoothness of the SPH. We stress that the above discussion is very informal and omits many details. For example, a verification key vk should be sent in the first message such that the client can generate a signature σ on the protocol transcripts in the third message (and thus the total communication cost is determined by $|\mathsf{hp}| + |c| + |vk| + |\sigma|$).

Clearly, a lattice-based PAKE is immediate if a PKE with associated SPH could be obtained from lattice assumptions. However, the literature [52] suggests that it is highly non-trivial, if not impossible, to instantiate SPH from lattices. Instead, Katz and Vaikuntanathan [38] provided a solution from a weaker notion of SPH—Approximate SPH (ASPH), which weakens both the correctness and smoothness properties of the SPH notion in [29]. First, ASPH only provides "approximate correctness" in the sense that $\mathsf{H}_{\mathsf{hk}}(c, pw)$ and $\mathsf{Hash}(\mathsf{hp}, (c, pw), r)$ may differ at a few positions when parsed as bit-strings. Second, the smoothness property of ASPH only holds for some (c, pw) that pw is not equal to the decryption of c, and hence leaves a gap that there exists (c, pw) for which ASPH provides neither correctness nor smoothness guarantee. This relaxation is necessary for instantiating ASPH on lattices, since in the lattice setting there is no clear boundary between "c is an encryption of pw" and "c is not an encryption of pw", which is actually one of the main difficulties for realizing SPH from lattices.

Thus, if one directly plugs ASPH into the GL-framework [29], neither the correctness nor the security of the resulting PAKE is guaranteed. Because both users may not compute the same session key, and the adversary may break the protocol by exploiting the (inherent) gap introduced by ASPH. The authors [38] fixed the issues by relying on error correcting codes (ECC) and the robustness of the GL-framework [29]. Specifically, in addition to sending a projection key hp_1, the client also randomly chooses a session key sk, computes $tk = \mathsf{H}_{\mathsf{hk}_1}(c_2, pw) \oplus \mathsf{Hash}(\mathsf{hp}_2, (c_1, pw), r_1)$, and appends $\Delta = tk \oplus \mathsf{ECC}(sk)$ to the third message (i.e., tk is used as a masking key to deliver sk to the server), where ECC and ECC^{-1} are the corresponding encoding and decoding algorithms. After receiving the third message, the server can compute the session key $sk' = \mathsf{ECC}^{-1}(tk' \oplus \Delta)$, where $tk' = \mathsf{Hash}(\mathsf{hp}_1, (c_2, pw), r_2) \oplus \mathsf{H}_{\mathsf{hk}_2}(c_1, pw)$. By the "approximate correctness" of the ASPH, we know that $tk' \oplus \Delta$ is not far from the codeword $\mathsf{ECC}(sk)$. Thus, both users can obtain the same session key $sk = sk'$ by the correctness of an appropriately chosen ECC, which finally allows [38] to obtain a three-round PAKE from PKE with associated ASPH.

However, the techniques of [38] are not enough to obtain a two-round PAKE (in particular, they cannot be applied into the PAKE framework [39]) due to the following two main reasons. First, the ASPH in [38] is *adaptive* (i.e., the projection key hp depends on the ciphertext c, and the smoothness only holds when c is independent of hp), which seems to inherently require at least three-round messages [29,39]. Second, the strategy of delivering a random session key to deal with the "approximate correctness" of ASPH can only be applied when one user (e.g., the client) obtained the masking key tk, and may be vulnerable to active attacks (e.g., modifications) because of the loose relation between the marking part (namely, Δ) and other protocol messages. This is not a problem for the GL-framework [29], since it had three-round messages and used one-time signatures, which allows the authors of [38] to simply send Δ in the third message and tightly bind it with other protocol messages by incorporating it into the one-time signature. Nevertheless, the above useful features are not available in the more efficient PAKE framework [39].

In order to get a two-round PAKE from PKE with associated ASPH, we strengthen the underlying primitive with several reasonable properties. First, we require that the ASPH is non-adaptive, i.e., the projection function only depends on the hash key, and the smoothness property holds even when the ciphertext c depends on hp. Second, we require that the underlying PKE is splittable. Informally, this property says that a ciphertext $c = (u, v)$ of the PKE scheme can be "independently" computed by two functions (f, g), where $u = f(pk, pw, \cdots)$ mainly takes a plaintext pw as input and plays the role of "encrypting" pw, while $v = g(pk, \mathsf{label}, \cdots)$ mainly takes a label as input and plays the role of providing non-malleability for CCA-security.[3] Third, we require that the hash value of the ASPH is determined by the hash key hk, the first part u of the ciphertext $c = (u, v)$, as well as the password pw. At a high level, the first enhancement allows us to safely compute the masking key tk after receiving the first message, while the second and third enhancements enable us to leverage the non-malleability of the underlying CCA-secure PKE scheme to tightly bind the masking part Δ with other protocol messages. Concretely, we let the client to send the projection hash key hp_1 together with the ciphertext c_1 in a single message, and let the server compute the masking key tk immediately after it has obtained the first part $u_2 = f(pk, pw, \cdots)$ of the ciphertext $c_2 = (u_2, v_2)$, and compute the second part $v_2 = g(pk, \mathsf{label}, \cdots)$ with a label consisting of $\mathsf{hp}_1, c_1, \mathsf{hp}_2, u_2$ and $\Delta = tk \oplus sk$ for some randomly chosen session key sk. The protocol ends with a message $(\mathsf{hp}_2, c_2, \Delta)$ sent by the server to the client. A high level overview of our two-round PAKE framework is given in Fig. 1.

Note that the PKEs with associated SPH in [39] can be used to instantiate our two-round PAKE framework, but the only known lattice-based PKE with associated ASPH [38] does not satisfy our requirements. Actually, it is highly non-trivial to realize non-adaptive ASPH from lattices. One of the main reason is that the smoothness should hold even when the ciphertext c is adversarially chosen and dependent on the projection key hp (and thus is stronger than that in [38]), which gives the adversary an ability to obtain non-trivial information about the secret hash key hk, and makes the above (inherent) gap introduced by the ASPH notion more problematic. In order to ensure the stronger smoothness property, we first develop an adaptive smoothing lemma for q-ary lattices, which may be of independent interest. Then, we combine it with several other techniques [30,38,47,49,55] to achieve our goal. As in [39], our PKE is computationally inefficient due to the use of simulation-sound non-interactive zero-knowledge (SS-NIZK) proofs. However, we can obtain an efficient SS-NIZK

---

[3] Similar properties were also considered for identity-based encryptions [4,60].

from lattices in the random oracle model, and finally get an efficient lattice-based PAKE. Despite the less message rounds, our PAKE (in the random oracle model) is also at least $O(\log n)$ times more efficient in the communication overhead than the one in [38], because they used the correlated products technique [54] (which introduces an expansion factor n w.r.t. the basic CPA-secure PKE scheme) and signatures (which usually consists of matrices on lattices [57]). In comparison, our framework does not use signatures, and the ciphertext of our PKE scheme is $n/\omega(\log n)$ times shorter than that of [38].

1.3 Other Related Work and Discussions

Gong et al. [33] first considered the problem of resisting off-line attacks in the "PKI model" where the server also has a public key in addition to a password. A formal treatment on this model was provided by Halevi and Krawczyk [36]. At CRYPTO 1993, Bellovin and Merritt [11] considered the setting where only a password is shared between users, and proposed a PAKE with heuristic security arguments. Formal security models for PAKE were provided in [9,16]. Goldreich and Lindell [32] showed a PAKE solution in the plain model, which does not support concurrent executions of the protocol by the same user. As a special case of secure multiparty computations, PAKEs supporting concurrent executions in the plain model were studied in [8,19,34]. All the protocols in [8,19,32,34] are inefficient in terms of both computation and communication. In the setting where all users share a common reference string, Katz et al. [37] provided a practical three-round PAKE based on the DDH assumption, which was later generalized and abstracted out by Gennaro and Lindell [29] to obtain a PAKE framework from PKE with associated SPH [22]. Canetti et al. [21] considered the security of PAKE within the framework of universal composability (UC) [18], and showed that an extension of the KOY/GL protocol was secure in the UC model.

Relations to [38,39]. The works [38,39] due to Katz and Vaikuntanathan are most related to our work. First, the ASPH notion in this paper is stronger than that in [38]. In particular, the PKE with associated ASPH in [38] cannot be used to instantiate our framework. Our PAKE framework with less message rounds is obtained by strengthening the underlying primitives with several useful and achievable features, which provide us a better way to handle lattice assumptions. Besides, our PKE with associated SPH can be used to instantiate the PAKE framework in [38] (with improved efficiency). Second, our ASPH notion is much weaker than the SPH in [39], which means that our PKE with associated ASPH cannot be used to instantiate the PAKE framework in [39]. In fact, it is still an open problem to construct PKE with associated SPH from lattices, and we still do not know how to instantiate the efficient one-round PAKE framework [39] with lattice assumptions (recall that our PAKE has two-round messages). Third, our PAKE framework is inspired by [38,39], and thus shares some similarities to the latter. However, as discussed above, there are technical differences among the underlying primitives used in the three papers, and several new ideas/techniques are needed to obtain a two-round PAKE from lattices.

1.4   Roadmap

After some preliminaries in Sect. 2, we propose a generic two-round PAKE from splittable PKE with associated ASPH in Sect. 3. In Sect. 4, we give some backgrounds together with a new technical lemma on lattices. We construct a concrete splittable PKE with associated ASPH from lattices in Sect. 5.

2   Preliminaries

2.1   Notation

Let κ be the natural security parameter. By $\log_2$ (resp. log) we denote the logarithm with base 2 (resp. the natural logarithm). A function $f(n)$ is negligible, denoted by $\mathrm{negl}(n)$, if for every positive c, we have $f(n) < n^{-c}$ for all sufficiently large n. A probability is said to be overwhelming if it is $1 - \mathrm{negl}(n)$. The notation $\leftarrow_r$ denotes randomly choosing elements from some distribution (or the uniform distribution over some finite set). If a random variable x follows some distribution D, we denote it by $x \backsim D$. For any strings $x, y \in \{0,1\}^\ell$, denote $\mathsf{Ham}(x,y)$ as the hamming distance of x and y.

By $\mathbb{R}$ (resp. $\mathbb{Z}$) we denote the set of real numbers (resp. integers). Vectors are used in the column form and denoted by bold lower-case letters (e.g., $\mathbf{x}$). Matrices are treated as the sets of column vectors and denoted by bold capital letters (e.g., $\mathbf{X}$). The concatenation of the columns of $\mathbf{X} \in \mathbb{R}^{n \times m}$ followed by the columns of $\mathbf{Y} \in \mathbb{R}^{n \times m'}$ is denoted as $(\mathbf{X} \| \mathbf{Y}) \in \mathbb{R}^{n \times (m+m')}$. By $\| \cdot \|$ and $\| \cdot \|_\infty$ we denote the l_2 and l_∞ norm, respectively. The largest singular value of a matrix $\mathbf{X}$ is $s_1(\mathbf{X}) = \max_{\mathbf{u}} \|\mathbf{X}\mathbf{u}\|$, where the maximum is taken over all unit vectors $\mathbf{u}$.

2.2   Security Model for PAKE

We recall the security model for password-based authenticated key exchange (PAKE) in [9,37,39]. Formally, the protocol relies on a setup assumption that a common reference string (CRS) and other public parameters are established (possibly by a trusted third party) before any execution of the protocol. Let $\mathcal{U}$ be the set of protocol users. For every distinct $A, B \in \mathcal{U}$, users A and B share a password $pw_{A,B}$. We assume that each $pw_{A,B}$ is chosen independently and uniformly from some dictionary set $\mathcal{D}$ for simplicity. Each user $A \in \mathcal{U}$ is allowed to execute the protocol multiple times with different partners, which is modeled by allowing A to have an unlimited number of *instances* with which to execute the protocol. Denote instance i of A as Π_A^i. An instance is for one-time use only and it is associated with the following variables that are initialized to $\bot$ or 0:

- $\mathsf{sid}_A^i, \mathsf{pid}_A^i$ and sk_A^i denote the *session id*, *parter id*, and *session key* for instance Π_A^i. The session *id* consists of the (ordered) concatenation of all messages sent and received by Π_A^i; while the partner *id* specifies the user with whom Π_A^i believes it is interacting;
- acc_A^i and term_A^i are boolean variables denoting whether instance Π_A^i has accepted or terminated, respectively.

For any user $A, B \in \mathcal{U}$, instances Π_A^i and Π_B^j are *partnered* if $\mathsf{sid}_A^i = \mathsf{sid}_B^j \neq \bot$, $\mathsf{pid}_A^i = B$ and $\mathsf{pid}_B^j = A$. We say that a PAKE protocol is *correct* if instances Π_A^i and Π_B^j are partnered, then we have that $\mathsf{acc}_A^i = \mathsf{acc}_B^j = 1$ and $\mathsf{sk}_A^i = \mathsf{sk}_B^j \neq \bot$ hold (with overwhelming probability).

Adversarial abilities. The adversary $\mathcal{A}$ is a probabilistic polynomial time (PPT) algorithm with full control over all communication channels between users. In particular, $\mathcal{A}$ can intercept all messages, read them all, and remove or modify any desired messages as well as inject its own messages. $\mathcal{A}$ is also allowed to obtain the session key of an instance, which models possible leakage of session keys. These abilities are formalized by allowing the adversary to interact with the various instances via access to the following oracles:

- $\mathsf{Send}(A, i, \mathsf{msg})$: This sends message msg to instance Π_A^i. After receiving msg, instance Π_A^i runs according to the protocol specification, and updates its states as appropriate. Finally, this oracle returns the message output by Π_A^i to the adversary. We stress that the adversary can prompt an unused instance Π_A^i to execute the protocol with partner B by querying $\mathsf{Send}(A, i, B)$, and obtain the first protocol message output by Π_A^i.
- $\mathsf{Execute}(A, i, B, j)$: If both instances Π_A^i and Π_B^j have not yet been used, this oracle executes the protocol between Π_A^i and Π_B^j, updates their states as appropriate, and returns the transcript of this execution to the adversary.
- $\mathsf{Reveal}(A, i)$: This oracle returns the session key sk_A^i to the adversary if it has been generated (i.e., $\mathsf{sk}_A^i \neq \bot$).
- $\mathsf{Test}(A, i)$: This oracle chooses a random bit $b \leftarrow_r \{0, 1\}$. If $b = 0$, it returns a key chosen uniformly at random; if $b = 1$, it returns the session key sk_A^i of instance Π_A^i. The adversary is only allowed to query this oracle once.

Definition 1 (Freshness). *We say that an instance Π_A^i is fresh if the following conditions hold:*

- *the adversary $\mathcal{A}$ did not make a $\mathsf{Reveal}(A, i)$ query to instance Π_A^i;*
- *the adversary $\mathcal{A}$ did not make a $\mathsf{Reveal}(B, j)$ query to instance Π_B^j, where instances Π_A^i and Π_B^j are partnered;*

Security Game. The security of a PAKE protocol is defined via the following game. The adversary $\mathcal{A}$ makes any sequence of queries to the oracles above, so long as only one $\mathsf{Test}(A, i)$ query is made to a fresh instance Π_A^i, with $\mathsf{acc}_A^i = 1$ at the time of this query. The game ends when $\mathcal{A}$ outputs a guess b' for b. We say $\mathcal{A}$ wins the game if its guess is correct, so that $b' = b$. The advantage $\mathbf{Adv}_{\Pi, \mathcal{A}}$ of adversary $\mathcal{A}$ in attacking a PAKE protocol Π is defined as $|2 \cdot \Pr[b' = b] - 1|$.

We say that an *on-line attack* happens when the adversary makes one of the following queries to some instance Π_A^i: $\mathsf{Send}(A, i, *)$, $\mathsf{Reveal}(A, i)$ or $\mathsf{Test}(A, i)$. In particular, the $\mathsf{Execute}$ queries are not counted as on-line attacks. Since the size of the password dictionary is small, a PPT adversary can always win by trying all password one-by-one in an on-line attack. The number $Q(\kappa)$ of on-line attacks represents a bound on the number of passwords the adversary could have tested in an on-line fashion. Informally, a PAKE protocol is secure if online password guessing attacks are already the best strategy (for all PPT adversaries).

Definition 2 (Security). *We say that a PAKE protocol Π is secure if for all dictionary $\mathcal{D}$ and for all PPT adversaries $\mathcal{A}$ making at most $Q(\kappa)$ on-line attacks, it holds that $\mathbf{Adv}_{\Pi,\mathcal{A}}(\kappa) \leq Q(\kappa)/|\mathcal{D}| + \mathrm{negl}(\kappa)$.*

3   PAKE from Splittable PKE with Associated ASPH

In this section, we give a new PAKE framework which only has two-round messages. We begin with the definition of splittable PKE with associated ASPH.

3.1   Public-Key Encryption

A (labeled) public-key encryption (PKE) with plaintext-space $\mathcal{P}$ consists of three PPT algorithms $\mathcal{PKE} = (\mathsf{KeyGen}, \mathsf{Enc}, \mathsf{Dec})$. The key generation algorithm KeyGen takes the security parameter κ as input, outputs a public key pk and a secret key sk, denoted as $(\mathsf{pk}, \mathsf{sk}) \leftarrow \mathsf{KeyGen}(1^{\kappa})$. The encryption algorithm Enc takes pk, a string $\mathsf{label} \in \{0,1\}^*$, and a plaintext $pw \in \mathcal{P}$ as inputs,[4] with an internal coin flipping r, outputs a ciphertext c, which is denoted as $c \leftarrow \mathsf{Enc}(\mathsf{pk}, \mathsf{label}, pw; r)$, or $c \leftarrow \mathsf{Enc}(\mathsf{pk}, \mathsf{label}, pw)$ in brief. The deterministic algorithm Dec takes sk and c as inputs, and produces as output a plaintext pw or $\bot$, which is denoted as $pw \leftarrow \mathsf{Dec}(\mathsf{sk}, \mathsf{label}, c)$.

For correctness, we require that for all $(\mathsf{pk}, \mathsf{sk}) \leftarrow \mathsf{KeyGen}(1^{\kappa})$, any $\mathsf{label} \in \{0,1\}^*$, any plaintext pw and $c \leftarrow \mathsf{Enc}(\mathsf{pk}, \mathsf{label}, pw)$, the equation $\mathsf{Dec}(\mathsf{sk}, \mathsf{label}, c) = pw$ holds with overwhelming probability. For security, consider the following game between a challenger $\mathcal{C}$ and an adversary $\mathcal{A}$.

Setup. The challenger $\mathcal{C}$ first computes $(\mathsf{pk}, \mathsf{sk}) \leftarrow \mathsf{KeyGen}(1^{\kappa})$. Then, it gives the public key pk to $\mathcal{A}$, and keeps the secret key sk to itself.
Phase 1. The adversary $\mathcal{A}$ can make a number of decryption queries on any pair (label, c), and $\mathcal{C}$ returns $pw \leftarrow \mathsf{Dec}(\mathsf{sk}, \mathsf{label}, c)$ to $\mathcal{A}$ accordingly.
Challenge. At some time, $\mathcal{A}$ outputs two equal-length plaintexts $pw_0, pw_1 \in \mathcal{P}$ and a $\mathsf{label}^* \in \{0,1\}^*$. The challenger $\mathcal{C}$ chooses a random bit $b^* \leftarrow_r \{0,1\}$, and returns the challenge ciphertext $c^* \leftarrow \mathsf{Enc}(\mathsf{pk}, \mathsf{label}^*, pw_{b^*})$ to $\mathcal{A}$.
Phase 2. $\mathcal{A}$ can make more decryption queries on any $(\mathsf{label}, c) \neq (\mathsf{label}^*, c^*)$, the challenger $\mathcal{C}$ responds as in Phase 1.
Guess. Finally, $\mathcal{A}$ outputs a guess $b \in \{0,1\}$.

The adversary $\mathcal{A}$ wins the game if $b = b^*$. The advantage of $\mathcal{A}$ in the above game is defined as $\mathrm{Adv}_{\mathcal{PKE},\mathcal{A}}^{\mathrm{ind\text{-}cca}}(1^{\kappa}) \overset{\mathrm{def}}{=} |\Pr[b = b^*] - \frac{1}{2}|$.

---

[4] The notation 'pw' stands for password, and we keep several other commonly used notations such as 'm' and 'w' for latter use on lattices.

Definition 3 (IND-CCA). *We say that a PKE scheme $\mathcal{PKE}$ is CCA-secure if for any PPT adversary $\mathcal{A}$, its advantage $\mathbf{Adv}^{ind\text{-}cca}_{\mathcal{PKE},\mathcal{A}}(1^{\kappa})$ is negligible in κ.*

Informally, the splittable property of a PKE scheme $\mathcal{PKE}$ requires that the encryption algorithm can be split into two functions.

Definition 4 (Splittable PKE). *A labeled CCA-secure PKE scheme $\mathcal{PKE} = (\mathsf{KeyGen}, \mathsf{Enc}, \mathsf{Dec})$ is splittable if there exists a pair of two efficiently computable functions (f, g) such that the followings hold:*

1. *for any $(\mathsf{pk}, \mathsf{sk}) \leftarrow \mathsf{KeyGen}(1^{\kappa})$, string $\mathsf{label} \in \{0,1\}^*$, plaintext $pw \in \mathcal{P}$ and randomness $r \in \{0,1\}^*$, we have $c = (u, v) = \mathsf{Enc}(\mathsf{pk}, \mathsf{label}, pw; r)$, where $u = f(\mathsf{pk}, pw, r)$ and $v = g(\mathsf{pk}, \mathsf{label}, pw, r)$. Moreover, the first part u of the ciphertext $c = (u, v)$ fixes the plaintext pw in the sense that for any v' and $\mathsf{label}' \in \{0,1\}^*$, the probability that $\mathsf{Dec}(\mathsf{sk}, \mathsf{label}', (u, v')) \notin \{\perp, pw\}$ is negligible in κ over the random choices of sk and r;*
2. *the security of $\mathcal{PKE}$ still holds in a CCA game with modified challenge phase: the adversary $\mathcal{A}$ first submits two equal-length plaintexts $pw_0, pw_1 \in \mathcal{P}$. Then, the challenger $\mathcal{C}$ chooses a random bit $b^* \leftarrow_r \{0,1\}$, randomness $r^* \leftarrow_r \{0,1\}^*$, and returns $u^* = f(\mathsf{pk}, pw_{b^*}, r^*)$ to $\mathcal{A}$. Upon receiving u^*, $\mathcal{A}$ outputs a string $\mathsf{label} \in \{0,1\}^*$. Finally, $\mathcal{C}$ computes $v^* = g(\mathsf{pk}, \mathsf{label}, pw_{b^*}, r^*)$, and returns the challenge ciphertext $c^* = (u^*, v^*)$ to $\mathcal{A}$;*

Definition 4 captures the "splittable" property in both the functionality and the security of the PKE scheme. In particular, the modified CCA game allows the adversary to see the first part u^* of c^* and then adaptively determine label to form the complete challenge ciphertext $c^* = (u^*, v^*)$. We note that similar properties had been used in the context of identity-based encryption (IBE) [4,60], where one part of the ciphertext is defined as a function of the plaintext, and the other part is a function of the user identity. By applying generic transformations such as the CHK technique [20] from IBE (with certain property) to PKE, it is promising to get a splittable PKE such that the g function simply outputs a tag or a signature which can be used to publicly verify the validity of the whole ciphertext. Finally, we stress that the notion of splittable PKE is not our main goal, but rather a crucial intermediate step to reaching two-round PAKE.

3.2 Approximate Smooth Projective Hash Functions

Smooth projective hash (SPH) functions were first introduced by Cramer and Shoup [22] for achieving CCA-secure PKEs. Later, several works [29,39] extended the notion for PAKE. Here, we tailor the definition of approximate SPH (ASPH) in [38] to our application. Formally, let $\mathcal{PKE} = (\mathsf{KeyGen}, \mathsf{Enc}, \mathsf{Dec})$ be a splittable PKE scheme with respect to functions (f, g), and let $\mathcal{P}$ be an efficiently recognizable plaintext space of $\mathcal{PKE}$. As in [38], we require that $\mathcal{PKE}$ defines a notion of ciphertext validity in the sense that the validity of a label-ciphertext pair (label, c) with respect to any public key pk can be efficiently determined using pk alone, and all honestly generated ciphertexts are valid. We also assume that given a valid ciphertext c, one can easily parse $c = (u, v)$ as the outputs of (f, g). Now, fix a key pair $(\mathsf{pk}, \mathsf{sk}) \leftarrow \mathsf{KeyGen}(1^\kappa)$, and let C_{pk} denote the set of valid label-ciphertexts with respect to pk. Define sets X, L and $\bar{L}$ as follows:

$$X = \{(\mathsf{label}, c, pw) \mid (\mathsf{label}, c) \in C_{\mathsf{pk}}; pw \in \mathcal{P}\}$$
$$L = \{(\mathsf{label}, c, pw) \in X \mid \mathsf{label} \in \{0, 1\}^*; c = \mathsf{Enc}(\mathsf{pk}, \mathsf{label}, pw)\}$$
$$\bar{L} = \{(\mathsf{label}, c, pw) \in X \mid \mathsf{label} \in \{0, 1\}^*; pw = \mathsf{Dec}(\mathsf{sk}, \mathsf{label}, c)\}$$

By the definitions, for any ciphertext c and label $\in \{0, 1\}^*$, there is at most a single plaintext $pw \in \mathcal{P}$ such that $(\mathsf{label}, c, pw) \in \bar{L}$.

Definition 5 (ϵ-approximate SPH). *An ϵ-approximate SPH function is defined by a sampling algorithm that, given a public key pk of $\mathcal{PKE}$, outputs $(K, \ell, \{\mathsf{H}_{\mathsf{hk}} : X \to \{0, 1\}^\ell\}_{\mathsf{hk} \in K}, S, \mathsf{Proj} : K \to S)$ such that*

- *There are efficient algorithms for (1) sampling a hash key $\mathsf{hk} \leftarrow_r K$, (2) computing $\mathsf{H}_{\mathsf{hk}}(x) = \mathsf{H}_{\mathsf{hk}}(u, pw)$ for all $\mathsf{hk} \in K$ and $x = (\mathsf{label}, (u, v), pw) \in X$,[5] and (3) computing $\mathsf{hp} = \mathsf{Proj}(\mathsf{hk})$ for all $\mathsf{hk} \in K$.*
- *For all $x = (\mathsf{label}, (u, v), pw) \in L$ and randomness r such that $u = f(\mathsf{pk}, pw, r)$ and $v = g(\mathsf{pk}, \mathsf{label}, pw, r)$, there exists an efficient algorithm computing the value $\mathsf{Hash}(\mathsf{hp}, x, r) = \mathsf{Hash}(\mathsf{hp}, (u, pw), r)$, and satisfies $\Pr[\mathsf{Ham}(\mathsf{H}_{\mathsf{hk}}(u, pw), \mathsf{Hash}(\mathsf{hp}, (u, pw), r)) \geq \epsilon \cdot \ell] = \mathsf{negl}(\kappa)$ over the choice of $\mathsf{hk} \leftarrow_r K$.*
- *For any (even unbounded) function $h : S \to X \backslash \bar{L}, \mathsf{hk} \leftarrow_r K, \mathsf{hp} = \mathsf{Proj}(\mathsf{hk}), x = h(\mathsf{hp})$ and $\rho \leftarrow_r \{0, 1\}^\ell$, the statistical distance between $(\mathsf{hp}, \mathsf{H}_{\mathsf{hk}}(x))$ and (hp, ρ) is negligible in the security parameter κ.*

Compared to the ASPH notion in [38], our ASPH notion in Definition 5 mainly has three modifications: (1) the projection function only depends on the hash key; (2) the value $\mathsf{H}_{\mathsf{hk}}(x) = \mathsf{H}_{\mathsf{hk}}(u, pw)$ is determined by the hash key hk, the first part u of the ciphertext $c = (u, v)$, as well as the plaintext pw (i.e., it is independent from the pair (label, v)); and (3) the smoothness property holds even for adaptive choice of $x = h(\mathsf{hp}) \notin \bar{L}$. Looking ahead, the first modification allows us to achieve PAKE with two-round messages, whereas the last two are needed for proving the security of the resulting PAKE. One can check that the PKEs with associated SPH (based on either DDH or decisional linear assumptions) in [39] satisfy Definition 5 with $\epsilon = 0$ (under certain choices of f and g). We will construct a splittable PKE with associated ASPH from lattices in Sect. 5.

3.3 A Framework for Two-Round PAKE

Let $\mathcal{PKE} = (\mathsf{KeyGen}, \mathsf{Enc}, \mathsf{Dec})$ be a splittable PKE scheme with respect to functions (f, g). Let $(K, \ell, \{\mathsf{H}_{\mathsf{hk}} : X \to \{0, 1\}^\ell\}_{\mathsf{hk} \in K}, S, \mathsf{Proj} : K \to S)$ be the associated ϵ-approximate SPH for some $\epsilon \in (0, 1/2)$. Let the session key space be $\{0, 1\}^\kappa$, where κ is the security parameter. Let $\mathsf{ECC} : \{0, 1\}^\kappa \to \{0, 1\}^\ell$ be

---

[5] For all $x = (\mathsf{label}, (u, v), pw) \in X$, we slightly abuse the notation $\mathsf{H}_{\mathsf{hk}}(x) = \mathsf{H}_{\mathsf{hk}}(u, pw)$ by omitting (label, v) from its inputs. Similarly, the notation $\mathsf{Hash}(\mathsf{hp}, x, r) = \mathsf{Hash}(\mathsf{hp}, (u, pw), r)$ will be used later.

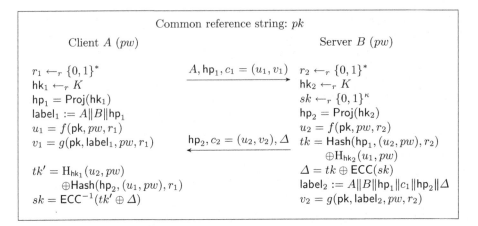

Fig. 1. PAKE from splittable PKE with ASPH

an error-correcting code which can correct 2ϵ-fraction of errors, and let ECC^{-1} : $\{0,1\}^\ell \to \{0,1\}^\kappa$ be the decoding algorithm. We assume that for uniformly distributed $\rho \in \{0,1\}^\ell$, the distribution of $w = \mathsf{ECC}^{-1}(\rho)$ conditioned on $w \neq \perp$ is uniform over $\{0,1\}^\kappa$. A high-level overview of our PAKE is given in Fig. 1.

Public Parameters. The public parameter consists of a public key pk of the scheme $\mathcal{PKE}$, which can be generated by a trusted third party using $\mathsf{KeyGen}(1^\kappa)$. No users in the system need to know the secret key corresponding to pk.

Protocol Execution. Consider an execution of the protocol between a client A and a server B holding a shared password $pw \in \mathcal{D} \subset \mathcal{P}$, where $\mathcal{D}$ is the set of valid passwords in the system. First, A chooses random coins $r_1 \leftarrow_r \{0,1\}^*$ for encryption, a hash key $\mathsf{hk}_1 \leftarrow_r K$ for the ASPH, and computes the projection key $\mathsf{hp}_1 = \mathsf{Proj}(\mathsf{hk}_1)$. Then, it defines $\mathsf{label}_1 := A\|B\|\mathsf{hp}_1$, and computes $(u_1, v_1) = \mathsf{Enc}(\mathsf{pk}, \mathsf{label}_1, pw; r_1)$, where $u_1 = f(\mathsf{pk}, pw, r_1)$ and $v_1 = g(\mathsf{pk}, \mathsf{label}_1, pw, r_1)$. Finally, A sends $(A, \mathsf{hp}_1, c_1 = (u_1, v_1))$ to the server B.

Upon receiving $(A, \mathsf{hp}_1, c_1 = (u_1, v_1))$ from the client A, the server B checks if c_1 is a valid ciphertext with respect to pk and $\mathsf{label}_1 := A\|B\|\mathsf{hp}_1$.[6] If not, B rejects and aborts. Otherwise, B chooses random coins $r_2 \leftarrow_r \{0,1\}^*$ for encryption, a hash key $\mathsf{hk}_2 \leftarrow_r K$ for the ASPH, and a random session key $sk \leftarrow_r \{0,1\}^\kappa$. Then, it computes $\mathsf{hp}_2 = \mathsf{Proj}(\mathsf{hk}_2), u_2 = f(\mathsf{pk}, pw, r_2), tk = \mathsf{Hash}(\mathsf{hp}_1, (u_2, pw), r_2) \oplus \mathsf{H}_{\mathsf{hk}_2}(u_1, pw)$, and $\Delta = tk \oplus \mathsf{ECC}(sk)$. Finally, let $\mathsf{label}_2 := A\|B\|\mathsf{hp}_1\|c_1\|\mathsf{hp}_2\|\Delta$, the server B computes $v_2 = g(\mathsf{pk}, \mathsf{label}_2, pw, r_2)$, and sends the message $(\mathsf{hp}_2, c_2 = (u_2, v_2), \Delta)$ to the client A.

After receiving $(\mathsf{hp}_2, c_2 = (u_2, v_2), \Delta)$ from the server B, the client A checks if c_2 is a valid ciphertext with respect to pk and $\mathsf{label}_2 := A\|B\|\mathsf{hp}_1\|c_1\|\mathsf{hp}_2\|\Delta$. If not, A rejects and aborts. Otherwise, A computes $tk' = \mathsf{H}_{\mathsf{hk}_1}(u_2, pw) \oplus \mathsf{Hash}(\mathsf{hp}_2, (u_1, pw), r_1)$, and decodes to obtain $sk = \mathsf{ECC}^{-1}(tk' \oplus \Delta)$. If $sk = \perp$ (i.e., an error

---

[6] Recall that the validity of a ciphertext can be efficiently determined using pk alone.

occurs during decoding), A rejects and aborts. Otherwise, A accepts $sk \in \{0,1\}^{\kappa}$ as the shared session key. This completes the description of our protocol.

In the following, we say that a user (or an instance of a user) accepts an incoming message msg as *a valid protocol message* if no abort happens during the computations after receiving msg. Note that a client/server will only obtain a session key when he accepts a received message as a valid protocol message.

Correctness. It suffices to show that honestly users can obtain the same session key $sk \in \{0,1\}^{\kappa}$ with overwhelming probability. First, all honestly generated ciphertexts are valid. Second, $\mathsf{H}_{hk_1}(u_2, pw) \oplus \mathsf{Hash}(hp_1, (u_2, pw), r_2) \in \{0,1\}^{\ell}$ has at most ϵ-fraction non-zeros by the ϵ-approximate correctness of the ASPH. Similarly, $\mathsf{Hash}(hp_2, (u_1, pw), r_1) \oplus \mathsf{H}_{hk_2}(u_1, pw) \in \{0,1\}^{\ell}$ has at most ϵ-fraction non-zeros. Thus, $tk' \oplus tk$ has at most 2ϵ-fraction non-zeros. Since ECC can correct 2ϵ-fraction of errors by assumption, we have that $sk = \mathsf{ECC}^{-1}(tk' \oplus tk \oplus \mathsf{ECC}(sk))$ holds. This completes the correctness argument.

Security. We now show that the above PAKE is secure. Formally,

Theorem 1. *If* $\mathcal{PKE} = (\mathsf{KeyGen}, \mathsf{Enc}, \mathsf{Dec})$ *is a splittable CCA-secure PKE scheme associated with an ϵ-approximate SPH* $(K, \ell, \{\mathsf{H}_{hk} : X \to \{0,1\}^{\ell}\}_{hk \in K}, S, \mathsf{Proj} : K \to S)$, *and* $\mathsf{ECC} : \{0,1\}^{\kappa} \to \{0,1\}^{\ell}$ *is an error-correcting code which can correct 2ϵ-fraction of errors, then the above protocol is a secure PAKE.*

Before giving the proof, we first give some intuitions. Without loss of generality we assume $0 \in \mathcal{P} \backslash \mathcal{D}$ (i.e., 0 is not a valid password in the system). First, by the CCA-security of the PKE scheme $\mathcal{PKE}$, the adversary cannot obtain any useful information of the real password pw via the Execute query (i.e., by eavesdropping on a protocol execution). In particular, it is computationally indistinguishable for the adversary if the encryption of pw is replaced by an encryption of 0 in answering the Execute queries. Since $0 \notin \mathcal{D}$, by the smoothness of the ASPH we have that the session keys corresponding to the instances used in the Execute queries are indistinguishable from uniform in the adversary's view.

Second, if the adversary simply relays the messages between honest instances, the proof is the same for the Execute queries. In case that the adversary modifies the message (i.e., the label-ciphertext pair) output by some instance, then one can use the decryption oracle provided by the CCA-security to decrypt the modified ciphertext, and check if the decrypted result pw' is equal to the real password pw. For $pw' = pw$ the attack is immediately considered successful (note that this will only increase the advantage of the adversary). By the CCA-security of $\mathcal{PKE}$ and the fact that pw is uniformly chosen from $\mathcal{D}$ at random, we have $\Pr[pw' = pw]$ is at most $1/|\mathcal{D}|$. Thus, for $Q(\kappa)$ times on-line attacks, this will only increase the adversary's advantage by at most $Q(\kappa)/|\mathcal{D}|$. Otherwise (i.e., $pw' \neq pw$) we again have that the corresponding session key is indistinguishable from uniform in the adversary's view by the smoothness of the ASPH.

Proof. We now formally prove Theorem 1 via a sequence of games from G_0 to G_{10}, where G_0 is the real security game, and G_{10} is a random game with uniformly chosen session keys. The security is established by showing that the adversary's advantage in game G_0 and G_{10} will differ at most $Q(\kappa)/|\mathcal{D}|+\mathrm{negl}(\kappa)$. Let $\mathbf{Adv}_{\mathcal{A},i}(\kappa)$ be the adversary $\mathcal{A}$'s advantage in game G_i.

Game G_0: This game is the real security game as defined in Sect. 2.2, where all the oracle queries are honestly answered following the protocol specification.

Game G_1: This game is similar to game G_0 except that in answering each Execute query the value tk' is directly computed using the corresponding hash keys hk_1 and hk_2, i.e., $tk' = \mathsf{H}_{\mathsf{hk}_1}(u_2, pw) \oplus \mathsf{H}_{\mathsf{hk}_2}(u_1, pw)$.

Lemma 1. *Let* $(K, \ell, \{\mathsf{H}_{\mathsf{hk}} : X \rightarrow \{0,1\}^\ell\}_{\mathsf{hk} \in K}, S, \mathsf{Proj} : K \rightarrow S)$ *be an ϵ-approximate SPH, and* $\mathsf{ECC} : \{0,1\}^\kappa \rightarrow \{0,1\}^\ell$ *be an error-correcting code which can correct 2ϵ-fraction of errors, then* $|\mathbf{Adv}_{\mathcal{A},1}(\kappa) - \mathbf{Adv}_{\mathcal{A},0}(\kappa)| \leq \mathrm{negl}(\kappa)$.

Proof. Since the simulator knows both hk_1 and hk_2, this lemma follows from the approximate correctness of the ASPH and the correctness of the ECC.   □

Game G_2: This game is similar to game G_1 except that the ciphertext c_1 is replaced with an encryption of $0 \notin \mathcal{D}$ in answering each Execute query.

Lemma 2. *If* $\mathcal{PKE} = (\mathsf{KeyGen}, \mathsf{Enc}, \mathsf{Dec})$ *is a CCA-secure scheme, then we have that* $|\mathbf{Adv}_{\mathcal{A},2}(\kappa) - \mathbf{Adv}_{\mathcal{A},1}(\kappa)| \leq \mathrm{negl}(\kappa)$.

Proof. Since the adversary $\mathcal{A}$ can only make polynomial times Execute queries, it is enough to consider that $\mathcal{A}$ only makes a single Execute query by a standard hybrid argument. In this case, the only difference between game G_1 and G_2 is that the encryption of pw is replaced by an encryption of $0 \notin \mathcal{D}$. We now show that any PPT adversary $\mathcal{A}$ that distinguishes the two games with non-negligible advantage can be directly transformed into an algorithm $\mathcal{B}$ that breaks the CCA-security of the underlying $\mathcal{PKE}$ scheme with the same advantage.

Formally, given a challenge public key pk, the algorithm $\mathcal{B}$ sets pk as the CRS of the protocol, and interacts with $\mathcal{A}$ as in game G_1. When $\mathcal{B}$ has to answer the adversary's Execute(A, i, B, j) query, it first randomly chooses a hash key $\mathsf{hk}_1 \leftarrow_r K$ for the ASPH, and computes the projection key $\mathsf{hp}_1 = \mathsf{Proj}(\mathsf{hk}_1)$. Then, $\mathcal{B}$ submits two plaintexts $(pw, 0)$ and $\mathsf{label}_1 := A\|B\|\mathsf{hp}_1$ to its own challenger, and obtains a challenge ciphertext c_1^*. Finally, $\mathcal{B}$ uses c_1^* to form the answer of the Execute(A, i, B, j) query, and returns whatever $\mathcal{A}$ outputs as its own guess.

Note that if c_1^* is an encryption of pw, then $\mathcal{B}$ exactly simulates the attack environment of game G_1 for adversary $\mathcal{A}$, else it simulates the attack environment of G_2 for $\mathcal{A}$. Thus, if $\mathcal{A}$ can distinguish G_1 and G_2 with non-negligible advantage, then $\mathcal{B}$ can break the CCA-security of $\mathcal{PKE}$ with the same advantage.   □

Game G_3: This game is similar to game G_2 except that in answering each Execute query: (1) the value tk is directly computed by using the corresponding hash keys hk_1 and hk_2, i.e., $tk = \mathsf{H}_{\mathsf{hk}_1}(u_2, pw) \oplus \mathsf{H}_{\mathsf{hk}_2}(u_1, pw)$; (2) the ciphertext c_2 is replaced with an encryption of $0 \notin \mathcal{D}$.

Lemma 3. *If* $\mathcal{PKE}$ = (KeyGen, Enc, Dec) *is a splittable CCA-secure scheme,* $(K, \ell, \{H_{hk} : X \to \{0,1\}^\ell\}_{hk \in K}, S, \text{Proj} : K \to S)$ *is an* ϵ-*approximate SPH, and* ECC : $\{0,1\}^\kappa \to \{0,1\}^\ell$ *is an error-correcting code which can correct* 2ϵ-*fraction of errors, then we have that* $|\textbf{Adv}_{\mathcal{A},3}(\kappa) - \textbf{Adv}_{\mathcal{A},2}(\kappa)| \leq \text{negl}(\kappa)$.

Proof. This lemma can be shown by using a sequence of games similar to that from G_0 to G_2 except the modified CCA-security game considered in Definition 4 is used instead of the standard CCA-security game, we omit the details. □

Game G_4: This game is similar to game G_3 except that a random session key $\text{sk}_A^i = \text{sk}_B^j$ is set for both Π_A^i and Π_B^j in answering each Execute(A, i, B, j) query.

Lemma 4. *If* $(K, \ell, \{H_{hk} : X \to \{0,1\}^\ell\}_{hk \in K}, S, \text{Proj} : K \to S)$ *is an* ϵ-*approximate SPH, then we have that* $|\textbf{Adv}_{\mathcal{A},4}(\kappa) - \textbf{Adv}_{\mathcal{A},3}(\kappa)| \leq \text{negl}(\kappa)$.

Proof. Since both ciphertexts $c_1 = (u_1, v_1)$ and $c_2 = (u_2, v_2)$ in answering each Execute(A, i, B, j) query are encryptions of $0 \notin \mathcal{D}$, the value $tk' = tk = H_{hk_1}(u_2, pw) \oplus H_{hk_2}(u_1, pw)$ is statistically close to uniform by the smoothness of the ASPH. Thus, the masking part $\Delta = tk \oplus \text{ECC}(sk)$ in answering each Execute(A, i, B, j) query statistically hides $sk \in \{0,1\}^\kappa$ from the adversary $\mathcal{A}$. Since $sk \in \{0,1\}^\kappa$ is uniformly random, the modification in game G_4 can only introduce a negligible statistical difference. Since $\mathcal{A}$ can only make polynomial times Execute queries, this lemma follows by a standard hybrid argument. □

Game G_5: This game is similar to game G_4 except that the simulator generates the CRS pk by running (pk, sk) $\leftarrow$ KeyGen(1^κ), and keeps sk private.

Lemma 5. $\textbf{Adv}_{\mathcal{A},5}(\kappa) = \textbf{Adv}_{\mathcal{A},4}(\kappa)$.

Proof. This lemma follows from the fact that the modification from game G_4 to G_5 is just conceptual. □

Before continuing, we divide the adversary's Send query into three types depending on the message which may be sent as part of the protocol:

- Send$_0(A, i, B)$: the adversary prompts an unused instance Π_A^i to execute the protocol with partner B. This oracle updates $\text{pid}_A^i = B$, and returns the message $\text{msg}_1 = (A, \text{hp}_1, c_1)$ output by Π_A^i to the adversary.
- Send$_1(B, j, (A, \text{hp}_1, c_1))$: the adversary sends message $\text{msg}_1 = (A, \text{hp}_1, c_1)$ to an unused instance Π_B^j. This oracle updates $(\text{pid}_B^j, \text{sk}_B^j, \text{acc}_B^j, \text{term}_B^j)$ as appropriate, and returns the message $\text{msg}_2 = (\text{hp}_2, c_2, \Delta)$ output by Π_B^j to the adversary (only if Π_B^j accepts msg_1 as a valid protocol message).
- Send$_2(A, i, (\text{hp}_2, c_2, \Delta))$: the adversary sends message $\text{msg}_2 = (\text{hp}_2, c_2, \Delta)$ to instance Π_A^i. This oracle updates $(\text{sk}_B^j, \text{acc}_B^j, \text{term}_B^j)$ as appropriate.

Game G_6: This game is similar to game G_5 except that each $\mathsf{Send}_1(B, j, \mathsf{msg}_1' = (A', \mathsf{hp}_1', c_1'))$ query is handled as follows:

- If msg_1' was output by a previous $\mathsf{Send}_0(A', *, B)$ query, the simulator $\mathcal{C}$ performs exactly as in game G_5;
- Otherwise, let $\mathsf{label}_1' := A' \| B \| \mathsf{hp}_1'$, and distinguish the following two cases:
  - If c_1' is not a valid ciphertext with respect to pk and label_1', the simulator $\mathcal{C}$ rejects this query;
  - Else, $\mathcal{C}$ decrypts $(\mathsf{label}_1', c_1')$ using the secret key sk corresponding to pk, and let pw' be the decryption result. If pw' is equal to the real password pw shared by A and B (i.e., $pw' = pw$), the simulator $\mathcal{C}$ declares that $\mathcal{A}$ succeeds, and terminates the experiment. Otherwise, $\mathcal{C}$ answers this query as in game G_5 but sets the session key sk_B^j for instance Π_B^j by using an independently and uniformly chosen element from $\{0,1\}^\kappa$.

Lemma 6. *If $(K, \ell, \{\mathsf{H}_{\mathsf{hk}} : X \to \{0,1\}^\ell\}_{\mathsf{hk} \in K}, S, \mathsf{Proj} : K \to S)$ is an ϵ-approximate SPH, then we have that $\mathbf{Adv}_{\mathcal{A},5}(\kappa) \leq \mathbf{Adv}_{\mathcal{A},6}(\kappa) + \mathrm{negl}(\kappa)$.*

Proof. We only have to consider the case that $\mathsf{msg}_1' = (A', \mathsf{hp}_1', c_1')$ was not output by any previous $\mathsf{Send}_0(A', *, B)$ query and c_1' is a valid ciphertext with respect to pk and label_1' (note that B will always reject invalid ciphertexts in the real run of the protocol). Since $\mathcal{C}$ knows the secret key sk corresponding to pk in both game G_5 and G_6, it can always decrypt $(\mathsf{label}_1', c_1')$ to obtain the decryption result pw'. Obviously, the modification for the case $pw' = pw$ can only increase the advantage of the adversary $\mathcal{A}$. As for the case $pw' \neq pw$, we have $(\mathsf{label}_1', c_1', pw) \notin \bar{L}$. By the smoothness of the underlying ASPH (in Definition 5), the masking part $\Delta = tk \oplus \mathsf{ECC}(sk)$ output by Π_B^j statistically hides $sk \in \{0,1\}^\kappa$ from the adversary $\mathcal{A}$ with knowledge of $\mathsf{hp}_2 = \mathsf{Proj}(\mathsf{hk}_2)$ (because tk has a term $\mathsf{H}_{\mathsf{hk}_2}(u_1', pw)$ for $c_1' = (u_1', v_1')$ and $\mathsf{hk}_2 \leftarrow_r K$). Using the fact that sk is essentially uniformly chosen from $\{0,1\}^\kappa$, we have that the modification for the case $pw' \neq pw$ in game G_6 can only introduce a negligible statistical difference. In all, we have that $\mathbf{Adv}_{\mathcal{A},5}(\kappa) \leq \mathbf{Adv}_{\mathcal{A},6}(\kappa) + \mathrm{negl}(\kappa)$. $\square$

Game G_7: This game is similar to game G_6 except that each $\mathsf{Send}_2(A, i, \mathsf{msg}_2' = (\mathsf{hp}_2', c_2', \Delta'))$ query is handled as follows: let $\mathsf{msg}_1 = (A, \mathsf{hp}_1, c_1)$ be the message output by a previous $\mathsf{Send}_0(A, i, B)$ query (note that such a query must exist),

- If msg_2' was output by a previous $\mathsf{Send}_1(B, j, \mathsf{msg}_1)$ query, the simulator $\mathcal{C}$ performs as in game G_6 except that $\mathcal{C}$ computes tk' directly using the corresponding hash keys hk_1 and hk_2, and sets the session key $\mathsf{sk}_A^i = \mathsf{sk}_B^j$;
- Otherwise, let $\mathsf{label}_2' := A \| B \| \mathsf{hp}_1 \| c_1 \| \mathsf{hp}_2' \| \Delta'$, and distinguish the following two cases:
  - If c_2' is not a valid ciphertext with respect to pk and label_2', the simulator $\mathcal{C}$ rejects this query;
  - Else, $\mathcal{C}$ decrypts $(\mathsf{label}_2', c_2')$ using the secret key sk corresponding to pk, and let pw' be the decryption result. If $pw' = pw$, the simulator $\mathcal{C}$ declares that $\mathcal{A}$ succeeds, and terminates the experiment. Otherwise, $\mathcal{C}$ performs

the computations on behalf of Π_A^i as in game G_6. If Π_A^i accepts msg_2' as a valid protocol message, $\mathcal{C}$ sets the session key sk_A^i for instance Π_A^i by using an independently and uniformly chosen element from $\{0,1\}^\kappa$ (note that Π_A^i might reject msg_2' if the decoding algorithm returns $\perp$, and thus no session key is generated in this case, i.e., $\mathsf{acc}_A^i = 0$ and $\mathsf{sk}_A^i = \perp$).

Lemma 7. *If* $(K, \ell, \{\mathsf{H}_{\mathsf{hk}} : X \rightarrow \{0,1\}^\ell\}_{\mathsf{hk} \in K}, S, \mathsf{Proj} : K \rightarrow S)$ *is an ϵ-approximate SPH, and* $\mathsf{ECC} : \{0,1\}^\kappa \rightarrow \{0,1\}^\ell$ *is an error-correcting code which can correct 2ϵ-fraction of errors, then* $\mathbf{Adv}_{\mathcal{A},6}(\kappa) \leq \mathbf{Adv}_{\mathcal{A},7}(\kappa) + \mathsf{negl}(\kappa)$.

Proof. First, if both msg_1 and msg_2' were output by previous oracle queries, then the simulator $\mathcal{C}$ knows the corresponding hash keys hk_1 and hk_2 needed for computing tk', and it is just a conceptual modification to compute tk' using $(\mathsf{hk}_1, \mathsf{hk}_2)$ and set $\mathsf{sk}_A^i = \mathsf{sk}_B^j$. Second, as discussed in the proof of Lemma 6, $\mathcal{C}$ knows the secret key sk corresponding to pk in both game G_6 and G_7, it can always decrypt $(\mathsf{label}_2', c_2')$ to obtain the decryption result pw'. Obviously, the modification for the case $pw' = pw$ can only increase the advantage of the adversary $\mathcal{A}$. Moreover, if $pw' \neq pw$, we have $(\mathsf{label}_2', c_2', pw) \notin \bar{L}$. By the smoothness of the ASPH, the value $tk' \in \{0,1\}^\ell$ computed by Π_A^i is statistically close to uniform over $\{0,1\}^\ell$ (because tk' has a term $\mathsf{H}_{\mathsf{hk}_1}(u_2', pw)$ for $c_2' = (u_2', v_2')$). By our assumption on ECC^{-1}, if $sk = \mathsf{ECC}^{-1}(tk' \oplus \Delta') \neq \perp$, then it is statistically close to uniform over $\{0,1\}^\kappa$. Thus, the modification for the case $pw' \neq pw$ in game G_6 can only introduce a negligible statistical difference. In all, we can have that $\mathbf{Adv}_{\mathcal{A},6}(\kappa) \leq \mathbf{Adv}_{\mathcal{A},7}(\kappa) + \mathsf{negl}(\kappa)$ holds. $\square$

Game G_8: This game is similar to game G_7 except that the ciphertext c_1 is replaced with an encryption of $0 \notin \mathcal{D}$ in answering each $\mathsf{Send}_0(A, i, B)$ query.

Lemma 8. *If* $\mathcal{PKE} = (\mathsf{KeyGen}, \mathsf{Enc}, \mathsf{Dec})$ *is a CCA-secure scheme, we have that* $|\mathbf{Adv}_{\mathcal{A},8}(\kappa) - \mathbf{Adv}_{\mathcal{A},7}(\kappa)| \leq \mathsf{negl}(\kappa)$.

Proof. By a standard hybrid argument, it is enough to consider that $\mathcal{A}$ only makes a single $\mathsf{Send}_0(A, i, B)$ query. In this case, the only difference between game G_8 and G_7 is that the encryption of pw is replaced with an encryption of $0 \notin \mathcal{D}$. We now show that any PPT adversary $\mathcal{A}$ that distinguishes the two games with non-negligible advantage can be directly transformed into an algorithm $\mathcal{B}$ that breaks the CCA-security of the underlying $\mathcal{PKE}$ scheme.

Formally, given a challenge public key pk, the algorithm $\mathcal{B}$ sets pk as the CRS of the protocol, and simulates the attack environment for $\mathcal{A}$ as in game G_7. When $\mathcal{B}$ has to answer the adversary's $\mathsf{Send}_0(A, i, B)$ query, it first randomly chooses a hash key $\mathsf{hk}_1 \leftarrow_r K$ for the ASPH, and computes the projection key $\mathsf{hp}_1 = \mathsf{Proj}(\mathsf{hk}_1)$. Then, $\mathcal{B}$ submits two plaintexts $(pw, 0)$ and $\mathsf{label}_1 := A\|B\|\mathsf{hp}_1$ to its own challenger, and obtains a challenge ciphertext c_1^*. Finally, $\mathcal{B}$ sends $(A, \mathsf{hp}_1, c_1^*)$ to the adversary $\mathcal{A}$. When $\mathcal{B}$ has to decrypt some valid label-ciphertext pair $(\mathsf{label}_1', c_1') \neq (\mathsf{label}_1, c_1^*)$, it submits $(\mathsf{label}_1', c_1')$ to its own CCA-security challenger for decryption. At some time, the adversary $\mathcal{A}$ outputs a bit $b \in \{0,1\}$, $\mathcal{B}$ outputs b as its own guess.

Note that if c_1^* is an encryption of pw, then $\mathcal{B}$ exactly simulates the attack environment of game G_7 for adversary $\mathcal{A}$, else it simulates the attack environment of game G_8 for $\mathcal{A}$. Thus, if $\mathcal{A}$ can distinguish game G_7 and G_8 with non-negligible advantage, then $\mathcal{B}$ can break the CCA-security of the PKE scheme $\mathcal{PKE}$ with the same advantage, which completes the proof.   □

Game G_9: This game is similar to game G_8 except that each $\mathsf{Send}_1(B, j, \mathsf{msg}_1' = (A', \mathsf{hp}_1', c_1'))$ query is handled as follows:

- If msg_1' was output by a previous $\mathsf{Send}_0(A', *, B)$ query, the simulator $\mathcal{C}$ performs as in game G_8 except that it computes tk directly using the corresponding hash keys $(\mathsf{hk}_1, \mathsf{hk}_2)$, and sets the session key sk_B^j for instance $\varPi_B^j$ by using an independently and uniformly chosen element from $\{0,1\}^\kappa$;
- Otherwise, $\mathcal{C}$ performs exactly as in game G_8.

Lemma 9. *If $(K, \ell, \{\mathsf{H}_{\mathsf{hk}} : X \to \{0,1\}^\ell\}_{\mathsf{hk} \in K}, S, \mathsf{Proj} : K \to S)$ is an ϵ-approximate SPH, and $\mathsf{ECC} : \{0,1\}^\kappa \to \{0,1\}^\ell$ is an error-correcting code which can correct 2ϵ-fraction of errors, then $|\mathbf{Adv}_{\mathcal{A},9}(\kappa) - \mathbf{Adv}_{\mathcal{A},8}(\kappa)| \leq \mathsf{negl}(\kappa)$.*

Proof. Note that if msg_1' was output by a previous $\mathsf{Send}_0(A', *, B)$ query, then we have that (1) the simulator $\mathcal{C}$ knows the corresponding hash keys $(\mathsf{hk}_1, \mathsf{hk}_2)$ and (2) $c_1' = (u_1', v_1')$ is an encryption of $0 \notin \mathcal{D}$. In other words, $\mathcal{C}$ can directly compute tk using $(\mathsf{hk}_1, \mathsf{hk}_2)$, and tk is statistically close to uniform (because $pw \neq 0$, and tk has a term $\mathsf{H}_{\mathsf{hk}_2}(u_1', pw)$ that is statistically close to uniform by the smoothness of the ASPH). Thus, the masking part $\Delta = tk \oplus \mathsf{ECC}(sk)$ output by $\varPi_B^j$ statistically hides $sk \in \{0,1\}^\kappa$ from the adversary $\mathcal{A}$. Since sk is essentially uniformly chosen from $\{0,1\}^\kappa$, we have that the modification in game G_9 can only introduce a negligible statistical difference, which means that $|\mathbf{Adv}_{\mathcal{A},9}(\kappa) - \mathbf{Adv}_{\mathcal{A},8}(\kappa)| \leq \mathsf{negl}(\kappa)$.   □

Game G_{10}: This game is similar to game G_9 except that each $\mathsf{Send}_1(B, j, \mathsf{msg}_1' = (A', \mathsf{hp}_1', c_1'))$ query is handled as follows:

- If msg_1' was output by a previous $\mathsf{Send}_0(A', *, B)$ query, the simulator $\mathcal{C}$ performs as in game G_9 except that the ciphertext c_2 is replaced with an encryption of $0 \notin \mathcal{D}$;
- Otherwise, $\mathcal{C}$ performs exactly as in game G_9.

Lemma 10. *If $\mathcal{PKE} = (\mathsf{KeyGen}, \mathsf{Enc}, \mathsf{Dec})$ is a splittable CCA-secure scheme, then $|\mathbf{Adv}_{\mathcal{A},10}(\kappa) - \mathbf{Adv}_{\mathcal{A},9}(\kappa)| \leq \mathsf{negl}(\kappa)$.*

Proof. As before, it is enough to consider that $\mathcal{A}$ only makes a single $\mathsf{Send}_1(B, j, \mathsf{msg}_1' = (A', \mathsf{hp}_1', c_1'))$ query with msg_1' output by some $\varPi_{A'}^i$. We now show that any PPT adversary $\mathcal{A}$ that distinguishes the two games with non-negligible advantage can be directly transformed into an algorithm $\mathcal{B}$ that breaks the modified CCA-security game of the underlying $\mathcal{PKE}$ scheme with the same advantage.

Formally, given a challenge public key pk, the algorithm $\mathcal{B}$ sets pk as the CRS of the protocol, and interacts with $\mathcal{A}$ as in game G_9. When $\mathcal{B}$ has to answer a $\mathsf{Send}_1(B, j, \mathsf{msg}_1' = (A', \mathsf{hp}_1', c_1'))$ query for some $c_1' = (u_1', v_1')$, it first randomly chooses a hash key $\mathsf{hk}_2 \leftarrow_r K$ for the ASPH, a random session key $sk \leftarrow_r \{0,1\}^\kappa$, and computes $\mathsf{hp}_2 = \mathsf{Proj}(\mathsf{hk}_2)$. Then, $\mathcal{B}$ submits two plaintexts $(pw, 0)$ to its own challenger. After obtaining u_2^*, $\mathcal{B}$ computes $tk = \mathsf{H}_{\mathsf{hk}_1}(u_2^*, pw) \oplus \mathsf{H}_{\mathsf{hk}_2}(u_1', pw)$, $\Delta = tk \oplus \mathsf{ECC}(sk)$, and submits $\mathsf{label}_2 := A'\|B\|\mathsf{hp}_1'\|c_1'\|\mathsf{hp}_2\|\Delta$ to its own modified CCA-security challenger to obtain the challenge ciphertext $c_2^* = (u_2^*, v_2^*)$. Finally, $\mathcal{B}$ sends $(\mathsf{hp}_2, c_2^*, \Delta)$ to the adversary $\mathcal{A}$. When $\mathcal{B}$ has to decrypt some valid label-ciphertext pair $(\mathsf{label}_2', c_2') \neq (\mathsf{label}_2, c_2^*)$, it submits $(\mathsf{label}_2', c_2')$ to its own challenger for decryption. At some time, the adversary $\mathcal{A}$ outputs a bit $b \in \{0, 1\}$, $\mathcal{B}$ outputs b as its own guess.

Note that if c_2^* is an encryption of pw, then $\mathcal{B}$ perfectly simulates the attack environment of game G_9 for adversary $\mathcal{A}$, else it simulates the attack environment of G_{10} for $\mathcal{A}$. Thus, if $\mathcal{A}$ can distinguish game G_9 and G_{10} with non-negligible advantage, then algorithm $\mathcal{B}$ can break the modified CCA-security of the PKE scheme $\mathcal{PKE}$ with the same advantage, which completes the proof. $\square$

Lemma 11. *If the adversary $\mathcal{A}$ only makes at most $Q(\kappa)$ times on-line attacks, then we have that $\mathbf{Adv}_{\mathcal{A},10}(\kappa) \leq Q(\kappa)/|\mathcal{D}| + \mathrm{negl}(\kappa)$.*

Proof. Let $\mathcal{E}$ be the event that $\mathcal{A}$ submits a ciphertext that decrypts to the real password pw. If $\mathcal{E}$ does not happen, we have that the advantage of $\mathcal{A}$ is negligible in κ (because all the session keys are uniformly chosen at random). Now, we estimate the probability that $\mathcal{E}$ happens. Since in game G_{10}, all the ciphertexts output by oracle queries are encryptions of $0 \notin \mathcal{D}$, the adversary cannot obtain useful information of the real password pw via the oracle queries. Thus, for any adversary $\mathcal{A}$ that makes at most $Q(\kappa)$ times on-line attacks, the probability that $\mathcal{E}$ happens is at most $Q(\kappa)/|\mathcal{D}|$, i.e., $\Pr[E] \leq Q(\kappa)/|\mathcal{D}|$. By a simple calculation, we have $\mathbf{Adv}_{\mathcal{A},10}(\kappa) \leq Q(\kappa)/|\mathcal{D}| + \mathrm{negl}(\kappa)$. $\square$

In all, we have that $\mathbf{Adv}_{\mathcal{A},0}(\kappa) \leq Q(\kappa)/|\mathcal{D}| + \mathrm{negl}(\kappa)$ by Lemmas 1–11. This completes the proof of Theorem 1. $\square$

4 Lattices

In this section, we first give some backgrounds on lattices. Then, we show an adaptive smoothing lemma for q-ary lattices, which was crucial for our concrete instantiation of two-round PAKE from lattices in Sect. 5.

4.1 Backgrounds on Lattices

An m-dimensional full-rank lattice $\mathbf{\Lambda} \subset \mathbb{R}^m$ is the set of all integral combinations of m linearly independent vectors $\mathbf{B} = (\mathbf{b}_1, \ldots, \mathbf{b}_m) \in \mathbb{R}^{m \times m}$, i.e., $\mathbf{\Lambda} = \mathcal{L}(\mathbf{B}) = \{\sum_{i=1}^m x_i \mathbf{b}_i : x_i \in \mathbb{Z}\}$. The dual lattice of $\mathbf{\Lambda}$, denote $\mathbf{\Lambda}^*$ is defined to be $\mathbf{\Lambda}^* = \{\mathbf{x} \in \mathbb{R}^m : \forall \mathbf{v} \in \mathbf{\Lambda}, \langle \mathbf{x}, \mathbf{v} \rangle \in \mathbb{Z}\}$. For $\mathbf{x} \in \mathbf{\Lambda}$, define the Gaussian function $\rho_{s,\mathbf{c}}(\mathbf{x})$ over $\mathbf{\Lambda} \subseteq \mathbb{Z}^m$ centered at $\mathbf{c} \in \mathbb{R}^m$ with parameter $s > 0$ as

$\rho_{s,\mathbf{c}}(\mathbf{x}) = \exp(-\pi \|\mathbf{x} - \mathbf{c}\|^2/s^2)$. Let $\rho_{s,\mathbf{c}}(\mathbf{\Lambda}) = \sum_{\mathbf{x} \in \mathbf{\Lambda}} \rho_{s,\mathbf{c}}(\mathbf{x})$, and define the discrete Gaussian distribution over $\mathbf{\Lambda}$ as $D_{\mathbf{\Lambda},s,\mathbf{c}}(\mathbf{y}) = \frac{\rho_{s,\mathbf{c}}(\mathbf{y})}{\rho_{s,\mathbf{c}}(\mathbf{\Lambda})}$, where $\mathbf{y} \in \mathbf{\Lambda}$. The subscripts s and $\mathbf{c}$ are taken to be 1 and $\mathbf{0}$ (resp.) when omitted.

Lemma 12 [48,51]. *For any positive integer $m \in \mathbb{Z}$, and large enough $s \geq \omega(\sqrt{\log m})$, we have $\Pr_{\mathbf{x} \leftarrow_r D_{\mathbb{Z}^m,s}}[\|\mathbf{x}\| > s\sqrt{m}] \leq 2^{-m+1}$.*

First introduced in [48], the smoothing parameter $\eta_\epsilon(\mathbf{\Lambda})$ for any real $\epsilon > 0$ is defined as the smallest real $s > 0$ s.t. $\rho_{1/s}(\mathbf{\Lambda}^* \backslash \{\mathbf{0}\}) \leq \epsilon$.

Lemma 13 [48]. *For any m-dimensional lattice $\mathbf{\Lambda}$, $\eta_\epsilon(\mathbf{\Lambda}) \leq \sqrt{m}/\lambda_1(\mathbf{\Lambda}^*)$, where $\epsilon = 2^{-m}$, and $\lambda_1(\mathbf{\Lambda}^*)$ is the length of the shortest vector in lattice $\mathbf{\Lambda}^*$.*

Lemma 14 [30]. *Let $\mathbf{\Lambda}, \mathbf{\Lambda}'$ be m-dimensional lattices, with $\mathbf{\Lambda}' \subseteq \mathbf{\Lambda}$. Then, for any $\epsilon \in (0, 1/2)$, any $s \geq \eta_\epsilon(\mathbf{\Lambda}')$, and any $\mathbf{c} \in \mathbb{R}^m$, the distribution of $(D_{\mathbf{\Lambda},s,\mathbf{c}} \mod \mathbf{\Lambda}')$ is within distance at most 2ϵ of uniform over $(\mathbf{\Lambda} \mod \mathbf{\Lambda}')$.*

Let $\mathbf{A} \in \mathbb{Z}_q^{n \times m}$, define lattices $\mathbf{\Lambda}_q^\perp(\mathbf{A}) = \{\mathbf{e} \in \mathbb{Z}^m \text{ s.t. } \mathbf{A}\mathbf{e} = 0 \mod q\}$ and $\mathbf{\Lambda}_q(\mathbf{A}) = \{\mathbf{y} \in \mathbb{Z}^m \text{ s.t. } \exists \mathbf{s} \in \mathbb{Z}^n, \mathbf{A}^t\mathbf{s} = \mathbf{y} \mod q\}$. We have the following facts.

Lemma 15 [30]. *Let integers $n, m \in \mathbb{Z}$ and prime q satisfy $m \geq 2n \log q$. Then, for all but an at most $2q^{-n}$ fraction of $\mathbf{A} \in \mathbb{Z}_q^{n \times m}$, we have that (1) the columns of $\mathbf{A}$ generate $\mathbb{Z}_q^n$, (2) $\lambda_1^\infty(\mathbf{\Lambda}_q(\mathbf{A})) \geq q/4$, and (3) the smoothing parameter $\eta_\epsilon(\mathbf{\Lambda}_q^\perp(\mathbf{A})) \leq \omega(\sqrt{\log m})$ for some $\epsilon = \mathrm{negl}(\kappa)$.*

Lemma 16 [30]. *Assume the columns of $\mathbf{A} \in \mathbb{Z}_q^{n \times m}$ generate $\mathbb{Z}_q^n$, and let $\epsilon \in (0, 1/2)$ and $s \geq \eta_\epsilon(\mathbf{\Lambda}_q^\perp(\mathbf{A}))$. Then for $\mathbf{e} \sim D_{\mathbb{Z}^m,s}$, the distribution of the syndrome $\mathbf{u} = \mathbf{A}\mathbf{e} \mod q$ is within statistical distance 2ϵ of uniform over $\mathbb{Z}_q^n$.*

Furthermore, fix $\mathbf{u} \in \mathbb{Z}_q^n$ and let $\mathbf{v} \in \mathbb{Z}^m$ be an arbitrary solution to $\mathbf{A}\mathbf{v} = \mathbf{u} \mod q$. Then the conditional distribution of $\mathbf{e} \sim D_{\mathbb{Z}^m,s}$ given $\mathbf{A}\mathbf{e} = \mathbf{u} \mod q$ is exactly $\mathbf{v} + D_{\mathbf{\Lambda}_q^\perp(\mathbf{A}),s,-\mathbf{v}}$.

There exist efficient algorithms [5,7,47] to generate almost uniform matrix $\mathbf{A}$ together with a trapdoor (or a short basis of $\mathbf{\Lambda}_q^\perp(\mathbf{A})$).

Proposition 1 [47]. *Given any integers $n \geq 1$, $q > 2$, sufficiently large $m = O(n \log q)$, and $k = \lceil \log_2 q \rceil$, there is an efficient algorithm $\mathsf{TrapGen}(1^n, 1^m, q)$ that outputs a matrix $\mathbf{A} \in \mathbb{Z}_q^{n \times m}$ and a trapdoor $\mathbf{R} \in \mathbb{Z}_q^{(m-nk) \times nk}$ such that $s_1(\mathbf{R}) \leq \sqrt{m} \cdot \omega(\sqrt{\log n})$, and $\mathbf{A}$ is $\mathrm{negl}(n)$-close to uniform.*

Moreover, given any $\mathbf{y} = \mathbf{A}^t\mathbf{s} + \mathbf{e} \in \mathbb{Z}_q^m$ satisfying $\|\mathbf{e}\| \leq \frac{q}{2\sqrt{5(s_1(\mathbf{R})^2+1)}}$, there exists an efficient algorithm $\mathsf{Solve}(\mathbf{A}, \mathbf{R}, \mathbf{y})$ that outputs the vector $\mathbf{s} \in \mathbb{Z}_q^n$.

Let $\mathrm{dist}(\mathbf{z}, \mathbf{\Lambda}_q(\mathbf{A}))$ be the distance of the vector $\mathbf{z}$ from the lattice $\mathbf{\Lambda}_q(\mathbf{A})$. For any $\mathbf{A} \in \mathbb{Z}_q^{n \times m}$, define $Y_{\mathbf{A}} = \{\tilde{\mathbf{y}} \in \mathbb{Z}_q^m : \forall a \in \mathbb{Z}_q \backslash \{0\}, \mathrm{dist}(a\tilde{\mathbf{y}}, \mathbf{\Lambda}_q(\mathbf{A})) \geq \sqrt{q}/4\}$.

Lemma 17 [30,38]. *Let integers n, m and prime q satisfy $m \geq 2n \log q$. Let $\gamma \geq \sqrt{q} \cdot \omega(\sqrt{\log n})$. Then, for all but a negligible fraction of $\mathbf{A} \in \mathbb{Z}_q^{n \times m}$, and for any $\mathbf{z} \in Y_A$, the distribution of $(\mathbf{A}\mathbf{e}, \mathbf{z}^t\mathbf{e})$ is statistically close to uniform over $\mathbb{Z}_q^n \times \mathbb{Z}_q$, where $\mathbf{e} \sim D_{\mathbb{Z}^m,\gamma}$.*

Lemma 18 [38]. *Let κ be the security parameter. Let intergers n_1, n_2, m and prime q satisfy $m \geq (n_1 + n_2 + 1) \log q$ and $n_1 = 2(n_2 + 1) + \omega(\log \kappa)$. Then, for all but a negligible fraction of $\mathbf{B} \in \mathbb{Z}_q^{m \times n_1}$, the probability that there exist numbers $a, a' \in \mathbb{Z}_q \backslash \{0\}$, vectors $\mathbf{w} \neq \mathbf{w}' \in \mathbb{Z}_q^{n_2}$, and a vector $\mathbf{c} \in \mathbb{Z}_q^m$, s.t.*

$$\mathsf{dist}(a\mathbf{y}, \Lambda_q(\mathbf{B}^t)) \leq \sqrt{q}/4 \text{ and } \mathsf{dist}(a'\mathbf{y}', \Lambda_q(\mathbf{B}^t)) \leq \sqrt{q}/4$$

is negligible in κ over the uniformly random choice of $\mathbf{U} \leftarrow_r \mathbb{Z}_q^{m \times (n_2+1)}$, where

$$\mathbf{y} = \mathbf{c} - \mathbf{U} \begin{pmatrix} 1 \\ \mathbf{w} \end{pmatrix} \text{ and } \mathbf{y}' = \mathbf{c} - \mathbf{U} \begin{pmatrix} 1 \\ \mathbf{w}' \end{pmatrix}.$$

Learning with Errors. For any positive integers $n, q \in \mathbb{Z}$, real $\alpha > 0$ and vector $\mathbf{s} \in \mathbb{Z}_q^n$, define the distribution $A_{\mathbf{s},\alpha} = \{(\mathbf{a}, \mathbf{a}^t\mathbf{s} + e \mod q) : \mathbf{a} \leftarrow_r \mathbb{Z}_q^n, e \leftarrow_r D_{\mathbb{Z},\alpha q}\}$. For any m independent samples $(\mathbf{a}_1, b_1), \ldots, (\mathbf{a}_m, b_m)$ from $A_{\mathbf{s},\alpha}$, we denote it in matrix form $(\mathbf{A}, \mathbf{b}) \in \mathbb{Z}_q^{n \times m} \times \mathbb{Z}_q^m$, where $\mathbf{A} = (\mathbf{a}_1, \ldots, \mathbf{a}_m)$ and $\mathbf{b} = (b_1, \ldots, b_m)^t$. We say that the $\mathsf{LWE}_{n,q,\alpha}$ problem is hard if, for uniformly random $\mathbf{s} \leftarrow_r \mathbb{Z}_q^n$ and given polynomially many samples, no PPT algorithm can recover $\mathbf{s}$ with non-negligible probability. The decisional LWE problem is asked to distinguish polynomially many samples from uniform. For certain parameters, the decisional LWE problem is polynomially equivalent to its search version, which is in turn known to be at least as hard as quantumly approximating SIVP on n-dimensional lattices to within polynomial factors in the worst case [53].

4.2   An Adaptive Smoothing Lemma for q-ary Lattices

Based on a good use of Lemma 17 from [30], the authors [38] constructed the first lattice-based ASPH with adaptive projection function [29,39] (i.e., the projection key is generated after given the input ciphertext). However, Lemma 17 is not enough to obtain a non-adaptive ASPH for constructing two-round PAKEs (where the ciphertext is chosen after seeing the projection key). Specifically, it provides no guarantee for the distribution of $\mathbf{z}^t\mathbf{e}$ when the choice of $\mathbf{z} \in Y_\mathbf{A}$ is dependent on $\mathbf{Ae} \in \mathbb{Z}_q^n$. In particular, it is possible that for each $\mathbf{z} \in Y_\mathbf{A}$, there is a negligible fraction of bad values $Bad_\mathbf{z} \subset \mathbb{Z}_q^n$ such that for all $\mathbf{Ae} \in Bad_\mathbf{z}$ the distribution of $\mathbf{z}^t\mathbf{e}$ is far from uniform (and thus given a fixed $\mathbf{u} = \mathbf{Ae} \in \mathbb{Z}_q^n$, the adversary may choose $\mathbf{z} \in Y_\mathbf{A}$ such that $\mathbf{u} \in Bad_\mathbf{z}$). Instead, we show a stronger result in Lemma 19, which is very crucial for our construction in Sect. 5.

Lemma 19. *Let positive integers $n, m \in \mathbb{Z}$ and prime q satisfy $m \geq 2n \log q$. Let $\gamma \geq 4\sqrt{mq}$. Then, for all but a negligible fraction of $\mathbf{A} \in \mathbb{Z}_q^{n \times m}$, and for any (even unbounded) function $h : \mathbb{Z}_q^n \rightarrow Y_\mathbf{A}$, the distribution of $(\mathbf{Ae}, \mathbf{z}^t\mathbf{e})$ is statistically close to uniform over $\mathbb{Z}_q^n \times \mathbb{Z}_q$, where $\mathbf{e} \sim D_{\mathbb{Z}^m, \gamma}$ and $\mathbf{z} = h(\mathbf{Ae})$.*

Proof. By Lemma 15, for all but a negligible fraction of $\mathbf{A} \in \mathbb{Z}_q^{n \times m}$, the columns of $\mathbf{A}$ generate $\mathbb{Z}_q^n$ and the length $\lambda_1(\Lambda_q(\mathbf{A}))$ (in the l_2 norm) of the shortest vector in $\Lambda_q(\mathbf{A})$ is at least $q/4$ (since $\lambda_1(\Lambda_q(\mathbf{A})) \geq \lambda_1^\infty(\Lambda_q(\mathbf{A})) \geq q/4$). Moreover, the smoothing parameter $\eta_\epsilon(\Lambda_q^\perp(\mathbf{A})) \leq \omega(\sqrt{\log m})$ for some negligible ϵ. In the following, we always assume that $\mathbf{A}$ satisfies the above properties. Since $\gamma \geq 4\sqrt{mq} > \eta_\epsilon(\Lambda_q^\perp(\mathbf{A}))$, by Lemma 16 the distribution of $\mathbf{Ae}$ mod q is within statistical distance 2ϵ of uniform over $\mathbb{Z}_q^n$, where $\mathbf{e} \sim D_{\mathbb{Z}^m,\gamma}$. Furthermore, fix $\mathbf{u} \in \mathbb{Z}_q^n$ and let $\mathbf{v}$ be an arbitrary solution to $\mathbf{Av} = \mathbf{u}$ mod q, the conditional distribution of $\mathbf{e} \sim D_{\mathbb{Z}^m,\gamma}$ given $\mathbf{Ae} = \mathbf{u}$ mod q is exactly $\mathbf{v} + D_{\Lambda_q^\perp(\mathbf{A}),\gamma,-\mathbf{v}}$. Thus, it is enough to show that for arbitrary $\mathbf{v} \in \mathbb{Z}^m$ and $\mathbf{z} = h(\mathbf{Av}) \in Y_{\mathbf{A}}$, the distribution $\mathbf{z}^t\mathbf{e}$ is statistically close to uniform over $\mathbb{Z}_q$, where $\mathbf{e} \sim D_{\Lambda_q^\perp(\mathbf{A}),\gamma,-\mathbf{v}}$.

Now, fix $\mathbf{v} \in \mathbb{Z}^m$ and $\mathbf{z} = h(\mathbf{Av}) \in Y_{\mathbf{A}}$, let $\mathbf{A}' = \begin{pmatrix} \mathbf{A} \\ \mathbf{z}^t \end{pmatrix} \in \mathbb{Z}_q^{(n+1)\times m}$. By the definition $Y_{\mathbf{A}} = \{\tilde{\mathbf{y}} \in \mathbb{Z}_q^m : \forall a \in \mathbb{Z}_q\backslash\{0\}, \mathsf{dist}(a\tilde{\mathbf{y}}, \Lambda_q(\mathbf{A})) \geq \sqrt{q}/4\}$, we have that the rows of $\mathbf{A}'$ are linearly independent over $\mathbb{Z}_q$. In other words, the columns of $\mathbf{A}'$ generate $\mathbb{Z}_q^{n+1}$. Let $\mathbf{x}$ be the shortest vector of $\Lambda_q(\mathbf{A}')$. Note that the lattice $\Lambda_q(\mathbf{A}')$ is obtained by adjoining the vector $\mathbf{z}$ to $\Lambda_q(\mathbf{A})$. Without loss of generality we assume $\mathbf{x} = \mathbf{y} + a\mathbf{z}$ for some $\mathbf{y} \in \Lambda_q(\mathbf{A})$ and $a \in \mathbb{Z}_q$. Then, if $a = 0$, we have $\|\mathbf{x}\| \geq q/4$ by the fact that $\lambda_1(\Lambda_q(\mathbf{A})) \geq q/4$. Otherwise, for any $a \in \mathbb{Z}_q\backslash\{0\}$, we have $\|\mathbf{x}\| \geq \mathsf{dist}(a\mathbf{z}, \Lambda_q(\mathbf{A})) \geq \sqrt{q}/4$. In all, we have that $\lambda_1(\Lambda_q(\mathbf{A}')) = \|\mathbf{x}\| \geq \sqrt{q}/4$. By Lemma 13 and the duality $\Lambda_q(\mathbf{A}') = q \cdot (\Lambda_q^\perp(\mathbf{A}'))^*$, we have $\eta_\epsilon(\Lambda_q^\perp(\mathbf{A}')) \leq 4\sqrt{mq} \leq \gamma$ for $\epsilon = 2^{-m}$.[7]

Since the columns of $\mathbf{A}' \in \mathbb{Z}_q^{(n+1)\times m}$ generate $\mathbb{Z}_q^{n+1}$, we have the set of syndromes $\{u = \mathbf{z}^t\mathbf{e} : \mathbf{e} \in \Lambda_q^\perp(\mathbf{A})\} = \mathbb{Z}_q$. By the fact $\Lambda_q^\perp(\mathbf{A}') = \Lambda_q^\perp(\mathbf{A}) \cap \Lambda_q^\perp(\mathbf{z}^t)$, the quotient group $(\Lambda_q^\perp(\mathbf{A})/\Lambda_q^\perp(\mathbf{A}'))$ is isomorphic to the set of syndromes $\mathbb{Z}_q$ via the mapping $\mathbf{e} + \Lambda_q^\perp(\mathbf{A}') \mapsto \mathbf{z}^t\mathbf{e}$ mod q. This means that computing $\mathbf{z}^t\mathbf{e}$ mod q for some $\mathbf{e} \in \Lambda_q^\perp(\mathbf{A})$ is equivalent to reducing $\mathbf{e}$ modulo the lattice $\Lambda_q^\perp(\mathbf{A}')$. By Lemma 14, for any $\epsilon = \mathsf{negl}(n)$, any $\gamma \geq \eta_\epsilon(\Lambda_q^\perp(\mathbf{A}'))$ and any $\mathbf{v} \in \mathbb{Z}^m$, the distribution of $D_{\Lambda_q^\perp(\mathbf{A}),\gamma,-\mathbf{v}}$ mod $\Lambda_q^\perp(\mathbf{A}')$ is within statistical distance at most 2ϵ of uniform over $(\Lambda_q^\perp(\mathbf{A})/\Lambda_q^\perp(\mathbf{A}'))$. Thus, the distribution $\mathbf{z}^t\mathbf{e}$ is statistically close to uniform over $\mathbb{Z}_q$, where $\mathbf{e} \sim D_{\Lambda_q^\perp(\mathbf{A}),\gamma,-\mathbf{v}}$. This completes the proof. □

5   Lattice-Based Splittable PKE with Associated ASPH

In order to construct a splittable PKE with associated ASPH from lattices, our basic idea is to incorporate the specific algebraic properties of lattices into the Naor-Yung paradigm [49,55], which is a generic construction of CCA-secure PKE scheme from any CPA-secure PKE scheme and simulation-sound non-interactive zero knowledge (NIZK) proof [55], and was used to achieve the first one-round PAKEs from DDH and decisional linear assumptions [39].

Looking ahead, we will use a CPA-secure PKE scheme from lattices and a simulation-sound NIZK proof for specific statements, so that we can freely apply Lemmas 18 and 19 to construct a non-adaptive approximate SPH and achieve the stronger smoothness property. Formally, we need a simulation-sound NIZK proof for the following relation:

---

[7] It is possible to set a smaller γ by a more careful analysis with $\epsilon = \mathsf{negl}(n)$.

$$R_{pke} := \left\{ \begin{array}{c} ((\mathbf{A}_0, \mathbf{A}_1, \mathbf{c}_0, \mathbf{c}_1, \beta), (\mathbf{s}_0, \mathbf{s}_1, \mathbf{w})) : \\[2mm] \left\| \mathbf{c}_0 - \mathbf{A}_0^t \begin{pmatrix} \mathbf{s}_0 \\ 1 \\ \mathbf{w} \end{pmatrix} \right\| \le \beta \wedge \left\| \mathbf{c}_1 - \mathbf{A}_1^t \begin{pmatrix} \mathbf{s}_1 \\ 1 \\ \mathbf{w} \end{pmatrix} \right\| \le \beta \end{array} \right\}$$

where $\mathbf{A}_0, \mathbf{A}_1 \in \mathbb{Z}_q^{n \times m}, \mathbf{c}_0, \mathbf{c}_1 \in \mathbb{Z}_q^m, \beta \in \mathbb{R}, \mathbf{s}_0, \mathbf{s}_1 \in \mathbb{Z}_q^{n_1}, \mathbf{w} \in \mathbb{Z}_q^{n_2}$ for some integers $n = n_1 + n_2 + 1, m, q \in \mathbb{Z}$. Note that under the existence of (enhanced) trapdoor permutations, there exist NIZK proofs with efficient prover for any NP relation [10,26,31]. Moreover, Sahai [55] showed that one can transform any general NIZK proof into a simulation-sound one. Thus, there exists a simulation-sound NIZK proof with efficient prover for the relation R_{pke}. In Sect. 5.3, we will also show how to directly construct an efficient one from lattices.

For our purpose, we require that the NIZK proof supports labels [1], which can be obtained from a normal NIZK proof by a standard way (e.g., appending the label to the statement [25,39]). Let (CRSGen, Prove, Verify) be a labeled NIZK proof for relation R_{pke}. The algorithm CRSGen(1^κ) takes a security parameter κ as input, outputs a common reference string crs, i.e., $crs \leftarrow$ CRSGen(1^κ). The algorithm Prove takes a pair $(x, wit) = ((\mathbf{A}_0, \mathbf{A}_1, \mathbf{c}_0, \mathbf{c}_1, \beta), (\mathbf{s}_0, \mathbf{s}_1, \mathbf{w})) \in R_{pke}$ and a label $\in \{0,1\}^*$ as inputs, outputs a proof π, i.e., $\pi \leftarrow$ Prove($crs, x, wit, $label). The algorithm Verify takes as inputs x, a proof π and a label $\in \{0,1\}^*$, outputs a bit $b \in \{0,1\}$ indicating whether π is valid or not, i.e., $b \leftarrow$ Verify($crs, x, \pi, $label). For completeness, we require that for any $(x, wit) \in R_{pke}$ and any label $\in \{0,1\}^*$, Verify($crs, x,$ Prove($crs, x, wit, $label), label) $= 1$. We defer more information of simulation-sound NIZK to the full version [58].

5.1 A Splittable PKE from Lattices

Let $n_1, n_2 \in \mathbb{Z}$ and prime q be polynomials in the security parameter κ. Let $n = n_1 + n_2 + 1, m = O(n \log q) \in \mathbb{Z}$, and $\alpha, \beta \in \mathbb{R}$ be the system parameters. Let $\mathcal{P} = \{-\alpha q + 1, \ldots, \alpha q - 1\}^{n_2}$ be the plaintext space. Let (CRSGen, Prove, Verify) be a simulation-sound NIZK proof for R_{pke}. Our PKE scheme $\mathcal{PKE} = $ (KeyGen, Enc, Dec) is defined as follows.

KeyGen(1^κ): Given the security parameter κ, compute $(\mathbf{A}_0, \mathbf{R}_0) \leftarrow$ TrapGen(1^n, $1^m, q$), $(\mathbf{A}_1, \mathbf{R}_1) \leftarrow$ TrapGen($1^n, 1^m, q$) and $crs \leftarrow$ CRSGen(1^κ). Return the public and secret key pair (pk, sk) $= ((\mathbf{A}_0, \mathbf{A}_1, crs), \mathbf{R}_0)$.

Enc(pk, label, $\mathbf{w} \in \mathcal{P}$): Given pk $= (\mathbf{A}_0, \mathbf{A}_1, crs)$, label $\in \{0,1\}^*$ and plaintext $\mathbf{w}$, randomly choose $\mathbf{s}_0, \mathbf{s}_1 \leftarrow_r \mathbb{Z}_q^{n_1}$, $\mathbf{e}_0, \mathbf{e}_1 \leftarrow_r D_{\mathbb{Z}^m, \alpha q}$. Finally, return the ciphertext $C = (\mathbf{c}_0, \mathbf{c}_1, \pi)$, where

$$\mathbf{c}_0 = \mathbf{A}_0^t \begin{pmatrix} \mathbf{s}_0 \\ 1 \\ \mathbf{w} \end{pmatrix} + \mathbf{e}_0, \quad \mathbf{c}_1 = \mathbf{A}_1^t \begin{pmatrix} \mathbf{s}_1 \\ 1 \\ \mathbf{w} \end{pmatrix} + \mathbf{e}_1,$$

and $\pi \leftarrow$ Prove($crs, (\mathbf{A}_0, \mathbf{A}_1, \mathbf{c}_0, \mathbf{c}_1, \beta), (\mathbf{s}_0, \mathbf{s}_1, \mathbf{w}), $label).

Dec(sk, label, C): Given $\mathsf{sk} = \mathbf{R}_0$, label $\in \{0,1\}^*$ and ciphertext $C = (\mathbf{c}_0, \mathbf{c}_1, \pi)$, if Verify($crs, (\mathbf{A}_0, \mathbf{A}_1, \mathbf{c}_0, \mathbf{c}_1, \beta), \pi,$ label) $= 0$, return $\perp$. Otherwise, compute

$$\mathbf{t} = \begin{pmatrix} \mathbf{s}_0 \\ 1 \\ \mathbf{w} \end{pmatrix} \leftarrow \mathsf{Solve}(\mathbf{A}_0, \mathbf{R}_0, \mathbf{c}_0),$$

and return $\mathbf{w} \in \mathbb{Z}_q^{n_2}$ (note that a valid π ensures that has the right form).

Correctness. By Lemma 12, we have that $\|\mathbf{e}_0\|, \|\mathbf{e}_1\| \leq \alpha q\sqrt{m}$ hold with overwhelming probability. Thus, it is enough to set $\beta \geq \alpha q\sqrt{m}$ for the NIZK proof to work. By Proposition 1, we have that $s_1(\mathbf{R}_0) \leq \sqrt{m} \cdot \omega(\sqrt{\log n})$, and the Solve algorithm can recover $\mathbf{t}$ from any $\mathbf{y} = \mathbf{A}_0^t \mathbf{t} + \mathbf{e}_0$ as long as $\|\mathbf{e}_0\| \cdot \sqrt{m} \cdot \omega(\sqrt{\log n}) \leq q$. Thus, we can set the parameters appropriately to satisfy the correctness. Besides, for the hardness of the LWE assumption, we need $\alpha q \geq 2\sqrt{n_1}$. In order to obtain an ϵ-approximate SPH function, we require $\beta \leq \sqrt{q}/4$, $\sqrt{mq}/4 \cdot \omega(\sqrt{\log n}) \leq q$ and $\alpha\gamma m < \epsilon/8$, where $\gamma \geq 4\sqrt{mq}$ is the parameter for ASPH in Sect. 5.2. In all, fix $\epsilon \in (0, 1/2)$, we can set the parameters $m, \alpha, \beta, q, \gamma$ as follows (where $c \geq 0$ is a real such that q is a prime) for both correctness and security[8]:

$$
\begin{aligned}
m &= O(n\log n), & \beta &> 16m\sqrt{mn}/\epsilon \\
q &= 16\beta^2 + c, & \alpha &= 2\sqrt{n}/q \\
\gamma &= 4\sqrt{mq}
\end{aligned}
\tag{1}
$$

Security. For any $C = (\mathbf{c}_0, \mathbf{c}_1, \pi) \leftarrow \mathsf{Enc}(\mathsf{pk}, \mathsf{label}, \mathbf{w})$, let r be the corresponding random coins which includes $(\mathbf{s}_0, \mathbf{s}_1, \mathbf{e}_0, \mathbf{e}_1)$ for generating $(\mathbf{c}_0, \mathbf{c}_1)$, and the randomness used for generating π. We define functions (f, g) as follows:

- The function f takes $(\mathsf{pk}, \mathbf{w}, r)$ as inputs, computes $(\mathbf{c}_0, \mathbf{c}_1)$ with random coins r, and returns $(\mathbf{c}_0, \mathbf{c}_1)$, i.e., $(\mathbf{c}_0, \mathbf{c}_1) = f(\mathsf{pk}, \mathbf{w}, r)$;
- The function g takes $(\mathsf{pk}, \mathsf{label}, \mathbf{w}, r)$ as inputs, computes the Prove algorithm with random coins r and returns the result π, i.e., $\pi = g(\mathsf{pk}, \mathsf{label}, \mathbf{w}, r)$.

We fix the two functions (f, g) in the rest of Sect. 5, and have the following theorem for security.

Theorem 2. *Let* $n = n_1 + n_2 + 1$, $m \in \mathbb{Z}$, $\alpha, \beta, \gamma \in \mathbb{R}$ *and prime* q *be as in Eq. (1). If* $\mathsf{LWE}_{n_1, q, \alpha}$ *is hard,* (CRSGen, Prove, Verify) *is a simulation-sound NIZK proof, then the scheme* $\mathcal{PKE}$ *is a splittable CCA-secure PKE scheme.*

Since $\mathcal{PKE}$ is essentially an instantiation of the Naor-Yung paradigm [49,55] using a special LWE-based CPA scheme (similar to the ones in [38,47]), and a SS-NIZK for a special relation R_{pke}, this theorem can be shown by adapting the proof techniques in [49,55]. We deter the proof to the full version [58].

---

[8] We defer the choices of concrete parameters to the full version [58].

5.2 An Associated Approximate SPH

Fix a public key $\mathsf{pk} = (\mathbf{A}_0, \mathbf{A}_1, crs)$ of the PKE scheme $\mathcal{PKE}$. Given any string label $\in \{0,1\}^*$ and $C = (\mathbf{c}_0, \mathbf{c}_1, \pi)$, we say that (label, C) is a valid label-ciphertext pair with respect to pk if $\mathsf{Verify}(crs, (\mathbf{A}_0, \mathbf{A}_1, \mathbf{c}_0, \mathbf{c}_1, \beta), \pi, \mathsf{label}) = 1$. Let sets X, L and $\bar{L}$ be defined as in Sect. 3.2. Define the associated ASPH function $(K, \ell, \{\mathsf{H}_{\mathsf{hk}} : X \to \{0,1\}^\ell\}_{\mathsf{hk} \in K}, S, \mathsf{Proj} : K \to S)$ for $\mathcal{PKE}$ as follows.

- The hash key is an ℓ-tuple of vectors $\mathsf{hk} = (\mathbf{x}_1, \ldots, \mathbf{x}_\ell)$, where $\mathbf{x}_i \sim D_{\mathbb{Z}^m, \gamma}$. Write $\mathbf{A}_0^t = (\mathbf{B} \| \mathbf{U}) \in \mathbb{Z}_q^{m \times n}$ such that $\mathbf{B} \in \mathbb{Z}_q^{m \times n_1}$ and $\mathbf{U} \in \mathbb{Z}_q^{m \times (n_2+1)}$. Define the projection key $\mathsf{hp} = \mathsf{Proj}(\mathsf{hk}) = (\mathbf{u}_1, \ldots, \mathbf{u}_\ell)$, where $\mathbf{u}_i = \mathbf{B}^t \mathbf{x}_i$.
- $\mathsf{H}_{\mathsf{hk}}(x) = \mathsf{H}_{\mathsf{hk}}((\mathbf{c}_0, \mathbf{c}_1), \mathbf{w})$: Given $\mathsf{hk} = (\mathbf{x}_1, \ldots, \mathbf{x}_\ell)$ and $x = (\mathsf{label}, C, \mathbf{w}) \in X$ for some $C = (\mathbf{c}_0, \mathbf{c}_1, \pi)$, compute $z_i = \mathbf{x}_i^t \left(\mathbf{c}_0 - \mathbf{U} \begin{pmatrix} 1 \\ \mathbf{w} \end{pmatrix} \right)$ for $i \in \{1, \ldots, \ell\}$. Then, treat each z_i as a number in $\{-(q-1)/2, \ldots, (q-1)/2\}$. If $z_i = 0$, then set $b_i \leftarrow_r \{0,1\}$. Else, set

$$b_i = \begin{cases} 0 \text{ if } z_i < 0 \\ 1 \text{ if } z_i > 0 \end{cases}.$$

Finally, return $\mathsf{H}_{\mathsf{hk}}((\mathbf{c}_0, \mathbf{c}_1), \mathbf{w}) = (b_1, \ldots, b_\ell)$.
- $\mathsf{Hash}(\mathsf{hp}, x, \mathbf{s}_0) = \mathsf{Hash}(\mathsf{hp}, ((\mathbf{c}_0, \mathbf{c}_1), \mathbf{w}), \mathbf{s}_0)$: Given $\mathsf{hp} = (\mathbf{u}_1, \ldots, \mathbf{u}_\ell), x = (\mathsf{label}, (\mathbf{c}_0, \mathbf{c}_1, \pi), \mathbf{w}) \in L$ and $\mathbf{s}_0 \in \mathbb{Z}_q^{n_1}$ such that $\mathbf{c}_0 = \mathbf{B}\mathbf{s}_0 + \mathbf{U} \begin{pmatrix} 1 \\ \mathbf{w} \end{pmatrix} + \mathbf{e}_0$ for some $\mathbf{e}_0 \leftarrow_r D_{\mathbb{Z}^m, \alpha q}$, compute $z_i' = \mathbf{u}_i^t \mathbf{s}_0$. Then, treat each z_i' as a number in $\{-(q-1)/2, \ldots, (q-1)/2\}$. If $z_i' = 0$, then set $b_i' \leftarrow_r \{0,1\}$. Else, set

$$b_i' = \begin{cases} 0 \text{ if } z_i < 0 \\ 1 \text{ if } z_i' > 0 \end{cases}.$$

Finally, return $\mathsf{Hash}(\mathsf{hp}, ((\mathbf{c}_0, \mathbf{c}_1), \mathbf{w}), \mathbf{s}_0) = (b_1', \ldots, b_\ell')$.

Theorem 3. *Let $\epsilon \in (0, 1/2)$, and let $n, m, q, \alpha, \beta, \gamma$ be as in Theorem 2. Let ℓ be polynomial in the security parameter κ. Then, $(K, \ell, \{\mathsf{H}_{\mathsf{hk}} : X \to \{0,1\}^\ell\}_{\mathsf{hk} \in K}, S, \mathsf{Proj} : K \times C_{\mathsf{pk}} \to S)$ is an ϵ-approximate SPH as in Definition 5.*

For space reason, we defer it to the full version [58].

5.3 Achieving Simulation-Sound NIZK for R_{pke} on Lattices

In this section, we will show how to construct an simulation-sound NIZK for R_{pke} from lattices in the random oracle model. Formally, let $n = n_1 + n_2 + 1, m, q \in \mathbb{Z}$ be defined as in Sect. 5.1. We begin by defining a variant relation R_{pke}' of R_{pke} (in the l_∞ form):

$$R_{pke}' := \left\{ ((\mathbf{A}_0, \mathbf{A}_1, \mathbf{c}_0, \mathbf{c}_1, \zeta), (\mathbf{s}_0, \mathbf{s}_1, \mathbf{w})) : \|\mathbf{w}\|_\infty \leq \zeta \wedge \left\| \mathbf{c}_0 - \mathbf{A}_0^t \begin{pmatrix} \mathbf{s}_0 \\ 1 \\ \mathbf{w} \end{pmatrix} \right\|_\infty \leq \zeta \wedge \left\| \mathbf{c}_1 - \mathbf{A}_1^t \begin{pmatrix} \mathbf{s}_1 \\ 1 \\ \mathbf{w} \end{pmatrix} \right\|_\infty \leq \zeta \right\},$$

where $\mathbf{A}_0, \mathbf{A}_1 \in \mathbb{Z}_q^{n \times m}, \mathbf{c}_0, \mathbf{c}_1 \in \mathbb{Z}_q^m, \zeta \in \mathbb{R}, \mathbf{s}_0, \mathbf{s}_1 \in \mathbb{Z}_q^{n_1}$ and $\mathbf{w} \in \mathbb{Z}_q^{n_2}$. Write $\mathbf{A}_0^t = (\mathbf{B}_0 \| \mathbf{U}_0) \in \mathbb{Z}_q^{m \times n_1} \times \mathbb{Z}_q^{m \times (n_2+1)}$. Note that for large enough $m = O(n_1 \log q)$, the rows of a uniformly random $\mathbf{B}_0 \in \mathbb{Z}_q^{m \times n_1}$ generate $\mathbb{Z}_q^{n_1}$ with overwhelming probability. By the duality [46], one can compute a parity check matrix $\mathbf{G}_0 \in \mathbb{Z}_q^{(m-n_1) \times m}$ such that (1) the columns of $\mathbf{G}_0$ generate $\mathbb{Z}_q^{m-n_1}$, and (2) $\mathbf{G}_0 \mathbf{B}_0 = \mathbf{0}$. Now, let vector $\mathbf{e}_0 \in \mathbb{Z}^m$ satisfy

$$\mathbf{c}_0 = \mathbf{A}_0^t \begin{pmatrix} \mathbf{s}_0 \\ 1 \\ \mathbf{w} \end{pmatrix} + \mathbf{e}_0 = \mathbf{B}_0 \mathbf{s}_0 + \mathbf{U}_0 \begin{pmatrix} 1 \\ \mathbf{w} \end{pmatrix} + \mathbf{e}_0. \tag{2}$$

By multiplying Eq. (2) with matrix $\mathbf{G}_0$ and rearranging the terms, we have the equation $\mathbf{D}_0 \mathbf{w} + \mathbf{G}_0 \mathbf{e}_0 = \mathbf{b}_0$, where $(\mathbf{a}_0 \| \mathbf{D}_0) = \mathbf{G}_0 \mathbf{U}_0 \in \mathbb{Z}_q^{(m-n_1) \times (1+n_2)}$, and $\mathbf{b}_0 = \mathbf{G}_0 \mathbf{c}_0 - \mathbf{a}_0 \in \mathbb{Z}_q^{m-n_1}$. Similarly, by letting $\mathbf{A}_1^t = (\mathbf{B}_1 \| \mathbf{U}_1)$ and $\mathbf{c}_1 = \mathbf{B}_1 \mathbf{s}_1 + \mathbf{U}_1 \begin{pmatrix} 1 \\ \mathbf{w} \end{pmatrix} + \mathbf{e}_1$, we can compute an equation $\mathbf{D}_1 \mathbf{w} + \mathbf{G}_1 \mathbf{e}_1 = \mathbf{b}_1$, where $\mathbf{G}_1 \in \mathbb{Z}_q^{(m-n_1) \times m}$ is a parity check matrix for $\mathbf{B}_1$, $(\mathbf{a}_1 \| \mathbf{D}_1) = \mathbf{G}_1 \mathbf{U}_1 \in \mathbb{Z}_q^{(m-n_1) \times (1+n_2)}$, and $\mathbf{b}_1 = \mathbf{G}_1 \mathbf{c}_1 - \mathbf{a}_1 \in \mathbb{Z}_q^{m-n_1}$. As in [40,44], in order to show $((\mathbf{A}_0, \mathbf{A}_1, \mathbf{c}_0, \mathbf{c}_1, \zeta), (\mathbf{s}_0, \mathbf{s}_1, \mathbf{w})) \in R'_{pke}$, it is enough to prove that there exists $(\mathbf{w}, \mathbf{e}_0, \mathbf{e}_1)$ such that $((\mathbf{D}_0, \mathbf{G}_0, \mathbf{D}_1, \mathbf{G}_1, \mathbf{b}_0, \mathbf{b}_1, \zeta), (\mathbf{w}, \mathbf{e}_0, \mathbf{e}_1)) \in \tilde{R}'_{pke}$:

$$\tilde{R}'_{pke} := \left\{ \begin{array}{c} ((\mathbf{D}_0, \mathbf{G}_0, \mathbf{D}_1, \mathbf{G}_1, \mathbf{b}_0, \mathbf{b}_1, \zeta), (\mathbf{w}, \mathbf{e}_0, \mathbf{e}_1)) : \\ \begin{pmatrix} \mathbf{D}_0 & \mathbf{G}_0 & \mathbf{0} \\ \mathbf{D}_1 & \mathbf{0} & \mathbf{G}_1 \end{pmatrix} \begin{pmatrix} \mathbf{w} \\ \mathbf{e}_0 \\ \mathbf{e}_1 \end{pmatrix} = \begin{pmatrix} \mathbf{b}_0 \\ \mathbf{b}_1 \end{pmatrix} \wedge \\ \|\mathbf{w}\|_\infty \leq \zeta \wedge \|\mathbf{e}_0\|_\infty \leq \zeta \wedge \|\mathbf{e}_1\|_\infty \leq \zeta \end{array} \right\},$$

which is essentially a special case of the ISIS relation R_{ISIS} (in the l_∞ norm):

$$R_{ISIS} := \{((\mathbf{M}, \mathbf{b}, \zeta), \mathbf{x}) : \mathbf{M}\mathbf{x} = \mathbf{b} \wedge \|\mathbf{x}\|_\infty \leq \zeta\}.$$

Notice that if there is a three-round public-coin honest-verifier zero-knowledge (HVZK) proof for the relation R_{ISIS}, one can obtain an NIZK proof for R_{ISIS} by applying the Fiat-Shamir transform [27] in the random oracle model [15]. Moreover, if the basic protocol additionally has the *quasi unique responses* property [13,24,28], the literature [13,14,24] shows that the resulting NIZK proof derived from the Fiat-Shamir transform meets the simulation-soundness needed for constructing CCA-secure PKE via the Naor-Yung paradigm [49,55]. Fortunately, we do have an efficient three-round public-coin HVZK proof with quasi unique responses in [42],[9] which is extended from the Stern protocol [56] and has the same structure as the latter. Specifically, the protocol [42] has three messages (a, e, z), where a consists of several commitments sent

---

[9] More precisely, the authors [42] showed a zero-knowledge proof of knowledge for R_{ISIS}, which has a constant soundness error about $2/3$. By repeating the basic protocol $t = \omega(\log n)$ times in parallel, we can obtain a desired public-coin HVZK proof for R_{ISIS} with negligible soundness error [41,43].

by the prover, e is the challenge sent by the verifier, and the third message z (i.e., the response) consists of the openings to the commitments specified by the challenge e.

Note that the quasi unique responses property [24, 28] essentially requires that it is computationally infeasible for an adversary to output (a, e, z) and (a, e, z') such that both (a, e, z) and (a, e, z') are valid. Thus, if, as is usually the case, the parameters of the commitment scheme are priorly fixed for all users, the protocol in [42] naturally has the quasi unique responses property by the binding property of the commitment scheme. In other words, the NIZK proof for R_{ISIS} [41, 43] (and thus for $\tilde{R}'_{pke}$) obtained by applying the Fiat-Shamir transform to the protocol in [42] suffices for our PKE scheme (where labels can be incorporated into the input of the hash function used for the transformation).

Finally, we clarify that the protocol [42] is designed for R_{ISIS} in the l_∞ norm, while the l_2 norm is used in Sect. 5.1. This problem can be easily fixed by setting $\zeta = \alpha q \cdot \omega(\sqrt{\log n})$ in the NIZK proof, and setting the parameter β in Eq. (1) such that $\beta \geq 2\zeta\sqrt{n}$ holds, since (1) for $\mathbf{e}_0, \mathbf{e}_1 \leftarrow_r D_{\mathbb{Z}^m, \alpha q}$, both $\Pr[\|\mathbf{e}_0\|_\infty \geq \zeta], \Pr[\|\mathbf{e}_1\|_\infty \geq \zeta]$ are negligible in n by [30, Lemma 4.2]; and (2) $\mathcal{P} = \{-\alpha q + 1, \ldots, \alpha q - 1\}^{n_2}$ in our PKE scheme $\mathcal{PKE}$. By [42], the resulting NIZK can be achieved with total communication cost $\log_2 \beta \cdot \tilde{O}(m \log q)$.

Acknowledgments. We would like to thank the anonymous reviewers for their helpful suggestions. Jiang Zhang is supported by the National Key Research and Development Program of China (Grant No. 2017YFB0802005), the National Grand Fundamental Research Program of China (Grant No. 2013CB338003), the National Natural Science Foundation of China (Grant Nos. 61602046, 61602045, U1536205), and the Young Elite Scientists Sponsorship Program by CAST (2016QNRC001). Yu Yu is supported by the National Natural Science Foundation of China (Grant Nos. 61472249, 61572192, 61572149), the National Cryptography Development Fund MMJJ20170209, and International Science & Technology Cooperation & Exchange Projects of Shaanxi Province (2016KW-038).

References

1. Abdalla, M., Benhamouda, F., MacKenzie, P.: Security of the J-PAKE password-authenticated key exchange protocol. In: IEEE S&P 2015, pp. 571–587, May 2015
2. Abdalla, M., Benhamouda, F., Pointcheval, D.: Disjunctions for hash proof systems: new constructions and applications. In: Oswald, E., Fischlin, M. (eds.) EUROCRYPT 2015. LNCS, vol. 9057, pp. 69–100. Springer, Heidelberg (2015). https://doi.org/10.1007/978-3-662-46803-6_3
3. Abdalla, M., Chevalier, C., Pointcheval, D.: Smooth projective hashing for conditionally extractable commitments. In: Halevi, S. (ed.) CRYPTO 2009. LNCS, vol. 5677, pp. 671–689. Springer, Heidelberg (2009). https://doi.org/10.1007/978-3-642-03356-8_39
4. Abe, M., Cui, Y., Imai, H., Kiltz, E.: Efficient hybrid encryption from ID-based encryption. Des. Codes Cryptogr. **54**(3), 205–240 (2010)
5. Ajtai, M.: Generating hard instances of the short basis problem. In: Wiedermann, J., van Emde Boas, P., Nielsen, M. (eds.) ICALP 1999. LNCS, vol. 1644, pp. 1–9. Springer, Heidelberg (1999). https://doi.org/10.1007/3-540-48523-6_1

6. Alkim, E., Ducas, L., Pöppelmann, T., Schwabe, P.: Post-quantum key exchange—a new hope. In: USENIX Security 2016, pp. 327–343. USENIX Association (2016)
7. Alwen, J., Peikert, C.: Generating shorter bases for hard random lattices. In: STACS, pp. 75–86 (2009)
8. Barak, B., Canetti, R., Lindell, Y., Pass, R., Rabin, T.: Secure computation without authentication. In: Shoup, V. (ed.) CRYPTO 2005. LNCS, vol. 3621, pp. 361–377. Springer, Heidelberg (2005). https://doi.org/10.1007/11535218_22
9. Bellare, M., Pointcheval, D., Rogaway, P.: Authenticated key exchange secure against dictionary attacks. In: Preneel, B. (ed.) EUROCRYPT 2000. LNCS, vol. 1807, pp. 139–155. Springer, Heidelberg (2000). https://doi.org/10.1007/3-540-45539-6_11
10. Bellare, M., Yung, M.: Certifying cryptographic tools: the case of trapdoor permutations. In: Brickell, E.F. (ed.) CRYPTO 1992. LNCS, vol. 740, pp. 442–460. Springer, Heidelberg (1993). https://doi.org/10.1007/3-540-48071-4_31
11. Bellovin, S.M., Merritt, M.: Encrypted key exchange: password-based protocols secure against dictionary attacks. In: Proceedings 1992 IEEE Computer Society Symposium on Research in Security and Privacy, pp. 72–84, May 1992
12. Benhamouda, F., Blazy, O., Chevalier, C., Pointcheval, D., Vergnaud, D.: New techniques for SPHFs and efficient one-round PAKE protocols. In: Canetti, R., Garay, J.A. (eds.) CRYPTO 2013. LNCS, vol. 8042, pp. 449–475. Springer, Heidelberg (2013). https://doi.org/10.1007/978-3-642-40041-4_25
13. Bernhard, D., Fischlin, M., Warinschi, B.: Adaptive proofs of knowledge in the random oracle model. In: Katz, J. (ed.) PKC 2015. LNCS, vol. 9020, pp. 629–649. Springer, Heidelberg (2015). https://doi.org/10.1007/978-3-662-46447-2_28
14. Bernhard, D., Pereira, O., Warinschi, B.: How not to prove yourself: pitfalls of the fiat-shamir heuristic and applications to helios. In: Wang, X., Sako, K. (eds.) ASIACRYPT 2012. LNCS, vol. 7658, pp. 626–643. Springer, Heidelberg (2012). https://doi.org/10.1007/978-3-642-34961-4_38
15. Blum, M., Feldman, P., Micali, S.: Non-interactive zero-knowledge and its applications. In: STOC 1988, pp. 103–112. ACM (1988)
16. Boyko, V., MacKenzie, P., Patel, S.: Provably secure password-authenticated key exchange using Diffie-Hellman. In: Preneel, B. (ed.) EUROCRYPT 2000. LNCS, vol. 1807, pp. 156–171. Springer, Heidelberg (2000). https://doi.org/10.1007/3-540-45539-6_12
17. Bresson, E., Chevassut, O., Pointcheval, D.: Security proofs for an efficient password-based key exchange. In: CCS 2003, pp. 241–250. ACM (2003)
18. Canetti, R.: Universally composable security: a new paradigm for cryptographic protocols. In: FOCS 2001, pp. 136–145. IEEE (2001)
19. Canetti, R., Goyal, V., Jain, A.: Concurrent secure computation with optimal query complexity. In: Gennaro, R., Robshaw, M. (eds.) CRYPTO 2015. LNCS, vol. 9216, pp. 43–62. Springer, Heidelberg (2015). https://doi.org/10.1007/978-3-662-48000-7_3
20. Canetti, R., Halevi, S., Katz, J.: Chosen-ciphertext security from identity-based encryption. In: Cachin, C., Camenisch, J.L. (eds.) EUROCRYPT 2004. LNCS, vol. 3027, pp. 207–222. Springer, Heidelberg (2004). https://doi.org/10.1007/978-3-540-24676-3_13
21. Canetti, R., Halevi, S., Katz, J., Lindell, Y., MacKenzie, P.: Universally composable password-based key exchange. In: Cramer, R. (ed.) EUROCRYPT 2005. LNCS, vol. 3494, pp. 404–421. Springer, Heidelberg (2005). https://doi.org/10.1007/11426639_24

22. Cramer, R., Shoup, V.: Universal hash proofs and a paradigm for adaptive chosen ciphertext secure public-key encryption. In: Knudsen, L.R. (ed.) EUROCRYPT 2002. LNCS, vol. 2332, pp. 45–64. Springer, Heidelberg (2002). https://doi.org/10.1007/3-540-46035-7_4

23. Diffie, W., Hellman, M.: New directions in cryptography. IEEE Trans. Inf. Theory 22(6), 644–654 (1976)

24. Faust, S., Kohlweiss, M., Marson, G.A., Venturi, D.: On the non-malleability of the fiat-shamir transform. In: Galbraith, S., Nandi, M. (eds.) INDOCRYPT 2012. LNCS, vol. 7668, pp. 60–79. Springer, Heidelberg (2012). https://doi.org/10.1007/978-3-642-34931-7_5

25. Faust, S., Mukherjee, P., Nielsen, J.B., Venturi, D.: Continuous non-malleable codes. In: Lindell, Y. (ed.) TCC 2014. LNCS, vol. 8349, pp. 465–488. Springer, Heidelberg (2014). https://doi.org/10.1007/978-3-642-54242-8_20

26. Feige, U., Lapidot, D., Shamir, A.: Multiple non-interactive zero knowledge proofs based on a single random string. In: FOCS 1990, pp. 308–317. IEEE (1990)

27. Fiat, A., Shamir, A.: How to prove yourself: practical solutions to identification and signature problems. In: Odlyzko, A.M. (ed.) CRYPTO 1986. LNCS, vol. 263, pp. 186–194. Springer, Heidelberg (1987). https://doi.org/10.1007/3-540-47721-7_12

28. Fischlin, M.: Communication-efficient non-interactive proofs of knowledge with online extractors. In: Shoup, V. (ed.) CRYPTO 2005. LNCS, vol. 3621, pp. 152–168. Springer, Heidelberg (2005). https://doi.org/10.1007/11535218_10

29. Gennaro, R., Lindell, Y.: A framework for password-based authenticated key exchange. In: Biham, E. (ed.) EUROCRYPT 2003. LNCS, vol. 2656, pp. 524–543. Springer, Heidelberg (2003). https://doi.org/10.1007/3-540-39200-9_33

30. Gentry, C., Peikert, C., Vaikuntanathan, V.: Trapdoors for hard lattices and new cryptographic constructions. In: STOC 2008, pp. 197–206. ACM (2008)

31. Goldreich, O.: Basing non-interactive zero-knowledge on (enhanced) trapdoor permutations: the state of the art. In: Goldreich, O. (ed.) Studies in Complexity and Cryptography. Miscellanea on the Interplay between Randomness and Computation. LNCS, vol. 6650, pp. 406–421. Springer, Heidelberg (2011). https://doi.org/10.1007/978-3-642-22670-0_28

32. Goldreich, O., Lindell, Y.: Session-key generation using human passwords only. In: Kilian, J. (ed.) CRYPTO 2001. LNCS, vol. 2139, pp. 408–432. Springer, Heidelberg (2001). https://doi.org/10.1007/3-540-44647-8_24

33. Gong, L., Lomas, M.A., Needham, R.M., Saltzer, J.H.: Protecting poorly chosen secrets from guessing attacks. IEEE J. Sel. Areas Commun. 11(5), 648–656 (1993)

34. Goyal, V., Jain, A., Ostrovsky, R.: Password-authenticated session-key generation on the internet in the plain model. In: Rabin, T. (ed.) CRYPTO 2010. LNCS, vol. 6223, pp. 277–294. Springer, Heidelberg (2010). https://doi.org/10.1007/978-3-642-14623-7_15

35. Groce, A., Katz, J.: A new framework for efficient password-based authenticated key exchange. In: CCS 2010, pp. 516–525. ACM (2010)

36. Halevi, S., Krawczyk, H.: Public-key cryptography and password protocols. ACM Trans. Inf. Syst. Secur. 2(3), 230–268 (1999)

37. Katz, J., Ostrovsky, R., Yung, M.: Efficient and secure authenticated key exchange using weak passwords. J. ACM 57(1), 3:1–3:39 (2009)

38. Katz, J., Vaikuntanathan, V.: Smooth projective hashing and password-based authenticated key exchange from lattices. In: Matsui, M. (ed.) ASIACRYPT 2009. LNCS, vol. 5912, pp. 636–652. Springer, Heidelberg (2009). https://doi.org/10.1007/978-3-642-10366-7_37

39. Katz, J., Vaikuntanathan, V.: Round-optimal password-based authenticated key exchange. In: Ishai, Y. (ed.) TCC 2011. LNCS, vol. 6597, pp. 293–310. Springer, Heidelberg (2011). https://doi.org/10.1007/978-3-642-19571-6_18

40. Laguillaumie, F., Langlois, A., Libert, B., Stehlé, D.: Lattice-based group signatures with logarithmic signature size. In: Sako, K., Sarkar, P. (eds.) ASIACRYPT 2013. LNCS, vol. 8270, pp. 41–61. Springer, Heidelberg (2013). https://doi.org/10.1007/978-3-642-42045-0_3

41. Libert, B., Mouhartem, F., Nguyen, K.: A lattice-based group signature scheme with message-dependent opening. In: Manulis, M., Sadeghi, A.-R., Schneider, S. (eds.) ACNS 2016. LNCS, vol. 9696, pp. 137–155. Springer, Cham (2016). https://doi.org/10.1007/978-3-319-39555-5_8

42. Ling, S., Nguyen, K., Stehlé, D., Wang, H.: Improved zero-knowledge proofs of knowledge for the ISIS problem, and applications. In: Kurosawa, K., Hanaoka, G. (eds.) PKC 2013. LNCS, vol. 7778, pp. 107–124. Springer, Heidelberg (2013). https://doi.org/10.1007/978-3-642-36362-7_8

43. Ling, S., Nguyen, K., Wang, H.: Group signatures from lattices: simpler, tighter, shorter, ring-based. In: Katz, J. (ed.) PKC 2015. LNCS, vol. 9020, pp. 427–449. Springer, Heidelberg (2015). https://doi.org/10.1007/978-3-662-46447-2_19

44. Lyubashevsky, V.: Lattice signatures without trapdoors. In: Pointcheval, D., Johansson, T. (eds.) EUROCRYPT 2012. LNCS, vol. 7237, pp. 738–755. Springer, Heidelberg (2012). https://doi.org/10.1007/978-3-642-29011-4_43

45. MacKenzie, P., Patel, S., Swaminathan, R.: Password-authenticated key exchange based on RSA. In: Okamoto, T. (ed.) ASIACRYPT 2000. LNCS, vol. 1976, pp. 599–613. Springer, Heidelberg (2000). https://doi.org/10.1007/3-540-44448-3_46

46. Micciancio, D., Mol, P.: Pseudorandom knapsacks and the sample complexity of LWE search-to-decision reductions. In: Rogaway, P. (ed.) CRYPTO 2011. LNCS, vol. 6841, pp. 465–484. Springer, Heidelberg (2011). https://doi.org/10.1007/978-3-642-22792-9_26

47. Micciancio, D., Peikert, C.: Trapdoors for lattices: simpler, tighter, faster, smaller. In: Pointcheval, D., Johansson, T. (eds.) EUROCRYPT 2012. LNCS, vol. 7237, pp. 700–718. Springer, Heidelberg (2012). https://doi.org/10.1007/978-3-642-29011-4_41

48. Micciancio, D., Regev, O.: Worst-case to average-case reductions based on gaussian measures. SIAM J. Comput. **37**, 267–302 (2007)

49. Naor, M., Yung, M.: Public-key cryptosystems provably secure against chosen ciphertext attacks. In: STOC 1990, pp. 427–437. ACM (1990)

50. Peikert, C.: Lattice cryptography for the internet. In: Mosca, M. (ed.) PQCrypto 2014. LNCS, vol. 8772, pp. 197–219. Springer, Cham (2014). https://doi.org/10.1007/978-3-319-11659-4_12

51. Peikert, C., Rosen, A.: Efficient collision-resistant hashing from worst-case assumptions on cyclic lattices. In: Halevi, S., Rabin, T. (eds.) TCC 2006. LNCS, vol. 3876, pp. 145–166. Springer, Heidelberg (2006). https://doi.org/10.1007/11681878_8

52. Peikert, C., Vaikuntanathan, V., Waters, B.: A framework for efficient and composable oblivious transfer. In: Wagner, D. (ed.) CRYPTO 2008. LNCS, vol. 5157, pp. 554–571. Springer, Heidelberg (2008). https://doi.org/10.1007/978-3-540-85174-5_31

53. Regev, O.: On lattices, learning with errors, random linear codes, and cryptography. In: STOC 2005, pp. 84–93. ACM (2005)

54. Rosen, A., Segev, G.: Chosen-ciphertext security via correlated products. In: Reingold, O. (ed.) TCC 2009. LNCS, vol. 5444, pp. 419–436. Springer, Heidelberg (2009). https://doi.org/10.1007/978-3-642-00457-5_25

55. Sahai, A.: Non-malleable non-interactive zero knowledge and adaptive chosen-ciphertext security. In: FOCS 1999, pp. 543–553. IEEE Computer Society (1999)
56. Stern, J.: A new paradigm for public key identification. IEEE Trans. Inf. Theory **42**(6), 1757–1768 (1996)
57. Zhang, J., Chen, Y., Zhang, Z.: Programmable hash functions from lattices: short signatures and IBEs with small key sizes. In: Robshaw, M., Katz, J. (eds.) CRYPTO 2016. LNCS, vol. 9816, pp. 303–332. Springer, Heidelberg (2016). https://doi.org/10.1007/978-3-662-53015-3_11
58. Zhang, J., Yu, Y.: Two-round PAKE from approximate SPH and instantiations from lattices. Cryptology ePrint Archive, Report 2017/838 (2017)
59. Zhang, J., Zhang, Z., Ding, J., Snook, M., Dagdelen, Ö.: Authenticated key exchange from ideal lattices. In: Oswald, E., Fischlin, M. (eds.) EUROCRYPT 2015. LNCS, vol. 9057, pp. 719–751. Springer, Heidelberg (2015). https://doi.org/10.1007/978-3-662-46803-6_24
60. Zhang, R.: Tweaking TBE/IBE to PKE transforms with chameleon hash functions. In: Katz, J., Yung, M. (eds.) ACNS 2007. LNCS, vol. 4521, pp. 323–339. Springer, Heidelberg (2007). https://doi.org/10.1007/978-3-540-72738-5_21

Tightly-Secure Signatures from Five-Move Identification Protocols

Eike Kiltz[1], Julian Loss[1], and Jiaxin Pan[2(✉)]

[1] Ruhr-Universität Bochum, Bochum, Germany
{eike.kiltz,julian.loss}@rub.de
[2] Karlsruher Institut für Technologie, Karlsruhe, Germany
jiaxin.pan@kit.edu

Abstract. We carry out a concrete security analysis of signature schemes obtained from five-move identification protocols via the Fiat-Shamir transform. Concretely, we obtain tightly-secure signatures based on the computational Diffie-Hellman (CDH), the short-exponent CDH, and the Factoring (FAC) assumptions. All our signature schemes have tight reductions to search problems, which is in stark contrast to all known signature schemes obtained from the classical Fiat-Shamir transform (based on three-move identification protocols), which either have a non-tight reduction to a search problem, or a tight reduction to a (potentially) stronger decisional problem. Surprisingly, our CDH-based scheme turns out to be (a slight simplification of) the Chevallier-Mames signature scheme (CRYPTO 05), thereby providing a theoretical explanation of its tight security proof via five-move identification protocols.

Keywords: Signatures · Five-move identification protocols · Fiat-Shamir · Tightness

1 Introduction

The security of public-key cryptographic primitives is commonly analyzed via a security reduction to a suitable cryptographic assumption (such as the factoring assumption). Concretely, a security reduction converts a successful adversary A against the cryptographic scheme's security into a successful solver B against the hardness of the underlying assumption. If the reduction provides the bound $\varepsilon_A \leq L \cdot \varepsilon_B$ (where ε_A is A's success probability and ε_B is B's success probability) then L is called the (multiplicative) security loss of the reduction.[1] Clearly, it is desirable to have the security loss L as small as a constant so that $\varepsilon_A \approx \varepsilon_B$. If furthermore the running times of A and B are approximately the same, then

---

E. Kiltz was partially supported by ERC Project ERCC (FP7/615074) and by DFG SPP 1736 Big Data.

J. Loss was supported by ERC Project ERCC (FP7/615074).

J. Pan was supported by the DFG grant HO 4534/4-1.

[1] We ignore the additive negligible terms to simplify our discussion.

© International Association for Cryptologic Research 2017
T. Takagi and T. Peyrin (Eds.): ASIACRYPT 2017, Part III, LNCS 10626, pp. 68–94, 2017.
https://doi.org/10.1007/978-3-319-70700-6_3

the reduction is said to be *tight*. Cryptographic schemes with tight reductions recently drew a large amount of attention (e.g., [4,12,13,18,29–31]) due to the fact that tightly-secure schemes do not need to compensate for the security loss with increased parameters.

Digital signature schemes are one of the most important public-key cryptographic primitives. They have numerous applications and often serve as a building block for advanced cryptographic protocols. Ideally, we desire to have signature schemes with short signature sizes, efficient signing and verification algorithms, and tight security based on weak, well studied assumptions.

TIGHTNESS, EFFICIENCY, AND DECISIONAL ASSUMPTIONS. We will focus on signature schemes in the random oracle model [8] which usually have better efficiency than the ones in the standard model. Even in the random oracle model, there seems to be a prevalence of efficient, yet tightly-secure signature schemes based on decisional rather than search assumptions [2,3,33,34]. Notable exceptions are the Rabin-Williams (RW) scheme from factoring (FAC) [9,10], the BLS and RSA-PSS variants with the "selector bit" technique [9,14,26,34], the Chevallier-Mames scheme (and its variants) (CM) from CDH [20,26], and the Micali-Reyzin scheme (MR) from FAC [7,39].

THE FIAT-SHAMIR TRANSFORM AND ITS TIGHTNESS. The Fiat-Shamir (FS) method [21] transforms a (canonical) three-move identification scheme ID into a digital signature scheme SIG[ID] using a hash function. A canonical identification scheme ID as formalized by [1] is a three-move public-key authentication protocol of a specific form. The prover (holding the secret-key) sends a commitment R to the verifier. The verifier (holding the public-key) returns a random challenge h, uniformly chosen from a set ChSet (of exponential size). The prover sends a response s. Finally, using the verification algorithm, the verifier publicly checks correctness of the transcript (R, h, s). There is a large number of canonical identification schemes known (e.g., [11,15,21,23,26,28,34,40–42,49], the most popular among them being the scheme by Schnorr [49].

As discussed above, obtaining tightly secure signatures with short parameters has been proven to be notoriously hard. In particular, schemes obtained via the classical FS transform are usually proven via the Forking Lemma [46] and therefore are not tightly secure. For example, the Schnorr signature is obtained from the Fiat-Shamir transform and has an *inherently* loose security reduction [22,35,50] to the discrete logarithm problem. This issue was addressed by [3,34] who showed how to improve the tightness of signature schemes obtained from the FS transform by basing their security on decisional assumptions such as DDH (and a short exponent variant thereof), quadratic residuosity, and (Ring)-LWE. However, to the best of our knowledge, there is currently no FS-derived signature scheme known which can be tightly proven secure under a *search assumption*. Moreover, there seems to be some evidence to support that this is impossible: the results of [22] show that the Schnorr scheme cannot be proven tightly secure under any non-interactive assumption. However, tight variants of the FS transform for three-move schemes may still exist if the scheme meets some additional requirements.

1.1 Our Contributions

In this work, we consider the Fiat-Shamir transform applied to a five-move identification scheme (rather than a three-move scheme). More precisely, we first formalize syntax and security of a five-move identification scheme and, following [35], provide a concrete and modular security analysis of the Fiat-Shamir transformed signature scheme. Next, we instantiate our framework to obtain schemes with security from search assumptions, such as the classical CDH and FAC assumptions. All our security reductions are tight.

FIVE-MOVE IDENTIFICATION SCHEMES. A five-move identification scheme ID is an extension of the three-move identification scheme, where there are two "commitment-challenge" rounds (compared to one), followed by a final response output by the prover. (Each round has two moves, so five moves in total.) Intuitively, the additional rounds give us the handle to tightly embed the challenge of a search assumption. Following [35], we consider PIMP-KOA security (parallel impersonation against key-only attacks) of identification schemes where the adversary, given the public-key, tries to impersonate a prover in one of many parallel "commitment-challenge" sessions.

FIAT-SHAMIR FOR FIVE-MOVE IDENTIFICATION SCHEMES. We consider two variants FS[ID] and OF[ID] of the five-move Fiat-Shamir transformation. Both have tight security reductions given that the identification scheme has honest-verifier zero-knowledge (HVZK) and is secure against parallel impersonation attacks.[2] The two variants come with different trade-offs. OF[ID] requires *special soundness* but results in an online/offline signature scheme [52], which allows to pre-compute most of the signature in an offline phase to have a computationally cheap online signing phase (that requires knowledge of the message to be signed). FS[ID] does not require special soundness but does not come with the online/offline property. Interestingly, we are able to explain the Chevallier-Mames signature scheme [20] in our framework and show that it can be obtained from a five-move identification scheme by applying OF[ID]. We now give some details of our obtained signature schemes. A detailed comparison of their properties is given in Table 1.

A NEW ONLINE/OFFLINE SCHEME WITH A TIGHT SECURITY REDUCTION. Using our OF[ID] transform, we present a modified version of the Girault-Poupard-Stern (GPS) signature scheme [24]. The main interest of this scheme lies within its online signing step which can be made extremely efficient, given that most of the work can be precomputed in the *offline-step*, i.e., before seeing the message m. Concretely, the scheme only performs arithmetics over the integers in its online step, thereby even getting rid of modular reductions. [24] provide a loose security reduction to the Short-Exponent Discrete Logarithm (SEDL) assumption [36] which states that the discrete logarithm problem remains hard

---

[2] We also consider identification schemes with correctness error and statistical HVZK and define the notion of non-aborting HVZK. In this section, we ignore these for simplicity.

even if the discrete logarithm is known to lie in some fixed interval. Subsequently, [3] proved a tight reduction for the GPS scheme to the decisional variant of the SEDL [36]. However, so far, there has been no known tight reduction of the GPS scheme to a *search assumption*. Our scheme resolves this issue by offering a tight reduction to the Short Exponent CDH (SCDH) assumption. The relation between these aforementioned problems is explained in [36] as follows. First, the SEDL assumption is (non-tightly) equivalent to its decisional version. Second, the SCDH assumption is (non-tightly) equivalent to the assumption that both the full length CDH problem *and* the SEDL problem are hard.

A TIGHTLY SECURE FACTORING-BASED SCHEME WITH EFFICIENT SIGNING. As an application of our second transform, FS[ID], we present a new signature scheme with a tight security reduction to factoring. While our signature generation step and size are not quite as efficient than the ones of the factoring-based schemes of [9,10,39], our scheme highlights the usefulness of our (generic) FS[ID] transform.

A TIGHTLY SECURE SCHEME FROM CDH. Our instantiation from CDH results in a slight simplification of the Chevallier-Mames signature scheme [20]. (Slight simplification in the sense that some inputs to the hash function can be left out in our scheme.) We believe that our framework provides interesting insights to the original scheme and underlines the usefulness of our (generic) OF[ID] transform.

Table 1. Comparison between some known tightly-secure signature schemes in the random oracle model. Top: schemes in a cyclic group $\mathbb{G}$ of prime order p. Bottom: schemes over $\mathbb{Z}_N$ for composite N. Elements of $\mathbb{G}$ have bit length G and k denotes the security parameter. $c < |p|$ is a parameter for the short Diffie-Hellman assumptions. Computational cost (x, y, z) denotes x modular exponentiations, y modular multiplications, and z hash operations, * indicates multiplication over integers, and const is a small constant.

Scheme	Origin	Approx. size	Off-line comp.	On-line comp.	Ass.	Search Ass.?
KW	[26]	$k + |p|$	$(2,0,0)$	$(0,1,1)$	DDH	–
GJKW	[26]	$G + k + |p|$	$(1,0,0)$	$(2,1,2)$	CDH	✓
FS$_{CDH}$	New	$G + k + |p|$	$(1,0,0)$	$(2,1,2)$	CDH	✓
OF$_{CDH}$	[20]	$G + k + |p|$	$(3,0,1)$	$(0,1,1)$	CDH	✓
AFLT	[3]	$2k + c$	$(1,0,0)$	$(0, 1^*, 1)$	DSDL	–
FS$_{SCDH}$	New	$G + 2k + c$	$(1,0,0)$	$(2, 1^*, 2)$	SCDH	✓
OF$_{SCDH}$	New	$G + 2k + c$	$(3,0,1)$	$(0, 1^*, 1)$	SCDH	✓
MR	[39, Sect. 4.3]	$k + |N|$	$(1, 0, 0)$	$(1, \text{const}, 1)$	FAC	✓
BR	[9]	$k + |N|$	$(0,0,0)$	$(0, \text{const}, 1)$	FAC	✓
FDH$_{RSA}$	[33]	$|N|$	$(0,0,0)$	$(1,0,1)$	ΦH	–
FS$_{FAC}$	New	$G + k + |N|$	$(1,0,0)$	$(2, 1, 2)$	FAC	✓

1.2   Related Work

There exists a large body of literature on variants and improvements for the Fiat-Shamir transform. [35] give a concrete and modularized security treatment for signatures obtained from identification schemes via the Fiat-Shamir transform in both the multi- and single user settings which yields optimal security parameters for a wide array of important signatures in the ROM, in particular for Schnorr signatures. Our work is based on their modular framework. More recently, [7] put forth a new framework which includes multiple transforms that allow to tightly convert an ID scheme satisfying certain requirements into a signature. Their framework captures some existing transformations that have so far not received any theoretical treatment. Most notably, they give a characterization of the 'swap' method used in [38] to obtain a signature based tightly on the factoring assumption. [2,3,26] propose methods to obtain tight security for FS-derived signature schemes, with two of the schemes in [3] satisfying (conjectured) *post-quantum security*.

Deriving signature schemes from five-move ID schemes (or more generally from schemes with $2n+1$ rounds) in an FS-like manner is a natural idea and thus has already been in the literature [5,16,17,19,44,45,48,51,53–55]. Surprisingly, many of them in the context of post-quantum security. However, none of these works proposes generic transforms for five-move ID schemes which makes the resulting proofs rather complex. Furthermore, none of the presented signature schemes admits a tight security reduction.

2   Preliminaries

2.1   Notations

We define $[N] := \{1, \ldots, N\}$ and $\mathbb{Z}_N := \mathbb{Z}/N\mathbb{Z}$ as the residual ring for an integer N. Let S be a finite set. $a \xleftarrow{\boxtimes} S$ denotes choosing an element a from S uniformly at random. Our algorithms are considered to be probabilistic polynomial time unless stated otherwise. If A is an algorithm, then $a \xleftarrow{\boxtimes} A(b)$ denotes the random variable which is defined as the output of A on input b. With $a \in A(b)$ we denote a possible output a of the execution of A on input b. When we want to make the randomness explicit, we use the notation $a := A(b; \rho)$ meaning that the algorithm is executed on input b and randomness ρ. Note that A's execution is now deterministic.

2.2   Digital Signatures

We begin by defining syntax and security of a (digital) signature scheme. Let par denote some common system parameters shared among all participants.

Definition 1 (Signature Scheme). *A digital signature scheme* SIG := (Gen, Sign, Ver) *is defined as follow.*

- *The key generation algorithm* Gen *takes system parameters* par *as inputs and returns the public and secret keys* (pk, sk). *We assume that* pk *implicitly defines a message space* $\mathcal{M}$ *and a signature space* Σ.
- *The signing algorithm* Sign *takes a secret key* sk *and a message* $m \in \mathcal{M}$ *as inputs and returns a signature* $\sigma \in \Sigma$.
- *The deterministic verification algorithm* Ver *takes a public key* pk, *a message* m *and a signature* σ *as inputs and returns 1 (accept) or 0 (reject).*

SIG *has correctness error* ρ *if for all* $(pk, sk) \in$ Gen(par) *and all messages* $m \in \mathcal{M}$, *with probability at least* $1 - \rho$ Sign(sk, m) *outputs a valid signature* σ *such that* Ver$(pk, m, \sigma) = 1$.

Definition 2 (UF-CMA Security). *A signature scheme* SIG *is said to satisfy* (t, ε, Q_s)-UF-CMA *security (unforgeability against chosen message attacks) if for all adversaries* A *running in time at most* t *and making at most* Q_s *queries to the signing oracle,*

$$\Pr \left[\begin{matrix} \text{Ver}(pk, m^*, \sigma^*) = 1 \\ \wedge\, m^* \notin \mathcal{M} \end{matrix} \,\middle|\, \begin{matrix} (pk, sk) \xleftarrow{\boxtimes} \text{Gen(par)} \\ (m^*, \sigma^*) \xleftarrow{\boxtimes} \text{A}^{\text{SIGN}(\cdot)}(pk) \end{matrix} \right] \leq \varepsilon,$$

where on query m *the signing oracle* SIGN *adds* m *to list* $\mathcal{M}$ *and returns* $\sigma \xleftarrow{\boxtimes}$ Sign(sk, m) *to* A, *i.e., a signature on message* m *under public-key* pk.

As a special case of UF-CMA security, we define (t, ε)-UF-KOA security (unforgeability against key-only attacks) as $(t, \varepsilon, 0)$-UF-CMA security, i.e. $Q_s = 0$. In other words, the adversary is not allowed to make any signing query in the UF-KOA security experiment.

SECURITY IN THE RANDOM ORACLE MODEL. A common approach to analyze the security of signature schemes that involve a hash function is to use the random oracle model [8] in which hash queries are answered by an oracle H. H is defined as follows. On input x, it first checks whether $H(x)$ has previously been defined. If so, it returns $H(x)$. Otherwise, it sets $H(x)$ to a uniformly random value in the codomain of H and then returns $H(x)$. This allows us to parametrize the maximal number of hash queries in our security notions. As an example, we define $(t, \varepsilon, Q_s, Q_h)$-UF-CMA as security against any adversary that makes at most Q_h queries to H in the UF-CMA game. Furthermore, we make the standard convention that any random oracle query that is asked as a result of a query to the signing oracle in the UF-CMA game is also counted as a query to the random oracle. This implies that $Q_s \leq Q_h$.

2.3 Identification Schemes

A five-move identification protocol of the form depicted in Fig. 1 is defined as follows.

Definition 3 (Five-move Identification Scheme). *A five-move identification scheme* ID $:=$ (IGen, P, ChSet$_1$, ChSet$_2$, V) *is defined as follow.*

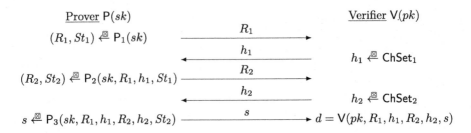

Fig. 1. A 5-move identification scheme and its transcript (R_1, h_1, R_2, h_2, s).

- *The key generation algorithm* IGen *takes system parameters* par *as input and returns a public key and a secret key* (pk, sk). *We assume that* pk *defines two challenge sets* ChSet$_1$ *and* ChSet$_2$.
- *The prover algorithm* P = (P$_1$, P$_2$, P$_3$) *is split into three algorithms.* P$_1$ *takes as input the secret key* sk *and returns a first-move commitment* R_1 *and a state* St_1; P$_2$ *takes as input the secret key* sk, *a first-move commitment* R_1, *a challenge* h_1, *and a state* St_1 *and returns a second-move commitment* R_2; P$_3$ *takes as input the secret key* sk, *a transcript* (R_1, h_1, R_2, h_2), *and a state* St_2 *and returns a response* s.
- *The deterministic verifier algorithm* V *takes the public key* pk *and the conversation transcript as input and outputs a* decision, *1 (acceptance) or 0 (rejection).*

We define some useful terms. A *transcript* for a canonical five-move identification scheme is of the form (R_1, h_1, R_2, h_2, s). A transcript (R_1, h_1, R_2, h_2, s) is *valid* (with respect to pk) if $V(pk, R_1, h_1, R_2, h_2, s) = 1$ and it is *real* if it is output by the following algorithm SKTran(sk). We elaborate further on the purpose of SKTran below when defining the notion of naHVZK (non-aborting honest-verifier zero-knowledge).

> SKTran(sk):
> $(R_1, St_1) \xleftarrow{\boxtimes} P_1(sk)$
> $h_1 \xleftarrow{\boxtimes}$ ChSet$_1$
> $(R_2, St_2) \xleftarrow{\boxtimes} P_2(sk, R_1, h_1, St_1)$
> $h_2 \xleftarrow{\boxtimes}$ ChSet$_2$
> $s \xleftarrow{\boxtimes} P_3(sk, R_1, h_1, R_2, h_2, St_2)$
> If $s = \bot$ then $T := (\bot, \bot, \bot, \bot, \bot)$
> Else $T := (R_1, h_1, R_2, h_2, s)$
> Return T

Definition 4 (Correctness error ρ). *We call an* ID *has correctness error* ρ *if, for all* $(pk, sk) \in$ IGen(par), *the following holds:*

- *For all* $(R_1, h_1, R_2, h_2, s) \xleftarrow{\boxtimes}$ SKTran(sk) *with* $s \neq \bot$, *we have* $V(pk, R_1, h_1, R_2, h_2, s) = 1$.

– A real transcript (R_1, h_1, R_2, h_2, s) contains $s = \perp$ with probability at most ρ, i.e., $\Pr[s = \perp \mid (R_1, h_1, R_2, h_2, s) \stackrel{\boxtimes}{\leftarrow} \mathsf{SKTran}(sk)] \leq \rho$.

Generalizing [35] we now define parallel impersonation against key-only attacks (KOA) for five-move identification schemes.

Definition 5 (Non-aborting (Parallel) Impersonation). *A five-move identification scheme* ID *is* $(t, \varepsilon, Q_{\mathrm{CH}_1}, Q_{\mathrm{CH}_2}, Q_{\mathrm{O}})$-naPIMP-ATK *secure (non-aborting parallel impersonation against* ATK *attacks,* ATK $\in \{\mathsf{KOA}, \mathsf{PA}\}$*) if for all adversaries* A *running in time at most* t *and making at most* Q_{CH_1} *queries to the challenge oracle* CH_1 *and* Q_{CH_2} *queries to oracle* CH_2*, we have*

$$\Pr\left[\begin{array}{c} \mathsf{V}(pk, R_1, h_1, R_2, h_2, s^*) = 1 \\ \wedge \; (R_1, h_1) \in \mathcal{L}_1 \wedge \; (R_2, h_2) \in \mathcal{L}_2 \end{array} \middle| \begin{array}{l} (pk, sk) \stackrel{\boxtimes}{\leftarrow} \mathsf{IGen}(par) \\ s^* \stackrel{\boxtimes}{\leftarrow} \mathsf{A}^{\mathrm{CH}_1(\cdot), \mathrm{CH}_2(\cdot)}(pk, St) \end{array} \right] \leq \varepsilon,$$

where the challenge oracles $\mathrm{CH}_i(R_i)$ *($i \in \{1, 2\}$) return* $h_i \stackrel{\boxtimes}{\leftarrow} \mathsf{ChSet}_i$ *to* A *and store* (R_i, h_i) *in set* $\mathcal{L}_i$.[3] *For different kinds of attacks, oracle* O *is defined as follows.*

– *If* ATK = KOA *(key-only attack), then* O *always returns* $\perp$.
– *If* ATK = PA *(passive attack), then* O := TRAN, *and the transcript oracle* TRAN() *returns a real transcript* $(R_1', h_1', R_2', h_2', s')$ *to* A, *i.e.,* $(R_1', h_1', R_2', h_2', s') \stackrel{\boxtimes}{\leftarrow} \mathsf{SKTran}(sk)$.

We do not use the parameter Q_{O} *for* ATK = KOA *and simply speak of* $(t, \varepsilon, Q_{\mathrm{CH}_1}, Q_{\mathrm{CH}_2})$-naPIMP-KOA. *Moreover,* $(t, \varepsilon, Q_{\mathrm{O}})$-naIMP-ATK *(impersonation against* ATK *attack) security is defined as* $(t, \varepsilon, 1, 1, Q_{\mathrm{O}})$-naPIMP-ATK *security, i.e., the adversary is only allowed* $Q_{\mathrm{CH}_1} = 1$ *query to the* CH_1 *oracle and* $Q_{\mathrm{CH}_2} = 1$ *to* CH_2.

Definition 6 (Special Soundness). *A five-move identification scheme* ID *is* SS *(special sound) if there exists an extractor* Ext *such that, for all* $(pk, sk) \in$ IGen(par), *given any two valid transcripts* (R_1, h_1, R_2, h_2, s) *and* $(R_1, h_1', R_2, h_2', s')$ *with* $h_2 \neq h_2'$, *it outputs a valid secret key* sk^* *such that* $(pk, sk^*) \in$ IGen(par), *i.e., we have* $\Pr[(sk^*, pk) \in \mathsf{IGen}(par) \mid sk^* \stackrel{\boxtimes}{\leftarrow} \mathsf{Ext}(pk, R_1, h_1, R_2, h_2, s, h_1', R_2', h_2', s')] = 1$. *The winning condition* $(pk, sk^*) \in$ IGen(par) *means that the tuple* (pk, sk^*) *is in the support of* IGen(par), *i.e., that* A *outputs a valid secret-key* sk^* *with respect to* pk.

We now introduce the notion of (statistical) *non-aborting honest-verifier zero-knowledge* [37], abbreviated as naHVZK. This notion generalizes the standard definition of honest-verifier zero-knowledge by including also identification schemes ID with correctness error ρ. Note that a real run of ID might produce a transcript of the form $(R_1, h_1, R_2, h_2, \perp)$ with probability ρ. Unfortunately, it

---

[3] On two queries $\mathrm{CH}_i(R_i)$ and $\mathrm{CH}_i(R_i')$ with the same input $R_i = R_i'$ the oracle returns two independent random challenges $h_i \stackrel{\boxtimes}{\leftarrow} \mathsf{ChSet}_i$ and $h_i' \stackrel{\boxtimes}{\leftarrow} \mathsf{ChSet}_i$.

might not be efficiently possible to simulate a correctly distributed transcript in this case. Therefore, we again make use of the algorithm SKTran which on input a secret key sk internally generates a valid transcript of ID (with respect to the matching public key pk), but outputs $\perp$ if the transcript matches an execution of ID in which $s = \perp$. We now require an efficient simulator Sim that produces transcripts which are statistically close in distribution to the ones output by SKTran.

Definition 7 (Δ-Statistical Non-Aborting Honest-Verifier Zero-Knowledge with α Bits Min-Entropy). *A five-move identification scheme* ID *is said to be* Δ-*statistically* naHVZK(non-aborting honest-verifier zero-knowledge) *with* α *bits min-entropy if there exists an algorithm* Sim *that, given a valid public key* pk, *outputs* (R_1, h_1, R_2, h_2, s) *such that the distribution of* (R_1, h_1, R_2, h_2, s) *has statistical distance at most* Δ *from the distribution of a transcript output by* SKTran *on input* sk *and if for all* $(pk, sk) \in \mathsf{IGen(par)}$ *and strings* R_1', R_2' *we have*

$$\Pr\left[R_1 = R_1' \text{ or } R_2 = R_2' \mid (R_1, h_1, R_2, h_2, s) \xleftarrow{\boxtimes} \mathsf{Sim}(pk) \right] \leq 2^{-\alpha}.$$

If $\Delta = 0$, *we say that* ID *is perfectly* naHVZK.

3   Signatures from Five-Move Identification Schemes

We extend the generalized Fiat-Shamir transform [6] to construct signatures for 5-move identification schemes. We also present an online/offline variant of our transformation, where parts of the computation can be performed off-line which leads to a better performance in the signing step, but requires special soundness for the underlying identification schemes. Furthermore, we give an alternative Fiat-Shamir transform, which outputs shorter signatures, but retains the same security.

3.1   The Fiat-Shamir Transform and Its Online/Offline Variant

Fix some system parameters par. Let $\mathsf{ID} := (\mathsf{IGen}, \mathsf{P}, \mathsf{ChSet}_1, \mathsf{ChSet}_2, \mathsf{V})$ be a five-move identification scheme and $H_1 : \{0,1\}^* \to \mathsf{ChSet}_1$ and $H_2 : \{0,1\}^* \to \mathsf{ChSet}_2$ be two hash functions. We also fix $\ell \in \mathbb{N}$ which controls the scheme's correctness. The signature scheme $\mathsf{FS}[\mathsf{ID}, H_1, H_2, \ell] := (\mathsf{Gen}, \mathsf{Sign}, \mathsf{Ver})$ from ID is defined as follows. Its online/offline variant $\mathsf{OF}[\mathsf{ID}, H_1, H_2, \ell] := (\mathsf{Gen}, \mathsf{Sign_o}, \mathsf{Ver_o})$ is defined with the boxed differences.

Gen(par):	Sign(sk, m), $\boxed{\text{Sign}_o(sk, m)}$:
$(pk, sk) \stackrel{\boxtimes}{\leftarrow} \text{IGen}(\text{par})$	$i := 0$
Return (pk, sk)	While $i \leq \ell$ and $s = \bot$:
	$\quad i := i + 1$
Ver(pk, m, σ), $\boxed{\text{Ver}_o(pk, m, \sigma)}$:	$\quad (R_1, St_1) \stackrel{\boxtimes}{\leftarrow} \text{P}_1(sk)$
Parse $\sigma = (R_1, R_2, s)$	$\quad h_1 = H_1(R_1, m)$
$h_1 = H_1(R_1, m)$	$\quad \boxed{h_1 = H_1(R_1)}$
$\boxed{h_1 = H_1(R_1)}$	$\quad (R_2, St_2) \stackrel{\boxtimes}{\leftarrow} \text{P}_2(sk, R_1, h_1, St_1)$
$h_2 = H_2(R_2, m)$	$\quad h_2 = H_2(R_2, m)$
Return $\text{V}(pk, R_1, h_1, R_2, h_2, s)$	$\quad s \stackrel{\boxtimes}{\leftarrow} \text{P}_2(sk, R_1, h_1, R_2, h_2, St_2)$
	If $s = \bot$ then $\sigma := \bot$
	Else $\sigma := (R_1, R_2, s)$
	Return σ

If ID has correctness error ρ, then both FS$[\text{ID}, H_1, H_2, \ell]$ and OF$[\text{ID}, H_1, H_2, \ell]$ are signature schemes with correctness error ρ^ℓ.

The following theorem states the security of FS$[\text{ID}, H_1, H_2, \ell]$ and OF$[\text{ID}, H_1, H_2, \ell]$.

Theorem 1 (Security of FS and OF). *Suppose that* ID *has Δ-statistical naHVZK with α bits min-entropy and is $(t, \varepsilon, Q_{\text{CH}_1}, Q_{\text{CH}_2})$-naPIMP-KOA secure. Then the signature scheme* FS$[\text{ID}, H_1, H_2, \ell]$ *is $(t', \varepsilon', Q_s, Q_1, Q_2)$-UF-CMA-secure in the random oracle model, where*

$$\varepsilon' \leq \varepsilon + \frac{(Q_1 + Q_2)Q_s}{2^\alpha} + Q_s \cdot \ell\Delta, \quad t \approx t', \quad Q_1 = Q_{\text{CH}_1} - 1, \quad Q_2 = Q_{\text{CH}_2} - 1,$$

and Q_{CH_1} and Q_{CH_2} are upper bounds on the number of CH_1 and CH_2 queries in the PIMP-KOA *experiment, respectively, and Q_s, Q_1, and Q_2 are upper bounds on the number of signing and random oracles H_1 and H_2 queries in the* UF-CMA *experiment, respectively. Moreover, if* ID *has special soundness (SS), then* OF$[\text{ID}, H_1, H_2, \ell]$ *is $(t', \varepsilon', Q_s, Q_1, Q_2)$-UF-CMA-secure in the random oracle model, where*

$$\varepsilon' \leq \varepsilon + \frac{(Q_1 + Q_2)Q_s}{2^\alpha} + Q_s \cdot \ell\Delta + \frac{1}{|\text{ChSet}_2|}, \quad t \approx t',$$
$$Q_1 = Q_{\text{CH}_1} - 2, \quad Q_2 = Q_{\text{CH}_2} - 2.$$

The proof of Theorem 1 is obtained by combining Lemmas 1 to 3.

ALTERNATIVE FIAT-SHAMIR TRANSFORM. We call ID *partially commitment-recoverable* if the second-move commitment R_2 can be partitioned into $R_2 = (R_L, R_R)$, a left part R_L and a right part R_R, and $\text{V}(pk, R_1, h_1, R_2, h_2, s)$ is such that it first recomputes $R_1' = \text{V}_1(pk, h_2, s)$ and $R_R' = \text{V}_2(pk, R_L, h_1, h_2, s)$ and then outputs 1 iff $(R_1', R_R') = (R_1, R_R)$. It is *fully commitment-recoverable* if $R_2 = R_R$ and R_L is the empty string. For commitment-recoverable ID, we can define an alternative Fiat-Shamir transformation FS$'[\text{ID}, H_1, H_2, \ell] :=$ (Gen, Sign$'$, Ver$'$), where Gen is as in FS$[\text{ID}, H_1, H_2, \ell]$. Algorithm Sign$'(sk, m)$ is defined as Sign(sk, m) with the modified output $\sigma' = (R_L, h_2, s)$. Algorithm Ver$'(pk, m, \sigma')$ first parses $\sigma' = (R_L, h_2, s)$, then recomputes $R_1' = \text{V}_1(pk, h_2, s)$

and $R'_R := V_2(pk, R_L, h_1, h_2, s)$, where $h_1 = H_1(R'_1)$, and finally returns 1 iff $H_2((R_L, R'_R), m) = h_2$. Its online/offline variant $OF'[ID, H_1, H_2, \ell] := (Gen, Sign_o', Ver_o')$ is defined in the similar manner with the boxed differences.

Gen(par):	Sign$'(sk, m)$, $\boxed{Sign_o'(sk, m)}$:
$(pk, sk) \overset{\boxtimes}{\leftarrow} IGen(par)$	$i := 0$
Return (pk, sk)	While $i \leq \ell$ and $s = \perp$:
	$\quad (R_1, St_1) \overset{\boxtimes}{\leftarrow} P_1(sk)$
Ver$'(pk, m, \sigma')$, $\boxed{Ver_o'(pk, m, \sigma')}$:	$\quad h_1 = H_1(R_1, m)$
Parse $\sigma' = (R_L, h_2, s)$	$\quad \boxed{h_1 = H_1(R_1)}$
$R_1 = V_1(pk, h_2, s)$	$\quad (R_2, St_2) \overset{\boxtimes}{\leftarrow} P_2(sk, R_1, h_1, St_1)$
$h_1 = H_1(R_1, m)$	$\quad h_2 = H_2(R_2, m)$
$\boxed{h_1 = H_1(R_1)}$	$\quad s \overset{\boxtimes}{\leftarrow} P_2(sk, R_1, h_1, R_2, h_2, St_2)$
$R_R = V_2(pk, R_L, h_1, h_2, s)$	If $s = \perp$ then $\sigma' := \perp$
$R_2 = (R_L, R_R)$	Else $\sigma' = (R_L, h_2, s)$
If $h_2 = H_2(R_2, m)$	Return σ'
$\quad$ then return 1	
Else return 0	

Since $\sigma = (R_1, R_2, s)$ can be publicly transformed into $\sigma' = (R_L, h_2, s)$ and vice versa, $FS[ID, H_1, H_2, \ell]$ and $FS'[ID, H_1, H_2, \ell]$ are equivalent in terms of security. The same argument holds for $OF[ID, H_1, H_2, \ell]$ and $OF'[ID, H_1, H_2, \ell]$. On the one hand, the alternative Fiat-Shamir transform yields shorter signatures if $h_2 \in ChSet_2$ has a smaller representation size than (R_1, R_R). On the other hand, signatures of the Fiat-Shamir transform maintain their algebraic structure which in some cases enables useful properties such as batch verification.

Lemma 1 (UF-KOA security of FS and OF). *Suppose that* ID *is* $(t, \varepsilon, Q_{CH_1}, Q_{CH_2})$-naPIMP-KOA-*secure. Then the signature schemes* $FS[ID, H_1, H_2, \ell]$ *and* $OF[ID, H_1, H_2, \ell]$ *are* $(t', \varepsilon', Q_1, Q_2)$-UF-KOA-*secure in the random oracle model, where*

$$\varepsilon = \varepsilon', t \approx t', Q_1 = Q_{CH_1} - 1, Q_2 = Q_{CH_2} - 1,$$

and Q_1, Q_2 *are upper bounds on the numbers of hash queries to* H_1 *and* H_2, *respectively.*

Proof. We prove the statement for $OF[ID, H_1, H_2, \ell]$; the proof for $FS[ID, H_1, H_2, \ell]$ is identical. Assume that an adversary A breaks the $(t', \varepsilon', Q_1, Q_2)$-UF-KOA-security of $OF[ID, H_1, H_2, \ell]$. We construct an adversary B that breaks the $(t, \varepsilon, Q_{CH_1}, Q_{CH_2})$-PIMP-KOA security of ID, with $(t, \varepsilon, Q_{CH_1}, Q_{CH_2})$ as claimed.

At the beginning, after obtaining pk from the PIMP-KOA experiment, B forwards it to A. If A makes a query R_1 to the random oracle H_1, B returns $H_1(R_1)$ if it is already defined, otherwise B makes a query $h_1 \overset{\boxtimes}{\leftarrow} CH_1(R_1)$ and programs $H_1(R_1) := h_1$. If A makes a query (R_2, m) to the random oracle H_2, B returns $H_2(R_2, m)$ if it is already defined, otherwise B makes a query $h_2 \overset{\boxtimes}{\leftarrow} CH(R_2)$ and programs the random oracle $H_2(R_2, m) := h_2$.

Eventually, A submits a forgery $(m, \sigma = (R_1, R_2, s))$, and terminates. We assume that $h_1 := H_1(R_1)$ and $h_2 := H_2(R_2, m)$ were already queried by A. (Otherwise, B queries $H_1(R_1)$ and $H_2(R_2, m)$ which are simulated as described above.) Hence, in total, there are $Q_{\mathrm{CH}_1} = Q_1 + 1$ and $Q_{\mathrm{CH}_2} = Q_2 + 1$ queries to H_1 and H_2, respectively. Adversary B outputs s and terminates. According to the simulations of H_1 and H_2, we have $(R_1, h_1) \in \mathcal{L}_1$ and $(R_2, h_2) \in \mathcal{L}_2$, and (R_1, h_1, R_2, h_2, s) is a valid transcript and hence breaks the PIMP-KOA security if A's forgery is valid. This establishes $\varepsilon = \varepsilon'$. The running time of B is roughly that of A, and thus $t' \approx t$.

Lemma 2 (UF-CMA security of FS). *If* ID *is* Δ-*statistically naHVZK with* α-*bits min-entropy and* FS[ID, H_1, H_2, ℓ] *is* $(t, \varepsilon, Q_1, Q_2)$-*UF-KOA secure, then* FS[ID, H_1, H_2, ℓ] *is* $(t', \varepsilon', Q_s, Q_1', Q_2')$-*UF-CMA secure in the random oracle model, where*

$$\varepsilon' \leq \varepsilon + \frac{(Q_1' + Q_2')Q_s}{2^\alpha} + Q_s \cdot \ell\Delta, \quad t' \approx t, \quad Q_1' = Q_1 \quad Q_2' = Q_2$$

and Q_1 *and* Q_2 *are upper bounds on the numbers of hash queries to* H_1 *and* H_2, *respectively, and* Q_s, Q_1' *and* Q_2' *are upper bounds on the number of signing and hash queries to* H_1' *and* H_2' *in the* UF-CMA *experiment, respectively.*

Proof Assume that an adversary A breaks $(t', \varepsilon', Q_s, Q_1', Q_2')$-UF-CMA security of FS[ID, H_1, H_2, ℓ]. We construct an adersary B invokes A and breaks $(t, \varepsilon, Q_1, Q_2)$-UF-KOA security of FS[ID, H_1, H_2, ℓ] with (t, ε) as stated in the lemma. Adversary B is executed in the UF-KOA experiment. It obtains public key pk and has access to random oracles H_1 and H_2.

Adversary B runs A on input pk answering hash queries to random oracles H_1' and H_2' and signing queries as follows.

SIMULATION OF HASH QUERIES. A hash query $H_1'(R, m)$ is answered by B by querying its own hash oracle $H_1(R, m)$ and storing its answer and returning it. H_2' is simulated in the same way by using B's own oracle H_2.

SIMULATION OF SIGNING QUERIES. On A's signature query m, B uses the naHVZK property of ID to generate a signature σ on message m. Concretely, B defines $i := 0$ and simulates the signing query as follows.

- While $i \leq \ell$ and $s = \perp$:
    - $(R_1, h_1, R_2, h_2, s) \xleftarrow{\boxtimes} \mathsf{Sim}(pk)$ and $i := i + 1$;
- If $s = \perp$, then return $\perp$;
- Else
    - If $H_1'(R_1, m) \neq \perp$ or $H_2'(R_2, m) \neq \perp$, then define INCON := 1 and return $\perp$
    - Else define
    $$H_1'(R_1, m) := h_1, \quad H_2'(R_2, m) := h_2 \tag{1}$$
    and return $\sigma := (R_1, R_2, s)$.

We note that, by Eq. (1), B makes the hash functions inconsistent, since $H_1(R_1, m) \neq h_1 =: H_1'(R_1, m)$ and $H_2(R_2, m) \neq h_2 =: H_2'(R_2, m)$ with high probability. Adversary A can detect this inconsistence if $\text{INCON} = 1$, namely, A queries the exact $H_1'(R_1, m)$ or $H_2'(R_2, m)$ before asking the signing query on m. For each signing query, this can be bounded (namely, B aborts because of $\text{INCON} = 1$) by probability at most $(Q_1' + Q_2')/2^\alpha$, since ID has α-bits min-entropy. Moreover, for each signing query, B runs Sim at most ℓ times and produces a real transcript from $\text{SKTran}(sk)$ oracle with statistical distance at most Δ in each of these runs. Since the number of signing queries is bounded by Q_s, the statistical distance between the real UF-CMA experiment and the simulated one is at most $Q_s \cdot ((Q_1' + Q_2')/2^\alpha + \ell\Delta)$.

FORGERY. Eventually, A submits its forgery $(m, \sigma := (R_1, R_2, s))$. We assume that it is a valid forgery in the UF-CMA experiment, namely, for $h_1 = H_1'(R_1, m)$ and $h_2 = H_2'(R_2, m)$ we have $\text{V}(pk, R_1, h_1, R_2, h_2, s) = 1$. Furthermore, it satisfies the freshness condition, i.e., $m \notin \mathcal{M}$. Note that by the freshness condition, we have $H_1(R_1, m) = H_1'(R_1, m) = h_1$ and $H_2(R_2, m) = H_2'(R_2, m) = h_2$ since $H_1'(R_1, m)$ and $H_2'(R_2, m)$ were not programmed via (1). After receiving A's forgery, B computes a forgery for the UF-KOA experiment as $\sigma = (R_1, R_2, s)$.

Overall, B returns a valid forgery of UF-KOA experiment with probability

$$\varepsilon \geq \varepsilon' - \frac{(Q_1' + Q_2')Q_s}{2^\alpha} - Q_s \cdot \ell\Delta.$$

The running time of B is that of A plus the Q_s executions of Sim. We write $t' \approx t$. This completes the proof.

Lemma 3 (UF-CMA-security of OF). *If* ID *is Δ-statistically* naHVZK *with α-bits min-entropy, has* SS, *and* OF$[\text{ID}, H_1, H_2, \ell]$ *is $(t, \varepsilon, Q_1, Q_2)$-UF-KOA secure, then* OF$[\text{ID}, H_1, H_2, \ell]$ *is $(t', \varepsilon', Q_s, Q_1', Q_2')$-UF-CMA secure in the random oracle model, where*

$$\varepsilon' \leq \varepsilon + \frac{(Q_1' + Q_2')Q_s}{2^\alpha} + Q_s \cdot \ell\Delta + \frac{1}{|\text{ChSet}_2|}, \quad t' \approx t, \quad Q_1' = Q_1 - 1 \quad Q_2' = Q_2 - 1$$

and Q_1 and Q_2 are upper bounds on the numbers of hash queries to H_1 and H_2, respectively, and Q_s, Q_1' and Q_2' are upper bounds on the number of signing and hash queries to H_1' and H_2' in the UF-CMA experiment, respectively.

Proof. Let A be an algorithm that breaks $(t', \varepsilon', Q_s, Q_1', Q_2')$-UF-CMA security of OF$[\text{ID}, H_1, H_2, \ell]$. We will describe an adversary B invoking A that breaks $(t, \varepsilon, Q_1, Q_2)$-UF-KOA security of OF$[\text{ID}, H_1, H_2, \ell]$ with (t, ε) as stated in the lemma. Adversary B is executed in the UF-KOA experiment and obtains public-key pk, and has access to random oracles H_1 and H_2.

Adversary B runs A on input pk answering hash queries to random oracles H_1' and H_2' and signing queries as follows.

SIMULATION OF HASH QUERIES. A hash query $H_1'(R)$ is answered by B by querying its own hash oracle $H_1(R)$ and storing its answer and returning it. H_2' is simulated in the same way by using B's own oracle H_2.

SIMULATION OF SIGNING QUERIES. The simulation here is similar to that in Lemma 2 except for the simulation of H_1'. For completeness, we present the details as follows.

On A's signature query m, B uses the naHVZK property of ID to generate a signature σ on message m. Concretely, B defines $i := 0$ and simulates the signing query as follows.

- While $i \leq \ell$ and $s = \bot$:
  - $(R_1, h_1, R_2, h_2, s) \overset{\boxtimes}{\leftarrow} \mathsf{Sim}(pk)$ and $i := i + 1$;
- If $s = \bot$, then return $\bot$;
- Else
  - If $H_1'(R_1) \neq \bot$ or $H_2'(R_2, m) \neq \bot$, then define INCON $:= 1$ and return $\bot$
  - Else define

$$H_1'(R_1) := h_1, \ H_2'(R_2, m) := h_2 \tag{2}$$

  and return $\sigma := (R_1, R_2, s)$.

We note that, by Eq. (2), B makes the hash functions inconsistent, since $H_1(R_1) \neq h_1 =: H_1'(R_1)$ and $H_2(R_2, m) \neq h_2 =: H_2'(R_2, m)$ with high probability. Adversary A can detect this inconsistence if INCON $= 1$, namely, A queries the exact $H_1'(R_1)$ or $H_2'(R_2, m)$ before asking the signing query on m. For each signing query, this can be bounded (namely, B aborts because of INCON $= 1$) by probability at most $(Q_1' + Q_2')/2^\alpha$, since ID has α-bits min-entropy. Moreover, for each signing query, B runs Sim at most ℓ times and produces a real transcript from $\mathsf{SKTran}(sk)$ oracle with statistical distance at most Δ in each of these runs. Since the number of signing queries is bounded by Q_s, the statistical distance between the real UF-CMA experiment and the simulated one is at most $Q_s \cdot ((Q_1' + Q_2')/2^\alpha + \ell\Delta)$.

FORGERY. Eventually, A will submit its forgery $(m, \sigma := (R_1, R_2, s))$. We assume that it is a valid forgery in the UF-CMA experiment, namely, for $h_1 = H_1'(R_1)$ and $h_2 = H_2'(R_2, m)$ we have $\mathsf{V}(pk, R_1, h_1, R_2, h_2, s) = 1$. Furthermore, it satisfies the freshness condition, i.e., $m \notin^* \mathcal{M}$. After receiving A's forgery, B computes a forgery for the UF-KOA experiment according to the following case distinction.

- Case 1: R_1 was defined in a signing query on some message m' via (2), i.e., $(R_1, h_1', R_2', h_2', s')$ was generated by using $\mathsf{Sim}(pk)$. The freshness condition implies $m' \neq m$ and hence $h_2 = H_2'(R_2, m) \neq H_2'(R_2', m') = h_2'$, except with probability $1/|\mathsf{ChSet}_2|$. In that case we have two valid transcripts (R_1, h_1, R_2, h_2, s) and $(R_1, h_1', R_2', h_2', s')$ with $h_2 \neq h_2'$. By the special soundness of ID, B extracts a valid sk^* by running Ext such that $(pk, sk^*) \in \mathsf{IGen}(par)$, and then B use sk^* to generate a fresh and valid UF-CMA forgery. In this case, $Q_1' = Q_1 + 1$ and $Q_2' = Q_2 + 1$.
- Case 2: R_1 was queried to the H_1' oracle, i.e., $H_1(R_1) = H_1'(R_1) = h_1$. By the freshness condition, $h_2 = H_2'(R_2, m)$ was defined in a hash query, but not in a signing query, i.e., $h_2 = H_2'(R_2, m) = H_2(R_2, m)$. B returns $\sigma = (R_1, R_2, s)$ as a valid forgery to its UF-CMA experiment.

Overall, B returns a valid forgery of UF-KOA experiment with probability

$$\varepsilon \geq \varepsilon' - \frac{(Q_1' + Q_2')Q_s}{2^\alpha} - Q_s \cdot \ell\Delta - \frac{1}{|\mathsf{ChSet}_2|}.$$

The running time of B is that of A plus the Q_s executions of Sim. We write $t' \approx t$. This completes the proof.

4   Instantiations

In the following, let $\mathsf{par} := (p, g, \mathbb{G})$ be a set of system parameters, where $\mathbb{G} = \langle g \rangle$ is a cyclic group of prime order p.

4.1   Instantiation from CDH

We briefly recall the CDH problem.

Definition 8. (Computation Diffie-Hellman Assumption). *The computational Diffie-Hellman problem* CDH *is* (t, ε)-*hard in* par *if for all adversaries* A *running in time at most* t,

$$\Pr\left[Z = g^{xy} \mid x, y \xleftarrow{\boxtimes} \mathbb{Z}_p; Z \xleftarrow{\boxtimes} \mathsf{A}(g^x, g^y)\right] \leq \varepsilon.$$

IDENTIFICATION SCHEME. The identification scheme $\mathsf{ID}_{\mathsf{CDH}} := (\mathsf{IGen}, \mathsf{P}, \mathsf{ChSet}_1, \mathsf{ChSet}_2, \mathsf{V})$ is defined as follows.

$\mathsf{IGen}(\mathsf{par})$:	$\mathsf{P}_1(sk)$:
$sk := x \xleftarrow{\boxtimes} \mathbb{Z}_p$	$St_1 := r \xleftarrow{\boxtimes} \mathbb{Z}_p; R_1 = g^r$
$pk := X = g^x$	Return (R_1, St_1)
$\mathsf{ChSet}_1 := \mathbb{G}; \mathsf{ChSet}_2 := \{0, ..., 2^k - 1\}$	
Return (pk, sk)	$\mathsf{P}_2(sk, R_1, h_1, St_1)$:
	Parse $St_1 := r$
$\mathsf{V}(pk, R_1, h_1, R_2, h_2, s)$:	$R_L := h_1^x; R_R := h_1^r$
Parse $R_2 := (R_L, R_R)$	Return $(R_2 := (R_L, R_R)), St_2 := r)$
If $R_1 = g^s \cdot X^{-h_2}$ and $R_R = h_1^s \cdot R_L^{-h_2}$	
then return 1	$\mathsf{P}_3(sk, R_1, h_1, R_2, h_2, St_2)$:
Else return 0	Parse $St_2 = r$
	Return $s = x \cdot h_2 + r \bmod p$

Lemma 4. $\mathsf{ID}_{\mathsf{CDH}}$ *is a perfectly correct five-move identification scheme and has perfect non-aborting honest-verifier zero-knowledge (naHVZK) with* $\alpha = \log p$ *bits min-entropy and special soundness (SS). Moreover, if* CDH *is* (t, ε)-*hard in* $\mathsf{par} = (p, g, \mathbb{G})$ *then* $\mathsf{ID}_{\mathsf{CDH}}$ *is* $(t', \varepsilon', Q_{\mathrm{CH}_1}, Q_{\mathrm{CH}_2})$-*PIMP-KOA secure, where* $t \approx t'$ *and* $\varepsilon \geq \varepsilon' - \frac{Q_{\mathrm{CH}_2}}{2^k}$.

Proof. The perfect correctness of $\mathsf{ID}_{\mathsf{CDH}}$ is straightforward to verify. We show the other properties as follows:

PERFECT NON-ABORTING HONEST-VERIFIER ZERO-KNOWLEDGE (naHVZK). Given public key $pk = X$, Sim first samples $s, w, h_2 \overset{\boxtimes}{\leftarrow} \mathbb{Z}_p$. It then computes $h_1 := g^w, R_1 := g^s X^{-h_2}, R_L := X^w$, and $R_R := R_1^w$, defines $R_2 := (R_L, R_R)$ and outputs the transcript (R_1, h_1, R_2, h_2, s). Clearly, (R_1, h_1, R_2, h_2, s) is distributed the same as the one from $\mathsf{SKTran}(sk)$, since s is uniformly random over $\mathbb{Z}_p$ and R_1, R_L and R_R satisfy $R_1 = g^s \cdot X^{-h_2}$ and $R_R = h_1^s \cdot R_L^{-h_2}$. Moreover, we note that the entropy of $(R_1, St_1 := r) \overset{\boxtimes}{\leftarrow} \mathsf{P}_1(sk)$ and $(R_2, St_2) \overset{\boxtimes}{\leftarrow} \mathsf{P}_2(sk, R_1, h_1)$ comes only from R_1, which is uniformly random over $\mathbb{G}$. Hence, since the outputs of $\mathsf{Sim}(pk)$ are identically distributed to the outputs of $\mathsf{SKTran}(sk)$, $\mathsf{ID}_{\mathsf{CDH}}$ has $\log |\mathbb{G}| = \log p$ bits min-entropy as claimed.

SPECIAL SOUNDNESS. Given two accepting transcripts (R_1, h_1, R_2, h_2, s) and $(R_1, h_1', R_2', h_2', s')$ with $h_2 \neq h_2'$, we define an extractor Ext with the property that $\mathsf{Ext}(pk, R_1, h_1, R_2, h_2, s, h_1', R_2', h_2', s')$ outputs $x^* := (s - s')/(h_2 - h_2') \bmod p$. We have $\Pr[g^{x^*} = X] = 1$ for all $(X := g^x, x) \in \mathsf{IGen}(par)$, since $R_1 = g^s \cdot X^{-h_2} = g^{s'} \cdot X^{-h_2'}$ and $X = g^{(s-s')/(h_2-h_2')}$.

PIMP-KOA-SECURITY. Let A be an attacker against the $(t', \varepsilon', Q_{\mathsf{CH}_1}, Q_{\mathsf{CH}_2})$-PIMP-KOA security of $\mathsf{ID}_{\mathsf{CDH}}$. We construct an attacker B that (t, ε)-breaks CDH.

CONSTRUCTION OF B. Let $(X := g^x, Y := g^y)$ denote the CDH instance. B runs A with input $pk := X$ and answers A's challenge queries as follows.

For A's CH_1 query on $R_1 \in \mathbb{G}$, B chooses $a \overset{\boxtimes}{\leftarrow} \mathbb{Z}_p$ and computes $h_1 = Y \cdot g^a$. For A's CH_2 query on $R_2 \in \mathbb{G} \times \mathbb{G}$, B chooses $h_2 \overset{\boxtimes}{\leftarrow} \mathbb{Z}_p$ and returns it to A. Clearly, B's simulation of the PIMP-KOA game is perfect, since both h_1 and h_2 are uniformly random over $\mathbb{G}$ and $\mathbb{Z}_p$, respectively.

Eventually, A returns its response s^* for the PIMP-KOA experiment. We assume that A's response is valid, i.e., there exist $(R_1, h_1) \in \mathcal{L}_1$ and $(R_2 := (R_L, R_R), h_2) \in \mathcal{L}_2$ such that $R_1 = g^{s^*} \cdot X^{-h_2}$ and $R_R = h_1^{s^*} \cdot R_L^{-h_2}$. We denote the discrete logarithm of R_L based on h_1 by $x' = \mathsf{DL}_{h_1}(R_L)$ and do the following cases distinction:

- <u>Case 1:</u> $x = x'$. By the simulation of CH_1, we have $R_L = h_1^x = (Yg^{a_i})^x = Y^x X^{a_i}$ for some $i \in [Q_{\mathsf{CH}_1}]$. Thus, B returns $Z := R_L \cdot X^{-a_i} = Y^x$ to break the CDH problem.
- <u>Case 2:</u> $x \neq x'$. We show in this case even an unbounded adversary A can only win with probability $Q_{\mathsf{CH}_2}/2^k$. For each index $i \in [Q_{\mathsf{CH}_2}]$, before receiving $h_{2,i}$, A first commits to some $R_1 = g^{r_1}, R_L = h_1^{x'}$ and $R_R = h_1^{r_2}$ (for arbitrary $r_1, r_2, x' \in \mathbb{Z}_p$ and $x' \neq x$) and there exists $(R_1, h_1) \in \mathcal{L}_1$. A can only win if there exists an $s_i \in \mathbb{Z}_p$ such that

$$r_1 + h_{2,i}x = s_i = r_2 + h_{2,i}x' \Leftrightarrow h_{2,i} = \tfrac{r_2 - r_1}{x - x'}.$$

where $h_{2,i} \overset{\boxtimes}{\leftarrow} \mathsf{ChSet}_2 := \{0, ..., 2^k - 1\}$ is chosen independently of r_1, r_2 and x'. This happens with probability at most $1/|\mathsf{ChSet}_2| = 2^{-k}$. By the union bound we obtain the bound $Q_{\mathsf{CH}_2}/2^k$ as claimed.

Overall, B returns a valid solution of the CDH challenge with probability $\varepsilon \geq \varepsilon' - \frac{Q_{\text{CH}_2}}{2^k}$. This completes the proof.

(ONLINE/OFFLINE) SIGNATURE. Let $H_1 : \{0,1\}^* \to \mathbb{G}$ and $H_2 : \{0,1\}^* \to \{0,\ldots,2^k - 1\}$ be two hash functions. As ID_{CDH} is perfectly correct and partially commitment-recoverable, we can use the alternative Fiat-Shamir transformation from Sect. 3.1 with $\ell := 1$ to obtain the signature scheme $\text{FS}_{\text{CDH}} := (\text{Gen}, \text{Sign}, \text{Ver})$ and its online/offline variant $\text{OF}_{\text{CDH}} := (\text{Gen}, \text{Sign}_\text{o}, \text{Ver}_\text{o})$. Here, FS_{CDH} does not include X, R_1 and (g, h_1) in the hash H_2, which is slightly simpler than the Chevallier-Mames scheme in [20].

Gen(par):	Sign(sk, m), $\boxed{\text{Sign}_\text{o}(sk, m)}$:	Ver(pk, m, σ), $\boxed{\text{Ver}_\text{o}(pk, m, \sigma)}$:
$sk := x \xleftarrow{\boxtimes} \mathbb{Z}_p$	$r \xleftarrow{\boxtimes} \mathbb{Z}_p$; $R_1 = g^r$	Parse $\sigma := (R_L, h_2, s)$
$pk := X = g^x$	$h_1 = H_1(R_1, m)$; $\boxed{h_1 = H_1(R_1)}$	$R_1 = g^s \cdot X^{-h_2}$
Return (pk, sk)	$R_L = h_1^x \in \mathbb{G}$; $R_R = h_1^r$	$h_1 = H_1(R_1, m)$; $\boxed{h_1 = H_1(R_1)}$
	$R_2 := (R_L, R_R)$	$R_R = h_1^s \cdot R_L^{-h_2}$
	$h_2 = H_2(R_2, m) \in \{0, ..., 2^k - 1\}$	$R_2 := (R_L, R_R)$
	$s = x \cdot h_2 + r \in \mathbb{Z}_p$	If $h_2 = H_2(R_2, m)$
	$\sigma = (R_L, h_2, s)$	then return 1
	Return σ	Else return 0.

By Lemma 4 and Theorem 1, we have

Theorem 2 (Security of FS_{CDH} and OF_{CDH}). *If* CDH *is (t, ε)-hard in* par $:= (p, g, \mathbb{G})$ *then scheme* FS_{CDH} *is $(t', \varepsilon', Q_s, Q_1, Q_2)$-UF-CMA secure and scheme* OF_{CDH} *is $(t'', \varepsilon'', Q_s, Q_1, Q_2)$-UF-CMA secure in the programmable random oracle model, where*

$$\varepsilon' \leq \varepsilon + \frac{Q_2 + 1}{2^n} + \frac{(Q_1 + Q_2)Q_s}{2^n}, \qquad t' \approx t,$$

$$\varepsilon'' \leq \varepsilon + \frac{Q_2 + 2}{2^n} + \frac{(Q_1 + Q_2)Q_s}{2^n} + \frac{1}{2^n}, \qquad t'' \approx t.$$

4.2 Instantiation from Short CDH

We recall the short exponent CDH assumption from [36].

Definition 9 (c-SCDH Assumption). *The c-short exponent computational Diffie-Hellman problem c-SCDH is (t, ε)-hard in* par *if for all adversaries* A *running in time at most t,*

$$\Pr\left[Z = g^{xy} \mid x, y \xleftarrow{\boxtimes} \{0, ..., 2^c - 1\}; Z \xleftarrow{\boxtimes} \text{A}(g^x, g^y)\right] \leq \varepsilon.$$

IDENTIFICATION SCHEME. Let par $:= (p, g, \mathbb{G}, k, k', c)$ be a set of system parameters, where $\mathbb{G} = \langle g \rangle$ is a cyclic group of prime order p with a hard c-SCDH problem and $k = \omega(\log p)$. The identification scheme $\text{ID}_{\text{SCDH}} := (\text{IGen}, \text{P}, \text{ChSet}_1, \text{ChSet}_2, \text{V})$ is defined as follows. Here the response s is computed over the integers (rather than over $\mathbb{Z}_p$).

IGen(par):	$P_1(sk)$:
$sk := x \xleftarrow{\boxtimes} \{0, ..., 2^c - 1\}$	$St_1 := r \xleftarrow{\boxtimes} \{0, ..., 2^{k+k'+c} - 1\};$
$pk := X = g^x$	$R_1 = g^r$
$\mathsf{ChSet}_1 := \mathbb{G}; \mathsf{ChSet}_2 := \{0, ..., 2^k - 1\}$	Return (R_1, St_1)
Return (pk, sk)	
	$P_2(sk, R_1, h_1, St_1)$:
$V(pk, R_1, h_1, R_2, h_2, s)$:	Parse $St_1 := r$
Parse $R_2 := (R_L, R_R)$	$R_L = h_1^x; R_R = h_1^r$
If $s \notin \{2^{k+c}, ..., 2^{k+k'+c} - 1\}$	$R_2 := (R_L, R_R); St_2 := St_1$
then return 0	Return (R_2, St_2)
If $R_1 = g^s X^{-h_2}$ and $R_R = h_1^s R_L^{-h_2}$	
then return 1	$P_3(sk, R_1, h_1, R_2, h_2, St_2)$:
Else return 0	Parse $St_2 := r$
	$s = x \cdot h_2 + r$
	If $s \notin \{2^{k+c}, ..., 2^{k+k'+c} - 1\}$
	then return $\bot$
	Else return s

Lemma 5 [3]. *Let $x \xleftarrow{\boxtimes} \{0, ...2^c - 1\}, h_2 \xleftarrow{\boxtimes} \{0, ..., 2^k - 1\}, r \xleftarrow{\boxtimes} \{0, ..., 2^{k'+k+c} - 1\}$. Then, $s := xh_2 + r \in \{2^{k+c}, ..., 2^{k'+k+c} - 1\}$ with probability $1 - 2^{-k'}$. Moreover, if $s \in \{2^{k+c}, ..., 2^{k+k+c} - 1\}$, then it is uniformly distributed in this interval.*

Lemma 6. $\mathsf{ID}_{\mathsf{SCDH}}$ *is a five-move identification scheme with correctness error $2^{-k'}$ and has perfect non-aborting honest-verifier zero-knowledge (naHVZK) with $\alpha = c$ bits min-entropy and special soundness. Moreover, if c-SCDH is (t, ε)-hard in $\mathsf{par} = (p, g, \mathbb{G})$ then $\mathsf{ID}_{\mathsf{SCDH}}$ is $(t', \varepsilon', Q_{\mathrm{CH}_1}, Q_{\mathrm{CH}_2})$-PIMP-KOA-secure, where $\varepsilon' \leq \varepsilon + Q_{\mathrm{CH}_2}/2^k$ and $t' \approx t$.*

Proof. By Lemma 5, $\mathsf{ID}_{\mathsf{SCDH}}$ has correctness error $2^{-k'}$. We note that the entropy of $R_1 = g^r$ and $R_2 := (h_1^x, h_1^r)$ comes only from r, which is chosen uniformly from $\{0, ..., 2^c - 1\}$. Hence $\mathsf{ID}_{\mathsf{SCDH}}$ has c bits min-entropy. We show the other properties as follows.

PERFECT naHVZK. Given public key $pk = X$, simulator Sim first samples $s \xleftarrow{\boxtimes} \{2^{k+c}, ..., 2^{k'+k+c} - 1\}, w \xleftarrow{\boxtimes} \mathbb{Z}_p$ and $h_2 \leftarrow \{0, ..., 2^k - 1\}$. It then computes $h_1 = g^w, R_1 = g^s X^{-h_2}, R_L = X^w$, and $R_R = R_1^w$, defines $R_2 := (R_L, R_R)$ and outputs the transcript (R_1, h_1, R_2, h_2, s) with probability $1 - 2^{-k'}$, or $(\bot, \bot, \bot, \bot, \bot)$ with probability $2^{-k'}$. Clearly, the output of $\mathsf{Sim}(pk)$ is identical to that of $\mathsf{SKTran}(sk)$. According to the simulation, with probability $1 - 2^{-k'}$, $\mathsf{Sim}(pk)$ will output a transcript (R_1, h_1, R_2, h_2, s), where s is uniformly random over the interval $\{2^{k+c}, ..., 2^{k'+k+c} - 1\}$ and R_1, R_L and R_R are values satisfying $R_1 = g^s \cdot X^{-h_2}$ and $R_R = h_1^s \cdot R_L^{-h_2}$. Due to Lemma 5, the probability that such a transcript is output by Sim is the same as that for SKTran, and, moreover, the probability that Sim outputs $(\bot, \bot, \bot, \bot, \bot)$ is also the same as that for SKTran. We note that the entropy of $R_1 = g^r$ and $R_2 := (h_1^x, h_1^r)$ comes only from r, which is chosen uniformly from $\{0, ..., 2^c - 1\}$. Since $\mathsf{Sim}(pk)$ outputs transcripts identically distributed to the ones output by $\mathsf{SKTran}(sk)$, $\mathsf{ID}_{\mathsf{SCDH}}$ has c bits min-entropy. Thus, $\mathsf{ID}_{\mathsf{SCDH}}$ is perfectly naHVZK.

SPECIAL SOUNDNESS. Given two accepting transcripts (R_1, h_1, R_2, h_2, s) and $(R_1, h_1', R_2', h_2', s')$ with $h_2 \neq h_2'$, we define an extractor Ext with the property that $\text{Ext}(pk, R_1, h_1, R_2, h_2, s, h_1', R_2', h_2', s')$ outputs $x^* := (s - s_i')/(h_2 - h_2')$. We note that $x^* \in \{0, ..., 2^c - 1\}$, since $s, s_i' \in \{2^{k+c}, ..., 2^{k+k'+c} - 1\}$ and $h_2, h_2' \in \{0, ..., 2^k\}$. Moreover, for all $(X := g^x, x) \in \text{IGen}(\text{par})$, we have $\Pr[X = g^{x^*}] = 1$, since $R_1 = g^s X^{-h_2} = g^{s'} X^{-h_2'}$ and then $X = g^{(s-s')/(h_2-h_2')}$.

PIMP-KOA-SECURITY. Let A be an attacker against the $(t', \varepsilon', Q_{\text{CH}_1}, Q_{\text{CH}_2})$-PIMP-KOA security of ID_{SCDH}. We construct an attacker B that (t, ε)-breaks c-SCDH.

CONSTRUCTION OF B. Let $(X := g^x, Y := g^y)$ denote the c-SCDH instance. B runs A with input $pk := X$ and answers A's challenge queries as follows.

For A's CH_1 query on $R_1 \in \mathbb{G}$, B chooses $a \stackrel{\boxtimes}{\leftarrow} \mathbb{Z}_p$ and computes $h_1 = Y \cdot g^a$. For A's CH_2 query on $R_2 \in \mathbb{G} \times \mathbb{G}$, B chooses $h_2 \stackrel{\boxtimes}{\leftarrow} \{0, ...2^k - 1\}$ and returns it to A. Clearly, B's simulation of the PIMP-KOA game is perfect, since both h_1 and h_2 are uniformly random over $\mathbb{G}$ and $\{0, ..., 2^k\}$, respectively.

Eventually, A returns its response s^* for the PIMP-KOA experiment. We assume that A's response is valid, i.e., $s^* \in \{2^{k+c}, ..., 2^{k+k'+c} - 1\}$ and there exist $(R_1, h_1) \in \mathcal{L}_1$ and $(R_2 := (R_L, R_R), h_2) \in \mathcal{L}_2$ such that $R_1 = g^{s^*} \cdot X^{-h_2}$ and $R_R = h_1^{s^*} \cdot R_L^{-h_2}$. We denote the discrete logarithm of R_L based on h_1 by $x' = \text{DL}_{h_1}(R_L)$ and do the following cases distinction:

- Case 1: $x = x'$. By the simulation of CH_1, we have $R_L = h_1^x = (Yg^{a_i})^x = Y^x X^{a_i}$ for some $i \in [Q_{\text{CH}_1}]$. Thus, B returns $Z := R_L \cdot X^{-a_i} = Y^x$ to break the c-SCDH problem.
- Case 2: $x \neq x'$. We show in this case even an unbounded adversary A can only win with probability $Q_{\text{CH}_2}/2^k$. For each index $i \in [Q_{\text{CH}_2}]$, before receiving $h_{2,i}$, A first commits to some $R_1 = g^{r_1}$, $R_L = h_1^{x'}$ and $R_R = h_1^{r_2}$ (for arbitrary $r_1, r_2, x' \in \mathbb{Z}_p$ and $x' \neq x$) and there exists $(R_1, h_1) \in \mathcal{L}_1$. A can only win if there exists an $s_i \in \{2^{k+c}, ..., 2^{k+k'+c} - 1\}$ such that

$$r_1 + h_{2,i} x = s_i = r_2 + h_{2,i} x' \Leftrightarrow h_{2,i} = \frac{r_2 - r_1}{x - x'},$$

where $h_{2,i} \stackrel{\boxtimes}{\leftarrow} \text{ChSet}_2 := \{0, ..., 2^k\}$ is chosen independently of r_1, r_2 and x'. This happens with probability at most $1/2^k$. By the union bound we obtain the bound $Q_{\text{CH}_2}/2^k$ as claimed.

Overall, B returns a valid solution of the c-SCDH challenge with probability $\varepsilon \geq \varepsilon' - \frac{Q_{\text{CH}_2}}{2^k}$. This completes the proof.

(ONLINE/OFFLINE) SIGNATURE. Let $H_1 : \{0,1\}^* \to \mathbb{G}$ and $H_2 : \{0,1\}^* \to \{0, ..., 2^k - 1\}$ be two hash functions. Since ID_{SCDH} has $2^{-k'}$ correctness error, given the required correctness of the signature, we fix a $\ell \in \mathbb{N}$, and as the scheme ID_{SCDH} is partially commitment recoverable, we can use the alternative Fiat-Shamir transform to obtain the signature scheme $\text{FS}_{\text{SCDH}} := (\text{Gen}, \text{Sign}, \text{Ver})$ and its online/offline variant $\text{OF}_{\text{SCDH}} := (\text{Gen}, \text{Sign}_o, \text{Ver}_o)$. For simplicity of notation, let $\mathcal{I} := \{2^{k+c}, ..., 2^{k+k'+c} - 1\}$.

Gen(par):	Sign(sk, m), Sign$_o(sk, m)$:	Ver(pk, m, σ), Ver$_o(pk, m, \sigma)$:
$x \xleftarrow{\boxtimes} \{0, ..., 2^c - 1\}$	$i := 0$	Parse $\sigma = (R_L, h_2, s)$
$pk := X = g^x$	While $i \leq \ell$ and $s = \bot$:	$R_1 = g^s \cdot X^{-h_2}$
$sk := x$	$\quad i := i + 1$	$h_1 = H_1(R_1, m)$
Return (pk, sk)	$\quad r \xleftarrow{\boxtimes} \{0, ..., 2^{k+k'+c} - 1\}$	$\boxed{h_1 = H_1(R_1)}$
	$\quad R_1 = g^r$	$R_R = h_1^s \cdot R_L^{-h_2}$
	$\quad h_1 = H_1(R_1, m) \in \mathbb{G}$	$R_2 := (R_L, R_R)$
	$\quad \boxed{h_1 = H_1(R_1) \in \mathbb{G}}$	If $s \in \mathcal{I} \wedge h_2 = H_2(R_2, m)$
	$\quad R_2 := (R_L, R_R) := (h_1^x, h_1^r)$	$\quad$ then return 1
	$\quad h_2 = H_2(R_2, m) \in \{0, ..., 2^k - 1\}$	Else return 0.
	$\quad s = x \cdot h_2 + r$	
	$\quad$ If $s \notin \mathcal{I}$ then $s := \bot$	
	If $s = \bot$ then $\sigma := \bot$	
	Else $\sigma := (R_L, h_2, s)$	
	Return σ	

By Lemma 6 and Theorem 1, we have

Theorem 3. (Security of FS$_{\mathsf{SCDH}}$ and OF$_{\mathsf{SCDH}}$) *If c-SCDH is (t, ε)-hard in* par $:= (p, g, \mathbb{G})$ *then* FS$_{\mathsf{SCDH}}$ *is* $(t', \varepsilon', Q_s, Q_1, Q_2)$-UF-CMA *secure and* OF$_{\mathsf{SCDH}}$ *is* $(t'', \varepsilon'', Q_s, Q_1, Q_2)$-UF-CMA *secure in the programmable random oracle model, where*

$$\varepsilon' \leq \varepsilon + \frac{Q_2 + 1}{2^k} + \frac{(Q_1 + Q_2)Q_s}{2^c}, \qquad t' \approx t,$$

$$\varepsilon'' \leq \varepsilon + \frac{Q_2 + 2}{2^k} + \frac{(Q_1 + Q_2)Q_s}{2^c} + \frac{1}{2^k}, \qquad t'' \approx t.$$

SIZE OF PARAMETERS. We follow the analyses provided in [20, 25, 35]. We set $t' = t = Q_s = Q_h := Q_1 + Q_2$. Let A be and adversary that runs in time t', makes at most Q_s signature queries, at most Q_h has queries and breaks the UF-CMA-security of FS$_{\mathsf{SCDH}}$ with probability at least ε' (the analysis applies also to OF$_{\mathsf{SCDH}}$). If A is run multiple times, the expected time to produce a forgery is t'/ε', so we are looking for a security parameter κ with $\varepsilon'/t' \leq 2^{-\kappa}$. Setting $\varepsilon'/t' \approx (\varepsilon + \frac{Q_2+1}{2^k} + \frac{(Q_1+Q_2)Q_s}{2^c})/t$, we can bound each of the additive terms separately by $2^{-\kappa}$ by choosing k and c accordingly. Concretely, we choose $c = 2\kappa$ and $k = \kappa$. Given that the Pollard lambda algorithm [47] running in time $O(\sqrt{2^c})$ is the best algorithm for solving the c-SCDH problem over $\mathbb{G}$, $\frac{\varepsilon}{t} \leq \frac{t}{2^c} \leq \frac{2^{c/2}}{2^c} = 2^{-\kappa}$. Also, by assumption, $\frac{Q_2+1}{t2^k} \leq \frac{t}{2^k t} = 2^{-\kappa}$ and $\frac{(Q_1+Q_2)Q_s}{t2^c} = \frac{t^2}{t2^c} = \frac{t}{2^c}$ which we have already shown to be bounded by $2^{-\kappa}$. Setting $k' = 8$, this gives a signature of size of about $\ell' + 4 \cdot \kappa + 8$, where ℓ' denotes the size of a group element.

4.3   Instantiation from Factoring

SIGNED QUADRATIC RESIDUES AND THE FACTORING ASSUMPTION. We begin by recalling the useful group of signed quadratic residues from [27, 32]. For $n \in \mathbb{N}$

we denote the set of all $n/2$-bit primes by $\mathbb{P}_{n/2}$ and $\mathsf{Blum}_n := \{N \mid N = (2p + 1)(2q + 1) \wedge (2p + 1), (2q + 1), p, q \in \mathbb{P}_{n/2} \wedge p \neq q\}$. Let $\varphi(N) = 4pq$ be Euler's totient function for $N \in \mathsf{Blum}_n$.

We define the factoring assumption as follows.

Definition 10 (Factoring Assumption). *The factoring problem* FAC *is* (t, ε)-*hard for* Blum_n *if for all adversaries* A *running in time at most* t,

$$\Pr\left[N = PQ \wedge P, Q \in \mathbb{P}_{n/2} \mid N \xleftarrow{\boxtimes} \mathsf{Blum}_n; (P, Q) \xleftarrow{\boxtimes} \mathsf{A}(N) \right] \leq \varepsilon.$$

For an element $a \in \mathbb{Z}_N$, we define the absolute value

$$|x| := \begin{cases} x & \text{if } x \leq (N - 1)/2 \\ -x & \text{otherwise} \end{cases}.$$

We define the group of signed quadratic residues as $\mathbb{QR}_N^+ := \{|x| : x \in \mathbb{QR}_N\}$. We have that $(\mathbb{QR}_N^+, \circ)$ is a cyclic group with order $|\mathbb{QR}_N^+| = \varphi(N)/4$, where, for all $a, b \in \mathbb{QR}_N^+$ and $x \in \mathbb{Z}_N$, group operations are defined as follows:

$$a \circ b := |a \cdot b \bmod N|, \quad a^x := \underbrace{a \circ a \circ \ldots \circ a}_{x \text{ times}} = |a^x \bmod N|, \quad a^{-1} := |a^{-1} \bmod N|.$$

By Theorem 2 of [32], we note that the factoring assumption tightly implies the CDH assumption over $\mathbb{QR}_N^+$ (henceforth denoted as CDH_N). Let $\mathsf{par}_1 := \mathsf{Blum}_n$ and $\mathsf{par}_2 := (N, g, \mathbb{QR}_N^+)$, where $N \xleftarrow{\boxtimes} \mathsf{Blum}_n$ and g is a random generator of $\mathbb{QR}_N^+$.

Corollary 1 (FAC $\rightarrow$ CDH$_N$). *If* FAC *is* (t, ε)-*hard in* par_1, *then* CDH_N *is* (t', ε')-*hard in* par_2, *where* $t' \approx t$ *and* $\varepsilon' \leq \varepsilon + 2^{-n/2}$.

IDENTIFICATION SCHEME. Let $\mathsf{par} := \mathbb{P}_{n/2}$. The identification scheme $\mathsf{ID}_{\mathsf{FAC}} := (\mathsf{IGen}, \mathsf{P}, \mathsf{ChSet}_1, \mathsf{ChSet}_2, \mathsf{V})$ is defined as follows.

$\mathsf{IGen}(\mathsf{par})$:	$\mathsf{P}_1(sk)$:
$p, q \xleftarrow{\boxtimes} \mathbb{P}_{n/2}$ s.t. $P = 2p + 1 \in \mathbb{P}_{n/2}$	$r \xleftarrow{\boxtimes} \mathbb{Z}_{N/4}$; $R_1 = g^r$; $St_1 := r$
and $Q = 2q + 1 \in \mathbb{P}_{n/2}$	Return (R_1, St_1)
$N = PQ$	
$x \xleftarrow{\boxtimes} \mathbb{Z}_{N/4}$; $X := g^x$	$\mathsf{P}_2(sk, R_1, h_1, St_1)$:
$sk := (x, p, q)$	Parse $St_1 := r$
$pk := (X, N)$	$R_L = h_1^x$; $R_R = h_1^r$
$\mathsf{ChSet}_1 := \mathbb{QR}_N^+$	$R_2 := (R_L, R_R)$; $St_2 := St_1$
$\mathsf{ChSet}_2 := \{0, \ldots, 2^k - 1\}$	Return (R_2, St_2)
Return (pk, sk)	
	$\mathsf{P}_3(sk, R_1, h_1, R_2, h_2, St_2)$:
$\mathsf{V}(pk, R_1, h_1, R_2, h_2, s)$:	Parse $St_2 := r$
Parse $R_2 := (R_L, R_R)$	$s = xh_2 + r \bmod (\varphi(N)/4)$
If $R_1 = g^s \circ X^{-h_2}$ and $R_R = h_1^s \circ R_L^{-h_2}$	Return s
then return 1	
Else return 0	

Lemma 7. *Let* $N' := \lceil N/4 \rceil$, $\mathbb{G} := \mathbb{QR}_N^+$, *and* $X \stackrel{\boxtimes}{\leftarrow} \mathbb{Z}_{N'}, Y \stackrel{\boxtimes}{\leftarrow} \mathbb{Z}_{|\mathbb{G}|}$. *Then the statistical distance* $D(X, Y)$ *satisfies* $D(X, Y) \leq \frac{2(P+Q)}{PQ}$.

Proof. We split the term $D(X, Y)$ as

$$D(X, Y) = \sum_{x \in \mathbb{Z}_{N'}} \left| \Pr[X = x] - \Pr[Y = x] \right|$$

$$= \sum_{x \in \mathbb{Z}_{|\mathbb{G}|}} \left| \Pr[X = x] - \Pr[Y = x] \right| + \sum_{x \in [N' - |\mathbb{G}|]} \left| \Pr[X = x] - \Pr[Y = x] \right|.$$

In the first term, $\Pr[X = x] = \frac{1}{N'} \leq \frac{4}{PQ}, \Pr[Y = x] = \frac{1}{|\mathbb{G}|} = \frac{4}{(P-1)(Q-1)}$. Therefore, the first summand is equal to $\left| \frac{1-P-Q}{PQ} \right| \leq \frac{P+Q}{PQ}$. Similarly, the second term can be bounded by $\frac{P+Q}{PQ}$ and thus, $D(X, Y) \leq \frac{2(P+Q)}{PQ}$.

Lemma 8. $\mathsf{ID}_{\mathsf{FAC}}$ *is a perfectly correct five-move identification scheme and has* $\frac{2(P+Q)}{PQ}$-*statistical non-aborting honest-verifier zero-knowledge* (naHVZK) *with* $\alpha = \log(\varphi(N)/4) - \frac{2(P+Q)}{PQ}$ *bits min-entropy. Moreover, if* FAC *is* (t, ε)-*hard then* $\mathsf{ID}_{\mathsf{FAC}}$ *is* $(t', \varepsilon', Q_{\mathrm{CH}_1}, Q_{\mathrm{CH}_2})$-$\mathsf{PIMP}$-$\mathsf{KOA}$-*secure, where* $\varepsilon' \leq \varepsilon + \frac{1}{2^k}$ *and* $t' \approx t$.

Proof. The perfect correctness of $\mathsf{ID}_{\mathsf{FAC}}$ is straightforward to verify, since $R_1 = g^s \circ X^{-h_2}$ and $R_R = h_1^s \circ R_L^{-h_2}$ hold if and only if $s = xh_2 + r \bmod \varphi(N)/4$ holds.

$\frac{2(P+Q)}{PQ}$-STATISTICAL naHVZK. Given public key $pk = X$, Sim first samples $s, w, h_2 \stackrel{\boxtimes}{\leftarrow} \mathbb{Z}_{N/4}$. It then computes $h_1 := g^w, R_1 := g^s \circ X^{-h_2}, R_L := R_1^w$, and $R_R := X^w$ and outputs the transcript $(R_1, h_1, R_L, R_R, h_2, s)$.

The simulated transcript (R_1, h_1, R_2, h_2, s) is close to the transcript output by $\mathsf{SKTran}(sk)$ with statistical distance $2(P+Q)/(PQ)$, since s has statistical distance at most $\frac{2(P+Q)}{PQ}$ from a uniformly random variable over $\mathbb{Z}_{|\mathbb{QR}_N^+|}$, according to Lemma 7 and R_1, R_L and R_R are values $R_1 = g^s \circ X^{-h_2}$ and $R_R = h_1^s \circ R_L^{-h_2}$. The entropy of $(R_1, St_1) \stackrel{\boxtimes}{\leftarrow} \mathsf{P}_1(sk)$ and $(R_2, St_2) \stackrel{\boxtimes}{\leftarrow} \mathsf{P}_2$ comes only from R_1, which is uniformly random from $\mathbb{QR}_N^+$. Since the transcripts output by $\mathsf{Sim}(pk)$ are statistically $\frac{2(P+Q)}{PQ}$ close to the ones produced by $\mathsf{SKTran}(sk)$, $\mathsf{ID}_{\mathsf{FAC}}$ has $\log |\mathbb{QR}_N^+| = \log(\varphi(N)/4) - \frac{2(P+Q)}{PQ}$ bits min-entropy.

PIMP-KOA-SECURITY. Let A be an attacker against the $(t', \varepsilon', Q_{\mathrm{CH}_1}, Q_{\mathrm{CH}_2})$-PIMP-KOA security of $\mathsf{ID}_{\mathsf{FAC}}$. We construct an attacker B that (t, ε)-breaks CDH over $\mathbb{QR}_N^+$.

CONSTRUCTION OF B. Let $(X := g^x, Y := g^y)$ denote the CDH instance. B runs A with input $pk := X$ and answers A's challenge queries as follows.

For A's CH_1 query on $R_1 \in \mathbb{QR}_N^+$, B chooses $a \stackrel{\boxtimes}{\leftarrow} \mathbb{Z}_{N/4}$ and computes $h_1 = Y \circ g^a$. For A's CH_2 query on $R_2 \in \mathbb{QR}_N^+ \times \mathbb{QR}_N^+$, B chooses $h_2 \stackrel{\boxtimes}{\leftarrow} \mathbb{Z}_{N/4}$ and returns it to A. Clearly, B's simulation of the PIMP-KOA game is perfect, since both h_1 and h_2 are uniformly random over $\mathbb{QR}_N^+$ and $\mathbb{Z}_{N/4}$, respectively.

Eventually, A returns its response s^* for the PIMP-KOA experiment. We assume that A's response is valid, i.e., there exist $(R_1, h_1) \in \mathcal{L}_1$ and $(R_2 := (R_L, R_R), h_2) \in \mathcal{L}_2$ such that $R_1 = g^{s^*} \circ X^{-h_2}$ and $R_R = h_1^{s^*} \circ R_L^{-h_2}$. We denote the discrete logarithm of R_L based on h_1 by $x' = \mathsf{DL}_{h_1}(R_L) \bmod (\varphi(N)/4)$ and do the following cases distinction:

- Case 1: $x = x' \bmod (\varphi(N)/4)$. By the simulation of CH_1, we have $R_L = h_1^x = (Y \circ g^{a_i})^x = Y^x \circ X^{a_i}$ for some $i \in [Q_{\mathsf{CH}_1}]$. Thus, B returns $Z := R_L \circ X^{-a_i} = Y^x$ to break the CDH problem over $\mathbb{QR}_N^+$.
- Case 2: $x \neq x' \bmod (\varphi(N)/4)$. We show in this case even an unbounded adversary A can only win with probability $Q_{\mathsf{CH}_2}/2^k$. For each index $i \in [Q_{\mathsf{CH}_2}]$, before receiving $h_{2,i}$, A first commits to some $R_1 = g^{r_1}$, $R_L = h_1^{x'}$ and $R_R = h_1^{r_2}$ (for arbitrary $r_1, r_2, x' \in \mathbb{Z}_{N/4}$ and $x' \neq x \bmod (\varphi(N)/4)$) and there exists $(R_1, h_1) \in \mathcal{L}_1$. A can only win if there exists an $s_i \in \mathbb{Z}_{N/4}$ such that

$$r_1 + h_{2,i}x = s_i = r_2 + h_{2,i}x' \bmod (\varphi(N)/4)$$
$$\Leftrightarrow h_{2,i} = \frac{r_2 - r_1}{x - x'} \bmod (\varphi(N)/4)$$

where $h_{2,i}$ is distributed uniformly over $\{0, ..., 2^k - 1\}$ and independently of r_1, r_2 and x'. The equation $h_{2,i} = \frac{r_2 - r_1}{x - x'} \bmod (\varphi(N)/4)$ holds with probability at most $1/2^k$. By the union bound we obtain the bound $Q_{\mathsf{CH}_2}/2^k$ as claimed.

Overall, B returns a valid solution of the CDH challenge with probability $\varepsilon \geq \varepsilon' - \frac{Q_{\mathsf{CH}_2}}{2^k}$. This completes the proof.

SIGNATURE SCHEME. Let $H_1 : \{0,1\}^* \to \mathbb{QR}_N^+$ and $H_2 : \{0,1\}^* \to \{0, ..., 2^k - 1\}$ be two hash functions. We note that H_1 has been used in [32,43]. As $\mathsf{ID}_{\mathsf{FAC}}$ is perfectly correct and partial commitment-recoverable, we can use the alternative Fiat-Shamir transformation from Sect. 3.1 with $\ell := 1$ to obtain the signature scheme $\mathsf{FS}_{\mathsf{FAC}} := (\mathsf{Gen}, \mathsf{Sign}, \mathsf{Ver})$.

Gen(par):	Sign(sk, m):	Ver(pk, m, σ):
$p, q \xleftarrow{\boxtimes} \mathbb{P}_{n/2}$ s.t	$r \xleftarrow{\boxtimes} \mathbb{Z}_{N/4}$; $R_1 = g^r$	Parse $\sigma = (R_L, h_2, s)$
$P = 2p + 1 \in \mathbb{P}_{n/2}$	$h_1 = H_1(R_1, m) \in \mathbb{QR}_N^+$;	$R_1 = g^s \circ X^{-h_2}$
$Q = 2q + 1 \in \mathbb{P}_{n/2}$	$R_L = h_1^x$; $R_R = h_1^r$	$h_1 = H_1(R_1, m)$;
$N = PQ$	$R_2 := (R_L, R_R)$	$R_R = h_1^s \circ R_L^{-h_2}$
$x \xleftarrow{\boxtimes} \mathbb{Z}_{N/4}$; $X := g^x$	$h_2 = H_2(R_2, m) \in \{0, ...2^k - 1\}$	$R_2 := (R_L, R_R)$
$sk := (x, p, q)$	$s = xh_2 + r \bmod (\varphi(N)/4)$	If $h_2 = H_2(R_2, m)$
$pk := (X, N)$	$\sigma = (R_L, h_2, s)$	then return 1
Return (pk, sk)	Return σ	Else return 0.

By Corollary 1, Lemma 8 and Theorem 1, we have

Theorem 4 (Security of $\mathsf{FS}_{\mathsf{FAC}}$). *If FAC is (t, ε)-hard in* par := Blum_n *then* $\mathsf{FS}_{\mathsf{FAC}}$ *is $(t', \varepsilon', Q_s, Q_1, Q_2)$-UF-CMA secure in the programmable random oracle model, where*

$$\varepsilon' \leq \varepsilon + \frac{1}{2^{n/2}} + \frac{Q_2 + 1}{2^k} + \frac{(Q_1 + Q_2)Q_s}{2^{n-3}} + \frac{Q_s(P + Q)}{2^{n-1}}, \qquad t' \approx t.$$

SIZE OF PARAMETERS. We argue along the lines of our analysis provided above. We again set $Q_s = Q_h = t' = t$. Here, we only need to bound the term $\frac{Q_2+1}{t2^k} \leq \frac{1}{2^k}$, as all terms in the above theorem vanish. Thus, we set again $k = \kappa$ to achieve a security of κ bits. This yields a signature of size $2 \cdot \ell' + \kappa$ bits where again ℓ' denotes the size of a group element of $\mathbb{G}$.

References

1. Abdalla, M., An, J.H., Bellare, M., Namprempre, C.: From identification to signatures via the Fiat-Shamir transform: minimizing assumptions for security and forward-security. In: Knudsen, L.R. (ed.) EUROCRYPT 2002. LNCS, vol. 2332, pp. 418–433. Springer, Heidelberg (2002). https://doi.org/10.1007/3-540-46035-7_28

2. Abdalla, M., Ben Hamouda, F., Pointcheval, D.: Tighter reductions for forward-secure signature schemes. In: Kurosawa, K., Hanaoka, G. (eds.) PKC 2013. LNCS, vol. 7778, pp. 292–311. Springer, Heidelberg (2013). https://doi.org/10.1007/978-3-642-36362-7_19

3. Abdalla, M., Fouque, P.-A., Lyubashevsky, V., Tibouchi, M.: Tightly-secure signatures from lossy identification schemes. In: Pointcheval, D., Johansson, T. (eds.) EUROCRYPT 2012. LNCS, vol. 7237, pp. 572–590. Springer, Heidelberg (2012). https://doi.org/10.1007/978-3-642-29011-4_34

4. Abe, M., Hofheinz, D., Nishimaki, R., Ohkubo, M., Pan, J.: Compact structure-preserving signatures with almost tight security. In: Katz, J., Shacham, H. (eds.) CRYPTO 2017. LNCS, vol. 10402, pp. 548–580. Springer, Cham (2017). https://doi.org/10.1007/978-3-319-63715-0_19

5. Yousfi Alaoui, S.M., Dagdelen, Ö., Véron, P., Galindo, D., Cayrel, P.-L.: Extended security arguments for signature schemes. In: Mitrokotsa, A., Vaudenay, S. (eds.) AFRICACRYPT 2012. LNCS, vol. 7374, pp. 19–34. Springer, Heidelberg (2012). https://doi.org/10.1007/978-3-642-31410-0_2

6. Bellare, M., Palacio, A.: GQ and Schnorr identification schemes: proofs of security against impersonation under active and concurrent attacks. In: Yung, M. (ed.) CRYPTO 2002. LNCS, vol. 2442, pp. 162–177. Springer, Heidelberg (2002). https://doi.org/10.1007/3-540-45708-9_11

7. Bellare, M., Poettering, B., Stebila, D.: From identification to signatures, tightly: a framework and generic transforms. In: Cheon, J.H., Takagi, T. (eds.) ASIACRYPT 2016. LNCS, vol. 10032, pp. 435–464. Springer, Heidelberg (2016). https://doi.org/10.1007/978-3-662-53890-6_15

8. Bellare, M., Rogaway, P.: Random oracles are practical: A paradigm for designing efficient protocols. In: Ashby, V. (ed.) ACM CCS 1993, pp. 62–73. ACM Press, November 1993

9. Bellare, M., Rogaway, P.: The exact security of digital signatures: how to sign with RSA and Rabin. In: Maurer, U. (ed.) EUROCRYPT 1996. LNCS, vol. 1070, pp. 399–416. Springer, Heidelberg (1996). https://doi.org/10.1007/3-540-68339-9_34

10. Bernstein, D.J.: Proving tight security for Rabin-Williams signatures. In: Smart, N. (ed.) EUROCRYPT 2008. LNCS, vol. 4965, pp. 70–87. Springer, Heidelberg (2008). https://doi.org/10.1007/978-3-540-78967-3_5

11. Beth, T.: Efficient zero-knowledge identification scheme for smart cards. In: Barstow, D., et al. (eds.) EUROCRYPT 1988. LNCS, vol. 330, pp. 77–84. Springer, Heidelberg (1988). https://doi.org/10.1007/3-540-45961-8_7

12. Blazy, O., Kakvi, S.A., Kiltz, E., Pan, J.: Tightly-secure signatures from chameleon hash functions. In: Katz, J. (ed.) PKC 2015. LNCS, vol. 9020, pp. 256–279. Springer, Heidelberg (2015). https://doi.org/10.1007/978-3-662-46447-2_12

13. Blazy, O., Kiltz, E., Pan, J.: (Hierarchical) identity-based encryption from affine message authentication. In: Garay, J.A., Gennaro, R. (eds.) CRYPTO 2014. LNCS, vol. 8616, pp. 408–425. Springer, Heidelberg (2014). https://doi.org/10.1007/978-3-662-44371-2_23

14. Boneh, D., Lynn, B., Shacham, H.: Short signatures from the Weil pairing. In: Boyd, C. (ed.) ASIACRYPT 2001. LNCS, vol. 2248, pp. 514–532. Springer, Heidelberg (2001). https://doi.org/10.1007/3-540-45682-1_30

15. Brickell, E.F., McCurley, K.S.: An interactive identification scheme based on discrete logarithms and factoring. In: Damgård, I.B. (ed.) EUROCRYPT 1990. LNCS, vol. 473, pp. 63–71. Springer, Heidelberg (1991). https://doi.org/10.1007/3-540-46877-3_6

16. Cayrel, P.-L., Lindner, R., Rückert, M., Silva, R.: Improved zero-knowledge identification with lattices. In: Heng, S.-H., Kurosawa, K. (eds.) ProvSec 2010. LNCS, vol. 6402, pp. 1–17. Springer, Heidelberg (2010). https://doi.org/10.1007/978-3-642-16280-0_1

17. Cayrel, P.-L., Véron, P., Yousfi Alaoui, S.M.: A zero-knowledge identification scheme based on the q-ary syndrome decoding problem. In: Biryukov, A., Gong, G., Stinson, D.R. (eds.) SAC 2010. LNCS, vol. 6544, pp. 171–186. Springer, Heidelberg (2011). https://doi.org/10.1007/978-3-642-19574-7_12

18. Chen, J., Wee, H.: Fully, (almost) tightly secure IBE and dual system groups. In: Canetti, R., Garay, J.A. (eds.) CRYPTO 2013. LNCS, vol. 8043, pp. 435–460. Springer, Heidelberg (2013). https://doi.org/10.1007/978-3-642-40084-1_25

19. Chen, M.-S., Hülsing, A., Rijneveld, J., Samardjiska, S., Schwabe, P.: From 5-pass $\mathcal{MQ}$-based identification to $\mathcal{MQ}$-based signatures. In: Cheon, J.H., Takagi, T. (eds.) ASIACRYPT 2016. LNCS, vol. 10032, pp. 135–165. Springer, Heidelberg (2016). https://doi.org/10.1007/978-3-662-53890-6_5

20. Chevallier-Mames, B.: An efficient CDH-based signature scheme with a tight security reduction. In: Shoup, V. (ed.) CRYPTO 2005. LNCS, vol. 3621, pp. 511–526. Springer, Heidelberg (2005). https://doi.org/10.1007/11535218_31

21. Fiat, A., Shamir, A.: How to prove yourself: practical solutions to identification and signature problems. In: Odlyzko, A.M. (ed.) CRYPTO 1986. LNCS, vol. 263, pp. 186–194. Springer, Heidelberg (1987). https://doi.org/10.1007/3-540-47721-7_12

22. Fleischhacker, N., Jager, T., Schröder, D.: On tight security proofs for schnorr signatures. In: Sarkar, P., Iwata, T. (eds.) ASIACRYPT 2014. LNCS, vol. 8873, pp. 512–531. Springer, Heidelberg (2014). https://doi.org/10.1007/978-3-662-45611-8_27

23. Girault, M.: An identity-based identification scheme based on discrete logarithms modulo a composite number. In: Damgård, I.B. (ed.) EUROCRYPT 1990. LNCS, vol. 473, pp. 481–486. Springer, Heidelberg (1991). https://doi.org/10.1007/3-540-46877-3_44

24. Girault, M., Poupard, G., Stern, J.: On the fly authentication and signature schemes based on groups of unknown order. J. Cryptol. 19(4), 463–487 (2006)

25. Goh, E.-J., Jarecki, S.: A signature scheme as secure as the Diffie-Hellman problem. In: Biham, E. (ed.) EUROCRYPT 2003. LNCS, vol. 2656, pp. 401–415. Springer, Heidelberg (2003). https://doi.org/10.1007/3-540-39200-9_25

26. Goh, E.-J., Jarecki, S., Katz, J., Wang, N.: Efficient signature schemes with tight reductions to the Diffie-Hellman problems. J. Cryptol. 20(4), 493–514 (2007)

27. Goldwasser, S., Micali, S., Rivest, R.L.: A digital signature scheme secure against adaptive chosen-message attacks. SIAM J. Comput. **17**(2), 281–308 (1988)

28. Guillou, L.C., Quisquater, J.-J.: A "paradoxical" indentity-based signature scheme resulting from zero-knowledge. In: Goldwasser, S. (ed.) CRYPTO 1988. LNCS, vol. 403, pp. 216–231. Springer, New York (1990). https://doi.org/10.1007/0-387-34799-2_16

29. Hofheinz, D.: Algebraic partitioning: Fully compact and (almost) tightly secure cryptography. In: Kushilevitz, E., Malkin, T. (eds.) TCC 2016. LNCS, vol. 9562, pp. 251–281. Springer, Heidelberg (2016). https://doi.org/10.1007/978-3-662-49096-9_11

30. Hofheinz, D.: Adaptive partitioning. In: Coron, J.-S., Nielsen, J.B. (eds.) EURO-CRYPT 2017. LNCS, vol. 10212, pp. 489–518. Springer, Cham (2017). https://doi.org/10.1007/978-3-319-56617-7_17

31. Hofheinz, D., Jager, T.: Tightly secure signatures and public-key encryption. In: Safavi-Naini, R., Canetti, R. (eds.) CRYPTO 2012. LNCS, vol. 7417, pp. 590–607. Springer, Heidelberg (2012). https://doi.org/10.1007/978-3-642-32009-5_35

32. Hofheinz, D., Kiltz, E.: The group of signed quadratic residues and applications. In: Halevi, S. (ed.) CRYPTO 2009. LNCS, vol. 5677, pp. 637–653. Springer, Heidelberg (2009). https://doi.org/10.1007/978-3-642-03356-8_37

33. Kakvi, S.A., Kiltz, E.: Optimal security proofs for full domain hash, revis-ited. In: Pointcheval, D., Johansson, T. (eds.) EUROCRYPT 2012. LNCS, vol. 7237, pp. 537–553. Springer, Heidelberg (2012). https://doi.org/10.1007/978-3-642-29011-4_32

34. Katz, J., Wang, N.: Efficiency improvements for signature schemes with tight secu-rity reductions. In: Jajodia, S., Atluri, V., Jaeger, T. (eds.) ACM CCS 2003, pp. 155–164. ACM Press, October 2003

35. Kiltz, E., Masny, D., Pan, J.: Optimal security proofs for signatures from identifica-tion schemes. In: Robshaw, M., Katz, J. (eds.) CRYPTO 2016. LNCS, vol. 9815, pp. 33–61. Springer, Heidelberg (2016). https://doi.org/10.1007/978-3-662-53008-5_2

36. Koshiba, T., Kurosawa, K.: Short exponent Diffie-Hellman problems. In: Bao, F., Deng, R., Zhou, J. (eds.) PKC 2004. LNCS, vol. 2947, pp. 173–186. Springer, Heidelberg (2004). https://doi.org/10.1007/978-3-540-24632-9_13

37. Lyubashevsky, V.: Fiat-Shamir with aborts: applications to lattice and factoring-based signatures. In: Matsui, M. (ed.) ASIACRYPT 2009. LNCS, vol. 5912, pp. 598–616. Springer, Heidelberg (2009). https://doi.org/10.1007/978-3-642-10366-7_35

38. Micali, S., Reyzin, L.: Improving the exact security of digital signature schemes. Cryptology ePrint Archive, Report 1999/020 (1999). http://eprint.iacr.org/1999/020

39. Micali, S., Reyzin, L.: Improving the exact security of digital signature schemes. J. Cryptol. **15**(1), 1–18 (2002)

40. Micali, S., Shamir, A.: An improvement of the Fiat-Shamir identification and signa-ture scheme. In: Goldwasser, S. (ed.) CRYPTO 1988. LNCS, vol. 403, pp. 244–247. Springer, New York (1990). https://doi.org/10.1007/0-387-34799-2_18

41. Okamoto, T.: Provably secure and practical identification schemes and correspond-ing signature schemes. In: Brickell, E.F. (ed.) CRYPTO 1992. LNCS, vol. 740, pp. 31–53. Springer, Heidelberg (1993). https://doi.org/10.1007/3-540-48071-4_3

42. Ong, H., Schnorr, C.P.: Fast signature generation with a Fiat-Shamir-like scheme. In: Damgård, I.B. (ed.) EUROCRYPT 1990. LNCS, vol. 473, pp. 432–440. Springer, Heidelberg (1991). https://doi.org/10.1007/3-540-46877-3_38

43. Poettering, B., Stebila, D.: Double-authentication-preventing signatures. In: Kutyłowski, M., Vaidya, J. (eds.) ESORICS 2014. LNCS, vol. 8712, pp. 436–453. Springer, Cham (2014). https://doi.org/10.1007/978-3-319-11203-9_25

44. Pointcheval, D.: A new identification scheme based on the perceptrons problem. In: Guillou, L.C., Quisquater, J.-J. (eds.) EUROCRYPT 1995. LNCS, vol. 921, pp. 319–328. Springer, Heidelberg (1995). https://doi.org/10.1007/3-540-49264-X_26

45. Pointcheval, D., Poupard, G.: A new NP-complete problem and public-key identification. Des. Codes Cryptogr. **28**(1), 5–31 (2003)

46. Pointcheval, D., Stern, J.: Security arguments for digital signatures and blind signatures. J. Cryptol. **13**(3), 361–396 (2000)

47. Pollard, J.M.: Monte Carlo methods for index computation mod p. Math. Comput. **32**, 918–924 (1978)

48. Sakumoto, K., Shirai, T., Hiwatari, H.: Public-key identification schemes based on multivariate quadratic polynomials. In: Rogaway, P. (ed.) CRYPTO 2011. LNCS, vol. 6841, pp. 706–723. Springer, Heidelberg (2011). https://doi.org/10.1007/978-3-642-22792-9_40

49. Schnorr, C.-P.: Efficient signature generation by smart cards. J. Cryptol. **4**(3), 161–174 (1991)

50. Seurin, Y.: On the exact security of Schnorr-type signatures in the random oracle model. In: Pointcheval, D., Johansson, T. (eds.) EUROCRYPT 2012. LNCS, vol. 7237, pp. 554–571. Springer, Heidelberg (2012). https://doi.org/10.1007/978-3-642-29011-4_33

51. Shamir, A.: An efficient identification scheme based on permuted kernels (extended abstract). In: Brassard, G. (ed.) CRYPTO 1989. LNCS, vol. 435, pp. 606–609. Springer, New York (1990). https://doi.org/10.1007/0-387-34805-0_54

52. Shamir, A., Tauman, Y.: Improved online/offline signature schemes. In: Kilian, J. (ed.) CRYPTO 2001. LNCS, vol. 2139, pp. 355–367. Springer, Heidelberg (2001). https://doi.org/10.1007/3-540-44647-8_21

53. Silva, R., Cayrel, P.-L., Lindner, R.: Zero-knowledge identification based on lattices with low communication costs. In: XI Simposio Brasileiro de Seguranca da Informacao e de Sistemas Computacionais, vol. 8, pp. 95–107 (2011)

54. Stern, J.: Designing identification schemes with keys of short size. In: Desmedt, Y.G. (ed.) CRYPTO 1994. LNCS, vol. 839, pp. 164–173. Springer, Heidelberg (1994). https://doi.org/10.1007/3-540-48658-5_18

55. Stern, J.: A new identification scheme based on syndrome decoding. In: Stinson, D.R. (ed.) CRYPTO 1993. LNCS, vol. 773, pp. 13–21. Springer, Heidelberg (1994). https://doi.org/10.1007/3-540-48329-2_2

On the Untapped Potential of Encoding Predicates by Arithmetic Circuits and Their Applications

Shuichi Katsumata[✉]

National Institute of Advanced Industrial Science and Technology (AIST),
The University of Tokyo, Tokyo, Japan
shuichi_katsumata@it.k.u-tokyo.ac.jp

Abstract. Predicates are used in cryptography as a fundamental tool to control the disclosure of secrets. However, how to embed a particular predicate into a cryptographic primitive is usually not given much attention. In this work, we formalize the idea of encoding predicates as arithmetic circuits and observe that choosing the right encoding of a predicate may lead to an improvement in many aspects such as the efficiency of a scheme or the required hardness assumption. In particular, we develop two predicate encoding schemes with different properties and construct cryptographic primitives that benefit from these: verifiable random functions (VRFs) and predicate encryption (PE) schemes.

- We propose two VRFs on bilinear maps. Both of our schemes are secure under a non-interactive Q-type assumption where Q is only poly-logarithmic in the security parameter, and they achieve either a poly-logarithmic verification key size or proof size. This is a significant improvement over prior works, where all previous schemes either require a strong hardness assumption or a large verification key and proof size.
- We propose a lattice-based PE scheme for the class of *multi-dimensional equality* (MultD-Eq) predicates. This class of predicate is expressive enough to capture many of the appealing applications that motivates PE schemes. Our scheme achieves the best in terms of the required approximation factor for LWE (we only require $\mathsf{poly}(\lambda)$) and the decryption time. In particular, all existing PE schemes that support the class of MultD-Eq predicates either require a subexponential LWE assumption or an exponential decryption time (in the dimension of the MultD-Eq predicates).

1 Introduction

A *predicate* is a function $P : \mathcal{X} \to \{0, 1\}$ that partitions an input domain $\mathcal{X}$ into two distinct sets according to some relation. Due to its natural compatibility with cryptographic primitives, predicates have been used in many scenarios to control the disclosure of secrets. This may either come up explicitly during construction (e.g., attribute-based encryptions [SW05, GPSW06], predicate encryptions

© International Association for Cryptologic Research 2017
T. Takagi and T. Peyrin (Eds.): ASIACRYPT 2017, Part III, LNCS 10626, pp. 95–125, 2017.
https://doi.org/10.1007/978-3-319-70700-6_4

[BW07, SBC+07, KSW08]) or implicitly during security proofs (e.g., in the form of programmable hashes [HK08, ZCZ16], admissible hashes [BB04a, CHKP10]). However, how to express predicates as (arithmetic) circuits is usually not given much attention in these works. Since the way we embed predicates into a cryptographic primitive has a direct effect on the concrete efficiency of the schemes, it is important to know how efficiently we can embed predicates. In this paper, we propose an efficient encoding for a specific class of predicates and focus on two primitives that benefit from this: verifiable random functions (VRFs) and predicate encryptions (PE) schemes.

Verifiable Random Functions. VRFs introduced by Micali, Rabin and Vadhan [MRV99] are a special form of pseudorandom functions (PRFs), which additionally enables a secret key holder to create a non-interactive and publicly verifiable proof that validates the output value. An attractive property for the VRF to have is the notion of *all the desired properties* coined by [HJ16], which captures the following features: an exponential-sized input space, adaptive pseudorandomness, and security under a non-interactive complexity assumption.

There currently exist two approaches for constructing VRFs with all the desired properties. The first approach is to use a specific number theory setting (mainly bilinear groups) to handcraft VRFs [HW10, BMR10, ACF14, Jag15, HJ16, Yam17], and the second approach is to use a more generic approach and build VRFs from general cryptographic primitives [GHKW17, Bit17, BGJS17]. While the second approach provides us with better insight on VRFs and allows us to base security on hardness assumptions other than bilinear map based ones, the major drawback is the need for large verification key/proof sizes or the need for strong hardness assumptions such as the subexponential Learning with Errors (LWE) assumption to instantiate the underlying primitives. Concretely, all generic approaches require general non-interactive witness indistinguishable proofs (NIWIs) and constrained PRFs for admissible hash friendly functions, which we currently do not know how to simultaneously construct compactly and base security under a weak hardness assumption.

The first approach is more successful overall in light of compactness and the required hardness assumptions, however, they come with their own shortcomings. Notably, [Yam17] presents three constructions where only $\omega(\log \lambda)$ group elements[1] are required for either the verification key or the proof. In particular, in one of their schemes, only sub-linear group elements are required for both verification key and proof. However, all three schemes require an L-DDH[2] assumption where $L = \tilde{\Omega}(\lambda)$. In contrast, [Jag15] presents a scheme secure under a much weaker L-DDH assumption where $L = O(\log \lambda)$ and [HJ16] under the DLIN assumption. However, these approaches require a linear number of group elements in the verification key and proof in the security parameter. Therefore,

---

[1] Here, $\omega(f(\lambda))$ denotes any function that grows asymptotically faster than $f(\lambda)$, e.g., $\log^2 \lambda = \omega(\log \lambda)$.

[2] The L-DDH problem is where we are given $(h, g, g^\alpha, \cdots, g^{\alpha^L}, \Psi)$ and have to decided whether $\Psi = e(g, h)^{1/\alpha}$ or a uniform random element.

we currently do not know how to construct VRFs that are both compact and secure under a weak hardness assumption.

Predicate Encryption. A predicate encryption (PE) scheme [BW07, SBC+07, KSW08] is a paradigm for public-key encryption that supports searching on encrypted data. In predicate encryption, ciphertexts are associated with some attribute X, secret keys are associated with some predicate P, and the decryption is successful if and only if $P(X) = 1$. The major difficulty of constructing predicate encryption schemes stems from the security requirement that enforces the privacy of the attribute X and the plaintext even amidst multiple secret key queries.

Some of the motivating applications for predicate encryption schemes that are often stated in the literatures are: inspection of recorded log files for network intrusions, credit card fraud investigation and conditional disclosure of patient records. Notably, all the above applications only require checking whether a subset or range conjunction predicate is satisfied. (For a more thorough discussion, see [BW07, SBC+07, KSW08].) Therefore, in some sense many of the applications that motivates for predicate encryption schemes can be implemented by predicate encryption schemes for the class of predicates that are expressive enough to support subset or range conjunctions.

On the surface, the present situation on lattice-based predicate encryption schemes seem bright. We have concrete constructions based on LWE for the class of predicates that supports equality [ABB10, CHKP10], inner-products [AFV11], multi-dimensional equality (MultD-Eq)[3] [GMW15], and all circuits [GVW15, GKW17, WZ17][4] Therefore, in theory, we can realize all the above applications in a secure manner, since subset or range conjunctions can be efficiently encoded by any predicate as expressive as the MultD-Eq predicate, i.e., the works of [GMW15, GVW15, GKW17, WZ17] are all sufficient for the above applications. However, all of these schemes may be too inefficient to use in real-life applications. Namely, the scheme of [GMW15] highly resembles the bilinear map based construction of [SBC+07] and inherits the same problem; it takes $\Omega(2^D)$ decryption time where D roughly corresponds to the number of set elements specifying the subset predicate or the number of conjunctions used in the range conjunction predicate. Further, the schemes of [GVW15, GKW17, WZ17] are powerful and elegant, albeit they all require subexponential LWE assumptions. Therefore, aiming at predicate encryption schemes with the above applications in mind, we currently do not have satisfactorily efficient lattice-based schemes. In particular, we do not know how to construct efficient lattice-based PE schemes for the class of MultD-Eq predicates. This is in sharp contrast with

---

[3] The precise definition and discussions of this predicate are given in Sect. 4.2. For the time being, it is enough to view it as a subset predicate.

[4] [GKW17, WZ17] give a generic conversion from ABEs to PEs that uses an obfuscation for a specific program proven secure under the subexponential LWE assumption. Therefore, we have provably secure lattice-based PEs for all circuits using the lattice-based ABE of [GVW13, BGG+14].

the bilinear map setting where we know how to obtain efficient schemes for the above applications [BW07].

1.1 Our Contributions

In this paper, we provide two results: a compact VRF under a weak assumption and an efficient lattice-based PE scheme for the class of MultD-Eq predicates. For the time being, it suffices to think of the MultD-Eq predicate as simply a predicate that supports the subset predicate. Here, although the two results may seem independent, they are in fact related by a common theme that they both implicitly or explicitly embed the subset predicates in their constructions.

Our idea is simple. We first detach predicates from cryptographic constructions, and view predicates simply as a function. Then, we introduce the notion of *predicate encoding schemes*[5], where we encode predicates as simple (arithmetic) circuits that have different properties fit for the underlying cryptographic applications. For example, we might not care that a predicate P outputs 0 or 1. We may only care that P behaves differently on satisfied/non-satisfied inputs, e.g., P outputs a value in S_0 when it is satisfied and S_1 otherwise, where S_0, S_1 are disjoint sets. In particular, we provide two predicate encoding schemes $\mathsf{PES_{FP}}$ and $\mathsf{PES_{Lin}}$ with different properties encoding the MultD-Eq predicates. Then, based on these encoded MultD-Eq predicates, we construct our VRFs, and PE schemes for the class of MultD-Eq predicates. The following is a summary of our two results.

VRF. We propose two VRFs with all the desired properties. The detailed comparison between the recent efficient VRF constructions are given in Table 1. Note that we exclude the recent VRF constructions of [Bit17, BGJS17, GHKW17] from the table, since their schemes cannot be instantiated efficiently due to the lack of efficient (general) NIWIs and constrained PRFs.

Our constructions are inspired by the bilinear map based VRFs of [Yam17], where they noticed that an admissible hash function [BB04b, CHKP10] can be represented much more compactly by using a subset predicate[6]. We improve their works by further noticing that subset predicates, when viewed as simply a function, can be encoded in various ways into a circuit. In particular, we propose a more efficient circuit encoding ($\mathsf{PES_{FP}}$) of the subset predicates that is compatible with the underlying algebraic structure of the VRF. We note that at the technical level the constructions are quite different; [Yam17] uses the inversion-based techniques [DY05, BMR10] whereas we do not. Here, simply using $\mathsf{PES_{FP}}$ already provides us with an improvement over previous schemes, however, by

---

[5] We note that the term "predicate encoding" has already been used in a completely different context by [Wee14]. See the section of related work for the differences.

[6] In particular, our idea is inspired by the VRFs based on the admissible hash function of [Yam17], Sect. 6. However, the construction is more similar to the VRF based on the variant of Water's hash in their Appendix C.

Table 1. Comparison of Recent VRFs with all the desired properties.

Schemes	$\|vk\|$ # $\mathbb{G}$	$\|\pi\|$ # $\mathbb{G}$	Assumption	Reduction cost
[Jag15]	$O(\lambda)$	$O(\lambda)$	$O(\log(Q/\epsilon))$-DDH	$O(\epsilon^{\nu+1}/Q^\nu)$
[HJ16]	$O(\lambda)$	$O(\lambda)$	DLIN	$O(\epsilon^{\nu+1}/\lambda Q^\nu)$
[Yam17]: Sect. 7.1	$\omega(\lambda\log\lambda)$	$\omega(\log\lambda)$	$\omega(\lambda\log\lambda)$-DDH	$O(\epsilon^{\nu+1}/Q^\nu)$
[Yam17]: Sect. 7.3	$\omega(\log\lambda)$	$\omega(\sqrt{\lambda}\log\lambda)$	$\omega(\lambda\log\lambda)$-DDH	$O(\epsilon^{\nu+1}/Q^\nu)$
[Yam17]: Appendix C.	$\omega(\log\lambda)$	$\mathsf{poly}(\lambda)$	$\mathsf{poly}(\lambda)$-DDH	$O(\epsilon^2/\lambda^2 Q)$
0.95,0.95,0.95 Ours: Sect. 5.1	$\omega(\log^2\lambda)$	$\omega(\lambda\log^2\lambda)$	$\omega(\log^2\lambda)$-DDH	$O(\epsilon^{\nu+1}/Q^\nu)$
0.95,0.95,0.95 Ours: Sect. 5.3	$\omega(\sqrt{\lambda}\log\lambda)$	$\omega(\log\lambda)$	$\omega(\log^2\lambda)$-DDH	$O(\epsilon^{\nu+1}/Q^\nu)$

To measure the verification key size $\|vk\|$ and proof size $\|\pi\|$, we count the number of group elements in $\mathbb{G}$. Q, ϵ denotes the number of adversarial queries and advantage, respectively. ν is a constant satisfying $c = 1 - 2^{-1/\nu}$, where c is the relative distance of the underlying error correcting code $C : \{0,1\}^n \to \{0,1\}^\ell$.

exploiting a special linear structure in $\mathsf{PES_{FP}}$, we can further improve the efficiency using an idea native to our scheme. Namely, we can skip some of the verification steps required to check the validity of the proof, hence, lowering the number of group elements in the verification key. Our schemes can be viewed as combining the best of [Jag15, Yam17]. In the following, to compare the efficiency, we count the number of group elements of the verification key and proof.

- In our first scheme, the verification key size is $\omega(\log^2\lambda)$, the proof size is $\omega(\lambda\log^2\lambda)$, and the scheme is proven secure under the L-DDH assumption with $L = \omega(\log^2\lambda)$. This is the first scheme that simultaneously achieves a small verification key size and security under an L-DDH assumption where L is poly-logarithm in the security parameter.
- Our second scheme is a modification of our first VRF with some additional ideas; the verification key size is $\omega(\sqrt{\lambda}\log\lambda)$, the proof size is $\omega(\log\lambda)$, and the scheme is proven secure under the L-DDH assumption with $L = \omega(\log^2\lambda)$. This achieves the smallest verification key and proof size among all the previous schemes while also reducing the underlying L of the L-DDH assumption significantly to poly-logarithm.

PE Schemes for the MultD-Eq **Predicates.** Based on the predicate encoding scheme $\mathsf{PES_{Lin}}$ for the MultD-Eq predicates, we propose a lattice-based PE scheme for the MultD-Eq predicates. Due to the symmetry of the MultD-Eq predicates, we obtain key-policy and ciphertext-policy predicate encryption schemes for the class of predicates that can be expressed as MultD-Eq, such as subset and range conjunction. The detailed overview and comparison are given in Table 2. We disclude the generic construction of [GVW15, GKW17, WZ17] from the table, since our primal goal was to compare the efficiency of the schemes. Our scheme achieves the best efficiency in terms of decryption time and the required modulus size q; [GMW15] requires to perform $\Omega(2^D)$ number of inner product operations (between secret key vectors and ciphertext vectors) to decrypt a

Table 2. Comparison of lattice PEs for MultD-Eq predicates (over $\mathbb{Z}_p^{D \times \ell}$).

Schemes	$|\mathsf{mpk}| \ \# \ \mathbb{Z}_q^{n \times m}$	$|\mathsf{sk}| \ \# \ \mathbb{Z}^{2m}$	$|\mathsf{ct}| \ \# \ \mathbb{Z}_q^m$	LWE param $1/\alpha$	Dec. time $\#$ IP
[GMW15]	$O(D\ell)$	$O(D\ell)$	$O(D\ell)$	$\tilde{O}(\sqrt{D} \cdot n^{1.5})^a$	$O(\ell^D)$
Ours: Sect. 6.2	$O(D\ell p)$	1	$O(D\ell p)$	$\tilde{O}(\max\{\frac{n^2}{\sqrt{D\ell p}}, \sqrt{D\ell p} \cdot n\})$	1

To compare (space) efficiency, we measure the master public key size $|\mathsf{mpk}|$, secret key size $|\mathsf{sk}|$ and ciphertext size $|\mathsf{ct}|$ by the required number of elements in $\mathbb{Z}_q^{n \times m}, \mathbb{Z}^{2m}, \mathbb{Z}_q^m$, respectively. We measure the decryption time as the number of inner products computed between vectors in $\mathbb{Z}_q^{2m}$.
[a] For fairness, we provided a more rigorous analysis for their parameter selections.

ciphertext, and [GVW15, GKW17, WZ17] require subexponential LWE for security. Our construction follows very naturally from the predicate encoding scheme $\mathsf{PES}_{\mathsf{Lin}}$ for the MultD-Eq predicates, and builds upon the proof techniques of [AFV11, BGG+14].

Other Applications. We also show how to make the identity-based encryption (IBE) scheme of [Yam17] more efficient by using our predicate encoding scheme for the MultD-Eq predicate. In particular, we are able to lower the approximation factor of the LWE problem from $\tilde{O}(n^{11})$ to $\tilde{O}(n^{5.5})$ (with some additional analysis). Furthermore, we are able to significantly reduce the parallel complexity of the required matrix multiplications during encryption and key generation. Notably, our construction does not rely on the heavy sequential matrix multiplication technique of [GV15] as the IBE scheme of [Yam17]. Finally, we note that the size of the public matrices and ciphertexts are unchanged. Details are provided the full version.

1.2 Related Work

The idea of encoding predicates to another form has already been implicitly or explicitly used in other works. The notion of randomized encoding [IK00, AIK04] (not specific to predicates) aims to trade the computation of a "complex" function $f(x)$ for the computation of a "simpler" randomized function $\hat{f}(x; r)$ whose output distribution on an input x encodes the value for $f(x)$. The notion of predicate encoding [Wee14, CGW15] (and also the related notion of pair encoding [Att14, Att16]) has already been used previously, in a completely different context, as a generic framework that abstracts the concept of dual system encryption techniques for bilinear maps, and not as a tool for lowering the circuit complexity of predicates.

2 Technical Overview

We now give a brief overview of our technical approaches. A formal treatment is given in the main body. We break our overview in two pieces. First, we give intuition for our notion of predicate encoding schemes PES and illustrate the significance of the MultD-Eq predicates. Then, we overview how the different types of PES schemes for the MultD-Eq predicates can be used to construct VRFs, and PE schemes for the MultD-Eq predicates.

Different Ways of Encoding Predicates. Predicates are often times implicit in cryptographic constructions and in some cases there lies an untapped potential. To highlight this, we recall the observation of [Yam17]. An admissible hash function is one of the central tools used to prove adaptive security (e.g., digital signatures, identity-based encryptions, verifiable random functions). At a high level, during the security proof, it allows the simulator to secretly partition the input space into two disjoint sets, so there is a noticeable probability that the input values submitted by the adversary as challenge queries fall inside the intended sets. Traditionally, the partition made by the admissible hash function is viewed as a bit-fixing predicate; a bit-fixing predicate is specified by a string $K \in \{0, 1, \perp\}^{\ell}$ where the number of non-$\perp$ symbols are $O(\log \lambda)$, and the input space $\{0,1\}^{\ell}$ is partitioned by the rule whether the string $x \in \{0,1\}^{\ell}$ matches the string K on all non-$\perp$ symbols.

[Yam17] observed that a bit-fixing predicate can be encoded as a subset predicate; an observation not made since the classical works of [BB04b, CHKP10]. In particular, Yamada observed that K has many meaningless $\perp$ symbols and only has $O(\log \lambda)$ meaningful non-$\perp$ symbols. Under this observation, he managed to encode K into a very small set T_K (e.g., $|\mathsf{T}_K| = O(\log^2 \ell)$) where each element indicates the position of the non-$\perp$ symbols. Now, the partition of the input space is done by checking whether the input includes the set T_K or not. Since admissible hash functions are implicitly embedded in the public parameters, this idea allowed them to significantly reduce the number of public parameters for identity-based encryption (IBE) schemes and the size of the verification key (or the proof size) for VRFs.

We take this observation one step further. A predicate defines a function, but often a function may be represented as a polynomial[7] in various ways depending on what kind of properties we require. This is easiest to explain through an example. Let us continue with the above example of the subset predicate used in [Yam17]: $P_{\mathsf{T}} : 2^{[2n]} \rightarrow \{0, 1\}$, where $P_{\mathsf{T}}(\mathsf{S}) = 1$ iff $\mathsf{T} \subseteq \mathsf{S}$. Here, assume $|\mathsf{T}| = m$ and all the inputs to P_{T} have cardinality n. One of the most natural ways to represent the subset predicate as a polynomial is by its boolean circuit representation:

$$\underbrace{\prod_{i=1}^{m}\left(1 - \prod_{j=1}^{n}\left(1 - \underbrace{\prod_{k=1}^{\zeta}\left(1 - (t_{i,k} - s_{j,k})^2\right)}_{\text{is } t_i = s_j?}\right)\right)}_{\text{is } t_i \in \mathsf{S}?} = \begin{cases} 1 & \text{if} \quad \mathsf{T} \subseteq \mathsf{S} \\ 0 & \text{if} \quad \mathsf{T} \not\subseteq \mathsf{S} \end{cases}, \quad (1)$$

where $\zeta = \lfloor \log 2n \rfloor + 1$, $\mathsf{T} = \{t_i\}_{i \in [m]}, \mathsf{S} = \{s_j\}_{j \in [n]} \subseteq [2n]$ and $t_{i,k}, s_{j,k}$ are the k-th bit of the binary representation of t_i, s_j. Here Eq. (1) is the polynomial

---

[7] It might be more precise to state that a predicate is represented by a circuit, however, in this section we adopt the view of polynomials to better convey the intuition.

representation of the boolean logic $\bigwedge_{i\in[m]} \bigvee_{j\in[n]} \bigwedge_{k\in[\zeta]}(t_{i,k} = s_{j,k})$. This is essentially what was used for the lattice-based IBE construction of [Yam17] with very short public parameters. Observe that this polynomial has degree $2mn\zeta$, which is $O(\lambda \log^3 \lambda)$ if we are considering the subset predicate specifying the admissible hash function, where we have $m = O(\log^2 \lambda), n = O(\lambda)$ and $\zeta = O(\log \lambda)$. However, in general, using a high degree polynomial may be undesirable for many reasons, even if it is only of degree linear in the security parameter. For the case of the IBE scheme of [Yam17], due to the highly multiplicative structure, the encryption and key generation algorithms require to rely on a linear number of heavy sequentialized matrix multiplication technique of [GV15]. Therefore, it is a natural question to ask whether we can embed a predicate into a polynomial with lower degree, and in some cases into a linear polynomial.

Indeed, we show that it is possible for the above predicate. Namely, we can do much better by noticing the extra structure of subset predicates; we know there exists at most one $j \in [n]$ that satisfies $t_i = s_j$. Therefore, we can equivalently express Eq. (1) as the following polynomial:

$$\prod_{i=1}^{m} \sum_{j=1}^{n} \prod_{k=1}^{\zeta} \left(1 - (t_{i,k} - s_{j,k})^2\right) = \begin{cases} 1 & \text{if } \mathsf{T} \subseteq \mathsf{S} \\ 0 & \text{if } \mathsf{T} \not\subseteq \mathsf{S} \end{cases} . \tag{2}$$

This polynomial is now down to degree $2m\zeta$. When this subset predicate specifies the admissible hash function, Eq. (2) significantly lowers the degree down to $O(\log^3 \lambda)$. Furthermore, if we do not require the output to be exactly 0 or 1, and only care that the predicate behaves differently on satisfied/non-satisfied inputs, we can further lower the degree down to 2ζ. In particular, consider the following polynomial:

$$m - \sum_{i=1}^{m} \sum_{j=1}^{n} \prod_{k=1}^{\zeta} \left(1 - (t_{i,k} - s_{j,k})^2\right) = \begin{cases} 0 & \text{if } \mathsf{T} \subseteq \mathsf{S} \\ \neq 0 & \text{if } \mathsf{T} \not\subseteq \mathsf{S} \end{cases} , \tag{3}$$

which follows from the observation that $|\mathsf{T}| = m$. Since, the output of the polynomial is different for the case $\mathsf{T} \subseteq \mathsf{S}$ and $\mathsf{T} \not\subseteq \mathsf{S}$, Eq. (3) indeed properly encodes the information of the subset predicate. Using this polynomial instead of Eq. (1) already allows us to significantly optimize the concrete parameters of the lattice-based IBE of [Yam17]. In fact, by encoding the inputs T, S in a different way and with some additional ideas, we can encode the subset predicate into a *linear* polynomial.

To summarize, depending on what we require for the encoding of a predicate (e.g., preserve the functionality, linearize the encoding) one has the freedom of choosing how to express a particular predicate. We formalize this idea of a "right encoding" by introducing the notion of *predicate encoding schemes*. In the above we used the subset predicate as an motivating example, however, in our work we focus on a wider class of predicates called the *multi-dimensional equality* MultD-Eq predicates, and propose two encoding schemes $\mathsf{PES_{FP}}$ and $\mathsf{PES_{Lin}}$ with different applications in mind.

Finally, we state two justifications for why we pursue the construction of predicate encoding schemes for the class of MultD-Eq predicates. First, the MultD-Eq predicates are expressive enough to encode many useful predicates that come up in cryptography (e.g., bit-fixing, subset conjunction, range conjunction predicates), that being for constructions of cryptographic primitives or for embedding secret information during in the security proof. Second, in spite of its expressiveness, the MultD-Eq predicates have a simple structure that we can exploit and offers us plenty of freedom on the types of predicate encoding schemes we can achieve. The definition and a more detailed discussion on the expressiveness of MultD-Eq is provided in Sects. 4.2 and 4.3.

Constructing VRFs. Similarly to many of the prior works [BMR10, ACF14, Jag15, Yam17] on VRFs with all the desired properties, we use admissible hash functions and base security on the L-DDH assumption, which states that given $(h, g, g^\alpha, \cdots, g^{\alpha^L}, \Psi)$ it is hard to distinguish whether $\Psi = e(g, h)^{1/\alpha}$ or a random element. Here, we briefly review the core idea used during the security proof of [Yam17] for the pseudorandomness property of the VRF. We note that many of the arguments made below are informal for the sake of intuition. Their observation was that the admissible hash function embedded during simulation can be stated in the following way using a subset predicate:

$$F_T(X) = \begin{cases} 0 & \text{if } T \subseteq S(X) \\ 1 & \text{if } T \not\subseteq S(X) \end{cases} \qquad \text{where} \quad S(X) = \{2i - C(X)_i \mid i \in [n]\}.$$

Here, $C(\cdot)$ is a public hash function that maps an input X (of the VRF) to a bit string $\{0, 1\}^n$, and $T \subseteq [2n]$ is a set defined as $T = \{2i - K_i \mid i \in [n], K_i \neq \bot\}$ where K is the secret string in $\{0, 1, \bot\}^n$ that specifies the partition made by the admissible hash. Since, the number of non-$\bot$ symbols in K are $O(\log^2 \lambda)$, the above function can be represented by a set T with cardinality $O(\log^2 \lambda)$. During security proof, by the property and definition of F_T, we have

$$\left(T \not\subseteq S(X^{(1)})\right) \wedge \cdots \wedge \left(T \not\subseteq S(X^{(Q)})\right) \wedge \left(T \subseteq S(X^*)\right),$$

with non-negligible probability, where X^* is the challenge input and $X^{(1)}, \cdots, X^{(Q)}$ are the inputs for which the adversary has made evaluation queries. The construction of [Yam17] is based on previous inversion-based VRFs [DY05, BMR10]. Here, we ignore the problem of how to add verifiability to the scheme and overview on how they prove pseudorandomness of the VRF evaluation. Informally, during simulation, the simulator uses the following polynomial to encode the admissible hash function:

$$Q(\alpha) \Big/ \left(\prod_{i=1}^m \prod_{j=1}^n (\alpha + t_i - s_j)\right) = \begin{cases} \frac{\text{const}}{\alpha} + \text{poly}(\alpha) & \text{if } T \subseteq S(X) \\ \text{poly}(\alpha) & \text{if } T \not\subseteq S(X) \end{cases}, \qquad (4)$$

where $Q(\alpha)$ is some fixed polynomial with degree roughly $4n$ independent of the input X. Here, recall $\alpha \in \mathbb{Z}_p$ is that of the L-DDH problem, and notice that in

Eq. (4) the polynomial will have α in the denominator if and only if $\mathsf{T} \subseteq \mathsf{S}(X)$. Although this may not seem quite like it, this polynomial is indeed an encoding of the subset predicate[8] since it acts differently depending on $\mathsf{T} \subseteq \mathsf{S}(X)$ and $\mathsf{T} \not\subseteq \mathsf{S}(X)$. Finally, we note that the output Y of the VRF is obtained by simply putting the above polynomial in the exponent of $e(g, h)$.

Now, if the simulator is given enough $(g^{\alpha^i})_{i \in [L]}$ as the L-DDH challenge, it can create a valid evaluation Y for inputs X such that $\mathsf{T} \not\subseteq \mathsf{S}(X)$, since it can compute terms of the form $e(g^{\mathsf{poly}(\alpha)}, h) = e(g, h)^{\mathsf{poly}(\alpha)}$. Furthermore, for the challenge query X^* it will use Ψ; if $\Psi = e(g, h)^{1/\alpha}$ it can correctly simulate for the case $\mathsf{T} \subseteq \mathsf{S}(X^*)$, otherwise the evaluation Y^* of the VRF is independent of X^*. Therefore, under the hardness of the L-DDH assumption, the output is proven pseudorandom. Observe that for the simulator to compute $e(g, h)^{\mathsf{poly}(\alpha)}$ from Eq. (4), it needs to have $(g^{\alpha^i})_{i \in [L]}$ where $L = O(n)$. Then, since $n = O(\lambda)$, we need to base this on an L-DDH assumption where $L = O(\lambda)$.[9] To reflect the above polynomial, the verification keys are set as $(h, \hat{g}, (W_i = \hat{g}^{w_i}))$ in the actual construction. During simulation the parameters are (roughly) set as $\hat{g} = g^{Q(\alpha)}$, $\hat{g}^{w_i} = \hat{g}^{\alpha + t_i}$.

The above construction is rather naive in that it checks whether $\mathsf{T} \subseteq \mathsf{S}(X)$ in a brute-force manner (as also noted in [Yam17]). Our idea is to instead use the polynomial from Eq. (2) to represent the admissible hash function. In other words, we embed the following polynomial during simulation:

$$\frac{1}{\alpha} \cdot \prod_{i=1}^{m} \sum_{j=1}^{n} \prod_{k=1}^{\zeta} \left(1 - (\alpha + t_{i,k} - s_{j,k})^2\right) = \begin{cases} \frac{1}{\alpha} + \mathsf{poly}(\alpha) & \text{if } \mathsf{T} \subseteq \mathsf{S}(X) \\ \mathsf{poly}(\alpha) & \text{if } \mathsf{T} \not\subseteq \mathsf{S}(X) \end{cases}. \tag{5}$$

We note that in our actual construction, we use an optimized version of Eq. (2) called $\mathsf{PES}_{\mathsf{FP}}$. Similarly to above, we put the above polynomial in the exponent of $e(g, h)$ for the VRF evaluation. The difference is that the degree of the polynomial in Eq. (5) is significantly lowered down to merely $2m\zeta$, which is $O(\log^3 \lambda)$. Therefore, when the simulator needs to compute $e(g, h)^{\mathsf{poly}(\alpha)}$ during simulation, we only require $(g^{\alpha^i})_{i \in [L]}$ for $L = O(\log^3 \lambda)$. Hence, we significantly reduced the required L of the L-DDH assumption to poly-logarithm. Note that we need to validate the output in a different way now, since the terms α, t_i, s_j that appear in the left-hand polynomial are not in the denominator as in Eq. (4). Now, to generate the proof, we take the so called "step ladder approach" [Lys02, ACF09, HW10], where we publish values of the form $(g^{\theta_{i'}})_{i' \in [m]}, (g^{\theta_{i,j,k'}})_{(i,j,k') \in [m] \times [n] \times [\zeta]}$ defined as follows:

$$\theta_{i'} = \prod_{i=1}^{i'} \sum_{j=1}^{n} \prod_{k=1}^{\zeta} \left(1 - (w_{i,k} - s_{j,k})^2\right), \quad \theta_{i,j,k'} = \prod_{k=1}^{k'} \left(1 - (w_{i,k} - s_{j,k})^2\right),$$

---

[8] To be strict, this does not exactly fit the definition of predicate encoding we define in Sect. 4. However, we can do so by appropriately arguing the size of α or by viewing α as an indeterminate.

[9] In the actual construction we require $L = \omega(\lambda \log \lambda)$, since we need to simulate a higher degree polynomial in the exponent.

where we (roughly) set $g^{w_{i,k}}$ as $g^{\alpha+t_{i,k}}$ during simulation. Although this scheme achieves a very short verification key, it comes at the cost of a rather long proof size of $O(mn\zeta) = O(\lambda \log^3 \lambda)$.

Finally, we describe how to make the proof much shorter, while still maintaining a sub-linear verification key size. As a first step, we can use the simple trick used in [Yam17] to make the proof much shorter. Namely, we add helper components to the verification key so that anyone can compute $(\theta_{i,j,k'})$ publicly. However, as in [Yam17], this leads to a long verification key with size $\tilde{\Omega}(\lambda)$. Interestingly, for our construction, we can do much better and shorten the verification key by a quadratic factor by in a sense *skipping* some ladders. The main observation is the additive structure in $(\theta_{i'})_{i'}$. In particular, if each $\theta_{i'}$ were simply a large product $\prod_{i,j,k} \left(1 - (w_{i,k} - s_{j,k})^2\right)$, we would have to prepare all the necessary helper components in the verification key that would allow to compute $g^{\theta_{i,j,\varsigma}}$. This is because in the step ladder approach, after computing $g^{\theta_{i,j,\varsigma}}$, we have to reuse this as an input to the bilinear map to validate the next term in the ladder. However, in our case, we only need the ability to publicly compute $e(g,g)^{\theta_{i,j,\varsigma}}$. Here, we crucially rely on the additive structure in $\theta_{i'}$ that allows us to compute $e(g,g)^{\sum_{j\in[n]} \theta_{i,j,\varsigma}}$ by ourselves; thus the notion of skipping some ladders. Note that we are not able to publicly compute $e(g,g)^{\prod_{j\in[n]} \theta_{i,j,\varsigma}}$. Finally, we continue with the step ladder approach for the outer $\prod_{i=1}^{i'}$ products. Therefore, since we only need the ability to generate $e(g,g)^{\theta_{i,j,\varsigma}}$ rather than $g^{\theta_{i,j,\varsigma}}$, we can reduce quadratically the number of helper components we have to publish in the verification key.

Constructing PE for the MultD-Eq Predicates. Our proposed predicate encryption scheme for the MultD-Eq predicates follows the general framework of [AFV11,BGG+14], which allows us to compute an inner product of a private attribute vector X associated to a ciphertext and a (public) predicate vector Y associated to a secret key. To accommodate this framework, we use our proposed *linear* predicate encoding scheme $\mathsf{PES_{Lin}}$ for the MultD-Eq predicates. In the overview, we continue with our examples with the subset predicate for simplicity. The core idea is the same for the MultD-Eq predicates. Essentially, $\mathsf{PES_{Lin}}$ will allow us to further modify Eq. (3), to the following linear polynomial:

$$\sum_{i=1}^{L} a_i \mathsf{X}_i = \begin{cases} 0 & \text{if } \mathsf{T} \subseteq \mathsf{S} \\ \neq 0 & \text{if } \mathsf{T} \nsubseteq \mathsf{S} \end{cases}, \tag{6}$$

where $(\mathsf{X}_i)_{i\in[L]}, (a_i)_{i\in[L]} \in \mathbb{Z}_q^L$ are encodings of the attribute set T and the predicate set S, respectively.

Following the general framework, the secret key for a user with predicate set S is a short vector $\mathbf{e}$ such that $[\mathbf{A}|\mathbf{B_S}]\mathbf{e} = \mathbf{u}$ for a random public vector $\mathbf{u}$, where $\mathbf{B_S}$ is defined as in Eq. (7) below. Furthermore, we privately embed an attribute set T into the ciphertext as

$$[\mathbf{c}_1^\top | \cdots | \mathbf{c}_L^\top] = \mathbf{s}^\top [\mathbf{B}_1 + \mathsf{X}_1 \mathbf{G} | \cdots | \mathbf{B}_L + \mathsf{X}_L \mathbf{G}] + [\mathbf{z}_1^\top | \cdots | \mathbf{z}_L^\top].$$

Using the gadget matrix $\mathbf{G}$ of [MP12], a user corresponding to the predicate set S can transform the ciphertext without knowledge of T as follows:

$$\sum_{i=1}^{L} \mathbf{c}_i^\top \mathbf{G}^{-1}(a_i\mathbf{G}) = \mathbf{s}^\top \Big(\underbrace{\sum_{i=1}^{L} \mathbf{B}_i \mathbf{G}^{-1}(a_i\mathbf{G}) + \sum_{i=1}^{L} a_i \mathsf{X}_i \cdot \mathbf{G}}_{= \,\mathbf{B}_\mathsf{S}} \Big) + \underbrace{\sum_{i=1}^{L} \mathbf{z}_i^\top \mathbf{G}^{-1}(a_i\mathbf{G})}_{= \,\mathbf{z} \text{ (noise term)}}.$$

$$(7)$$

Observe the matrix $\mathbf{B}_\mathsf{S}$ is defined independently of X (i.e., the attribute set S). By Eq. (6) and the correctness of the predicate encoding scheme $\mathsf{PES}_{\mathsf{Lin}}$, we have $\sum_{i\in[L]} a_i\mathsf{X}_i = 0$ when the subset predicate is satisfied, as required for decryption. To prove security, we set the matrices $(\mathbf{B}_i)_{i\in[L]}$ as $\mathbf{B}_i = \mathbf{A}\mathbf{R}_i - \mathsf{X}_i^* \cdot \mathbf{G}$, where $\mathbf{A}$ is from the problem instance of LWE, $\mathbf{R}_i$ is a random matrix with small coefficients and $(\mathsf{X}_i^*)_{i\in[L]}$ is the encoding of the challenge attribute set T^*. During simulation we have

$$\mathbf{B}_\mathsf{S} = \mathbf{A}\mathbf{R}_\mathsf{S} - \sum_{i=1}^{L} a_i\mathsf{X}^* \cdot \mathbf{G}, \quad \text{where} \quad \mathbf{R}_\mathsf{S} = \sum_{i=1}^{L} \mathbf{R}_i \mathbf{G}^{-1}(a_i\mathbf{G}).$$

for any set S. Here, we have $\sum_{i\in[L]} a_i\mathsf{X}^* \neq 0$ iff $\mathsf{T}^* \not\subseteq \mathsf{S}$. Therefore, for the key extraction queries for S such that $\mathsf{T}^* \not\subseteq \mathsf{S}$, we can use $\mathbf{R}_\mathsf{S}$ as the $\mathbf{G}$-trapdoor [MP12] for the matrix $[\mathbf{A} \,|\, \mathbf{B}_\mathsf{S}]$ to simulate the secret keys. We are able to generate the challenge ciphertext for the subset T^* by computing

$$\underbrace{(\mathbf{s}^\top\mathbf{A} + \mathbf{z}'^\top)}_{\text{LWE Problem}}[\mathbf{I} \,|\, \mathbf{R}_1 | \cdots | \mathbf{R}_L] = \mathbf{s}^\top[\mathbf{A} \,|\, \mathbf{B}_1 + \mathsf{X}_1^*\mathbf{G} | \cdots | \mathbf{B}_L + \mathsf{X}_L^*\mathbf{G}] + \underbrace{\mathbf{z}'^\top[\mathbf{I} \,|\, \mathbf{R}_1 | \cdots | \mathbf{R}_L]}_{\text{simulation noise term}}$$

A subtle point here is that the simulation noise term is not distributed correctly as in Eq. (7). However, this can be resolved by the noise rerandomization technique of [KY16].

Finally, we propose a technique to finer analyze the growth of the noise term $\mathbf{z} = \sum_{i\in[L]} \mathbf{z}_i^\top \mathbf{G}^{-1}(a_i\mathbf{G})$ and the $\mathbf{G}$-trapdoor $\mathbf{R}_\mathsf{S} = \sum_{i\in[L]} \mathbf{R}_i \mathbf{G}^{-1}(a_i\mathbf{G})$ used during simulation. This allows us to choose narrower Gaussian parameters and let us base security on a weaker LWE assumption. The main observation is that $\mathbf{G}^{-1}(a_i\mathbf{G}) \in \{0,1\}^{nk\times nk}$ is a block-diagonal matrix with n square matrices with size k along its diagonals where $n = O(\lambda)$ and $k = O(\log\lambda)$. Exploiting this additional block-diagonal structure, we are able to finer control the growth of $\|\mathbf{v}\|_2$ and $s_1(\mathbf{R}_\mathsf{S})$ (i.e., the largest singular value of $\mathbf{R}_\mathsf{S}$). We believe this technique to be useful for obtaining tighter analysis on other lattice-based constructions.

3 Preliminaries

Notation. We use $\{\cdot\}$ to denote sets and use $(\cdot)$ to denote a finite ordered list of elements. When we use notations such as $(w_{i,j})_{(i,j)\in[n]\times[m]}$ for $n, m \in \mathbb{N}$, we assume the elements are sorted in the lexicographical order. For $n, m \in \mathbb{N}$ with $n \leq m$, denote $[n]$ as the set $\{1, \cdots, n\}$ and $[n, m]$ as the set $\{n, \cdots, m-1, m\}$.

3.1   Verifiable Random Functions

We define a verifiable random function $\mathsf{VRF} = (\mathsf{Gen}, \mathsf{Eval}, \mathsf{Verify})$ as a tuple of three probabilistic polynomial time algorithms [MRV99].

$\mathsf{Gen}(1^\lambda) \to (\mathsf{vk}, \mathsf{sk})$: The key generation algorithm takes as input the security parameter 1^λ and outputs a verification key vk and a secret key sk.

$\mathsf{Eval}(\mathsf{sk}, X) \to (Y, \pi)$: The evaluation algorithm takes as input the secret key sk and an input $X \in \{0,1\}^n$, and outputs a value $Y \in \mathcal{Y}$ and a proof π, where $\mathcal{Y}$ is some finite set.

$\mathsf{Verify}(\mathsf{vk}, X, (Y, \pi)) \to 0/1$: The verification algorithm takes as input the verification key vk, $X \in \{0,1\}^n$, $Y \in \mathcal{Y}$ and a proof π, and outputs a bit.

Definition 1. *We say a tuple of polynomial time algorithms* $\mathsf{VRF} = (\mathsf{Gen},$ *Eval,Verify) is a verifiable random function if all of the following requirements hold:*

Correctness. *For all* $\lambda \in \mathbb{N}$, *all* $(\mathsf{vk}, \mathsf{sk}) \leftarrow \mathsf{Gen}(1^\lambda)$ *and all* $X \in \{0,1\}^n$, *if* $(Y, \pi) \leftarrow \mathsf{Eval}(\mathsf{sk}, X)$ *then* $\mathsf{Verify}(\mathsf{vk}, X, (Y, \pi))$.

Uniqueness. *For an arbitrary string* $\mathsf{vk} \in \{0,1\}^*$ *(not necessarily generated by* Gen*) and all* $X \in \{0,1\}^n$, *there exists at most a single* $Y \in \mathcal{Y}$ *for which there exists an accepting proof* π.

Pseudorandomness. *This security notion is defined by the following game between a challenger and an adversary* $\mathcal{A}$.

  Setup. *The challenger runs* $(\mathsf{vk}, \mathsf{sk}) \leftarrow \mathsf{Gen}(1^\lambda)$ *and gives* vk *to* $\mathcal{A}$.
  Phase 1. $\mathcal{A}$ *adaptively submits an evaluation query* $X \in \{0,1\}^n$ *to the challenger, and the challenger returns* $(Y, \pi) \leftarrow \mathsf{Eval}(\mathsf{sk}, X)$.
  Challenge Query. *At any point,* $\mathcal{A}$ *may submit a challenge input* $X^* \in \{0,1\}^n$. *Here, we require that* $\mathcal{A}$ *has not submitted* X^* *as an evaluation query in Phase 1. The challenger picks a random coin* $\mathsf{coin} \leftarrow \{0,1\}$. *Then it runs* $(Y_0^*, \pi_0^*) \leftarrow \mathsf{Eval}(\mathsf{sk}, X^*)$ *and picks* $Y_1^* \leftarrow \mathcal{Y}$. *Finally it returns* Y_{coin}^* *to* $\mathcal{A}$.
  Phase 2. $\mathcal{A}$ *may continue on submitting evaluation queries as in Phase 1 with the added restriction that* $X \neq X^*$.
  Guess. *Finally,* $\mathcal{A}$ *outputs a guess* $\widehat{\mathsf{coin}}$ *for* coin.

The advantage of $\mathcal{A}$ *is defined as* $|\Pr[\widehat{\mathsf{coin}} = \mathsf{coin}] - \frac{1}{2}|$. *We say that the* VRF *satisfies (adaptive) pseudorandomness if the advantage of any probabilistic polynomial time algorithm* $\mathcal{A}$ *is negligible.*

3.2   Predicate Encryption

We use the standard syntax of predicate encryption (PE) schemes [BW07, KSW08, AFV11], where $P(X) = 1$ signifies the ability to decrypt. We briefly recall the security notion of PE schemes and refer the exact definition to the full version. In our paper, we define the notion of *selectively secure* and *weakly attribute hiding* using a standard game-based security formalization. The former notion requires the challenge ciphertext to leak no information on the message, given the challenge attribute at the outset of the game. The latter notion also requires that the challenge ciphertext leaks no information on the attribute, if the adversary is only allowed to obtain secret keys that do no decrypt the challenge ciphertext.

3.3   Background on Lattices

For an integer $m > 0$, let $D_{\mathbb{Z}^m, \sigma}$ be the discrete Gaussian distribution over $\mathbb{Z}^m$ with parameter $\sigma > 0$. Other lattice notions are defined in the standard way.

Hardness Assumption. We define the Learning with Errors (LWE) problem introduced by Regev [Reg05].

Definition 2 (Learning with Errors). *For integers n, m, a prime $q > 2$, an error distribution over χ over $\mathbb{Z}$, and a PPT algorithm $\mathcal{A}$, an advantage for the learning with errors problem $\mathsf{LWE}_{n,m,q,\chi}$ of $\mathcal{A}$ is defined as follows:*

$$\mathsf{Adv}_{\mathcal{A}}^{\mathsf{LWE}_{n,m,q,\chi}} = \left| \Pr\left[\mathcal{A}(\mathbf{A}, \mathbf{A}^\top \mathbf{s} + \mathbf{z}) = 1 \right] - \Pr\left[\mathcal{A}(\mathbf{A}, \mathbf{w} + \mathbf{z}) = 1 \right] \right|$$

where $\mathbf{A} \leftarrow \mathbb{Z}_q^{n \times m}$, $\mathbf{s} \leftarrow \mathbb{Z}_q^n$, $\mathbf{w} \leftarrow \mathbb{Z}_q^m$, $\mathbf{z} \leftarrow \chi$. We say that the LWE assumption holds if $\mathsf{Adv}_{\mathcal{A}}^{\mathsf{LWE}_{n,m,q,\chi}}$ is negligible for all PPT $\mathcal{A}$.

The (decisional) $\mathsf{LWE}_{n,m,q,D_{\mathbb{Z},\alpha q}}$ for $\alpha q > 2\sqrt{n}$ has been shown by Regev [Reg05] to be as hard as approximating the worst-case SIVP and GapSVP problems to within $\tilde{O}(n/\alpha)$ factors in the ℓ_2-norm in the worst case. In the subsequent works, (partial) dequantumization of the reduction were achieved [Pei09, BLP+13].

Gadget Matrix. We use the gadget matrix $\mathbf{G} \in \mathbb{Z}_q^{n \times m}$ defined in [MP12]. Here, $\mathbf{G}$ is a full rank matrix such that the lattice $\Lambda^\perp(\mathbf{G})$ has a publicly known basis $\mathbf{T_G}$ with $\|\mathbf{T_G}\|_{\mathsf{GS}} \leq \sqrt{5}$. Further properties on $\mathbf{G}$ can be found in [MP12] or the full version.

Sampling Algorithms. The following lemma states useful algorithms for sampling short vectors from lattices.

Lemma 1 [GPV08, ABB10, CHKP10, MP12]. *Let $n, m, q > 0$ be integers with $m > 2n\lceil \log q \rceil$.*

- *TrapGen$(1^n, 1^m, q) \to (\mathbf{A}, \mathbf{T_A})$: There exists a randomized algorithm that outputs a matrix $\mathbf{A} \in \mathbb{Z}_q^{n \times m}$ and a full-rank matrix $\mathbf{T_A} \in \mathbb{Z}^{m \times m}$, where $\mathbf{T_A}$ is a basis for $\Lambda^\perp(\mathbf{A})$, $\mathbf{A}$ is statistically close to uniform and $\|\mathbf{T_A}\|_{\mathsf{GS}} = O(\sqrt{n \log q})$.*
- *SampleLeft$(\mathbf{A}, \mathbf{B}, \mathbf{u}, \mathbf{T_A}, \sigma) \to \mathbf{e}$: There exists a randomized algorithm that, given matrices $\mathbf{A}, \mathbf{B} \in \mathbb{Z}_q^{n \times m}$, a vector $\mathbf{u} \in \mathbb{Z}_q^n$, a basis $\mathbf{T_A} \in \mathbb{Z}^{m \times m}$ for $\Lambda^\perp(\mathbf{A})$, and a Gaussian parameter $\sigma > \|\mathbf{T_A}\|_{\mathsf{GS}} \cdot \omega(\sqrt{\log m})$, outputs a vector $\mathbf{e} \in \mathbb{Z}^{2m}$ sampled from a distribution which is $\mathsf{negl}(n)$-close to $D_{\Lambda_{\mathbf{u}}^\perp([\mathbf{A}|\mathbf{B}]), \sigma}$.*

3.4 Background on Bilinear Maps

We define certified bilinear group generators following [HJ16]. We require that there is an efficient bilinear group generator algorithm GrpGen that on input 1^λ outputs a description Π of bilinear groups $\mathbb{G}, \mathbb{G}_T$ with prime order p and a map $e : \mathbb{G} \times \mathbb{G} \to \mathbb{G}_T$. We require GrpGen to be certified in the sense that there is an efficient algorithm GrpVfy that on input a description of the bilinear groups, outputs the validity of the description. Furthermore, we require that each group element has a unique encoding, which can be efficiently recognized. For the precise definition, we refer [HJ16]. The following is the hardness assumption we use in our scheme.

Definition 3 (L-Diffie-Hellman Assumption). *For a PPT algorithm $\mathcal{A}$, an advantage for the decisional L-Diffie-Hellman problem L-DDH of $\mathcal{A}$ with respect to GrpGen is defined as follows:*

$$\mathsf{Adv}_{\mathcal{A}}^{L\text{-}DDH} = |\Pr[\mathcal{A}(\Pi, g, h, g^{\alpha}, g^{\alpha^2}, \cdots, g^{\alpha^L}, \Psi_0) \to 1]$$

$$- \Pr[\mathcal{A}(\Pi, g, h, g^{\alpha}, g^{\alpha^2}, \cdots, g^{\alpha^L}, \Psi_1) \to 1]|,$$

where $\Pi \leftarrow \mathsf{GrpGen}(1^{\lambda}), \alpha \leftarrow \mathbb{Z}_p^*, g, h \leftarrow \mathbb{G}, \Psi_0 = e(g, h)^{1/\alpha}$ and $\Psi_1 \leftarrow \mathbb{G}_T$. We say that L-DDH assumption holds if $\mathsf{Adv}_{\mathcal{A}}^{L\text{-}DDH}$ is negligible for all PPT $\mathcal{A}$.

4 Encoding Predicates with Arithmetic Circuits

In this section, we formalize the intuition outlined in the introduction on how to encode predicates as circuits. Here, we view predicates as simply a function $P : \mathcal{X} \to \{0, 1\}$ over some domain $\mathcal{X}$ with image $\{0, 1\}$. Furthermore, to capture the algebraic properties of arithmetic circuits, we adapt the view of treating circuits as polynomials and vice versa.

4.1 Predicate Encoding Scheme

We formalize our main tool: predicate encoding scheme.

Definition 4 (Predicate Encoding Scheme). *Let $\mathcal{P} = \{\mathcal{P}_\lambda\}_{\lambda \in \mathbb{N}}$ be a family of set of efficiently computable predicates where $\mathcal{P}_\lambda$ is a set of predicates of the form $P : \mathcal{X}_\lambda \to \{0, 1\}$ for some input space $\mathcal{X}_\lambda$, and let $\mathcal{R} = \{\mathcal{R}_\lambda\}_{\lambda \in \mathbb{N}}$ be a family of rings. We define a predicate encoding scheme over a family of rings $\mathcal{R}$ for a family of set of predicates $\mathcal{P}$, as a tuple of deterministic polynomial time algorithms PES = (EncInpt, EncPred) such that*

- EncInpt$(1^\lambda, x) \to \hat{\boldsymbol{x}}$: *The input encoding algorithm takes as inputs the security parameter 1^λ and input $x \in \mathcal{X}_\lambda$, and outputs an encoding $\hat{\boldsymbol{x}} \in \{0_{\mathcal{R}_\lambda}, 1_{\mathcal{R}_\lambda}\}^t \subseteq \mathcal{R}_\lambda^t$, where $t = t(\lambda)$ is an integer valued polynomial and $0_{\mathcal{R}_\lambda}, 1_{\mathcal{R}_\lambda}$ denote the zero and identity element of the ring $\mathcal{R}_\lambda$, respectively.*
- EncPred$(1^\lambda, P) \to \hat{C}$: *The predicate encoding algorithm takes as inputs the security parameter 1^λ and a predicate $P \in \mathcal{P}_\lambda$, and outputs a polynomial representation of an arithmetic circuit $\hat{C} : \mathcal{R}_\lambda^t \to \mathcal{R}_\lambda$. We denote $\hat{\mathcal{C}}_\lambda$ as the set of arithmetic circuits $\{\hat{C} \mid \hat{C} \leftarrow \mathsf{EncPred}(1^\lambda, P), \forall P \in \mathcal{P}_\lambda\}$.*

Correctness. *We require a predicate encoding scheme over a family of rings $\mathcal{R}$ for a family of set of predicates $\mathcal{P}$ to satisfy the following: for all $\lambda \in \mathbb{N}$ there exist disjoint subsets $S_{\lambda,0}, S_{\lambda,1} \subset \mathcal{R}_\lambda$ (i.e., $S_{\lambda,0} \cap S_{\lambda,1} = \phi$), such that for all predicates $P \in \mathcal{P}_\lambda$, all inputs $x \in \mathcal{X}_\lambda$ if $P(\boldsymbol{x}) = b$ then $\hat{C}(\hat{\boldsymbol{x}}) \in S_{\lambda,b}$, where $\hat{\boldsymbol{x}} \leftarrow \mathsf{EncInpt}(1^\lambda, \boldsymbol{x}), \hat{C} \leftarrow \mathsf{EncPred}(1^\lambda, P)$, and $b \in \{0, 1\}$.*

Degree. *We say that a predicate encoding scheme PES is of degree $d = d(\lambda)$ if the maximal degree of the circuits in $\hat{\mathcal{C}}_\lambda$ (in their polynomial representation) is d. In case $d = 1$, we say PES is linear.*

In the following, we will be more loose in our use of notations. For simplicity, we omit the subscripts expressing the domain or the security parameter such as $0_\mathcal{R}, S_{\lambda,b}, \mathcal{R}_\ell$ when it is clear from context. We also omit the expression family and simply state that it is a predicate encoding scheme over a ring $\mathcal{R}$ for a set of predicates P. Finally, in the following we assume that the algorithms $\mathsf{EncInpt}(1^\lambda, \cdot), \mathsf{EncPred}(1^\lambda, \cdot)$ will implicitly take the security parameter 1^λ as input and omit it unless stated otherwise.

4.2 Encoding Multi-dimensional Equality Predicates

Here, we propose two predicate encoding schemes for the *multi-dimensional equality predicate*[10] (MultD-Eq) whose constructions are motivated by different applications. As we show later, the multi-dimensional equality predicate is expressive enough to encode many useful predicates that come up in cryptography (e.g., bit-fixing, subset conjunction, range conjunction predicates), that being for constructions of cryptographic primitives or for embedding secret information during in the security proof.

We first define the domains on which the multi-dimensional equality predicates MultD-Eq are defined over, and then formally define what they are.

Definition 5 (Compatible Domains for MultD-Eq). *Let p, D, ℓ be positive integers. We call a pair of domains $(\mathcal{X}, \mathcal{Y}) \subseteq \mathbb{Z}_p^{D \times \ell} \times \mathbb{Z}_p^{D \times \ell}$ to be compatible with the multi-dimensional equality predicates if it satisfies the following:*

For all $\mathsf{X} \in \mathcal{X}, \mathsf{Y} \in \mathcal{Y}$ and for all $i \in [D]$, there exists at most one $j \in [\ell]$ such that $\mathsf{X}_{i,j} = \mathsf{Y}_{i,j}$, where $\mathsf{X}_{i,j}$ and $\mathsf{Y}_{i,j}$ denote the (i,j)-th element of X and Y respectively.

Definition 6 (MultD-Eq Predicates). *Let p, D, ℓ be positive integers and let $(\mathcal{X}, \mathcal{Y}) \subseteq \mathbb{Z}_p^{D \times \ell} \times \mathbb{Z}_p^{D \times \ell}$ be any compatible domains for MultD-Eq. Then, for all $\mathsf{Y} \in \mathcal{Y}$, the multi-dimensional equality predicate $\mathsf{MultD\text{-}Eq_Y} : \mathcal{X} \rightarrow \{0, 1\}$ is defined as follows:*

---

[10] This predicate is presented in the works of [GMW15] as the AND-OR-EQ predicate satisfying the so called "at most one" promise. The conceptual differences between their formalization and ours is that, they view predicates as functions on both variables X and Y, whereas we view only X as a variable and treat Y as a constant. (Compare [GMW15] Sect. 3.1 and our Definition 6).

$$\text{MultD-Eq}_Y(X) = \begin{cases} 1 & if \quad \forall i \in [D], \ \exists unique \ j \in [\ell] \ suchthat \ X_{i,j} = Y_{i,j} \\ 0 & otherwise \end{cases},$$

where $X_{i,j}$ and $Y_{i,j}$ denote the (i,j)-th element of X and Y respectively.

Note that $\text{MultD-Eq}_Y(X)$ is satisfied only if for each $i \in [D]$, there exists exactly one $j \in [\ell]$ such that $X_{i,j} = Y_{i,j}$. Furthermore, since we restrict (X,Y) to be over the compatible domains $(\mathcal{X}, \mathcal{Y})$ for MultD-Eq, for all $i \in [D]$ we will never have $X_{i,j} = Y_{i,j}$ and $X_{i,j'} = Y_{i,j'}$ for distinct $j, j' \in [\ell]$. This restriction may appear contrived and inflexible at first, however, this proves to be very useful for constructing predicate encoding schemes with nice qualities, and in fact does not seem to lose much generality in light of expressiveness of the predicate. In particular, by appropriately instantiating the compatible domains, we can embed many useful predicates into the MultD-Eq predicate. Further discussions are given in Sect. 4.3.

We now present two types of predicate encoding schemes for the MultD-Eq predicate. The correctness of the two schemes are provided in the full version.

Functionality Preserving Encoding Scheme PES$_{\text{FP}}$. Our first predicate encoding scheme preserves the functionality of the multi-dimensional equality predicate and can be viewed as an efficient polynomial representation of the circuit computing MultD-Eq_Y. This encoding scheme will be used for our VRF construction in Sect. 5.

Lemma 2. *Let $q = q(\lambda), p = p(\lambda), D = D(\lambda), \ell = \ell(\lambda)$ be positive integers and let $(\mathcal{X}, \mathcal{Y}) \subseteq \mathbb{Z}_p^{D \times \ell} \times \mathbb{Z}_p^{D \times \ell}$ be any compatible domains for the MultD-Eq predicate. Further, let $\mathcal{P} = \{\text{MultD-Eq}_Y : \mathcal{X} \to \{0,1\} \mid Y \in \mathcal{Y}\}$ be a set of MultD-Eq predicates. Then the following algorithms $\text{PES}_{\text{FP}} = (\text{EncInpt}_{\text{FP}}, \text{EncPred}_{\text{FP}})$ is a predicate encoding scheme over the ring $\mathbb{Z}_q$ with degree $d = D\zeta$ where $\zeta = \lfloor \log p \rfloor + 1$:*

- $\text{EncInpt}_{\text{FP}}(X) \to \hat{X}$: *It takes as input $X \in \mathcal{X}$, and outputs an encoding $\hat{X} \in \{0,1\}^{D\ell\zeta}$ as follows:*

$$\hat{X} = (X_{i,j,k})_{(i,j,k) \in [D] \times [\ell] \times [\zeta]},$$

where $X_{i,j,k}$ is the k-th bit of the binary representation of the (i,j)-th element of X. Here, the output tuple $(X_{i,j,k})$ is sorted in the lexicographical order.
- $\text{EncPred}_{\text{FP}}(\text{MultD-Eq}_Y) \to \hat{C}_Y$: *It takes as input a predicate $\text{MultD-Eq}_Y \in \mathcal{P}$, and outputs the following polynomial representation of an arithmetic circuit $\hat{C}_Y : \mathbb{Z}_q^{D\ell\zeta} \to \mathbb{Z}_q$:*

$$\hat{C}_Y(\hat{X}) = \prod_{i=1}^{D} \sum_{j=1}^{\ell} \prod_{k=1}^{\zeta} \left((1 - \hat{Y}_{i,j,k}) + (-1 + 2\hat{Y}_{i,j,k}) \cdot \hat{X}_{i,j,k} \right),$$

where $\hat{X}, \hat{Y} \in \{0,1\}^{D\ell\zeta}$ are encodings of X, Y respectively.

The correctness of $\mathsf{PES_{FP}}$ *holds for the two disjoint subsets* $S_0 = \{0\}$, $S_1 = \{1\} \subset \mathbb{Z}_q$.

Linear Encoding Scheme. $\mathsf{PES_{Lin}}$. Our second construction is a linear predicate encoding scheme. It achieves linearity by increasing the length of the encoded input $\hat{\mathsf{X}}$ and takes advantage of the fact that we can change the functionality of the encoded arithmetic circuit $\hat{C}$; the output of $\hat{C}$ can be values other than 0 or 1, whereas outputs of predicates are defined to be in $\{0, 1\}$. This encoding scheme will be used for our lattice-based PE scheme for the MultD-Eq predicate in Sect. 6.

Lemma 3. *Let* $q = q(\lambda), p = p(\lambda), D = D(\lambda), \ell = \ell(\lambda)$ *be positive integers such that* $q > D$ *and let* $(\mathcal{X}, \mathcal{Y}) \subseteq \mathbb{Z}_p^{D \times \ell} \times \mathbb{Z}_p^{D \times \ell}$ *be any compatible domains for the* MultD-Eq *predicate. Further, let* $\mathcal{P} = \{\mathsf{MultD\text{-}Eq_Y} : \mathcal{X} \to \{0, 1\} \mid \mathsf{Y} \in \mathcal{Y}\}$ *be a set of* MultD-Eq *predicates. Then the following algorithms* $\mathsf{PES_{Lin}} = (\mathsf{EncInpt_{Lin}}, \mathsf{EncPred_{Lin}})$ *is a predicate encoding scheme over the ring* $\mathbb{Z}_q$ *with degree* $d = 1$, *i.e., a linear scheme, where we set* $L = 2^\varsigma$ *and* $\varsigma = \lfloor \log p \rfloor + 1$ *below.*

- $\mathsf{EncInpt_{Lin}}(\mathsf{X}) \to \hat{\mathsf{X}}$: *It takes as input* $\mathsf{X} \in \mathcal{X}$, *and outputs an encoding* $\hat{\mathsf{X}} \in \{0, 1\}^{D\ell L}$ *defined as follows:*

$$\hat{\mathsf{X}} = \left(\prod_{k=1}^{\varsigma} (\mathsf{X}_{i,j,k})^{w_k} \right)_{(i,j,w) \in [D] \times [\ell] \times [L]},$$

  where w_k *and* $\mathsf{X}_{i,j,k}$ *is the* k-*th bit of the binary representation of* $w - 1$[11] *and the* (i, j)-*th element of* X, *respectively. In case* $\mathsf{X}_{i,j,k} = w_k = 0$, *we define* $(\mathsf{X}_{i,j,k})^{w_k}$ *to be 1.*
- $\mathsf{EncPred_{Lin}}(\mathsf{MultD\text{-}Eq_Y}) \to \hat{C}_\mathsf{Y}$: *It takes as input a predicate* $\mathsf{MultD\text{-}Eq_Y} \in \mathcal{P}$, *and outputs the following polynomial representation of an arithmetic circuit* $\hat{C}_\mathsf{Y} : \mathbb{Z}_q^{D\ell L} \to \mathbb{Z}_q$:

$$\hat{C}_\mathsf{Y}(\hat{\mathsf{X}}) = D - \sum_{i=1}^{D} \sum_{j=1}^{\ell} \sum_{w=1}^{L} a_{i,j,w} \cdot \hat{\mathsf{X}}_{i,j,w},$$

  where $a_{i,j,w} \in \{-1, 0, 1\} \subset \mathbb{Z}_q$ *is the coefficient for the term* $\hat{\mathsf{X}}_{i,j,w} = \prod_{k=1}^{\varsigma} (\mathsf{X}_{i,j,k})^{w_k}$ *of the polynomial*

$$\prod_{k=1}^{\varsigma} \left((1 - \mathsf{Y}_{i,j,k}) + (-1 + 2\mathsf{Y}_{i,j,k}) \cdot \mathsf{X}_{i,j,k} \right).$$

  Here we treat Y *as a constant.*

---

[11] This inconvenient notion is due to the fact that the bit length of p and L may differ by one in case $p = 2^n - 1$ for $n \in \mathbb{N}$.

The correctness of $\mathsf{PES_{Lin}}$ holds for the two disjoint subsets $S_0 = \{1, \cdots, D\}, S_1 = \{0\} \subset \mathbb{Z}_q$.

Remark 1. In some applications, the compatible domains $(\mathcal{X}, \mathcal{Y})$ for MultD-Eq will have some additional structures that we can exploit to obtain more efficient encoding schemes. For example, in some case for all $\mathsf{X} \in \mathcal{X}$, all of the rows of X will be equal, i.e., $\mathsf{X}_i = \mathsf{X}_{i'}$ for all $i, i' \in [D]$ where X_i denotes the i-th row of X. In this case, we can reduce the output length of EncInpt by a factor of D by discarding the redundant terms.

4.3 Expressiveness of Multi-dimensional Equality Predicates

Here we comment on the expressiveness of the multi-dimensional equality predicates MultD-Eq. Notably, many predicates that come up in cryptography (e.g., bit-fixing, subset conjunction, range conjunction predicates) can be expressed as the multi-dimensional equality predicate instantiated with appropriate compatible domains $(\mathcal{X}, \mathcal{Y})$. Combining this with the result of the previous section, we obtain a functionality preserving ($\mathsf{PES_{FP}}$) or a linear ($\mathsf{PES_{Lin}}$) encoding scheme for all those predicates. We provide a thorough discussion in the full version.

5 Verifiable Random Functions

Modified Admissible Hash Functions. In this work, we use the *modified admissible hash function* of [Yam17] to prove security of our VRF. This allows us to use the same techniques employed by admissible hash functions, while providing for a more compact representation. The following is obtained by the results of [Jag15, Yam17].

Definition 7 *(Modified Admissible Hash Function). Let $n = n(\lambda), \ell = \ell(\lambda)$ and $\eta = \eta(\lambda)$ be an integer-valued function of λ such that $n, \ell = \Theta(\lambda)$ and $\eta = \omega(\log \lambda)$, and $\{C_n : \{0,1\}^n \to \{0,1\}^\ell\}_{n \in \mathbb{N}}$ be a family of error correcting codes with minimal distance $c \cdot \ell$ for a constant $c \in (0, 1/2)$. Let*

$$\mathcal{K}_{\mathsf{MAH}} = \{\mathsf{T} \subseteq [2\ell] \mid |\mathsf{T}| < \eta\} \quad and \quad \mathcal{X}_{\mathsf{MAH}} = \{0,1\}^n.$$

Then, we define the modified admissible hash function $\mathsf{F_{MAH}} : \mathcal{K}_{\mathsf{MAH}} \times \mathcal{X}_{\mathsf{MAH}} \to \{0,1\}$ *as*

$$\mathsf{F_{MAH}}(\mathsf{T}, X) = \begin{cases} 0, & if \ \mathsf{T} \subseteq \mathsf{S}(X) \\ 1, & otherwise \end{cases} \quad where \quad \mathsf{S}(X) = \{2i - C(X)_i \mid i \in [\ell]\}.$$
(8)

In the above, $C(X)_i$ is the i-th bit of $C(X) \in \{0,1\}^\ell$.

We also need the notion of *partitioning functions* as introduced in [Yam17] to prove security of our VRF. Informally, there exists a PPT algorithm PrtSmp called the partitioning function that given some polynomial function $Q(\lambda)$ and a noticeable function $\epsilon_0(\lambda)$, outputs a set $\mathsf{T} \in \mathcal{K}_{\mathsf{MAH}}$ such that for all $X^*, \{X_i\}_{i=1}^Q \in \mathcal{X}_{\mathsf{MAH}}$ the probability of $\mathsf{F_{MAH}}(\mathsf{T}, X^*) = 0 \wedge \bigwedge_{i=1}^Q \mathsf{F_{MAH}}(\mathsf{T}, X^{(i)}) = 1$ is noticeable. The concrete definition can be found in [Yam17] or in the full version.

5.1   Construction

Below, $n, \ell, \eta, \mathsf{S}(\cdot)$ are the parameters and function specified by the modified admissible hash function and ζ is set as $\lfloor \log p \rfloor + 1$. Note that $n, \ell = \Theta(\lambda)$ and $\eta = \omega(\log \lambda)$.

$\mathsf{Gen}(1^\lambda)$: On input 1^λ, it runs $\Pi \leftarrow \mathsf{GrpGen}(1^\lambda)$ to obtain a group description. It then chooses random generators $g, h \leftarrow \mathbb{G}^*$ and $w_0, w_{i,k} \leftarrow \mathbb{Z}_p$ for $(i, k) \in [\eta] \times [\zeta]$. Finally, it outputs

$$\mathsf{vk} = \left(\Pi, g, h, g_0 = g^{w_0}, (g_{i,k} = g^{w_{i,k}})_{(i,k) \in [\eta] \times [\zeta]} \right),$$

$$\mathsf{sk} = \left(w_0, (w_{i,k})_{(i,k) \in [\eta] \times [\zeta]} \right).$$

$\mathsf{Eval}(\mathsf{sk}, X)$: On input $X \in \{0, 1\}^n$, it first computes $\mathsf{S}(X) = \{s_1, \cdots, s_\ell\} \in [2\ell]$. In the following, let $s_{j,k}$ be the k-th bit of the binary representation of s_j, where $k \in [\zeta]$. It then computes

$$\begin{cases} \theta_{i'} = \prod_{i=1}^{i'} \sum_{j=1}^{\ell} \prod_{k=1}^{\zeta} \left((1 - s_{j,k}) + (-1 + 2s_{j,k}) \cdot w_{i,k} \right) \\ \theta_{i,j,k'} = \prod_{k=1}^{k'} \left((1 - s_{j,k}) + (-1 + 2s_{j,k}) \cdot w_{i,k} \right) \end{cases},$$

for $i' \in [\eta]$ and $(i, j, k') \in [\eta] \times [\ell] \times [\zeta]$, and defines $\theta := \theta_\eta$. Finally, it outputs

$$Y = e(g, h)^{\theta/w_0}$$

$$\pi = \left(\pi_0 := g^{\theta/w_0}, (\pi_{i'} := g^{\theta_{i'}})_{i' \in [\eta]}, (\pi_{i,j,k'} := g^{\theta_{i,j,k'}})_{(i,j,k') \in [\eta] \times [\ell] \times [\zeta]} \right).$$

$\mathsf{Verify}(\mathsf{vk}, X, (Y, \pi))$: First, it checks the validity of vk. It outputs 0 if any of the following properties are not satisfied.

1. vk is of the form $\left(\Pi, g, h, g_0, (g_{i,k})_{(i,k) \in [\eta] \times [\zeta]} \right)$.
2. $\mathsf{GrpVfy}(\Pi) = 1$ and $\mathsf{GrpVfy}(\Pi, s) = 1$ for all $s \in (g, h, g_0) \cup (g_{i,k})_{(i,k) \in [\eta] \times [\zeta]}$. Then, it checks the validity of X, Y and π. In doing so, it first prepares the terms $\Phi_{i'}, \bar{g}_{i,j,k'}$ for all $i' \in [\eta], (i, j, k') \in [\eta] \times [\ell] \times [\zeta]$ defined as

$$\Phi_{i'} := \prod_{j=1}^{\ell} \pi_{i',j,\zeta}, \quad \text{and} \quad \bar{g}_{i,j,k'} := g^{1 - s_{j,k'}} \cdot (g_{i,k'})^{-1 + 2s_{j,k'}}.$$

It outputs 0 if any of the following properties are not satisfied.
3. $X \in \{0,1\}^n, Y \in \mathbb{G}_T, \pi$ is of the above form
4. It holds that for all $i' \in [\eta - 1]$ and $(i, j, k') \in [\eta] \times [\ell] \times [\zeta - 1]$,

$$e(\pi_1, g) = e(\Phi_1, g), \qquad\qquad e(\pi_{i,j,1}, g) = e(\bar{g}_{i,j,1}, g),$$
$$e(\pi_{i'+1}, g) = e(\Phi_{i'+1}, \pi_{i'}), \qquad e(\pi_{i,j,k'+1}, g) = e(\bar{g}_{i,j,k'+1}, \pi_{i,j,k'}).$$

5. It holds that $e(\pi_\eta, g) = e(\pi_0, g_0)$ and $e(\pi_0, h) = Y$.

If all the above checks are passed, it outputs 1.

5.2   Correctness, Unique Provability, and Pseudorandomness

Correctness and unique provability for the above scheme can be shown by simple calculation. The proof is provided in the full version. The following theorem addresses the pseudorandomness of the scheme.

Theorem 1 *(Pseudorandomness).*   *Our scheme satisfies pseudorandomness assuming L-DDH with $L = \eta\zeta = \omega(\log^2 \lambda)$.*

Proof. Let $\mathcal{A}$ be a PPT adversary that breaks the pseudorandomness of the scheme with non-negligible advantage. Let $\epsilon = \epsilon(\lambda)$ be its advantage and $Q = Q(\lambda)$ be the upper bound on the number of evaluation queries it makes. Here, since $\mathcal{A}$ is a valid adversary, Q is a polynomially bounded function and there exists a noticeable function $\epsilon_0 = \epsilon_0(\lambda)$ such that $\epsilon(\lambda) \geq \epsilon_0(\lambda)$ holds for infinitely many λ. Then, by the definition of partitioning functions for the admissible hash function, if we run $\mathsf{T} \leftarrow \mathsf{PrtSmp}_{\mathsf{MAH}}(1^\lambda, Q(\lambda), \epsilon_0(\lambda))$, we have $\mathsf{T} \subseteq [2\ell]$ and $|\mathsf{T}| < \eta$ with probability 1 for all sufficiently large λ. Therefore, in the following, we assume this condition always holds. We show security of the scheme through a sequence of games. In each game, a value $\mathsf{coin}' \in \{0, 1\}$ is defined. While it is set $\mathsf{coin}' = \widehat{\mathsf{coin}}$ in the first game, these values may be different in the later games. In the following we define E_i to be the event that $\mathsf{coin}' = \mathsf{coin}$ in Game_i.

Game_0: This is the actual security game. Since $\mathcal{Y} = \mathbb{G}_T$, when $\mathsf{coin} = 1$, a random element $Y_1^* \leftarrow \mathbb{G}_T$ is returned to $\mathcal{A}$ as the challenge query. At the end of the game, $\mathcal{A}$ outputs a guess $\widehat{\mathsf{coin}}$ for coin. Finally, the challenger sets $\mathsf{coin}' = \widehat{\mathsf{coin}}$. By assumption on the adversary $\mathcal{A}$, we have $\left|\Pr[\mathsf{E}_0] - \frac{1}{2}\right| = \left|\Pr[\mathsf{coin}' = \mathsf{coin}] - \frac{1}{2}\right| = \left|\Pr[\widehat{\mathsf{coin}} = \mathsf{coin}] - \frac{1}{2}\right| = \epsilon$.

Game_1: In this game, we change Game_0 so that the challenger performs an additional step at the end of the game. Namely, the challenger first runs the partitioning function $\mathsf{T} \leftarrow \mathsf{PrtSmp}_{\mathsf{MAH}}(1^\lambda, Q(\lambda), \epsilon_0(\lambda))$. As noted earlier, we have $|\mathsf{T}| \subseteq [2\ell]$ and $|\mathsf{T}| < \eta$. Then, it checks whether the following condition holds:

$$\mathsf{F}_{\mathsf{MAH}}(\mathsf{T}, X^{(1)}) = 1 \ \wedge \ \cdots = \ \wedge \ \mathsf{F}_{\mathsf{MAH}}(\mathsf{T}, X^{(Q)}) = 1 \ \wedge \ \mathsf{F}_{\mathsf{MAH}}(\mathsf{T}, X^*) = 0$$
$$\Longleftrightarrow \ \left(\mathsf{T} \not\subseteq \mathsf{S}(X^{(1)})\right) \wedge \cdots \wedge \left(\mathsf{T} \not\subseteq \mathsf{S}(X^{(Q)})\right) \wedge \left(\mathsf{T} \subseteq \mathsf{S}(X^*)\right) \qquad (9)$$

where X^* is the challenge input and $\{X^{(i)}\}_{i\in[Q]}$ are the inputs for which $\mathcal{A}$ has queried the evaluation of the function. If it does not hold, the challenger ignores the output $\widehat{\mathsf{coin}}$ of $\mathcal{A}$ and sets $\mathsf{coin}' \leftarrow \{0, 1\}$. In this case, we say that the challenger aborts. If condition (9) holds, the challenger sets $\mathsf{coin}' = \widehat{\mathsf{coin}}$. By the property of the partitioning function we have $|\Pr[\mathsf{E}_1] - 1/2| \geq \tau$ for infinitely many λ, where $\tau = \tau(\lambda)$ is a noticeable function. See the full version for a formal treatment concerning the partitioning function.

Game_2 : In this game, we change the way $w_0, (w_{i,k})_{(i,k)\in[\eta]\times[\zeta]}$ are chosen. First, at the beginning of the game, the challenger picks $\mathsf{T} \leftarrow \mathsf{PrtSmp}_{\mathsf{MAH}}(1^\lambda, Q(\lambda),$

$\epsilon_0(\lambda))$ and parses it as $\mathsf{T} = \{t_1, \cdots, t_{\eta'}\} \subset [2\ell]$. Note that changing the time on which the adversary runs the algorithm is only conceptual. Now, recalling that by our assumption $\eta' < \eta$, it sets $t_i = 0$ for $i \in [\eta'+1, \eta]$. Next, it samples $\alpha \leftarrow \mathbb{Z}_p^*$ and $\tilde{w}_0, \tilde{w}_{i,k} \leftarrow \mathbb{Z}_p$ for $(i,k) \in [\eta] \times [\zeta]$. Finally, the challenger sets

$$w_0 = \tilde{w}_0 \cdot \alpha, \quad w_{i,k} = \tilde{w}_{i,k} \cdot \alpha + t_{i,k} \quad \text{for} \quad (i,k) \in [\eta] \times [\zeta], \qquad (10)$$

where $t_{i,k}$ is the k-th bit of the binary representation of t_i. The rest of the game is identical to Game_1. Here, the statistical distance of the distributions of $w_0, (w_{i,k})_{(i,k)\in[\eta]\times[\zeta]}$ in Game_1 and Game_2 is at most $(\eta\zeta + 1)/p$, which is negligible. Therefore, we have $|\Pr[\mathsf{E}_1] - \Pr[\mathsf{E}_2]| = \mathsf{negl}(\lambda)$.

Before, getting into Game_3, we introduce polynomials (associated with each input X) that implicitly embeds the information on the partitioning function $\mathsf{F}_{\mathsf{MAH}}(\mathsf{T}, X)$, i.e., the form of the polynomials depend on whether $\mathsf{T} \subseteq \mathsf{S}(X)$ or not. For any $\mathsf{T} \subseteq [2\ell]$ with $|\mathsf{T}| = \eta' < \eta$ and $X \in \{0,1\}^n$ (i.e., for any $\mathsf{S}(X)$), we define the polynomial $\mathsf{P}_{\mathsf{T}\subseteq\mathsf{S}(X)}(\mathsf{Z}) : \mathbb{Z}_p \to \mathbb{Z}_p$ as

$$\mathsf{P}_{\mathsf{T}\subseteq\mathsf{S}(X)}(\mathsf{Z}) = \prod_{i=1}^{\eta}\sum_{j=1}^{\ell}\prod_{k=1}^{\zeta}\left((1 - s_{j,k}) + (-1 + 2s_{j,k}) \cdot (\tilde{w}_{i,k}\mathsf{Z} + t_{i,k})\right), \qquad (11)$$

where $\{s_{j,k}\}_{(j,k)\in[\ell]\times[\zeta]}$ and $\{t_{i,k}\}_{(i,k)\in[\eta]\times[\zeta]}$ are defined as in Game_2. Note that $\mathsf{P}_{\mathsf{T}\subseteq\mathsf{S}(X)}(\alpha) = \theta$. Our security proof is built upon the following lemma on the partitioning function.

Lemma 4. *There exists* $\mathsf{R}_{\mathsf{T}\subseteq\mathsf{S}(X)}(\mathsf{Z}) : \mathbb{Z}_p \to \mathbb{Z}_p$ *such that*

$$\mathsf{P}_{\mathsf{T}\subseteq\mathsf{S}(X)}(\mathsf{Z}) = \begin{cases} 1 + \mathsf{Z} \cdot \mathsf{R}_{\mathsf{T}\subseteq\mathsf{S}(X)}(\mathsf{Z}), & \text{if } \mathsf{F}_{\mathsf{MAH}}(\mathsf{T}, X) = 0 \\ \mathsf{Z} \cdot \mathsf{R}_{\mathsf{T}\subseteq\mathsf{S}(X)}(\mathsf{Z}), & \text{if } \mathsf{F}_{\mathsf{MAH}}(\mathsf{T}, X) = 1 \end{cases}.$$

In other words, $\mathsf{P}_{\mathsf{T}\subseteq\mathsf{S}(X)}(\mathsf{Z})$ *is not divisible by* Z *if and only if* $\mathsf{T} \subseteq \mathsf{S}(X)$.

This can be checked by the property of the functionality preserving encoding scheme $\mathsf{PES}_{\mathsf{FP}}$ scheme. We omit the proof of this lemma to the full version. With an abuse of notation, for all $X \in \{0,1\}^n$, we define the following polynomials that map $\mathbb{Z}_p$ to $\mathbb{Z}_p$, which are defined analogously to the values computed during the Eval algorithm:

$$\begin{cases} \theta_{i'}^X(\mathsf{Z}) = \prod_{i=1}^{i'}\sum_{j=1}^{\ell}\prod_{k=1}^{\zeta}\left((1 - s_{j,k}) + (-1 + 2s_{j,k})(\tilde{w}_{i,k}\mathsf{Z} + t_{i,k})\right) \\ \theta_{i,j,k'}^X(\mathsf{Z}) = \prod_{k=1}^{k'}\left((1 - s_{j,k}) + (-1 + 2s_{j,k})(\tilde{w}_{i,k}\mathsf{Z} + t_{i,k})\right) \end{cases},$$

for $i' \in [\eta]$ and $(i, j, k') \in [\eta] \times [\ell] \times [\zeta]$, and define $\theta^X(\mathsf{Z}) := \theta_\eta^X(\mathsf{Z})$. Note that we have $\mathsf{P}_{\mathsf{T}\subseteq\mathsf{S}(X)}(\mathsf{Z}) = \theta^X(\mathsf{Z}), \theta_{i'} = \theta_{i'}^X(\alpha), \theta_{i,j,k'} = \theta_{i,j,k'}^X(\alpha)$, and $\theta = \theta^X(\alpha)$.

Game$_3$: Recall that in the previous game, the challenger aborts at the end of the game if condition (9) is not satisfied. In this game, we change the game so that the challenger aborts as soon as the abort condition becomes true. Since this is only a conceptual change, we have $\Pr[E_2] = \Pr[E_3]$.

Game$_4$: In this game, we change the way the evaluation queries are answered. When the adversary $\mathcal{A}$ queries an input X to be evaluated, it first checks whether $\mathsf{F_{MAH}}(T, X) = 1$, i.e., it checks if condition (9) is satisfied. If it does not hold, it aborts as in Game$_3$. Otherwise, it computes the polynomial $\mathsf{R}_{T \subseteq S(X)}(Z) \in \mathbb{Z}_p[Z]$ such that $\mathsf{P}_{T \subseteq S(X)}(Z) = Z \cdot \mathsf{R}_{T \subseteq S(X)}(Z)$, and returns

$$Y = e(g^{\mathsf{R}_{T \subseteq S(X)}(\alpha)/\tilde{w}_0}, h),$$

$$\pi = \left(\pi_0 = g^{\mathsf{R}_{T \subseteq S(X)}(\alpha)/\tilde{w}_0}, \right.$$

$$\left. \left(\pi_{i'} = g^{\theta_{i'}^X(\alpha)} \right)_{i' \in [\eta]}, \left(\pi_{i,j,k'} = g^{\theta_{i,j,k'}^X(\alpha)} \right)_{(i,j,k') \in [\eta] \times [\ell] \times [\zeta]} \right).$$

Note that existence of such a polynomial $\mathsf{P}_{T \subseteq S(X)}(Z)$ is guaranteed by Lemma 4. By the definition of $\theta_{i'}^X(Z)$ and $\theta_{i,j,k'}^X(Z)$, the components $\pi_{i'}$ and $\pi_{i,j,k'}$ are correctly generated. Furthermore, we have

$$\frac{\mathsf{R}_{T \subseteq S(X)}(\alpha)}{\tilde{w}_0} = \frac{\alpha \cdot \mathsf{R}_{T \subseteq S(X)}(\alpha)}{\alpha \cdot \tilde{w}_0} = \frac{\mathsf{P}_{T \subseteq S(X)}(\alpha)}{w_0} = \frac{\theta}{w_0}.$$

Therefore, Y and π_0 are also correctly generated, and the challenger simulates the evaluation queries perfectly. Hence, $\Pr[E_3] = \Pr[E_4]$.

Game$_5$: In this game, we change the way the challenge ciphertext is created when coin $= 0$. Recall in the previous games when coin $= 0$, we created a valid $Y_0^* = \mathsf{Eval}(\mathsf{sk}, X^*)$ as in the real scheme. If coin $= 0$ and $\mathsf{F_{MAH}}(X^*) = 0$ (i.e., if it does not abort), to create Y_0^*, the challenger first computes the polynomial $\mathsf{R}_{T \subseteq S(X^*)}(Z) \in \mathbb{Z}_p[X]$ such that $\mathsf{P}_{T \subseteq S(X^*)}(Z) = 1 + Z \cdot \mathsf{R}_{T \subseteq S(X^*)}(Z)$, whose existence is guaranteed by Lemma 4. It then sets,

$$Y_0^* = \left(e(g, h)^{1/\alpha} \cdot e(g, h)^{\mathsf{R}_{T \subseteq S(X^*)}(\alpha)} \right)^{1/\tilde{w}_0}$$

and returns it to $\mathcal{A}$. Here, the above term can be written equivalently as

$$\left(e(g, h)^{1/\alpha} \cdot e(g, h)^{\mathsf{R}_{T \subseteq S(X^*)}(\alpha)} \right)^{1/\tilde{w}_0} = e(g^{(1 + \alpha \mathsf{R}_{T \subseteq S(X^*)}(\alpha))/\alpha \tilde{w}_0}, h)$$

$$= e(g^{\mathsf{P}_{T \subseteq S(X^*)}(\alpha)/w_0}, h) = e(g^{\theta/w_0}, h).$$

Therefore, the view of the adversary in unchanged. Hence, $\Pr[E_4] = \Pr[E_5]$.

Game$_6$: In this game, we change the challenge value to be a random value in $\mathbb{G}_T$ regardless of whether coin $= 0$ of coin $= 1$. Namely, the challenger sets $Y^* \leftarrow \mathbb{G}_T$. We show in the full version that assuming $L\text{-}DDH$ is hard for $L = \eta\zeta$, we have $|\Pr[E_5] = \Pr[E_6]| = \mathsf{negl}(\lambda)$.

Analysis. From the above, we have $|\Pr[\mathsf{E}_6] - 1/2| = |\Pr[\mathsf{E}_1] - 1/2 + \sum_{i=1}^{5}(\Pr[\mathsf{E}_{i+1}] - \Pr[\mathsf{E}_i])| \geq |\Pr[\mathsf{E}_1] - 1/2| - \sum_{i=1}^{5}|\Pr[\mathsf{E}_{i+1}] - \Pr[\mathsf{E}_i]| \geq \tau(\lambda) - \mathsf{negl}(\lambda)$, for infinitely many λ. Since $\Pr[\mathsf{E}_6] = 1/2$, this implies $\tau(\lambda) \leq \mathsf{negl}(\lambda)$ for infinitely many λ, which is a contradiction.

5.3   Achieving Smaller Proof Size

In this section, we propose a variant of the VRF presented in Sect. 5.1 with a much shorter proof size. In particular, using the idea outlined in the technical overview, we obtain a VRF with proof size $|\pi| = \omega(\log \lambda)$ and verification key size $|\mathsf{vk}| = \omega(\sqrt{\lambda} \log \lambda)$.

Preparation. To make the presentation more clean, we define the notion of *power tuples*. We define *power tuples* $\mathcal{P}(W)$ for a tuple W, analogously to power sets. Namely, we create a tuple that contains all the subsequence of W in lexicographical order, i.e., $\mathcal{P}(W) = (w_1, w_2, w_3, w_1w_2, w_1w_3, w_2w_3, w_1w_2w_3)$ for $W = (w_1, w_2, w_3)$. Here, we do not consider the empty string as a subsequence of W. For a group element $g \in \mathbb{G}$ or $\mathbb{G}_T$ and a tuple W with elements in $\mathbb{Z}_p$, we denote $g^{\mathcal{P}(W)}$ as the tuple $(g^w \mid w \in \mathcal{P}(W))$. Furthermore, for tuples W, W' with elements in $\mathbb{Z}_p$ we define $e(g^{\mathcal{P}(W)}, g^{\mathcal{P}(W')})$ to be the tuple $(e(g,g)^{ww'} \mid w \in W, w' \in W')$. Assume all the tuples are sorted in the lexicographical order.

Construction. Below, we provide a VRF with small proof size.

$\mathsf{Gen}(1^\lambda)$: On input 1^λ, it runs $\Pi \leftarrow \mathsf{GrpGen}(1^\lambda)$ to obtain a group description. It then chooses random generators $g, h \leftarrow \mathbb{G}^*$, $w_0, w_{i,k} \leftarrow \mathbb{Z}_p$ for $(i, k) \in [\eta] \times [\zeta]$ and sets $L_i = (w_{i,k})_{k \in [\lfloor \zeta/2 \rfloor]}$ and $R_i = (w_{i,k})_{k \in [\lfloor \zeta/2 \rfloor + 1, \zeta]}$. Finally, it outputs

$$\mathsf{vk} = \left(\Pi, g, h, g_0 := g^{w_0}, (g^{\mathcal{P}(L_i)}, g^{\mathcal{P}(R_i)})_{i \in [\eta]}\right), \quad \mathsf{sk} = \left(w_0, (w_{i,k})_{(i,k) \in [\eta] \times [\zeta]}\right).$$

Note that we have $e(g^{\mathcal{P}(L_i)}, g^{\mathcal{P}(R_i)}) = e(g, g)^{\mathcal{P}(W_i)}$ where $W_i = (w_{i,k})_{k \in [\zeta]}$.

$\mathsf{Eval}(\mathsf{sk}, X)$: On input $X \in \{0,1\}^n$, it first computes $\mathsf{S}(X) = \{s_1, \cdots, s_\ell\} \in [2\ell]$. In the following, let $s_{j,k}$ be the k-th bit of the binary representation of s_j, where $k \in [\zeta]$. It then computes

$$\begin{cases} \theta_i = \sum_{j=1}^{\ell} \prod_{k=1}^{\zeta} \left((1 - s_{j,k}) + (-1 + 2s_{j,k}) \cdot w_{i,k}\right) \\ \theta_{[1:i']} = \prod_{i=1}^{i'} \sum_{j=1}^{\ell} \prod_{k=1}^{\zeta} \left((1 - s_{j,k}) + (-1 + 2s_{j,k}) \cdot w_{i,k}\right) \end{cases},$$

for $i \in [\eta], i' \in [2, \eta]$ and sets $\theta := \theta_{[1:\eta]}$. Note that we do not require $i' = 1$ since $\theta_1 = \theta_{[1:1]}$. Finally, it outputs

$$Y = e(g, h)^{\theta/w_0}, \quad \pi = \left(\pi_0 := g^{\theta/w_0}, (\pi_i := g^{\theta_i})_{i \in [\eta]}, (\pi_{[1:i']} := g^{\theta_{[1:i']}})_{i' \in [2,\eta]}\right).$$

Verify(vk, X, (Y, π)): First, it checks the validity of vk. It outputs 0 if any of the following properties are not satisfied.

1. vk is of the form $\left(\Pi, g, h, g_0, (g^{\mathcal{P}(L_i)}, g^{\mathcal{P}(R_i)})_{i \in [\eta]}\right)$.
2. $\mathsf{GrpVfy}(\Pi) = \mathsf{GrpVfy}(\Pi, s) = 1$ for all $s \in (g, h, g_0) \cup (g^{\mathcal{P}(L_i)}, g^{\mathcal{P}(R_i)})_{i \in [\eta]}$.

Then, it checks the validity of X, Y and π. In doing so, it first computes the coefficients $(\alpha_S)_{S \subseteq [\zeta]}$ of the multi-variate polynomial

$$p(\mathsf{Z}_1, \cdots, \mathsf{Z}_\zeta) = \sum_{j=1}^{\ell} \prod_{k=1}^{\zeta} \left((1 - s_{j,k}) + (-1 + 2s_{j,k}) \cdot \mathsf{Z}_k\right) = \sum_{S \subseteq [\zeta]} \alpha_S \prod_{k \in S} \mathsf{Z}_k.$$

Next, for all $i \in [\eta]$ and $S \subseteq [\zeta]$, it sets $L_S = S \cap [\lfloor \zeta/2 \rfloor]$ and $R_S = S \cap [\lfloor \zeta/2 \rfloor + 1, \zeta]$, and computes $\Phi_{i,S}$ as

$$\Phi_{i,S} = e(g^{\prod_{k \in L_S} w_{i,k}}, g^{\prod_{k \in R_S} w_{i,k}}).$$

Here, in case $L_S = \phi$ (resp. $R_S = \phi$), we define $\prod_{k \in L_S} w_{i,k}$ (resp. $\prod_{k \in R_S} w_{i,k}$) to be 1. Note that these values can be computed efficiently, since $g^{\mathcal{P}(L_i)}, g^{\mathcal{P}(R_i)}$ are given as part of the verification key. It outputs 0 if any of the following properties are not satisfied.

3. $X \in \{0, 1\}^n, Y \in \mathbb{G}_T, \pi$ is of the form $\pi = (\pi_0, (\pi_i)_{i \in [\eta]}, (\pi_{[1:i']})_{i' \in [2,\eta]})$.
4. It holds that for all $i \in [\eta]$ and $i' \in [3, \eta]$,

$$e(\pi_i, g) = \prod_{S \subseteq [\zeta]} \Phi_{i,S}^{\alpha_S}, \quad e(\pi_{[1:2]}, g) = e(\pi_1, \pi_2), \quad e(\pi_{[1:i']}, g) = e(\pi_{[1:i'-1]}, \pi_{i'}).$$

5. It holds that $e(\pi_{[1:\eta]}, g) = e(\pi_0, g_0)$ and $e(\pi_0, h) = Y$.

If all the above checks are passed, it outputs 1.

The correctness, unique provability and pseudorandomness of the above VRF can be proven in a similar manner to the VRF in Sect. 5.1. The proof is provided in the full version.

6 Predicate Encryption for MultD-Eq Predicates

In this section, we show how to construct a predicate encryption scheme for the multi-dimensional equality predicates MultD-Eq. This directly yields predicate encryption schemes for all the predicates presented in Sect. 4.3. Due to the symmetry of the MultD-Eq predicate and the compatible domains $(\mathcal{X}, \mathcal{Y})$, we obtain both key-policy and ciphertext-policy predicate encryption schemes.

6.1   Embedding Predicate Encoding Schemes into Matrices

The following definition gives a sufficient condition for constructing predicate encryption schemes. For discussions and comparisons with the related definition of [BGG+14] for attribute-based encryption schemes are given in the full version.

Definition 8. *We say the* deterministic *algorithms* $(\mathsf{Eval}_{\mathrm{pk}}, \mathsf{Eval}_{\mathrm{ct\text{-}priv}}, \mathsf{Eval}_{\mathrm{sim}})$ *are* $\alpha_{\mathcal{C}}$*-predicate encryption (PE)* enabling *for a family of arithmetic circuits* $\mathcal{C} = \{C : \mathbb{Z}_q^t \to \mathbb{Z}_q\}$ *if they are efficient and satisfy the following properties:*

- $\mathsf{Eval}_{\mathrm{pk}}\big(C \in \mathcal{C},\ \mathbf{B}_0, (\mathbf{B}_i)_{i \in [t]} \in \mathbb{Z}_q^{n \times m}\big) \to \mathbf{B}_C \in \mathbb{Z}_q^{n \times m}$
- $\mathsf{Eval}_{\mathrm{ct\text{-}priv}}\big(C \in \mathcal{C},\ \mathbf{c}_0, (\mathbf{c}_i)_{i \in [t]} \in \mathbb{Z}_q^n\big) \to \mathbf{c}_C \in \mathbb{Z}_q^m$
- $\mathsf{Eval}_{\mathrm{sim}}\big(C \in \mathcal{C},\ \mathbf{R}_0, (\mathbf{R}_i)_{i \in [t]} \in \mathbb{Z}^{m \times m}\big) \to \mathbf{R}_C \in \mathbb{Z}^{m \times m}$

We further require that the following holds:

1. $\mathsf{Eval}_{\mathrm{pk}}(C, (\mathbf{AR}_0 - \mathbf{G}), (\mathbf{AR}_i - x_i\mathbf{G})_{i \in [t]}) = \mathbf{A} \cdot \mathsf{Eval}_{\mathrm{sim}}(C, \mathbf{R}_0, (\mathbf{R}_i)_{i \in [t]}) - C(\mathbf{x})\mathbf{G}$ *for any* $\mathbf{x} = (x_1, \cdots, x_t) \in \{0,1\}^t$.
2. *If* $\mathbf{c}_0 = (\mathbf{B}_0 + \mathbf{G})^\top \mathbf{s} + \mathbf{z}_0$ *and* $\mathbf{c}_i = (\mathbf{B}_i + x_i\mathbf{G})^\top \mathbf{s} + \mathbf{z}_i$ *for some* $\mathbf{s} \in \mathbb{Z}_q^n$, *and* $\mathbf{z}_0, \mathbf{z}_i \leftarrow D_{\mathbb{Z}^m, \beta}, x_i \in \{0,1\}$ *for all* $i \in [t]$, *then* $\|\mathbf{c}_C - (\mathbf{B}_C + C(\mathbf{x})\mathbf{G})^\top \mathbf{s}\|_2 < \alpha_{\mathcal{C}} \cdot \beta\sqrt{m}$ *with all but negligible probability.*
3. *If* $\mathbf{R}_i \leftarrow \{-1,1\}^{m \times m}$ *for all* $i \in [0,t]$, *then* $s_1(\mathbf{R}_C) < \alpha_{\mathcal{C}}$ *with all but negligible probability.*

The linear predicate encoding scheme $\mathsf{PES}_{\mathsf{Lin}}$ for the $\mathsf{MultD\text{-}Eq}$ predicates (Sect. 4.2, Lemma 3) provides us with a family of arithmetic circuits $\hat{\mathcal{C}}$ that allows for $\alpha_{\hat{\mathcal{C}}}$-PE enabling algorithms $(\mathsf{Eval}_{\mathrm{pk}}, \mathsf{Eval}_{\mathrm{ct\text{-}priv}}, \mathsf{Eval}_{\mathrm{sim}})$. In particular we have the following lemma, which we provide the proof in the full version.

Lemma 5. *There exist* $\alpha_{\hat{\mathcal{C}}}$*-PE enabling algorithms for the family of arithmetic circuits* $\hat{\mathcal{C}}$ *defined by the predicate encoding scheme* $\mathsf{PES}_{\mathsf{Lin}}$ *for the* $\mathsf{MultD\text{-}Eq}$ *predicates defined over* $\mathbb{Z}_p^{D \times \ell}$, *where* $\alpha_{\hat{\mathcal{C}}} = C \cdot \max\{m\sqrt{m/n}, \sqrt{D\ell pm}\}$ *for some absolute constant* $C > 0$.

6.2   Construction

Given $\alpha_{\hat{\mathcal{C}}}$-PE enabling algorithms $(\mathsf{Eval}_{\mathrm{pk}}, \mathsf{Eval}_{\mathrm{ct\text{-}priv}}, \mathsf{Eval}_{\mathrm{sim}})$ for a family of arithmetic circuits defined by the predicate encoding scheme $\mathsf{PES}_{\mathsf{Lin}} = (\mathsf{EncInpt}_{\mathsf{Lin}}, \mathsf{EncPred}_{\mathsf{Lin}})$ for the $\mathsf{MultD\text{-}Eq}$ predicates with compatible domains $(\mathcal{X}, \mathcal{Y})$, we build a predicate encryption scheme for the same family of predicates.

Parameters. In the following, let n, m, q, p, D, ℓ be positive integers such that q is a prime and $q > D$, and let σ, α, α' be positive reals denoting the Gaussian parameters. Furthermore, let $(\mathcal{X}, \mathcal{Y}) \in \mathbb{Z}_p^{D \times \ell} \times \mathbb{Z}_p^{D \times \ell}$ be any compatible domains for the $\mathsf{MultD\text{-}Eq}$ predicates, let $\mathcal{P} = \{\mathsf{MultD\text{-}Eq}_{\mathsf{Y}} : \mathcal{X} \to \{0,1\} \mid \mathsf{Y} \in \mathcal{Y}\}$ be the set of multi-dimensional predicates and $\hat{\mathcal{C}} = \{\hat{C}_{\mathsf{Y}} \mid \hat{C}_{\mathsf{Y}} \leftarrow \mathsf{EncPred}$

$(\mathsf{MultD\text{-}Eq_Y}), \forall \mathsf{MultD\text{-}Eq_Y} \in \mathcal{P}\}$ be the set of polynomials representing the multidimensional predicates. Finally, let $\zeta = \lfloor \log p \rfloor + 1$ and $L = 2^\zeta$. Here, we assume that all of the parameters are a function of the security parameter $\lambda \in \mathbb{N}$. We provide a concrete parameter selection of the scheme in the full version. The following is our PE scheme.

$\mathsf{Setup}(1^\lambda)$: It first runs $(\mathbf{A}, \mathbf{T_A}) \leftarrow \mathsf{TrapGen}(1^n, 1^m, q)$ to obtain $\mathbf{A} \in \mathbb{Z}_q^{n \times m}$ and $\mathbf{T_A} \in \mathbb{Z}^{m \times m}$. It also picks $\mathbf{u} \leftarrow \mathbb{Z}_q^n$, $\mathbf{B}_0, \mathbf{B}_{i,j,w} \leftarrow \mathbb{Z}_q^{n \times m}$ for $(i, j, w) \in [D] \times [\ell] \times [L]$ and outputs

$$\mathsf{mpk} = \left(\mathbf{A}, \mathbf{B}_0, (\mathbf{B}_{i,j,w})_{(i,j,w) \in [D] \times [\ell] \times [L]}, \mathbf{u} \right) \quad \text{and} \quad \mathsf{msk} = \mathbf{T_A}.$$

$\mathsf{KeyGen}(\mathsf{mpk}, \mathsf{msk}, \mathsf{MultD\text{-}Eq_Y})$: Given a predicate $\mathsf{MultD\text{-}Eq_Y} \in \mathcal{P}$ for $\mathsf{Y} \in \mathbb{Z}_p^{D \times \ell}$ as input, it runs $\hat{C}_\mathsf{Y} \leftarrow \mathsf{EncPred}_{\mathsf{Lin}}(\mathsf{MultD\text{-}Eq_Y})$ and computes

$$\mathsf{Eval}_{\mathrm{pk}} \left(\hat{C}_\mathsf{Y}, \mathbf{B}_0, (\mathbf{B}_{i,j,w})_{(i,j,w) \in [D] \times [\ell] \times [L]} \right) \to \mathbf{B}_\mathsf{Y} \in \mathbb{Z}_q^{n \times m}.$$

Then, it runs $\mathsf{SampleLeft}(\mathbf{A}, \mathbf{B}_\mathsf{Y}, \mathbf{u}, \mathbf{T_A}, \sigma) \to \mathbf{e}$, where $[\mathbf{A}|\mathbf{B}_\mathsf{Y}]\mathbf{e} = \mathbf{u} \mod q$, and finally returns $\mathsf{sk}_\mathsf{Y} = \mathbf{e} \in \mathbb{Z}^{2m}$.

$\mathsf{Enc}(\mathsf{mpk}, \mathsf{X}, \mathsf{M})$: Given an attribute $\mathsf{X} \in \mathbb{Z}_p^{D \times \ell}$ as input, it first runs $\hat{\mathsf{X}} \leftarrow \mathsf{EncInpt}_{\mathsf{Lin}}(\mathsf{X})$ where $\hat{\mathsf{X}} \in \{0, 1\}^{D\ell L}$. Then it samples $\mathbf{s} \leftarrow \mathbb{Z}_q^n$, $z \leftarrow D_{\mathbb{Z}, \alpha q}$, $\mathbf{z}, \mathbf{z}_0, \mathbf{z}_{i,j,w} \leftarrow D_{\mathbb{Z}^m, \alpha' q}$ for $(i, j, w) \in [D] \times [\ell] \times [L]$, and computes

$$\mathbf{c}_\mathsf{X} = \begin{cases} c & = \mathbf{u}^\top \mathbf{s} + z + \mathsf{M} \cdot \lfloor q/2 \rceil, \\ \mathbf{c} & = \mathbf{A}^\top \mathbf{s} + \mathbf{z}, \\ \mathbf{c}_0 & = (\mathbf{B}_0 + \mathbf{G})^\top \mathbf{s} + \mathbf{z}_0, \\ \mathbf{c}_{i,j,w} & = (\mathbf{B}_{i,j,w} + \hat{\mathsf{X}}_{i,j,w} \mathbf{G})^\top \mathbf{s} + \mathbf{z}_{i,j,w} \quad \text{for} \quad (i, j, w) \in [D] \times [\ell] \times [L], \end{cases}$$

where $\hat{\mathsf{X}}_{i,j,w}$ is the (i, j, w)-th element of $\hat{\mathsf{X}}$. Finally, it returns the ciphertext $\mathbf{c}_\mathsf{X} \in \mathbb{Z}_q \times (\mathbb{Z}_q^m)^{D\ell L + 2}$.

$\mathsf{Dec}(\mathsf{mpk}, (\hat{C}_\mathsf{Y}, \mathsf{sk}_\mathsf{Y}), \mathbf{c}_\mathsf{X})$: To decrypt the ciphertext $\mathbf{c}_\mathsf{X} = (c, \mathbf{c}, \mathbf{c}_0, (\mathbf{c}_{i,j,w}))$ given a predicate and a secret key $(\hat{C}_\mathsf{Y}, \mathsf{sk}_\mathsf{Y})$, it computes

$$\mathsf{Eval}_{\mathrm{ct\text{-}priv}} \left(\hat{C}_\mathsf{Y}, \mathbf{c}_0, (\mathbf{c}_{i,j,w})_{(i,j,w) \in [D] \times [\ell] \times [L]} \right) \to \bar{\mathbf{c}} \in \mathbb{Z}_q^m.$$

Then using the secret key $\mathsf{sk}_\mathsf{Y} = \mathbf{e} \in \mathbb{Z}^{2m}$, it computes $d = c - [\mathbf{c}^\top | \bar{\mathbf{c}}^\top]^\top \mathbf{e} \in \mathbb{Z}_q$. Finally, it returns $|d - \lfloor q/2 \rceil| < q/4$ and 0 otherwise.

Correctness and Parameter Selection. We omit the correctness of our scheme and a candidate parameter selection to the full version. We note that we can chose the modulus size as small as $q = \sqrt{m} \cdot (\sqrt{D\ell p})^{-1} \cdot \alpha_{\hat{C}}^2 \cdot \omega(\log m)$. In particular, we can base security on the polynomial LWE assumption.

Security Proof. The following theorem addresses the security of the scheme.

Theorem 2. *Given PE enabling algorithms* $(\mathsf{Eval}_{pk}, \mathsf{Eval}_{ct\text{-}priv}, \mathsf{Eval}_{sim})$ *for the family of arithmetic circuits* $\hat{\mathcal{C}}$ *defined above, our predicate encryption scheme is selectively secure and weakly attribute hiding with respect to the* MultD-Eq *predicates, assuming the hardness of* $\mathsf{LWE}_{n,m+1,q,D_{\mathbb{Z},\alpha q}}$.

Acknowledgement. We would like to thank the anonymous reviewers of Asiacrypt 2016 for insightful comments. In particular, we are grateful for Takahiro Matsuda and Shota Yamada for precious comments on the earlier version of this work. We also thank Atsushi Takayasu, Jacob Schuldt and Nuttapong Attrapadung for helpful comments on the draft. The research was partially supported by JST CREST Grant Number JPMJCR1302 and JSPS KAKENHI Grant Number 17J05603.

References

[ABB10] Agrawal, S., Boneh, D., Boyen, X.: Efficient lattice (H)IBE in the standard model. In: Gilbert, H. (ed.) EUROCRYPT 2010. LNCS, vol. 6110, pp. 553–572. Springer, Heidelberg (2010). doi:10.1007/978-3-642-13190-5_28

[ACF09] Abdalla, M., Catalano, D., Fiore, D.: Verifiable random functions from identity-based key encapsulation. In: Joux, A. (ed.) EUROCRYPT 2009. LNCS, vol. 5479, pp. 554–571. Springer, Heidelberg (2009). doi:10.1007/978-3-642-01001-9_32

[ACF14] Abdalla, M., Catalano, D., Fiore, D.: Verifiable random functions: relations to identity-based key encapsulation and new constructions. J. Cryptol. **27**(3), 544–593 (2014)

[AFV11] Agrawal, S., Freeman, D.M., Vaikuntanathan, V.: Functional encryption for inner product predicates from learning with errors. In: Lee, D.H., Wang, X. (eds.) ASIACRYPT 2011. LNCS, vol. 7073, pp. 21–40. Springer, Heidelberg (2011). doi:10.1007/978-3-642-25385-0_2

[AIK04] Applebaum, B., Ishai, Y., Kushilevitz, E.: Cryptography in NC^0. In: FOCS, pp. 166–175 (2004)

[Att14] Attrapadung, N.: Dual system encryption via doubly selective security: framework, fully secure functional encryption for regular languages, and more. In: Nguyen, P.Q., Oswald, E. (eds.) EUROCRYPT 2014. LNCS, vol. 8441, pp. 557–577. Springer, Heidelberg (2014). doi:10.1007/978-3-642-55220-5_31

[Att16] Attrapadung, N.: Dual system encryption framework in prime-order groups via computational pair encodings. In: Cheon, J.H., Takagi, T. (eds.) ASIACRYPT 2016. LNCS, vol. 10032, pp. 591–623. Springer, Heidelberg (2016). doi:10.1007/978-3-662-53890-6_20

[BB04a] Boneh, D., Boyen, X.: Efficient selective-ID secure identity-based encryption without random oracles. In: Cachin, C., Camenisch, J.L. (eds.) EUROCRYPT 2004. LNCS, vol. 3027, pp. 223–238. Springer, Heidelberg (2004). doi:10.1007/978-3-540-24676-3_14

[BB04b] Boneh, D., Boyen, X.: Secure identity based encryption without random oracles. In: Franklin, M. (ed.) CRYPTO 2004. LNCS, vol. 3152, pp. 443–459. Springer, Heidelberg (2004). doi:10.1007/978-3-540-28628-8_27

[BGG+14]   Boneh, D., Gentry, C., Gorbunov, S., Halevi, S., Nikolaenko, V., Segev, G., Vaikuntanathan, V., Vinayagamurthy, D.: Fully key-homomorphic encryption, arithmetic circuit abe and compact garbled circuits. In: Nguyen, P.Q., Oswald, E. (eds.) EUROCRYPT 2014. LNCS, vol. 8441, pp. 533–556. Springer, Heidelberg (2014). doi:10.1007/978-3-642-55220-5_30

[BGJS17]   Badrinarayanan, S., Goyal, V., Jain, A., Sahai, A.: A note on VRFs from verifiable functional encryption. Cryptology ePrint Archive, Report 2017/051 (2017). https://eprint.iacr.org/2017/051.pdf

[Bit17]   Bitansky, N.: Verifiable random functions from non-interactive witness-indistinguishable proofs. Cryptology ePrint Archive, Report 2017/18 (2017). https://eprint.iacr.org/2017/018.pdf

[BLP+13]   Brakerski, Z., Langlois, A., Peikert, C., Regev, O., Stehlé, D.: Classical hardness of learning with errors. In: STOC, pp. 575–584 (2013)

[BMR10]   Boneh, D., Montgomery, H.W., Raghunathan, A.: Algebraic pseudorandom functions with improved efficiency from the augmented cascade. In: CCS, pp. 131–140. ACM (2010)

[BW07]   Boneh, D., Waters, B.: Conjunctive, subset, and range queries on encrypted data. In: Vadhan, S.P. (ed.) TCC 2007. LNCS, vol. 4392, pp. 535–554. Springer, Heidelberg (2007). doi:10.1007/978-3-540-70936-7_29

[CGW15]   Chen, J., Gay, R., Wee, H.: Improved dual system ABE in prime-order groups via predicate encodings. In: Oswald, E., Fischlin, M. (eds.) EUROCRYPT 2015. LNCS, vol. 9057, pp. 595–624. Springer, Heidelberg (2015). doi:10.1007/978-3-662-46803-6_20

[CHKP10]   Cash, D., Hofheinz, D., Kiltz, E., Peikert, C.: Bonsai trees, or how to delegate a lattice basis. In: Gilbert, H. (ed.) EUROCRYPT 2010. LNCS, vol. 6110, pp. 523–552. Springer, Heidelberg (2010). doi:10.1007/978-3-642-13190-5_27

[DY05]   Dodis, Y., Yampolskiy, A.: A verifiable random function with short proofs and keys. In: Vaudenay, S. (ed.) PKC 2005. LNCS, vol. 3386, pp. 416–431. Springer, Heidelberg (2005). doi:10.1007/978-3-540-30580-4_28

[GHKW17]   Goyal, R., Hohenberger, S., Koppula, V., Waters, B.: A generic approach to constructing and proving verifiable random functions. Cryptology ePrint Archive, Report 2017/021 (2017). https://eprint.iacr.org/2017/021.pdf

[GKW17]   Goyal, R., Koppula, V., Waters, B.: Lockable obfuscation. Cryptology ePrint Archive, Report 2017/274, to appear in FOCS 2017. https://eprint.iacr.org/2017/274.pdf

[GMW15]   Gay, R., Méaux, P., Wee, H.: Predicate encryption for multi-dimensional range queries from lattices. In: Katz, J. (ed.) PKC 2015. LNCS, vol. 9020, pp. 752–776. Springer, Heidelberg (2015). doi:10.1007/978-3-662-46447-2_34

[GPSW06]   Goyal, V., Pandey, O., Sahai, A., Waters, B.: Attribute-based encryption for fine-grained access control of encrypted data. In: CCS, pp. 89–98. ACM (2006)

[GPV08]   Gentry, C., Peikert, C., Vaikuntanathan, V.: Trapdoors for hard lattices and new cryptographic constructions. In: STOC, pp. 197–206. ACM (2008)

[GV15]   Gorbunov, S., Vinayagamurthy, D.: Riding on asymmetry: efficient ABE for branching programs. In: Iwata, T., Cheon, J.H. (eds.) ASIACRYPT 2015. LNCS, vol. 9452, pp. 550–574. Springer, Heidelberg (2015). doi:10.1007/978-3-662-48797-6_23

[GVW13]   Gorbunov, S., Vaikuntanathan, V., Wee, H.: Attribute-based encryption for circuits. In: STOC, pp. 545–554. ACM (2013)

[GVW15] Gorbunov, S., Vaikuntanathan, V., Wee, H.: Predicate encryption for circuits from LWE. In: Gennaro, R., Robshaw, M. (eds.) CRYPTO 2015. LNCS, vol. 9216, pp. 503–523. Springer, Heidelberg (2015). doi:10.1007/978-3-662-48000-7_25

[HJ16] Hofheinz, D., Jager, T.: Verifiable random functions from standard assumptions. In: Kushilevitz, E., Malkin, T. (eds.) TCC 2016. LNCS, vol. 9562, pp. 336–362. Springer, Heidelberg (2016). doi:10.1007/978-3-662-49096-9_14

[HK08] Hofheinz, D., Kiltz, E.: Programmable hash functions and their applications. In: Wagner, D. (ed.) CRYPTO 2008. LNCS, vol. 5157, pp. 21–38. Springer, Heidelberg (2008). doi:10.1007/978-3-540-85174-5_2

[HW10] Hohenberger, S., Waters, B.: Constructing verifiable random functions with large input spaces. In: Gilbert, H. (ed.) EUROCRYPT 2010. LNCS, vol. 6110, pp. 656–672. Springer, Heidelberg (2010). doi:10.1007/978-3-642-13190-5_33

[IK00] Ishai, Y., Kushilevitz, E.: Randomizing polynomials: a new representation with applications to round-efficient secure computation. In: FOCS (2000)

[Jag15] Jager, T.: Verifiable random functions from weaker assumptions. In: Dodis, Y., Nielsen, J.B. (eds.) TCC 2015. LNCS, vol. 9015, pp. 121–143. Springer, Heidelberg (2015). doi:10.1007/978-3-662-46497-7_5

[KSW08] Katz, J., Sahai, A., Waters, B.: Predicate encryption supporting disjunctions, polynomial equations, and inner products. In: Smart, N. (ed.) EUROCRYPT 2008. LNCS, vol. 4965, pp. 146–162. Springer, Heidelberg (2008). doi:10.1007/978-3-540-78967-3_9

[KY16] Katsumata, S., Yamada, S.: Partitioning via non-linear polynomial functions: more compact IBEs from ideal lattices and bilinear maps. In: Cheon, J.H., Takagi, T. (eds.) ASIACRYPT 2016. LNCS, vol. 10032, pp. 682–712. Springer, Heidelberg (2016). doi:10.1007/978-3-662-53890-6_23

[Lys02] Lysyanskaya, A.: Unique signatures and verifiable random functions from the DH-DDH separation. In: Yung, M. (ed.) CRYPTO 2002. LNCS, vol. 2442, pp. 597–612. Springer, Heidelberg (2002). doi:10.1007/3-540-45708-9_38

[MP12] Micciancio, D., Peikert, C.: Trapdoors for lattices: simpler, tighter, faster, smaller. In: Pointcheval, D., Johansson, T. (eds.) EUROCRYPT 2012. LNCS, vol. 7237, pp. 700–718. Springer, Heidelberg (2012). doi:10.1007/978-3-642-29011-4_41

[MRV99] Micali, S., Rabin, M., Vadhan, S.: Verifiable random functions. In: FOCS, pp. 120–130. IEEE (1999)

[Pei09] Peikert, C.: Public-key cryptosystems from the worst-case shortest vector problem. In: STOC, pp. 333–342. ACM (2009)

[Reg05] Regev, O.: On lattices, learning with errors, random linear codes, and cryptography. In: STOC, pp. 84–93. ACM Press (2005)

[SBC+07] Shi, E., Bethencourt, J., Chan, T.H.H., Song, D., Perrig, A.: Multi-dimensional range query over encrypted data. In: S&P, pp. 350–364. IEEE (2007)

[SW05] Sahai, A., Waters, B.: Fuzzy identity-based encryption. In: Cramer, R. (ed.) EUROCRYPT 2005. LNCS, vol. 3494, pp. 457–473. Springer, Heidelberg (2005). doi:10.1007/11426639_27

[Wee14] Wee, H.: Dual system encryption via predicate encodings. In: Lindell, Y. (ed.) TCC 2014. LNCS, vol. 8349, pp. 616–637. Springer, Heidelberg (2014). doi:10.1007/978-3-642-54242-8_26

[WZ17]   Wichs, D., Zirdelis, G.: Obfuscating compute-and-compare programs under LWE. Cryptology ePrint Archive, Report 2017/276, to appear in FOCS 2017. http://eprint.iacr.org/2017/276

[Yam17] Yamada, S.: Asymptotically compact adaptively secure lattice IBEs and verifiable random functions via generalized partitioning techniques. In: Katz, J., Shacham, H. (eds.) CRYPTO 2017. LNCS, vol. 10403, pp. 161–193. Springer, Cham (2017). doi:10.1007/978-3-319-63697-9_6

[ZCZ16] Zhang, J., Chen, Y., Zhang, Z.: Programmable hash functions from lattices: short signatures and IBEs with small key sizes. In: Robshaw, M., Katz, J. (eds.) CRYPTO 2016. LNCS, vol. 9816, pp. 303–332. Springer, Heidelberg (2016). doi:10.1007/978-3-662-53015-3_11

The Minimum Number of Cards in Practical Card-Based Protocols

Julia Kastner[1], Alexander Koch[1(✉)], Stefan Walzer[2], Daiki Miyahara[3],
Yu-ichi Hayashi[4], Takaaki Mizuki[3], and Hideaki Sone[3]

[1] Karlsruhe Institute of Technology (KIT), Karlsruhe, Germany
julia.kastner@student.kit.edu, alexander.koch@kit.edu
[2] Technische Universität Ilmenau, Ilmenau, Germany
stefan.walzer@tu-ilmenau.de
[3] Tohoku University, Sendai, Japan
daiki.miyahara.q4@dc.tohoku.ac.jp,
tm-paper+cardcopy@g-mail.tohoku-university.jp
[4] Nara Institute of Science and Technology, Ikoma, Japan

Abstract. The elegant "five-card trick" of den Boer (EUROCRYPT 1989) allows two players to securely compute a logical AND of two private bits, using five playing cards of symbols $\heartsuit$ and $\clubsuit$. Since then, card-based protocols have been successfully put to use in classroom environments, vividly illustrating secure multiparty computation – and evoked research on the minimum number of cards needed for several functionalities.

Securely computing arbitrary circuits needs protocols for negation, AND and bit copy in committed-format, where outputs are commitments again. Negation just swaps the bit's cards, computing AND and copying a bit n times can be done with six and $2n + 2$ cards, respectively, using the simple protocols of Mizuki and Sone (FAW 2009).

Koch et al. (ASIACRYPT 2015) showed that five cards suffice for computing AND in finite runtime, albeit using relatively complex and unpractical shuffle operations. In this paper, we show that if we restrict shuffling to closed permutation sets, the six-card protocol is optimal in the finite-runtime setting. If we additionally assume a uniform distribution on the permutations in a shuffle, we show that restart-free four-card AND protocols are impossible. These shuffles are easy to perform even in an actively secure manner (Koch and Walzer, ePrint 2017).

For copying bit commitments, the protocol of Nishimura et al. (ePrint 2017) needs only $2n + 1$ cards, but performs a number of complex shuffling steps that is only finite in expectation. We show that it is impossible to go with less cards. If we require an a priori bound on the runtime, we show that the $(2n + 2)$-card protocol is card-minimal.

Keywords: Card-based protocols · Committed format · Boolean AND · COPY · Secure computation · Cryptography without computers

---

This article is the result of a merge. The main contribution by the first three authors is in Sects. 3 to 7, the one of the last four authors is in Sects. 8 and 9.

© International Association for Cryptologic Research 2017
T. Takagi and T. Peyrin (Eds.): ASIACRYPT 2017, Part III, LNCS 10626, pp. 126–155, 2017.
https://doi.org/10.1007/978-3-319-70700-6_5

1   Introduction

Card-based cryptography is best illustrated by example. Let us begin by giving a concise and graphical description of the six-card AND protocol of Mizuki and Sone [MS09]. This protocol enables two players, Alice and Bob, to compute the AND of their private bits. For instance, they may wish to determine whether they both like to watch a particular movie, without giving away their (possibly embarrassing) preference if there is no match. Using card-based cryptography, this is possible without computers – making the security tangible and eliminating the danger of malware.[1]

For this, we use a deck of six cards with indistinguishable backs and either ♡ or ♣ on the front. Each player is handed one card of each symbol and is asked to enter his or her bit by arranging the cards in one of two ways.

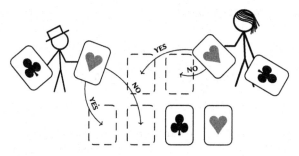

Bob (on the left) inputs "yes", by placing his ♡-card in the first position, and "no" by placing it in the second position; he places his ♣-card in the unused position. Alice encodes her input bit in a similar manner in the first row. We employ two additional cards, encoding "no" (♣♡) in the lower right part of the arrangement. Of course, as the players want their input bit to be secret, their cards are put face-down on the table, hence, making it impossible for the other party to observe the bit at this point. (The extra cards encoding "no" can be put publicly on the table and are turned face-down at the start of the protocol.)

This puts the protocol in one of the following (hidden) configurations:

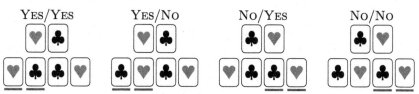

Observe that the *correct result* in the above encoding, (♣♡ = "no", and ♡♣ = "yes"), is *on the side of the heart* in the upper row. This stays invariant, if we split the cards in the middle of the arrangement and exchange both sides. This property is crucial for the protocol, as in the following we want to randomly exchange the two halves of the arrangement to obscure the input order of Alice's

---

[1] Imagine a setting where Alice asks Bob to enter his bit into an app on a smartphone, which might well raise concerns, even if Bob has the app's source code.

cards (they will be inverted with probability one half). For this, it is suggested to split the cards as discussed and put them into two indistinguishable envelopes:

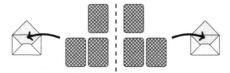

Next, we *shuffle* the envelopes, such that they changed places with probability one half and no player was able to keep track.

We extract the cards again and put them back into the geometric arrangement as before. (This assumes that they have been carefully put into the envelope so that it is still clear which card to place where.)

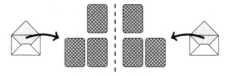

As discussed before, the upper two cards do not give away Alice's bit any longer, so they can be safely turned over. The invariant ensures that the result is still on the side with the heart:

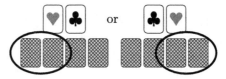

In total, we have performed an AND protocol in *committed format*, as the output are two face-down cards encoding the result. From observing the protocol (as an outsider) we did not learn anything about the order of these cards in the process.

Protocols, which are *not* in committed format, are already well-understood in terms of the minimum number of cards, cf. [MKS12, MWS15, Miz16], and they have the important disadvantage, that they do not allow to use the output directly in subsequent calculations, such as in a three-player AND. Hence, the quest for card-minimal committed format protocols has emerged as a central research task in card-based cryptography.

Besides committed-format AND and negation (inverting the cards), there is one more ingredient necessary for computing arbitrary boolean circuits: commitment copy. A circuit may contain forking wires which enter two or more gates. To see this, consider the three-input majority function as an example:

$$\mathsf{maj}(x_1, x_2, x_3) = (x_1 \wedge x_2) \vee (x_2 \wedge x_3) \vee (x_3 \wedge x_1).$$

Here, before computing $x_1 \wedge x_2$ using an AND protocol, we need to duplicate the input commitment of x_1 and x_2, as they are used in the other clauses as well. (Trusting a user to just input the same bit again is not an option as he or she might deviate and cause wrong outputs, but also because the inputs might not be known to any user if they are the output of some previously run protocol.)

Current Protocols and Their Practicality. Using the very general computational model of Mizuki and Shizuya [MS14, MS17], it was shown in [KWH15] that there are four-card AND protocols in committed format, albeit with a running time that is finite only in expectation (a Las Vegas protocol). Moreover, they showed that if a finite runtime is required, five cards are necessary and sufficient for computing AND. While on the first glance the task of finding card-minimal protocols seems to be settled, a closer look reveals that there is much more to be hoped for. The most pressing point here is not the protocols' increased complexity, but a certain type of card shuffling which is much more difficult to perform than in previous protocols. The authors identified two properties whose violation seems to be a cause of the difficulty in their implementation, namely:

Closedness. The set of possible permutations in a card shuffle is invariant under repetition, i.e., a subgroup of the respective permutation group.

Uniform Probability. Every possible permutation in a shuffle action has the same probability.

These *uniform closed shuffles* can be easily implemented in the honest-but-curious setting by two parties taking turns in performing a random permutation from the specified permutation set, while the other party is not looking. More importantly, there is a recent *actively secure* implementation of uniform closed shuffles [KW17], removing the assumption that players do only permutations from the allowed set, even when not under surveillance of the other party. (The non-uniform shuffle and/or non-closed shuffles in [KWH15] have been implemented in [NHMS16] using sliding cover boxes, but such an approach requires extra tools and security is not achieved against attackers that reorder the boxes in a illicit way.)

The most extreme example for the power of non-closed shuffles seems to be the $2k$-card Las Vegas protocol of [KWH15, Sect. 7] which computes an arbitrary k-ary boolean function in three steps: we shuffle, turn and either output a result or restart. Here, all the work in computing the function is done by the complex and (in general) non-closed shuffle – suggesting that *non-closed shuffles in general* are too broad a shuffle class to consider. Besides these shuffle restrictions, there is one additional central parameter for *practical* card-based protocols, namely runtime behavior. We consider the following practical:

Finite Runtime. This guarantees an a priori bound on the runtime and allow to precisely predict how long the protocol will run. All tree diagrams (introduced in Sect. 3) will be finite. We regard this as the most practical.

Restart-Free Las Vegas (LV). While the runtime of the protocol is finite only in expectation, it is usually just a small constant. When running these protocols we may run in cycles but exit these cycles at least with some constant probability in every run. More importantly we do not end in a failure state where we have to restart the whole protocol and query players for their inputs again.

Protocols which do not belong to these classes are called *restarting LV* protocols. Note that running a COPY protocol on the input bits before the protocol itself requires already at least five cards for the copy process of a single input bit, as shown in Sect. 8. In total this strategy likely needs more cards than just going for restart-free protocols instead.

The aim of this paper is to derive tight lower bounds on the number of cards for protocols using practical – namely (uniform) closed – shuffles and/or a practical runtime, namely finite runtime or restart-free LV, for the two central ingredients of boolean circuits: AND protocols and COPY protocols. Our results are given Table 1, which includes a survey on current bounds relative to certain restrictions on the operations.

Table 1. Minimum number of cards required by committed format AND and n-COPY protocols, subject to the requirements specified in the first two columns. The second column specifies shuffle restrictions. See also Fig. 6.

Runtime	Shuffle Restr.	#Cards	Reference
AND PROTOCOLS:			
Restarting LV	Uniform	4	[KWH15, Sect. 7]
Restart-free LV	Closed	4	[KWH15, Sect. 4]
Restart-free LV	Uniform closed	≥ 5	Theorem 2
Finite runtime	–	5	[KWH15, Sects. 5 and 6]
Finite runtime	Closed	$\Big\} \; 6$	Theorem 1, [MS09]
Finite runtime	Uniform closed		
COPY PROTOCOLS:			
Restarting LV	–	$\Big\} \; 2n + 1$	Theorem 3, [NNH+17]
Restart-free LV	Uniform		
Finite runtime	–	$\Big\} \; 2n + 2$	Theorem 4, [MS09]
Finite runtime	Uniform closed		

Contribution. Regarding committed-format AND protocols, we

– show that five-card finite-runtime protocols are impossible if restricted to closed shuffle operations. This identifies the six-card protocol of [MS09] as card-minimal w.r.t. finite-runtime closed-shuffle protocols using two colors.

- introduce an analysis tool, namely orbit partitions of shuffles, which might be of independent interest for finding protocols using only closed shuffles.
- show that four-card Las Vegas protocols are impossible, if they are restricted to uniform closed shuffle operations and may not use restart operations.

Regarding n-COPY protocols, which produce n copies of a commitment, we

- show that at least $2n + 1$ cards are necessary, even for protocols that may restart. This identifies the $(2n + 1)$-card protocol of [NNH+17] as card-minimal.
- show that finite-runtime protocols need at least $2n + 2$ cards. This identifies the $(2n+2)$-card protocol of [MS09] as a card-minimal finite-runtime protocol. These proofs are simple enough to additionally serve as impossibility results that may be presented in didactic contexts, e.g., to high-school students.
- give simple state trees of the protocols from the literature, cf. Figs. 2 and 3.

Moreover, we show that public randomness does not help in secure protocols. This simplifies future attempts at inventing protocols or showing their impossibility.

Outline. We give necessary preliminaries for card-based protocols in Sect. 2. We introduce [KWH15]'s tree-based protocol notation, argue formally for its usefulness and collect interesting properties in Sects. 3 to 5. Section 6 gives the main result, namely that six cards are necessary for computing AND when restricted to closed shuffle operations. A lower bound of 5 for the number of cards in restart-free Las Vegas protocols using only uniform closed shuffles is given in Sect. 7. Moreover, tight lower bounds for the number of cards in Las Vegas and finite-runtime COPY protocols are proven in Sects. 8 and 9, respectively.

Related Work. Except [FAN+16], which extends the work of [KWH15] to two-bit output functionalities, no other (tight) lower bounds on the number of cards have been obtained so far. Researchers seem to have concentrated on how to implement non-uniform or non-closed shuffles, which are usually more complex or require other tools, such as sliding cover boxes, cf. [NHMS16].

Notation. In the paper we use the following notation.

- *Cycle Decomposition.* For elements $x_1, \ldots, x_k$ the *cycle* $(x_1 \ x_2 \ \ldots \ x_k)$ denotes the *cyclic* permutation π with $\pi(x_i) = x_{i+1}$ for $1 \leq i < k$, $\pi(x_k) = x_1$ and $\pi(x) = x$ for all x not occurring in the cycle. If several cycles act on pairwise disjoint sets, we write them next to one another to denote their composition. For instance $(1\ 2)(3\ 4\ 5)$ denotes a permutation with mappings $\{1 \mapsto 2, 2 \mapsto 1, 3 \mapsto 4, 4 \mapsto 5, 5 \mapsto 3\}$. Every permutation can be written in such a *cycle decomposition*. S_n denotes the *symmetric group* on $\{1, \ldots, n\}$.
- *Entries in Sequences.* Given a sequence $x = (\alpha_1, \ldots, \alpha_l)$ and $i \in \{1, \ldots, l\}$, $x[i]$ denotes the i-th entry of the sequence, namely α_i.
- *Cyclic Group.* Let Π be a group (usually a subgroup of S_n) and $\pi \in \Pi$. Then $\langle \pi \rangle := \{\pi^k \mid k \in \mathbb{Z}\}$ is the *cyclic* subgroup of Π *generated by* π.

2   Card-Based Protocols

We introduce a restricted version of the computational model for card-based protocols as introduced in [MS14,MS17]. We later argue that the liberties we take are essentially cosmetic in Sect. 4.

A *deck* $\mathcal{D}$ is a finite multiset of *symbols* – in this paper only the symbols $\heartsuit$, $\clubsuit$ and (rarely) $\diamondsuit$ are used. For a symbol $c \in \mathcal{D}$, $\frac{c}{?}$ denotes a *face-up card* and $\frac{?}{c}$ a *face-down card*, respectively. Here, '?' is a back symbol, which is not in $\mathcal{D}$. The deck is lying on the table in a *sequence* of cards that contains each symbol of the deck either as a face-up or face-down card. For a face-up or face-down card α, $\mathsf{top}(\alpha)$ and $\mathsf{symb}(\alpha)$ denote the symbol in the "numerator" and the symbol which is not '?', respectively. These definitions are canonically extended to map card sequences to symbol sequences. We call $\mathsf{top}(\Gamma)$ the *visible sequence* of a card sequence Γ and denote by $\mathsf{Vis}^{\mathcal{D}}$ the set of visible sequences on a deck $\mathcal{D}$.

A *protocol* $\mathcal{P}$ is a quadruple $(\mathcal{D}, U, Q, A)$, where $\mathcal{D}$ is a deck, U is a set of input sequences over $\mathcal{D}$, Q is a set of states with two distinguished states q_0 and q_f, being the initial and the final state. Moreover, we have a (partial) action function $A \colon (Q \setminus \{q_f\}) \times \mathsf{Vis}^{\mathcal{D}} \to Q \times \mathsf{Action}$, depending on the current state and visible sequence, which specifies the next state and an operation on the sequence. The following actions exist:

– (turn, T) for $T \subseteq \{1, \ldots, |\mathcal{D}|\}$. This turns the cards at all positions in T, i.e., it transforms $\Gamma = (\alpha_1, \ldots, \alpha_{|\mathcal{D}|})$ into

$$\mathsf{turn}_T(\Gamma) := (\beta_1, \ldots, \beta_{|\mathcal{D}|}), \text{ where } \beta_j = \begin{cases} \mathsf{swap}(\alpha_j), & \text{if } j \in T, \\ \alpha_j, & \text{otherwise.} \end{cases}$$

Here, $\mathsf{swap}(\frac{c}{?}) := \frac{?}{c}$ and $\mathsf{swap}(\frac{?}{c}) := \frac{c}{?}$, for $c \in \mathcal{D}$.

– (perm, π) for a permutation $\pi \in S_{|\mathcal{D}|}$ from the symmetric group $S_{|\mathcal{D}|}$. This applies π to the sequence, i.e., if $\Gamma = (\alpha_1, \ldots, \alpha_{|\mathcal{D}|})$, the resulting sequence is

$$\mathsf{perm}_\pi(\Gamma) := (\alpha_{\pi^{-1}(1)}, \ldots, \alpha_{\pi^{-1}(|\mathcal{D}|)}).$$

– $(\mathsf{shuffle}, \Pi, \mathcal{F})$ for a probability distribution $\mathcal{F}$ on $S_{|\mathcal{D}|}$ with support Π. This transforms a sequence Γ into

$$\mathsf{shuffle}_{\Pi, \mathcal{F}}(\Gamma) := \mathsf{perm}_\pi(\Gamma), \text{ for } \pi \text{ drawn from } \mathcal{F},$$

i.e., $\pi \in \Pi$ is drawn according to $\mathcal{F}$ and then applied to Γ. This is done in such a way that no player/bystander can observe anything about π in the process. If $\mathcal{F}$ is the uniform distribution on Π, we just write $(\mathsf{shuffle}, \Pi)$.

– $(\mathsf{restart})$. This transforms a sequence into the start sequence. In that case the first component of A's output must be q_0. This allows for Las Vegas protocols to start over and cannot be used in finite-runtime protocols.

– $(\mathsf{result}, p_1, \ldots, p_r)$ declares the ordered sequence of cards in distinct positions $p_1, \ldots, p_r \in \{1, \ldots, |\mathcal{D}|\}$ as the *output* of the protocol and halts. This special operation occurs if and only if the first component of A's output is q_f.

A *sequence trace* of a finite protocol run is a tuple $(\Gamma_0, \Gamma_1, \ldots, \Gamma_t)$ of sequences such that $\Gamma_0 \in U$ and Γ_{i+1} arises from Γ_i by an operation as specified by the action function. Moreover, $(\mathsf{top}(\Gamma_0), \mathsf{top}(\Gamma_1), \ldots, \mathsf{top}(\Gamma_t))$ is a corresponding *visible sequence trace*.

Security of card-based protocols intuitively means that input and output are perfectly hidden, i.e., from the outside the execution of a protocol looks the same (has the same distribution), regardless of what input and output are.

Definition 1 (Security, cf. [KWH15,KW17]). *Let $\mathcal{P} = (\mathcal{D}, U, Q, A)$ be a protocol. It is* (input- and output-)secure *if for any random variable I with values in the set of input sequences U, the following holds. A random protocol run starting with $\Gamma_0 = I$, terminates almost surely. Moreover, if V and O are random variables denoting the visible sequence trace and the output of the run, then the pair (I, O) is stochastically independent of V.*

3 State Trees and Their Properties

The authors of [KWH15] introduce a representation of secure protocols as a tree-like diagram, we call it a *state tree*. In it, all protocol runs are captured simultaneously in a way that makes output behavior and security of the protocol directly recognizable.

An assumption that significantly reduces notational complexity is that cards are essentially always face-down. This relates to all input and output sequences[2], but moreover, turn actions occur in pairs, such that cards that are turned face-up by the first turn are directly turned face-down again afterwards. After revealing the symbols, there is no reason to keep the cards face-up any longer. Corollary 2 of Sect. 4 is a formal version of this argument.

3.1 Constructing State Trees from Protocols

We describe how to construct state trees from protocol descriptions. See Fig. 1 for a reference. In the following v is always the visible sequence trace of the prefix of a protocol run.[3] For each such v, the state tree contains a state μ_v. In our model, each state is associated with a unique action that the protocol prescribes for this situation. We annotate the state (essentially its outgoing edge) with this action. If the action may extend v to v' by appending a visible sequence v^+, then $\mu_{v'}$ is a *child* of its *parent* μ_v. A turn action may result in several children which are *siblings* of each other and the edge to each child is additionally annotated with v^+. If the action is a result action, then μ_v is a *leaf*.

From the perspective of an observer of the protocol who does not know the input I or the permutations chosen during shuffle actions, the *actual sequence* S_v

---

[2] For positions in the input and output sequence that are not constant, security requires cards to be face-down anyway.

[3] By our assumption that cards are usually face-down, many visible sequences are trivial, i.e., are $(?\ldots?)$, with the sole exception of turn actions themselves.

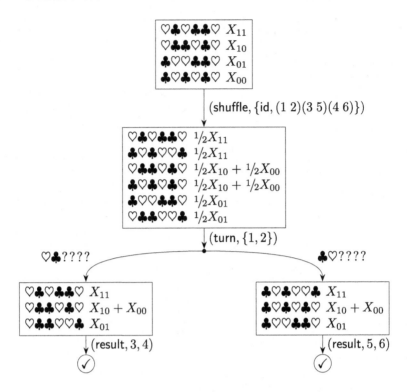

Fig. 1. The six-card AND protocol of [MS09].

lying on the table in a particular run when the state μ_v is reached, is unknown. We annotate μ_v with the corresponding probabilities, i.e. with $\mu_v(s) := \Pr[S_v = s \mid v]$ where s is any sequence of symbols and v stands for the event that v is a prefix of the visible sequence trace of the complete protocol run. We can rewrite this as:

$$\mu_v(s) = \sum_{\Gamma \in U} \Pr[S_v = s, I = \Gamma \mid v] = \sum_{\Gamma \in U} \Pr[S_v = s \mid I = \Gamma, v] \cdot \Pr[I = \Gamma \mid v]$$

$$= \sum_{\Gamma \in U} \underbrace{\Pr[S_v = s \mid I = \Gamma, v]}_{=:p_{v,\Gamma,s}} \cdot \underbrace{\Pr[I = \Gamma]}_{=:X_\Gamma} = \sum_{\Gamma \in U} p_{v,\Gamma,s} X_\Gamma.$$

At the line-break, we exploited the independence of visible sequence trace and input in secure protocols and introduced two abbreviations. Note that $p_{v,\Gamma,s}$ is constant, but X_Γ is a variable since we have not actually defined a specific probability distribution on the inputs (nor will we). We treat the result as a formal sum, making μ_v a map from sequences of symbols to polynomials.

We say a state μ_v *contains* a sequence of symbols s (or s *is in* μ_v for short) if $\mu_v(s)$ is not the zero polynomial, and write $\text{supp}(\mu_v)$ (as in *support*) for the set of all such sequences. Let $|\mu| := |\text{supp}(\mu_v)|$. We now describe how the tree of all states can be computed for a given secure protocol inductively starting from the root.

Start state. First note that the start state is unique: Regardless of the (face-down) input sequence, the visible sequence trace is $v_0 = ((?, \ldots, ?))$ only containing one trivial visible sequence. The distribution of S_{v_0} is the input distribution, i.e.,

$$\mu_{v_0}(s) = \begin{cases} X_\Gamma, & \text{if } s = \mathsf{symb}(\Gamma) \text{ for } \Gamma \in U, \\ 0, & \text{otherwise.} \end{cases}$$

Shuffle action. Assume for a state μ_v the action (shuffle, $\Pi, \mathcal{F}$) is prescribed. No non-trivial information is revealed when shuffling, so obtaining v' by appending a trivial visible sequence to v, $\mu_{v'}$ is the unique child of μ_v, fulfilling

$$\mu_{v'}(s') = \Pr[S_{v'} = s' \mid v']$$
$$= \sum_{\pi \in \Pi} \mathcal{F}(\pi) \cdot \Pr[S_v = \pi^{-1}(s') \mid v] = \sum_{\pi \in \Pi} \mathcal{F}(\pi) \cdot \mu_v(\pi^{-1}(s')).$$

This just takes into account all sequences $\pi^{-1}(s')$ in μ from which s' may have originated from via some π as well as corresponding probabilities.

Note that perm actions are a special case and need not be described separately.

Restart action. States in which a restart action happens have a single child. The subtree at this child is equal to the entire tree.

Turn action. Let μ_v be a state with a turn action that possibly results in the visible sequence v^+, which appended to v yields v'. The child $\mu_{v'}$ of μ_v contains those sequences from μ_v that are compatible with v^+, i.e., equal to v^+ in the positions i with $v^+[i] \neq ?$. Moreover, the probability to reach $\mu_{v'}$ from μ_v is:

$$\Pr[v' \mid v] = \Pr[S_v \in \mathrm{supp}(\mu_{v'}) \mid v] = \sum_{s \in \mathrm{supp}(\mu_{v'})} \Pr[S_v = s \mid v] = \sum_{s \in \mathrm{supp}(\mu_{v'})} \mu_v(s).$$

Recall that the right hand side is a polynomial of the form $\sum_{\Gamma \in U} a_\Gamma X_\Gamma$ where X_Γ is a placeholder for the input probability $\Pr[I = \Gamma]$. By security, input and visible sequence trace are independent, in particular no matter how the variables $(X_\Gamma)_{\Gamma \in U}$ are initialized, the polynomial $\Pr[v' \mid v]$ evaluates to the same real number. Thus, all a_Γ are equal to a constant $\lambda_{v'} \in [0, 1]$ and using $\sum_{\Gamma \in U} X_\Gamma = 1$ we have:

$$\sum_{s \in \mathrm{supp}(\mu_{v'})} \mu_v(s) = \lambda_{v'} \sum_{\Gamma \in U} X_\Gamma = \lambda_{v'}.$$

Moreover, for $s \in \mathrm{supp}(\mu_{v'})$ we have:

$$\mu_{v'}(s) = \Pr[S_{v'} = s \mid v'] = \frac{\Pr[S_{v'} = s \mid v]}{\Pr[v' \mid v]} = \frac{\Pr[S_v = s \mid v]}{\lambda_{v'}} = \frac{\mu_v(s)}{\lambda_{v'}}.$$

Hence, $\mu_{v'}$ is a restriction of μ_v to sequences compatible with v', scaled by $\frac{1}{\lambda_{v'}}$. Capturing the observation we made for secure protocols, we define

$$i \text{ is } turnable \text{ in } \mu :\Leftrightarrow \forall c \in \mathcal{D}: \exists \lambda \in [0,1]: \sum_{s \text{ with } s[i]=c} \mu(s) = \lambda \sum_{\Gamma \in U} X_\Gamma. \qquad (1)$$

Output behavior. Consider any leaf state μ_v with action (result, p_1, $p_2, \ldots, p_r$). The output $O = (S_v)_{p_1,\ldots,p_r}$ is the projection of S_v to the components $p_1, \ldots, p_r$. We can easily obtain from μ_v the probabilities $\Pr[O = o \mid v]$ for any sequence of symbols o. By security, the visible sequence trace v is irrelevant, which implies that we obtain the same polynomial $\Pr[O = o]$ for a fixed o, regardless of which leaf state we examine. If the protocol computes a deterministic function (see below), this polynomial is the sum of all X_Γ for which Γ evaluates to o.

3.2 The Utility of State Trees

In the following proposition we argue that state trees are a superior way to represent secure protocols, as security and correctness are more tangible.

Proposition 1. *Every secure protocol $\mathcal{P}$ has a unique state tree that can be obtained as described above. Moreover, local checks suffice to verify that a tree of states describes a secure protocol with a desired output behavior.*

Proof. The proof is given in the full version. By local checks we mean checking that each state arises from its parent in the way we discussed. In particular, for turns, the probability of sequences compatible with an outcome must sum to a constant and all leaf states must provide the same output distribution.   □

State trees can be infinite while state machines are typically required to have a finite number of states. This is mostly beside the point for our purposes, but the claim should be understood with the following qualification. Either a countably infinite number of states is permitted for protocols or only certain self-similar state trees can be transformed into protocols (see Fig. 3 for such a state tree).

Runtime of a Protocol. A secure protocol *terminates* when entering the final state q_f. It has *finite-runtime*, if there is an a priori upper bound on the number of steps. It is *Las Vegas* if it terminates almost surely in a number of steps that is finite in expectation, but unbounded.

We restate these definitions with respect to state trees. A *path in a state tree p* starts at the root and descends according to a visible sequence trace v. If v is finite, p ends in a leaf. The *probability* of p is $\Pr[v]$, which is a constant (independent of the distribution of inputs) by our discussion of security before. We say that a protocol is *finite-runtime* if there is no infinite path in its state tree. If a protocol is not finite-runtime but the expected length of any path in its state tree is finite, we call it a *Las Vegas protocol*.

3.3 Protocols Computing a Boolean Function

The way we defined it, a protocol takes an unstructured input and computes an unstructured output that is in general a randomized function of the input. While randomization (as in privately computing a fixed-point-free permutation [CK93]) and unstructured (non-committed) outputs (as in the "five card trick" [dBoe89]) appear in the literature, the most important use case is arguably the computation of a deterministic Boolean function in committed format. In this section we provide useful terminology and some simple insights for this setting.

The Setting. A commitment to a bit value $b \in \{0,1\}$ is represented by two cards, with their symbols arranged as $\heartsuit \clubsuit$ if $b = 1$ and as $\clubsuit \heartsuit$ otherwise. Extend this in the natural way, we say a sequence of symbols (or cards) $(\alpha_1, \ldots, \alpha_{2k})$ *encodes* a sequence $b \in \{0,1\}^k$ of bits, if $\alpha_{2i-1}\alpha_{2i}$ (or $\mathsf{top}(\alpha_{2i-1})\mathsf{top}(\alpha_{2i})$) represents $b[i]$ for $1 \leq i \leq k$. We say that a protocol $\mathcal{P} = (\mathcal{D}, U, Q, A)$ *computes a function* $f\colon \{0,1\}^k \to \{0,1\}^\ell$, if the following holds:

- The deck $\mathcal{D}$ contains at least $\max(k, \ell)$ cards of each symbol,
- There is an input sequence Γ_b for each $b \in \{0,1\}^k$, the first $2k$ cards of which encode b. The remaining $|\mathcal{D}| - 2k$ cards are independent of b.
  Favoring light notation, we use b instead of Γ_b as index to refer to variables from the family $(X_b)_{b \in \{0,1\}^k} = (X_\Gamma)_{\Gamma \in U}$ from now on.
- Correctness: A protocol run starting with Γ_b almost surely terminates with an output encoding $f(b)$.

We call a protocol computing $f_\wedge\colon \{0,1\}^2 \to \{0,1\}$, with $(a,b) \mapsto a \wedge b$ an *AND protocol*. A protocols computing $f_{\mathrm{copy}}\colon \{0,1\} \to \{0,1\}^n$, with $a \mapsto a^n = (a, \ldots, a)$ and $n \geq 2$ is called an *n-COPY* or just *COPY protocol*.

Figures 1, 2 and 3 depict state trees of current AND and COPY protocols, which are identified as card-minimal with respect to certain restrictions, in this paper. See the full version for a discussion of the security and correctness of Fig. 1.

3.4 Reduced State Trees and Possibilistic Security

When trying to prove that no secure protocol with certain properties exists, the space of possible states can seem vast. We therefore opt to show stronger results, namely the non-existence of protocols with the weakened requirement of possibilistic security, defined next. This allows to consider a projection of the state space which is finite in size.

Definition 2. *A protocol* $\mathcal{P} = (\mathcal{D}, U, Q, A)$ *computing a function* $f\colon \{0,1\}^k \to \{0,1\}^\ell$ *has* possibilistic input security (possibilistic output security) *if it is correct, i.e.,* $O = f(I)$ *almost surely and for uniformly*[4] *random input* $I \in U$ *and any visible sequence trace* v *with* $\Pr[v] > 0$ *as well as any input* $i \in \{0,1\}^k$ *(any output* $o \in \{0,1\}^\ell$*) we have* $\Pr[v \mid I = i] > 0$ *(*$\Pr[v \mid f(I) = o]$*).*

---

[4] Actually, the distribution does not matter, as long as $\Pr[I = i] > 0$ for all $i \in \{0,1\}^k$.

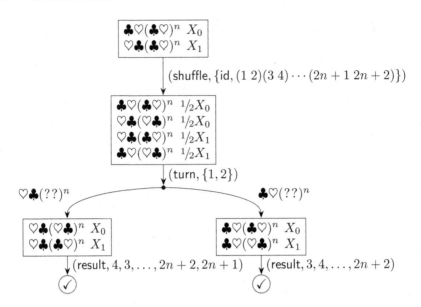

Fig. 2. The state tree of the $(2n+2)$-card COPY protocol [MS09]. As in the six-card AND protocol, the shuffle consists only of the identity and a permutation that is a cycle decomposition into (disjoint) 2-cycles. Hence, it can be implemented by a random bisection cut. See [CHL13, Sect. 6] for a more elaborate explanation.

In other words, from observing a visible sequence trace it is never possible to exclude an input (or output) with certainty. Clearly, security implies possibilistic input and output security. To decide whether a protocol has possibilistic output security, it suffices to consider projections of states in the following sense.

Definition 3 (adapted from [KWH15]). *Let $\mathcal{P} = (\mathcal{D}, U, Q, A)$ be a protocol computing a function $f \colon \{0,1\}^k \to \{0,1\}^\ell$. If μ is a state in the state tree, then the reduced state $\hat{\mu}$ has the same sequences as μ with simpler annotations. Assume $\mu(s)$ is a polynomial with positive coefficients for the variables $X_{b_1}, \ldots, X_{b_i}$ ($i \geq 1$). Then we set $\hat{\mu}(s) = o \in \{0,1\}^\ell$ if $o = f(b_1) = f(b_2) = \ldots = f(b_i)$. If not all b_i evaluate to the same output under f, set $\hat{\mu}(s) = \perp$. Accordingly, sequences in $\hat{\mu}$ are called o-sequences or $\perp$-sequences. We also say they are of type o or of type $\perp$, respectively. For COPY protocols, we call 1^ℓ- and 0^ℓ-sequences just 1- and 0-sequences, respectively.*

Note that if a state μ has children $\mu_1, \ldots, \mu_i$ in the state tree, then the reduced states $\hat{\mu}_1, \ldots, \hat{\mu}_i$ can be computed from the reduced state $\hat{\mu}$. In particular, it makes sense to define *reduced state trees* as projections of state trees and aim to prove the non-existence of certain state trees by proving the non-existence of the corresponding reduced state trees.

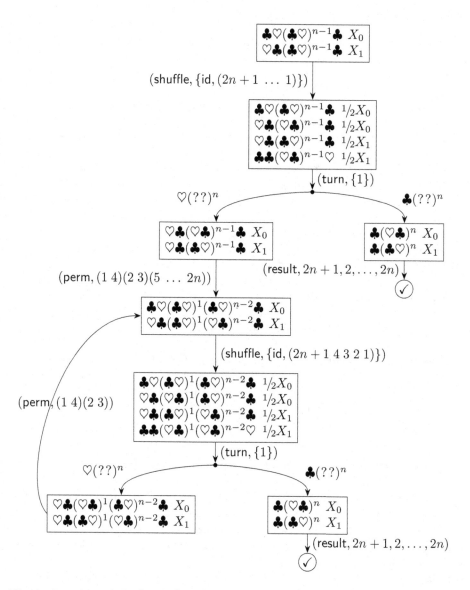

Fig. 3. A variant of the $(2n + 1)$-card COPY protocol of Nishimura et al. [NNH+17, Sect. 5], with less permuting. After the permutation in the first $\heartsuit$-branch, the protocol resembles a 2-COPY protocol [NNH+15] on the first four and the last card. The parentheses $(\cdot)^1$ are to emphasize this symmetry. The shuffle steps in this protocol are uniform, but non-closed. As they consist of the identity and exactly one odd-length cycle, they can be performed using an "unequal division shuffle". A proposed implementation of these shuffles using sliding cover boxes or envelopes can be found in [NNH+17, Sect. 6]. Although this diagram includes a backwards edge, this is for presentation only. We regard it as an infinite (self-similar) tree.

3.5 Important Properties of (Reduced) States

For most of our arguments, reduced states offer a sufficient granularity of detail and the following observations and definitions for reduced states will prove useful. Clearly, they also apply to the richer, non-reduced states which we will briefly require for a more subtle argument in Sect. 7.

Protocols computing a function f that have a $\perp$-sequence in a reduced state cannot be restart-free, and hence also not finite-runtime: once the $\perp$-sequence is actually on the table, it does not contain sufficient information to deduce the unique correct output, making a restart necessary. The protocol might then repeat the unfortunate path of execution an arbitrary number of times.

We call two reduced states *similar*, if one is just a permuted version of the other, i.e. interpreting a state as a matrix of symbols with annotated rows, there is a permutation on the columns mapping one state to the other.

In restart-free protocols with two outputs (say 1 and 0) any reduced state $\hat{\mu}$ is composed of some number i of 0-sequences and some number j of 1-sequences with $|\hat{\mu}| = i + j$. We call μ and $\hat{\mu}$ an i/j-state. They are *final* if they admit a correct output action ($\mathsf{result}, m_1, n_1 \ldots, m_\ell, n_\ell$), i.e. they does not contain a $\perp$-sequence and for all outputs $r \in \{0,1\}^\ell$ there is at least one r-sequence and r-sequences have a $\heartsuit$ at position m_i if $r[i] = 1$, or a $\clubsuit$ otherwise, for all $1 \leq i \leq \ell$, and the respective other symbol at position n_i.

Turnable Positions. Recall the definition of a turnable position for a state μ from (1). For a reduced state $\hat{\mu}$ this simplifies to: For each symbol c occurring in column i, among the sequences with $s[i] = c$, there is either a $\perp$-sequence or an r-sequence for each $r \in \{0,1\}^\ell$. If some position in a (reduced) state is turnable, we call the state turnable, otherwise unturnable.

4 Simplifications to the Computational Model

In the following, we argue that public randomness does not provide any benefit in secure protocols and hence can be safely excluded from the computational model. For this assume there was an additional class of actions in protocols, that produces public randomness: The action $\mathsf{rand}(p_1, \ldots, p_i)$ for real numbers $p_1, \ldots, p_i \in (0,1)$ summing to 1 appends a value $x \in \{0, 1, \ldots, i\}$ to the visible sequence trace with $\Pr[x = j] = p_j$. This extends the computational model[5]. An example would be a protocol that receives a bit in committed format, and outputs either this bit unchanged with probability $1/3$ or its negation with probability $2/3$. This can be implemented by a $\mathsf{rand}(1/3, 2/3)$ action and, depending on the outcome a permutation or not, followed by a result action. This is also possible with a shuffle, but requires a non-uniform probability distribution on the permutations, something we are inclined to prohibit later. However, this protocol is *not secure* since the result of the random experiment is part of the

---

[5] Consider a protocol that receives a single card $\Gamma = \frac{?}{\heartsuit}$ as input and outputs $\frac{?}{\heartsuit}$ with probability $1/2$ and $\frac{\heartsuit}{?}$ with probability $1/2$, which was not possible before.

visible sequence trace (it is *public*) which introduces a non-trivial relationship between output and visible sequence trace precluding independence. Indeed we show:

Proposition 2 (Public randomness is unnecessary). *Any* $\mathsf{rand}(p_1, \ldots, p_i)$ *action can be removed from a secure protocol $\mathcal{P}$ without affecting security or the output distribution and without increasing (worst-case or expected) runtime.*

Proof. Consider i copies $\mathcal{P}_1, \ldots, \mathcal{P}_i$ of $\mathcal{P}$, where in $\mathcal{P}_j$, if the rand action is encountered, execution continues deterministically as if j had been chosen. The distribution of input output and visible sequence trace (I_j, O_j, V_j) for $\mathcal{P}_j$ equals the distribution of (I, O, V) for $\mathcal{P}$, except conditioned under a certain event E_j regarding V. In particular, independence of (I, O) and V implies independences of (I_j, O_j) and V_j and (I_j, O_j) is distributed like (I, O) (security and output behavior are unaffected). Thus, choosing j such that V_j has least (worst-case or expected) length gives the desired modification $\mathcal{P}_j$ of $\mathcal{P}$.   □

Mizuki and Shizuya [MS14] introduced a randomized action called rflip that works just like a turn-action, except the set of cards T that is turned is chosen randomly according to some distribution. We call it *randomized flip*. As the random choice is clearly public we can directly derive the following:

Corollary 1. *The randomized flip can be removed from a secure protocol without negative effects as in Proposition 2.*

When introducing state trees, we assumed that cards are always face-down, in particular during shuffle operations. This hides a complication: Shuffling with face-up cards, i.e., with the current visible sequence v being non-trivial, may result in several visible sequences v', depending on how the face-up cards are permuted. But similar to rflip, such a *branching shuffle* can be simulated with a rand action, determining which v' is obtained, followed by a shuffle action restricted to those permutations that transform v into v'. From the discussion above, we can derive the following corollary.

Corollary 2 (We can assume all cards are face down). *It suffices to look at protocols which start with face-down cards and which, after turning any card, directly turns them down again.*

In Lemma 4 we argue that this still works for uniform and/or closed shuffles, i.e. if the original branching shuffle was closed (uniform), the restricted shuffles can be implemented using closed (uniform) shuffles as well.

5   Properties of Restricted-Shuffle Protocols

As discussed before, it is natural to restrict card-based protocols to closed shuffles. The fact that a closed permutation set Π is a subgroup of the symmetric group $S_{|\mathcal{D}|}$ implies strong structural properties for the corresponding shuffle action on states. This structure is captured by the orbits associated with the group action of Π on sequences on symbols, introduced in the following.

Definition 4 (Group action, e.g. [DM96]). *Let X be a nonempty set, G a group, and $\varphi\colon G \times X \to X$ a function written as $g(x) := \varphi(g, x)$ for $g \in G, x \in X$. We say that G acts on X if*

 - *$e(x) = x$ for all $x \in X$, where e denotes the identity element in G,*
 - *$(g \circ h)(x) = g(h(x))$ for all $x \in X$ and all $g, h \in G$.*

In card-based protocols, the permutation group $S_{|\mathcal{D}|}$ acts on a set of sequences of a deck $\mathcal{D}$ in the natural way, by reordering the cards in a sequence according to the permutation.

For a group G acting on a set X the *orbit of an element* $x \in X$ is $G(x) = \{g(x)\colon g \in G\}$. Any subgroup of $S_{|\mathcal{D}|}$ also acts on X in the same way. This is interesting for closed shuffle operations. The orbit of an element x under a subgroup Π is $\Pi(x) = \{\pi(x)\colon \pi \in \Pi\}$. It is easy to see that the orbits of two elements are either the same or disjoint. Therefore, Π induces a partition of the set of possible sequences into disjoint orbits. We call this partition the *orbit partition* of the sequences through Π. For a shuffle with permutation set Π we also say that the orbit partition of Π is the orbit partition of the shuffle.

5.1   Properties of (Uniform) Closed Shuffles

For our proof we will need the orbit partitions of closed shuffles on sequences of five cards with three $\heartsuit$ and two $\clubsuit$. A display of all relevant orbit partitions can be seen in Fig. 4. Before we proceed with the main theorem in Sect. 7, it is beneficial to state some observations about orbit partitions and closed shuffles.

Lemma 1. *Assume we shuffle a state μ into a state μ' using a closed permutation set Π. If there is $\pi \in \Pi$ with $\pi(i) = j$, then in μ' the columns i and j contain the same multiset of symbols.*

Proof. Assume column i has k $\clubsuit$ in μ'. Because Π is closed, shuffling again with Π yields μ' again. In particular if s' is contained in μ', then so is $\pi(s')$. Since π is a bijection on the set of all sequences, the sequences with $\clubsuit$ in position i are mapped to distinct sequences with $\clubsuit$ in position j. In particular column j contains at least as many $\clubsuit$ as column i in μ'. Since the same argument works for all other symbols and the total number of sequences did not increase, the numbers of $\clubsuit$ (and other symbols) in column i and j coincides. Since $\clubsuit$ was arbitrary, the claim follows.   □

Observe that Lemma 1 is false, if Π is not closed. Take for instance an action (perm, π) for $\pi \neq$ id. It is a non-closed shuffle clearly lacking this property.

Lemma 2 (Shuffling and Orbit Partitions). *Let μ be transformed into μ' via some shuffle with closed permutation set Π. Let O be an orbit of Π.*

1. If $\mu \cap O = \emptyset$, then[6] $\mu' \cap O = \emptyset$

---

[6] We slightly abuse notation here, using μ for the set of sequences contained in μ.

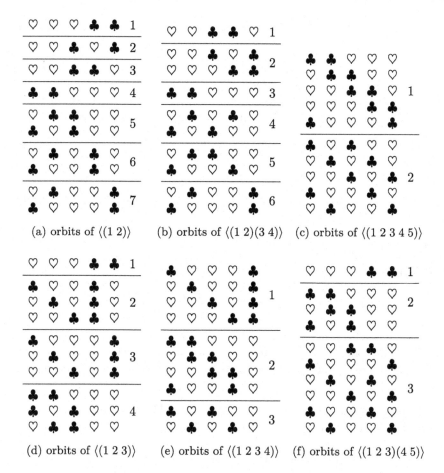

Fig. 4. Orbits of different closed (cyclic) shuffle operations on the card sequences.

2. If $\mu \cap O$ contains a sequence of type r and no other sequences of type $r' \neq r$, then $\mu' \cap O = O$ contains only r-sequences.

3. Otherwise, $\mu' \cap O = O$ contains only $\perp$-sequences.

Proof. Note that for any pair s_1, s_2 of sequences, there is some $\pi \in \Pi$ with $\pi(s_1) = s_2$ precisely if they are in the same orbit of Π. Thus when shuffling with Π, the type $\mu(s_1)$ directly "infects" precisely the entire orbit of s_1. With $\perp$ indicating several types jumbled together, the three cases are easily checked. □

For the more restricted case of uniform closed shuffles, let us state the following simple but interesting observation: all sequences in an orbit have the same symbolic probability after the shuffle.

Lemma 3. *Let* act $= (\text{shuffle}, \Pi)$ *be a uniform closed shuffle and* μ *a state. Let* μ' *be the state arising from* μ *through* act *and* s'_1 *and* s'_2 *two sequences of symbols* in μ'. If $s_1' = \pi(s_2')$ for $\pi \in \Pi$, then $\mu'(s_1') = \mu'(s_2')$, i.e., the sequences have the same probability.

Finally, let us note that it is not possible to perform shuffles that are more powerful than closed shuffles indirectly by using a closed shuffle while cards are face-up (branching shuffles). This re-vindicates an assumption made in Sect. 4. Taking into account Corollary 2, we need only prove the following:

Lemma 4. *Let Π be any closed permutation set and v and v' visible sequences. Assume $\Pi_{v \to v'} := \{\pi \in \Pi : \pi(v) = v'\}$ is non-empty. Then for any $\pi_{v \to v'} \in \Pi_{v \to v'}$, the set $\Pi_v := \Pi_{v \to v'} \circ \pi_{v \to v'}^{-1}$ is closed.*

In particular, we can implement (shuffle, $\Pi_{v \to v'}$) using (perm, $\pi_{v \to v'}$) followed by the closed shuffle (shuffle, Π_v).

Proof. Let $x = \varphi_1 \circ \pi_{v \to v'}^{-1}$ and $y = \varphi_2 \circ \pi_{v \to v'}^{-1}$ be two elements of Π_v, in particular $\varphi_1, \varphi_2 \in \Pi_{v \to v'}$. We have $(\varphi_1 \circ \pi_{v \to v'}^{-1} \circ \varphi_2)(v) = v'$, so $\varphi_1 \circ \pi_{v \to v'}^{-1} \circ \varphi_2 \in \Pi_{v \to v'}$. This implies:

$$x \circ y = \varphi_1 \circ \pi_{v \to v'}^{-1} \circ \varphi_2 \circ \pi_{v \to v'}^{-1} \in \Pi_{v \to v'} \circ \pi_{v \to v'}^{-1} = \Pi_v. \qquad \square$$

For uniform shuffles, the analogous claim is even easier, as a uniformly random variable conditioned to be contained in some subset is still uniform on that subset. For uniform *and* closed shuffles these arguments can be combined.

6   Impossibility of Five-Card Finite-Runtime AND Protocols with Closed Shuffles

In this section, we prove that AND protocols which are restricted to closed shuffle operations, require six cards. The proof uses a similar technique as in [KWH15, Sect. 6]. While in the latter there was a set of good states including the final states which you never enter with all branches of a branching point in the protocol, here we found that it was easier to start the other way round, namely to define a set of bad states, about which we prove that starting from these there is always a path in the tree which will enter a bad state again and this path does not contain any final states. Here the situation is more complex as there are many more possible states and we needed to derive new tools (orbit partitions) to make use of the structure of the restricted permutation sets. For this we will make heavy use of Lemmas 1 and 2, because they enable us to exploit the rich structure of closed shuffles. Let us begin by stating our theorem.

Theorem 1. *Let $\mathcal{P}$ be a (possibilistically) secure protocol computing AND in committed format using only closed shuffles with five cards of two symbols.[7] Then $\mathcal{P}$ is not finite-runtime.*

---

[7] Whether a five-card protocol is possible using a deck of three colors, i.e. $\mathcal{D} = [\heartsuit, \heartsuit, \clubsuit, \clubsuit, \diamondsuit]$, is an interesting open question.

Proof Outline. We define a set of *bad states*, such that the start state is one of them. We then show that in any protocol from each bad state there is a path into another bad state. In particular, none of the bad states and none of the states on the paths between them are final. This implies that there is an infinite path starting from the start state precluding finite runtime.

Without loss of generality, the protocols we consider have the following properties, since each protocol that does not have some of these properties can be transformed into an equivalent protocol that does.

- $\mathcal{P}$ does not use operations that transform a (reduced) state into any similar state. These operations are clearly not necessary, when arguing about possibilistic security, which is sufficient in our case.
- $\mathcal{P}$ does not use shuffle operations while cards are lying face up. These are unnecessary by Corollary 2 and Lemma 4.
- each shuffle set Π is a cyclic subgroup of S_5. This is because each subgroup Π of S_5 can be written as the product $\Pi = \prod_{\pi \in \Pi} \langle \pi \rangle$, implying that doing the cyclic shuffles $\langle \pi \rangle$ one after the other gives the same set of permutations that can happen in total as in Π itself.
- The deck is $\mathcal{D} = [\heartsuit, \heartsuit, \heartsuit, \clubsuit, \clubsuit]$. We need two $\clubsuit$ and two $\heartsuit$ for the inputs. Our arguments work regardless of whether the fifth card is $\clubsuit$ or $\heartsuit$.

Definition of Bad States. Any state μ with one of the following properties is bad:

$\mathcal{B}_{4*}$: μ has four sequences of the same type. In particular, this includes all states with seven or more sequences,

$\mathcal{B}_{3\clubsuit}$: μ has a constant $\clubsuit$-column and three or more sequences,

$\mathcal{B}_{5\heartsuit}$: μ has a constant $\heartsuit$-column and five or more sequences,

$\mathcal{B}_{3\heartsuit\heartsuit}$: μ has two constant $\heartsuit$-columns and three sequences,

$\mathcal{B}_{\heartsuit 3/1}$: μ has a constant $\heartsuit$-column and is of type 3/1 or 1/3.

To see that states of these types are all non-final, first note that any final state is, up to reordering, a (not necessarily proper) subset of the sequences of the following state including at least one 0-sequence and at least one 1-sequence. Hence, we will call it the *maximal final state*:

$$
\begin{array}{ccccccc}
\heartsuit & \clubsuit & \clubsuit & \heartsuit & \heartsuit & 1 \\
\heartsuit & \clubsuit & \heartsuit & \clubsuit & \heartsuit & 1 \\
\heartsuit & \clubsuit & \heartsuit & \heartsuit & \clubsuit & 1 \\
\clubsuit & \heartsuit & \clubsuit & \heartsuit & \heartsuit & 0 \\
\clubsuit & \heartsuit & \heartsuit & \clubsuit & \heartsuit & 0 \\
\clubsuit & \heartsuit & \heartsuit & \heartsuit & \clubsuit & 0
\end{array}
$$

The bad states do not "fit" into this state, since

$\mathcal{B}_{4*}$: the maximal final state has only three sequences of each type.

$\mathcal{B}_{3\clubsuit}$: the only two columns in the maximal final state with three $\clubsuit$ are the first two. But a restriction to the sequences with $\clubsuit$ in the first (or second) position contains 0-sequences (1-sequences) only.

$\mathcal{B}_{5\heartsuit}$: all columns of the maximal final state contain at most four $\heartsuit$.

$\mathcal{B}_{3\heartsuit\heartsuit}$: no two columns in the maximal final state have three $\heartsuit$ in the same positions.

$\mathcal{B}_{\heartsuit3/1}$: any admissible restriction of the maximal final state with four sequences and a constant $\heartsuit$-column is of type $2/2$.

The start state is bad as required as it falls into category $\mathcal{B}_{\heartsuit3/1}$. We begin by making a few observations that greatly simplify the proof.

Claim (Restriction to Simple Shuffles). We need to only consider the three shuffle sets $\langle(1\,2)\rangle$, $\langle(1\,2\,3)\rangle$ and $\langle(1\,2)(3\,4)\rangle$ in the main proof.

Proof. As discussed before, we only consider cyclic shuffles, i.e. shuffles of the form $\Pi = \langle\pi\rangle$ for $\pi \in S_{|\mathcal{D}|}$. Unless $\pi = \mathsf{id}$, the cycle decomposition of π can have either one cycle of length 2, 3, 4 or 5 or two cycles of length 2 or one cycle of length 2 and one cycle of length 3. We treat states that are equal up to similarity (reordering) as equivalent; it therefore suffices to consider one shuffle of each type. Among them, $\langle(1\,2)(3\,4\,5)\rangle$ can be decomposed[8] as $\langle(1\,2)\rangle \circ \langle(3\,4\,5)\rangle$ and we handle those factors anyway. In the cases $\langle(1\,2\,3\,4\,5)\rangle$ and $\langle(1\,2\,3\,4)\rangle$ there is no choice of two orbits such that both contain less than four sequences, so any shuffle that does not produce $\bot$-sequences will produce at least four sequences of the same type, yielding a bad state of type $\mathcal{B}_{4*}$.   □

Claim (Criteria for Dead Columns). The following criteria for columns ensure that if the next turn in the protocol occurs in this column, then we are either already bad or this turn entails a bad successor state. We say the column is dead.[9] In particular, if all columns are dead, we know that after the next turn, we get a bad state.

$\mathcal{D}_{3\clubsuit}$: The column contains 3 $\clubsuit$.

$\mathcal{D}_{5\heartsuit}$: The column contains 5 $\heartsuit$.

$\mathcal{D}_{2*\clubsuit}$: The column contains 2 $\clubsuit$ belonging to sequences of the same type.

$\mathcal{D}_{3*\heartsuit}$: The column contains 3 $\heartsuit$ belonging to sequences of the same type.

Proof. – If a column contains three or more $\clubsuit$, turning this column yields a bad state with a constant $\clubsuit$-column and three or more sequences.

- If a column contains five or more $\heartsuit$, turning this column yields a bad state with a constant $\heartsuit$-column and five or more sequences.

- if a column contains two $\clubsuit$ belonging to sequences of the same type, an additional sequence of the other type with $\clubsuit$ must be added in this position to make it turnable. This leads to a column with three $\clubsuit$, and turning there yields a $\mathcal{B}_{3\clubsuit}$.

- If three sequences of the same type all have $\heartsuit$ in a column, there needs to be an additional sequence of this type with $\clubsuit$ in this column to make it turnable. Adding such a sequence yields a bad state of type $\mathcal{B}_{4*}$.   □

---

[8] The cycles have co-prime length, as opposed to the case $\langle(1\,2)(3\,4)\rangle$, which we handle explicitly in the proof of the theorem.

[9] If the column is unturnable then any method (not involving a turn) to make it turnable first before turning it in this column, will retain/ensure this deadness property.

Claim (Death is Contagious). Consider a dead column with index i in a state μ and a closed shuffle act with permutation set Π such that $\pi(i) = j$ for a $\pi \in \Pi$. Then the column with index j is dead as well after applying act to μ.

Proof. – By Lemma 1, the number of ♣ must be the same in columns i and j after the shuffle. Therefore $\mathcal{D}_{3♣}$ spreads.
- This case is completely analogous to the previous, with $\mathcal{D}_{5♡}$ instead of $\mathcal{D}_{3♣}$.
- By Lemma 1 column j must have at least two ♣ in the same type of sequences after the shuffle.
- For any shuffle that does not create ⊥-sequences, by Lemma 1 there must be three sequences of the same type that have ♡ in column j. Therefore j is dead after the shuffle.   □

Proof (of Theorem 1). We show that from each bad state there is a path into another bad state by case analysis.

States with four sequences of the same type. Any non-trivial shuffle not producing ⊥-sequences retains the four sequences of the same type.
Assume w.l.o.g. that there are four 0-sequences and consider turn operations. This requires at least two 1-sequences. If there are two sequences of type 1, we have six sequences in total. Turning either yields two states of at least three sequences, in particular one with constant ♣, a $\mathcal{B}_{3♣}$, or a 3/1 state with constant ♡, a $\mathcal{B}_{♡3/1}$. If there are more that two 1-sequences, there are at least seven sequences in total. A turn yields two successor states – one with a constant ♡-column and k sequences, and one with a constant ♣-column with ℓ sequences and $k + \ell \geq 7$. So we have $k \geq 5$ ($\mathcal{B}_{5♡}$) or $\ell \geq 3$ ($\mathcal{B}_{3♣}$).

States with a constant ♣-column and four sequences. Up to reordering, the state looks like

$$
\begin{array}{ccccc}
♣ & ♣ & ♡ & ♡ & ♡ \\
♣ & ♡ & ♣ & ♡ & ♡ \\
♣ & ♡ & ♡ & ♣ & ♡ \\
♣ & ♡ & ♡ & ♡ & ♣
\end{array}
$$

This state admits no non-trivial turn operation. Any shuffle operation that does not involve the first column produces a similar state. Since any column other than the first contains three ♡, any other shuffle produces three additional ♡ in the first column by Lemma 1, resulting in a state of type $\mathcal{B}_{4*}$.

States with a constant ♣-column and three sequences. Without loss of generality, these states are of type 2/1. They are non-turnable. Shuffles that do not involve the constant column will retain the property of being constant ♣ in that column and three or more sequences. Hence, in the following we consider shuffles involving the constant column.
To keep the proof simple, an important tool are the orbit partitions of each of the three equivalence classes of shuffles, as in Fig. 4. We try to place the sequences into the orbits, s.t. completing these does not yield a bad state. W.l.o.g. we choose sequences, s.t. the constant ♣-column is the first column.
   Case 1: (1 2). The orbit partition looks as in Fig. 4a. The first three orbits contain no ♣ there and are out. No orbits contain two ♣ in the first

column, so both 0-sequences must be placed in distinct orbits. If both orbits are of size 2, shuffling yields four 0-sequences giving a $\mathcal{B}_{4*}$-state. Otherwise, one of the 0-sequences is ♣♣♡♡♡. The other 0-sequence and the 1-sequence must be placed into orbits of size 2. All of them have ♣ only in one column out of the columns 3, 4, 5, and ♣♣♡♡♡ has ♣ in none of them, so for all choices we end up in a $\mathcal{B}_{5\heartsuit}$-state.

Case 2: $(1\,2)(3\,4)$. Similarly to before, we need to choose one 0-sequence as ♣♣♡♡♡ and need to place the remaining two sequences into the last three orbits of Fig. 4b. Choosing orbit 4 and 5 yields a $\mathcal{B}_{5\heartsuit}$-state.

If we choose orbits 4 (or 5) and 6 then the first two columns are dead ($\mathcal{D}_{3♣}$) and the fifth column is dead as well ($\mathcal{D}_{2*♣}$). If we choose the second 0-sequence within orbit 6, then columns 3 and 4 are dead ($\mathcal{D}_{3*\heartsuit}$).

If we choose the 0-sequence within orbit 4 (or 5) and the 1-sequence within orbit 6, there are two living non-turnable columns. Any shuffle that contains only columns 3 and 4 in one cycle does not produce a 1-sequence with ♣ in those columns, so they remain non-turnable. Any shuffle that makes column 3 or 4 turnable by shuffling them with at least one of the dead columns kills the column in the process.

Case 3: $(1\,2\,3)$. If we place the three sequences in the pairwise distinct orbits of Fig. 4d, we end up with nine sequences after the shuffle. Otherwise, the two 0-sequences share the bottommost orbit and the 1-sequence must be in the second or third orbit, and we get a $\mathcal{B}_{5\heartsuit}$-state.

States with a constant ♡-column and five or more sequences. W.l.o.g. the first column is constant ♡. Consider a turn operation on column $i \neq 1$. Column i has either three ♡ and turning therefore leads to a $\mathcal{B}_{3\heartsuit\heartsuit}$-state, or column i has three ♣ and turning leads to a $\mathcal{B}_{3♣}$-state.

Any shuffle not involving the first column keeps it constant ♡, and therefore bad. Consider any shuffle involving the first column, say π is a possible permutation in the permutation set of the shuffle with $\pi(i) = 1$. Column i contains at least two ♣ and by Lemma 1 shuffling yields two new sequences with ♣ in position 1, giving seven or more sequences – a $\mathcal{B}_{4*}$-state.

States with two constant ♡-columns and three sequences. No turn operation is possible as the state is of type 1/2 or 2/1. Shuffles involving none of the constant columns keep them constant. Shuffles involving only one of them, produce a state with two ♣ in that column, so five sequences in total, where the other ♡-column stays constant. This is a $\mathcal{B}_{5\heartsuit}$-state.

For the interesting case of shuffles involving both constant ♡-columns, we again try to place the sequences into the orbits of the different types of shuffles such that completing these orbits does not yield two or more additional 0-sequences, as this would lead to a 4/2 state or to seven or more sequences.

Case 1: $(1\,2)$. As both constant columns have to be involved in the shuffle, they have to be in positions 1 and 2. This leaves the state constant.

Case 2: $(1\,2)(3\,4)$. If the constant ♡-columns are both in the same cycle, we do not get additional sequences. Otherwise, say they are in positions 2 and 3, the only three sequences are ♡♡♡♣♣ from orbit 2, ♣♡♡♣♡ from orbit 5, and ♣♡♡♡♣ from orbit 6, and shuffling results in a $\mathcal{B}_{4*}$-state.

Case 3: (1 2 3). W.l.o.g. the constant columns are 1 and 2. The sequences with $\heartsuit$ in those positions are all in distinct orbits with a combined size of 7, which is a $\mathcal{B}_{4*}$-state.

3/1-states with a constant $\heartsuit$-column. This state is not turnable as it has only one 1-sequence. A non-trivial shuffle not involving the constant column yields a state with five or more sequences and a constant $\heartsuit$-column. For shuffles involving the constant column we try to place the sequences into the orbits of the different classes of shuffles such that completing these orbits does not yield any additional 0-sequences.

Case 1: (1 2). We have to involve the constant $\heartsuit$-column, which is w.l.o.g. in position 1. The only orbits with constant $\heartsuit$ are 1, 2 and 3. To not produce additional 0-sequences, those need to be the ones occupied by the 0-sequences. We need to place the 1-sequence in orbits 5, 6 or 7, and choose w.l.o.g. 5. Then, the first two columns are dead via $\mathcal{D}_{3*\heartsuit}$. The third column is dead via $\mathcal{D}_{3\clubsuit}$. Columns 4 and 5 are dead via $\mathcal{D}_{2*\clubsuit}$.

Case 2: (1 2)(3 4). Regardless, the first two columns are dead via $\mathcal{D}_{3*\heartsuit}$ and the other columns are dead via $\mathcal{D}_{2*\clubsuit}$

Case 3: (1 2 3). The constant $\heartsuit$-column is w.l.o.g. the first. No orbit contains three $\heartsuit$ in column 1, so the 0-sequences are spread over at least two orbits. Any choice of two orbits contains four or more sequences, so we have four sequences of the same type after the shuffle.

This concludes the proof. □

7 Impossibility of Four-Card Restart-Free AND Protocols with Uniform Closed Shuffles

As discussed on Sect. 1, if we are willing to discard the finite-runtime requirement, there is an intermediate property, namely restart-freeness. This section focuses on restart-free AND protocols. Our proof is similar to the setting of [KWH15] in that we start from a (slightly enlarged) set of good states, but instead of showing that there is an infinite path of non-good states, we show the stronger property that no path contains a good state.

Note that for protocols with only closed, but possibly non-uniform shuffles, the four-card AND protocol of [KWH15] provides an already card-minimal solution. For protocols which are additionally restricted to uniform and closed shuffles, we show in the following that five cards are necessary to compute AND in committed format without restarts.

The following argument speaks about non-reduced states. We still use the terminology for reduced states, but the precise notion of turnability from Eq. (1).

Theorem 2. *There is no secure restart-free Las Vegas four-card AND protocol in committed format if shuffling is restricted to uniform closed shuffles.*

Proof. The proof is similar to the proof of [KWH15, Theorem 3]. Let $\mathcal{P}$ be a secure protocol computing AND with four cards using only uniform closed shuffles and no restart actions. We define a set $\mathcal{G}$ of good states that contains all final states but not the start state. We then show that we cannot get into the set of good states from a non-good state with a turn or a uniform closed shuffle that does not create $\bot$-sequences. A state is *good* if it is

- a state of type 1/1 or 2/2,
- a state of type 1/2 or 2/1 without a constant column,
- a turnable state of type 2/3 or 3/2.

All other states are *bad*. The main difference to [KWH15, Theorem 3] is that we need to consider states where turning yields at least one good state instead of only the ones that yield only good states. In particular, the additional good states are the turnable 2/3- and 3/2-states.

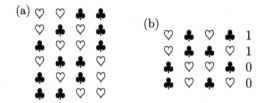

Fig. 5. (a) All sequences with four cards. (b) A sequence-maximal final state with four cards w.r.t. (result, 1, 2).

Note that as our deck is $[\heartsuit, \heartsuit, \clubsuit, \clubsuit]$, we can form at most 6 sequences shown in Fig. 5a. The start state is bad because it is of type 3/1. The final states are amongst the good states because they are of type i/j with $i, j \leq 2$. If they are of type 2/1 or 1/2 they do not have a constant column as one can easily see when looking at the sequence-maximal final state for four cards (Fig. 5b).

Turns. Only states of type i/j with $i, j \geq 2$ are turnable. The only bad states that fit this criterion are of type 3/3 or 4/2 (2/4). The maximum state with a constant column has three sequences, therefore these states can only be turned into two states both of which are of types 1/2 or 2/1 with a constant column.

Shuffles. We cannot shuffle a bad state μ into a good state μ' of type i/j where both $i, j \leq 2$. Assume we could. This would require μ to be a bad 1/2-state, thus μ has a constant column and three sequences. Any shuffle that adds sequences involves the constant column. Any other column of μ contains the symbol of the constant column only once, hence, by Lemma 1 it must contain it at least three times after the column has been shuffled with the constant column. This adds at least two sequences, contradicting our assumption on the type of μ'. Therefore we only need to look at shuffles that yield a turnable 2/3 or 3/2-state.

Assume that an action (shuffle, Π) transforms a bad μ into a turnable 3/2 state μ'. Without loss of generality, we assume position 1 is turnable in μ' and the sequence not contained in μ' is $s = \heartsuit\heartsuit\clubsuit\clubsuit$.[10] By Lemma 2, $\{s\}$ is an orbit of Π of size 1, i.e. s is invariant under Π. In particular, Π is a non-trivial subgroup of $\{\mathrm{id}, (1\ 2), (3\ 4), (1\ 2)(3\ 4)\} \subseteq S_4$.

If we had $(3\ 4) \in \Pi$, then the only two sequences of μ' with $\heartsuit$ in position 1, namely $\heartsuit\clubsuit\heartsuit\clubsuit$ and $\heartsuit\clubsuit\clubsuit\heartsuit$, would share an orbit of Π, and therefore have the same type in μ', contradicting turnability of the first column of μ'. Thus, either $(1\ 2)$ or $(1\ 2)(3\ 4)$ is in Π (but not both). Assume $(1\ 2) \in \Pi$, the other case is analogous. By Lemma 3, it holds that

$$\mu'(\heartsuit\clubsuit\heartsuit\clubsuit) = \mu'(\clubsuit\heartsuit\heartsuit\clubsuit) \text{ and } \mu'(\heartsuit\clubsuit\clubsuit\heartsuit) = \mu'(\clubsuit\heartsuit\clubsuit\heartsuit), \qquad (*)$$

because these each share an orbit. As position 1 of μ' is turnable, we know that

$$\mu'(\heartsuit\clubsuit\heartsuit\clubsuit) + \mu'(\heartsuit\clubsuit\clubsuit\heartsuit) = p \in (0, 1), \text{ and}$$
$$\mu'(\clubsuit\clubsuit\heartsuit\heartsuit) + \mu'(\clubsuit\heartsuit\heartsuit\clubsuit) + \mu'(\clubsuit\heartsuit\clubsuit\heartsuit) = q \in (0, 1),$$

i.e., constant. Using this and $(*)$, we obtain $\mu'(\clubsuit\clubsuit\heartsuit\heartsuit) = q-p$, which is constant. It is non-zero, so each of the monomials $X_{00}, X_{01}, X_{10}, X_{11}$ occurs with a positive coefficient, meaning $\clubsuit\clubsuit\heartsuit\heartsuit$ is a $\perp$-sequence of μ' – a contradiction. □

8   Impossibility of n-COPY with $2n$ Cards

In this section, we prove that any protocol that produces n copies of a commitment needs at least $2n + 1$ cards, showing that we cannot improve w.r.t. to the number of cards on the protocol in [NNH+17], cf. Fig. 3. We start with a lemma.

Lemma 5. *Let $\mathcal{P}$ be a secure protocol computing a function f. Assume that a reduced state μ of $\mathcal{P}$ is transformed to a non-similar state μ' by an action. Then,*

(a) the action cannot be of type perm.
(b) if the action is a turn, μ' has a sibling state.
(c) if it is a shuffle, either $|\mu| = |\mu'|$ and μ' has a $\perp$-sequence, or $|\mu| < |\mu'|$.

Proof.

(a) A permutation can only produce similar states by definition.
(b) As μ' and μ are not similar, μ' contains a proper subset of the sequences of μ. The sequences which were removed from μ' are not compatible with the visible sequence annotation to μ' and hence it needs to have a sibling state.
(c) Clearly, a (shuffle, $\Pi, \mathcal{F}$) action cannot reduce the number of sequences. Assume that $|\mu| = |\mu'|$ and μ' contains no $\perp$-sequences. Let $s_1, s_2, \ldots, s_i$ be the sequences in μ. For any $\pi \in \Pi$, μ' includes all sequences $\pi(s_1), \pi(s_2), \ldots, \pi(s_i)$. Because $|\mu| = |\mu'|$, μ' cannot have any other sequences. Thus, μ and μ' are similar via π, a contradiction. □

---

[10] If this is not the case we can apply a permutation such that the constant sequence either is $\heartsuit\heartsuit\clubsuit\clubsuit$ or $\clubsuit\clubsuit\heartsuit\heartsuit$ and use the symmetry of exchanging $\heartsuit$ and $\clubsuit$.

Our impossibility proof first assumes the existence of a $2n$-card COPY protocol which is minimal in the sense that it admits the shortest run, i.e. there is no protocol with the same functionality where there is a leaf state that is less deep.[11] We call it *run-minimal*. We then derive a contradiction by showing that the leaf state of the shortest run cannot be reached by one of the admissible actions.

Theorem 3. *There is no (possibilistically) secure $2n$-card n-COPY protocol.*

Proof. Suppose for a contradiction that there are $2n$-card n-COPY protocols, and let $\mathcal{P} = (\mathcal{D}, U, Q, A)$ be a run-minimal one. Let μ' be a leaf state of the state tree of $\mathcal{P}$ of minimum depth. As μ' is final, it is similar to the state

$$(\clubsuit \heartsuit)^n \quad 0$$
$$(\heartsuit \clubsuit)^n \quad 1,$$

meaning μ' contains exactly two sequences and the distance between them is $2n$. Let μ be the parent state of μ'. Note that μ and μ' are not similar, as otherwise the shortest run would obviously not be minimal, as we could remove μ' from the tree. Hence, by Lemma 5 we only need to consider the following actions:

- (turn, T): Sequences in μ' coincide at positions in the (non-empty) turn set T. But then μ' would not contain two sequences of distance $2n$, a contradiction.
- $(\mathsf{shuffle}, \Pi, \mathcal{F})$: The only way for μ to have only one sequence is if it is a $\perp$-sequence. Note that from that situation a shuffle cannot produce any 0- or 1-sequences and hence cannot end in μ'. Therefore, assume $|\mu| \geq 2$. Then, by Lemma 5(c) there are two possibilities. If $|\mu'| = |\mu|$ we would have introduced a $\perp$-sequence, which is not present in μ'. Therefore, we have $|\mu'| > |\mu| \geq 2$, which is a contradiction to $|\mu'| = 2$.
- $(\mathsf{restart})$: μ' would be the start state, where there are exactly two sequences of distance 2, contrary to the distance of $2n$, with $n > 1$. $\square$

By Theorem 3 and the existing $(2n+1)$-card COPY protocol [NNH+17], $2n+1$ is the necessary and sufficient number of cards for making n copied commitments. In the next section, we restrict our attention to finite-runtime protocols.

9   Impossibility of $2n + 1$-Card Finite-Runtime n-COPY

In Sect. 8, we showed that $2n + 1$ cards are minimal for COPY computations. Note that the existing $(2n + 1)$-card COPY protocol [NNH+17] has a runtime that is finite only in expectation. In this section, we show that if we are restricted to a finite runtime, there cannot be a protocol using only $2n + 1$ cards. For this, we look at the finite state tree of an assumed minimal protocol and consider a leaf state with a position of largest depth in the tree (i.e. there is no leaf state which is even deeper). By contradiction, we show that no such leaf state can exist, as it would have a sibling which is not yet a leaf state.

---

[11] This is to avoid intricacies with definitions such as: "we cannot delete a leaf while retaining functionality" (used in Sect. 9) for *infinite trees*.

Theorem 4. *There is no (possibilistically) secure $(2n + 1)$-card finite-runtime n-COPY protocol.*

Proof. Suppose for a contradiction that there exist $(2n + 1)$-card finite-runtime n-COPY protocols and let $\mathcal{P} = (\mathcal{D}, U, Q, A)$ be a minimal one, in the sense that we cannot remove a leaf while retaining functionality. The deck $\mathcal{D}$ includes at least n ♣s and n ♡s, and let remaining card be of symbol ♣ or of third color, say ♢. Because of the finite runtime, the height of the state tree of $\mathcal{P}$ is finite, and hence, there must be a deepest leaf state, i.e. a leaf state with no other leaf state being on a level below it. We call it μ'. As it is a final state, it needs to look like this, up to reordering:

$$♣(♣♡)^n \quad 0 \qquad\qquad ♢(♣♡)^n \quad 0$$
$$♣(♡♣)^n \quad 1, \qquad \text{or} \qquad ♢(♡♣)^n \quad 1.$$

The distance between the two sequences is $2n$, and one column is constant. Let μ be the parent state of μ'. Similarly to the proof of Theorem 3, perm and shuffle cannot have transformed μ to μ'. Hence, from the form of μ' the action must be (turn, $\{1\}$). By the minimality of $\mathcal{P}$, μ and μ' are not similar. Therefore, Lemma 5(b) implies that there is a sibling state μ'' of μ'. This sibling state has a constant ♡-column (or possibly a ♣-column, if we have a ♢) in the first position, as it is the only way for the visible sequences of the turn to differ. In this case, we cannot construct two sequences whose distance is $2n$ with the remaining $n - 1$ ♡s and $n + 1$ ♣s (or n ♣s and a ♢) in μ''. Therefore, μ'' cannot be a leaf state of $\mathcal{P}$, and there would be a deeper leaf than μ', contrary to our assumption. □

10   Conclusion

We extended the analysis on the necessary and sufficient number of cards in card-based protocols computing an AND in committed format to certain plausible restrictions on the operations that can be performed during a protocol run. These are restrictions to certain forms of shuffling, namely closed and/or uniform shuffles and whether running in loops or restarting is allowed. This focus allows to get a clearer view on how many cards are necessary in protocols with *favorable* properties, such as finite-runtime or easy-to-do/actively secure shuffling. It is for example useful to now be aware that a search for five-card finite-runtime protocols using only closed shuffles will be fruitless – and thereby identifying the six-card protocol by [MS09] as optimal w.r.t. closed-shuffle protocols. In the process, we highlight interesting properties from which the orbit partitions are the most useful. Furthermore, we extended the four-card impossibility result of [KWH15] to the case of uniform closed shuffles for restart-free Las Vegas protocol.

For bit copy, we proved that the $(2n + 1)$-card COPY protocol of [NNH+17] is card-minimal, and the $(2n + 2)$-card COPY protocol of [MS09] is card-minimal w.r.t. finite-runtime protocols. Figure 6 summarizes our results and surveys current bounds on the number of cards for all combinations of restrictions.

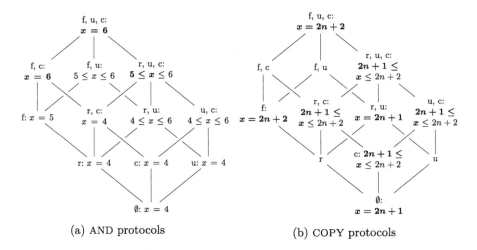

(a) AND protocols   (b) COPY protocols

Fig. 6. Currently known bounds on the numbers of cards needed for committed format protocols in form of a Hasse diagram. A line between two settings in the lattice describes that the configuration above the other has more restrictions and hence needs as least as many cards as the dominated configuration. The new results are in bold. Here, f means finite-runtime (which includes restart-freeness), r is restart-freeness, c is a restriction to closed shuffles, and u for uniform shuffles. The values of corners which only have a label are completely determined by the surrounding nodes and hence omitted for brevity.

Open Problems. We find it interesting whether our impossibility result concerning closed-shuffle finite-runtime five-card AND protocols carries over to the setting where we use a helping card of a third color, say $\diamondsuit$. A second open problem is related to the analysis of restart-free Las Vegas protocols from Sect. 7: As we show that a lower bound is five cards, we are keen to find out whether this bound is tight, i.e. whether there really is a (restart-free) five-card protocol which uses only uniform closed shuffles. In addition to that it would be very interesting to extend the analysis with respect to restricted shuffling to COPY protocols. For example, does there exist a $(2n + 1)$-card COPY protocol using only (uniform) closed shuffles?

Acknowledgments. We would like to thank our reviewers for their valuable comments. This work was supported by JSPS KAKENHI Grant Number 17K00001.

References

[CHL13] Cheung, E., Hawthorne, C., Lee, P.: CS 758 Project: Secure Computation with Playing Cards (2013). https://csclub.uwaterloo.ca/~cdchawth/files/papers/secure_laying_cards.pdf

[CK93] Crépeau, C., Kilian, J.: Discreet solitary games. In: Stinson, D.R. (ed.) CRYPTO 1993. LNCS, vol. 773, pp. 319–330. Springer, Heidelberg (1994). https://doi.org/10.1007/3-540-48329-2_27

[dBoe89] den Boer, B.: More efficient match-making and satisfiability The Five Card Trick. In: Quisquater, J.-J., Vandewalle, J. (eds.) EUROCRYPT 1989. LNCS, vol. 434, pp. 208–217. Springer, Heidelberg (1990). https://doi.org/ 10.1007/3-540-46885-4_23

[DM96] Dixon, J.D., Mortimer, B.: Permutation Groups. Graduate Texts in Mathematics, vol. 163. Springer, New York (1996). https://doi.org/10.1007/ 978-1-4612-0731-3

[FAN+16] Francis, D., Aljunid, S.R., Nishida, T., Hayashi, Y., Mizuki, T., Sone, H.: Necessary and sufficient numbers of cards for securely computing two-bit output functions. In: Phan, R.C.-W., Yung, M. (eds.) Mycrypt 2016. LNCS, vol. 10311, pp. 193–211. Springer, Cham (2017). https://doi.org/10.1007/ 978-3-319-61273-7_10

[KW17] Koch, A., Walzer, S.: Foundations for actively secure card-based cryptography. Cryptology ePrint Archive, Report 2017/423 (2017). https://eprint. iacr.org/2017/423

[KWH15] Koch, A., Walzer, S., Härtel, K.: Card-based cryptographic protocols using a minimal number of cards. In: Iwata, T., Cheon, J.H. (eds.) ASIACRYPT 2015. LNCS, vol. 9452, pp. 783–807. Springer, Heidelberg (2015). https:// doi.org/10.1007/978-3-662-48797-6_32

[Miz16] Mizuki, T.: Card-based protocols for securely computing the conjunction of multiple variables. Theoret. Comput. Sci. 622, 34–44 (2016). https:// doi.org/10.1016/j.tcs.2016.01.039

[MKS12] Mizuki, T., Kumamoto, M., Sone, H.: The five-card trick can be done with four cards. In: Wang, X., Sako, K. (eds.) ASIACRYPT 2012. LNCS, vol. 7658, pp. 598–606. Springer, Heidelberg (2012). https://doi.org/10.1007/ 978-3-642-34961-4_36

[MS09] Mizuki, T., Sone, H.: Six-card secure AND and four-card secure XOR. In: Deng, X., Hopcroft, J.E., Xue, J. (eds.) FAW 2009. LNCS, vol. 5598, pp. 358–369. Springer, Heidelberg (2009). https://doi.org/10.1007/ 978-3-642-02270-8_36

[MS14] Mizuki, T., Shizuya, H.: A formalization of card-based cryptographic protocols via abstract machine. Int. J. Inf. Secur. 13(1), 15–23 (2014). https:// doi.org/10.1007/s10207-013-0219-4

[MS17] Mizuki, T., Shizuya, H.: Computational model of card-based cryptographic protocols and its applications. In: IEICE Transactions, vol. 100-A.1, pp. 3–11 (2017). http://search.ieice.org/bin/summary.php?id=e100-a_1_3

[MWS15] Marcedone, A., Wen, Z., Shi, E.: Secure dating with four or fewer cards. Cryptology ePrint Archive, Report 2015/1031 (2015)

[NHMS16] Nishimura, A., Hayashi, Y.-I., Mizuki, T., Sone, H.: An implementation of non-uniform shuffle for secure multi-party computation. In: Workshop on ASIA Public-Key Cryptography, Proceedings. AsiaPKC 2016, pp. 49–55. ACM, New York (2016). https://doi.org/10.1145/2898420.2898425

[NNH+15] Nishimura, A., Nishida, T., Hayashi, Y., Mizuki, T., Sone, H.: Five-card secure computations using unequal division shuffle. In: Dediu, A.-H., Magdalena, L., Martín-Vide, C. (eds.) TPNC 2015. LNCS, vol. 9477, pp. 109–120. Springer, Cham (2015). https://doi.org/10.1007/978-3-319-26841-5_9

[NNH+17] Nishimura, A., Nishida, T., Hayashi, Y., Mizuki, T., Sone, H.: Card-based protocols using unequal division shuffles. Soft Computing (to appear). https://doi.org/10.1007/s00500-017-2858-2

Foundations

Succinct Spooky Free Compilers Are Not Black Box Sound

Zvika Brakerski[1(✉)], Yael Tauman Kalai[2], and Renen Perlman[1]

[1] Weizmann Institute of Science, Rehovot, Israel
zvika.brakerski@weizmann.ac.il, renenp@gmail.com
[2] Microsoft Research and MIT, Cambridge, USA
yaelism@gmail.com

Abstract. It is tempting to think that if we encrypt a sequence of messages $\{x_i\}$ using a semantically secure encryption scheme, such that each x_i is encrypted with its own independently generated public key pk_i, then even if the scheme is malleable (or homomorphic) then malleability is limited to acting on each x_i independently. However, it is known that this is not the case, and in fact even non-local malleability might be possible. This phenomenon is known as *spooky interactions*.

We formally define the notion of *spooky free compilers* that has been implicit in the delegation of computation literature. A spooky free compiler allows to encode a sequence of queries to a multi-prover interactive proof system (MIP) in a way that allows to apply the MIP prover algorithm on the encoded values on one hand, and prevents spooky interactions on the other. In our definition, the compiler is allowed to be tailored to a specific MIP and does not need to support any other operation.

We show that (under a plausible complexity assumption) spooky free compilers that are sufficiently succinct to imply delegation schemes for NP with communication n^α (for any constant $\alpha < 1$) cannot be proven secure via black-box reduction to a falsifiable assumption. On the other hand, we show that it is possible to construct *non-succinct spooky free fully homomorphic encryption*, the strongest conceivable flavor of spooky free compiler, in a straightforward way from any fully homomorphic encryption scheme.

Our impossibility result relies on adapting the techniques of Gentry and Wichs (2011) which rule out succinct adaptively sound delegation protocols. We note that spooky free compilers are only known to imply *non-adaptive* delegation, so the aforementioned result cannot be applied directly. Interestingly, we are still unable to show that spooky free compilers imply adaptive delegation, nor can we apply our techniques directly to rule out arbitrary non-adaptive NP-delegation.

1 Introduction

The PCP Theorem [AS98, ALM+98] is one of the most formidable achievements of computer science in the last decades. Probabilistically Checkable Proofs (PCPs)

Z. Brakerski and R. Perlman—Supported by the Israel Science Foundation (Grant No. 468/14) and Binational Science Foundation (Grants Nos. 2016726, 2014276).

© International Association for Cryptologic Research 2017
T. Takagi and T. Peyrin (Eds.): ASIACRYPT 2017, Part III, LNCS 10626, pp. 159–180, 2019.
https://doi.org/10.1007/978-3-319-70700-6_6

and Multi-Prover Interactive Proofs (MIPs) allow to reduce the communication complexity of verifying an NP statement to logarithmic in the input length (and linear in the security parameter), in a single round of communication. However, they require sending multiple queries to isolated non-colluding provers.[1] It is impossible (under plausible complexity assumptions) to achieve the same communication complexity with a single computationally unbounded prover. However, if we only require computational soundness this may be possible.

Indeed, it has been shown by Micali [Mic94] and Damgård et al. and Bitansky et al. [DFH12,BCCT12,BCCT13,BCC+14] that in the random oracle model or relying on knowledge assumptions, it is indeed possible. However, in the standard model and under standard hardness assumptions (in particular falsifiable [Nao03]), this is not known. Gentry and Wichs [GW11] showed that if adaptive security is sought, i.e. if the adversary is allowed to choose the NP instance after seeing the challenge message from the verifier, then soundness cannot be proved under any falsifiable assumption, so long as the security reduction uses the adversary as a black-box, and relying on the existence of sufficiently hard languages in NP. This still leaves open the possibility of non-adaptive protocols which seems to be beyond the reach of the techniques of [GW11].[2]

A notable attempt was made by Biehl et al. [BMW98], and by Aiello et al. [ABOR00]. They suggested to generate MIP queries and encode them using independent instances of a private information retrieval (PIR) scheme. Intuitively, since each query is encoded separately, it should be impossible to use the content of one encoding to effect another. However, as Dwork et al. [DLN+01] showed, the provable guarantees of PIR (or semantically secure encryption) are insufficient to imply the required soundness. They showed that semantic security does not preclude non-local *spooky interactions* which cannot be simulated by independent provers.

Dodis et al. [DHRW16] recently showed that there exist explicit secure PIR schemes (under widely believed cryptographic assumptions) that actually exhibit spooky interactions, and thus fail the [BMW98, ABOR00] approach. They complemented this negative result with a construction of a *spooky free* fully homomorphic encryption (FHE) scheme, which is an FHE scheme with the additional guarantee that if multiple inputs are encrypted using independently generated public keys, then any operation on the collection of ciphertexts can be simulated by independent processes applied to each encrypted message separately. In particular, a spooky free FHE has strong enough security guarantees to allow proving the [BMW98, ABOR00] approach, since a single computationally bounded prover "has no choice" but to behave like a collection of isolated provers as is

---

[1] We purposely refrain from distinguishing between a PCP, where multiple queries are made to a fixed proof string, and a single round MIP, where there are multiple provers. The difference is insignificant for the purpose of our exposition and the two forms are often equivalent.

[2] Extending the black-box impossibility to non-adaptive delegation is a well motivated goal by itself and has additional implications, e.g. for the study of program obfuscation.

required for MIP soundness. However, the spooky free encryption scheme constructed by Dodis et al. relies on knowledge assumptions, the same knowledge assumptions that imply short computationally sound proofs (and in fact uses them as building blocks).

Our Results. In this work, we notice that spooky free FHE is a flavor of a more general notion that we call *spooky free compiler*. This notion has been implicit in previous works since [BMW98, ABOR00]. A spooky free compiler provides a way to encode and decode a set of queries in such a way that any operation on an encoded set, followed by decoding, is equivalent to performing an independent process on each of the queries separately. In addition, for functionality purposes, it should be possible to apply the MIP prover algorithm on encoded queries. This notion generalizes much of the research efforts in providing a proof for [BMW98, ABOR00]-style protocols. In particular, spooky free FHE can be viewed as a *universal* spooky free compiler that is applicable to all MIPs.

We show that spooky free compilers cannot have succinct encodings if they are proven based on a falsifiable hardness assumption using a reduction that uses the adversary as black-box. Our negative result holds for any compiler where the encoding is succinct enough to imply a delegation scheme with sub-linear communication complexity. We note that this does not follow from [GW11] since spooky free compilers are only known to imply non-adaptive delegation protocols whereas [GW11] only rules out adaptive protocols.

On the other hand, we show that if succinctness is not imposed, then it is straightforward to achieve spooky free FHE based on the existence of any FHE scheme. Namely, spooky free compilation in its strongest sense becomes trivial. Specifically, we present a scheme where the encoding size corresponds to the size of the query space for the MIP, i.e. the length of the truth table of the MIP provers.

Other Related Works. Kalai et al. [KRR13, KRR14] showed that the [BMW98, ABOR00] approach is in fact applicable and sound when using *no signaling MIP*. These are proof systems that remain sound even when spooky interactions are allowed. However, such MIPs can only be used to prove statements for P and not for all of NP unless NP=P.

1.1   Overview of Our Techniques

We provide an overview of our techniques. For this outline we only require an intuitive understanding of the notion of spooky free compiler as we tried to convey above. The formal definition appears in Sect. 3.

Ruling Out Succinct Compilers. Our method for ruling out succinct compilers draws from the [GW11] technique for showing the impossibility of reductions for adaptively secure delegation schemes, i.e. ones where the instance x can be chosen after the encoded MIP queries are received. At a high level, [GW11] produce an adversary that chooses instances x that are not in the NP language in question, but are computationally indistinguishable from ones that are in the language. This allows to simulate accepting short delegation responses for those x's using a brute force process, since the complexity of the exhaustive process is still insufficient to distinguish whether x is in the language or not (this argument makes use of the dense model theorem [DP08, RTTV08, VZ13]). The crucial property that is required is that each x is only used once, since otherwise the combined complexity of applying the brute force process many times will not allow us to rely on the computational indistinguishability. The adaptive setting allows to choose a new x for each query, and thus to apply this argument.

We notice that a spooky free compiler is similar to an adaptive delegation protocol, since it does not preclude the adversary from using a fresh x for each set of queries. We will consider an adversary that samples x not in the language similarly to [GW11], but instead of performing the MIP evaluation on the encoded queries it uses the dense model theorem to produce an accepting response.

We would like to then argue that this adversary breaks the spooky-freeness, since it cannot be simulated by a sequence of local operations on the queries due to the unconditional soundness of the MIP. However, we need to be rather careful here, since an attempt to simulate will only fail w.r.t. a distinguisher who knows x (otherwise the soundness of the MIP is meaningless). It may seem that this can be handled by giving x to the distinguisher together with the MIP answers, e.g. by considering an additional "dummy MIP prover" that always returns x, so that x is now sent together with the MIP answers. Alas, this approach seems to fail, since a simulator can simulate the adversary by using x in the language, and answering the queries locally. The dense model theorem implies that the two views are indistinguishable, which in turn implies that this adversary does not break the spooky freeness.

We overcome this obstacle by confining the adversary to choose x from a small bank $\overline{X}$ of randomly chosen x's that are not in the language, and are a priori sampled and hardwired to the adversary's code. We consider a distinguisher that also has this bank $\overline{X}$ hardwired into its code, and will output 1 if and only if the answers are accepting with respect to *some* $x \in \overline{X}$. We denote this adversary and distinguisher pair by $(\overline{\mathcal{A}}, \overline{\Psi})$, and use the soundness of the MIP to argue that the distinguisher $\overline{\Psi}$ can distinguish between the adversary $\overline{\mathcal{A}}$ and any local process, which implies that $(\overline{\mathcal{A}}, \overline{\Psi})$ break the spooky freeness.

The fact that $(\overline{\mathcal{A}}, \overline{\Psi})$ break spooky freeness implies that the black-box reduction breaks the assumption given oracle access to $(\overline{\mathcal{A}}, \overline{\Psi})$.[3] We reach a contradiction by showing efficient (probabilistic polynomial time) algorithms $(\mathcal{A}, \Psi)$ which are indistinguishable from $(\overline{\mathcal{A}}, \overline{\Psi})$ in the eyes of the reduction, which implies that the underlying assumption is in fact solvable in probabilistic polynomial time.

---

[3] In fact, the situation is more delicate since $(\overline{\mathcal{A}}, \overline{\Psi})$ is actually a *distribution* over adversaries and distinguishers, where the distribution is over the choice of the bank $\overline{X}$. We argue that almost all $(\overline{\mathcal{A}}, \overline{\Psi})$ break the spooky freeness, and then prove that the average advantage is also non-negligible (see Lemma 16 in Sect. 2).

See Sect. 5 for the full details of this negative result.

Straightforward Non-Succinct Spooky Free FHE. We show that any FHE scheme with message space Σ, implies a spooky free FHE scheme with message space Σ and ciphertext size $\approx |\Sigma|$. We explain the construction for $\Sigma = \{0, 1\}$, the extension to the general case is fairly straightforward, and we refer the reader to Sect. 4 for the full details.

Our starting point is an FHE scheme with message space $\{0, 1\}$. Our spooky free scheme is essentially an *equivocable* variant of the FHE scheme, namely one where there is a special ciphertext that can be explained as either an encryption of 0 or an encryption of 1 given an appropriate secret key. Formally, the spooky free key generation generates two key sets for the FHE scheme: $(\mathsf{fhepk}_0, \mathsf{fhesk}_0)$, $(\mathsf{fhepk}_1, \mathsf{fhesk}_1)$, it also flips a coin $b \xleftarrow{\$} \{0, 1\}$. Finally it outputs the spooky free key pair: $\mathsf{sfpk} = (\mathsf{fhepk}_0, \mathsf{fhepk}_1)$ and $\mathsf{sfsk} = (b, \mathsf{fhesk}_b)$. To encrypt, encrypt the same message with both fhepk's to obtain $c' = (c_0, c_1)$. Homomorphic evaluation can be performed on c_0, c_1 independently, and since both components of the ciphertext will always encrypt the same value, then decrypting with fhesk_b will be correct regardless of the value of b. Note that the size of the ciphertext blew up by a factor of $|\Sigma| = 2$.

To show that the scheme is spooky free, we notice that it is possible to generate an equivocable ciphertext $c^* = (\mathsf{Enc}_{\mathsf{fhepk}_0}(\beta), \mathsf{Enc}_{\mathsf{fhepk}_1}(\bar{\beta}))$, for a random $\beta \xleftarrow{\$} \{0, 1\}$. Note that for $b = \beta \oplus x$, it holds that $\mathsf{sfsk}_x = (b, \mathsf{fhesk}_b)$ decrypts c^* to the value x, and furthermore, the joint distribution $(\mathsf{sfpk}, \mathsf{sfsk}_x, c^*)$ is computationally indistinguishable from the case where b was chosen randomly and c^* was a proper encryption of x.

To see why this scheme is spooky free, we consider an adversary that receives a number of ciphertexts under independently generated sfpk's and attempts to perform some non-local spooky interaction. Namely, the adversary takes $\{\mathsf{sfpk}_i, c'_i = \mathsf{Enc}_{\mathsf{sfpk}_i}(x_i)\}_i$, performs some operation to produce $\{\tilde{c}_i\}_i$ s.t. when decrypting $y_i = \mathsf{Dec}_{\mathsf{sfsk}_i}(\tilde{c}_i)$, the entries y_i should be distributed in a way that cannot be simulated locally by operating on each x_i independently. We will show that this is impossible and in fact there is a local way to generate the y_i values, up to computational indistinguishability.

To this end, we first consider a setting where instead of c'_i, we feed the adversary with the equivocable ciphertext c^*_i. Recall that the value x_i that c^*_i encrypts is determined by sfsk and not by c^* itself. Still, as we explained above, the distribution of (public key, secret key, ciphertext) is indistinguishable from the previous one. Therefore, in this experiment the adversary should return a computationally indistinguishable distribution over the y_i's as it did before. However, notice that now the adversary's operation does not depend on the x_i's at all. Namely, it is possible to decide on the value of x_i only at decryption time and not at encryption time, and it is possible to do so for each i independently (by selecting an appropriate value for b in the i'th instantiation of the scheme). It follows that the distribution of y_i in this experiment, which is computationally indistinguishable from the original one, is spooky free in the sense that it can be generated by executing a local process on each x_i to compute y_i.[4]

Tightness. Similarly to the [GW11] argument, our black-box impossibility result shows there is no spooky free compiler where the length of the evaluated answers is less than $|x|^\alpha$ for a constant $\alpha > 0$ which is determined by the hardness of the NP language used in the proof. Assuming that we use an MIP with a small (polylogarithmic) number of queries and small (polynomial) query alphabet Σ,[5] we get that the average encoded answer size is lower bounded by $|\Sigma|^{\Omega(1)}$. However, as we showed above, we can construct spooky free FHE with ciphertext size $\approx |\Sigma|$, which matches the lower bound up to a polynomial.

2   Preliminaries

Definition 1. *Two distributions $\mathcal{X}, \mathcal{Y}$ are said to be $(\epsilon(\lambda), s(\lambda))$-indistinguishable if for every distinguisher Ψ of size $\mathrm{poly}(s(\lambda))$ it holds that*

$$|Pr\left[\Psi(\mathcal{X}) = 1\right] - Pr\left[\Psi(\mathcal{Y}) = 1\right]| \le \epsilon(\lambda) .$$

We say that the distributions $\mathcal{X}, \mathcal{Y}$ are α-sub-exponentially indistinguishable if they are $(2^{-n^\alpha}, 2^{n^\alpha})$-indistinguishable.

Lemma 2 (Borel-Cantelli). *For any sequence of events $\{E_\lambda\}_{\lambda \in \mathbb{N}}$, if the sum of the probabilities of E_λ is finite, i.e. $\sum_{\lambda \in \mathbb{N}} \Pr[E_\lambda] < \infty$, then the probability that infinitely many of them occur is 0.*

Definition 3 (One-Round Multi-Prover Interactive Proofs (MIP)). *Let R be an NP relation, and let L be the induced language. A one-round p-prover interactive proof for L is a triplet of PPT algorithms $\Pi = (\mathcal{G}, (\mathcal{P}_1, \dots, \mathcal{P}_p), \mathcal{V})$ as follows:*

- ***Query Generation** $\vec{q} \leftarrow \mathcal{G}(1^\kappa)$* : *Outputs a set of queries $\vec{q} = (q_1, \dots, q_p)$ for the provers.*
- ***Provers** $a_i \leftarrow \mathcal{P}_i(q_i, x, w)$* : *Given the query corresponding to the i'th prover, outputs an answer a_i for x using the query q_i, the instance x and its witness w.*
- ***Verifier** $b \leftarrow \mathcal{V}(\vec{q}, \vec{a}, x)$* : *Using the set of queries $\vec{q}$ with matching answers $\vec{a}$ and the instance x outputs a bit b.*

We require that there is a soundness parameter $\sigma > 0$ such that $\sigma(\kappa) < 1 - 1/\mathrm{poly}(\kappa)$, for which the following two properties hold:

---

[4] A meticulous reader may have noticed that it is required that for all i the local process uses the same sequence of c_i^*. Indeed the definition of spooky freeness allows the provers to pre-share a joint state.

[5] We note that all "natural" MIPs that we are aware of have this property. In particular, any MIP that is constructed from a poly-size PCP with polylogarithmically many queries, has this property.

- **Completeness:** For every $(x, w) \in R$ such that $x \in \{0, 1\}^{\leq 2^{\kappa}}$,

$$\Pr[\mathcal{V}(\vec{q}, \vec{a}, x) = 1] = 1 ,$$

where $\vec{q} \leftarrow \mathcal{G}(1^{\kappa})$, $\vec{a} = (a_1, \ldots, a_p)$ and $a_i \leftarrow \mathcal{P}_i(q_i, x, w)$ for every $i \in [p]$.
- **Soundness:** For every $x \in \{0, 1\}^{\leq 2^{\kappa}} \setminus L$ and for every (not necessarily efficient) cheating provers $\mathcal{P}'_1, \ldots, \mathcal{P}'_p$ the following holds:

$$\Pr\left[\mathcal{V}(\vec{q}, \vec{a}', x) = 1\right] < \sigma(\kappa) ,$$

where $\vec{q} \leftarrow \mathcal{G}(1^{\kappa})$, $\vec{a}' = (a'_1, \ldots, a'_p)$ and $a'_i \leftarrow \mathcal{P}'_i(q_i, x)$ for every $i \in [p]$.

Definition 4 An $\mathcal{NP}$ language $L \subset \{0, 1\}^*$, is said to have sub-exponentially hard subset-membership problem $(\mathcal{L}, \overline{\mathcal{L}}, \mathsf{Sam})$ if the following holds:

- $\mathcal{L} = \{\mathcal{L}_n\}_{n \in \mathbb{N}}$ is a PPT distribution ensemble, each over $L \cap \{0, 1\}^n$.
- $\overline{\mathcal{L}} = \{\overline{\mathcal{L}}_n\}_{n \in \mathbb{N}}$ is a PPT distribution ensemble, each over $\overline{L} \cap \{0, 1\}^n = \{0, 1\}^n \setminus L$.
- Sam is a PPT algorithm, that on input 1^n outputs a tuple $(x, w) \in R_L$ where x is distributed as in $\mathcal{L}_n$.
- $\mathcal{L}, \overline{\mathcal{L}}$ are $(\epsilon(n), s(n))$-indistinguishable for $\epsilon(n) = 1/2^{n^{\alpha}}$, $s(n) = 2^{n^{\alpha}}$, where $\alpha > 0$ is some constant referred to the **hardness-parameter**.

In such case we will say that $(\mathcal{L}, \overline{\mathcal{L}}, \mathsf{Sam})$ is α-sub-exponentially hard.

Lemma 5 If $(\mathcal{L}, \overline{\mathcal{L}}, \mathsf{Sam})$ is α-sub-exponentially hard, then $H_{\infty}(\mathcal{L}), H_{\infty}(\overline{\mathcal{L}}) \geq n^{\alpha}$.

Proof Let δ be the probability of x^*, the maximum likelihood element in the support of $\mathcal{L}$. Then there is a constant size distinguisher between $\mathcal{L}, \overline{\mathcal{L}}$ that succeeds with probability δ. On input x, output 1 if and only if $x = x^*$. It follows that $\delta \leq 2^{-n^{\alpha}}$, and a symmetric argument holds for $\overline{\mathcal{L}}$ as well.

Theorem 6 (Dense Model Theorem [VZ13, Lemma 6.9]**).** *There exists a fixed polynomial p such that the following holds:Let $\mathcal{X}$ and $\mathcal{Y}$ be two $(\epsilon(\lambda), s(\lambda))$-indistinguishable distributions. Let $\mathcal{A}$ be a distribution over $\{0, 1\}^{\ell}$ jointly distributed with $\mathcal{X}$. Then there exists a (probabilistic) function $h : \mathcal{Y} \rightarrow \{0, 1\}^{\ell}$ such that $(\mathcal{X}, \mathcal{A})$ and $(\mathcal{Y}, h(\mathcal{Y}))$ are $(\epsilon^*(\lambda), s^*(\lambda))$-indistinguishable, where $\epsilon^*(\lambda) = 2 \cdot \epsilon(\lambda)$ and $s^*(\lambda) = s(\lambda) \cdot p(\epsilon(\lambda), 1/2^{\ell(\lambda)})$.*

Corollary 7. *Let $(\mathcal{X}, \mathcal{A})$ be a joint distribution s.t. $\mathcal{A}$ is supported over $\{0, 1\}^{\ell}$ for $\ell = O(n^{\alpha})$, and let $\mathcal{Y}$ be a distribution such that $\mathcal{X}$ and $\mathcal{Y}$ are α-sub-exponentially indistinguishable. Then there exists a probabilistic function h s.t. $(\mathcal{X}, \mathcal{A})$ and $(\mathcal{Y}, h(\mathcal{Y}))$ are $(2 \cdot 2^{-n^{\alpha}}, 2^{n^{\alpha}})$ indistinguishable.*

Proof. Let $\epsilon(n) = 2^{-n^{\alpha}}$, $s(n) = 2^{n^{\alpha}}$ be such that $\mathcal{X}, \mathcal{Y}$ are $(\epsilon(n), s(n))$-indistinguishable. Then it follows from Definition 1 that they are also $(\epsilon(n), s'(n))$-indistinguishable for any $s'(n) = \text{poly}(s(n))$, in particular let $s'(n) = s(n)/p(\epsilon, 1/2^{\ell}) = 2^{O(n^{\alpha})} = \text{poly}(s(n))$. Theorem 6 implies that there exists a probabilistic function h s.t. $(\mathcal{X}, \mathcal{A})$ and $(\mathcal{Y}, h(\mathcal{Y}))$ are $(2\epsilon(n), s(n))$-indistinguishable. $\qquad\square$

Definition 8 (fully-homomorphic encryption). *A fully-homomorphic (public-key) encryption scheme* FHE = (FHE.Keygen, FHE.Enc, FHE.Dec, FHE.Eval) *is a 4-tuple of* PPT *algorithms as follows (λ is the security parameter):*

- **Key generation** (pk, sk)←FHE.Keygen(1^λ): *Outputs a public encryption key* pk *and a secret decryption key* sk.
- **Encryption** c←FHE.Enc(pk, μ): *Using the public key* pk, *encrypts a single bit message* $\mu \in \{0, 1\}$ *into a ciphertext* c.
- **Decryption** μ←FHE.Dec(sk, c): *Using the secret key* sk, *decrypts a ciphertext* c *to recover the message* $\mu \in \{0, 1\}$.
- **Homomorphic evaluation** $\widehat{c}$←FHE.Eval($\mathcal{C}$, $(c_1, \ldots, c_\ell)$, pk): *Using the public key* pk, *applies a boolean circuit* $\mathcal{C} : \{0, 1\}^\ell \to \{0, 1\}$ *to* $c_1, \ldots, c_\ell$, *and outputs a ciphertext* $\widehat{c}$.

A homomorphic encryption scheme is said to be secure if it is semantically secure. A scheme FHE *is fully homomorphic, if for any circuit* $\mathcal{C}$ *and any set of inputs* $\mu_1, \ldots, \mu_\ell$, *letting* (pk, sk)←FHE.Keygen(1^λ) *and* c_i←FHE.Enc(pk, μ_i), *it holds that*

$$\Pr\left[\text{FHE.Dec}(\text{sk}, \text{FHE.Eval}(\mathcal{C}, (c_1, \ldots, c_\ell), \text{pk})) \neq \mathcal{C}(\mu_1, \ldots, \mu_\ell)\right] = \text{negl}(\lambda) \ ,$$

A fully homomorphic encryption scheme is compact if the output length of FHE.Eval *is a fixed polynomial in* λ *(and does not depend on the length of* $\mathcal{C}$*).*

2.1   Spooky-Free Encryption

Let PKE = (PKE.KeyGen, PKE.Enc, PKE.Dec) be a public-key encryption scheme. Let $\mathcal{D}$ be some distribution and let $\mathcal{A}$ and $\mathcal{S}$ be some algorithms. Consider the following experiments:

REAL$_{\mathcal{D},\mathcal{A}}(1^\kappa)$

1. Sample messages and auxiliary information $(\vec{m}, \alpha) = (m_1, \ldots, m_n, \alpha)$←$\mathcal{D}(1^\kappa)$.
2. Generate keys and encryptions for every $i \in [n]$ (pk$_i$, sk$_i$) ← PKE.KeyGen(1^κ), c_i ← PKE.Enc (pk$_i$, m_i).
3. Evaluate $\vec{c'}$ ← $\mathcal{A}(1^\kappa, \vec{\text{pk}}, \vec{c})$
4. Decrypt each evaluated ciphertext m_i':=PKE.Dec(sk$_i$, c_i').
5. Output $(\vec{m}, \vec{m'}, \alpha)$.

SIM$_{\mathcal{D},\mathcal{S}}(1^\kappa)$

1. Sample messages and auxiliary information $(\vec{m}, \alpha) = (m_1, \ldots, m_n, \alpha)$←$\mathcal{D}(1^\kappa)$.
2. Sample random coins r for the simulator $\mathcal{S}$, and evaluate for every $i \in [n]$ m_i'←$\mathcal{S}(1^\kappa, 1^n, i, m_i; r)$.
3. Output $(\vec{m}, \vec{m'}, \alpha)$.

Definition 9. *Let* PKE = (PKE.KeyGen, PKE.Enc, PKE.Enc) *be a public-key encryption scheme. We say that* PKE *is* **strongly spooky − free** *if there exists* *a* PPT *simulator* $\mathcal{S}$ *such that for every* PPT *adversary* $\mathcal{A}$, *distribution* $\mathcal{D}$ *and distinguisher* Ψ, *the following holds:*

$$\left| \Pr\left[\Psi(\vec{m}, \vec{m'}, \alpha) = 1 \mid (\vec{m}, \vec{m'}, \alpha) \leftarrow \boldsymbol{REAL}_{\mathcal{D},\mathcal{A}}(1^{\kappa})\right] - \right.$$

$$\left. \Pr\left[\Psi(\vec{m}, \vec{m'}, \alpha) = 1 \mid (\vec{m}, \vec{m'}, \alpha) \leftarrow \boldsymbol{SIM}_{\mathcal{D},\mathcal{S}^{\mathcal{A}}}(1^{\kappa})\right] \right| = \mathrm{negl}(\kappa)$$

We say that PKE *is* weakly spooky − free *if the simulator can be chosen after the adversary, the distribution and the distinguisher have been set. Similarly, we say that* PKE *is* strongly spooky − free without auxiliary information *(weakly spooky-free without auxiliary information), if it is strongly spooky-free (weakly spooky-free), and the distribution* $\mathcal{D}$ *must output* $\alpha = \bot$.

For our negative result, we prove the impossibility with respect to the *weak* definition *without auxiliary information*, thus strengthening the impossibility result. On the other hand, for the positive result we construct a *strongly spooky-free (with auxiliary information)* scheme. We note that in the original definition in [DHRW16] the order of quantifiers was somewhere "in between" our two definitions. They allowed the simulator to be chosen after seeing the adversary $\mathcal{A}$, but before seeing the distribution $\mathcal{D}$ and the distinguisher Ψ.

2.2   Falsifiable Assumptions and Black-Box Reductions

In what follows, we recall the notion of falsifiable assumptions as defined by Naor [Nao03]. We follow the formalization of Gentry and Wichs [GW11].

Definition 10 (falsifiable assumption). *A* falsifiable assumption *consists of a* PPT *interactive challenger* $\mathcal{C}(1^{\lambda})$ *that runs in time* $\mathrm{poly}(\lambda)$ *and a constant* $\eta \in [0, 1)$. *The challenger* $\mathcal{C}$ *interacts with a machine* $\mathcal{A}$ *and may output a special symbol* win. *If this occurs,* $\mathcal{A}$ *is said to win* $\mathcal{C}$. *For any adversary* $\mathcal{A}$, *the advantage of* $\mathcal{A}$ *over* $\mathcal{C}$ *is defined as:*

$$\mathsf{Adv}_{\mathcal{A}}^{(\mathcal{C},\eta)}(1^{\lambda}) = \Pr[\mathcal{A}(1^{\lambda}) \text{ wins } \mathcal{C}(1^{\lambda})] - \eta,$$

where the probability is taken over the random coins of $\mathcal{A}$ *and* $\mathcal{C}$. *The assumption associated with the tuple* $(\mathcal{C}, \eta)$ *states that for every (non-uniform) adversary* $\mathcal{A}(1^{\lambda})$ *running in polynomial time,*

$$\mathsf{Adv}_{\mathcal{A}}^{(\mathcal{C},\eta)}(1^{\lambda}) = \mathrm{negl}(\lambda).$$

If the advantage of $\mathcal{A}$ *is non-negligible in* λ *then* $\mathcal{A}$ *is said to* break *the assumption.*

Definition 11. *A falsifiable assumption* $(\mathcal{C}_1, \eta_1)$ *is* black-box stronger *than a falsifiable assumption* $(\mathcal{C}_2, \eta_2)$, *denoted* $(\mathcal{C}_1, \eta_1) \geq (\mathcal{C}_2, \eta_2)$ *if there exists a reduction* $\mathcal{R}$ *such that for every adversary* $\mathcal{A}$ *with non-negligible advantage against* $(\mathcal{C}_1, \eta_1)$, *it holds that* $\mathcal{R}^{\mathcal{A}}$ *has non-negligible advantage against* $(\mathcal{C}_2, \eta_2)$.

We say that $(\mathcal{C}_1, \eta_1)$ and $(\mathcal{C}_2, \eta_2)$ are black-box equivalent, *denoted $(\mathcal{C}_1, \eta_1) \equiv (\mathcal{C}_2, \eta_2)$ if $(\mathcal{C}_1, \eta_1) \geq (\mathcal{C}_2, \eta_2)$ and $(\mathcal{C}_2, \eta_2) \geq (\mathcal{C}_1, \eta_1)$.*

Definition 12. *Let $(\mathcal{C}, \eta)$ be a falsifiable assumption, and define the challenger $\mathcal{C}_\eta^\otimes$ that interacts with an adversary $\mathcal{A}$ as follows. First $\mathcal{A}$ sends a polynomially bounded unary number 1^t to the challenger. Then the challenger executes the $\mathcal{C}$ game with $\mathcal{A}$ sequentially and independently t times. Finally $\mathcal{C}_\eta^\otimes$ declares that $\mathcal{A}$ won if and only if $\mathcal{A}$ won in at least $\lceil \eta t \rceil + 1$ of the games.*

Lemma 13. *For any falsifiable assumption $(\mathcal{C}, \eta)$ it holds that $(\mathcal{C}, \eta) \equiv (\mathcal{C}_\eta^\otimes, 0)$.*

Proof. Let $\mathcal{A}$ be an adversary with non-negligible advantage δ in $(\mathcal{C}, \eta)$. Then $\mathcal{R}^\mathcal{A}(1^\lambda)$ is an adversary against $\mathcal{C}_\eta^\otimes$ as follows. It starts by sending 1^t for $t = \lceil \lambda/\delta \rceil$ in the first message. Then for every iteration it simply executes $\mathcal{A}$. By definition, the expected number of wins is at least $\lfloor \eta t + \lambda \rfloor > \lceil \eta t \rceil + 1 + \lambda/2$. By a Chernoff argument the probability to win against $\mathcal{C}_\eta^\otimes$ is at least $1 - \mathrm{negl}(\lambda)$.[6]

Now let $\mathcal{A}$ be an adversary with non-negligible advantage δ against $(\mathcal{C}_\eta^\otimes, 0)$. Then $\mathcal{R}^\mathcal{A}(1^\lambda)$ is an adversary against $(\mathcal{C}, \eta)$ as follows. It simulates $\mathcal{C}_\eta^\otimes$ for $\mathcal{A}$ by first reading 1^t, then sampling $i^* \xleftarrow{\$} [t]$, simulating $\mathcal{C}$ in all iterations except i^*, and in iteration i^* forward messages back and forth to the real challenger. By definition the advantage of $\mathcal{R}^\mathcal{A}(1^\lambda)$ is at least $1/t$ which is noticeable. $\square$

Definition 14 (black box reduction). *We say that the security of a scheme Π can be proven via a* black-box reduction *to a falsifiable assumption, if there is an oracle-access machine $\mathcal{R}$ such that for every (possibly inefficient) adversary $\mathcal{A}$ that breaks the security of Π, the oracle machine $\mathcal{R}^\mathcal{A}$ runs in time $\mathrm{poly}(\lambda)$ and breaks the assumption.*

Corollary 15. *If Π can be proven via a black-box reduction to a falsifiable assumption $(\mathcal{C}, \eta)$ then it can also be proven via a black-box reduction to a falsifiable assumption $(\mathcal{C}', 0)$, and furthermore if $(\mathcal{C}, \eta)$ is hard for all polynomial adversaries then so is $(\mathcal{C}', 0)$.*

Proof. Letting $\mathcal{C}' = \mathcal{C}_\eta^\otimes$, the corollary directly follows from Lemma 13 and Definition 14. $\square$

Lemma 16. *Let Π be a scheme whose security can be proven via a black-box reduction $\mathcal{R}$ to a falsifiable assumption $(\mathcal{C}, 0)$ (note that $\eta = 0$). Let $\widetilde{\mathcal{A}}$ be a distribution on adversaries such that with probability 1, $\mathcal{A} \xleftarrow{\$} \widetilde{\mathcal{A}}$ breaks the security of Π. Then there exists a non-negligible δ such that*

$$\Pr_{\mathcal{A}, \mathcal{R}, \mathcal{C}}[\mathcal{R}^\mathcal{A}(1^\lambda) \text{ wins } \mathcal{C}(1^\lambda)] \geq \delta(\lambda) .$$

Namely, the expected advantage of $\mathcal{R}$ against $(\mathcal{C}, 0)$ is non-negligible.

---

[6] We assumed that δ is known to the reduction, which could be viewed as non-black-box access. However, note that δ can be estimated by running the oracle many times, simulating $\mathcal{C}$.

Proof. For every $\mathcal{A}$ denote:

$$\tilde{\delta}_\mathcal{A}(\lambda) = \Pr_{\mathcal{R},\mathcal{C}}[\mathcal{R}^\mathcal{A}(1^\lambda) \text{ wins } \mathcal{C}(1^\lambda)] .$$

By the correctness of the reduction $\mathcal{R}$ we are guaranteed that with probability 1 over $\mathcal{A} \xleftarrow{\$} \widetilde{\mathcal{A}}$, it holds that $\tilde{\delta}_\mathcal{A}$ is a non-negligible function. Furthermore, notice that by definition

$$\Pr_{\mathcal{A},\mathcal{R},\mathcal{C}}[\mathcal{R}^\mathcal{A}(1^\lambda) \text{ wins } \mathcal{C}(1^\lambda)] = \mathbb{E}_\mathcal{A}[\tilde{\delta}_\mathcal{A}(\lambda)] ,$$

and our goal therefore is to prove that $\mathbb{E}_\mathcal{A}[\tilde{\delta}_\mathcal{A}(\lambda)]$ is non-negligible.

Let us consider a random $\mathcal{A}^* \xleftarrow{\$} \widetilde{\mathcal{A}}$ and define $\tilde{\delta}^*(\lambda) = \tilde{\delta}_{\mathcal{A}^*}(\lambda)$. We define a sequence of events $\{E_\lambda\}_{\lambda \in \mathbb{N}}$, where E_λ is the event that

$$\Pr_\mathcal{A}[\tilde{\delta}_{\mathcal{A}^*}(\lambda) \leq \tilde{\delta}_\mathcal{A}(\lambda)] \leq 1/\lambda^2 .$$

Trivially, $\Pr[E_\lambda] \leq 1/\lambda^2$. Therefore, by the Borel-Cantelli Lemma, with probability 1 on the choice of $\mathcal{A}^*$ it holds that only finitely many of the events E_λ can occur.

Let us consider some value of λ for which E_λ does not hold (as explained above, this includes all but finitely many λ values). That is, where

$$\Pr_\mathcal{A}[\tilde{\delta}_{\mathcal{A}^*}(\lambda) \leq \tilde{\delta}_\mathcal{A}(\lambda)] > 1/\lambda^2 .$$

By definition, for these values, we can apply the Markov inequality

$$\mathbb{E}_\mathcal{A}[\tilde{\delta}_\mathcal{A}(\lambda)] \geq \Pr_\mathcal{A}[\tilde{\delta}_{\mathcal{A}^*}(\lambda) \leq \tilde{\delta}_\mathcal{A}(\lambda)] \cdot \tilde{\delta}_{\mathcal{A}^*}(\lambda) > \tilde{\delta}_{\mathcal{A}^*}(\lambda)/\lambda^2 .$$

Since with probability 1 it holds that both $\tilde{\delta}_{\mathcal{A}^*}(\lambda)$ is noticeable and that only finitely many of the E_λ can occur, then obviously there exists $\mathcal{A}^*$ for which both hold, which implies that indeed $\Pr_{\mathcal{A},\mathcal{R},\mathcal{C}}[\mathcal{R}^\mathcal{A}(1^\lambda) \text{ wins } \mathcal{C}(1^\lambda)]$ is non-negligible. $\square$

3   Spooky-Free Compiler

Definition 17 (Spooky-Free Compiler). *Let $\Pi = (\mathcal{G}, \vec{\mathcal{P}}, \mathcal{V})$ be a p-provers, one-round MIP with soundness σ for an* **NP** *language L with an induced relation R. A* Spooky-Free Compiler *for Π is a triplet of* PPT *algorithms* SFC = (SFC.Enc, SFC.Dec, SFC.Eval) *as follows:*

- **Encoding** $(e, \mathsf{dk}) \leftarrow \mathsf{SFC.Enc}(\vec{q})$: *Outputs an encoding of the queries, and a decoding-key.*
- **Evaluation** $e' \leftarrow \mathsf{SFC.Eval}(e, x, w)$: *Evaluates the MIP answers on the encoded queries, instance x, and witness w.*
- **Decoding** $\vec{a} \leftarrow \mathsf{SFC.Dec}(e', \mathsf{dk})$: *Decodes the evaluated queries using the decoding-key.*

We require the following properties:

- **Completeness** *For every* $(x, w) \in R$ *such that* $x \in \{0, 1\}^{\leq 2^\kappa}$, *the following holds: Sample queries* $\vec{q} \leftarrow \mathcal{G}(1^\kappa)$ *and encode* $(e, \mathsf{dk}) \leftarrow \mathsf{SFC.Enc}(\vec{q})$. *Evaluate* $e' \leftarrow \mathsf{SFC.Eval}(e, x, w)$, *and decode* $\vec{a} \leftarrow \mathsf{SFC.Dec}(e', \mathsf{dk})$. *Then,*

$$\Pr[\mathcal{V}(\vec{q}, \vec{a}, x) = 1] = 1.$$

- **Spooky-Freeness** *Define the following experiments:*

$\boldsymbol{REAL}_{\mathcal{A}=(\mathcal{A}_1, \mathcal{A}_2)}(1^\kappa)$

1. *Sample input* $x \leftarrow \mathcal{A}_1(1^\kappa)$.
2. *Sample queries* $\vec{q} \leftarrow \mathcal{G}(1^\kappa)$.
3. *Encode* $(e, \mathsf{dk}) \leftarrow \mathsf{SFC.Enc}(\vec{q})$.
4. *Evaluate* $e' \leftarrow \mathcal{A}_2(1^\kappa, x, e)$.
5. *Decode* $\vec{a} \leftarrow \mathsf{SFC.Dec}(e', \mathsf{dk})$.
6. *Output* $(x, \vec{q}, \vec{a})$.

$\boldsymbol{SIM}_\mathcal{S}(1^\kappa)$

1. *Sample random coins* r, *and using these coins sample input* $x \leftarrow \mathcal{S}(1^\kappa, 1^p, 0, 0, r)$.
2. *Sample queries* $\vec{q} \leftarrow \mathcal{G}(1^\kappa)$.
3. *For all* $i \in [p]$ *compute the value* $a_i \leftarrow \mathcal{S}(1^\kappa, 1^p, i, q_i, r)$.
4. *Output* $(x, \vec{q}, \vec{a})$.

We say that SFC *is* strongly spooky-free *if there exists a* PPT *simulator* $\mathcal{S}$ *such that for every* PPT *adversary* $\mathcal{A}$ *the experiments* $\boldsymbol{REAL}_\mathcal{A}(1^\kappa)$ *and* $\boldsymbol{SIMS}^\mathcal{A}(1^\kappa)$ *are computationally-indistinguishable. Similarly, we say that* SFC *is* weakly spooky-free *if the simulator can be chosen after the adversary and the distinguisher have been set.*

On Black-Box Reductions of Spooky Free Compilers. Let us explicitly instantiate the definition of black box reductions (Definition 14 above) in the context of (weak) spooky free compilers. This is the definition that will be used to prove our main technical result in Theorem 20.

Consider a candidate spooky free compiler as in Definition 17 above. Then a pair of (not necessarily efficient) algorithms $(\mathcal{A}, \Psi)$ breaks weak spooky freeness if for any simulator $\mathcal{S}$ (possibly dependent on $\mathcal{A}, \Psi$) allowed to run in time $\mathrm{poly}(\mathsf{time}(\mathcal{A}), \mathsf{time}(\Psi))$, it holds that Ψ can distinguish between the distributions $\boldsymbol{REAL}_\mathcal{A}$ and $\sim \mathcal{S}$ with non-negligible probability (we refer to this as "breaking spooky freeness").

A black-box reduction from a falsifiable assumption $(\mathcal{C}, \eta)$ to a weakly spooky free compiler is an oracle machine $\mathcal{R}$ that, given oracle access to a pair of machines $(\mathcal{A}, \Psi)$ that break weak spooky freeness as defined above, $\mathcal{R}^{(\mathcal{A}, \Psi)}$ has non-negligible advantage against $(\mathcal{C}, \eta)$. We note that we will prove an even stronger result that places no computational restrictions at all on the running time of $\mathcal{S}$.

4   Non-Succinct Spooky Freeness Is Trivial

In this section, we construct a non-succinct spooky free FHE, where the length of each ciphertext and the length of each public-key is exponential in the length of the messages. Specifically, we show how to convert any FHE scheme into a spooky-free FHE scheme such that the length of each ciphertext and each public key is $2^k \cdot \mathrm{poly}(\lambda)$, where k is the length of the messages.

We note that a spooky-free FHE is stronger than a spooky-free compiler, since the latter is tied to a specific MIP whereas the former is "universal".

Theorem 18. *There exists an efficient generic transformation from any fully-homomorphic encryption scheme* FHE = (FHE.Keygen, FHE.Enc, FHE.Dec, FHE.Eval) *into a scheme* FHE′ = (FHE.Keygen′, FHE.Enc′, FHE.Dec′, FHE.Eval′) *that is fully-homomorphic and* **strongly spooky-free.** *The length of each ciphertext generated by* FHE.Enc′ *and the length of each public-key generated by* FHE.Keygen′ *is* $2^k \cdot \mathrm{poly}(\lambda)$, *where k is the length of each message.*

Proof Overview. We transform the scheme to have equivocal properties. Specifically, the transformed scheme's ciphertexts can be replaced with ones that can be decrypted to any value using different pre-computed secret-keys. The joint distribution of each secret-key and the special ciphertext are indistinguishable from a properly generated secret-key and ciphertext. This allows us to define a simulator that precomputes those secret-keys and queries the adversary using an equivocable ciphertext. Then, it decrypts with the secret-key corresponding to the given message to extract the adversary's answers. By indistinguishability, this is the same answer that would be produced by querying the adversary, if it was queried with an encryption of that message.

We achieve this property by simply generating independently 2^k public-keys, whereas the secret-key corresponds only to one of the public-keys. Each ciphertext is 2^k encryptions, under each public-key. The equivocable ciphertext is produced by encrypting each of the 2^k possible messages under some public-key, in a randomly chosen order. Indistinguishability follows from the semantic-security of the original scheme.

Remark 19. We assume, without the loss of generality, that the length of an encryption of k bits is bounded by $k \cdot \mathrm{poly}(\lambda)$, since we can always encrypt bit-by-bit while preserving security and homomorphism.

Proof. Let $k = k(\lambda)$ be an upper-bound on the length of the messages $|m_i| \leq k$, where $(m_1, \ldots, m_n, \alpha) \leftarrow \mathcal{D}$. We define the scheme FHE′ as follows:

- FHE.Keygen′(1^λ): Generate 2^k pairs of keys $(\mathsf{pk}_i, \mathsf{sk}_i) \leftarrow$ FHE.Keygen(1^λ). Then, choose uniformly at random a n index $j \xleftarrow{\$} [2^k]$, and output $(\vec{\mathsf{pk}}, (\mathsf{sk}_j, j))$.
- FHE.Enc′($\vec{\mathsf{pk}}, \mu$): Encrypt the message under each public-key $c_i \leftarrow$ FHE.Enc(pk_i, μ), then output $\vec{c}$.
- FHE.Dec′($(\mathsf{sk}, j), \hat{\vec{c}}$): Decrypt according to the indexed secret-key and output $\mu' := $FHE.Dec($\mathsf{sk}, \hat{c}_j$).
- FHE.Eval′($\vec{\mathsf{pk}}, \vec{c}, \mathcal{C}$): For every $i \in [2^k]$ compute $\hat{c}_i \leftarrow$ FHE.Eval($\mathsf{pk}_i, c_i, \mathcal{C}$) and output $\hat{\vec{c}}$.

Clearly FHE$'$ is a fully-homomorphic encryption scheme. It is thus left to prove that it is strongly spooky-free.

The simulator $\mathcal{S}^{\mathcal{A}}(1^\lambda, 1^n, i, m_i; r)$ First, the simulator uses its randomness r to sample $n \cdot 2^k$ pairs of keys $(\mathsf{pk}_{\ell,j}, \mathsf{sk}_{\ell,j}) \leftarrow \mathsf{FHE.Keygen}(1^\lambda), \ell \in [n], j \in [2^k]$. Then, for every 2^k-tuple $\vec{\mathsf{pk}}_\ell = (\mathsf{pk}_{\ell,1}, \ldots, \mathsf{pk}_{\ell,2^k})$, it chooses a permutation $\pi_\ell : [2^k] \rightarrow [2^k]$, encrypts the message $\pi_\ell^{-1}(j)$ under the public-key $\mathsf{pk}_{\ell,j}$, for every $j \in \{0,1\}^k$

$$c_{\ell,j} = \mathsf{FHE.Enc}(\mathsf{pk}_{\ell,j}, \pi_\ell^{-1}(j)) \ .$$

Next, it sets $\vec{c}_\ell = (c_{\ell,1}, \ldots, c_{\ell,2^k})$ and queries the adversary to get $(\vec{c'}_1, \ldots, \vec{c'}_n)$ $\leftarrow \mathcal{A}((\vec{\mathsf{pk}}_1, \ldots, \vec{\mathsf{pk}}_n), (\vec{c}_1, \ldots, \vec{c}_n))$. Finally it outputs $m_i' := \mathsf{FHE.Dec}(\mathsf{sk}_{i,\pi_\ell(m_i)}, c'_{i,\pi_\ell(m_i)})$.

Claim 1. *For every* PPT *adversary* $\mathcal{A}$ *and distribution* $\mathcal{D}$, *the experiments* $\boldsymbol{REAL}_{\mathcal{D},\mathcal{A}}$ *and* $\boldsymbol{SIM}_{\mathcal{D},\mathcal{S}^{\mathcal{A}}}$ *are computationally indistinguishable.*

Proof. We prove using a sequence of hybrids.

- $\mathcal{H}_0$: This is simply the distribution $\boldsymbol{REAL}_{\mathcal{D},\mathcal{A}}$.
- $\mathcal{H}_{1,i}$ ($i \in [n]$): In these hybrids we modify the key generation step in $\boldsymbol{REAL}_{\mathcal{D},\mathcal{A}}$: Instead of choosing $j_i \xleftarrow{\$} [2^k]$ uniformly at random, we choose uniformly at random a permutation $\pi_i : [2^k] \rightarrow [2^k]$ and set $j_i = \pi_i(m_i)$. These hybrids are identically distributed, since the π_i's are random permutations, so each j_i is distributed uniformly over $[2^k]$.
- $\mathcal{H}_{2,i,j}$ ($i \in [n], j \in [2^k]$): In these hybrids we modify the encryption step in $\boldsymbol{REAL}_{\mathcal{D},\mathcal{A}}$: Instead of letting $c_{i,j} \leftarrow \mathsf{FHE.Enc}(\mathsf{pk}_{i,j}, m_i)$, set $c_{i,\pi_i(j)} \leftarrow \mathsf{FHE.Enc}(\mathsf{pk}_{i,\pi_i(j)}, j)$, where π_i is the permutation from the previous hybrids. These hybrids are computationally indistinguishable by the semantic security of FHE.

Finally, note that $\mathcal{H}_{2,n,2^k}$ is actually $\boldsymbol{SIM}_{\mathcal{D},\mathcal{S}^{\mathcal{A}}}$. This is since for every $i \in [n]$, the simulator queries the adversary the same query every time, and that query is distributed as the one in $\mathcal{H}_{2,n,2^k}$. Moreover, the adversary's answer is decrypted in the same manner both in $\boldsymbol{SIM}_{\mathcal{D},\mathcal{S}^{\mathcal{A}}}$ and $\mathcal{H}_{2,n,2^k}$. Thus $\boldsymbol{REAL}_{\mathcal{D},\mathcal{A}} \stackrel{c}{\approx} \boldsymbol{SIM}_{\mathcal{D},\mathcal{S}^{\mathcal{A}}}$, as desired.

5   Succinct Spooky Freeness Cannot Be Proven Using a Black-Box Reduction

We state and prove our main theorem.

Theorem 20. *Let* $\Pi = (\mathcal{G}, \vec{\mathcal{P}}, \mathcal{V})$ *be a succinct one-round MIP for L. Let* $\mathsf{SFC} = (\mathsf{SFC.Enc}, \mathsf{SFC.Dec}, \mathsf{SFC.Eval})$ *be a spooky-free compiler for* Π, *with* $|e'| = \mathrm{poly}(\kappa) \cdot |x|^{\alpha'}$ *for some* $\alpha' < 1$. *Finally, let* $(\mathcal{C}, \eta)$ *be a falsifiable assumption.*

Then, assuming the existence of a language L with a sub-exponentially hard subset-membership problem $(\mathcal{L}, \overline{\mathcal{L}}, \mathsf{Sam})$ *with hardness parameter* $\alpha > \alpha'$, *there is no black-box reduction showing the weakly spooky-freeness of* SFC *based on the assumption* $(\mathcal{C}, \eta)$, *unless* $(\mathcal{C}, \eta)$ *is polynomially solvable.*

Proof Overview. We start by defining an inefficient adversary $(\overline{\mathcal{A}}, \overline{\Psi})$ against SFC, or more precisely a distribution over adversaries specified by a family of sets $\overline{X}$. These sets contain, for each value of the security parameter 1^κ a large number of inputs from $\overline{\mathcal{L}}$ of length $n = \mathrm{poly}(\kappa)$ for a sufficiently large polynomial to make $|e'|$ bounded by n^α. The adversary $\overline{\mathcal{A}}$ picks a random x from the respective set and generates a response e' as follows. As a thought experiment, if it was the case that $x \in \mathcal{L}$, then SFC allows us to generate e' that will be accepted by the MIP verifier. Therefore, the Dense Model Theorem states that it is possible to generate a computationally indistinguishable e' also for $x \in \overline{\mathcal{L}}$. The distinguisher $\overline{\Psi}$ will check that x is indeed in the respective set of $\overline{X}$ and if so, it will apply the MIP verifier. The soundness of MIP guarantees that this distribution cannot be simulated by independent provers. Note that the phase where $\overline{\Psi}$ checks that indeed $x \in \overline{X}$ is critical since otherwise the simulator could produce $x \in \mathcal{L}$ which will cause $\overline{\Psi}$ to accept! The use of a common $\overline{X}$ allows $\overline{\mathcal{A}}$ and $\overline{\Psi}$ to share a set of inputs for which they know the simulator cannot work.

Since $(\overline{\mathcal{A}}, \overline{\Psi})$ is successful against SFC, it means that the reduction breaks the assumption given oracle access to $(\overline{\mathcal{A}}, \overline{\Psi})$.[7] Our goal now is to show an efficient procedure $(\mathcal{A}, \Psi)$ which is indistinguishable from $(\overline{\mathcal{A}}, \overline{\Psi})$ in the eyes of the reduction. This will show that the underlying assumption is in fact polynomially solvable.

To do this, we notice that the reduction can only ever see polynomially many x's, so there is no need to sample a huge set $\overline{X}$, and an appropriately defined polynomial subset would be sufficient. Furthermore, instead of sampling from $\overline{\mathcal{L}}$, we can sample from $\mathcal{L}$ together with a witness, and compute e' as a legitimate SFC.Eval response. The Dense Model Theorem ensures that this strategy will be indistinguishable to the reduction, and therefore it should still be successful in breaking the assumption. Note that we have to be careful since the reduction might query its oracle on tiny security parameter values for which n is not large enough to apply the Dense Model Theorem. For those small values we create a hard-coded table of adversary responses (since these are tiny values, the table is still not too large).

Finally, we see that our simulated adversary runs in polynomial time since it only needs to sample from $\mathcal{L}$, which is efficient using Sam, and use the witness to compute e' via SFC.Eval. We conclude that we have a polynomial time algorithm that succeeds in breaking the assumption, as required in the theorem.

---

[7] In fact, the situation is more delicate since as explained above $(\overline{\mathcal{A}}, \overline{\Psi})$ is a *distribution over adversaries*, and while almost all adversaries in the support succeed against SFC, it still requires quite a bit of work to prove that the average advantage is also non-negligible (see Lemma 16 in Sect. 2).

Proof. We proceed as in the sketch above. By the properties of SFC as stated in the theorem, there exist constants $\beta_1, \beta_2, \beta_3 > 0$ such that $\beta_1 = \alpha - \alpha'$, $|e'| \leq O(\kappa^{\beta_2} \cdot |x|^{\alpha - \beta_1})$, $|e| \leq \kappa^{\beta_3}$. We define

$$n(\kappa) \triangleq \kappa^{\max\{2\beta_2/\beta_1, \beta_3/\alpha'\}} ,$$

and note that $|e|, |e'| = o(n^\alpha)$, when $|x| = n(\kappa)$.

Proofs Can Be Spoofed. We start by showing how to inefficiently spoof SFC answers for non-accepting inputs. Consider an encoded query e for SFC w.r.t. security parameter 1^κ, and define the distribution $(\mathcal{L}, \mathcal{E})$ as follows:

1. Sample $(x, w) \leftarrow \mathsf{Sam}(1^{n(\kappa)})$.
2. Evaluate $e' \leftarrow \mathsf{SFC.Eval}(e, x, w)$.
3. Output (x, e').

The following claim shows that it is possible to sample from a distribution that is computationally indistinguishable distribution from $(\mathcal{L}, \mathcal{E})$, but where the first component comes from $\overline{\mathcal{L}}$.

Claim 2. *For every e, there exists a randomized function $h = h_e$ such that the distributions $(\mathcal{L}, \mathcal{E})$ and $(\overline{\mathcal{L}}, h(\overline{\mathcal{L}}))$ are $(2 \cdot 2^{-n^\alpha}, 2^{n^\alpha})$ indistinguishable.*

Proof. Follows from Corollary 7 since $\mathcal{L}_n$ and $\overline{\mathcal{L}}_n$ are α-sub-exponentially indistinguishable. ∎

Constructing a Spooky Adversary. We define an adversary $\overline{\mathcal{A}}$, along with a distinguisher $\overline{\Psi}$ for the spooky-free experiment in SFC. We note that both $\overline{\mathcal{A}}$ and $\overline{\Psi}$ are *inefficient* algorithms, and more precisely, they are *distributions* over algorithms.

For every value of κ, define $\nu(\kappa) = 2^{0.1 \cdot n(\kappa)^\alpha}$. Define a vector $\overline{X}_{n(\kappa)} \overset{\$}{\leftarrow} \overline{\mathcal{L}}_{n(\kappa)}^{\otimes \nu(\kappa)}$, i.e. a sequence of independent samples from $\overline{\mathcal{L}}_n$. The functionality of $\overline{\mathcal{A}}$ and $\overline{\Psi}$ is as follows:

- $\overline{\mathcal{A}}(1^\kappa, e)$: Samples $i \in [\nu(\kappa)]$, sets $\bar{x} \overset{\$}{\leftarrow} \overline{X}_{n(\kappa)}[i]$ (i.e. the i'th element in the vector), and outputs $h(\bar{x})$.
- $\overline{\Psi}(1^\kappa, x, \vec{q}, \vec{a})$: Outputs 1 if and only if $x \in \overline{X}_{n(\kappa)}$ and $\mathcal{V}(\vec{q}, \vec{a}, x) = 1$.

The following claim asserts that the adversary $\overline{\mathcal{A}}$ and the distinguisher $\overline{\Psi}$ win the spooky-freeness game for the compiler SFC with probability 1 over the choice of the respective $\overline{X} = \{\overline{X}_{n(\kappa)}\}_\kappa$.

Claim 3. *With probability 1 over the choice of $\overline{X} = \{\overline{X}_{n(\kappa)}\}_\kappa$ it holds that $(\overline{\mathcal{A}}, \overline{\Psi})$ has non-negligible advantage in the spooky freeness game against SFC with any (possibly computationally unbounded) simulator.*

Proof. Let σ denote the soundness of the underlying MIP system. According to the definition of an MIP (see Definition 3), the soundness gap, $\sigma_{\mathrm{gap}} \triangleq 1 - \sigma$, is non-negligible.

We start by showing that for all $\overline{X}$, any value of κ, and any (possibly unbounded) spooky-free simulator $\mathcal{S}$ for the compiler SFC, it holds that

$$\Pr[\overline{\Psi}(\mathbf{SIM}_{\mathcal{S}}(1^{\kappa})) = 1 | \overline{X}] \leq \sigma(\kappa).$$

This follows since by the definition of the simulator, each value of its random string r defines an input x and induces a sequence of algorithms $\vec{\mathcal{S}}$ where

$$(\mathcal{S}_1(q_1), \ldots, \mathcal{S}_p(q_p)) = \vec{\mathcal{S}}(\vec{q}) \ .$$

If the induced $x \notin \overline{X}_{n(\kappa)}$ then $\overline{\Psi}$ will output 0. If $x \in \overline{X}_{n(\kappa)}$ then by the soundness of Π, the probability that the verifier $\mathcal{V}$ accepts answers generated by $\vec{\mathcal{S}}$ is at most $\sigma(\kappa)$, and thus $\overline{\Psi}$ outputs 1 with probability at most $\sigma(\kappa)$.

Next, we turn to show that $\Pr[\overline{\Psi}(\mathbf{REAL}_{\overline{\mathcal{A}}}(1^{\kappa})) = 1 | \overline{X}]$ is bounded away from $\sigma(\kappa)$ with probability 1 on $\overline{X}$. To this end, we define a sequence of events $\{E_{\kappa}\}_{\kappa \in \mathbb{N}}$, where E_{κ} is the event that

$$\Pr[\overline{\Psi}(\mathbf{REAL}_{\overline{\mathcal{A}}}(1^{\kappa})) = 1 | \overline{X}] \leq 1 - \sigma_{\mathrm{gap}}(\kappa)/2 \ ,$$

where the probability is over everything except the choice of $\overline{X}$. We show that with probability 1 over the choice of $\overline{X}$, only finitely many of the events E_{κ} occur.

To see this, fix queries $\vec{q} \leftarrow \mathcal{G}(1^{\kappa})$ and encoding $(e, \mathrm{dk}) \leftarrow \mathrm{SFC.Enc}(\vec{q})$ for the experiment $\mathbf{REAL}_{\overline{\mathcal{A}}}$. Note that since the compiler's decoding algorithm SFC.Dec can be described by a poly(κ) sized circuit, then we can describe the MIP's verifier $\mathcal{V}$ as a poly(κ) sized circuit that takes inputs from $(\overline{\mathcal{L}}, h(\overline{\mathcal{L}}))$.

Recall that by Claim 2, the distributions $(\mathcal{L}, \mathcal{E})$ and $(\overline{\mathcal{L}}, h(\overline{\mathcal{L}}))$ are $(2 \cdot 2^{-n^{\alpha}}, 2^{n^{\alpha}})$-indistinguishable. Moreover, by the completeness of the MIP, $\mathcal{V}$ outputs 1 with probability 1 on inputs from $(\mathcal{L}, \mathcal{E})$. We conclude that $\mathcal{V}$ accepts inputs from $(\overline{\mathcal{L}}, h(\overline{\mathcal{L}}))$ with overwhelming probability, and therefore $\overline{\Psi}$ also accepts with overwhelming probability inputs from $\mathbf{REAL}_{\overline{\mathcal{A}}}$. In other words, we have that

$$\underset{\overline{X}}{\mathbb{E}} \left[\Pr[\overline{\Psi}(\mathbf{REAL}_{\overline{\mathcal{A}}}(1^{\kappa})) = 1 | \overline{X}]] \right] \geq 1 - \mathrm{negl}(\kappa) \ .$$

By a Markov argument this implies that

$$\underset{\overline{X}}{\Pr} \left[\Pr[\overline{\Psi}(\mathbf{REAL}_{\overline{\mathcal{A}}}(1^{\kappa})) = 1 | \overline{X}] \leq 1 - \sigma_{\mathrm{gap}}(\kappa)/2 \right] \leq \mathrm{negl}(\kappa) \ .$$

Finally, we apply the Borel-Cantelli Lemma to conclude that with probability 1 over the choice of $\overline{X}$, only finitely many of the events E_{κ} occur, as desired.

Thus, with probability 1 (over the choice of $\overline{X}$), it holds that $(\overline{\mathcal{A}}, \overline{\Psi})$ has advantage at least $\sigma_{\mathrm{gap}}/2$ in the spooky free game. This completes the proof of the claim. ∎

Fooling the Reduction. We now notice that by Corollary 15, it is sufficient to prove the theorem for $\eta = 0$. Assume that there exists a black-box reduction $\mathcal{R}$ as in the theorem statement, and we will prove that $(\mathcal{C}, 0)$ is solvable in polynomial time. We notice that since $(\overline{\mathcal{A}}, \overline{\Psi})$ break spooky freeness with probability 1, it follows from Lemma 16 that

$$\delta(\lambda) = \Pr[\mathcal{R}^{(\overline{\mathcal{A}}, \overline{\Psi})}(1^\lambda) \text{ wins } \mathcal{C}(1^\lambda)]$$

is noticeable, where the probability is taken over the randomness of sampling $(\overline{\mathcal{A}}, \overline{\Psi})$, the randomness of the reduction and the randomness of $\mathcal{C}$.

We turn to define another adversary $\mathcal{A}$ and distinguisher Ψ, by modifying $\overline{\mathcal{A}}$ and $\overline{\Psi}$ in a sequence of changes. Our goal is to finally design $\mathcal{A}, \Psi$ computable in poly(λ) time, while ensuring that $\mathcal{R}^{(\mathcal{A}, \Psi)}$ still has advantage $\Omega(\delta)$.

Hybrid $\mathcal{H}_0$. In this hybrid we execute $\mathcal{R}^{(\overline{\mathcal{A}}, \overline{\Psi})}$ as defined.

$$\delta(\lambda) = \Pr_{\mathcal{H}_0}[\mathcal{R}^{(\overline{\mathcal{A}}, \overline{\Psi})}(1^\lambda) \text{ wins } \mathcal{C}(1^\lambda)] .$$

Hybrid $\mathcal{H}_1$. Let $\kappa_{\max} = \kappa_{\max}(\lambda) = \text{poly}(\lambda)$ be a bound on the size of the security parameters that the reduction $\mathcal{R}$ uses when interacting with its oracle. Note that $\kappa_{\max}$ is bounded by the runtime of the reduction, which in turn is bounded by some fixed polynomial (in λ). In this step, we remove all the sets relative to $\kappa > \kappa_{\max}$ from the ensemble $\{\overline{X}_{n(\kappa)}\}$. That is, now $\{\overline{X}_{n(\kappa)}\}$ only contains finite (specifically poly(λ)) many sets. Since by definition $\mathcal{R}$ cannot query on such large values of κ this step does not affect the advantage of $\mathcal{R}$.

$$\left| \Pr_{\mathcal{H}_1}[\mathcal{R}^{(\overline{\mathcal{A}}, \overline{\Psi})}(1^\lambda) \text{ wins } \mathcal{C}(1^\lambda)] - \Pr_{\mathcal{H}_0}[\mathcal{R}^{(\overline{\mathcal{A}}, \overline{\Psi})}(1^\lambda) \text{ wins } \mathcal{C}(1^\lambda)] \right| = 0 .$$

Hybrid $\mathcal{H}_2$. Let $\kappa_{\min} = \kappa(\lambda)$ be the maximal κ such that $2^{n(\kappa)^\alpha} \leq \lambda^c$ for a constant c to be selected large enough to satisfy constraints that will be explained below. Note that for all $\kappa \leq \kappa_{\min}$ it holds that $\nu(\kappa) = |\overline{X}_{n(\kappa)}| = 2^{O(n^\alpha)} = \text{poly}(\lambda)$ and for all $\kappa > \kappa_{\min}$ it holds that $\nu(\kappa) = 2^{0.1 \cdot n^\alpha} \geq \lambda^{0.1c}$.

From here on, we will focus on $\kappa \in (\kappa_{\min}, \kappa_{\max})$, since in the other regimes we can indeed execute $\overline{\mathcal{A}}, \overline{\Psi}$ efficiently. We call this *the relevant domain*.

We now change $\overline{\mathcal{A}}$ and make it *stateful*. Specifically, for all $\kappa \in (\kappa_{\min}, \kappa_{\max})$, instead of randomly selecting an $\bar{x} \in \overline{X}_{n(\kappa)}$ for every invocation, we go over the elements of $\overline{X}_{n(\kappa)}$ in order.

Claim 4. *If c is chosen so that $t(\lambda)^2 / \lambda^{0.1c} \leq \delta/10$ then*

$$\left| \Pr_{\mathcal{H}_2}[\mathcal{R}^{(\overline{\mathcal{A}}, \overline{\Psi})}(1^\lambda) \text{ wins } \mathcal{C}(1^\lambda)] - \Pr_{\mathcal{H}_1}[\mathcal{R}^{(\overline{\mathcal{A}}, \overline{\Psi})}(1^\lambda) \text{ wins } \mathcal{C}(1^\lambda)] \right| \leq \delta/10 .$$

Proof. The view of $\mathcal{R}$ can only change in the case where the same index i was selected more than once throughout the executions of $\overline{\mathcal{A}}$ on $\kappa > \kappa_{\min}$. Since $\mathcal{R}$ makes at most t queries, this event happens with probability at most $t^2 / \nu(\kappa_{\min}) \leq t(\lambda)^2 / \lambda^{c \cdot \gamma}$. If we choose c as in the claim statement, the probabilistic distance follows. ∎

Hybrid $\mathcal{H}_3$. We now change $\overline{\varPsi}$ to also be stateful, and in fact its state is joint with $\overline{\mathcal{A}}$. Specifically for the relevant domain $\kappa \in (\kappa_{\min}, \kappa_{\max})$, instead of checking whether $x \in \overline{X}_{n(\kappa)}$, it only checks whether x is in the prefix of $\overline{X}_{n(\kappa)}$ that had been used by $\overline{\mathcal{A}}$ so far.

Claim 5. *If c is chosen so that* $t(\lambda) \cdot \lambda^{-0.9c} \leq \delta/10$ *then*

$$\left| \Pr_{\mathcal{H}_3}[\mathcal{R}^{(\overline{\mathcal{A}},\overline{\varPsi})}(1^\lambda) \text{ wins } \mathcal{C}(1^\lambda)] - \Pr_{\mathcal{H}_2}[\mathcal{R}^{(\overline{\mathcal{A}},\overline{\varPsi})}(1^\lambda) \text{ wins } \mathcal{C}(1^\lambda)] \right| \leq \delta/10 .$$

Proof. We note that if $\mathcal{R}$ only makes $\overline{\varPsi}$ queries for these κ values with x inputs that are either in the prefix of $\overline{X}_{n(\kappa)}$ or not in $\overline{X}_{n(\kappa)}$ at all, then its view does not change. Let us consider the first $\overline{\varPsi}$ query of $\mathcal{R}$ that violates the above. Since this is the first such query, the view of $\mathcal{R}$ so far only depends on the relevant prefixes of $\overline{X}_{n(\kappa)}$'s. Let x be the value queried by $\mathcal{R}$ and let us bound the probability that x is in the suffix of $\overline{X}_{n(\kappa)}$ for the respective κ. Recall that the length of the suffix is at most $\nu(\kappa)$.

Since the view of $\mathcal{R}$ so far, and therefore x, is independent of this suffix, we can consider a given x and bound the probability that some entry in the suffix hits x. By Lemma 5, the entropy of each entry in $\overline{X}_{n(\kappa)}$ is at least n^α, which means that each value hits x with probability at most 2^{-n^α}. Applying the union bound we get a total probability of at most $\nu(\kappa) \cdot 2^{-n(\kappa)^\alpha} = 2^{-0.9n(\kappa)^\alpha}$. Since $\kappa > \kappa_{\min}$ we get that this probability is at most $\lambda^{-0.9c}$.

Applying the union bound over all at most t queries of $\mathcal{R}$, the probability that the above is violated for any of them is at most $t \cdot \lambda^{-0.9c}$ and the claim follows. ∎

Hybrid $\mathcal{H}_4$. We now change $\overline{\mathcal{A}}, \overline{\varPsi}$ in the relevant domain to sample the values of $\bar{x}$ *on the fly* rather than have them predetermined ahead of time. Specifically, $\overline{X}_{n(\kappa)}$ is initialized as an empty vector in the relevant domain. Whenever a query to $\overline{\mathcal{A}}$ is made relative to such a κ, $\overline{\mathcal{A}}$ samples a fresh $\bar{x} \xleftarrow{\$} \overline{\mathcal{L}}_{n(\kappa)}$, and applies h_e to it to compute the response e'. The sampled $\bar{x}$ is then appended to $\overline{X}_{n(\kappa)}$. When the distinguisher $\overline{\varPsi}$ is called, it uses the current value of $\overline{X}_{n(\kappa)}$ for its execution. Note that this change does not change the view of $\mathcal{R}$ at all.

$$\left| \Pr_{\mathcal{H}_4}[\mathcal{R}^{(\overline{\mathcal{A}},\overline{\varPsi})}(1^\lambda) \text{ wins } \mathcal{C}(1^\lambda)] - \Pr_{\mathcal{H}_3}[\mathcal{R}^{(\overline{\mathcal{A}},\overline{\varPsi})}(1^\lambda) \text{ wins } \mathcal{C}(1^\lambda)] \right| = 0 .$$

Hybrid $\mathcal{H}_5$. Now instead of sampling $\bar{x} \xleftarrow{\$} \overline{\mathcal{L}}_n$ and evaluating h_e on it, $\overline{\mathcal{A}}$ instead samples $(x, w) \xleftarrow{\$} \mathsf{Sam}(1^n)$ and computes e' by running $e' \leftarrow \mathsf{SFC.Eval}(e, x, w)$. The value x is appended to $\overline{X}_{n(\kappa)}$ just as before and the behavior of $\overline{\varPsi}$ does not change.

We would like to prove that this does not change the winning probability using a hybrid. Specifically, go over all samples of $\bar{x}$ and apply Corollary 7 to argue indistinguishability, but we need to be careful since indistinguishability only holds against adversaries of size $2^{O(n^\alpha)}$ but for the smaller values of κ in the relevant domain this is not necessarily the case. We will therefore need the following claim.

Claim 6. *Let* $\hat{\kappa} \in (\kappa_{\min}, \kappa_{\max})$ *and let* $\hat{n} = n(\hat{\kappa})$. *Then the functionality of* $(\overline{\mathcal{A}}, \overline{\Psi})_{\mathcal{H}_4}$ *on all* $\kappa \leq \hat{\kappa}$ *is computable by a size* $2^{O(\hat{n}^\alpha)}$ *circuit.*

Proof. We note that for all κ, $\overline{\mathcal{A}}$ takes inputs e of length at most $o(n^\alpha)$ and outputs e' of length at most $o(n^\alpha)$. We can therefore completely define its functionality using a table of size $2^{o(n^\alpha)}$. In addition to this truth table, we can also pre-sample $\overline{X}_{n(\kappa)}$. For $\kappa \leq \kappa_{\min}$, the set $\overline{X}_{n(\kappa)}$ contains at most $\nu(\kappa) \leq \nu(\hat{\kappa}) = 2^{0.1 \cdot \hat{n}^\alpha}$ samples, and for $\kappa > \kappa_{\min}$ it contains at most $t(\lambda) = \text{poly}(\lambda) \leq 2^{(1/c) \cdot O(\hat{n}^\alpha)}$ samples. Using these sets, we can simulate on-line sampling by going over these samples one by one. Taking the sum of table sizes for all $\kappa \leq \hat{\kappa}$, the claim follows. ∎

This will allow us to prove a bound on the difference between the hybrids.

Claim 7. *If c is chosen so that $t(\lambda)^2 \cdot \lambda^{-c} \leq \delta/10$ then*

$$\left| \Pr_{\mathcal{H}_5}[\mathcal{R}^{(\overline{\mathcal{A}}, \overline{\Psi})}(1^\lambda) \text{ wins } \mathcal{C}(1^\lambda)] - \Pr_{\mathcal{H}_4}[\mathcal{R}^{(\overline{\mathcal{A}}, \overline{\Psi})}(1^\lambda) \text{ wins } \mathcal{C}(1^\lambda)] \right| \leq \delta/10 \ .$$

Proof. The proof uses a hybrid argument. In hybrid (j_1, j_2), we construct the circuit from Claim 6 with respect to $\hat{\kappa} = \kappa_{\max} - j_1$, whose size is $2^{O(\hat{n}^\alpha)}$. We use this circuit to answer all queries for $\kappa < \hat{\kappa}$, as well as the first $t - j_2$ queries for $\kappa = \hat{\kappa}$. The rest of the queries are answered as in $\mathcal{H}_5$ by sampling x, w. The Distinguisher answers consistently with the x's that were used by $\overline{\mathcal{A}}$.

One can see that taking $j_1 = 0$, $j_2 = 0$, we get a functionality that is identical to $\mathcal{H}_4$, and taking $j_1 = \kappa_{\max} - \kappa_{\min}$, $j_2 = 0$ we get a functionality identical to $\mathcal{H}_5$. Furthermore, the functionality with (j_1, t) is identical to the functionality with $(j_1+1, 0)$. Now consider the difference between hybrids (j_1, j_2) and $(j_1, j_2 + 1)$. The only difference is whether (x, e') is generated by sampling $x \stackrel{\$}{\leftarrow} \overline{\mathcal{L}}$ and $e' = h_e(x)$, or whether $(x, w) \stackrel{\$}{\leftarrow} \text{Sam}(1^n)$ and $e' \leftarrow \text{SFC.Eval}(e, x, w)$. Furthermore, in this hybrid $\mathcal{R}^{(\overline{\mathcal{A}}, \overline{\Psi})}$ can be computed by a size $2^{O(n^\alpha)}$ circuit. Corollary 7 implies that $\Pr[\mathcal{R}^{(\overline{\mathcal{A}}, \overline{\Psi})}(1^\lambda) \text{ wins } \mathcal{C}(1^\lambda)]$ changes by at most $2 \cdot 2^{-n^\alpha} \leq 2 \cdot 2^{-n(\kappa_{\min})^\alpha} \leq \lambda^{-c}$. The total number of hybrids is at most $\kappa_{\max}(\lambda) \cdot t(\lambda)$, and recalling that $\kappa_{\max}(\lambda) \leq t(\lambda)$ the claim follows. ∎

We note that if we choose c to be an appropriately large constant, the resulting $\overline{\mathcal{A}}, \overline{\Psi}$ run in $\text{poly}(\lambda)$ time and only use x values in $\mathcal{L}$. We therefore denote them by $\mathcal{A}, \Psi$. This is formally stated below.

Claim 8. *There exist $(\mathcal{A}, \Psi)$ computable by $\text{poly}(\lambda)$ circuit that implements an identical functionality to $(\overline{\mathcal{A}}, \overline{\Psi})_{\mathcal{H}_5}$.*

Proof. We define $(\mathcal{A}, \Psi)$ as follows. Consider the circuit described in Claim 6 for $\hat{\kappa} = \kappa_{\min}$ and use it to answer queries with $\kappa \leq \kappa_{\min}$. Note that the circuit size is poly(λ). Queries with $\kappa > \kappa_{\max}$ don't need to be answered by $(\overline{\mathcal{A}}, \overline{\Psi})_{\mathcal{H}_5}$. As for queries in the relevant domain, the computation of $(\overline{\mathcal{A}}, \overline{\Psi})_{\mathcal{H}_5}$ for these values of κ runs in polynomial time in κ and therefore also in λ. $\qquad\square$

Conclusion. Combining the hybrids above, we get that $\mathcal{R}^{(\mathcal{A},\Psi)}$ is a poly(λ)-time algorithm with advantage

$$\mathsf{Adv}^{(\mathcal{C},0)}_{\mathcal{R}^{(\mathcal{A},\Psi)}}(1^\lambda) \geq \delta - 3 \cdot \delta/10 = \Omega(\delta) \ .$$

That is, $\mathcal{R}^{(\mathcal{A},\Psi)}$ is a polynomial time algorithm that breaks the assumption $(\mathcal{C},0)$ as required. $\qquad\square$

Acknowledgments. We wish to thank the Asiacrypt reviewers for the extremely thorough review process, and for their useful and enlightening comments that helped improve this manuscript significantly.

References

[ABOR00] Aiello, W., Bhatt, S., Ostrovsky, R., Rajagopalan, S.R.: Fast verification of any remote procedure call: short witness-indistinguishable one-round proofs for NP. In: Montanari, U., Rolim, J.D.P., Welzl, E. (eds.) ICALP 2000. LNCS, vol. 1853, pp. 463–474. Springer, Heidelberg (2000). doi:10.1007/3-540-45022-X_39

[ALM+98] Arora, S., Lund, C., Motwani, R., Sudan, M., Szegedy, M.: Proof verification and the hardness of approximation problems. J. ACM **45**(3), 501–555 (1998)

[AS98] Arora, S., Safra, S.: Probabilistic checking of proofs: a new characterization of NP. J. ACM **45**(1), 70–122 (1998)

[BCC+14] Bitansky, N., Canetti, R., Chiesa, A., Goldwasser, S., Lin, H., Rubinstein, A., Tromer, E.: The hunting of the SNARK. IACR Cryptology ePrint Archive, 2014:580 (2014)

[BCCT12] Bitansky, N., Canetti, R., Chiesa, A., Tromer, E.: From extractable collision resistance to succinct non-interactive arguments of knowledge, and back again. In: Proceedings of the 3rd Innovations in Theoretical Computer Science Conference, ITCS 2012, pp. 326–349 (2012)

[BCCT13] Bitansky, N., Canetti, R., Chiesa, A., Tromer, E.: Recursive composition and bootstrapping for SNARKS and proof-carrying data. In: STOC, pp. 111–120. ACM (2013)

[BFL91] Babai, L., Fortnow, L., Lund, C.: Non-deterministic exponential time has two-prover interactive protocols. Comput. Complex. **1**(1), 3–40 (1991)

[BMW98] Biehl, I., Meyer, B., Wetzel, S.: Ensuring the integrity of agent-based computations by short proofs. In: Rothermel, K., Hohl, F. (eds.) MA 1998. LNCS, vol. 1477, pp. 183–194. Springer, Heidelberg (1998). doi:10.1007/BFb0057658

[DBL08] 49th Annual IEEE Symposium on Foundations of Computer Science, FOCS 2008, 25–28 October 2008, Philadelphia, PA. IEEE Computer Society, USA (2008)

[DFH12]   Damgård, I., Faust, S., Hazay, C.: Secure two-party computation with low communication. In: Cramer, R. (ed.) TCC 2012. LNCS, vol. 7194, pp. 54–74. Springer, Heidelberg (2012). doi:10.1007/978-3-642-28914-9_4

[DHRW16]   Dodis, Y., Halevi, S., Rothblum, R.D., Wichs, D.: Spooky encryption and its applications. In: Robshaw, M., Katz, J. (eds.) CRYPTO 2016. LNCS, vol. 9816, pp. 93–122. Springer, Heidelberg (2016). doi:10.1007/978-3-662-53015-3_4

[DLN+01]   Dwork, C., Langberg, M., Naor, M., Nissim, K., Reingold, O.: Succinct proofs for NP and spooky interactions. Unpublished manuscript (2001)

[DP08]   Dziembowski, S., Pietrzak, K.: Leakage-resilient cryptography. In: 49th Annual IEEE Symposium on Foundations of Computer Science, FOCS 2008, 25–28 October 2008, Philadelphia, PA, USA [DBL08], pp. 293–302 (2008)

[GW11]   Gentry, C., Wichs, D.: Separating succinct non-interactive arguments from all falsifiable assumptions. In: Proceedings of the Forty-third Annual ACM Symposium on Theory Of Computing, pp. 99–108. ACM (2011)

[KRR13]   Kalai, Y.T., Raz, R., Rothblum, R.D.: Delegation for bounded space. In: Boneh, D., Roughgarden, T., Feigenbaum, J., (eds) Symposium on Theory of Computing Conference, STOC 2013, Palo Alto, CA, USA, 1–4 June 2013, pp. 565–574. ACM (2013)

[KRR14]   Kalai, Y.T., Raz, R., Rothblum, R.D.: How to delegate computations: the power of no-signaling proofs. In: STOC, pp. 485–494. ACM (2014)

[Mic94]   Micali, S.: CS proofs (extended abstracts). In: 35th Annual Symposium on Foundations of Computer Science, Santa Fe, New Mexico, USA, 20–22 November 1994, pp. 436–453. IEEE Computer Society (1994)

[Nao03]   Naor, M.: On cryptographic assumptions and challenges. In: Boneh, D. (ed.) CRYPTO 2003. LNCS, vol. 2729, pp. 96–109. Springer, Heidelberg (2003). doi:10.1007/978-3-540-45146-4_6

[RTTV08]   Reingold, O., Trevisan, L., Tulsiani, M., Vadhan, S.P.: Dense subsets of pseudorandom sets. In: 49th Annual IEEE Symposium on Foundations of Computer Science, FOCS 2008, 25–28 October 2008, Philadelphia, PA, USA [DBL08], pp. 76–85 (2008)

[VZ13]   Vadhan, S., Zheng, C.J.: A uniform min-max theorem with applications in cryptography. In: Canetti, R., Garay, J.A. (eds.) CRYPTO 2013. LNCS, vol. 8042, pp. 93–110. Springer, Heidelberg (2013). doi:10.1007/978-3-642-40041-4_6

Non-Interactive Multiparty Computation
Without Correlated Randomness

Shai Halevi[1]([✉]), Yuval Ishai[2], Abhishek Jain[3], Ilan Komargodski[4],
Amit Sahai[5], and Eylon Yogev[6]

[1] IBM Research, New York, USA
shaih@alum.mit.edu
[2] Technion and UCLA, Haifa, Israel
yuvali@cs.technion.ac.il
[3] Johns Hopkins University, Baltimore, USA
abhishek@cs.jhu.edu
[4] Cornell Tech, New York, USA
komargodski@cornell.edu
[5] UCLA, Los Angeles, USA
sahai@cs.ucla.edu
[6] Weizmann Institute, Rehovot, Israel
eylon.yogev@weizmann.ac.il

Abstract. We study the problem of non-interactive multiparty computation (NI-MPC) where a group of completely asynchronous parties can evaluate a function over their joint inputs by sending a *single* message to an evaluator who computes the output. Previously, the only general solutions to this problem that resisted collusions between the evaluator and a set of parties were based on multi-input functional encryption and required the use of complex correlated randomness setup.

In this work, we present a new solution for NI-MPC against arbitrary collusions using a public-key infrastructure (PKI) setup supplemented with a common random string. A PKI is, in fact, the minimal setup that one can hope for in this model in order to achieve a meaningful "best possible" notion of security, namely, that an adversary that corrupts the evaluator and an arbitrary set of parties only learns the residual function obtained by restricting the function to the inputs of the uncorrupted parties. Our solution is based on indistinguishability obfuscation and DDH both with sub-exponential security. We extend this main result to the case of *general interaction patterns*, providing the above best possible security that is achievable for the given interaction.

Our main result gives rise to a novel notion of *(public-key) multiparty obfuscation*, where n parties can *independently* obfuscate program modules M_i such that the obfuscated modules, when put together, exhibit the functionality of the program obtained by "combining" the underlying modules M_i. This notion may be of independent interest.

© International Association for Cryptologic Research 2017
T. Takagi and T. Peyrin (Eds.): ASIACRYPT 2017, Part III, LNCS 10626, pp. 181–211, 2017.
https://doi.org/10.1007/978-3-319-70700-6_7

1   Introduction

The recent breakthrough on general-purpose program obfuscation [16] has spurred a large body of research that studies different flavors of obfuscation and their applications. Much of this research is driven by the goal of *minimizing interaction*. In this work we revisit the question of minimizing interaction in the broad context of secure multiparty computation (MPC). This question is independently motivated by the goal of obtaining useful multi-party variants of obfuscation and related primitives such as functional encryption [5,28,29].

Non-Interactive MPC. Consider the following *non-interactive* model of computation: there are n parties who wish to securely evaluate a function f on their joint inputs. For instance, the parties may wish to compute a majority of their $0/1$ inputs for the purpose of voting. The parties are completely asynchronous. At any point, a party can "wake up" and submit a *single* message to a central server, also referred to as the *evaluator*. Upon receiving all the messages, the evaluator computes the output.

The above model represents an example of MPC with limited interaction, and contrasts with traditional MPC models [18,31] where the parties freely interact with each other. Indeed, there are many scenarios where more limited forms of interaction are desirable, e.g., when the parties cannot be simultaneously available due to physical constraints or due to efficiency considerations.

The non-interactive model of computation was first proposed by Feige, Kilian and Naor (FKN) [13]. In their model, they allow the messages of the parties to depend on secret randomness that is unknown to the evaluator. As a result, a major disadvantage of their model is that it does not provide any security in case the evaluator colludes with one or more parties.[1]

The study of *collusion-resistant* MPC protocols with restricted interaction was initiated by Halevi et al. (HLP) [24]. They considered a variation of the non-interactive model in which the parties *sequentially* interact with the evaluator, where this interaction can depend on a public-key infrastructure (PKI). Subsequent to their work, Beimel et al. [3] and Halevi et al. [23] considered a model (similar to FKN) where each of the parties can *independently* interact with the evaluator by sending it a single message. (It was also implicitly considered in the context of multi-input functional encryption by Goldwasser et al. [19].) Crucially, in this model, unlike the FKN model, security does not completely break down in the presence of collusion attacks. We refer to MPC in this model as *non-interactive MPC*, or NI-MPC for short.

Best-Possible Security for NI-MPC. An NI-MPC protocol should provide the "best-possible security" for any given set of corrupted parties. The latter can be formalized via the following notion of *residual function* [24]: when the evaluator colludes with a set T of corrupted parties, it can inevitably learn

---

[1] Indeed, it is not hard to see that in any protocol for a non-trivial function in this model, all inputs can be recovered from the n messages and the common randomness. This makes such protocols inherently vulnerable to collusions.

the value of the original function f on the honest parties' inputs coupled with *every possible choice* of inputs from T. Thus, the adversary effectively has oracle access to the residual function obtained by restricting f to the honest parties' inputs and allowing a free choice of the corrupted parties' inputs. The security requirement of NI-MPC informally asserts that the adversary learns no more than the residual function. This gives a meaningful security guarantee in many natural cases. For instance, for the above mentioned example of voting or, more generally, in the case of symmetric functions, the residual function still hides the sensitive information about the inputs of the uncorrupted parties. See [3] for a more detailed discussion.

The strongest formalization of the above security requirement is via efficient simulation. However, this cannot be achieved in general. Indeed, NI-MPC for general functions can be easily seen to imply "virtual black-box" obfuscation for general functions [3,19], which was ruled out in [2]. Instead, we consider a natural relaxation that replaces efficient simulation by unbounded simulation, or equivalently *indistinguishability* of different sets of inputs that induce the same residual function. Roughly, the indistinguishability-based security definition is described as follows – for simplicity, we state it for the three-party case where the last party is corrupted and is colluding with the evaluator: for any input pairs (x, y) and (x', y') for the honest parties such that $f(x, y, \cdot) \equiv f(x', y', \cdot)$, the adversary's view is indistinguishable when the inputs of the honest parties to the computation are (x, y) or (x', y'). We refer the reader to the technical sections for the general definition.

Necessity of Setup and iO. We next turn our attention to the necessity of setup for NI-MPC. Unlike the FKN model, collusion-resistant NI-MPC cannot be realized with a single common source of randomness that is shared by all clients. Instead, NI-MPC requires a setup that allows authentication of the messages of parties, while remaining robust to collusion. This is because otherwise, an adversarial evaluator can spoof an honest party in order to learn private information about the inputs of the other honest parties: consider NI-MPC for three parties where the last party and the evaluator are corrupted. The security definition for this case was described above. However, suppose there exists a "splitting input" y^* for f s.t. $f(x, y^*, z) \neq f(x', y^*, z)$, where z is some third party input. In the absence of authentication, the adversary can prepare a message $\widehat{y}^*$ on its own using y^* and then evaluate it together with the first and third party messages to distinguish whether the first honest party's input to the computation is x or x'.

Further, as already noted in [3,19,23], NI-MPC for general functions implies indistinguishability obfuscation (iO) [2,15]: recall the security definition of three-party NI-MPC as described above, where the last party and the evaluator are corrupted. Substituting x and x' with two circuits C_0 and C_1, y and y' with null inputs and f with a universal circuit such that $f(C, \bot, z) = C(z)$, we can recover the security definition of iO. We stress that this implication holds regardless of the choice of setup used in the NI-MPC protocol.

Our Question: NI-MPC with minimal setup? Known n-party NI-MPC protocols [3,19,23] require intricate n-wise correlated randomness setups where the *private* randomness r_i input to each party i is computed as $G(r) = (r_1, \ldots, r_n)$ for some function G. In particular, such a correlated randomness setup can only be realized by using a trusted dealer with private channels to each protocol participant, or alternatively a fully interactive MPC protocol that is executed during an offline phase. The main question we ask is whether the use of such *general* correlated randomness is necessary for NI-MPC.

Indeed, secure protocols in the HLP model or even the FKN model can be realized using a standard PKI setup.[2] A PKI setup can be viewed as a special form of "pairwise" correlation that can be realized with no direct interaction between the clients: it suffices for every client to post its own public key and read all of the other clients' public keys. Given that some form of setup is necessary to prevent the aforementioned spoofing attacks in NI-MPC, using a PKI setup is the best one could hope for.[3]

1.1 Our Results

I. NI-MPC without Correlated Randomness Setup. We construct the first collusion-resistant NI-MPC protocols without general correlated randomness. Instead, we use a PKI and a common *random* string (CRS).[4] The security of our construction is proven against malicious adversaries assuming iO and DDH, both with sub-exponential security.

Theorem 1 (Informal). *Assuming indistinguishability obfuscation and DDH, both with sub-exponential security, there exists a collusion-resistant non-interactive multi-party computation protocol that achieves security in the malicious model given a CRS and a PKI.*

As explained earlier, iO is a necessary assumption for NI-MPC for general functions.

II. Multi-party Indistinguishability Obfuscation. Our main result gives rise to a new notion of multiparty obfuscation (and iO) that may be of independent interest. We start by explaining this notion via a motivating example: consider a scenario where there are n parties, each holding a private program module M_i. Suppose there is a public Combine algorithm that combines $M_1, \ldots, M_n$ to

---

[2] Protocols in the FKN model can be realized using standard PKI by combining a non-interactive 2-party key agreement protocol, which can be based on the DDH assumption, with garbled circuits. The idea is to have every party P_i agree with the last party P_n on the input keys of P_i, and have P_n generate and send the garbled circuit based on all keys.

[3] In particular, note that a common random string setup cannot prevent spoofing attacks.

[4] We note that even a construction with a PKI and a common *reference* string was unknown prior to this work.

yield a program M. The parties wish to jointly obfuscate M; however, due to proprietary reasons, they cannot share their modules with each other.

One option for the parties is to simply run a standard MPC protocol using their modules as their respective private inputs to jointly compute an obfuscation of M. We ask whether it is possible for the parties to jointly obfuscate M in a non-interactive manner. In particular, we require that each party i can individually obfuscate its module M_i s.t. the obfuscated modules $\overline{M_1}, \ldots, \overline{M_n}$ put together, exhibit the functionality of M while still hiding the individual modules.

We refer to this as multiparty obfuscation, and in the context of iO, as multiparty iO. Roughly speaking, the security of multiparty iO says that for any pair of tuples $(M_1^0, \ldots, M_n^0)$ and $(M_1^1, \ldots, M_n^1)$ s.t. $\mathsf{Combine}(M_1^0, \ldots, M_n^0)$ and $\mathsf{Combine}(M_1^1, \ldots, M_n^1)$ are functionally equivalent, the pair of tuples $(\overline{M_1^0}, \ldots, \overline{M_n^0})$ and $(\overline{M_1^1}, \ldots, \overline{M_n^1})$ are computationally indistinguishable. Here $\overline{M_i^b}$ denotes the obfuscation of M_i^b computed by the i'th party.

We note that if we set $n = 1$ and the $\mathsf{Combine}$ function to be the identity function, then we recover the standard notion of iO from the above definition. Further, while the above definition considers the case where all the parties are honest and only the obfuscation evaluator is dishonest, it can be easily modified to allow for dishonest parties. However, in this case, the functional equivalence requirement on the modules must be suitably modified, as in the case of NI-MPC.

Multiparty iO from NI-MPC. We now explain how our NI-MPC can be used to obtain multiparty iO. The setup in multiparty iO is the same as in the NI-MPC, namely, a PKI and a CRS. We consider NI-MPC between $(n + 1)$ parties. The inputs of the first n parties are set to be their respective modules $M_1, \ldots, M_n$. The $(n + 1)$th party is the evaluator which holds an input x for the "combined" program $M = \mathsf{Combine}(M_1, \ldots, M_n)$. The function computed by the parties contains the $\mathsf{Combine}$ algorithm hardwired in its description. On input $(M_1, \ldots, M_n, x)$, it computes $M = \mathsf{Combine}(M_1, \ldots, M_n)$ and outputs $M(x)$.

Now, suppose that only the $(n+1)$th party and the evaluator are corrupted. Then, this case exactly corresponds to multiparty iO where all the parties (a.k.a., obfuscators) are honest. The case of more general corruptions is defined similarly.

III. General interaction patterns. As observed by Halevi et al. [23], the problem of NI-MPC can be viewed as secure computation on a *star* interaction pattern. Towards a generalization, Halevi et al. also considered the problem of secure multiparty computation with *general* interaction patterns. They presented a result for the same assuming (sub-exponentially secure) indistinguishability obfuscation by using a complex correlated randomness setup.

We improve upon their result by providing a construction in the PKI model assuming (sub-exponentially secure) indistinguishability obfuscation and DDH. In particular, our main protocol for NI-MPC can be extended to handle more general interaction patterns described by directed acyclic graphs, where each node represents a party who expects to receive messages from all of its parents and can then send messages to all of its children, and where the sinks of the graph compute the output.

We note that any interaction pattern gives rise to a "best-possible" notion of security, as formalized in [24] for the case of a "chain" pattern and in [23, Definition 6] for general patterns. In [23], it was shown that the star pattern is "complete" in the sense that any pattern can be reduced to it with no additional cryptographic assumptions. However, their reduction inherently requires a correlated randomness setup and thus is not applicable in our setting.

Instead, we show how to directly modify our scheme to guarantee best-possible security for any interaction pattern. A representative example is that of a chain pattern considered in [20,24], namely a simple directed path traversing all nodes. The "best-possible" security notion that can be achieved in this case is typically stronger than the security notion for NI-MPC which corresponds to a star pattern. To see this, notice that in the star pattern the adversary can reset the input of each party it controls, whereas in the chain pattern he cannot reset the input of a malicious party that is followed by an honest party along the chain. Our results for the chain pattern provide the first extension of the positive results in the HLP model [24] to general functions with a similar PKI setup, at the (necessary) price of relying on stronger assumptions.[5]

1.2   Our Techniques

We now explain the main ideas in our construction of collusion-resistant NI-MPC. As explained earlier, this suffices to obtain multiparty iO.

Initial Challenges. Recall that in an NI-MPC protocol, each party sends a *single* message to the evaluator. Then, a starting idea to construct an NI-MPC protocol is to "compile" an interactive MPC protocol into a non-interactive one using program obfuscation. That is, let f be the function that the parties wish to compute and let Π be an interactive MPC protocol for computing f. Then, each party i simply computes an obfuscation $\overline{\mathsf{NMF}_i}$ of its next-message function NMF_i in Π, where NMF_i contains its input x_i and random tape r_i hardwired. It then sends $\overline{\mathsf{NMF}_i}$ to the evaluator. Upon receiving all the obfuscated programs, the evaluator computes a transcript of Π by acting as the communication channel between the "virtual parties" $\overline{\mathsf{NMF}_1}, \ldots, \overline{\mathsf{NMF}_n}$. At the end, it obtains the output $f(x_1, \ldots, x_n)$.

The main problem with the above approach is that an adversarial evaluator (that colludes with one or more parties) can perform resetting attacks to break the security of the underlying MPC protocol Π. Indeed, since the obfuscated programs $\overline{\mathsf{NMF}_i}$ are computing a "reactive" functionality, an evaluator can simply reset them to an earlier state and feed them different messages. Since the input of each honest party i is fixed in the obfuscated program $\overline{\mathsf{NMF}_i}$, this means that the adversary can now execute multiple "sessions" of Π w.r.t. the same fixed inputs of the honest parties. The security of standard MPC protocols completely breaks down in such a case.

---

[5] Indeed, MPC for computing general functions on a chain implies indistinguishability obfuscation (iO). The positive results of [20,24] based on weaker assumptions are only for weaker classes of functions where positive results do not imply (general) iO.

To address this problem, a natural idea is to replace Π with a resettably secure MPC protocol [8,21,22]. Roughly speaking, resettable MPC guarantees that the only advantage an adversary can gain by performing a resetting attack is to change its input. As such, it can learn multiple function outputs w.r.t. a fixed input vector of the honest parties. No information beyond this set of outputs is leaked to the adversary. While resettably secure MPC for general functions is impossible in the plain model [21], it can be realized in the CRS model by compiling a UC-secure MPC protocol [10] using pseudorandom functions à la [8].

The security guarantee of resettable MPC coincides with our definition of NI-MPC, with the only difference that our definition is indistinguishability-based while resettable MPC security is defined w.r.t. simulation. Nevertheless, it is not immediately clear how to compile a resettably secure MPC protocol into NI-MPC when using iO. In particular, note that the natural approach to argue security in this context is to hardwire the simulated transcript in the obfuscated programs so as to "force" the simulated execution on the adversary. This strategy was previously used in [14] for constructing two-round MPC protocols. In our context, however, since the adversary can legitimately perform resetting attacks, the number of possible transcripts we may need to hardwire is *unbounded*.

A Starting Point: Obfuscation Combiners. In order to find a path to a solution, let us first consider a weaker problem: suppose we are given N candidates for program obfuscation, many of which are possibly "bad", i.e., either they do not hide the program or do not preserve correctness. We do not know which of the candidates are bad. The goal is to use these candidates to obfuscate a function in a secure manner. This problem is referred to as obfuscation combiner [1].

To see why this problem is relevant to us, note that the "bad" candidate obfuscations can be thought of as the corrupted parties in our setting. The role of the evaluator is the same. Furthermore, in this setting, similar to ours, resetting attacks are unavoidable. The key difference, however, is that while our setting is inherently distributed, the above setting is centralized, in that a common entity performs all the obfuscations.

Nevertheless, we use obfuscation combiner as a starting point for our construction. Specifically, we build on ideas used in the obfuscation combiner of Ananth et al. [1]. In their construction, they use a special-purpose MPC protocol [27] based on multi-key fully homomorphic encryption [11,12,26,27] (see Sect. 2 for the definition). Our solution also uses this MPC scheme as a starting basis. In particular, our construction implicitly compiles the MPC scheme of [27] into a resettably secure one in an "iO-friendly" manner. In order to develop a full solution for our setting, however, we have to address several additional challenges that do not arise in the setting of obfuscation combiners. Below, we elaborate on the details.

Our Approach. We first recall the notion of a multi-key FHE. A multi-key FHE allows for generating individual encryption/decryption-key pairs ek_i, dk_i such that they can be later combined to obtain a joint public key ek. To be more precise, given a ciphertext with respect to ek_i, there is an Expand operation that transforms it into a ciphertext with respect to a joint public key ek. Once this done, the resulting ciphertext can be homomorphically evaluated just like any FHE scheme. The resulting ciphertexts can then be individually decrypted using the dk_i's to obtain partial decryptions. Finally, there is a mechanism to combine the partial decryptions to obtain the final output.

Given a multi-key FHE, a first idea for a non-interactive MPC protocol for a function f is described below.[6]

1. **Setup of party i**: Sample an encryption/decryption key pair ek_i, dk_i and sets its private-key to be $\mathsf{SK}_i = (dk_i)$ and published the public-key $\mathsf{PK}_i = (ek_i)$.
2. **Encryption of party i on input x_i**: Compute an encryption of x_i with respect to the key ek_i: $\mathsf{CT}_i = \mathsf{Enc}(ek_i, x_i)$ and compute an obfuscation $\overline{G_i}$ of the circuit given in Fig. 1. The message of this party is the pair $(\overline{G_i}, \mathsf{CT}_i)$.
3. **Evaluation given the messages of the parties**: Given $(\overline{G_i}, \mathsf{CT}_i)$ for each $i \in [n]$, the evaluator executes each $\overline{G_i}$ on $\mathsf{CT}_1, \ldots, \mathsf{CT}_n$ to get p_i. Then, it combines all partial decryptions $p_1, \ldots, p_n$ to get y.

Correctness of the protocol is immediate by the correctness of the underlying obfuscator and the correctness of the multi-key FHE scheme. However, proving security of this scheme runs into several obstacles. First, we have to deal with the fact that the Eval operation is randomized. Second, there seems to be a very simple attack for the evaluator stemming from the fact that the encryption keys can be used to encrypt arbitrary values. Namely, the evaluator, given ek_i, can simulate ciphertexts generated by the i-th party and potentially fully recover all inputs.

The first issue is (by now) quite standard in the iO literature and we overcome it by using a puncturable PRF family. The second issue is more complicated to handle. Our idea is to augment the setup procedure to generate a signature/verification key pair and let each party sign its input. The circuit that

$$G_i\left[dk_i, \{ek_j\}_{j\in[n]}\right]$$

Hardwired: the decryption key of party i, dk_i, and all the encryption keys $\{ek_j\}_{j\in[n]}$.
Input: n ciphertexts: $\mathsf{CT}_1, \ldots, \mathsf{CT}_n$.

- For all $j \in [n]$ compute $\widehat{\mathsf{CT}}_j \leftarrow \mathsf{Expand}((ek_1, \ldots, ek_n), j, \mathsf{CT}_j)$.
- Perform $\widehat{\mathsf{CT}_{\mathsf{out}}} \leftarrow \mathsf{Eval}\left(f, \widehat{\mathsf{CT}}_1, \ldots, \widehat{\mathsf{CT}}_n\right)$.
- Output $p_i \leftarrow \mathsf{PartDec}\left(\widehat{\mathsf{CT}_{\mathsf{out}}}, ek_1, \ldots, ek_n, i, dk_i\right)$.

Fig. 1. The circuit G_i. The parameters in the square brackets are hardwired values and the inputs are specified separately (this is the convention throughout the paper).

---

[6] For simplicity of notation we assume a single public function.

each party will obfuscate will verify each such signature to make sure it was honestly generated. Intuitively, it seems to solve the problem, but formally arguing security using iO and an arbitrary signature scheme seems to be problematic.[7] To overcome this, we construct a special kind of signatures with which we are able to complete the proof. Specifically, our signatures have a property that allows us to replace the verification key with a "punctured verification key" in an indistinguishable way such that the punctured key (statistically) verifies *only a single signature*.

Having these signatures, the proof proceeds by a sequence of hybrid experiments in which we slowly replace each partial decryption with a simulation which does not require the decryption key. The number of hybrids is proportional to the input domain of the circuit G_i. Throughout the proof, we extensively use the probabilistic iO of [9] and the partition-programming technique of [1].

The malicious case. Notice that in the malicious case, since our protocol is non-interactive, the evaluator can plug in any value it wishes in the corrupted parties and obtain a value for the function f. Thus, in this case, if the evaluator colludes with a subset $A \subset [n]$ of the parties (i.e. $H = [n] \setminus A$ is the honest set of parties), then we could hope to get security only for functions f and challenge $(\{x_i^0\}_{i \in H}, \{x_i^1\}_{i \in H})$ that satisfy

$$f(\langle \{x_i^0\}_{i \in H}, \{x_i\}_{i \in A}\rangle) = f(\langle \{x_i^1\}_{i \in H}, \{x_i\}_{i \in A}\rangle)$$

for every $\{x_i\}_{i \in A}$ and where $\langle \cdot \rangle$ sorts the inputs according to the index i.

The proof and construction above do not guarantee security in the case where the evaluator colludes with a subset of the parties. Specifically, there is no guarantee as to what happens when one of the (malicious) parties sends a malformed ciphertext CT_i. There, the partial decryption would still output some string, but we have no guarantee on whether the simulation of this partial decryption is going to be indistinguishable.

To overcome this we use non-interactive zero-knowledge proofs (NIZKs). Specifically, during encryption of the value x_i, party i will also provide (in addition to the obfuscated circuit, the ciphertext and the signature) a NIZK proof asserting that the ciphertext is a legal well-formed ciphertext. As in most works that combine NIZKs and iO, we need a special kind of NIZKs called statistically-simulation-sound NIZKs (SSS-NIZKs) [15]. However, this is still not enough for us. The reason is that for the partial-decryption simulation we need to know the final output, but we have no control over the inputs coming from the malicious parties (recall that they are encrypted!). To solve this we use statistical-simulation-extractable zero-knowledge proofs (SSE-NIZKs) which are a special kind of non-interactive zero-knowledge proofs of knowledge (NIWI-POKs) which allow us to *extract* from the proof of the malicious parties their decryption key, decrypt their ciphertexts and compute the expected final output of the

---

[7] One standard way to handle this is to assume a differing-input obfuscator (diO) rather than plain iO. A diO guarantees that if an adversary distinguishes two obfuscated circuits, then it must know a differing input.

computation. We construct such SSE-NIZKs in the CRS model starting with any NIWI-POK and any one-way permutation.

Reusable PKI with a session ID. The solution we described above only gives a non-reusable scheme. Namely, the PKI cannot be used for more than one encryption. We consider a stronger model where the PKI can be reused but this requires a more delicate security definition since the evaluator can mix-and-match ciphertexts corresponding to honest parties which makes the security definition much less meaningful (as it applies to a much smaller set of functions). We consider a hybrid security definition in which we support a reusable PKI but introduce session IDs and require correctness only for ciphertexts that are generated with the same session ID.

To support such functionality we construct puncturable signatures that can be punctured such that only a single message with a specific prefix is allowed. The session ID is used as a prefix and after puncturing no signature exists for messages agreeing on the prefix, except the one generated by the honest party. Our construction uses NIZKs and statistically-binding commitments.

1.3 Related Work

The problem of devising non-interactive protocols was first addressed in [13], where it was shown how to compute any function assuming that all parties share common private randomness which is unknown to the adversary.

As observed in [23], collusion-resistant non-interactive MPC in the correlated randomness model follows from multi-input functional encryption [19]. Thus, the MIFE scheme of Goldwasser et al. [19] gives an NI-MPC protocol in the correlated randomness model assuming iO and one-way functions. As noted already in [19], this is in fact tight since MIFE implies iO.

Non-interactive MPC protocols in the information-theoretic setting for a simple class of functions were constructed by [3] (see also improvements in [23]), again using correlated randomness.

2   Preliminaries

We use the following primitives in our constructions.

1. We use the notion of **indistinguishability obfuscation** (iO), as defined in [2,15].
2. We use the notion of **puncturable PRFs**, as defined in [6,7,25,30]. As observed in these works, the GGM construction [17] of PRFs from one-way functions yields puncturable PRFs.
3. We will also use the notion of **threshold multi-key FHE** [11,26,27]. The initial constructions of threshold multi-key FHE based on learning with errors assumption relied on a common random string. Recently, Dodis et al. [12] constructed a multi-key FHE scheme based on iO and DDH. We will use their scheme in our constructions.

The definitions of the above primitives are provided in the full version.

3   Building Blocks

3.1   Statistical Simulation-Extractable Zero Knowledge Proofs

Let $\mathcal{R}$ be an efficiently computable relation that consists of pairs (x, w), where x is called the statement and w is the witness. Let L denote the language consisting of statements in $\mathcal{R}$. A statistical simulation-extractable non-interactive zero-knowledge (SSE-NIZK) proof system for a language L consists of a tuple of algorithms $(\mathcal{K}, \mathcal{P}, \mathcal{V}, \mathcal{S}_1, \mathcal{S}_2, \mathcal{E})$. We start by describing the basic algorithms $\mathcal{K}, \mathcal{P}, \mathcal{V}$ below:

- $\sigma \leftarrow \mathcal{K}(1^\lambda)$: On input the security parameter, it outputs a common random string (CRS) σ.
- $\pi \leftarrow \mathcal{P}(\sigma, x, w)$: On input a CRS σ, a statement x and a witness w s.t. $(x, w) \in \mathcal{R}$, it outputs a proof string π.
- $b \leftarrow \mathcal{V}(\sigma, x, \pi)$: On input a CRS σ, a statement x and a proof π, it outputs 1 or 0, denoting accept or reject.

Perfect Completeness. A non-interactive proof system is complete if an honest prover with a valid witness for a statement can convince a verifier of the validity of the statement. Formally, for every $(x, w) \in \mathcal{R}$,

$$\Pr[\sigma \leftarrow \mathcal{K}(1^\lambda); \pi \leftarrow \mathcal{P}(\sigma, x, w) : \mathcal{V}(\sigma, x, \pi) = 1] = 1$$

Statistical Soundness. A non-interactive proof system is sound if it is infeasible to convince a verifier if the statement is false. Formally, for all (possibly unbounded) adversaries $\mathcal{A}$,

$$\Pr[\sigma \leftarrow \mathcal{K}(1^k); (x, \pi) \leftarrow \mathcal{A}(\sigma) : \mathcal{V}(\sigma, x, \pi) = 1 : x \notin L] \leq \mathsf{negl}(\lambda)$$

Computational Zero Knowledge. A non-interactive proof system is computational zero knowledge if the proof does not reveal any information about the witness to the adversary. Formally, we require that for all non-uniform PPT adversaries $\mathcal{A}$, for all $(x, w) \in \mathcal{R}$,

$$\Pr[\sigma \leftarrow \mathcal{K}(1^\lambda) : \pi \leftarrow \mathcal{P}(\sigma, x, w) : \mathcal{A}(\sigma, x, \pi) = 1] \approx$$
$$\Pr[(\sigma, \tau) \leftarrow \mathcal{S}_1(1^\lambda, x) : \pi \leftarrow \mathcal{S}_2(\sigma, \tau, x) : \mathcal{A}(\sigma, x, \pi) = 1]$$

Statistical Simulation-Extractability. A NIZK proof system is statistical simulation-extractable if under a simulated CRS, no proof for false statement exists, except for simulated proof for a fixed statement fed into $\mathcal{S}_1$ to generate the simulated CRS. Furthermore, using an efficient extractor algorithm $\mathcal{E}$, it is possible to extract a witness from any accepting proof generated by an unbounded adversary using the simulated CRS. Formally, for all statements x and all unbounded adversaries $\mathcal{A}$,

$$\Pr[(\sigma, \tau) \leftarrow \mathcal{S}_1(1^\lambda, x) : \pi \leftarrow \mathcal{S}_2(\sigma, \tau, x) : (x^*, \pi^*) \leftarrow \mathcal{A}(\sigma, x, \pi) : (x^* \neq x) :$$
$$1 \leftarrow \mathcal{V}(\sigma, x^*, \pi^*) : w^* \leftarrow \mathcal{E}(\sigma, \tau, x^*, \pi^*) : (x^*, w^*) \in \mathcal{R}] \geq 1 - \mathsf{negl}(\lambda)$$

The Construction

Let $(\mathcal{K}, \mathcal{P}, \mathcal{V})$ be a non-interactive witness-indistinguishable proof of knowledge (NIWI-POK) system in the common random string model. Let $\mathsf{Com}(\cdot, \cdot)$ be a non-interactive perfectly binding string commitment scheme with pseudorandom commitments. Using these ingredients, we will construct an SSE-NIZK proof system $(\mathcal{K}', \mathcal{P}', \mathcal{V}')$.

Let ℓ be the upper bound on the length of the statements to be proven and let $L = L(\ell)$ denote the length of commitments output by Com.

- $\mathcal{K}'(1^\lambda)$: Generate $\sigma \leftarrow \mathcal{K}(1^\lambda)$ and sample random strings $c_1, \ldots, c_\lambda \stackrel{\$}{\leftarrow} \{0,1\}^L$. Output the common random string as $\sigma' = (\sigma, c_1, \ldots, c_\lambda)$ (notice that σ' is a uniformly random string).
- $\mathcal{P}'(\sigma', x, w)$: Parse $\sigma' = (\sigma, c_1, \ldots, c_\lambda)$ and generate $\pi' \leftarrow \mathcal{P}(\sigma, x', w)$ where x' is the statement:

$$\exists \tilde{w}, r_1, \ldots, r_\lambda : (x, \tilde{w}) \in \mathcal{R} \vee \left(c_1 = \mathsf{Com}\,(x; r_1) \wedge \ldots \wedge c_\lambda = \mathsf{Com}\,(x; r_\lambda) \right). \quad (1)$$

- $\mathcal{V}'(\sigma', x, \pi')$: Parse $\sigma' = (\sigma, c_1, \ldots, c_\lambda)$ and output $\mathcal{V}(\sigma, x', \pi')$, where x' is as defined in Eq. (1).

Completeness. The completeness property of the scheme $(\mathcal{K}', \mathcal{P}', \mathcal{V}')$ follows directly from the completeness property of the underlying NIWI-POK scheme $(\mathcal{K}, \mathcal{P}, \mathcal{V})$.

Statistical Soundness. Let $\sigma' = (\sigma, c_1, \ldots, c_\lambda)$ be a CRS sampled at random by $\mathcal{K}$. With overwhelming probability, there does not exist $x, r_1, \ldots, r_\lambda$ s.t. for every i, $c_i = \mathsf{Com}(x; r_i)$.

Now, for any $x \notin L$, let x' be the corresponding statement as defined in Eq. (1). It follows from above that the second part of the statement x' is false. Then, from the statistical soundness of $(\mathcal{K}, \mathcal{P}, \mathcal{V})$, it follows that there does not exist any accepting proof for x'.

Zero-Knowledge and Statistical Simulation-Extractability. We first describe the simulator and extractor algorithms $(\mathcal{S}_1', \mathcal{S}_2', \mathcal{E}')$ below. Let $\mathcal{E} = (\mathcal{E}_1, \mathcal{E}_2)$ denote the extractor for the NIWI-POK scheme $(\mathcal{K}, \mathcal{P}, \mathcal{V})$.

- $\mathcal{S}_1'(1^\lambda, x)$: On input a statement x, it first computes $(\sigma, \tau) \leftarrow \mathcal{E}_1(1^\lambda)$. Next, for every $i \in [\lambda]$, it samples a random string r_i and computes $c_i \leftarrow \mathsf{Com}(x; r_i)$. It sets $\sigma' = (\sigma, c_1, \ldots, c_\lambda)$, $\tau_1' = (r_1, \ldots, r_\lambda)$, $\tau_2' = \tau$ and outputs $(\sigma', \tau_1', \tau_2')$.
- $\mathcal{S}_2'(\sigma', \tau_1', x)$: It sets $w = (r_1, \ldots, r_\lambda)$ and computes $\pi \leftarrow \mathcal{P}(\sigma', x', w)$ where x' is as defined in Eq. (1). Note that here the honest prover algorithm uses w to prove the second part of x'.
- $\mathcal{E}'(\sigma', \tau_2', x^*, \pi^*)$: It parses $\sigma' = (\sigma, c_1, \ldots, c_\lambda)$ and outputs the value returned by the extractor $\mathcal{E}_2(\sigma, \tau_2', x^*, \pi^*)$.

We first argue computational zero-knowledge property of our scheme. Let $x \in L$ be any statement. Consider the following sequence of hybrid experiments:

- H_0: In this experiment, the CRS $\sigma' = (\sigma, c_1, \ldots, c_\lambda)$ is honestly generated and we compute a proof π' for x using the honest prover algorithm.
- H_1: Same as above, except that the CRS $\sigma' = (\sigma, c_1, \ldots, c_\lambda)$ is computed using the simulator algorithm $\mathcal{S}'_1(1^\lambda, x)$ as described above. Let (τ'_1, τ'_2) be the trapdoor computed by $\mathcal{S}'_1$.
- H_2: Same as above, except that we now compute π' using the simulator algorithm $\mathcal{S}'_2(\sigma', x, \tau'_1)$.

In order to prove the zero knowledge property of $(\mathcal{K}', \mathcal{P}', \mathcal{V}')$, it suffices to show that H_0 and H_2 are computationally indistinguishable. The indistinguishability of H_0 and H_1 follows immediately from the hiding property of the commitment scheme Com. Further, the indistinguishability of H_1 and H_2 follows from the witness indistinguishability property of the underlying NIWI-POK $(\mathcal{K}, \mathcal{P}, \mathcal{V})$. Put together, we have that H_0 and H_2 are computationally indistinguishable.

We now argue statistical simulation-extractability. We first note that if $x \notin L$, then in experiment H_2 described above, x is the only false statement for which an accepting proof exists. This follows from the perfectly binding property of Com and the statistical soundness property of the scheme. Now, let (x^*, π^*) be a statement and proof output by an unbounded adversary $\mathcal{A}$ who is given σ' s.t. $\mathcal{V}(\sigma', x^*, \pi^*) = 1$. Since $x^* \neq x$, it follows from above that $x^* \in L$. We now run the extractor $\mathcal{E}'(\sigma', \tau'_2, x^*, \pi^*)$ to compute w^*. From the proof of knowledge property of the underlying NIWI-POK $(\mathcal{K}, \mathcal{P}, \mathcal{V})$, it follows that w^* is a valid witness for x^*, except with negligible probability.

3.2 Puncturable Signatures

We define a special kind of signature scheme which is puncturable at any prefix, such that after puncturing no signature exists for messages agreeing on the prefix. A puncturable signature scheme is a tuple of efficient algorithms PuncSig = (KeyGen, Sign, Verify, Puncture) described as follows:

- $(sk, vk) \leftarrow$ KeyGen(1^λ): is a randomized algorithm which takes as input the security parameter and outputs a key pair.
- $\sigma \leftarrow$ Sign(sk, m): is a randomized algorithm which takes as input the signing key sk, some message m and outputs a signature σ.
- $b \leftarrow$ Verify(vk, m, σ): is a deterministic algorithm which takes as input the verification key, a message m, and a signature σ and outputs a bit b.
- $vk_{m,s} \leftarrow$ Puncture(sk, m, s): is a randomized algorithm which takes as input the signing key, a message m, a prefix s of m, and outputs a punctured verification key $vk_{m,s}$.

We require following properties from the scheme.

1. **Correctness:** For any message $m \in \{0, 1\}^\lambda$ it holds that

$$\Pr[\mathsf{Verify}(vk, m, \sigma) = 1 : (sk, vk) \leftarrow \mathsf{KeyGen}(1^\lambda), \sigma \leftarrow \mathsf{Sign}(sk, m)] = 1$$

2. **Security:** For any message m, and prefix s of m it holds that:

$$\{vk_{m,s}, m, \sigma\} \approx_c \{vk, m, \sigma\},$$

where $(sk, vk) \leftarrow \mathsf{KeyGen}(1^\lambda)$, $\sigma \leftarrow \mathsf{Sign}(sk, m)$, $vk_{m,s} \leftarrow \mathsf{Puncture}(sk, m, s)$.

3. **Punctured functionality:** There exist a negligible function $\mathsf{negl}(\cdot)$ such that for any message m and for any $m' \neq m$ such that s is a prefix of m' it holds that

$$\Pr[\exists \sigma : \mathsf{Verify}(vk_{m,s}, m', \sigma) = 1] \leq \mathsf{negl}(\lambda),$$

where $(sk, vk) \leftarrow \mathsf{KeyGen}(1^\lambda)$ and $vk_{m,s} \leftarrow \mathsf{Puncture}(sk, m, s)$.

The Construction

We show how to construct a puncturable signature scheme from NIZK proofs and a statistically binding commitment scheme. The construction is as follows:

- $\mathsf{KeyGen}(1^\lambda)$: Sample a PRF key K, sample a crs, compute $c \leftarrow \mathsf{Com}(C_K; r)$ where C is a circuit that on input x outputs $\mathsf{PRF}_K(x)$ (padded to be large enough). Set $sk = C_K, r$ and $vk = c, \mathsf{crs}$.
- $\mathsf{Sign}(sk, m)$: Compute $y = C_K(m)$ and a proof π that y is indeed that output of the circuit committed in c on the input m.
- $\mathsf{Verify}(vk, m, \sigma)$: Parse σ as y and π. Verify that proof π to the instance (m, y) and that $y \neq \bot$.
- $\mathsf{Puncture}(sk, m, s)$: Compute PRF key K^* that is punctured at the prefix s except the message m. That is, using K^* one can compute the PRF value on m and on all inputs that do not begin with s. In the verification key, replace the PRF key in the committed circuit to $C_{K^*, s, m}$. In the secret key, replace K with K^*.

Correctness is immediate from the construction. Security follows from the hiding property of the commitment scheme. Punctured security follows from the statistical soundness of the NIZK proof. After the circuit C is altered to output $\bot$ on messages with prefix s, we know that (with high probability over the crs) no valid proof exists for messages with prefix s, and thus no valid signatures.

4   Non-Interactive Multiparty Computation

A non-interactive multiparty computation protocol (NI-MPC) for a function $f \colon \mathcal{X}^n \to \mathcal{Y}$ is a protocol between n parties and a single evaluator. Each party holds an input x_i and sends exactly one message to an evaluator, who computes the output. The correctness requirement of the protocol is that the evaluator, given one message from each party, is able to compute the value $y = f(x_1, \ldots, x_n)$ correctly.

Our security guarantee is formalized as an indistinguishability game in which the evaluator commits on a set of parties with whom he is colluding. Specifically, since our setting is non-interactive, the evaluator can use the controlled parties to make resetting attacks that allow him to evaluate the function on different inputs of his choice (but he has no control over the inputs of the honest parties). Thus, our security is assuming that the "residual" function determined after fixing the challenge inputs of the honest parties is functionally equivalent in the view of the evaluator [24].

Definition 1 (Admissible inputs). *Let $f\colon \mathcal{X}^n \to \mathcal{Y}$ be a function. We say that $H \subseteq [n]$, $\{x_i^0\}_{i \in H}$ and $\{x_i^1\}_{i \in H}$ are **admissible** for f if for any $\{x_i\}_{i \notin H}$ it holds that*

$$f(\langle \{x_i^0\}_{i \in H}, \{x_i\}_{i \notin H}\rangle) = f(\langle \{x_i^1\}_{i \in H}, \{x_i\}_{i \notin H}\rangle),$$

where $\langle \cdot, \ldots, \cdot \rangle$ sorts the inputs according to the index i.

Definition 2 (Non-interactive multiparty computation). *A non-interactive multiparty computation protocol Π for the function $f\colon \mathcal{X}^n \to \mathcal{Y}$ consists of a probabilistic setup procedure Setup, a probabilistic encryption procedure Enc and a probabilistic evaluation procedure Eval that satisfy the following requirements:*

1. $\mathsf{Setup}(1^\lambda, \mathsf{crs}, i)$ *takes as input a security parameter λ, a common random string crs, a party index $i \in [n]$, computes a private key SK_i and outputs a public key PK_i.*
2. $\mathsf{Enc}(1^\lambda, \mathsf{crs}, x_i, \mathsf{PK}_1, \ldots, \mathsf{PK}_n, \mathsf{SK}_i, i)$ *takes as input a security parameter λ, a common random string crs, an input $x_i \in \mathcal{X}$, public keys $\mathsf{PK}_1, \ldots, \mathsf{PK}_n$, a secret key SK_i and a party index $i \in [n]$, and outputs a ciphertext $\widehat{x}_i$.*
3. $\mathsf{Eval}(1^\lambda, \mathsf{crs}, \widehat{x_1}, \ldots, \widehat{x_n}, \mathsf{PK}_1, \ldots, \mathsf{PK}_n)$ *takes as input a security parameter λ, a common random string crs, n ciphertexts $\widehat{x_1}, \ldots, \widehat{x_n}$, n public keys $\mathsf{PK}_1 \ldots, \mathsf{PK}_n$, and outputs a value $y \in \mathcal{Y}$.*
4. **Correctness:** *This property states that the evaluation on n ciphertexts of $x_1, \ldots, x_n$ gives $f(x_1, \ldots, x_n)$. Specifically, for every $\lambda \in \mathbb{N}$ and every $x_1, \ldots, x_n \in \mathcal{X}$, it holds that*

$$\Pr\Big[y = f(x_1, \ldots, x_n);$$
$$y = \mathsf{Eval}(1^\lambda, \mathsf{crs}, \widehat{x_1}, \ldots, \widehat{x_n}, \mathsf{PK}_1, \ldots, \mathsf{PK}_n),$$
$$\forall i \in [n]\colon \widehat{x}_i \leftarrow \mathsf{Enc}(1^\lambda, \mathsf{crs}, x_i, \mathsf{PK}_1, \ldots, \mathsf{PK}_n, \mathsf{SK}_i, i),$$
$$\forall i \in [n]\colon (\mathsf{PK}_i, \mathsf{SK}_i) \leftarrow \mathsf{Setup}(1^\lambda, \mathsf{crs}, i),$$
$$\mathsf{crs} \leftarrow \{0,1\}^\lambda \Big] = 1,$$

where the probability is over the internal randomness of Setup, Enc and Eval.

5. **Security:** *Informally speaking, this property states that an adversary that controls a subset of parties does not learn anything about the underlying plaintexts from the ciphertexts of the remaining parties, besides the output of the function. Specifically, for any probabilistic polynomial-time adversary $\mathcal{A} = (\mathcal{A}_1, \mathcal{A}_2)$, any admissible $H \subseteq [n]$, $\boldsymbol{x}^0 = \{x_i^0\}_{i \in H}$ and $\boldsymbol{x}^1 = \{x_i^1\}_{i \in H}$, there exists a negligible function $\mathsf{negl}(\lambda)$ such that*

$$\mathsf{Adv}_{\Pi, \mathcal{A}, H, \boldsymbol{x}^0, \boldsymbol{x}^1}^{\mathsf{NI\text{-}MPC}}(\lambda) \overset{\mathrm{def}}{=} \left| \Pr\Big[\mathsf{Exp}_{\Pi, \mathcal{A}, H, \boldsymbol{x}^0, \boldsymbol{x}^1}^{\mathsf{NI\text{-}MPC}}(\lambda) = 1 \Big] - \frac{1}{2} \right| \le \mathsf{negl}(\lambda),$$

for all sufficiently large $\lambda \in \mathbb{N}$, *where the random variable* $\mathsf{Exp}_{\Pi,\mathcal{A},H,x^0,x^1}^{\text{NI-MPC}}(\lambda)$
is defined via the following experiment:

1. *Sample* $\mathsf{crs} \leftarrow \{0,1\}^{\mathsf{poly}(\lambda)}$ *and* $b \leftarrow \{0,1\}$ *at random.*
2. *For* $i \in H$: $(\mathsf{PK}_i, \mathsf{SK}_i) \leftarrow \mathsf{Setup}(1^\lambda, \mathsf{crs}, i)$
3. $(\{\mathsf{PK}_i\}_{i \notin H}, \mathsf{st}) \leftarrow \mathcal{A}_1(1^\lambda, \mathsf{crs}, \{\mathsf{PK}_i\}_{i \in H})$
4. *For* $i \in H$: $\widehat{x}_i \leftarrow \mathsf{Enc}(1^\lambda, \mathsf{crs}, x_i^b, \mathsf{PK}_1, \ldots, \mathsf{PK}_n, \mathsf{SK}_i, i).$
5. $b' \leftarrow \mathcal{A}_2(1^\lambda, \mathsf{crs}, \{\mathsf{PK}_i\}_{i \in H}, \{\widehat{x}_i\}_{i \in H}, \mathsf{st}).$
6. *If* $b' = b$ *then output 1, and otherwise output 0.*

A Reusable PKI and Session IDs. The above definition is not reusable. Namely, it cannot be used across multiple ciphertexts generated by the honest parties. A definition that support multiple encryption sessions is given below. We remark that in our non-interactive setting, session IDs are required to prevent an adversary for combining ciphertexts from one session with another (otherwise, the definition of an admissible function becomes very weak and makes sense only for a very restricted set of functions). Our construction and proof given in Sect. 5 satisfy the weaker definition above but we sketch in Sect. 5.2 how to prove that our construction satisfies the stronger reusable-PKI with session IDs notion as well.

Definition 3 (Non-interactive multiparty computation with session IDs). *A non-interactive multiparty computation protocol* Π *for the function* $f \colon \mathcal{X}^n \to \mathcal{Y}$ *consists of a probabilistic setup procedure* Setup, *a probabilistic encryption procedure* Enc *and a probabilistic evaluation procedure* Eval. *The setup procedure and evaluation procedure have the same syntax as in Definition 2 but* Enc *also received a session ID. Moreover, the correctness and security are modified as follows:*

1. $\mathsf{Enc}(1^\lambda, \mathsf{crs}, x_i, \mathsf{PK}_1, \ldots, \mathsf{PK}_n, \mathsf{SK}_i, i, \textsf{session})$ *takes as input a security parameter* λ, *a common random string* crs, *an input* $x_i \in \mathcal{X}$, *public keys* $\mathsf{PK}_1, \ldots,$ PK_n, *a secret key* SK_i, *a party index* $i \in [n]$, *and a session ID* $\textsf{session} \in \{0,1\}^\lambda$, *and outputs a ciphertext* $\widehat{x}_i$.
2. **Correctness:** *This property states that the evaluation on* n *ciphertexts of* $x_1, \ldots, x_n$ *encrypted in the same session gives* $f(x_1, \ldots, x_n)$. *Specifically, for every* $\lambda \in \mathbb{N}$, *every session ID* $\textsf{session} \in \{0,1\}^\lambda$, *and every* $x_1, \ldots, x_n \in \mathcal{X}$, *it holds that*

$$\Pr\big[y = f(x_1, \ldots, x_n);$$
$$y = \mathsf{Eval}(1^\lambda, \mathsf{crs}, \widehat{x_1}, \ldots, \widehat{x_n}, \mathsf{PK}_1, \ldots, \mathsf{PK}_n),$$
$$\forall i \in [n]: \widehat{x}_i \leftarrow \mathsf{Enc}(1^\lambda, \mathsf{crs}, x_i, \mathsf{PK}_1, \ldots, \mathsf{PK}_n, \mathsf{SK}_i, i, \textsf{session}),$$
$$\forall i \in [n]: (\mathsf{PK}_i, \mathsf{SK}_i) \leftarrow \mathsf{Setup}(1^\lambda, \mathsf{crs}, i),$$
$$\mathsf{crs} \leftarrow \{0,1\}^\lambda\big] = 1,$$

where the probability is over the internal randomness of $\mathsf{Setup}, \mathsf{Enc}$ *and* Eval.

3. **Security:** *For any probabilistic polynomial-time adversary* $\mathcal{A} = (\mathcal{A}_1, \mathcal{A}_2)$, *any admissible* $H \subseteq [n]$, $\boldsymbol{x}^0 = \{x_i^0\}_{i \in H}$ *and* $\boldsymbol{x}^1 = \{x_i^1\}_{i \in H}$, *and any session ID* session $\in \{0,1\}^\lambda$, *there exists a negligible function* $\mathsf{negl}(\lambda)$ *such that*

$$\mathsf{Adv}^{\mathsf{NI\text{-}MPC}}_{\Pi,\mathcal{A},H,\boldsymbol{x}^1,session}(\lambda) \stackrel{\mathsf{def}}{=} \left| \Pr\left[\mathsf{Exp}^{\mathsf{NI\text{-}MPC}}_{\Pi,\mathcal{A},H,\boldsymbol{x}^0,\boldsymbol{x}^1,session}(\lambda) = 1 \right] - \frac{1}{2} \right|$$

$$\leq \mathsf{negl}(\lambda),$$

for all sufficiently large $\lambda \in \mathbb{N}$, *where* $\mathsf{Exp}^{\mathsf{NI\text{-}MPC}}_{\Pi,\mathcal{A},H,\boldsymbol{x}^0,\boldsymbol{x}^1,session}(\lambda)$ *is a random variable defined via the following experiment:*

1. *Sample* crs $\leftarrow \{0,1\}^{\mathsf{poly}(\lambda)}$ *and* $b \leftarrow \{0,1\}$ *at random.*
2. *For* $i \in H$: $(\mathsf{PK}_i, \mathsf{SK}_i) \leftarrow \mathsf{Setup}(1^\lambda, \mathsf{crs}, i)$
3. $(\{\mathsf{PK}_i\}_{i \notin H}, \mathsf{st}) \leftarrow \mathcal{A}_1(1^\lambda, \mathsf{crs}, \{\mathsf{PK}_i\}_{i \in H})$
4. *For* $i \in H$: $\widehat{x}_i \leftarrow \mathsf{Enc}(1^\lambda, \mathsf{crs}, x_i^b, \mathsf{PK}_1, \ldots, \mathsf{PK}_n, \mathsf{SK}_i, i, session)$.
5. $b' \leftarrow \mathcal{A}_2^{\mathsf{Enc}(\cdot,\cdot,\cdot)}(1^\lambda, \mathsf{crs}, \{\mathsf{PK}_i\}_{i \in H}, \{\widehat{x}_i\}_{i \in H}, \mathsf{st})$, *where the encryption oracle* $\mathsf{Enc}(\cdot,\cdot,\cdot)$ *on input triple* $(x_i, i, session')$ *produces an encryption of* x_i *with respect to party* $i \in H$ *as long as* $session' \neq session$ *(by running* $\mathsf{Enc}(1^\lambda, \mathsf{crs}, x_i, \mathsf{PK}_1, \ldots, \mathsf{PK}_n, \mathsf{SK}_i, i, session'))$.
6. *If* $b' = b$ *then output 1, and otherwise output 0.*

The fully honest case. A particularly interesting case is when $H = [n]$ and the evaluator does not collude with any party. In this case, which we call the fully honest case, we can assume without loss of generality that one of the parties will also generate the CRS as part of the PKI and thus we get a construction in the PKI model.

Selective vs. adaptive security. Our security definitions are selective in the sense that the adversary commits on the challenge before the challenger published the PKI. By standard "random guessing" we can turn any selectively secure scheme into an adaptively secure one. This is done by guessing ahead of time all the adaptive choices made by the adversary throughout the game and reducing to the selective case. By setting the security parameter to be large enough and using the sub-exponential security of the scheme, we can tolerate this exponential loss. (Recall that our main result assumes sub-exponentially secure primitives to begin with so we can get adaptive security for free.)

4.1 Communication on a Chain

We follow Halevi et al. [23] and consider the case of more general interaction patterns described by a directed acyclic graph (DAG), where each node represents a party who expects to receive messages from all of its parents and can then send messages to all of its children, and where the sinks of the graph compute outputs. The setting of Definitions 2 and 3 is a special case of the above where the communication pattern is a *star*, a graph connecting all nodes to a single central node. Another special case is a chain, a simple directed path traversing all nodes.

Here, we focus on a chain and define NI-MPC for this pattern. We assume that there are n parties with corresponding inputs x_i. The first party is the source of the chain and party n is the last party in the chain which sends its message to the evaluator.

Definition 4 (Chain admissible inputs). *Let $f\colon \mathcal{X}^n \to \mathcal{Y}$ be a function. Let $H \subseteq [n]$ with maximal index i^*. We say that H, $\{x_i^0\}_{i \in H}$, $\{x_i^1\}_{i \in H}$, and $\{x_i\}_{i \notin H \wedge i \leq i^*}$ are chain-admissible for f if for any $\{x_i\}_{i \notin H \wedge i > i^*}$ it holds that*

$$f(\langle \{x_i^0\}_{i \in H}, \{x_i\}_{i \notin H} \rangle) = f(\langle \{x_i^1\}_{i \in H}, \{x_i\}_{i \notin H} \rangle),$$

where $\langle \cdot, \ldots, \cdot \rangle$ sorts the inputs according to the index i.

Definition 5 (Non-interactive multiparty computation on a chain). *A non-interactive multiparty computation protocol Π for the function $f\colon \mathcal{X}^n \to \mathcal{Y}$ on a chain consists of a probabilistic setup procedure Setup, a probabilistic encryption procedure Enc and a probabilistic evaluation procedure Eval that satisfy the following requirements:*

1. Setup$(1^\lambda, \mathsf{crs}, i)$ *takes as input a security parameter λ, a common random string crs, a party index $i \in [n]$, computes a private key SK_i and outputs a public key PK_i.*
2. Enc$(1^\lambda, \mathsf{crs}, \widehat{x_{i-1}}, x_i, \mathsf{PK}_1, \ldots, \mathsf{PK}_n, \mathsf{SK}_i, i)$ *takes as input a security parameter λ, a common random string crs, the message sent by party $i - 1$ an input $x_i \in \mathcal{X}$, public keys $\mathsf{PK}_1, \ldots, \mathsf{PK}_n$, a secret key SK_i and a party index $i \in [n]$, and outputs a ciphertext $\widehat{x_i}$. We assume that $\widehat{x_0} = \bot$.*
3. Eval$(1^\lambda, \mathsf{crs}, \widehat{x_n}, \mathsf{PK}_1, \ldots, \mathsf{PK}_n)$ *takes as input a security parameter λ, a common random string crs, the ciphertexts of the last party $\widehat{x_n}$, n public keys $\mathsf{PK}_1 \ldots, \mathsf{PK}_n$, and outputs a value $y \in \mathcal{Y}$.*
4. **Correctness:** *This property states that the evaluation on n ciphertexts of $x_1, \ldots, x_n$ gives $f(x_1, \ldots, x_n)$. Specifically, for every $\lambda \in \mathbb{N}$ and every $x_1, \ldots, x_n \in \mathcal{X}$, it holds that*

$$\begin{aligned} \Pr \big[& y = f(x_1, \ldots, x_n); \\ & y = \mathsf{Eval}(1^\lambda, \mathsf{crs}, \widehat{x_n}, \mathsf{PK}_1, \ldots, \mathsf{PK}_n), \\ & \forall i \in [n] \colon \widehat{x_i} \leftarrow \mathsf{Enc}(1^\lambda, \mathsf{crs}, \widehat{x_{i-1}}, x_i, \mathsf{PK}_1, \ldots, \mathsf{PK}_n, \mathsf{SK}_i, i), \\ & \forall i \in [n] \colon (\mathsf{PK}_i, \mathsf{SK}_i) \leftarrow \mathsf{Setup}(1^\lambda, \mathsf{crs}, i), \\ & \mathsf{crs} \leftarrow \{0, 1\}^\lambda \big] = 1, \end{aligned}$$

   where the probability is over the internal randomness of Setup, Enc and Eval.
5. **Security:** *For any probabilistic polynomial-time adversary $\mathcal{A} = (\mathcal{A}_1, \mathcal{A}_2)$, any chain-admissible $H \subseteq [n]$, $\boldsymbol{x}^0 = \{x_i^0\}_{i \in H}$, $\boldsymbol{x}^1 = \{x_i^1\}_{i \in H}$, and $\boldsymbol{x} = \{x_i\}_{i \notin H \wedge i < i^*}$, where i^* is the maximal index in H, there exists a negligible function $\mathsf{negl}(\lambda)$ such that*

$$\mathsf{Adv}^{\mathsf{NI\text{-}MPC}}_{\Pi, \mathcal{A}, H, \boldsymbol{x}^0, \boldsymbol{x}^1, \boldsymbol{x}}(\lambda) \overset{\mathsf{def}}{=} \left| \Pr\left[\mathsf{Exp}^{\mathsf{NI\text{-}MPC}}_{\Pi, \mathcal{A}, H, \boldsymbol{x}^0, \boldsymbol{x}^1, \boldsymbol{x}}(\lambda) = 1\right] - \frac{1}{2} \right| \leq \mathsf{negl}(\lambda),$$

for all sufficiently large $\lambda \in \mathbb{N}$, *where* $\mathsf{Exp}^{\mathsf{NI\text{-}MPC}}_{\Pi,\mathcal{A},H,\boldsymbol{x}^0,\boldsymbol{x}^1,\boldsymbol{x}}(\lambda)$ *is a random variable defined via the following experiment:*

1. *Sample* $\mathsf{crs} \leftarrow \{0,1\}^{\mathsf{poly}(\lambda)}$ *and* $b \leftarrow \{0,1\}$ *at random.*
2. *For* $i \in H$: $(\mathsf{PK}_i, \mathsf{SK}_i) \leftarrow \mathsf{Setup}(1^\lambda, \mathsf{crs}, i)$
3. $(\{\mathsf{PK}_i\}_{i \notin H}, \mathsf{st}) \leftarrow \mathcal{A}_1(1^\lambda, \mathsf{crs}, \{\mathsf{PK}_i\}_{i \in H})$
4. *For* $i = 1 \ldots i^*$:
   i. *If* $i \in H$: $\widehat{x}_i \leftarrow \mathsf{Enc}(1^\lambda, \mathsf{crs}, \widehat{x_{i-1}}, x_i^b, \mathsf{PK}_1, \ldots, \mathsf{PK}_n, \mathsf{SK}_i, i)$.
   ii. *If* $i \notin H$: $\widehat{x}_i \leftarrow \mathsf{Enc}(1^\lambda, \mathsf{crs}, \widehat{x_{i-1}}, x_i, \mathsf{PK}_1, \ldots, \mathsf{PK}_n, \mathsf{SK}_i, i)$
5. $b' \leftarrow \mathcal{A}_2(1^\lambda, \mathsf{crs}, \{\mathsf{PK}_i\}_{i \in H}, \widehat{x_{i^*}}, \mathsf{st})$.
6. *If* $b' = b$, *then output* 1, *and otherwise, output* 0.

5   Our Construction

We now present our construction of NI-MPC as in Definition 2. The main ingredients in our construction are:

- A threshold multi-key FHE scheme $\mathsf{MFHE} = (\mathsf{KeyGen}, \mathsf{Enc}, \mathsf{Expand}, \mathsf{Eval}, \mathsf{Dec}, \mathsf{PartDec}, \mathsf{FinDec})$.
- An SSE-NIZK proof system $\mathsf{NIZK} = (\mathcal{K}, \mathcal{P}, \mathcal{V}, \mathcal{S}_1, \mathcal{S}_2, \mathcal{E})$ for **NP**.
- A puncturable PRF family $\mathcal{F}$.
- A puncturable signature scheme $\mathsf{PuncSig} = (\mathsf{KeyGen}, \mathsf{Sign}, \mathsf{Verify}, \mathsf{Puncture})$
- An indistinguishability obfuscator iO for general circuits.

Let $f: \mathcal{X}^n \to \mathcal{Y}$ be a deterministic function that takes n inputs $x_1, \ldots, x_n \in \mathcal{X}$ and outputs $y \in \mathcal{Y}$. We construct a non-interactive multiparty computation protocol $\Pi = (\mathsf{Setup}, \mathsf{Enc}, \mathsf{Eval})$ for f below.

$\underline{\mathsf{Setup}(1^\lambda, \mathsf{crs}, i)}$: It takes as input security parameter λ, a common random string crs, and does the following.

1. Execute the key generation procedure of the signature scheme, $(sk_i, vk_i) \leftarrow \mathsf{PuncSig.KeyGen}(1^\lambda)$.
2. Execute $\{(dk_i, ek_i) \leftarrow \mathsf{MFHE.KeyGen}(1^\lambda)\}_{i \in [n]}$.
3. Output public key $\mathsf{PK}_i = (vk_i, ek_i)$ and private key $\mathsf{SK}_i = (sk_i, dk_i)$.

$\underline{\mathsf{Enc}(1^\lambda, \mathsf{crs}, x_i, \mathsf{PK}_1, \ldots, \mathsf{PK}_n, \mathsf{SK}_i, i)}$: It takes as input security parameter λ, a common random string crs, an input $x_i \in \mathcal{X}$, a list of n public keys $\mathsf{PK}_1 \ldots, \mathsf{PK}_n$, one private key SK_i, an index $i \in [n]$, and does the following.

1. For every $j \in [n]$, parse $\mathsf{PK}_j = (vk_j, ek_j)$ and let $\mathsf{SK}_i = (sk_i, dk_i)$.
2. Encrypt the input x_i using ek_i: $\mathsf{CT}_i \leftarrow \mathsf{MFHE.Enc}(ek_i, x_i)$.
3. Sign on (CT_i, i) using sk_i: $\psi_i \leftarrow \mathsf{PuncSig.Sign}(sk_i, (\mathsf{CT}_i, i))$.
4. Sample puncturable PRF key $K_i \xleftarrow{\$} \{0,1\}^\lambda$.
5. Generate a proof π_i that CT_i and ek_i are valid ciphertext and encryption key, respectively. Namely, compute $\pi_i \leftarrow \mathsf{NIZK.P}(\mathsf{crs}, (\mathsf{CT}_i, ek_i), w = (r, r', x_i, dk_i))$, where r and r' are the randomness used for the computation of CT_i and (dk_i, ek_i), respectively. The exact statement is given in Fig. 2.

$$G\left[i, \mathsf{crs}, K_i, dk_i, \{ek_j\}_{j \in [n]}, \{vk_j\}_{j \in [n]}\right]$$

Input: $(\mathsf{CT}_1, \psi_1, \pi_1), \ldots, (\mathsf{CT}_n, \psi_n, \pi_n)$.

- If $\mathsf{PuncSig.Verify}(vk_j, (\mathsf{CT}_j, j), \psi_j)$ is false for some $j \in [n]$ then output $\bot$.
- For every $j \in [n]$, verify $\mathsf{NIZK.}\mathcal{V}(\mathsf{crs}, (ek_j, \mathsf{CT}_j), \pi_j)$ for the statement:

$$\exists r, r', M_j, dk_j \colon \mathsf{CT}_j = \mathsf{MFHE.Enc}(ek_j, M_j; r) \wedge (ek_j, dk_j) = \mathsf{MFHE.KeyGen}(1^\lambda; r').$$

   If this check fails for some $j \in [n]$ then output $\bot$.
- For all $j \in [n]$ compute $\widehat{\mathsf{CT}}_j \leftarrow \mathsf{MFHE.Expand}((ek_1, \ldots, ek_n), j, \mathsf{CT}_j)$.
- Perform $\widehat{\mathsf{CT}}_\mathsf{out} \leftarrow \mathsf{MFHE.Eval}\left(f, \widehat{\mathsf{CT}}_1, \ldots, \widehat{\mathsf{CT}}_n\right)$.
- $r_i \leftarrow \mathsf{PRF}_{K_i}(\mathsf{CT}_1\|\pi_1\|\ldots\|\mathsf{CT}_n\|\pi_n)$.
- $p_i \leftarrow \mathsf{MFHE.PartDec}\left(\widehat{\mathsf{CT}}_\mathsf{out}, ek_1, \ldots, ek_n, i, dk_i; r_i\right)$.
- Output p_i.

Fig. 2. The circuit G.

6. Obfuscate the circuit $\overline{G_i} \leftarrow \mathsf{iO}\left(1^\lambda, G\left[i, \mathsf{crs}, K_i, dk_i, \{ek_j\}_{j \in [n]}, \{vk_j\}_{j \in [n]}\right]\right)$ as described in Fig. 2.
7. Output the tuple $\widehat{x}_i = (\overline{G_i}, \mathsf{CT}_i, \psi_i, \pi_i)$.

$\underline{\mathsf{Eval}(1^\lambda, \mathsf{crs}, \widehat{x_1}, \ldots, \widehat{x_n}, \mathsf{PK}_1, \ldots, \mathsf{PK}_n)}$: It takes as input security parameter λ, a common random string crs, n strings $\widehat{x_1}, \ldots, \widehat{x_n}$, a list of n public keys $\mathsf{PK}_1 \ldots, \mathsf{PK}_n$, and does the following:

1. Parse each $\widehat{x}_i = (\overline{G_i}, \mathsf{CT}_i, \psi_i, \pi_i)$.
2. Evaluate each obfuscation $\overline{G_i}$ on the input $\left((\widehat{\mathsf{CT}}_1, \psi_1, \pi_1), \ldots, (\widehat{\mathsf{CT}}_n, \psi_n, \pi_n)\right)$ to get the partial decryption p_i.
3. Execute the final decryption algorithm, $y \leftarrow \mathsf{MFHE.FinDec}(p_1, \ldots, p_n)$ and output y.

We proceed with the proof of correctness and security of the scheme. Correctness follows by the correctness of the underlying building blocks. Specifically, let $x_1, \ldots, x_n \in \mathcal{X}$ be inputs such that party i holds x_i. By the correctness of the signature scheme and the threshold multi-key FHE scheme, each obfuscated circuit will output a partial decryption p_i such that $\mathsf{MFHE.FinDec}(p_1, \ldots, p_n)$ must be equal to $f(x_1, \ldots, x_n)$.

5.1 Proof of Security

We show that our scheme is secure for a restricted set of input vector pairs.

Lemma 1. *Assume the existence of a indistinguishability obfuscator, a threshold multi-key FHE scheme, an SSE-NIZK in the common random string model, a* *puncturable PRF, and a puncturable signature, all of which are sub-exponentially secure. Then, for any $H \subseteq [n]$, any $\mathsf{ind} \in H$ and any $\boldsymbol{x}^0 = \{x_i\}_{i \in H}$, $\boldsymbol{x}^1 = \{x_i\}_{i \in H \setminus \{\mathsf{ind}\}} \cup \{x'_{\mathsf{ind}}\}$, such that $H, \boldsymbol{x}^0, \boldsymbol{x}^1$ are admissible, it holds that*

$$\mathsf{Adv}^{\mathsf{NI\text{-}MPC}}_{\Pi, f, \mathcal{A}, H, \boldsymbol{x}^0, \boldsymbol{x}^1}(\lambda) \leq \mathsf{negl}(\lambda).$$

Given this lemma, we can construct a scheme that is secure as in Definition 2.

Lemma 2. *Assume the existence of a indistinguishability obfuscator, a threshold multi-key FHE scheme, an SSE-NIZK in the common random string model, a puncturable PRF, and a puncturable signature, all of which are sub-exponentially secure. Then, there exists a secure non-interactive multiparty computation scheme for any efficiently computable function f.*

The proof of Lemma 2 (given Lemma 1) is given in the full version of the paper.

Threshold multi-key FHE scheme exists based on the Learning with Errors assumption [27] in the CRS model or on indistinguishability obfuscation and DDH [12]. In Sect. 3.1 we constructed an SSE-NIZK scheme in the common random string model assuming any NIWI-POK and one-way permutations. A NIWI-POK can be constructed from any NIWI together with a standard encryption scheme. NIWI and one-way permutations exist based on iO and DDH [4]. Finally, we constructed a puncturable signature based on NIZKs and statistically-binding commitments. In total, we can instantiate our construction based on indistinguishability obfuscation and DDH, both with sub-exponential security.

Proof (of Lemma 1). Fix $H \subseteq [n]$, $\mathsf{ind} \in H$, $\boldsymbol{x}^0 = \{x_i\}_{i \in H}$, and $\boldsymbol{x}^1 = \{x_i\}_{i \in H \setminus \{\mathsf{ind}\}} \cup \{x'_{\mathsf{ind}}\}$, such that $H, \boldsymbol{x}^0, \boldsymbol{x}^1$ are admissible. The lemma is proved by a sequence of hybrid experiments. Denote by $\ell(\lambda)$ a polynomial bounding the total length of n ciphertexts and n NIZK proofs computed with security parameter λ.

$\underline{\mathsf{Hyb}_1}$: This experiment corresponds to the original experiment $\mathsf{Exp}^{\mathsf{NI\text{-}MPC}}_{\Pi, f, \mathcal{A}, H, \boldsymbol{x}^0, \boldsymbol{x}^1}(\lambda)$:

1. Sample a random bit $b \in \{0, 1\}$.
2. Do the following for party $i \in H$:
   2.1. Sample $(dk_i, ek_i) \leftarrow \mathsf{MFHE.KeyGen}(1^\lambda; r)$, where r is a random string.
   2.2. $\mathsf{CT}_i \leftarrow \mathsf{MFHE.Enc}(ek_i, x_i^b; r')$, where r' is a random string.
   2.3. Sample $(sk_i, vk_i) \leftarrow \mathsf{PuncSig.KeyGen}(1^\lambda)$.
   2.4. Set $\mathsf{PK}_i = (vk_i, ek_i)$ and $\mathsf{SK}_i = (sk_i, dk_i)$.
3. The adversary, given $\{\mathsf{PK}_i\}_{i \in H}$ and crs, publishes a public key PK_i for every $i \notin H$ of the form $\mathsf{PK}_i = (vk_i, ek_i)$.
4. Do the following for party $i \in H$:
   4.1 $\psi_i \leftarrow \mathsf{PuncSig.Sign}(sk_i, (\mathsf{CT}_i, i))$.
   4.2 Sample a PRF key K_i.
   4.3 Compute $\pi_i \leftarrow \mathsf{NIZK.P}(\mathsf{crs}, (\mathsf{CT}_i, ek_i), w = (r, r', x_i^b, dk_i))$.

4.4 Compute $\overline{G_i}$ as the obfuscation of the following circuit defined in Fig. 2:

$$G\left[i, \mathsf{crs}, K_i, dk_i, \{ek_j\}_{j \in [n]}, \{vk_j\}_{j \in [n]}\right].$$

4.5. Let $\widehat{x_i} = (\overline{G_i}, \mathsf{CT}_i, \psi_i, \pi_i)$.

5. The challenge: $b' \leftarrow \mathcal{A}(\mathsf{crs}, \{\mathsf{PK}_i\}_{i \in H}, \{\widehat{x_i}\}_{i \in H})$ and output b'.

Hyb_2: This experiment corresponds to experiment Hyb_1 except that now the published verification keys of parties in H are modified to verify only one pre-computed message and nothing else. Specifically, now between Item 2.3 and Item 2.4 we puncture the verification key of each $i \in H$ and set

$$vk_i \leftarrow \mathsf{PuncSig.Puncture}(sk_i, (\mathsf{CT}_i, i), \perp),$$

where the $\perp$ corresponds to an empty prefix. That is, the new verification key accepts the precisely one message which is (CT_i, i). The indistinguishability of this experiment and experiment Hyb_1 follows directly from the security of the signature scheme (see Item 2 at Sect. 3.2).

Hyb_3: This experiment corresponds to experiment Hyb_2 except that now the NIZK proof of the honest party ind is generated via the zero-knowledge simulator.

1. Sample $(dk_{\mathsf{ind}}, ek_{\mathsf{ind}}) \leftarrow \mathsf{MFHE.KeyGen}(1^\lambda; r)$, where r is a random string.
2. $\mathsf{CT}_{\mathsf{ind}} \leftarrow \mathsf{MFHE.Enc}(ek_{\mathsf{ind}}, x_{\mathsf{ind}}^b; r')$, where r' is a random string.
3. Sample $(\mathsf{crs}, \tau) \leftarrow \mathsf{NIZK}.\mathcal{S}_1(1^\lambda, (\mathsf{CT}_{\mathsf{ind}}, ek_{\mathsf{ind}}))$.
4. Sample random a bit $b \in \{0, 1\}$.
5. Proceed as before (Item 2) for party $i \in H \setminus \{\mathsf{ind}\}$.
6. Do the following for party ind:
   6.1. Sample $(sk_{\mathsf{ind}}, vk_{\mathsf{ind}}) \leftarrow \mathsf{PuncSig.KeyGen}(1^\lambda)$.
   6.2. $vk_{\mathsf{ind}} \leftarrow \mathsf{PuncSig.Puncture}(sk_{\mathsf{ind}}, (\mathsf{CT}_{\mathsf{ind}}, i), \perp)$.
   6.3. Set $\mathsf{PK}_i = (vk_{\mathsf{ind}}, ek_{\mathsf{ind}})$ and $\mathsf{SK}_{\mathsf{ind}} = (sk_{\mathsf{ind}}, dk_{\mathsf{ind}})$.
7. The adversary, given $\{\mathsf{PK}_i\}_{i \in H}$ and crs, publishes a public key PK_i for every $i \notin H$ of the form $\mathsf{PK}_i = (vk_i, ek_i)$.
8. Proceed as before (Item 4) for party $i \in H \setminus \{\mathsf{ind}\}$.
9. Do the following for party ind:
   9.1. Sample a PRF key K_{ind}.
   9.2. $\pi_{\mathsf{ind}} \leftarrow \mathsf{NIZK}.\mathcal{S}_2(\mathsf{crs}, \tau, (\mathsf{CT}_{\mathsf{ind}}, ek_{\mathsf{ind}}))$.
   9.3. Compute $\overline{G_{\mathsf{ind}}}$ as the obfuscation of the following circuit defined in Fig. 2:

$$G\left[\mathsf{ind}, \mathsf{crs}, K_{\mathsf{ind}}, dk_{\mathsf{ind}}, \{ek_j\}_{j \in [n]}, \{vk_j\}_{j \in H}\right].$$

   9.4. Let $\widehat{x_{\mathsf{ind}}} = (\overline{G_{\mathsf{ind}}}, \mathsf{CT}_{\mathsf{ind}}, \psi_{\mathsf{ind}}, \pi_{\mathsf{ind}})$.
10. The challenge: $b' \leftarrow \mathcal{A}(\mathsf{crs}, \{\mathsf{PK}_i\}_{i \in H}, \{\widehat{x_i}\}_{i \in H})$ and output b'.

The indistinguishability of this experiment and experiment Hyb_2 follows directly from the zero-knowledge property of the NIZK (see definition in Sect. 3.1), since the adversary does not get the trapdoor τ as input. The proof here relies on the fact that the adversary has to commit to his challenge input vectors before seeing the common random string.

$\underline{\mathsf{Hyb}_{4,1,\{\mathsf{CT}_i^*,\pi_i^*\}_{i\notin H}}}$ for $\{\mathsf{CT}_i^*,\pi_i^*\}_{i\notin H} \in \{0,1\}^{\ell(\lambda)}$:

1. Sample $(dk_{\mathsf{ind}}, ek_{\mathsf{ind}}) \leftarrow \mathsf{MFHE.KeyGen}(1^\lambda; r)$, where r is a random string.
2. $\mathsf{CT}_{\mathsf{ind}} \leftarrow \mathsf{MFHE.Enc}(ek_{\mathsf{ind}}, x_{\mathsf{ind}}^b; r')$, where r' is a random string.
3. Sample $(\mathsf{crs}, \tau) \leftarrow \mathsf{NIZK}.\mathcal{S}_1(1^\lambda, (\mathsf{CT}_{\mathsf{ind}}, ek_{\mathsf{ind}}))$.
4. Sample random a bit $b \in \{0,1\}$.
5. Proceed as before (Item 2) for party $i \in H \setminus \{\mathsf{ind}\}$.
6. Do the following for party ind:
   6.1. Sample $(sk_{\mathsf{ind}}, vk_{\mathsf{ind}}) \leftarrow \mathsf{PuncSig.KeyGen}(1^\lambda)$.
   6.2. $vk_{\mathsf{ind}} \leftarrow \mathsf{PuncSig.Puncture}(sk_{\mathsf{ind}}, (\mathsf{CT}_{\mathsf{ind}}, i), \bot)$.
   6.3. Set $\mathsf{PK}_{\mathsf{ind}} = (vk_{\mathsf{ind}}, ek_{\mathsf{ind}})$ and $\mathsf{SK}_{\mathsf{ind}} = (sk_{\mathsf{ind}}, dk_{\mathsf{ind}})$.
7. The adversary, given $\{\mathsf{PK}_i\}_{i\in H}$ and crs, publishes a public key PK_i for every $i \notin H$ of the form $\mathsf{PK}_i = (vk_i, ek_i)$.
8. Proceed as before (Item 4) for party $i \in H \setminus \{\mathsf{ind}\}$.
9. Do the following for party ind:
   9.1. $\psi_{\mathsf{ind}} \leftarrow \mathsf{PuncSig.Sign}(sk_{\mathsf{ind}}, (\widehat{\mathsf{CT}}_{\mathsf{ind}}, i))$.
   9.2. Sample a PRF key K_{ind}.
   9.3. $\pi_{\mathsf{ind}} \leftarrow \mathsf{NIZK}.\mathcal{S}_2(\mathsf{crs}, \tau, (\mathsf{CT}_{\mathsf{ind}}, ek_{\mathsf{ind}}))$.
   9.4. Compute $G_{\mathsf{ind}}^{(2)}$ as the obfuscation of the circuit defined in Fig. 3 :

$$G^{(2)}\Big[\mathsf{ind}, \mathsf{crs}, \{\tau^j\}_{j\notin H}, K_{\mathsf{ind}}, \{dk_j\}_{j\in H}, \{ek_j\}_{j\in[n]}, \{vk_j\}_{j\in[n]}, \{x_j^b\}_{j\in H},$$
$$\{\mathsf{CT}_j^*,\pi_j^*\}_{j\notin H}\Big]$$

   9.5. Let $\widehat{x_{\mathsf{ind}}} = (\overline{G_{\mathsf{ind}}}, \mathsf{CT}_{\mathsf{ind}}, \psi_{\mathsf{ind}}, \pi_{\mathsf{ind}})$.
10. The challenge: $b' \leftarrow \mathcal{A}(\mathsf{crs}, \{\mathsf{PK}_i\}_{i\in H}, \{\widehat{x}_i\}_{i\in H})$ and output b'.

Notice that when $\{\mathsf{CT}_i^*,\pi_i^*\}_{i\notin H}$ are all equal to the all zero string, then this hybrid is identical to Hyb_3.

$\underline{\mathsf{Hyb}_{4,2,\{\mathsf{CT}_i^*,\pi_i^*\}_{i\notin H}}}$ for $\{\mathsf{CT}_i^*,\pi_i^*\}_{i\notin H} \in \{0,1\}^{\ell(\lambda)}$: Proceed as in the previous hybrid, except that now hardwire in the circuit $G_i^{(2)}$ a punctured PRF key $K^* = \mathsf{Puncture}(K, \{\mathsf{CT}_i\}_{i\in H}\|\{\mathsf{CT}_i^*\}_{i\notin H})$, sample r_i^* at random instead of via a PRF.

1. Repeat Steps 1–9.1. from the previous experiment.
2. Modify Steps 9.2.–9.4. by doing the following for party ind:
   2.1. Compute punctured key $K_{\mathsf{ind}}^* = \mathsf{PRF.Puncture}(K, \{\mathsf{CT}_i\}_{i\in H}\|\{\mathsf{CT}_i^*\}_{i\notin H})$.
   2.2. For $j \notin H$, compute $\widehat{\mathsf{CT}}_j^* = \mathsf{MFHE.Expand}((ek_1,\ldots,ek_n), j, \mathsf{CT}_j^*)$.
   2.3. Compute $\widehat{\mathsf{CT}}_{\mathsf{out}}^* \leftarrow \mathsf{MFHE.Eval}\left(f, \{\widehat{\mathsf{CT}}_j\}_{j\in H}, \{\widehat{\mathsf{CT}}_j^*\}_{j\in H}\right)$.
   2.4. $r_{\mathsf{ind}} \leftarrow \mathsf{PRF}_{K_{\mathsf{ind}}}(\mathsf{CT}_1\|\pi_1\|\ldots\|\mathsf{CT}_n\|\pi_n)$.
   2.5. Compute $p_{\mathsf{ind}}^* \leftarrow \mathsf{MFHE.PartDec}\left(\widehat{\mathsf{CT}}_{\mathsf{out}}^*, ek_1,\ldots,ek_n, \mathsf{ind}, dk_{\mathsf{ind}}; r_{\mathsf{ind}}\right)$.
   2.6 $\pi_{\mathsf{ind}} \leftarrow \mathsf{NIZK}.\mathcal{S}_2(\mathsf{crs}, \tau, (\mathsf{CT}_{\mathsf{ind}}, ek_{\mathsf{ind}}))$.

$$G^{(2)}\left[i, \mathsf{crs}, \tau, K_i, \{dk_j\}_{j\in H}, \{ek_j\}_{j\in[n]}, \{vk_j\}_{j\in[n]}, \{x_j\}_{j\in H}, \underline{\{\mathsf{CT}_j^*, \pi_j^*\}_{j\notin H}}\right]$$

Input: $(\mathsf{CT}_1, \psi_1, \pi_1), \dots, (\mathsf{CT}_n, \psi_n, \pi_n)$.

- If $\mathsf{PuncSig.Verify}(vk_j, (\mathsf{CT}_j, j), \psi_j)$ is false for some $j \in [n]$ then output $\bot$.
- For every $j \in [n]$, verify $\mathsf{NIZK}.\mathcal{V}(\mathsf{crs}, (ek_j, \mathsf{CT}_j), \pi_j)$ for the statement:

$$\exists r, r', x_j, dk_j : \mathsf{CT}_j = \mathsf{MFHE.Enc}(ek_j, x_j; r) \ \wedge \ (ek_j, dk_j) = \mathsf{MFHE.KeyGen}(1^\lambda; r').$$

  If this check fails for some $j \in [n]$ then output $\bot$.
- For all $j \in [n]$ compute $\widehat{\mathsf{CT}_j} \leftarrow \mathsf{MFHE.Expand}((ek_1, \dots, ek_n), j, \mathsf{CT}_j)$.
- Perform $\widehat{\mathsf{CT}_{\mathsf{out}}} \leftarrow \mathsf{MFHE.Eval}\left(f, \widehat{\mathsf{CT}_1}, \dots, \widehat{\mathsf{CT}_n}\right)$.
- $r_i \leftarrow \mathsf{PRF}_{K_i}(\mathsf{CT}_1\|\pi_1\|\dots\|\mathsf{CT}_n\|\pi_n)$.
- If $\{\mathsf{CT}_i, \pi_i\}_{i\in H} < \{\mathsf{CT}_i^*, \pi_i^*\}_{i\in H}$:
    - For all $j \notin H$, let $(r, r', x_j, dk_j) \leftarrow \mathsf{NIZK}.\mathcal{E}(\mathsf{crs}, \tau, (\mathsf{CT}_j, ek_j), \pi_j)$.
    - $p_i \leftarrow \mathsf{MFHE.Sim}\left(y, \widehat{\mathsf{CT}_{\mathsf{out}}}, \{dk_j\}_{j\in[n]\setminus\{i\}}; r_i\right)$, where $y = f(x_1, \dots, x_n)$.
- Otherwise:
    - $p_i \leftarrow \mathsf{MFHE.PartDec}\left(\widehat{\mathsf{CT}_{\mathsf{out}}}, ek_1, \dots, ek_n, i, dk_i; r_i\right)$
- Output p_i.

Fig. 3. The circuit $G^{(2)}$

2.7 Compute $\overline{G_{\mathsf{ind}}^{(3)}}$ as the obfuscation of the circuit defined in Fig. 4 :

$$G^{(3)}\left[\mathsf{ind}, \mathsf{crs}, \{\tau^j\}_{j\notin H}, \underline{K_{\mathsf{ind}}^*}, \{dk_j\}_{j\in H}, \{ek_j\}_{j\in[n]}, \{vk_j\}_{j\in[n]}, \{x_j^b\}_{j\in H},\right.$$
$$\left.\{\mathsf{CT}_j^*, \pi_j^*\}_{j\notin H}, \underline{p_{\mathsf{ind}}^*}\right]$$

2.8 Let $\widehat{x_{\mathsf{ind}}} = (\overline{G_{\mathsf{ind}}^{(3)}}, \widehat{\mathsf{CT}_{\mathsf{ind}}}, \psi_{\mathsf{ind}}, \pi_{\mathsf{ind}})$.
3. The challenge: $b' \leftarrow \mathcal{A}(\mathsf{crs}, \{\mathsf{PK}_i\}_{i\in H}, \{\widehat{x}_i\}_{i\in H})$ and output b'.

This hybrid is indistinguishable from $\mathsf{Hyb}_{4,1,\{\mathsf{CT}_i^*, \pi_i^*\}}$ since the circuits $G^{(2)}$ and $G^{(3)}$ (with all the hardwired values) are functionally equivalent, and therefore, their obfuscation is indistinguishable.

$\mathsf{Hyb}_{4,3,\{\mathsf{CT}_i^*, \pi_i^*\}_{i\notin H}}$ for $\{\mathsf{CT}_i^*, \pi_i^*\}_{i\notin H} \in \{0,1\}^{\ell(\lambda)}$: Proceed as in the previous hybrid, except that in Line 2.5., sample r_{ind}^* at random instead of via a PRF. This hybrid is indistinguishable from $\mathsf{Hyb}_{4,2,\{\mathsf{CT}_i^*, \pi_i^*\}}$ from the security of the puncturable PRF.

$\mathsf{Hyb}_{4,4,\{\mathsf{CT}_i^*, \pi_i^*\}_{i\notin H}}$ for $\{\mathsf{CT}_i^*, \pi_i^*\}_{i\notin H} \in \{0,1\}^{\ell(\lambda)}$: Proceed as in the previous hybrid, except that in Line 2.5., p_{ind}^* is computed as follows:

- For all $j \notin H$, compute $(r, r', x_j, dk_j) \leftarrow \mathsf{NIZK}.\mathcal{E}(\mathsf{crs}, \tau, (\mathsf{CT}_j^*, ek_j), \pi_j^*)$.
- $p_{\mathsf{ind}}^* \leftarrow \mathsf{MFHE.Sim}\left(y, \widehat{\mathsf{CT}_{\mathsf{out}}^*}, \{dk_i\}_{i\neq\mathsf{ind}}\right)$ for $y = f(\{x_j^b\}_{j\in H}, \{x_j\}_{j\notin H})$.

$G^{(3)} \left[i, \mathsf{crs}, \tau, K_i, \{dk_j\}_{j \in H}, \{ek_j\}_{j \in [n]}, \{vk_j\}_{j \in [n]}, \{x_j\}_{j \in H}, \{\mathsf{CT}_j^*, \pi_j^*\}_{j \notin H}, \underline{p_i^*} \right]$

Input: $(\mathsf{CT}_1, \psi_1, \pi_1), \ldots, (\mathsf{CT}_n, \psi_n, \pi_n)$.

- If $\mathsf{PuncSig.Verify}(vk_j, (\mathsf{CT}_j, j), \psi_j)$ is false for some $j \in [n]$ then output $\bot$.
- For every $j \in [n]$, verify $\mathsf{NIZK}.\mathcal{V}(\mathsf{crs}, (ek_j, \mathsf{CT}_j), \pi_j)$ for the statement:

$$\exists r, r', x_j, dk_j : \mathsf{CT}_j = \mathsf{MFHE.Enc}(ek_j, x_j; r) \land (ek_j, dk_j) = \mathsf{MFHE.KeyGen}(1^\lambda; r').$$

  If this check fails for some $j \in [n]$ then output $\bot$.
- For all $j \in [n]$ compute $\widehat{\mathsf{CT}}_j \leftarrow \mathsf{MFHE.Expand}((ek_1, \ldots, ek_n), j, \mathsf{CT}_j)$.
- Perform $\widehat{\mathsf{CT}_{\mathsf{out}}} \leftarrow \mathsf{MFHE.Eval}\left(f, \widehat{\mathsf{CT}}_1, \ldots, \widehat{\mathsf{CT}}_n\right)$.
- $r_i \leftarrow \mathsf{PRF}_{K_i}(\mathsf{CT}_1 \| \pi_1 \| \ldots \| \mathsf{CT}_n \| \pi_n)$.
- If $\{\mathsf{CT}_i, \pi_i\}_{i \notin H} < \{\mathsf{CT}_i^*, \pi_i^*\}_{i \notin H}$ and $i = \mathsf{ind}$:
    - For all $j \notin H$, let $(r, r', x_j, dk_j) \leftarrow \mathsf{NIZK}.\mathcal{E}(\mathsf{crs}, \tau, (\mathsf{CT}_j, ek_j), \pi_j)$.
    - $p_i \leftarrow \mathsf{MFHE.Sim}\left(y, \widehat{\mathsf{CT}_{\mathsf{out}}}, \{dk_j\}_{j \in [n] \setminus \{\mathsf{ind}\}}; r_i\right)$, where $y = f(x_1, \ldots, x_n)$.
- If $\{\mathsf{CT}_i, \pi_i\}_{i \notin H} = \{\mathsf{CT}_i^*, \pi_i^*\}_{i \notin H}$, output p_i^*.
- Otherwise:
    - $p_i \leftarrow \mathsf{MFHE.PartDec}\left(\widehat{\mathsf{CT}_{\mathsf{out}}}, ek_1, \ldots, ek_n, i, dk_i; r_i\right)$.
- Output p_i.

Fig. 4. The circuit $G^{(3)}$

This hybrid is indistinguishable from $\mathsf{Hyb}_{4,3,\{\mathsf{CT}_i^*, \pi_i^*\}}$ from the simulation security of the MFHE scheme. For the simulation security to hold, we need to show that the value y actually corresponds to the message underlying the ciphertext $\widehat{\mathsf{CT}_{\mathsf{out}}^*}$ with respect to the public keys $ek_1, \ldots, ek_n$. Notice that y is computed as $y = f(\{x_j^b\}_{j \in H}, \{x_j\}_{j \notin H})$. For all $j \in H$ we have that the signature ψ_j is punctured such that it can verify only CT_j, which is indeed an encryption of x_j^b. For all $j \notin H$ by the statistical simulation extractability of the NIZK scheme, we get the corresponding secret key and message x_j for the ciphertext CT_j, and thus indeed it is a valid encryption of x_j. Altogether, we get that y is the correct value underlying the expanded ciphertext $\mathsf{CT}_{\mathsf{out}}^*$ and thus the security of the simulation holds.

$\underline{\mathsf{Hyb}_{4,5,\{\mathsf{CT}_i^*, \pi_i^*\}_{i \notin H}}}$ for $\{\mathsf{CT}_i^*, \pi_i^*\}_{i \notin H} \in \{0,1\}^{\ell(\lambda)}$: Proceed as in the previous hybrid, except that in Line 2.5., sample r_i^* via the PRF instead of uniformly at random. Again, this hybrid is indistinguishable from the previous one from the security of the puncturable PRF.

$\underline{\mathsf{Hyb}_5}$: This experiment corresponds to experiment $\mathsf{Hyb}_{4,1,1^{\ell(\lambda)}}$ except that now in Step 2.7. we obfuscate the circuit with the values $\{x_j^0\}_{j \in H}$ hardwired instead of $\{x_j^b\}_{j \in H}$.

Since f is admissible (see Definition 1), we get that the two circuits are equivalent, and thus the indistinguishability of this experiment and the previous one follows from the security of the obfuscation scheme.

Hyb_6: This experiment corresponds to experiment Hyb_5 except that now in Step 2 we encrypt x_{ind}^0 in $\mathsf{CT}_{\mathsf{ind}}$ and not x_{ind}^b. Notice that this experiment is independent of the bit b and thus the probability of any adversary in guessing b is $1/2$.

The indistinguishability of this experiment and the previous one follows from the semantic security of the MFHE scheme, since dk_{ind} is not used in this hybrid any more.

In conclusion, notice that each two experiments are indistinguishable and there are in total $5 \cdot 2^{\ell(\lambda)} + 3$ hybrids, where in each such hybrid we lose the security of the primitive in hand. Thus, if we start with sub-exponentially secure schemes and initialize the underlying primitives with a large enough security parameter $\mathsf{poly}(\lambda)$, our resulting scheme is secure, as required.

5.2 Reusable PKI

As discussed, the security we defined for the MPC protocol is a single challenge, for a single PKI instantiation. A protocol that works for many sessions using the same PKI instantiation can be achieved by using session ids for each computation. The only modification needed is to sign the ciphertexts with a prefix of the session id. The puncturable signature scheme is constructed to support creating verification keys that do not accept any message within a given prefix except one. Thus, the only change in the proof would be to puncture the signature at the given session id, instead of the prefix $\bot$ as currently performed.

6   The Chain Construction

We now present our construction of NI-MPC for the chain graph communication pattern. The construction is rather similar to the one of the start pattern, and share the same building blocks. The setup procedure is the same, the main difference is the encryption procedure. The obfuscated circuit will take as input only the future inputs on the chain, and the rest will be hard-wired in the circuit.

$\mathsf{Setup}(1^\lambda, \mathsf{crs}, i)$: It takes as input security parameter λ, a common random string crs, and does the following.

1. Execute the key generation procedure of the signature scheme, $(sk_i, vk_i) \leftarrow \mathsf{PuncSig.KeyGen}(1^\lambda)$.
2. Execute $\{(dk_i, ek_i) \leftarrow \mathsf{MFHE.KeyGen}(1^\lambda)\}_{i \in [n]}$.
3. Output public key $\mathsf{PK}_i = (vk_i, ek_i)$ and private key $\mathsf{SK}_i = (sk_i, dk_i)$.

$\mathsf{Enc}(1^\lambda, \mathsf{crs}, \widehat{x_{i-1}}, x_i, \mathsf{PK}_1, \ldots, \mathsf{PK}_n, \mathsf{SK}_i, i)$: It takes as input security parameter λ, a common random string crs, an input $x_i \in \mathcal{X}$, a list of n public keys $\mathsf{PK}_1 \ldots, \mathsf{PK}_n$, one private key SK_i, an index $i \in [n]$, the previous message m, and does the following.

1. For every $j \in [n]$, parse $\mathsf{PK}_j = (vk_j, ek_j)$ and let $\mathsf{SK}_i = (sk_i, dk_i)$.
2. Parse $\widehat{x_{i-1}} = (\overline{G_1}, \mathsf{CT}_1, \psi_1, \pi_1), \ldots, (\overline{G_i}, \mathsf{CT}_{i-1}, \psi_{i-1}, \pi_{i-1})$, and let
   $\sigma = (\mathsf{CT}_1, \psi_1, \pi_1), \ldots, (\mathsf{CT}_{i-1}, \psi_{i-1}, \pi_{i-1})$.
3. Encrypt the input x_i using ek_i: $\mathsf{CT}_i \leftarrow \mathsf{MFHE.Enc}(ek_i, x_i)$.
4. Sign on (CT_i, i) using sk_i: $\psi_i \leftarrow \mathsf{PuncSig.Sign}(sk_i, (\mathsf{CT}_i, i))$.
5. Sample puncturable PRF key $K_i \xleftarrow{\$} \{0,1\}^\lambda$.
6. Generate a proof π_i that CT_i and ek_i are valid ciphertext and encryption key, respectively. Namely, compute $\pi_i \leftarrow \mathsf{NIZK}.\mathcal{P}(\mathsf{crs}, (\mathsf{CT}_i, ek_i), w = (r, r', x_i, dk_i))$, where r and r' are the randomness used for the computation of CT_i and (dk_i, ek_i), respectively. The exact statement is given in Fig. 5.
7. Obfuscate the circuit $\overline{G_i} \leftarrow \mathsf{iO}\big(1^\lambda, G\big[i, \mathsf{crs}, K_i, dk_i, \{ek_j\}_{j \in [n]}, \{vk_j\}_{j \in [n]}, \sigma\big]\big)$ as described in Fig. 5.
8. Output $\widehat{x_i} = \widehat{x_{i-1}}, (\overline{G_i}, \mathsf{CT}_i, \psi_i, \pi_i)$.

$$G\big[i, \mathsf{crs}, K_i, dk_i, \{ek_j\}_{j \in [n]}, \{vk_j\}_{j \in [n]}, \sigma\big]$$

Input: $(\mathsf{CT}_i, \psi_i, \pi_i), \ldots, (\mathsf{CT}_n, \psi_n, \pi_n)$.

- Parse $\sigma = (\mathsf{CT}_1, \psi_1, \pi_1), \ldots, (\mathsf{CT}_{i-1}, \psi_{i-1}, \pi_{i-1})$.
- If $\mathsf{PuncSig.Verify}(vk_j, (\mathsf{CT}_j, j), \psi_j)$ is false for some $j \in [n]$ then output $\bot$.
- For every $j \in [n]$, verify $\mathsf{NIZK}.\mathcal{V}(\mathsf{crs}, (ek_j, \mathsf{CT}_j), \pi_j)$ for the statement:

  $$\exists r, r', M_j, dk_j : \mathsf{CT}_j = \mathsf{MFHE.Enc}(ek_j, M_j; r) \wedge (ek_j, dk_j) = \mathsf{MFHE.KeyGen}(1^\lambda; r').$$

  If this check fails for some $j \in [n]$ then output $\bot$.
- For all $j \in [n]$ compute $\widehat{\mathsf{CT}}_j \leftarrow \mathsf{MFHE.Expand}((ek_1, \ldots, ek_n), j, \mathsf{CT}_j)$.
- Perform $\widehat{\mathsf{CT}_{\mathsf{out}}} \leftarrow \mathsf{MFHE.Eval}\big(f, \widehat{\mathsf{CT}}_1, \ldots, \widehat{\mathsf{CT}}_n\big)$.
- $r_i \leftarrow \mathsf{PRF}_{K_i}(\mathsf{CT}_1 \| \pi_1 \| \ldots \| \mathsf{CT}_n \| \pi_n)$.
- $p_i \leftarrow \mathsf{MFHE.PartDec}\big(\widehat{\mathsf{CT}_{\mathsf{out}}}, ek_1, \ldots, ek_n, i, dk_i; r_i\big)$.
- Output p_i.

Fig. 5. The circuit G.

$\underline{\mathsf{Eval}(1^\lambda, \mathsf{crs}, \widehat{x_n}, \mathsf{PK}_1, \ldots, \mathsf{PK}_n)}$: It takes as input security parameter λ, a common random string crs, a string $\widehat{x_n}$, a list of n public keys $\mathsf{PK}_1 \ldots, \mathsf{PK}_n$, and performs:

1. Parse $\widehat{x_n} = (\overline{G_1}, \mathsf{CT}_1, \psi_1, \pi_1), \ldots, (\overline{G_n}, \mathsf{CT}_n, \psi_n, \pi_n)$.
2. Evaluate each obfuscation $\overline{G_i}$ on the input $\big((\widehat{\mathsf{CT}}_i, \psi_i, \pi_i), \ldots, (\widehat{\mathsf{CT}}_n, \psi_n, \pi_n)\big)$ to get the partial decryption p_i.
3. Execute the final decryption, $y \leftarrow \mathsf{MFHE.FinDec}(p_1, \ldots, p_n)$ and output y.

The correctness of the scheme is immediate and follows by the correctness of the underlying building blocks. Specifically, let $x_1, \ldots, x_n \in \mathcal{X}$ be inputs such that party i holds x_i. By the correctness of the signature scheme and the threshold multi-key FHE scheme, each obfuscated circuit will output a partial decryption p_i such that $\mathsf{MFHE.FinDec}(p_1, \ldots, p_n)$ must be equal to $f(x_1, \ldots, x_n)$.

For security, we use the same overall strategy we used in the star case in Sect. 5. Note however that the admissibility condition in the case of a chain (see Definition 4) is *weaker* than the admissibility condition in the case of a star. Indeed, in the chain case, the "free" inputs for which the adversary can reset their values are only those appearing after the last honest party, whereas in the star case he can reset the input of any dishonest party.

Consider first a function f that is admissible for a star (and thus also chain admissible). Every (obfuscated) circuit in the chain construction is exactly like the circuit in the star construction except that it has hardwired a subset of the inputs (appearing earlier in the chain). Then, the same sequence of hybrids from the star case applies and proves the security for f.

For the general case, where f is admissible for a chain pattern, we follow the same sequence of hybrids, but make appropriate modifications. Instead of hardwiring the inputs of the honest parties $\{x_i^b\}_{i \in H}$, we hardwire the whole set of inputs of parties up to i^* (where i^* is the index of the last honest party), namely, $\{x_i^b\}_{i \in H} \cup \{x_i\}_{[i^*] \setminus H}$. Notice that we know the inputs $\{x_i\}_{[i^*] \setminus H}$ in advance. Then, the loop of hybrids from $\mathsf{Hyb}_{4,1,\{\mathsf{CT}_i^*, \pi_i^*\}_{i > i^*}}$ to $\mathsf{Hyb}_{4,5,\{\mathsf{CT}_i^*, \pi_i^*\}_{i > i^*}}$ is only over the inputs of parties whose index is larger than i^* (rather than $i \notin H$).

Notice that in the original proof the only hybrid that uses the fact that f is admissible (for a star) is for indistinguishability (based on iO) between $\mathsf{Hyb}_{4,1,1^{\ell(\lambda)}}$ (the last hybrid in the loop) and Hyb_5. Thus, here we use a similar argument for the indistinguishability based on iO and the fact that f is admissible. Since , we know the input of all parties before i^* and since f is chain admissible (with the hardwired inputs of the parties before i^*), we get that the circuit with $\{x_i^b\}_{i \in H} \cup \{x_i\}_{[i^*] \setminus H}$ hardwired is functionally equivalent to the circuit with $\{x_i^0\}_{i \in H} \cup \{x_i\}_{[i^*] \setminus H}$ hardwired. Thus, indistinguishability of the above hybrids follows from the security of iO.

General interaction patterns. One can generalize the above idea, support arbitrary interaction patterns and achieve the "best-possible" security. The formalization of the "best-possible" security per interaction pattern is given in [23, 24]. The modification of our construction is: each party will forward all its input messages to the next party in the DAG in addition to its own message that includes an obfuscated circuit and another ciphertext. The new obfuscated circuit, as in the chain pattern, will have hardwired all ciphertexts it received from previous parties in the chain. The contruction, other than this change, remains the same.

The security proof is also easily modified. Specifically, the only part in the proof that relies on the interaction pattern is the part where we use the admissibility of the function we compute (notice that the definition of admissible functions varies per interaction pattern). We follow the modification we did for the chain case and obtain a security proof for every pattern.

Acknowledgments. Shai Halevi was supported by the Defense Advanced Research Projects Agency (DARPA) and Army Research Office (ARO) under Contract No. W911NF-15-C-0236. Yuval Ishai was supported in part by NSF-BSF grant 2015782, BSF grant 2012366, ISF grant 1709/14, ERC grant 742754, DARPA/ARL SAFEWARE award, NSF Frontier Award 1413955, NSF grants 1619348, 1228984, 1136174, and 1065276, a Xerox Faculty Research Award, a Google Faculty Research Award, an equipment grant from Intel, and an Okawa Foundation Research Grant. This material is based upon work supported by the DARPA through the ARL under Contract W911NF-15-C-0205. Abhishek Jain was supported in part by a DARPA/ARL Safeware Grant W911NF-15-C-0213 and a sub-award from NSF CNS-1414023. Ilan Komargodski is supported in part by Elaine Shi's Packard Foundation Fellowship. Most of this work done while he was a Ph.D student at the Weizmann Institute of Science, supported in part by grants from the Israel Science Foundation and by a Levzion Fellowship. Amit Sahai was supported in part from a DARPA/ARL SAFEWARE award, NSF Frontier Award 1413955, NSF grants 1619348, 1228984, 1136174, and 1065276, BSF grant 2012378, a Xerox Faculty Research Award, a Google Faculty Research Award, an equipment grant from Intel, and an Okawa Foundation Research Grant. This material is based upon work supported by the Defense Advanced Research Projects Agency through the ARL under Contract W911NF-15-C-0205. Eylon Yogev is supported in part by a grant from the Israel Science Foundation. The views expressed are those of the authors and do not reflect the official policy or position of the Department of Defense, the National Science Foundation, or the U.S. Government.

References

1. Ananth, P., Jain, A., Naor, M., Sahai, A., Yogev, E.: Universal constructions and robust combiners for indistinguishability obfuscation and witness encryption. In: Robshaw, M., Katz, J. (eds.) CRYPTO 2016. LNCS, vol. 9815, pp. 491–520. Springer, Heidelberg (2016). doi:10.1007/978-3-662-53008-5_17
2. Barak, B., Goldreich, O., Impagliazzo, R., Rudich, S., Sahai, A., Vadhan, S., Yang, K.: On the (im)possibility of obfuscating programs. In: Kilian, J. (ed.) CRYPTO 2001. LNCS, vol. 2139, pp. 1–18. Springer, Heidelberg (2001). doi:10.1007/3-540-44647-8_1
3. Beimel, A., Gabizon, A., Ishai, Y., Kushilevitz, E., Meldgaard, S., Paskin-Cherniavsky, A.: Non-interactive secure multiparty computation. In: Garay, J.A., Gennaro, R. (eds.) CRYPTO 2014. LNCS, vol. 8617, pp. 387–404. Springer, Heidelberg (2014). doi:10.1007/978-3-662-44381-1_22
4. Bitansky, N., Paneth, O.: ZAPs and non-interactive witness indistinguishability from indistinguishability obfuscation. In: Dodis, Y., Nielsen, J.B. (eds.) TCC 2015. LNCS, vol. 9015, pp. 401–427. Springer, Heidelberg (2015). doi:10.1007/978-3-662-46497-7_16
5. Boneh, D., Sahai, A., Waters, B.: Functional encryption: definitions and challenges. In: Ishai, Y. (ed.) TCC 2011. LNCS, vol. 6597, pp. 253–273. Springer, Heidelberg (2011). doi:10.1007/978-3-642-19571-6_16
6. Boneh, D., Waters, B.: Constrained pseudorandom functions and their applications. In: Sako, K., Sarkar, P. (eds.) ASIACRYPT 2013. LNCS, vol. 8270, pp. 280–300. Springer, Heidelberg (2013). doi:10.1007/978-3-642-42045-0_15
7. Boyle, E., Goldwasser, S., Ivan, I.: Functional signatures and pseudorandom functions. In: Krawczyk, H. (ed.) PKC 2014. LNCS, vol. 8383, pp. 501–519. Springer, Heidelberg (2014). doi:10.1007/978-3-642-54631-0_29

8. Canetti, R., Goldreich, O., Goldwasser, S., Micali, S.: Resettable zero-knowledge (extended abstract). In: STOC, pp. 235–244 (2000)
9. Canetti, R., Lin, H., Tessaro, S., Vaikuntanathan, V.: Obfuscation of probabilistic circuits and applications. In: Dodis, Y., Nielsen, J.B. (eds.) TCC 2015. LNCS, vol. 9015, pp. 468–497. Springer, Heidelberg (2015). doi:10.1007/978-3-662-46497-7_19
10. Canetti, R., Lindell, Y., Ostrovsky, R., Sahai, A.: Universally composable two-party and multi-party secure computation. In: STOC, pp. 494–503 (2002)
11. Clear, M., McGoldrick, C.: Multi-identity and multi-key leveled FHE from learning with errors. In: Gennaro, R., Robshaw, M. (eds.) CRYPTO 2015. LNCS, vol. 9216, pp. 630–656. Springer, Heidelberg (2015). doi:10.1007/978-3-662-48000-7_31
12. Dodis, Y., Halevi, S., Rothblum, R.D., Wichs, D.: Spooky encryption and its applications. In: Robshaw, M., Katz, J. (eds.) CRYPTO 2016. LNCS, vol. 9816, pp. 93–122. Springer, Heidelberg (2016). doi:10.1007/978-3-662-53015-3_4
13. Feige, U., Kilian, J., Naor, M.: A minimal model for secure computation (extended abstract). In: STOC (1994)
14. Garg, S., Gentry, C., Halevi, S., Raykova, M.: Two-round secure MPC from indistinguishability obfuscation. In: TCC (2014)
15. Garg, S., Gentry, C., Halevi, S., Raykova, M., Sahai, A., Waters, B.: Candidate indistinguishability obfuscation and functional encryption for all circuits. In: FOCS (2013)
16. Garg, S., Gentry, C., Halevi, S., Raykova, M., Sahai, A., Waters, B.: Candidate indistinguishability obfuscation and functional encryption for all circuits. SIAM J. Comput. 45(3), 882–929 (2016)
17. Goldreich, O., Goldwasser, S., Micali, S.: How to construct random functions. J. ACM (JACM) (1986)
18. Goldreich, O., Micali, S., Wigderson, A.: How to play any mental game. In: STOC (1987)
19. Goldwasser, S., et al.: Multi-input functional encryption. In: Nguyen, P.Q., Oswald, E. (eds.) EUROCRYPT 2014. LNCS, vol. 8441, pp. 578–602. Springer, Heidelberg (2014). doi:10.1007/978-3-642-55220-5_32
20. Gordon, S.D., Malkin, T., Rosulek, M., Wee, H.: Multi-party computation of polynomials and branching programs without simultaneous interaction. In: Johansson, T., Nguyen, P.Q. (eds.) EUROCRYPT 2013. LNCS, vol. 7881, pp. 575–591. Springer, Heidelberg (2013). doi:10.1007/978-3-642-38348-9_34
21. Goyal, V., Maji, H.K.: Stateless cryptographic protocols. In: FOCS, pp. 678–687 (2011)
22. Goyal, V., Sahai, A.: Resettably secure computation. In: Joux, A. (ed.) EUROCRYPT 2009. LNCS, vol. 5479, pp. 54–71. Springer, Heidelberg (2009). doi:10.1007/978-3-642-01001-9_3
23. Halevi, S., Ishai, Y., Jain, A., Kushilevitz, E., Rabin, T.: Secure multiparty computation with general interaction patterns. In: ITCS, pp. 157–168 (2016)
24. Halevi, S., Lindell, Y., Pinkas, B.: Secure computation on the web: computing without simultaneous interaction. In: Rogaway, P. (ed.) CRYPTO 2011. LNCS, vol. 6841, pp. 132–150. Springer, Heidelberg (2011). doi:10.1007/978-3-642-22792-9_8
25. Kiayias, A., Papadopoulos, S., Triandopoulos, N., Zacharias, T.: Delegatable pseudorandom functions and applications. In: ACM SIGSAC (2013)
26. López-Alt, A., Tromer, E., Vaikuntanathan, V.: On-the-fly multiparty computation on the cloud via multikey fully homomorphic encryption. In: STOC (2012)
27. Mukherjee, P., Wichs, D.: Two round multiparty computation via multi-key FHE. In: Fischlin, M., Coron, J.-S. (eds.) EUROCRYPT 2016. LNCS, vol. 9666, pp. 735–763. Springer, Heidelberg (2016). doi:10.1007/978-3-662-49896-5_26

28. O'Neill, A.: Definitional issues in functional encryption. IACR Cryptology ePrint Archive 2010, 556 (2010)
29. Sahai, A., Waters, B.: Slides on functional encryption (2008). http://www.cs.utexas.edu/~bwaters/presentations/files/functional.ppt
30. Sahai, A., Waters, B.: How to use indistinguishability obfuscation: deniable encryption, and more. In: STOC (2014)
31. Yao, A.C.C.: How to generate and exchange secrets (extended abstract). In: FOCS (1986)

Optimal-Rate Non-Committing Encryption

Ran Canetti[1,2]([✉]), Oxana Poburinnaya[2], and Mariana Raykova[3]

[1] Tel Aviv University, Tel Aviv, Israel
canetti@bu.edu
[2] Boston University, Boston, USA
oxanapob@bu.edu
[3] Yale University, New Haven, USA
mariana.raykova@yale.edu

Abstract. Non-committing encryption (NCE) was introduced in order
to implement secure channels under adaptive corruptions in situations
when data erasures are not trustworthy. In this paper we are interested
in the *rate* of NCE, i.e. in how many bits the sender and receiver need
to send per plaintext bit.

In initial constructions the length of both the receiver message,
namely the public key, and the sender message, namely the ciphertext, is
$m \cdot \text{poly}(\lambda)$ for an m-bit message, where λ is the security parameter. Subsequent work improve efficiency significantly, achieving rate $\text{poly} \log(\lambda)$.

We show the first construction of a constant-rate NCE. In fact, our
scheme has rate $1 + o(1)$, which is comparable to the rate of plain semantically secure encryption. Our scheme operates in the common reference
string (CRS) model. Our CRS has size $\text{poly}(m \cdot \lambda)$, but it is reusable for
an arbitrary polynomial number of m-bit messages. In addition, ours is
the first NCE construction with perfect correctness. We assume one way
functions and indistinguishability obfuscation for circuits.

Keywords: Adaptive security · Non-committing encryption

1 Introduction

Informally, *non-committing*, or *adaptively secure*, encryption (NCE) is an encryption scheme for which it is possible to generate a dummy ciphertext which

This work was done [in part] while the authors were visiting the Simons Institute for the Theory of Computing, supported by the Simons Foundation and by
the DIMACS/Simons Collaboration in Cryptography through NSF grant #CNS-
1523467.

R. Canetti—Supported in addition by the NSF MACS project and ISF Grant
1523/14. The author is a member of Check Point Institute for Information Security.

O. Poburinnaya—Supported in addition by the NSF MACS project and NSF grant
1421102.

M. Raykova—Supported by NSF grants 1421102, 1633282, 1562888, DARPA
W911NF-15-C-0236, W911NF-16-1-0389.

© International Association for Cryptologic Research 2017
T. Takagi and T. Peyrin (Eds.): ASIACRYPT 2017, Part III, LNCS 10626, pp. 212–241, 2017.
https://doi.org/10.1007/978-3-319-70700-6_8

is indistinguishable from a real one, but can later be opened to any message [CFGN96]. This primitive is a central tool in building adaptively secure protocols: one can take an adaptively secure protocol in secure channels setting and convert it into adaptively secure protocol in computational setting by encrypting communications using NCE. In particular, NCE schemes are secure under selective-opening attacks [DNRS99].

This additional property of being able to open dummy ciphertexts to any message has its price in efficiency: while for plain semantically secure encryption we have constructions with $O(\lambda)$-size, reusable public and secret keys for security parameter λ, and $m+\text{poly}(\lambda)$ size ciphertext for m-bit messages, non-committing encryption has been far from being that efficient. Some justification for this state of affairs is the lower bound of Nielsen [Nie02], which shows that the secret key of any NCE has to be at least m where m is the overall number of bits decrypted with this key. Still, no bound is known on the size of the public key or the ciphertext.

In this paper we focus on building NCE with better efficiency: specifically, we optimize the *rate* of NCE, i.e. the total amount of communication sent per single bit of a plaintext.

1.1 Prior Work

The first construction of adaptively secure encryption, presented by Beaver and Haber [BH92], is interactive (3 rounds) and relies on the ability of parties to reliably erase parts of their internal state. An adaptively secure encryption that does not rely on secure erasures, or *non-committing encryption,* is presented in [CFGN96]. The scheme requires only two messages, just like standard encryption, and is based on joint-domain trapdoor permutations. It requires both the sender and the receiver to send $\Theta(\lambda^2)$ bits per each bit of a plaintext. Subsequent work has focused on reducing rate and number of rounds. Beaver [Bea97] and Damgård and Nielsen [DN00] propose a 3-round NCE protocol from, respectively, DDH and a *simulatable PKE* (which again can be built from similar assumptions to those of [CFGN96]) with $m \cdot \Theta(\lambda^2)$ bits overall communication for m bit messages, but only $m \cdot \Theta(\lambda)$ bits from sender to receiver. These results were improved by Choi et al. [CDMW09] who reduce the number of rounds to two, which matches optimal number of rounds since non-interactive NCE is impossible [Nie02]. Also they reduced simulatable PKE assumption to a weaker *trapdoor simulatable PKE assumption*; such a primitive can be constructed from factoring. A recent work of Hemenway et al. [HOR15] presented a two-round NCE construction based on the Φ-hiding assumption which has $\Theta(m \log m) + \text{poly}(\lambda)$ ciphertext size and $m \cdot \Theta(\lambda)$ communication from receiver to sender. In a concurrent work, Hemenway et al. [HORR16] show how to build NCE with rate $\text{poly} \log(\lambda)$ under the ring-LWE assumption.

We remark that the recent results on adaptively secure multiparty computation (MPC) from indistinguishability obfuscation in the common reference string (CRS) model [CGP15, GP15, DKR15] do not provide an improvement of NCE rate. Specifically, [CGP15, DKR15] already use NCE as a building block in their constructions, and the resulting NCE is as inefficient as underlying NCE. The scheme by Garg and Polychroniadou [GP15] does not use NCE, but their second message is of size $\text{poly}(m\lambda)$ due to the statistically sound non-interactive zero knowledge proof involved.

Another line of work focuses on achieving better parameters for weaker notions of NCE where the adversary sees the internal state of only one of the parties (receiver or sender). Jarecki and Lysyanskaya [JL00] propose a scheme which is non-committing for the receiver only, which has two rounds and ciphertext expansion factor 3 (i.e., the ciphertext size is $3m + \text{poly}(\lambda)$), under DDH assumption. Furthermore, their public key is also short and thus their scheme achieves rate 4. Hazay and Patra [HP14] build a constant-rate NCE which is secure as long as only one party is corrupted, which was later modified by [HLP15] to obtain a constant-rate NCE in the partial erasure model, meaning that security would hold even with both parties corrupted, as long as one party is allowed to erase. Canetti et al. [CHK05] construct a constant-rate NCE with erasures, meaning that the sender has to erase encryption randomness, and the receiver has to erase the randomness used for the initial key generation. Their NCE construction has rate 13.

1.2   Our Results

We present two NCE schemes with constant-rate in the programmable CRS model. We first present a simpler construction which gives us rate 13, and then, using more sophisticated techniques, we construct the second scheme with rate $1 + o(1)$.

Our first construction is given by a rate-preserving transformation from any NCE with erasures to full NCE, assuming indistinguishability obfuscation ($i\mathcal{O}$) and one way functions (OWFs). The known construction of constant-rate NCE with erasures [CHK05] requires decisional composite residuosity assumption and has rate 13.

Our second construction assumes only $i\mathcal{O}$ and OWFs and achieves rate $1 + o(1)$. To be more precise, the public key, which is the first protocol message in our scheme, has the size $O(\lambda)$. The ciphertext, which is the second message, has the size $O(\lambda) + |m|$. The CRS size is $O(\text{poly}(m\lambda))$, but the CRS is reusable for any polynomially-many executions without an a priori bound on the number of executions. Thus when the length $|m|$ of a plaintext is large, the scheme has overall rate that approaches 1.

In addition, this NCE scheme is the first to guarantee perfect correctness. Note that NCE in the plain model cannot be perfectly correct, and therefore some setup assumption is necessary to achieve this property.

1.3   Construction and Proof Techniques

Definition of NCE. Before describing our construction, we recall what a non-committing encryption is in more detail. Such a scheme consists of algorithms (Gen, Enc, Dec, Sim), which satisfy usual correctness and security requirements.

Additionally, the scheme should remain secure even if the adversary first decides to see the communications in the protocol and later corrupt the parties. This means that the simulator should be able to generate a dummy ciphertext c_f (without knowing which message it encrypts). Later, upon corruption of the parties, the simulator learns a message m, and it should generate internal state of the parties consistent with m and c_f - namely, encryption randomness of the sender and generation randomness of the receiver.

First attempts and our first construction. Recall that the recent puncturing technique adds a special trapdoor to a program, which allows to "explain" any input-output behavior of a program, i.e. to generate randomness consistent with a given input-output pair [SW14, DKR15]. Given such a technique, we could try to build NCE as follows. Start from any rate-efficient non-committing encryption scheme in a model with erasures. Obfuscate key generation algorithm Gen and put it in the CRS. The protocol then proceeds as follows: the receiver runs Gen, obtains (pk, sk), sends pk to the sender, gets back c and decrypts it with sk. In order to allow simulation of the receiver, augment Gen with a trapdoor which allows a simulator to come up with randomness for Gen consistent with (pk, sk). However, this approach doesn't allow to simulate the sender.

One natural way to allow simulation of the sender is to modify Gen: instead of outputting pk, it should output an obfuscated encryption algorithm $E = i\mathcal{O}(\mathsf{Enc}[pk])$ with the public key hardwired, and the receiver should send E (instead of pk) to the sender in round 1. In the simulation $\mathsf{Enc}[pk]$ can be augmented with a trapdoor, thus allowing to simulate the sender. The problem is that this scheme is no longer efficient: in all known constructions the trapdoor (and therefore the whole program E) has the size of at least $\lambda|m|$, meaning that the rate is at least λ (this is due to the fact that this trapdoor uses a punctured PRF applied to the message m, and, to the best of our knowledge, in all known constructions of PPRFs the size of a punctured key is at least $\lambda|m|$).

Another attempt to allow simulation of the sender is to add to the CRS an obfuscated encryption program $E' = i\mathcal{O}(\mathsf{Enc}(pk, m, r))$, augmented with a trapdoor in the simulation. Just like in the initial scheme, the receiver should send pk to the sender; however, instead of computing c directly using pk, the sender should run obfuscated program E' on pk, m and r. This scheme allows to simulate both the sender and the receiver, and at the same time keeps communication as short as in the original PKE. However, we can only prove *selective* security, meaning that the adversary has to commit to the challenge message m *before* it sees the CRS. This is a limitation of the puncturing technique being used: in the security proof the input to the program Enc, including message m, has to be hardwired into the program.

We get around this issue by using another level of indirection, similar to the approach taken by [KSW14] to obtain adaptive security. Instead of publishing $E' = i\mathcal{O}(\mathsf{Enc}(pk, m, r))$ in the CRS, we publish a program GenEnc which generates E' and outputs it. The protocol works as follows: the receiver uses Gen to generate (pk, sk) and sends pk to the sender. The sender runs GenEnc and obtains E', and then executes $E'(pk, m, r) \to c$ and sends c back to the receiver.

Note that GenEnc doesn't take m as input, therefore there is no need to hardwire m into CRS and in particular there is no need to know m at the CRS generation step.

When this scheme uses [CHK05] as underlying NCE with erasures, it has rate 13. The scheme from [CHK05] additionally requires the decisional composite residuosity assumption.

Our second construction. We give another construction of NCE, which achieves nearly optimal rate. That is, the amount of bits sent is $|m| + \text{poly}(\lambda)$, and by setting m to be long enough, we can achieve rate close to 1. The new scheme assumes only indistinguishability obfuscation and one-way functions; there is no need for composite residuosity, used in our previous scheme.

Our construction proceeds in two steps. We first construct a primitive which we call *same-public-key non-committing encryption with erasures*, or *seNCE* for short; essentially this is a non-committing encryption secure with erasures, but there is an additional technical requirement on public keys. Our seNCE scheme will have short ciphertexts, i.e. ciphertext size is $m+\text{poly}(\lambda)$. However, the public keys will still be long, namely $\text{poly}(m\lambda)$.

The second step in our construction is to transform any seNCE into a full NCE scheme such that the ciphertext size is preserved and the public key size depends only on security parameter. We achieve this at the cost of adding a CRS.

Same-public-key NCE with erasures (seNCE). As a first step we construct a special type of non-committing encryption which we can realize in the standard model (without a CRS). This NCE scheme has the following additional properties:

- *security with erasures:* the receiver is allowed to erase its generation randomness (but not sk); the sender is allowed to erase its encryption randomness. (This means that sk is the only information the adversary expects to see upon corrupting both parties.)
- *same public key:* the generation and simulation algorithms executed on the same input r produce the same public keys.

Construction of seNCE. The starting point for our seNCE construction is the PKE construction from $i\mathcal{O}$ by Sahai and Waters [SW14]. Similarly to that approach we set our public key to be an obfuscated program with a key k inside, which takes as inputs message m and randomness r and outputs a ciphertext $c = (c_1, c_2) = (\text{prg}(r), \mathsf{F}_k(\text{prg}(r)) \oplus m)$, where F is a pseudorandom function (PRF). However, instead of setting k to be a secret key, we set the secret key to be an obfuscated program (with k hardwired) which takes an input $c = (c_1, c_2)$ and outputs $\mathsf{F}_k(c_1) \oplus c_2$. Once the encryption and decryption programs are generated, the key k and the randomness used for the obfuscations are erased, and the only thing the receiver keeps is its secret key. Note that ciphertexts in the above scheme have length $m + \text{poly}(\lambda)$.

To see that this construction is secure with erasures, consider the simulator that sets a dummy ciphertext c_f to be a random value. To generate a fake decryption key sk_f, which behaves like a real secret key except that it decrypts c_f to a challenge message m, the simulator obfuscates a program (with m, c_f, k hardwired) that takes as input (c_1, c_2) and does the following: if $c_1 = c_{f1}$ then the program outputs $c_{f2} \oplus c_2 \oplus m$, otherwise the output is $F_k(c_1) \oplus c_2$. Encryption randomness of the sender, as well as k and obfuscation randomness of the receiver, are erased and do not need to be simulated. (Note that the simulated secret key is larger than the real secret key. So, to make sure that the programs have the same size, the real secret key has to be padded appropriately.)

Furthermore, the scheme has the same-public-key property: The simulated encryption key is generated in exactly the same way as the honest encryption key.

Note that this scheme has perfect correctness.

From seNCE to full NCE. Our first step is to enhance the given seNCE scheme, such that the scheme remains secure even when the sender is not allowed to erase its encryption randomness. Specifically, following ideas from the deniable encryption of Sahai and Waters [SW14], we add a trapdoor branch to the encryption program, i.e. the public key. This allows the simulator to create fake randomness $r_{f,\mathsf{Enc}}$, which activates this trapdoor branch and makes the program output c_f on input m. In order to create such randomness, the simulator generates $r_{f,\mathsf{Enc}}$ as an encryption (using a scheme with pseudorandom ciphertexts[1]) of an instruction for the program to output c_f. The program will first try to decrypt $r_{f,\mathsf{Enc}}$ and check whether it should output c_f via trapdoor branch, or execute a normal branch instead.

The above construction of enhanced seNCE still has the following shortcomings. First, its public key (recall that it is an encryption program) is long: the program has to be padded to be at least of size $\mathrm{poly}(\lambda) \cdot |m|$, since in the proof the keys for the trapdoor branch are punctured and have an increased size, and therefore the size of an obfuscated program is $\mathrm{poly}(m\lambda)$.[2] Second, the simulator still cannot simulate the randomness which the receiver used to generate its public key, e.g. keys for the trapdoor branch and randomness for obfuscation. Third, the scheme is only selectively secure, meaning that the adversary has to fix the message *before* it sees a public key. This is due to the fact that our way for explaining a given output (i.e. trapdoor branch mechanism) requires hardwiring the message inside the encryption program in the proof.

We resolve these issues by adding another "level of indirection" for the generation of obfuscated programs. Specifically, we introduce a common reference string that will contain two obfuscated programs, called GenEnc and GenDec, which are generated independently of the actual communication of the protocol and can be reused for unboundedly many messages. The CRS allows the

---

[1] For this purpose we use a *puncturable deterministic encryption scheme* (PDE), since it is $i\mathcal{O}$-friendly and has pseudorandom ciphertexts.

[2] To the best of our knowledge, in all known puncturable PRFs the size of a punctured key applied to m is at least $\lambda|m|$.

sender and the receiver to locally and independently generate their long public and private keys for the underlying enhanced seNCE while communicating only a short token. Furthermore, we will only need to puncture these programs at points which are unrelated to the actual encrypted and decrypted messages. The protocol proceeds as follows.

Description of our protocol. The receiver chooses randomness r_{GenDec} and runs a CRS program $GenDec(r_{GenDec})$. This program uses r_{GenDec} to sample a short token t. Next the program uses this token t to internally compute a secret generation randomness r_{seNCE}, from which it derives (pk, sk) pair for underlying seNCE scheme. Finally, the program outputs (t, sk). In round 1 the receiver sends the token t (which therefore is a short public key of the overall NCE scheme) to the sender.

The sender generates its own randomness r_{GenEnc} and runs a CRS program $GenEnc(t, r_{GenEnc})$. GenEnc, in the same manner as GenDec, first uses t to generate secret r_{seNCE} and sample (the same) key pair (pk, sk) for the seNCE scheme. Further, GenEnc generates trapdoor keys and obfuscation randomness, which it uses to compute a public key program $P_{Enc}[pk]$ of enhanced seNCE, which extends the underlying seNCE public key with a trapdoor as described above. $P_{Enc}[pk]$ is the output of GenEnc. After obtaining P_{Enc}, the sender chooses encryption randomness r_{Enc} and runs $c \leftarrow P_{Enc}[pk](m, r_{Enc})$. In its response message, the sender sends c to the receiver, who decrypts it using sk.

Correctness of this scheme follows from correctness of the seNCE scheme, since at the end a message is being encrypted and decrypted using the seNCE scheme. To get some idea of why security holds, note that the seNCE generation randomness r_{seNCE} is only computed internally by the programs. This value is never revealed to the adversary, and therefore can be thought of as being "erased". In particular, if we had a VBB obfuscation, we could almost immediately reduce security of our scheme to security of seNCE. Due to the fact that we only have $i\mathcal{O}$, the actual security proof becomes way more intricate.

To see how we resolved the three issues from above (namely, with the length of the public key, with simulating the receiver, and with selective security), note:

(a) The only information communicated between sender and receiver is the short token t which depends only on the security parameter, and the ciphertext c which has size $poly(\lambda) + |m|$. Thus the total communication is $poly(\lambda) + |m|$.

(b) The simulator will show slightly modified programs with trapdoor branches inside; they allow the simulator to "explain" the randomness for any desired output, thus allowing it to simulate internal state of both parties.

(c) We no longer need to hardwire message-dependent values into the programs in the CRS, which previously made security only selective. Indeed, in a real world the inputs and outputs of these programs no longer depend on the message sent. They still do depend on the message in the ideal world (for instance, the output of GenDec is sk_m); however, due to the trapdoor branches in the programs it is possible for the simulator to encode sk_m into randomness r_{GenDec} rather than the program GenDec itself. Therefore m can be chosen adaptively after seeing the CRS (and the public key).

To give more details about adaptivity issues which come up in the analysis of the simulator, let us look closely at the following three parts of the proof:

- Starting from a real execution, the first step is to switch real sender generation randomness r_{GenEnc} to fake randomness $r_{f,\mathsf{GenEnc}}$ (which "explains" a real output $\mathsf{P_{Enc}}$). During this step we need to hardwire $\mathsf{P_{Enc}}$ inside GenEnc, which can be done, since $\mathsf{P_{Enc}}$ doesn't depend on m yet.
- Later in the proof we need to switch real encryption randomness r_{Enc} to fake randomness $r_{f,\mathsf{Enc}}$. During this step we need to hardwire m into $\mathsf{P_{Enc}}$. However, at this point $\mathsf{P_{Enc}}$ is not hardwired into the program GenEnc; instead it is being encoded into randomness r_{GenEnc}, and therefore it needs to be generated only when the sender is corrupted (which means that the simulator learns m and can create $\mathsf{P_{Enc}}$ with m hardwired).
- Eventually we need to switch real seNCE values pk, c, sk to simulated pk, c_f, sk_f. Before we can do this, we have to hardwire pk into GenEnc. Luckily, in the underlying seNCE game the adversary is allowed to choose m after it sees pk, and therefore the requirement to hardwire pk into the CRS program doesn't violate adaptive security.

In the proof of our NCE we crucially use the same public-key property of underlying seNCE: Our programs use the master secret key MSK to compute the generation randomness r_{seNCE} from token t, and then sample seNCE keys (pk, sk) using this randomness. In the proof we hardwire pk in the CRS, then puncture MSK and choose r_{seNCE} at random. Next we switch the seNCE values, including the public key pk, to simulated ones. Then we choose r_{seNCE} as a result of a PRF, and unhardwire pk. In order to unhardwire (now simulated) pk from the program and compute $(pk, sk) = \mathsf{F_{MSK}}(r_{\mathsf{seNCE}})$ instead, simulated pk generated from r_{seNCE} should be exactly the same as the real public key pk which the program normally produces by running $\mathsf{seNCE.Gen}(r_{\mathsf{seNCE}})$. This ensures that the programs with and without pk hardwired have the same functionality, and thus security holds by $i\mathcal{O}$.

An additional interesting property of this transformation is that it preserves the correctness of underlying seNCE scheme, meaning that if seNCE is computationally (statistically, perfectly) correct, then the resulting NCE is also computationally (statistically, perfectly) correct. Therefore, when instantiated with our perfectly correct seNCE scheme presented earlier, the resulting NCE achieves perfect correctness. To the best of our knowledge, this is the first NCE scheme with such property.

Shrinking the secret key. The secret key in the above scheme consists of an obfuscated program D, where D is the secret key (i.e. decryption program) for the seNCE scheme, together with some padding that will leave room to "hardwire", in the hybrid distributions in the proof of security, the $|m|$-bit plaintext m into D. Overall, the description size of D is $|m| + O(\lambda)$; when using standard IO, this means that the obfuscated version of D is of size $\mathrm{poly}(|m|\lambda)$.

Still, using the succinct Turing and RAM machine obfuscation of [KLW15, CHJV15, BGL+15, CH15] it is possible to obtain program obfuscation where the size of the obfuscated program is the size of the original program plus a polynomial in the security parameter. This can be done in a number of ways. One simple way is to generate the following (short) obfuscated TM machine OU: The input is expected to contain a description of a program that is one-time-padded, and then authenticated using a signed accumulator as in [KLW15], all with keys expanded from an internally known short key. The machine decrypts, authenticates, and then runs the input circuit. Now, to obfuscate a program, simply one time pad the program, authenticate it, and present it alongside machine OU with the authentication information and keys hardwired.

Augmented explainability compiler. In order to implement the trapdoor branch in the proof of our NCE scheme, we use among other things the "hidden sparse triggers" method of Sahai and Waters [SW14]. This method proved to be useful in other applications as well, and Dachman-Soled et al. [DKR15] abstracted it into a primitive called "explainability compiler". Roughly speaking, explainability compiler turns a randomized program into its "trapdoored" version, such that it becomes possible, for those who know faking keys, to create fake randomness which is consistent with a given input-output pair.

We use a slightly modified version of this primitive, which we call an *augmented explainability compiler*. The difference here is that we can use the original (unmodified) program in the protocol, and only in the proof replace it with its trapdoor version. This is important for perfect correctness of NCE: none of the programs GenEnc, GenDec, and Enc in the real world contain trapdoor branches (indeed, if there was a trapdoor branch in, say, encryption program Enc, it would be possible that an honest sender accidentally chooses randomness which contains an instruction to output an encryption of 0, making the program output this encryption of 0 instead of an encryption of m).

Organization. In Sect. 2 we define the different variants of non-committing encryption, as well as other primitives we use. In Sect. 3 we define and construct an augmented explainability compiler, used in the construction of our NCE. Our optimal-rate NCE and a sketch of security proof are described in Sect. 4.

2   Preliminaries

2.1   Non-committing Encryption and Its Variants

Non-committing encryption. Non-committing encryption is an adaptively secure encryption scheme, i.e., it remains secure even if the adversary decides to see the ciphertext first and only later corrupt parties. This means that the simulator should be able to first present a "dummy" ciphertext without knowing what the real message m is. Later, when parties are corrupted and the simulator learns m, the simulator should be able to present receiver decryption key (or receiver randomness) which decrypts dummy c to m and sender randomness under which m is encrypted to c.

Definition 1. *A non-committing encryption scheme for a message space* $M = \{0,1\}^l$ *is a tuple of algorithms* $(\mathsf{Gen}, \mathsf{Enc}, \mathsf{Dec}, \mathsf{Sim})$, *such that correctness and security hold:*

- **Correctness:** *For all* $m \in M$ $\Pr \left[m = m' \left| \begin{array}{l} (pk, sk) \leftarrow \mathsf{Gen}(1^\lambda, r_{\mathsf{Gen}}); \\ c \leftarrow \mathsf{Enc}(m, r_{\mathsf{Enc}}); \\ m' \leftarrow \mathsf{Dec}(c) \end{array} \right. \right] \geq 1 - \mathsf{negl}(\lambda).$

- **Security:** *An adversary cannot distinguish between real and simulated ciphertexts and internal state even if it chooses message m adaptively depending on the public key pk. More concretely, no PPT adversary* $\mathcal{A}$ *can win the following game with more than negligible advantage:*

A challenger chooses random $b \in \{0,1\}$. *If* $b = 0$, *it runs the following experiment (real):*

1. *It chooses randomness* r_{Gen} *and creates* $(pk, sk) \leftarrow \mathsf{Gen}(1^\lambda, r_{\mathsf{Gen}})$. *It shows pk to the adversary.*
2. *The adversary chooses message m.*
3. *The challenger chooses randomness* r_{Enc} *and creates* $c \leftarrow \mathsf{Enc}(pk, m; r_{\mathsf{Enc}})$. *It shows* $(c, r_{\mathsf{Enc}}, r_{\mathsf{Gen}})$ *to the adversary.*

If $b = 1$, *the challenger runs the following experiment (simulated):*

1. *It runs* $(pk^s, c^s) \leftarrow \mathsf{Sim}(1^\lambda)$. *It shows* pk^s *to the adversary.*
2. *The adversary chooses message m.*
3. *The challenger runs* $(r^s_{\mathsf{Enc}}, r^s_{\mathsf{Gen}}) \leftarrow \mathsf{Sim}(m)$ *and shows* $(c^s, r^s_{\mathsf{Enc}}, r^s_{\mathsf{Gen}})$ *to the adversary.*

The adversary outputs a guess b' *and wins if* $b = b'$.

Note that we allow Sim to be interactive, and in addition we omit its random coins.

In this definition we only spell out the case where *both* parties are corrupted, and all corruptions happen *after* the execution and *simultaneously*. Indeed, if any of the parties is corrupted *before* the ciphertext is sent, then the simulator learns m and can present honest execution of the protocol; therefore we concentrate on the case where corruptions happen afterwards. Next, m is the only information the simulator needs, and after learning it (regardless of which party was corrupted) the simulator can already simulate *both* parties; thus we assume that corruptions of parties happen simultaneously. Finally, without loss of generality we assume that *both* parties are corrupted: if only one or no party is corrupted, then the adversary sees strictly less information in the experiment, and therefore cannot distinguish between real execution and simulation, as long as the scheme is secure under our definition.

Note that this definition only allows parties to encrypt a single message under a given public key. This is due to impossibility result of Nielsen [Nie02], who showed that a secret key of any NCE can support only bounded number of ciphertexts. If one needs to send many messages, it can run several instances of a protocol (each with a fresh pair of keys). Security for this case can be shown via a simple hybrid argument.

Non-committing encryption in a programmable common reference string model. In this work we build NCE in a CRS model, meaning that both parties and the adversary are given access to a CRS, and the simulator, in addition to simulating communications and parties' internal state, also has to simulate the CRS. Before giving a formal definition, we briefly discuss possible variants of this definition.

Programmable CRS. One option is to consider a *global* (non-programmable) CRS model, where the CRS is given to the simulator, or *local* (programmable) CRS model, where the simulator is allowed to generate a CRS. The first variant is stronger and more preferable, but in our construction the simulator needs to know underlying trapdoors and we therefore focus on a weaker definition.

Reusable CRS. Given the fact that in a non-committing encryption a public key can be used to send only bounded number of bits, a bounded-use CRS would force parties to reestablish CRS after sending each block of messages. Since sampling a CRS is usually an expensive operation, it is good to be able to generate a CRS which can be reused for any number of times set a priori. It is even better to have a CRS which can be reused any polynomially many times without any a priori bound. In our definition we ask a CRS to be reusable in this stronger sense.

Security of multiple executions. Unlike NCE in the standard model, in the CRS model single-execution security of NCE does not immediately imply multi-execution security. Indeed, in a reduction to a single-execution security we would have to, given a challenge and a CRS, simulate other executions. But we cannot do this since we didn't generate this CRS ourselves and do not know trapdoors. Therefore in our definition we explicitly require that the protocol remains secure even when the adversary sees many executions with the same CRS.

Definition 2. *An NCE scheme for a message space* $M = \{0,1\}^l$ *in a common reference string model is a tuple of algorithms* (GenCRS, Gen, Enc, Dec, Sim) *which satisfy correctness and security.*

Correctness: *For all* $m \in M$
$$\Pr\left[m = m' \middle| \begin{array}{l} \mathsf{CRS} \leftarrow \mathsf{GenCRS}(1^\lambda); \\ (pk, sk) \leftarrow \mathsf{Gen}(1^\lambda, \mathsf{CRS}; r_{\mathsf{Gen}}); \\ c \leftarrow \mathsf{Enc}(m, \mathsf{CRS}; r_{\mathsf{Enc}}); \\ m' \leftarrow \mathsf{Dec}(\mathsf{CRS}, c) \end{array} \right] \geq$$
$1 - \mathsf{negl}(\lambda)$.

If this probability is equal to 1, then we say that the scheme is perfectly correct.[3]

Security: *For any PPT adversary* $\mathcal{A}$, *advantage of* $\mathcal{A}$ *in distinguishing the following two cases is negligible:*

A challenger chooses random $b \in \{0,1\}$. *If* $b = 0$, *it runs the following experiment (real):*

---

[3] Note that this definition implies that there are no decryption errors for *any* CRS.

First it generates a CRS as CRS $\leftarrow$ GenCRS($1^\lambda, l$). *CRS is given to the adversary. Next the challenger does the following, depending on the adversary's request:*

- *On a request to initiate a protocol instance with session ID* id, *the challenger chooses randomness* $r_{\mathsf{Gen},\mathsf{id}}$ *and creates* $(pk_{\mathsf{id}}, sk_{\mathsf{id}}) \leftarrow$ Gen(1^λ, CRS, $r_{\mathsf{Gen},\mathsf{id}}$). *It shows* pk_{id} *to the adversary.*
- *On a request to encrypt a message* m_{id} *in a protocol instance with session ID* id, *the challenger chooses randomness* $r_{\mathsf{Enc},\mathsf{id}}$ *and creates* $c_{\mathsf{id}} \leftarrow$ Enc($pk_{\mathsf{id}}, m_{\mathsf{id}}; r_{\mathsf{Enc},\mathsf{id}}$). *It shows* c_{id} *to the adversary.*
- *On a request to corrupt the sender of a protocol instance with ID* id, *the challenger shows* $r_{\mathsf{Enc},\mathsf{id}}$ *to the adversary.*
- *On a request to corrupt the receiver of a protocol instance with ID* id, *the challenger shows* $r_{\mathsf{Gen},\mathsf{id}}$ *to the adversary.*

If $b = 1$, it runs the following experiment (simulated):

First it generates a CRS as CRS^s $\leftarrow$ Sim($1^\lambda, l$). *CRS^s is given to the adversary. Next the challenger does the following, depending on the adversary's request:*

- *On a request to initiate a protocol instance with session ID* id, *the challenger runs* $(pk_{\mathsf{id}}^s, c_{\mathsf{id}}^s) \leftarrow$ Sim(1^λ) *and shows* pk_{id}^s *to the adversary.*
- *On a request to encrypt a message* m_{id} *in a protocol instance with session ID* id, *the challenger shows* c_{id}^s *to the adversary.*
- *On a request to corrupt the sender of a protocol instance with ID* id, *the challenger shows* $r_{\mathsf{Enc},\mathsf{id}}^s \leftarrow$ Sim(m_{id}) *to the adversary.*
- *On a request to corrupt the receiver of a protocol instance with ID* id, *the challenger shows* $r_{\mathsf{Gen},\mathsf{id}}^s \leftarrow$ Sim(m_{id}) *to the adversary.*

The adversary outputs a guess b' and wins if $b = b'$.

Constant rate NCE. The rate of an NCE scheme is how many bits the sender and receiver need to communicate in order to transmit a single bit of a plaintext: NCE scheme for a message space $M = \{0,1\}^l$ has rate $f(l, \lambda)$, if $(|pk| + |c|)/l = f(l, \lambda)$. If $f(l, \lambda)$ is a constant, the scheme is said to have constant rate.

Same-public-key non-committing encryption with erasures (seNCE). Here we define a different notion of NCE which we call *same-public-key non-committing encryption with erasures (seNCE)*. First, such a scheme allows parties to erase unnecessary information: the sender is allowed to erase its encryption randomness, and the receiver is allowed to erase its generation randomness r_{Gen} (but not its public or secret key). Furthermore, this scheme should have "the same public key" property, which says that both real generation and simulated generation algorithms should output *exactly the same* public key pk, if they are executed with the same random coins.

Definition 3. *The same-public-key non-committing encryption with erasures (seNCE) for a message space $M = \{0,1\}^l$ is a tuple of algorithms* (Gen, Enc, Dec, Sim), *such that correctness, security, and the same-public-key property hold:*

- **Correctness:** *For all* $m \in M$ $\Pr \left[m = m' \left| \begin{array}{l} (pk, sk) \leftarrow \mathsf{Gen}(1^\lambda, r_{Gen}); \\ c \leftarrow \mathsf{Enc}(m, r_{\mathsf{Enc}}); \\ m' \leftarrow \mathsf{Dec}(c) \end{array} \right. \right] \geq 1 -$
negl(λ).
- **Security with erasures:** *No PPT adversary $\mathcal{A}$ can win the following game with more than negligible advantage:*
  A challenger chooses random $b \in \{0,1\}$. If $b = 0$, it runs a real experiment:
  1. *The challenger chooses randomness r_{Gen} and creates $(pk, sk) \leftarrow \mathsf{Gen}(1^\lambda, r_{\mathsf{Gen}})$. It shows pk to the adversary.*
  2. *The adversary chooses a message m.*
  3. *The challenger chooses randomness r_{Enc} and creates $c \leftarrow \mathsf{Enc}(pk, m; r_{\mathsf{Enc}})$. It shows c to the adversary.*
  4. *Upon corruption request, the challenger shows to the adversary the secret key sk.*
  If $b = 1$, the challenger runs a simulated experiment:
  1. *A challenger generates simulated public key and ciphertext $(pk^s, c^s) \leftarrow \mathsf{Sim}(1^\lambda)^4$. It shows pk^s to the adversary.*
  2. *The adversary chooses a message m.*
  3. *The challenger shows the ciphertext c^s to the adversary.*
  4. *Upon corruption request, the challenger runs $sk^s \leftarrow \mathsf{Sim}(m)$ and shows to the adversary simulated secret key sk^s.*
  The adversary outputs a guess b' and wins if $b = b'$.
- **The same public key:** *For any r if $\mathsf{Gen}(1^\lambda, r) = (pk, sk); \mathsf{Sim}(1^\lambda, r) = (pk_f, c_f)$, then $pk = pk_f$.*

2.2   Puncturable Pseudorandom Functions and Their Variants

Puncturable PRFs. In puncrurable PRFs it is possible to create a key that is punctured at a set S of polynomial size. A key k punctured at S (denoted $k\{S\}$) allows evaluating the PRF at all points not in S. Furthermore, the function values at points in S remain pseudorandom even given $k\{S\}$.

Definition 4. *A puncturable pseudorandom function family for input size $n(\lambda)$ and output size $m(\lambda)$ is a tuple of algorithms {Sample, Puncture, Eval} such that the following properties hold:*

- **Functionality preserved under puncturing:** *For any PPT adversary A which outputs a set $S \subset \{0,1\}^n$, for any $x \notin S$,*

$$\Pr[F_k(x) = F_{k\{S\}}(x) : k \leftarrow \mathsf{Sample}(1^\lambda), k\{S\} \leftarrow \mathsf{Puncture}(k, S)] = 1.$$

- **Pseudorandomness at punctured points:** *For any PPT adversaries A_1, A_2, define a set S and state* state *as $(S, \mathsf{state}) \leftarrow A_1(1^\lambda)$. Then*

$$\Pr[A_2(\mathsf{state}, S, k\{S\}, F_k(S))] - \Pr[A_2(\mathsf{state}, S, k\{S\}, U_{|S| \cdot m(\lambda)})] < \mathsf{negl}(\lambda),$$

where $F_k(S)$ denotes concatenated PRF values on inputs from S, i.e. $F_k(S) = \{F_k(x_i) : x_i \in S\}$.

The GGM PRF [GGM84] satisfies this definition.

---

[4] We omit the random coins and state of Sim.

Statistically injective puncturable PRFs. Such PRFs are injective with overwhelming probability over the choice of a key. Sahai and Waters [SW14] show that if F is a puncturable PRF where the output length is large enough compared to the input length, and h is 2-universal hash function, then $F'_{k,h} = F_k(x) \oplus h(x)$ is a statistically injective puncturable PRF.

Extracting puncturable PRFs. Such PRFs have a property of a strong extractor: even when a full key is known, the output of the PRF is statistically close to uniform, as long as there is enough min-entropy in the input. Sahai and Waters [SW14] showed that if the input length is large enough compared to the output length, then such PRF can be constructed from any puncturable PRF F as $F'_{k,h} = h(F_k(x))$, where h is 2-universal hash function.

3   Augmented Explainability Compiler

In this section we describe a variant of an explainability compiler of [DKR15]. This compiler is used in our construction of NCE, as discussed in the introduction.

Roughly speaking, explainability compiler modifies a randomized program such that it becomes possible, for those who know faking keys, to create fake randomness r_f which is consistent with a given input-output pair. Explainability techniques were first introduced by Sahai and Waters [SW14] as a method to obtain deniability for encryption (there they were called "a hidden sparse trigger meachanism"). Later Dachman-Soled, Katz and Rao [DKR15] generalized these ideas and introduced a notion of explainability compiler.

We modify this primitive for our construction and call it an "augmented explainability compiler". Before giving a formal definition, we briefly describe it here. Such a compiler Comp takes a randomized algorithm $\mathsf{Alg}(input; u)$ with input *input* and randomness u and outputs three new algorithms:

- Comp.Rerand(Alg) outputs a new algorithm $\mathsf{Alg}'(input; r)$ which is a "rerandomized" version of Alg. Namely, Alg' first creates fresh randomness u using a PRF on input $(input, r)$ and then runs Alg with this fresh randomness u.
- Comp.Trapdoor(Alg) outputs a new algorithm $\mathsf{Alg}''(input; r)$ which is a "trapdoored" version of Alg', which allows to create randomness consistent with a given *output*: namely, before executing Alg', Alg'' interprets its randomness r as a ciphertext and tries to decrypt it using internal key. If it succeeds and r encrypts an instruction to output *output*, then Alg'' complies. Otherwise it runs Alg'.
- Comp.Explain(Alg) outputs a new algorithm $\mathsf{Explain}(input, output)$ which outputs randomness for algorithm Alg'' consistent with given *input* and *output*. It uses an internal key to encrypt an instruction to output *output* on an input *input*, and outputs the resulting ciphertext.

Definition 5. *An augmented explainability compiler* Comp *is an algorithm which takes as input algorithm* Alg *and randomness and outputs programs* $\mathsf{P_{Rerand}}, \mathsf{P_{Trapdoor}}, \mathsf{P_{Explain}}$, *such that the following properties hold:*

– **Indistinguishability of the source of the output.** *For any input it holds that*
$$\{(\mathsf{P}_{\mathsf{Trapdoor}}, \mathsf{P}_{\mathsf{Explain}}, output) : r \leftarrow U, output \leftarrow \mathsf{Alg}(input; r)\}$$
and
$$\{(\mathsf{P}_{\mathsf{Trapdoor}}, \mathsf{P}_{\mathsf{Explain}}, output) : r \leftarrow U, output \leftarrow \mathsf{P}_{\mathsf{Trapdoor}}(input; r)\}$$
are indistinguishable.

– **Indistinguishability of programs with and without a trapdoor.** $\mathsf{P}_{\mathsf{Rerand}}$ *and* $\mathsf{P}_{\mathsf{Trapdoor}}$ *are indistinguishable.*

– **Selective explainability.** *Any PPT adversary has only negligible advantage in winning the following game:*
  1. *Adv fixes an input* $input^*$;
  2. *The challenger runs* $\mathsf{P}_{\mathsf{Rerand}}, \mathsf{P}_{\mathsf{Trapdoor}}, \mathsf{P}_{\mathsf{Explain}} \leftarrow \mathsf{Comp}(\mathsf{Alg})$;
  3. *It chooses random* r^* *and computes* $output^* \leftarrow \mathsf{P}_{\mathsf{Trapdoor}}(input^*; r^*)$;
  4. *It chooses random* ρ *and computes fake* $r_f^* \leftarrow \mathsf{P}_{\mathsf{Explain}}(input^*, output^*; \rho)$
  5. *It chooses random bit* b. *If* $b = 0$, *it shows* $(\mathsf{P}_{\mathsf{Trapdoor}}, \mathsf{P}_{\mathsf{Explain}}, output^*, r^*)$, *else it shows* $(\mathsf{P}_{\mathsf{Trapdoor}}, \mathsf{P}_{\mathsf{Explain}}, output^*, r_f^*)$
  6. *Adv outputs* b' *and wins if* $b = b'$.

Differences between [DKR15] *compiler and our construction.* For the reader familiar with [SW14, DKR15], we briefly describe the differences.

First, we split compiling procedure into two parts: the first part, rerandomization, adds a PRF to the program Alg, such that the program uses randomness $F(input, r)$ instead of r. The second part adds a trapdoor branch to rerandomized program. This is done for a cleaner presentation of the proof.

Second, we slightly change a trapdoor branch activation mechanism: together with faking keys we hardwire an image S of a pseudorandom generator into the program. Whenever this program decrypts fake r, it follows instructions inside r only if these instructions contain a correct preimage of S. This trick allows us to first change S to random and then to indistinguishably "delete" the whole trapdoor branch from the program. Thus it becomes possible to use a program without a trapdoor in the protocol (and only in the proof change it to its trapdoor version), which is crucial for achieving perfect correctness.

Construction. Our explainability compiler is described in Fig. 1. It takes as input algorithm Alg and randomness r. It uses r to sample keys Ext (for an extracting PRF), f (for a special encryption scheme called puncturable deterministic encryption, or PDE[SW14]), as well as random s, and randomness for $i\mathcal{O}$. It sets $S = \mathsf{prg}(s)$. Then it obfuscates programs $\mathsf{Rerand}[\mathsf{Alg}, \mathsf{Ext}]$, $\mathsf{Trapdoor}[\mathsf{Alg}, \mathsf{Ext}, f, S]$, and $\mathsf{Explain}[f, s]$. It outputs these programs.

Theorem 1. *Algorithm* Comp *presented in Fig. 1 is an augmented explainability compiler.*

The proof of security can be found in the full version of the paper.

Explainability compiler Comp.

Program Comp(Alg; r)
Inputs: Algorithm Alg, randomness r

1. Use r to sample keys Ext (for extracting PRF), f (for PDE), as well as random s and randomness for $i\mathcal{O}$ r_1, r_2, r_3.
2. Set $S \leftarrow \mathsf{prg}(s)$;
3. Set $\mathsf{P_{Rerand}} \leftarrow i\mathcal{O}(\mathsf{Rerand}[\mathsf{Alg}, \mathsf{Ext}]; r_1)$, $\mathsf{P}_v \leftarrow i\mathcal{O}(\mathsf{Trapdoor}[\mathsf{Alg}, \mathsf{Ext}, f, S]; r_2)$, and $\mathsf{P_{Explain}} \leftarrow i\mathcal{O}(\mathsf{Explain}[f, s]; r_3)$.
4. Output $\mathsf{P_{Rerand}}$, $\mathsf{P_{Trapdoor}}$, and $\mathsf{P_{Explain}}$.

Program Rerand

Program Rerand[Alg, Ext](**input**; **r**)
Constants: underlying randomized algorithm Alg($input$; u), a key for extracting prf Ext
Inputs: input $input$, randomness r

1. Create randomness $u \leftarrow \mathsf{F_{Ext}}(input, r)$;
2. output $output \leftarrow \mathsf{Alg}(input; u)$

Program Trapdoor

Program Trapdoor[Alg, **f**, Ext, S](**input**; **r**)
Constants: underlying randomized algorithm Alg($input$; u), a faking key f, a key for extracting prf Ext, prg image S
Inputs: input $input$, randomness r

1. **Trapdoor branch:**
   (a) decode out $\leftarrow \mathsf{PDE.Dec}_f(r)$; if out $= \perp$ then goto normal branch;
   (b) parse out as $(input', output', s', \bar{\rho})$. If $input = input'$ and $\mathsf{prg}(s') = S$ then output $output'$ and halt, else goto normal branch;
2. **Normal branch:**
   (a) Create randomness $u \leftarrow \mathsf{F_{Ext}}(input, r)$;
   (b) output $output \leftarrow \mathsf{Alg}(input; u)$

Program Explain

Program Explain[**f**, **s**](**input**, **output**; ρ)
Constants: a faking key f, secret s, which is a prg preimage of S
Inputs: input and output ($input$, $output$), randomness ρ

1. output $r \leftarrow \mathsf{PDE.Enc}_f(input, output, s, \mathsf{prg}(\rho))$

Fig. 1. Explainability compiler and programs used.

4 Optimal-Rate Non-committing Encryption in the CRS Model

In this section we show how to construct a fully non-committing encryption with rate $1+o(1)$. A crucial part of our protocol is the underlying seNCE scheme with short ciphertexts, which we will transform into a full NCE in Sect. 4.2.

4.1 Same-Public-Key Non-committing Encryption with Erasures

In this section we present our construction of the same-public-key non-committing encryption with erasures (seNCE for short) (defined in Sect. 2, Definition 3), which is a building block in our construction of a full fledged NCE.

The seNCE Protocol:

Inputs: sender's message m

- **Round 1.** The receiver chooses randomness r_{Gen} and generates keys $(\mathsf{P_{Enc}}, \mathsf{P_{Dec}}) \leftarrow \mathsf{Gen}(r_{\mathsf{Gen}})$. It sends $\mathsf{P_{Enc}}$ to the sender and erases r_{Gen}.
- **Round 2.** The sender chooses randomness r_{Enc} and generates a ciphertext $c \leftarrow \mathsf{P_{Enc}}(m; r_{\mathsf{Enc}})$. It sends c to the receiver and erases r_{Enc}.
- The receiver decrypts $m' \leftarrow \mathsf{P_{Dec}}(c)$ and outputs m'.

Program Gen(r)

Inputs: randomness r which consists of three parts $r = (r_1, r_2, r_3)$

(a) Set $k \leftarrow r_1$ and generate $\mathsf{P_{Enc}} \leftarrow i\mathcal{O}(\mathsf{Enc}[k]; r_2)$ and $\mathsf{P_{Dec}} \leftarrow i\mathcal{O}(\mathsf{Dec}[k]; r_3)$.

(b) Output $(\mathsf{P_{Enc}}, \mathsf{P_{Dec}})$

Program Enc[k](m, r)   // hardcoded PRF key k

Inputs: message m, randomness r

Program Size: this program is padded to be of the maximum size of Enc and Enc:1

(a) Set $c_1 \leftarrow \mathsf{prg}(r)$ and $c_2 \leftarrow \mathsf{F}_k(c_1) \oplus m$.

(b) output $c = (c_1, c_2)$

Program Dec[k](c)   // hardcoded PRF key k

Inputs: ciphertext c consisting of two parts (c_1, c_2)

Program Size: this program is padded to be of the maximum size of Dec and SimDec.

(a) Output $\mathsf{F}_k(c_1) \oplus c_2$.

Fig. 2. seNCE protocol

Inspired by Sahai and Waters [SW14] way of converting a secret key encryption scheme into a public-key encryption, we set our public key to be an obfuscated encryption algorithm $pk = i\mathcal{O}(\mathsf{Enc}[k])$ (see Fig. 2). To allow the simulator to generate a fake secret key, we apply the same trick to the secret key: we set the secret key to be an obfuscated decryption algorithm with hardcoded PRF key, namely $sk = i\mathcal{O}(\mathsf{Dec}[k])$. In other words, the seNCE protocol proceeds as follows: the receiver generates the obfuscated programs pk, sk and then erases generation randomness, including the key k. Then it sends pk to the sender; the sender encrypts its message m, erases his encryption randomness, and sends back the resulting ciphertext c, which the receiver decrypts with sk. We present the detailed description of the seNCE protocol in Fig. 2.

Theorem 2. *The scheme given on Fig. 2 is the same-public-key non-committing encryption scheme with erasures, assuming indistinguishability obfuscation for circuits and one way functions. In addition, it has ciphertexts (the second message in the protocol) of size* $\text{poly}(\lambda) + |m|$. *The protocol is also perfectly correct.*

Proof. We show that the scheme from Fig. 2. is a seNCE and has short ciphertexts.

Perfect correctness. The underlying secret key encryption scheme is perfectly correct, since $\text{Dec}(\text{Enc}(m, r)) = F_k(c_1) \oplus (F_k(c_1) \oplus m) = m$. Due to perfect correctness of $i\mathcal{O}$, our seNCE protocol is also perfectly correct.

Security with erasures: We need to show that real and simulated pk, c, sk are indistinguishable, even when the adversary can choose m adaptively after seeing pk.

1. **Real experiment.** In this experiment P_{Enc} and P_{Dec} are generated honestly using Gen, c^* is a ciphertext encrypting m^* with randomness r^*, i.e. $c_1^* = \text{prg}(r^*)$, $c_2^* = F_k(c_1^*) \oplus m^*$.
2. **Hybrid 1.** In this experiment c_1^* is generated at random instead of $\text{prg}(r^*)$. Indistinguishability from the previous hybrid follows by security of the PRG.
3. **Hybrid 2.** In this experiment we puncture key k in both programs Enc and Dec, more specifically, we obfuscate programs $P_{\text{Enc}} = i\mathcal{O}(\text{Enc:1}[k\{c_1^*\}])$, $P_{\text{Dec}} = i\mathcal{O}(\text{SimDec}[k\{c_1^*\}, c^*, m^*])$. We claim that functionality of these programs is the same as that of Enc and Dec:
   Indeed, in Enc:1 (defined in Fig. 4), c_1^* is random and thus with high probability it is outside the image of the PRG; therefore no input r results in evaluating F at the punctured point c_1^*, and we can puncture safely. In SimDec (defined in Fig. 3), if $c_1 \neq c_1^*$, then the program behaves exactly like the original one (i.e. computes $F_k(c_1) \oplus c_2$); if $c_1 = c_1^*$, then SimDec outputs $c_2^* \oplus c_2 \oplus m = (F_k(c_1^*) \oplus m) \oplus c_2 \oplus m = F_k(c_1^*) \oplus c_2$, which is exactly what Dec outputs when $c_1 = c_1^*$. Note that c_1^* is random (and thus independent of m), therefore $pk = \text{Enc:1}[k\{c_1^*\}]$ can be generated *before* the message m^* is fixed. Indistinguishability from the previous hybrid follows by the security of $i\mathcal{O}$.
4. **Hybrid 3.** In this hybrid we switch c_2^* from $F_k(c_1^*) \oplus m^*$ to random. This hybrid relies on the indistinguishability between punctured value $F_k(c_1^*)$ and a truly random value, even given a punctured key $k\{c_1^*\}$.
   Indeed, to reconstruct this hybrid, first choose random c_1^* and get $k\{c_1^*\}$ and val^* (which is either random or $F_k(c_1^*)$) from the PPRF challenger. Show obfuscated $\text{Enc} : 1[k\{c_1^*\}]$ as a public key. When the adversary fixes message m^*, set $c_2^* = val^* \oplus m^*$ and upon corruption show obfuscated

SimDec$[k\{c_1^*\}, c^*, m^*]$. If val^* is truly random, then $c_2^* = val^* \oplus m^*$ is distributed uniformly and thus we are in hybrid 3. If val^* is the actual PRF value, then $c_2^* = F_k(c_1^*) \oplus m^*$ and we are in hybrid 2.

Indistinguishability holds by security of a punctured PRF.

5. **Hybrid 4 (Simulation).** In this hybrid we unpuncture the key k in both programs and show $\mathsf{P_{Enc}} \leftarrow i\mathcal{O}(\mathsf{Enc}[k])$, $\mathsf{P_{Dec}} \leftarrow i\mathcal{O}(\mathsf{SimDec}[k, c^*, m^*])$.

   This is without changing the functionality of the programs: Indeed, in Enc no random input r results in $\mathsf{prg}(r) = c_1^*$, thus we can remove the puncturing. In Dec:1 due to preceding "if" no input c causes evaluation of $F_{k\{c_1^*\}}$, thus we can unpuncture it as well.

   The indistinguishability from the previous hybrid follows by the security if the $i\mathcal{O}$.

We observe that the last hybrid is indeed the simulation experiment described in Fig. 3: c^* is a simulated ciphertext since c_1^* is random, $c_2^* = \mathsf{F}_k(c_1^*)$, $\mathsf{P_{Enc}}$ is honestly generated, and $\mathsf{P_{Dec}}$ is a simulated key SimDec$[k, c^*, m^*]$, which decrypts c^* to m^*. Thus, we have shown that this scheme is non-committing with erasures.

The same public key. Both real generation algorithm Gen and the simulator on randomness $r_{\mathsf{Gen}} = (r_1, r_2, r_3)$ produce exactly the same public key $pk = i\mathcal{O}(Enc[r_1]; r_2)$.

Simulation:

(a) Generate a simulated public key $\mathsf{P_{Enc}}$ as follows: choose a random PRF key k and randomness r, set
$\mathsf{P_{Enc}} \leftarrow i\mathcal{O}(\mathsf{Enc}[k]; r)$.

(b) Generate a simulated ciphertext $c^* = (c_1^*, c_2^*)$ for random c_1^*, c_2^*.

(c) Generate a simulated receiver's internal state $\mathsf{P_{Dec}}$ for message m^* as follows:
$\mathsf{P_{Dec}} \leftarrow i\mathcal{O}(\mathsf{SimDec}[k, c^*, m^*])$.

Program SimDec[k, c*, m*](c)   // hardcoded PRF key k, dummy ciphertext c^*, challenge message m^*

Inputs: ciphertext c which consists of two parts (c_1, c_2)

Program Size: this program is padded to be of the maximum size of Dec and SimDec.

(a) If $c_1 = c_1^*$, output $c_2^* \oplus c_2 \oplus m^*$. Otherwise, output $\mathsf{F}_k(c_1) \oplus c_2$.

Fig. 3. seNCE simulator.

Program Enc:1[k$\{c_1^*\}$](m, r)   // hardcoded punctured PRF key $k\{c_1^*\}$

Inputs: message m, randomness r

Program Size: this program is padded to be of the maximum size of Enc and Enc:1

(a) Set $c_1 \leftarrow \mathsf{prg}(r)$ and $c_2 \leftarrow \mathsf{F}_{k\{c_1^*\}}(c_1) \oplus m$.

(b) Output $c = (c_1, c_2)$.

Fig. 4. Program Enc:1 used in the proof.

Efficiency: Our PRG should be length-doubling to ensure that its image is sparse. Thus $|c_1| = 2\lambda$, and $|c_2| = |m|$. Thus the size of our ciphertext is $2\lambda + |m|$.

4.2   From seNCE to Full NCE

In this section we show how to transform any seNCE (for instance, seNCE constructed in Sect. 4.1) into full non-committing encryption in the CRS model. We start with a brief overview of the construction:

Construction. Our CRS contains algorithms Comp.Rerand(GenEnc) and Comp.Rerand (GenDec) which share master secret key MSK. Both programs can internally generate the parameters for the underlying seNCE scheme using their MSK and then output an encryption program or a decryption key. More specifically, GenDec takes a random input, produces generation token t and then uses this token and MSK to generate randomness r_{NCE} for seNCE.Gen. Then the program samples seNCE keys pk, sk from r_{NCE}. It outputs the token t and the generated decryption key sk for a seNCE scheme. The receiver keeps sk for itself and sends the token t to the sender.

GenEnc, given a token t, can produce (the same) pair (pk, sk) and outputs an algorithm Comp.Rerand(Enc_{pk}), which has pk hardwired. This algorithm takes a message m and outputs its encryption c, which the sender sends back to the receiver. Then receiver decrypts it using sk.

We present our full NCE protocol and its building block functions GenEnc, GenDec, Enc in Fig. 5.

Theorem 3. *Assuming* Comp *is a secure explainability compiler,* seNCE *is a secure same-public-key NCE with erasures with a ciphertext size* $O(\text{poly}(\lambda)) + m$, *and assuming one-way functions, the described construction is a constant-rate non-committing public key encryption scheme in a common reference string model. Assuming perfect correctness of underlying* seNCE *and* Comp, *our NCE scheme is also perfectly correct.*

4.3   Proof of the Theorem 3

Proof. We first show correctness of the scheme. Next we present a simulator and argue that the scheme is secure. Finally we argue that the scheme is constant-rate.

Correctness. The presented scheme is perfectly correct, as long as the underlying seNCE and Comp are perfectly correct: First, due to perfect correctness of Comp, using compiled versions Comp.Rerand(GenEnc), Comp.Rerand(GenDec), Comp.Rerand(Enc) is as good as the using original programs. Next, both the sender and receiver generate public and secret seNCE keys as $(pk, sk) \leftarrow$ seNCE.Gen($F_{MSK}(t)$). The sender also generates c, which is an encryption of m under pk, which is decrypted under sk by receiver. Thus the scheme is as correct as the underlying seNCE scheme is.

The NCE Protocol

CRS: programs P_{GenEnc} and P_{GenDec}, where
$P_{\mathsf{GenEnc}} = \mathsf{Comp.Rerand}(\mathsf{GenEnc}[\mathsf{MSK}])$, $P_{\mathsf{GenDec}} = \mathsf{Comp.Rerand}(\mathsf{GenDec}[\mathsf{MSK}])$
Inputs: sender's message m

1. **Round 1.** The receiver chooses randomness r_{GenDec} and generates $(t, sk) \leftarrow P_{\mathsf{GenDec}}(r_{\mathsf{GenDec}})$. The receiver sends t to the sender.
2. **Round 2.** The sender chooses randomness r_{GenEnc} and generates $P_{\mathsf{Enc}} \leftarrow P_{\mathsf{GenEnc}}(t; r_{\mathsf{GenEnc}})$. Then the sender chooses randomness r_{Enc} and encrypts $c \leftarrow P_{\mathsf{Enc}}(m; r_{\mathsf{Enc}})$. The sender sends c to the receiver.
3. The receiver decrypts $m' \leftarrow \mathsf{seNCE.Dec}_{sk}(c)$ and outputs m'

Program GenEnc[MSK]$(t; e)$
// hardcoded values: master key MSK
Inputs: token t, randomness e

1. Set the randomness $r_{\mathsf{NCE}} \leftarrow F_{\mathsf{MSK}}(t)$, run $(pk, sk) \leftarrow \mathsf{seNCE.Gen}(r_{\mathsf{NCE}})$.
2. Generate $P_{\mathsf{Enc}} \leftarrow \mathsf{Comp.Rerand}(\mathsf{Enc}[pk]; e)$.
3. Output the program P_{Enc}.

Program Enc[pk](m; u)
// hardcoded values: seNCE public key pk
Inputs: message m, randomness u

1. Output ciphertext $c \leftarrow \mathsf{seNCE.Enc}_{pk}(m; u)$.

Program GenDec[MSK](w)
// hardcoded values: master key MSK
Inputs: randomness w

1. Generate token $t \leftarrow \mathsf{prg}(w)$.
2. Set the randomness $r_{\mathsf{NCE}} \leftarrow F_{\mathsf{MSK}}(t)$, run $pk, sk \leftarrow \mathsf{seNCE.Gen}(r_{\mathsf{NCE}})$.
3. Output (t, sk).

Fig. 5. The NCE protocol.

Since the protocol for seNCE which we give in Sect. 4.1 has perfect correctness, the overall NCE scheme, when instantiated with our seNCE protocol from Sect. 4.1, also achieves perfect correctness.

Description of the Simulator. In this subsection we first explain which variables the adversary sees and then describe our simulator.

The view of the adversary. The view of the adversary consists of the CRS (programs P^*_{GenEnc}, P^*_{GenDec}), as well as the communication and the internal states of several protocol instances. Namely, for each protocol instance the adversary sees the following variables:

1. The first protocol message t^*, after which the adversary assigns an input m for this protocol instance;
2. The second protocol message c^*;
3. The sender internal state $r^*_{\mathsf{Enc}}, r^*_{\mathsf{GenEnc}}$;
4. The receiver internal states r^*_{GenDec}.

Other values, such as $\mathsf{P}^*_{\mathsf{Enc}}$ and sk^*, can be obtained by the adversary by running programs in the CRS: $\mathsf{P}^*_{\mathsf{Enc}} \leftarrow \mathsf{P}^*_{\mathsf{GenEnc}}(t^*, r^*_{\mathsf{GenEnc}})$, $(sk^*, t^*) \leftarrow \mathsf{P}^*_{\mathsf{GenDec}}(r^*_{\mathsf{GenDec}})$.

Simulation. The simulator runs the compiler Comp on programs Enc, GenEnc, GenDec and sets a simulated CRS to be a description of programs Comp.Trapdoor(GenEnc), Comp.Trapdoor(GenDec). The difference from the real-world CRS is that these simulated programs have a trapdoor branch inside them, which allows the simulator to produce randomness such that a program outputs a desired output on this randomness.

The simulator keeps programs $\mathsf{Expl}_{\mathsf{Enc}} = $ Comp.Explain(Enc), $\mathsf{Expl}_{\mathsf{GenEnc}} = $ Comp.Explain(GenEnc), $\mathsf{Expl}_{\mathsf{GenDec}} = $ Comp.Explain(GenDec) for later use.

- **CRS generation.** The simulator sets the CRS to be a description of programs $\mathsf{P}^*_{\mathsf{GenEnc}} = $ Comp.Trapdoor(GenEnc), $\mathsf{P}^*_{\mathsf{GenDec}} = $ Comp.Trapdoor(GenDec).

  Next the simulator responds to requests of the adversary. The adversary can interactively ask to setup a new execution of the protocol (where the input m can be chosen based on what the adversary has already learn from other executions), or ask to deliver messages or corrupt parties in protocols which are already being executed. Below we describe what our simulator does in each case:

- **Simulation of the first message.** If the receiver is already corrupted, then the simulator generates the first message by choosing random r^*_{GenDec} and running $(t^*, sk^*) \leftarrow \mathsf{P}^*_{\mathsf{GenDec}}(r^*_{\mathsf{GenDec}})$. Otherwise the simulator chooses random t^* as the first message.

- **Simulation of the second message.** If either the sender or the receiver is already corrupted, then the simulator learns m and therefore can generate the second message honestly. If neither the sender nor the receiver in this execution are corrupted by this moment, the simulator runs $(pk^*_f, c^*_f) \leftarrow $ seNCE.Sim($\mathsf{F}_{\mathsf{MSK}}(t^*)$) and gives c^*_f to the adversary as the second message.

- **Simulation of the sender internal state.** If either the sender or the receiver had been corrupted before the second message was sent, then the simulator has generated the second message honestly and can thus show true sender randomness.

  Otherwise it first generates a program $\mathsf{P}^*_{\mathsf{Enc}} = $ Comp.Trapdoor(Enc[pk^*_f]) with simulated pk^*_f hardwired inside. Next it encodes m^*, c^*_f into sender encryption randomness, i.e. sets $r^*_{f,\mathsf{Enc}} \leftarrow \mathsf{Expl}_{\mathsf{Enc}}(m^*, c^*_f; \rho_3)$ for random ρ_3; so that $\mathsf{P}^*_{\mathsf{Enc}}$ on input $(m^*, r^*_{f,\mathsf{Enc}})$ outputs c^*_f.

Simulation

1. **Generate a CRS:**
   (a) Choose a PRF key MSK and randomness ρ_{GenEnc}, ρ_{GenDec}
   (b) Compute $\mathsf{P}^*_{\mathsf{GenEnc}} \leftarrow \mathsf{Comp.Trapdoor}(\mathsf{GenEnc}[\mathsf{MSK}]; \rho_{\mathsf{GenEnc}})$,
       $\mathsf{Expl}_{\mathsf{GenEnc}} \leftarrow \mathsf{Comp.Explain}(\mathsf{GenEnc}[\mathsf{MSK}]; \rho_{\mathsf{GenEnc}})$.
   (c) Compute $\mathsf{P}^*_{\mathsf{GenDec}} \leftarrow \mathsf{Comp.Trapdoor}(\mathsf{GenDec}[\mathsf{MSK}]; \rho_{\mathsf{GenDec}})$,
       $\mathsf{Expl}_{\mathsf{GenDec}} \leftarrow \mathsf{Comp.Explain}(\mathsf{GenDec}[\mathsf{MSK}]; \rho_{\mathsf{GenDec}})$.
   (d) Set the CRS to be $(\mathsf{P}^*_{\mathsf{GenEnc}}, \mathsf{P}^*_{\mathsf{GenDec}})$. Publish the CRS.
2. **Generate communications in the protocol:**
   (a) Choose a random t^* and generate $r^*_{\mathsf{NCE}} \leftarrow F_{\mathsf{MSK}}(t^*)$. Show t^* as the first message in the protocol.
   (b) Run the seNCE simulator to simulate the public key $\mathsf{pk}^*_f \leftarrow \mathsf{seNCE.Sim}(r^*_{\mathsf{NCE}})$
   (c) After the adversary decides on the message m^*, run the seNCE simulator $c^*_f \leftarrow \mathsf{seNCE.Sim}(r^*_{\mathsf{NCE}})$ to generate a simulated ciphertext.
   (d) Show c^*_f as the second message in the protocol.
3. **Generate parties' internal state consistent with message m^* and communications:**
   (a) Run the seNCE simulator to create a simulated secret key: $sk^*_f \leftarrow \mathsf{seNCE.Sim}(st, m^*)$
   (b) Set the receiver's randomness $r^*_{f,\mathsf{GenDec}} \leftarrow \mathsf{Expl}_{\mathsf{GenDec}}(t^*, sk^*_f; \rho_1)$ for random ρ_1.
   (c) Compute $\mathsf{P}^*_{Enc} \leftarrow \mathsf{Comp.Trapdoor}(\mathsf{Enc}[\mathsf{pk}_f]; \rho_{\mathsf{Enc}})$,
       $\mathsf{Expl}_{\mathsf{Enc}} \leftarrow \mathsf{Comp.Explain}(\mathsf{Enc}[\mathsf{pk}_f]; \rho_{\mathsf{Enc}})$.
   (d) Set the sender's generation randomness $r^*_{f,\mathsf{GenEnc}} \leftarrow \mathsf{Expl}_{\mathsf{GenEnc}}(t^*, \mathsf{P}^*_{Enc}; \rho_2)$ for random ρ_2.
   (e) Set the sender's encryption randomness $r^*_{f,\mathsf{Enc}} \leftarrow \mathsf{Expl}_{\mathsf{Enc}}(m^*, c^*_f; \rho_3)$ for random ρ_3.
   (f) Show $(r^*_{f,\mathsf{GenEnc}}, r^*_{f,\mathsf{Enc}})$ as the sender's internal state and $r^*_{f,\mathsf{GenDec}}$ as receiver's internal state.

Fig. 6. Simulation.

Finally, it encodes $\mathsf{P}^*_{\mathsf{Enc}}$ into $r^*_{f,\mathsf{GenEnc}}$, i.e. sets the sender's generation randomness $r^*_{f,\mathsf{GenEnc}} \leftarrow \mathsf{Expl}_{\mathsf{GenEnc}}(t^*, \mathsf{P}^*_{\mathsf{Enc}}; \rho_2)$ for random ρ_2, so that $\mathsf{P}^*_{\mathsf{GenEnc}}$ outputs $\mathsf{P}^*_{\mathsf{Enc}}$ on input $(t^*, r^*_{f,\mathsf{GenEnc}})$.

The pair $(r^*_{f,\mathsf{GenEnc}}, r^*_{f,\mathsf{Enc}})$ is set to be the sender internal state.

- **Simulation of the receiver internal state.** If the corruption happens before the first message is sent, then the simulator has generated the first message honestly and thus can show true receiver internal state.

  If corruption happens after the first message, but before the second, then the first message t^* was generated at random. In this case the simulator computes $sk^* \leftarrow \mathsf{seNCE.Gen}(F_{\mathsf{MSK}}(t^*))$. It encodes (t^*, sk^*) into receiver randomness, i.e. sets $r^*_{f,\mathsf{GenDec}} \leftarrow \mathsf{Expl}_{\mathsf{GenDec}}(t^*, sk^*; \rho_1)$ for random ρ_1, so that P^*_{GenDec} on input $r^*_{f,\mathsf{GenDec}}$ outputs (t^*, sk^*).

  If corruption happens after the second message, then the simulator runs seNCE simulator and gets fake secret key sk^*_f which decrypts dummy c^*_f to m^*, chosen by the adversary. Next it encodes (t^*, sk^*_f) into receiver randomness, i.e. sets $r^*_{f,\mathsf{GenDec}} \leftarrow \mathsf{Expl}_{\mathsf{GenDec}}(t^*, sk^*_f; \rho_1)$ for random ρ_1, so that P^*_{GenDec} on input $r^*_{f,\mathsf{GenDec}}$ outputs (t^*, sk^*_f).

Note that simulation of each protocol instance is independent of simulation of other protocol instances (except for the fact that they share the same CRS).

Therefore in order to keep the description of the simulator simple enough, in Fig. 6 we present a detailed description of the simulator for a single execution only; it can be trivially generalized to a multiple-execution case according to what is written above. In addition, the simulator is presented for a difficult case, i.e. when nobody is corrupted by the time the ciphertext is sent, and therefore the simulator has to present a dummy c and later open it to a correct m.

Next we outline the intuition for the security proof and after that provide the detailed description of the hybrids.

Overview of the Analysis of the Simulator. Before presenting hybrids, let us give a roadmap of the proof: Starting from the real execution, we first switch the programs in the CRS: instead of compiling them with Comp.Rerand, we compile them using Comp.Trapdoor; in other words, we add trapdoor branches to the programs in the CRS, in order to allow creating fake randomness which explains a given output. Next we change what the simulator shows as internal states of the parties: instead of showing their real randomness, the simulator shows fake randomness (which explains outputs of programs from a real execution, i.e. this randomness explains honestly generated sk^*, c^*, and P^*_{Enc}). Our next step is to puncture the key $\mathsf{MSK}\{t^*\}$ in both CRS programs. This allows us to switch seNCE generation randomness r^*_{NCE} from $\mathsf{F}_{\mathsf{MSK}}(t^*)$ to a random value; this means that seNCE parameters (pk^*, sk^*) are now freshly generated and do not depend on the rest of an experiment anymore. Therefore we can use security of seNCE and switch seNCE values (pk^*, c^*, sk^*) from real to simulated (in particular, the simulator hardwires these simulated c^*_f, sk^*_f into fake randomness, instead of hardwiring real-execution c^*, sk^*). Next we undo previous hybrids: we set r^*_{NCE} as the result of $\mathsf{F}_{\mathsf{MSK}}(t^*)$, and then unpuncture $\mathsf{MSK}\{t^*\}$ in both CRS programs.

In security proof we will be using the following properties of explainability compiler Comp for any algorithm Alg:

1. *Indistinguishability of programs with and without trapdoor branch;*
   Comp.Rerand(Alg) $\approx$ Comp.Trapdoor(Alg).
2. *Indistinguishability of explanations:*
   given programs $\mathsf{P}(x; r) = $ Comp.Trapdoor(Alg) and Expl $= $ Comp.Explain(Alg), it is impossible to distinguish between real randomness and input (x, r) and fake randomness $(x, r_f \leftarrow $ Comp.Expl$(x, \mathsf{P}(x, r))$. In particular, evaluating $\mathsf{P}(x; r_f)$ results in $\mathsf{P}(x, r)$, with the only difference that the computation $\mathsf{P}(x; r_f)$ uses the trapdoor branch, which is however undetectable.
3. *Indistinguishability of source of the output:*
   given programs $\mathsf{P}(x; r) = $ Comp.Trapdoor(Alg) and Expl $= $ Comp.Explain(Alg), it is infeasible to tell whether a given output y was obtained by running original program Alg or its compiled version Comp.Trapdoor(Alg).

We omit proofs of these statements, since they generally follow the proofs of explainability compiler in previous works [DKR15], with some adaptations for our scenario (such as added indistinguishablity of programs with and without a trapdoor). Formal proofs appear in the full version of our paper.

We now briefly describe each hybrid. The full description with detailed security reductions is given in the full version of the paper.

- **Hybrid 0.** We start with a real execution of the protocol.
- **Hybrids 1a–1b.** We change how we generate the CRS programs: instead of obtaining them as Comp.Rerand(GenEnc) and Comp.Rerand(GenDec), we generate them as Comp.Trapdoor(GenEnc) and Comp.Trapdoor(GenDec). Security holds by indistinguishability of programs with and without trapdoor branch. **Next for every execution i, in which the receiver is corrupted between the first and the second messages, we run hybrids $2_i - 3_i$.**
  - **Hybrid 2_i.** Instead of showing the real randomness r^*_{GenDec}, the simulator shows fake $r^*_{f,\mathsf{GenDec}}$, which encodes t^*, sk^*. These experiments are indistinguishable because of the indistinguishability of explanation: indeed, $\mathsf{P}^*_{\mathsf{GenDec}}$ on both inputs r^*_{GenDec} and $r^*_{f,\mathsf{GenDec}}$ outputs t^*, sk^*, therefore true randomness r^*_{GenDec} is indistinguishable from randomness $r^*_{f,\mathsf{GenDec}}$, which explains the output t^*, sk^*.
    Note that since there is no non-random input to our program $\mathsf{P}_{\mathsf{GenDec}}$, it is enough to use the selective indistinguishability of explanation.
  - **Hybrid 3_i.** We set $t^* = \mathsf{prg}(w^*)$ for random w^* and then compute sk^* as $(pk^*, sk^*) \leftarrow \mathsf{seNCE.Gen}(F_{\mathsf{MSK}}(t^*))$. In other words, we compute (t^*, sk^*) as the result of running GenDec instead of Comp.Trapdoor(GenDec). Indistinguishability holds by indistinguishability of the source of the output for the compiler Comp and program GenDec.
  - **Hybrid 4_i.** Finally we set t^* to be randomly chosen instead of being the result of $\mathsf{prg}(w^*)$. Security follows from security of the prg.
    This is the simulation for the case when the receiver is corrupted between the first and the second message.

  For every execution i, in which both corruptions happen after the second message is sent, we run hybrids $2_i - 5h_i$.
- **Hybrid 2_i.** Instead of showing the real randomness r^*_{GenEnc}, the simulator shows fake $r^*_{f,\mathsf{GenEnc}}$, which encodes $t^*, \mathsf{P}^*_{\mathsf{Enc}}$. These experiments are indistinguishable because of the indistinguishability of explanation: indeed, $\mathsf{P}^*_{\mathsf{GenEnc}}$ on both inputs $t^*, r^*_{\mathsf{GenEnc}}$ and $t^*, r^*_{f,\mathsf{GenEnc}}$ outputs $\mathsf{P}^*_{\mathsf{Enc}}$, and by the theorem true randomness r^*_{GenEnc} is indistinguishable from fake randomness $r^*_{f,\mathsf{GenEnc}}$ which explains $\mathsf{P}^*_{\mathsf{Enc}}$ on input t^*. Note that non-random input to our program $\mathsf{P}_{\mathsf{GenEnc}}$ is t^*, obtained by running $t^* \leftarrow P^*_{\mathsf{GenDec}}(r^*_{\mathsf{GenDec}})$ for random r^*_{GenDec}, i.e., it can be generated before a CRS is shown to the adversary. Thus it is enough to use the selective indistinguishability of explanation.
- **Hybrid 3_i.** In the next step instead of showing the real r^*_{GenDec}, the simulator shows fake $r^*_{f,\mathsf{GenDec}}$, which encodes t^*, sk^*. These experiments are indistinguishable because of the indistinguishability of explanation: indeed, $\mathsf{P}^*_{\mathsf{GenDec}}$ on both inputs r^*_{GenDec} and $r^*_{f,\mathsf{GenDec}}$ outputs t^*, sk^*, therefore true randomness r^*_{GenDec} is indistinguishable from randomness $r^*_{f,\mathsf{GenDec}}$, which explains the output t^*, sk^* on empty non-random input.
  Note that since there is no non-random input to our program $\mathsf{P}_{\mathsf{GenDec}}$, it is enough to use the selective indistinguishability of explanation.

- **Step 4$_i$.** Next global step is to switch random r^*_{Enc} to fake $r^*_{f,\mathsf{Enc}}$ which encodes (m^*, c^*). We do this in several steps:
  - **Hybrid 4a$_i$.** We obtain t^*, sk^* by running GenDec on random w^* instead of running $\mathsf{P}^*_{\mathsf{GenDec}} = \mathsf{Comp.Trapdoor}(\mathsf{GenDec})$ on r^*_{GenDec}. Indistinguishability holds by indistinguishability of a source of the output for programs GenDec and Comp. Trapdoor(GenDec).
  - **Hybrid 4b$_i$.** We choose t^* at random instead of choosing it as $\mathsf{prg}(w^*)$ for random w^*. (pk^*, sk^*) are then obtained from $r^*_{\mathsf{NCE}} = \mathsf{F}_{\mathsf{MSK}}(t^*)$. Indistinguishability holds by security of a prg.
  - **Hybrid 4c$_i$.** We generate $\mathsf{P}^*_{\mathsf{Enc}}$ by running GenEnc on t^* and random e^*, instead of running $\mathsf{P}^*_{\mathsf{GenEnc}} = \mathsf{Comp.Trapdoor}(\mathsf{GenEnc})$ on $(t^*, r^*_{\mathsf{GenEnc}})$. Security holds by indistinguishability of source of the output for programs GenEnc and Comp.Trapdoor(GenEnc).
  - **Hybrid 4d$_i$.** We generate the program $\mathsf{P}^*_{\mathsf{Enc}} \leftarrow \mathsf{Comp.Trapdoor}(\mathsf{Enc}[pk^*])$ instead of $\mathsf{Comp.Rerand}(\mathsf{Enc}[pk^*])$. Security holds by indistinguishability of programs with and without trapdoor branch for program Enc.
  - **Hybrid 4e$_i$.** In this step we finally change r^*_{Enc} to $r^*_{f,\mathsf{Enc}}$ as follows: we first create a CRS and give it to the adversary. Then we generate random t^* and show t^* to the adversary as the first message in the protocol. Next the adversary fixes an input m^*. Then we generate pk^*, sk^* as $\mathsf{seNCE.Gen}(F_{\mathsf{MSK}}(t^*))$ and give $\mathsf{Enc}() = \mathsf{seNCE.Enc}_{pk^*}()$ to the explainability challenger as the underlying program. The challenger chooses random e^*, runs $\mathsf{Comp}(\mathsf{Enc}; e^*)$ and gives us either $(r^*_{\mathsf{Enc}}, m^*, c^*, \mathsf{P}^*_{\mathsf{Enc}})$ or $(r^*_{f,\mathsf{Enc}}, m^*, c^*, \mathsf{P}^*_{\mathsf{Enc}})$, where r^*_{Enc} is random, $\mathsf{P}^*_{\mathsf{Enc}} = \mathsf{Comp.Trapdoor}(\mathsf{Enc}; e^*)$, $c^* = \mathsf{P}^*_{\mathsf{Enc}}(m^*; r^*_{\mathsf{Enc}})$, and $r^*_{f,\mathsf{Enc}}$ encodes m^*, c^*. We show the given c^* as the second message in the protocol. Once asked to open the internal state, we present the given r^*_{Enc} or $r^*_{f,\mathsf{Enc}}$, generate r^*_{GenEnc} explaining the given $\mathsf{P}^*_{\mathsf{Enc}}$, and generate r^*_{GenDec} explaining (t^*, sk^*). We can rely on the selective indistinguishability of explanation for program $\mathsf{Comp.Trapdoor}(\mathsf{Enc})$ since at the moment when we need to see the challenge in explanation game (i.e., when we need to show c^* to the adversary), $\mathsf{P}^*_{\mathsf{Enc}}$'s input m^* is already fixed.
- **Step 5$_i$.** Our next global step is to change the underlying seNCE values to simulated. We proceed in several steps:
  - **Hybrids 5a$_i$–5b$_i$.** We puncture MSK at t^*. In $\mathsf{P}^*_{\mathsf{GenDec}}$ we can puncture immediately, since due to the sparseness of the length-doubling prg, t^* lies outside of the prg image and therefore $\mathsf{F}_{\mathsf{MSK}}$ is never called at t^*. In $\mathsf{P}^*_{\mathsf{GenEnc}}$ we hardwire pk^* and use it whenever $t = t^*$; otherwise, we use the punctured key $\mathsf{MSK}\{t^*\}$ to generate r_{NCE} and then sample pk.
  - **Hybrid 5c$_i$.** Once $\mathsf{MSK}\{t^*\}$ is punctured, we can choose the generation randomness for underlying seNCE scheme r^*_{NCE} at random.
  - **Hybrids 5d$_i$.** We generate c^* as a result of running Enc on m^* and random u^* instead of running $\mathsf{P}^*_{\mathsf{Enc}} = \mathsf{Comp.Trapdoor}(\mathsf{Enc})$ on $(m^*; r^*_{\mathsf{Enc}})$. We rely on indistinguishability of the source of the output for program Enc.

- **Hybrid 5e$_i$.** Next we switch the seNCE values from real to simulated: namely, c_f^* is now simulated and sk_f^* is now a simulated key decrypting c_f^* to m^*. We rely on the security of the underlying seNCE. Here we crucially use the fact that in the underlying NCE scheme pk^* is shown *before* the adversary chooses a message, since we hardwire this pk^* into the CRS (in $\mathsf{P}^*_{\mathsf{GenEnc}}$).
- **Hybrid 5f$_i$.** We switch back r^*_{NCE} to be the result of $\mathsf{F}_{\mathsf{MSK}}(t^*)$.
- **Hybrid 5g$_i$-5h$_i$.** We unpuncture $\mathsf{MSK}\{t^*\}$ in $\mathsf{P}^*_{\mathsf{GenEnc}}$ and $\mathsf{P}^*_{\mathsf{GenDec}}$ and remove the hardwired pk^* from $\mathsf{P}^*_{\mathsf{GenEnc}}$. To remove hardwired pk^*, we crucially use the fact that pk^*, although simulated, is the same as real pk^*, generated from randomness $\mathsf{F}_{\mathsf{MSK}}(t^*)$, which is guaranteed by the same-public-key property of seNCE.

This concludes the overview of hybrids. For the detailed description of the hybrids with security reductions, see the full version of the paper.

Sizes in Our Construction. Our construction has a lot of size dependencies. We present a size diagram on Fig. 7, assuming our implementation of explainability compiler based on $i\mathcal{O}$ and puncturable deterministic encryption (PDE). There all sizes are grouped in "complexity classes". Here we outline several main dependencies:

- if a fake randomness has values encoded, it should be longer than these values, but not much longer. Namely, if underlying encoded message has size l, then the size of the plaintext for PDE (which consists of encoded message, secret s and $\mathsf{prg}(\rho)$) is $l + 3\lambda$, and the size of PDE ciphertext should be at least 4 times bigger (the latter is because explainability compiler uses statistically injective PRF). Therefore randomness and encoded value are in the same "complexity class".
- if a key is punctured on some input, its size is at least $\lambda|input|$.
- size of an obfuscated program is significantly larger than the size of the original program (polynomial in original size s and security parameter λ).

Note that all dependencies in the graph are due to the "hardwired values", i.e. due to the fact that some values should be hardcoded into programs, or messages should be encrypted into ciphertexts. In particular, the same length restrictions remain even when succinct $i\mathcal{O}$ for TM or RAM [CHJV15, CH15, KLW15] is used.

Note that the dependency graph is acyclic, and variables which we actually send over the channel - t and c - are in the very top of the graph. This means that we can set length of t and m to be a security parameter, and then set lengths of other variables as large as needed by following edges in dependency graph.

Acknowledgements. We thank anonymous ASIACRYPT reviewers for pointing out that explainability compiler can be used in a black box manner, which greatly simplified the presentation of the results.

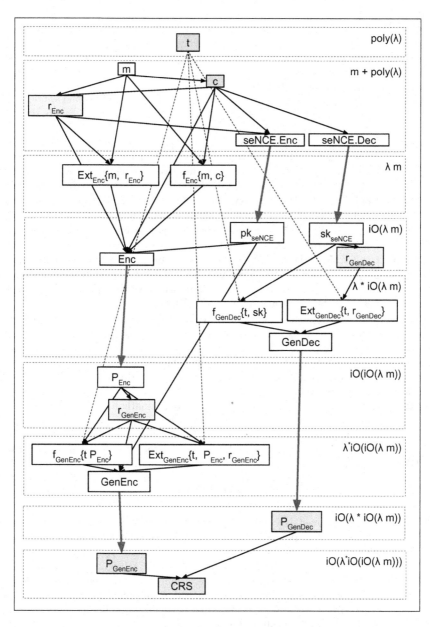

Fig. 7. Size dependency graph between different variables, when underlying seNCE is instantiated with our construction from Sect. 4.1. Notation: $i\mathcal{O}(s)$ for size s means the resulting size of an obfuscated program of the initial approximate size s. Dependencies due to obfuscation are drawn as fat blue arrows. Green boxes mark CRS, yellow boxes mark randomness used for extracting PRF, and blue denotes variables which are sent in the protocol. Arrows for t are shown dashed for easier tracking. Red dashed rectangles with size in the top right corner denote a "size group", e.g. any variable inside $i\mathcal{O}(\lambda m)$ box is as large as an obfuscated program of initial size λm. (Color figure online)

References

[Bea97] Beaver, D.: Plug and play encryption. In: Kaliski, B.S. (ed.) CRYPTO 1997. LNCS, vol. 1294, pp. 75–89. Springer, Heidelberg (1997). https://doi.org/10.1007/BFb0052228

[BGL+15] Bitansky, N., Garg, S., Lin, H., Pass, R., Telang, S.: Succinct randomized encodings and their applications. In: Rubinfeld, R. (ed.) Symposium on the Theory of Computing (STOC) (2015)

[BH92] Beaver, D., Haber, S.: Cryptographic protocols provably secure against dynamic adversaries. In: Rueppel, R.A. (ed.) EUROCRYPT 1992. LNCS, vol. 658, pp. 307–323. Springer, Heidelberg (1993). https://doi.org/10.1007/3-540-47555-9_26

[CDMW09] Choi, S.G., Dachman-Soled, D., Malkin, T., Wee, H.: Simple, black-box constructions of adaptively secure protocols. In: Reingold, O. (ed.) TCC 2009. LNCS, vol. 5444, pp. 387–402. Springer, Heidelberg (2009). https://doi.org/10.1007/978-3-642-00457-5_23

[CFGN96] Canetti, R., Friege, U., Goldreich, O., Naor, M.: Adaptively secure multiparty computation. In: Proceedings of the Twenty-Eighth Annual ACM Symposium on the Theory of Computing, Philadelphia, Pennsylvania, USA, 22–24 May 1996, pp. 639–648 (1996)

[CGP15] Canetti, R., Goldwasser, S., Poburinnaya, O.: Adaptively secure two-party computation from indistinguishability obfuscation. In: Dodis, Y., Nielsen, J.B. (eds.) TCC 2015. LNCS, vol. 9015, pp. 557–585. Springer, Heidelberg (2015). https://doi.org/10.1007/978-3-662-46497-7_22

[CH15] Canetti, R., Holmgren, J.: Fully succinct garbled RAM. IACR Cryptology ePrint Archive 2015, 388 (2015)

[CHJV15] Canetti, R., Holmgren, J., Jain, A., Vaikuntanathan, V.: Succinct garbling and indistinguishability obfuscation for RAM programs. In: Proceedings of the Forty-Seventh Annual ACM on Symposium on Theory of Computing, STOC 2015, Portland, OR, USA, 14–17 June 2015, pp. 429–437 (2015)

[CHK05] Canetti, R., Halevi, S., Katz, J.: Adaptively-secure, non-interactive public-key encryption. In: Kilian, J. (ed.) TCC 2005. LNCS, vol. 3378, pp. 150–168. Springer, Heidelberg (2005). https://doi.org/10.1007/978-3-540-30576-7_9

[DKR15] Dachman-Soled, D., Katz, J., Rao, V.: Adaptively secure, universally composable, multiparty computation in constant rounds. In: Dodis, Y., Nielsen, J.B. (eds.) TCC 2015. LNCS, vol. 9015, pp. 586–613. Springer, Heidelberg (2015). https://doi.org/10.1007/978-3-662-46497-7_23

[DN00] Damgård, I., Nielsen, J.B.: Improved non-committing encryption schemes based on a general complexity assumption. In: Bellare, M. (ed.) CRYPTO 2000. LNCS, vol. 1880, pp. 432–450. Springer, Heidelberg (2000). https://doi.org/10.1007/3-540-44598-6_27

[DNRS99] Dwork, C., Naor, M., Reingold, O., Stockmeyer, L.J.: Magic functions. In: 40th Annual Symposium on Foundations of Computer Science, FOCS 1999, New York, NY, USA, 17–18 October 1999, pp. 523–534 (1999)

[GGM84] Goldreich, O., Goldwasser, S., Micali, S.: How to construct random functions (extended abstract). In: 25th Annual Symposium on Foundations of Computer Science, West Palm Beach, Florida, USA, 24–26 October 1984, pp. 464–479 (1984)

[GP15]   Garg, S., Polychroniadou, A.: Two-round adaptively secure MPC from indistinguishability obfuscation. In: Dodis, Y., Nielsen, J.B. (eds.) TCC 2015. LNCS, vol. 9015, pp. 614–637. Springer, Heidelberg (2015). https://doi.org/10.1007/978-3-662-46497-7_24

[HLP15] Hazay, C., Lindell, Y., Patra, A.: Adaptively secure computation with partial erasures. In: Proceedings of the 2015 ACM Symposium on Principles of Distributed Computing, PODC 2015, Donostia-San Sebastián, Spain, 21–23 July 2015, pp. 291–300 (2015)

[HOR15] Hemenway, B., Ostrovsky, R., Rosen, A.: Non-committing encryption from Φ-hiding. In: Proceedings of Theory of Cryptography - 12th Theory of Cryptography Conference, TCC 2015, Warsaw, Poland, 23–25 March 2015, Part I, pp. 591–608 (2015)

[HORR16] Hemenway, B., Ostrovsky, R., Richelson, S., Rosen, A.: Adaptive security with quasi-optimal rate. In: Kushilevitz, E., Malkin, T. (eds.) TCC 2016. LNCS, vol. 9562, pp. 525–541. Springer, Heidelberg (2016). https://doi.org/10.1007/978-3-662-49096-9_22

[HP14]   Hazay, C., Patra, A.: One-sided adaptively secure two-party computation. In: Lindell, Y. (ed.) TCC 2014. LNCS, vol. 8349, pp. 368–393. Springer, Heidelberg (2014). https://doi.org/10.1007/978-3-642-54242-8_16

[JL00]   Jarecki, S., Lysyanskaya, A.: Adaptively secure threshold cryptography: introducing concurrency, removing erasures. In: Preneel, B. (ed.) EUROCRYPT 2000. LNCS, vol. 1807, pp. 221–242. Springer, Heidelberg (2000). https://doi.org/10.1007/3-540-45539-6_16

[KLW15] Koppula, V., Lewko, A.B., Waters, B.: Indistinguishability obfuscation for turing machines with unbounded memory. In: Rubinfeld, R. (ed.) Symposium on the Theory of Computing (STOC) (2015)

[KSW14] Khurana, D., Sahai, A., Waters, B.: How to generate and use universal parameters. IACR Cryptology ePrint Archive 2014, 507 (2014)

[Nie02] Nielsen, J.B.: Separating random oracle proofs from complexity theoretic proofs: the non-committing encryption case. In: Yung, M. (ed.) CRYPTO 2002. LNCS, vol. 2442, pp. 111–126. Springer, Heidelberg (2002). https://doi.org/10.1007/3-540-45708-9_8

[SW14]   Sahai, A., Waters, B.: How to use indistinguishability obfuscation: deniable encryption, and more. In: Symposium on Theory of Computing, STOC 2014, New York, NY, USA, 31 May–03 June 2014, pp. 475–484 (2014)

Preventing CLT Attacks on Obfuscation with Linear Overhead

Rex Fernando$^{(\boxtimes)}$, Peter M.R. Rasmussen, and Amit Sahai

Center for Encrypted Functionalities, UCLA, Los Angeles, USA
{rex,rasmussen,sahai}@cs.ucla.edu

Abstract. We describe a defense against zeroizing attacks on indistinguishability obfuscation (iO) over the CLT13 multilinear map construction that only causes an additive blowup in the size of the branching program. This defense even applies to the most recent extension of the attack by Coron *et al.* (PKC 2017), under which a much larger class of branching programs is vulnerable. To accomplish this, we describe an attack model for the current attacks on iO over CLT13 by distilling an essential common component of all previous attacks.

This essential component is a constraint on the function being obfuscated. We say the function needs to be *input partionable*, meaning that the bits of the function's input can be partitioned into somewhat independent subsets. This notion constitutes an attack model which we show captures all known attacks on obfuscation over CLT13. We find a way to thwart these attacks by requiring a "stamp" to be added to the input of every function. The stamp is a function of the original input and eliminates the possibility of finding the independent subsets of the input necessary for a zeroizing attack. We give three different constructions of such "stamping functions" and prove formally that they each prevent any input partition.

We also give details on how to instantiate one of the three functions efficiently in order to secure any branching program against this type of attack. The technique presented alters any branching program obfuscated over CLT13 to be secure against zeroizing attacks with only an additive blowup of the size of the branching program that is linear in the input size and security parameter.

We can also apply our defense to a recent extension of annihilation attacks by Chen *et al.* (EUROCRYPT 2017) on obfuscation over the GGH13 multilinear map construction.

Keywords: Obfuscation · Zeroizing attacks

Research supported in part from a DARPA/ARL SAFEWARE award, NSF Frontier Award 1413955, NSF grants 1619348, 1228984, 1136174, and 1065276, BSF grant 2012378, a Xerox Faculty Research Award, a Google Faculty Research Award, an equipment grant from Intel, and an Okawa Foundation Research Grant. This material is based upon work supported by the Defense Advanced Research Projects Agency through the ARL under Contract W911NF-15-C-0205. The views expressed are those of the authors and do not reflect the official policy or position of the Department of Defense, the National Science Foundation, or the U.S. Government.

T. Takagi and T. Peyrin (Eds.): ASIACRYPT 2017, Part III, LNCS 10626, pp. 242–271, 2017.
https://doi.org/10.1007/978-3-319-70700-6_9

1 Introduction

Indistinguishability obfuscation (iO) has so far relied on multilinear maps for instantiation (e.g. [GGH+13b]) and viable candidates for such are sparse. On top of that, the few that exist [GGH13a, CLT13, GGH15] have all been shown to suffer from significant vulnerabilities. However, not all attacks against these multilinear maps can be applied to iO. The very particular structure that most iO candidates induce puts numerous constraints on the way the encoded values can be combined, thus often not allowing the flexible treatment needed to mount an attack. Attacks on iO schemes have nonetheless been found for obfuscation of increasingly general families of functions.

In this paper, we focus on the Coron-Lepoint-Tibouchi (CLT13) multilinear maps [CLT13]. The known attacks over CLT13 are called *zeroizing attacks* [CHL+15, CGH+15, CLLT17]. To be carried out, they require multiple zero encodings that are the result of multiplications of elements that satisfy a certain structure. Since obfuscations of matrix branching programs only produce zeroes when evaluated in a very specific manner, setting up such a zeroizing attack on an obfuscated branching program is rather non-trivial.

Because of this, the first paper applying zeroizing attacks over CLT13 to iO only showed how to apply the attack to very simple branching programs [CGH+15], and attacking more realistic targets seemed out of reach of this technique. However, a very recent work by Coron et al. [CLLT17] introduced a simple method that can transform a much larger class of branching programs into ones that have this very specific structure. As such, zeroizing attacks appear much more threatening to the security of iO over CLT13 than previously thought.

1.1 The Story So Far: Branching Programs and Zeroizing Attacks

This section will serve as a light introduction to the terminology and concepts at work in this paper.

Branching Programs. The "traditional" method of obfuscation works with matrix branching programs that encode boolean functions. A (single input) matrix branching program BP is specified by the following information. It has a length ℓ, input size n, input function $\mathsf{inp}: [\ell] \to [n]$, square matrices $\{A_{i,b}\}_{i \in [\ell], b \in \{0,1\}}$, and bookend vectors A_0 and $A_{\ell+1}$. An evaluation of the branching program BP on input $x \in \{0,1\}^n$ is carried out by computing

$$A_0 \times \prod_{i=1}^{\ell} A_{i,x_{\mathsf{inp}(i)}} \times A_{\ell+1}.$$

If the product is zero then $\mathsf{BP}(x) = 0$ and otherwise, $\mathsf{BP}(x) = 1$.

Multilinear Maps and Obfuscating Branching Programs. Current instantiations of iO are based on graded multilinear maps [GGH+13b, BGK+14]. This primitive allows values $\{a_i\}$ to be encoded to $\{[a_i]\}$ in such a manner that they are hidden. The multilinear map allows evaluation of a very restricted class of polynomials over these encoded values. Moreover, evaluating a polynomial, p, over the encodings in this way should only yield one bit of information: whether or not the result, $p(\{a_i\})$, is zero.

To obfuscate a branching program, we first randomize the matrices and then encode the entries of the matrix using a multilinear map. (See for example [BGK+14] for details on how the matrices are randomized.) The hope is that the multilinear map will allow evaluations of the branching program but will not allow other malicious polynomials over the encodings that would violate indistinguishability. In fact, Barak *et al.* [BGK+14] show that if zero testing the result of evaluations over the multilinear map truly only reveals whether or not the evaluation is zero and does not leak anything else then this scheme is provably secure.

Zeroizing Attacks on Obfuscated Branching Programs. Unfortunately, the assumption that zero-testing does not leak any information is unrealistic. In particular, the zeroizing attacks over the CLT13 multilinear map work by exploiting the information leaked during successful zero tests to obtain the secret parameters.

Before discussing the zeroizing attacks on iO, we first consider how the attacks work over raw encodings. Each of the known zeroizing attacks requires sets of encodings that satisfy a certain structure. We describe here the structure required for the simplest attack. This version of the attack was first presented in [CHL+15]. Namely, to attack a CLT instance of dimension n, an adversary needs three sets of encodings $\{B_i\}_{i \in [n]}$, $\{C_0, C_1\}$, and $\{D_j\}_{j \in [n]}$ such that for every $i, j \in [n]$ and $\sigma \in \{0, 1\}$, $B_i C_\sigma D_j$ is a top-level encoding of zero. In other words, we must be able to vary the choice of encoding in each set independently of the other choices and always produce an encoding of zero. If an adversary is able to obtain such sets, then the adversary is able to factor the modulus of the ciphertext ring, completely breaking the CLT instance. There are several variants of this attack, some requiring sets of vectors or matrices of encodings instead of plain encodings. But all have the requirement of obtaining three sets which we can combine in some way to achieve encodings of zero, and where we can vary the choice from each set independently of the others. We give more details about these attacks in Sect. 3.

We show (Theorem 2, Sect. 3) that all currently known zeroizing attacks over CLT13 to branching program obfuscation give rise to a constraint on the function being obfuscated, which we call an *input partition*, described below. In fact, Theorem 2 shows that this applies not only to all current zeroizing attacks over CLT13 but to a broader class of zeroizing attacks, as we discuss further below.

Input Partitioning. Let $f: \{0,1\}^n \to \{0,1\}$ be a function to be obfuscated. We say that there is an *input partitioning* of f if there exist sets $A \subseteq \{0,1\}^k, B \subseteq \{0,1\}^l, k+l=n$ and a permutation $\pi \in S_n$ such that $|A|, |B| > 1$ and for every $a \in A$ and $b \in B$, $f(\pi(a||b)) = 0$, where π acts on the bit-string $a||b$ by permuting its bits. In words, the function f can be input partitioned if the indices of the input can be partitioned into two sets such that varying the bits of the partitions independently within certain configurations will always yield zero as the output.

All current attacks on obfuscation over CLT13 require an input partition, and we will describe later how there is strong evidence that more generally any zeroizing attack will also require it. In fact, the current attacks need a partition into three parts to succeed, but since any input partition into three parts can be viewed as an input partition into two parts, we treat that more general case instead.

It is worth noting that zeroizing attacks, in fact, require a stronger condition on the branching program in order to succeed; the matrices of the obfuscated branching program must be organized in a specific way in relation to the three sets of inputs. But preventing an input partition necessarily prevents this stronger condition.

1.2 Our Contributions

Our aim in this paper is to provide a robust defense against the known classes of zeroizing attacks for iO over CLT13 and against potential future extensions of these attacks. Further, we want the defense to have a minimal impact on the efficiency of the obfuscated program. In this section, we describe how we achieve such a defense that only incurs an additive linear blowup in the multilinearity.[1]

Attack Model. Our defense is built on an attack model based on the input partitions described above. Previous authors [CGH+15, CLLT17] have considered the stronger condition mentioned at the end of the section on input partitioning above as a requirement for their attacks, but we are the first to consider the input partition of a boolean function as the basis of a formal attack model.

In Sect. 4.1 we define this attack model formally. Before this, we show in Sect. 3.3 that the model captures all current zeroizing attacks on obfuscation over CLT13. We also argue that the model broadly captures any new attack which uses the general strategy of these attacks.

There is a simple intuition behind why the attack model is sufficient for capturing any zeroizing attack on obfuscation over CLT13. Obfuscation schemes are designed so that the ways in which an adversary can construct encodings of zero are severely restricted. Intuitively, the sets of polynomials over the encodings that an adversary should be allowed to compute should only be honest evaluations of branching programs, or something very close to this. In fact, [BMSZ16]

---

[1] An important practical caveat is that we will work with the variant of CLT13 described in [GLW14], which avoids a vulnerability by increasing the dimension of the CLT13 instance. We give more details on this in Sect. 5.3.

prove that the standard obfuscation scheme used in this paper has a property very close to this intuition, namely that the only successful evaluations of polynomials of the encodings are given by linear combinations of honest evaluations of the obfuscated program. Recall that zeroizing attacks over CLT13 require three sets of encodings; the result of [BMSZ16] shows the adversary has very few degrees of freedom in constructing the three sets other than varying the inputs to the branching program.

Given this model, we construct a procedure which takes an input partitionable function $f\colon \{0,1\}^n \to \{0,1\}$ and produces a function g with the same functionality, but on which no input partitions exist. The existence of an input partition depends on which inputs cause g to output zero; note that a branching program is defined to output zero if the result of the multiplications of the matrices is zero and one if the result is any other value. So we will think of g as being a function $\{0,1\}^{n'} \to \{0,\perp\}$.

Input Stamping. The idea behind such a procedure is to append a "stamp" to the end of the input of the function f. The stamp is designed to not allow the input as a whole to be input partitioned. More specifically, we will construct a function $h\colon \{0,1\}^n \to \{0,1\}^m$ such that given any $f\colon \{0,1\}^n \to \{0,1\}$ we can construct a new program $g\colon \{0,1\}^{n+m} \to \{0,\perp\}$ from f such that $g(s)$ outputs 0 if and only if the input is of the form $s = x\|h(x)$ and $f(x) = 0$ and otherwise outputs $\perp$. Note that the original $\{0,1\}$-output of f is recoverable from g as long as the evaluation took place with the correct stamp appended to the input.

Our main theoretical result is to find a necessary and sufficient condition on h such that g cannot be input partitioned. If this is the case then we say h *secures* f. We state a sufficient condition below as Theorem 1. In Sect. 4.3 we restate Theorem 1 with both necessary and sufficient conditions after introducing some preliminaries which are required for the stronger version of the theorem.

Theorem 1 (Weakened). *Let* $h\colon \{0,1\}^n \to \{0,1\}^m$. *Let* $x_{1,1}, x_{1,2}, x_{2,1}, x_{2,2} \in \{0,1\}^n$ *be treated as integers. If whenever*

$$x_{1,1} - x_{1,2} = x_{2,1} - x_{2,2}$$
$$h(x_{1,1}) - h(x_{1,2}) = h(x_{2,1}) - h(x_{2,2})$$

it is the case that $x_{1,1} = x_{1,2}$ *or* $x_{1,1} = x_{2,1}$, *then* h *secures all functions* $f\colon \{0,1\}^n \to \{0,1\}$.

With this theorem, two questions arise: whether is feasible to construct such an h, and how efficiently we can construct the modified g to be. The second question is relevant with respect to the work in [AGIS14, BISW17] on improving the efficiency of obfuscation of branching programs. It is especially relevant to [BISW17] since this paper uses CLT13 to achieve a significant speedup factor from previous constructions. Thus, establishing the security of obfuscation over CLT13 with minimal overhead is pertinent.

With regards to efficiency, the size of the image of h becomes important. Using an h that has an output size m which is large relative to n will necessarily affect the efficiency of the resulting g. In Sect. 4.4 we explore the minimum value of m necessary in order for h to be secure. We show that m must be at least linear in terms of n.

Constructions. The first two instantiations we present for h address the question of the feasibility of constructing such a function. They are both number-theoretic and follow from the fact that Theorem 1 can be interpreted as a sort of nonlinearity property. We show that squaring and exponentiation modulo a large enough prime satisfy this property and thus secure any function f. We stress that we do not rely on any number-theoretic assumptions in the proofs that these functions satisfy Theorem 1.

The third instantiation is a combinatorial function and is motivated by the desire for efficiency. To that end, instead of defining a single function which is guaranteed to have the property specified above, we define a family of very simple functions where the probability of a random choice from this family is very likely to have the property.

We define h as follows. Let k and t be parameters set beforehand. For each $i \in [t]$ and $j \in [n]$, choose $\pi_{i,j,0}$ and $\pi_{i,j,1}$ at random from the set of permutations acting on k elements. For an input $x \in \{0,1\}^n$, define

$$h_i(x) = (\pi_{i,1,x_1} \circ \pi_{i,2,x_2} \circ \cdots \circ \pi_{i,n,x_n})(1).$$

Then $h(x) = h_1(x)||h_2(x)|| \ldots ||h_t(x)$.

We give a combinatorial probabilistic argument that with $k = O(1)$ and $t = O(n + \lambda)$ the choice of h secures all functions $f \colon \{0,1\}^n \to \{0,1\}$ with overwhelming probability as a function of λ.

Parallel Initialization. Since this construction for h is defined in terms of permutations and processes the input in the same way that a branching program does, constructing a branching program that computes such an h and subsequently modifying a branching program for f to create a branching program for the corresponding g is fairly straightforward. While this is already vastly more efficient than implementing the first two instantiations of h using a matrix branching program, running the functions h_i in sequence would cause a linear blowup in the size of the branching program. We do much better than this and achieve a constant blowup factor with the following trick. Unlike the GGH13 multilinear map construction, CLT13 allows for a composite ring size. We use this fact to evaluate all the h_i in parallel. This achieves a constant factor overhead in the branching program size.[2] (This technique was used, for example, in [AS17, GLSW15], albeit for different purposes.)

Perspectives. We remark that the attacks in [CLLT17] still do not apply to obfuscations of all branching programs. Specifically, if the branching program

---

[2] In fact, our actual overhead is additive and linear in terms of the input size of f, not the size of its branching program. See Sect. 5.3 for details.

is too long compared to the input size then there is a blowup associated with the transformation in [CLLT17] which becomes infeasible. Also, it is not yet known how to apply the attack to dual-input branching programs, due to a similar blowup in complexity. Although this is the case, it is definitely possible that future work will extend zeroizing attacks to longer branching programs and dual-input branching programs. Our defense hedges against these possible future attacks, because it defends against any attack which requires an input partition.

It is noteworthy to contrast this line of work with the recent attacks on iO over the GGH13 [GGH13a] multilinear maps construction. In [MSZ16] Miles et al. implement the first known such attacks, which they call *annihilating attacks*. A follow-up paper [GMM+16] gives a weakened multilinear map model and an obfuscation construction in this model which is safe against all annihilating attacks. We stress that the attacks over CLT13 are not related to these annihilating attacks, which are not known to work over CLT13. However, a recent paper attacking obfuscation over GGH13 [CGH17], in which the authors extend annihilating attacks to the original GGHRSW construction of iO, does use an input partition as part of their attack. They do this as a first step in order to recover a basis for the ideal which defines the plaintext space. Our defense applies to this step of their attack.

As a final note, our defense does not operate in a weak multilinear map model, in contrast to the one defined in [GMM+16]. We leave it as an important open question to develop such a weak multilinear map model for CLT13.

Organization. In Sect. 3 we discuss the attacks on obfuscation over CLT13 in more detail. In Sect. 4 we define formally what it means for a function to be input partitionable, and give a necessary and sufficient condition for any h to secure a function. We also give our lower bound on the size of the image of h. Finally, in Sect. 5 we define and prove the correctness of our instantiations of h.

2 Notation

We first introduce some notation for our exposition.

Definition 1. *For any positive integer $k \in \mathbb{N}$ we denote by $\mathbb{Z}_k$ the set $\mathbb{Z}/k\mathbb{Z}$.*

Definition 2. *Let $n \in \mathbb{N}$ be a positive integer and $\boldsymbol{v} \in \mathbb{N}^n$ a vector. We will denote by $\mathbb{Z}_{\boldsymbol{v}}$ the set*

$$\mathbb{Z}_{v_1} \times \mathbb{Z}_{v_2} \times \cdots \times \mathbb{Z}_{v_n}.$$

In this and following sections we will consider functions f that we want to secure and input stamping functions h. We will consider such functions as having domains and/or codomains of the form $\mathbb{Z}_{\boldsymbol{v}}$. Note that if we define $\boldsymbol{v} = (2, 2, \ldots, 2)$, then $\mathbb{Z}_{\boldsymbol{v}} = \{0, 1\}^n$, so this is a generalization of binary functions. We do this because in the instantiations section we will define an h which needs this generalized input format. Thus we state all theorems using this more general format to accommodate such instantiations.

Definition 3. *For a positive integer $n \in \mathbb{N}$ we denote by $[n]$ the set $\{1, 2, \ldots, n\}$.*

Definition 4. *For a positive integer $t \in \mathbb{N}$, we denote by S_t the set of permutations of the set $[t]$.*

Definition 5. *For two vectors or strings a and b let $a \| b$ denote their concatenation.*

3 Attack Model for Zeroizing Attacks on Obfuscation

In this section, we give a high-level overview of the new attack by Coron, Lee, Lepoint and Tibouchi [CLLT17]. We start by reviewing the older attacks in [CHL+15, CGH+15] which this attack is based on.

3.1 CLT13 Zeroizing Attacks

The basic idea behind all the zeroizing attacks over CLT13 is to exploit the specific structure of the zero-test of CLT13, which differs from the other multilinear map constructions. CLT13 works over a ring $\mathbb{Z}_{x_0} \cong \bigoplus_{i=1}^{k} \mathbb{Z}_{p_i}$ where each p_i is a large prime and zero-testing of an encoding works by multiplying the encoding by a zero-test parameter $\mathbf{p}_{zt}$, checking that the result is "small" in $\mathbb{Z}_{x_0}$. The simplest version of the attack adheres to the following outline: the adversary finds three sets of CLT13 encodings $\{a_i\}_{i \in [n]}$, $\{b_0, b_1\}$, and $\{c_j\}_{j \in [n]}$ such that for every $i, j \in [n]$ and $\sigma \in \{0, 1\}$, $a_i b_\sigma c_j$ is a top-level encoding of zero. Now, define the matrices $W_\sigma, \sigma \in \{0, 1\}$ by

$$W_\sigma[i, j] = \mathbf{p}_{zt} a_i b_\sigma c_j$$

where $W_\sigma[i, j]$ denotes the entry of W in the ith row and jth column. Since each W_σ is an outer product of elements of $\mathbb{Z}_{x_0}$ it can never be full-rank over $\mathbb{Z}_{x_0}$. A key point in the zeroizing attacks, however, is that the way the zero-test works, the matrix W_σ will be invertible over $\mathbb{Q}$ when the right encodings $\{a_i\}_{i \in [n]}$, $\{b_0, b_1\}$, and $\{c_j\}_{j \in [n]}$ are chosen. Write b_σ as its decomposition in $\mathbb{Z}_{x_0}$ by the Chinese Remainder Theorem, $b_\sigma = (b_{\sigma,1}, \ldots, b_{\sigma,k})$, where for each i, $b_{\sigma,i} \equiv b_\sigma \pmod{p_i}$. Then computing $W_0 W_1^{-1}$ and finding the eigenvalues (over the rationals), one obtains the rational ratios $b_{0,i}/b_{1,i}$ for each i. This leads to a factorization of x_0.

More general attacks allow that the encodings $\{a_i\}_{i \in [n]}$, $\{b_0, b_1\}$, and $\{c_j\}_{j \in [n]}$ be replaced with matrices of encodings $\{A_i\}_{i \in [n]}$, $\{B_0, B_1\}$, $\{C_j\}_{j \in [n]} \in \mathbb{Z}_{x_0}^{d \times d}$ together with vectors $s \in \mathbb{Z}_{x_0}^{d \times 1}, t \in \mathbb{Z}_{x_0}^{1 \times d}$. In this case it is required that for every $i, j \in [n], \sigma \in \{0, 1\}$, $s \times A_i \times B_\sigma \times C_j \times t$ is an encoding of zero and the matrices W_0, W_1 are defined by

$$W_\sigma[i, j] = \mathbf{p}_{zt}(s \times A_i \times B_\sigma \times C_j \times t).$$

We again require that W_1 is invertible. Write B_σ as its decomposition in $\mathbb{Z}_{x_0}$ by the Chinese Remainder Theorem $B_\sigma = (B_{\sigma,1}, \ldots, B_{\sigma,k})$ where each $B_{\sigma,i}$ is a matrix. Analyzing the characteristic polynomial of $W_0 W_1^{-1}$ now yields the characteristic polynomials of $B_{0,i} B_{1,i}^{-1}$ for every i. This again leads to a factorization of x_0.

We refer to [CGH+15] for further details.

3.2 Zeroizing Attacks on Obfuscation over CLT13

All known attacks on obfuscation over CLT13 have proceeded in a very similar manner to the method just described: since the evaluation of a branching program is a product of matrices over encodings in a multilinear map, they divide the steps of the branching program into three parts corresponding to the sets of encodings above such that these three parts can be varied independently of the others.

To see what we mean by this, let

$$M(x) = \widehat{M_0} \times \prod_{i=1}^{r} \widehat{M}_{i, x_{\mathsf{inp}(i)}} \times \widehat{M}_{r+1}, x \in \{0, 1\}^t$$

be an obfuscation of a matrix branching program. We try to find $B_x = \widehat{M_0} \times \prod_{i=1}^{a} \widehat{M}_{i, x_{\mathsf{inp}(i)}}$, $C_x = \prod_{i=a+1}^{b} \widehat{M}_{i, x_{\mathsf{inp}(i)}}$, and $D_x = \prod_{i=b+1}^{r} \widehat{M}_{i, x_{\mathsf{inp}(i)}} \times M_{i,r+1}$ such that we can partition the input bits as $\mathcal{B} \cup \mathcal{C} \cup \mathcal{D} = [t]$ and the value of B_x, C_x, and D_x rely only on $\mathcal{B}$, $\mathcal{C}$, and $\mathcal{D}$, respectively. Write $M(bcd)$ to mean the evaluation of M where b specifies the bits with positions in $\mathcal{B}$, and with c, d likewise with $\mathcal{C}, \mathcal{D}$. We further try to find sets of bit strings $\mathscr{B}, \mathscr{C}, \mathscr{D}$ where $\mathscr{B}, \mathscr{D}$ are large and $\mathscr{C}$ is at least of size two and for all $b \in \mathscr{B}, c \in \mathscr{C}, d \in \mathscr{D}$, $M(bcd) = 0$. If we can do all this, then the corresponding products of matrices form products of zero which decompose in a similar manner to the products of encodings used for the previous attack, and can similarly be used to mount an attack on the CLT13 instance used.

The problem with using this attack directly is that only the very simplest of branching programs can be decomposed in this way. The three pieces of the branching program B_x, C_x and D_x which correspond to the three sets of encodings in the zeroizing attack must be consecutively arranged in the branching program. In particular, this rules out attacks on any branching program that makes several passes over its input.

The modified attack in [CLLT17] overcomes this limitation with a matrix identity which allows a rearranging of the matrix product corresponding to a branching program execution. The identity is as follows:

$$\mathsf{vec}(A \cdot B \cdot C) = (C^T \otimes A) \cdot \mathsf{vec}(B),$$

where vec is the function sending a matrix $D = [d_1, d_2, \ldots, d_n]$ for column vectors d_i to the vector

$$\mathbf{vec}(D) = \begin{bmatrix} d_1 \\ d_2 \\ \vdots \\ d_n \end{bmatrix}.$$

Using this identity, [CLLT17] shows how to attack a branching program with input function $\mathbf{inp}(i) = \min(i, 2t + 1 - i)$ for $1 \leq i \leq 2t + 1$. Note that any branching program with this function does not satisfy the property above which was required for the earlier CLT13 attacks, since every input bit except for the t-th bit controls two nonconsecutive positions in the branching program. This input function was originally used in the branching programs which were attacked in [MSZ16]. We can write such a program evalution as

$$A(x) = B(x)C(x)D(x)C'(x)B'(x) \times p_{zt} \pmod{x_0}$$

where $B(x)$ and $B'(x)$ are both controlled by the same inputs, and likewise for $C(x)$ and $C'(x)$. [CLLT17] show that this can be rewritten as

$$(B'(x)^T \otimes B(x)) \times (C'(x)^T \otimes C(x)) \times \mathbf{vec}(D(x)) \times p_{zt} \pmod{x_0}$$

where now the three sets of inputs control consecutive pieces of the product. They then show how to use a modification of the original attack on this product, factoring x_0.

3.3   Attack Model

To defend against zeroizing attacks as described above and other attacks following a similar tangent, we distil an attack model. The model we will be employing is fairly general and introduces the notion of an input partition, which will be formally defined in the following section. We here bring an informal version of the definition. Note that this definition only deals with functions with binary strings as input.

Definition 6 (Input Partition – Informal). *Let $f \colon \{0,1\}^n \to \{0, \bot\}$ be a function. An* input partition *for f is a tuple*

$$(\sigma \in S_n; a_1, a_2 \in \{0,1\}^{n_1}; c_1, c_2 \in \{0,1\}^{n_2})$$

such that $n_1 + n_2 = n$ and for every choice of $i, j \in \{1,2\}$,

$$f(\sigma(a_i || c_j)) = 0,$$

where σ permutes the string $a_i || c_j$ by permuting its indices.

Intuitively, the above definition describes that the input bits of the function f can be partitioned into two parts that can be varied independently while f still evaluates to 0. In the following we will demonstrate that the currently known zeroizing attacks against obfuscation and the most natural derivatives of such all require an input partition of the obfuscated function to exist.

Now, suppose that we were given an obfuscated branching program

$$M(x) = \widehat{M_0} \times \prod_{i=1}^{r} \widehat{M_{i,x_{\mathsf{inp}(i)}}} \times \widehat{M_{r+1}}, x \in \{0,1\}^t$$

as above, which we wish to use the technique of zeroizing attacks against. We will think of the entries of the matrices $\widehat{M_0}, \{\widehat{M_{i,b}}\}_{i \in [r]}, \widehat{M_{r+1}}$ as indeterminates since they are given to us as encoded values. Thus, an encoding from one of the matrices will uniquely identify that matrix. We need to find the partitioned set of encodings which is necessary to perform a zeroizing attack using the encodings of the branching program. We elaborate as follows. We are given CLT13 encodings for each element of the matrices $\widehat{M_{i,b}}$ and the vectors $\widehat{M_0}$ and $\widehat{M_{r+1}}$. We need to combine these in some way to obtain encodings a_i^s, b_σ^s, c_j^s indexed by $i, j \in [n], \sigma \in \{0,1\}$, and $s \in I$ for an index set I such that for every choice of i, j, σ, $\sum_{s \in I} a_i^s b_\sigma^s c_j^s$ is an encoding of zero. Note that the matrix products of the more general attack can also be written out in this manner.

In every known zeroizing attack against an obfuscated program M, the zeroizing attack employed is such that $\sum_{s \in I} a_i^s b_\sigma^s c_j^s$ is an evaluation of M at a point $x^{i,\sigma,j} \in \{0,1\}^n$. Specifically, the index set I and the a_i^s, b_σ^s, c_j^s are such that the a_i^s, b_σ^s, c_j^s correspond to entries in the $M_{i,b}$ of the branching program and

$$M(x^{i,\sigma,j}) = \widehat{M_0} \times \prod_{l=1}^{r} \widehat{M_{l,x_{\mathsf{inp}(l)}^{i,\sigma,j}}} \times \widehat{M_{r+1}} = \sum_{s \in I} a_i^s b_\sigma^s c_j^s.$$

This is even the case with the new attacks in [CLLT17]; they show that each evaluation can be *conceptually* rewritten as some other product, but they still use plain evaluations of a branching program.

Setting each element in the zeroizing attack matrix to be a branching program evaluation is obviously the most natural way to try to apply zeroizing attacks to obfuscations of branching programs, and there is strong evidence that it is the only way. This is because [BMSZ16] show that given an obfuscated branching program encoded in a multilinear map the only way to obtain a top-level zero is by taking a linear combination of honest branching program evaluations over inputs that evaluate to zero. In other words, for any zeroizing attack $\{a_i^s\}_{i,s}, \{b_\sigma^s\}_{\sigma,s}, \{c_j^s\}_{j,s}, i, j \in [n], \sigma \in \{0,1\}, s \in I$, we have that for all i, j, σ,

$$\sum_{s \in I} \alpha_s a_i^s b_\sigma^s c_j^s = \sum_{x \in \chi_{i,\sigma,j}} \alpha_x M(x)$$

This is nearly the condition in Theorem 2. In all current attacks, in fact, only a single $M(x)$ is computed for each i, σ, j, and so we restrict our analysis to this setting.

We denote by f_x the zero-tested encoding $\mathbf{p}_{zt} \cdot M(x)$. In our analysis, we make the following simplifying assumptions about the encodings of a zeroizing attack. We stress that these assumptions are in line with every known zeroizing attack on obfuscation.

Assumption 1: Suppose that $\sum_{s \in I} a_i^s b_\sigma^s c_j^s = M(x)$ and define the set

$$\mathcal{E} = \{d \mid \exists i \text{ such that } d \text{ occurs in } \widehat{M}_{i, x_{\mathsf{inp}(i)}}\}.$$

We assume that for every $s \in I$, a_i^s, b_σ^s, and c_j^s is a product of encodings from $\mathcal{E}$.

Assumption 2: The sum $\sum_{s \in I} a_i^s b_\sigma^s c_j^s$ is equal to the evaluation of M on a single input x.

We now show that all zeroizing attacks which follow this pattern must yield an input partition in the function being obfuscated, which constitutes a strong justification for our model's usefulness.

Theorem 2. *Let a valid zeroizing attack against an obfuscated program M using the CLT13 encodings*

$$\{a_i^s\}_{i,s}, \{b_\sigma^s\}_{\sigma,s}, \{c_j^s\}_{j,s}, i, j \in [n], \sigma \in \{0, 1\}, s \in I$$

be given. Assume that the zeroizing attack satisfies Assumptions 1 and 2 and that for a family of inputs $\{x^{i,\sigma,j}\}$,

$$\sum_{s \in I} a_i^s b_\sigma^s c_j^s = M(x^{i,\sigma,j})$$

Then there must exist an input partition for the function encoded by the branching program.

Proof. For each encoding d of the obfuscated branching program M, define the *origin* of d_i as $\mathbf{or}(d) = (i, b)$ where $M_{i,b}$ is the matrix in which d occurs. For a product $\prod_{i=1}^k d_i$ of encodings, denote by $\mathbf{input}\left(\prod_{i=1}^k d_i\right)$ the set

$$\mathbf{input}\left(\prod_{i=1}^k d_i\right) = \bigcup_{i=1}^k \{(\mathbf{inp}(i), b) \mid \mathbf{or}(d_i) = (i, b)\}.$$

In words, considering $\prod_{i=1}^k d_i$ as part of an evaluation of the obfuscated branching program on some input x, $\mathbf{input}\left(\prod_{i=1}^k d_i\right)$ specifies which bits of x are determined to be what by the product. I.e. if $(i, b) \in \mathbf{input}\left(\prod_{i=1}^k d_i\right)$ then somewhere in the (partial) evaluation of the branching program, $x_i = b$ was used to determine the choice of matrix.

Now, assume for contradiction that we have a valid zeroizing attack as above for an obfuscated program M that has no input partition. We have that for every $i, j \in [n], \sigma \in \{0, 1\}$,

$$\sum_{s \in I} a_i^s b_\sigma^s c_j^s = M(x^{i,\sigma,j}).$$

Since every monomial $a_i^s b_\sigma^s c_j^s$ must produce a top-level encoding and because of Assumption 1,

$$\mathsf{input}(a_i^s) \cup \mathsf{input}(b_\sigma^s) \cup \mathsf{input}(c_j^s) = \{(k, x_k^{i,\sigma,j}) \mid k \in [t]\}$$

Define the sets

$$A = \{\mathsf{input}(a_i^s) \mid i \in [n], s \in I\},$$
$$B = \{\mathsf{input}(b_\sigma^s) \mid \sigma \in \{0, 1\}, s \in I\},$$
$$C = \{\mathsf{input}(c_j^s) \mid j \in [n], s \in I\}.$$

We say that A (resp. B, C) contains a *switch of input* if there exists $k \in [t]$ such that $\{(k, 0), (k, 1)\} \in A$ (resp. $\{(k, 0), (k, 1)\} \in B$, $\{(k, 0), (k, 1)\} \in C$). In that case we say that k is a bit position of A that *switches*. If A contains a switch of input, $\{(k, 0), (k, 1)\} \in A$, it means that there exists $i_1, i_2 \in [n]$ such that $x^{i_1,\sigma,j}$ and $x^{i_2,\sigma,j}$ differ in bit k. Note that since $\{(k, x_k^{i,\sigma,j}) \mid k \in [t]\}$ never contains a switch of input, it must be the case that $B \cap \{(k, 0), (k, 1)\} = \emptyset$ and $C \cap \{(k, 0), (k, 1)\} = \emptyset$ as for every s,

$$\mathsf{input}(b_\sigma^s) \cup \mathsf{input}(c_j^s) \subseteq \{(k, x_k^{i_1,\sigma,j}) \mid k \in [t]\}$$
$$\mathsf{input}(b_\sigma^s) \cup \mathsf{input}(c_j^s) \subseteq \{(k, x_k^{i_2,\sigma,j}) \mid k \in [t]\}$$

and the two sets on the right of the inclusions differ in bit k while the two sets on the left of the inclusions are the same.

Assume for contradiction that any two of A, B, C contain a switch of input and assume without loss of generality that it is A and B. Let $k_1, \ldots, k_{m_1}$ be the bit positions of A that switch and $l_1, \ldots, l_{m_2}$ be the bit positions of B that switch. Then by the above argument, there is no pair (k_s, b) contained in $B \cup C$ and no pair (l_s, b) contained in $A \cup C$. Let i_1, i_2 be given such that $x^{i_1,\sigma,j}$ and $x^{i_2,\sigma,j}$ differ in some bits and similarly let j_1, j_2 be such that x^{i,σ,j_1} and x^{i,σ,j_2} differ in some bits. Then there is an input partition given by the $x^{i_1,\sigma,j_1}, x^{i_1,\sigma,j_2}, x^{i_2,\sigma,j_1}, x^{i_2,\sigma,j_2}$. To see that this is an input partition, note that x^{i_1,σ,j_1} and x^{i_1,σ,j_2} differ in the same bit positions as x^{i_2,σ,j_1} and x^{i_2,σ,j_2} and that x^{i_1,σ,j_1} and x^{i_2,σ,j_1} differ in the same bit positions as x^{i_1,σ,j_2} and x^{i_2,σ,j_2}. In the first case, these bit positions are contained in the set $\{l_1, \ldots, l_{m_2}\}$ and in the second case in the set $\{k_1, \ldots, k_{m_1}\}$. This is a contradiction since we assumed that no input partition existed for the obfuscated function.

Suppose instead that there are no two of A, B, C that contain a switch of input. Recall from Sect. 3.1 that for a zeroizing attack to be successful, the matrix W_1 must be invertible and further, $W_0 W_1^{-1}$ must yield information about

b_σ^s encodings. If B does not contain an input switch then $x^{i,\sigma,j}$ does not depend on σ and for the zeroizing attack, we construct the matrices

$$W_0 = \mathbf{p}_{zt} \begin{pmatrix} M(x^{1,0,1}) & \cdots & M(x^{1,0,n}) \\ \vdots & & \vdots \\ M(x^{n,0,1}) & \cdots & M(x^{n,0,n}) \end{pmatrix} = \mathbf{p}_{zt} \begin{pmatrix} M(x^{1,1,1}) & \cdots & M(x^{1,1,n}) \\ \vdots & & \vdots \\ M(x^{n,1,1}) & \cdots & M(x^{n,1,n}) \end{pmatrix} = W_1.$$

Thus, $W_0 = W_1$ and we get nothing out of computing $W_0 W_1^{-1}$. If instead neither A nor C contain an input switch then $x^{i,\sigma,j}$ depends on neither i nor j. In that case we get the matrices

$$W_0 = \mathbf{p}_{zt} \begin{pmatrix} M(x^{1,0,1}) & \cdots & M(x^{1,0,n}) \\ \vdots & & \vdots \\ M(x^{n,0,1}) & \cdots & M(x^{n,0,n}) \end{pmatrix} = \begin{pmatrix} f_{x^{1,0,1}} & \cdots & f_{x^{1,0,1}} \\ \vdots & & \vdots \\ f_{x^{1,0,1}} & \cdots & f_{x^{1,0,1}} \end{pmatrix}$$

$$W_1 = \mathbf{p}_{zt} \begin{pmatrix} M(x^{1,1,1}) & \cdots & M(x^{1,1,n}) \\ \vdots & & \vdots \\ M(x^{n,1,1}) & \cdots & M(x^{n,1,n}) \end{pmatrix} = \begin{pmatrix} f_{x^{1,1,1}} & \cdots & f_{x^{1,1,1}} \\ \vdots & & \vdots \\ f_{x^{1,1,1}} & \cdots & f_{x^{1,1,1}} \end{pmatrix}.$$

These two matrices both have rank 1 and are thus not invertible. Therefore, a zeroizing attack cannot be carried out, a contradiction.   □

4 Securing Functions Against Partition Attacks

4.1 Input Partition Attacks

In this section, we define formally the notion of an input partition attack. We also define what it means for a function to be hard or impossible to partition. Since in this section we only are concerned with whether a function outputs zero or not, we consider functions with codomain $\{0, \perp\}$, where $\perp$ represents any non-zero branching program output.

Definition 7 (Input Partition). *Let $\boldsymbol{v} \in \mathbb{N}^t$ be a vector and $f \colon \mathbb{Z}_{\boldsymbol{v}} \to \{0, \perp\}$ be a function. An* input partition *for f of degree k is a tuple*

$$\mathcal{I}_f^k = \left(\sigma \in S_t, \ \{a_i\}_{i \in [k]} \subseteq \mathbb{Z}_{\boldsymbol{u}_1}, \ \{c_j\}_{j \in [k]} \subseteq \mathbb{Z}_{\boldsymbol{u}_2} \right)$$

satisfying $a_i \neq a_j$ and $c_i \neq c_j$ for all $i, j \in [k]$ with $i \neq j$ and $\sigma(\boldsymbol{u}_1 \| \boldsymbol{u}_2) = \boldsymbol{v}$ such that for all $i, j \in [k]$,

$$f(\sigma(a_i \| c_j)) = 0.$$

Definition 8 (Input Partition Attack). *For each $t \in \mathbb{N}$ let $\boldsymbol{v}_t \in \mathbb{N}^t$ be a vector, $\mathcal{F}_t$ be a family of functions $f \colon \mathbb{Z}_{\boldsymbol{v}_t} \to \{0, \perp\}$, and let $\mathcal{F} = \{\mathcal{F}_t\}_{t \in \mathbb{N}}$. We say that a PPT adversary $\mathcal{A}$ performs an* input partition attack *of degree k on $\mathcal{F}$ if for a non-negligible function ϵ,*

$$\Pr_{w, f \leftarrow \mathcal{F}_t} \left[\mathcal{A}(f) = \mathcal{I}_f^k \text{ is an input partition of } f \text{ of degree } k \right] > \epsilon(t),$$

where the probability is taken over the randomness w of $\mathcal{A}$ and a uniform choice of f from $\mathcal{F}_t$.

Turning the above definition around, we can ensure security against input partition attacks if the function we obfuscate satisfies the following.

Definition 9 (Input Partition Resistance). *For each $t \in \mathbb{N}$ let $v_t \in \mathbb{N}^t$ be a vector, $\mathcal{F}_t$ be a family of functions $f\colon \mathbb{Z}_{v_t} \to \{0, \perp\}$, and let $\mathcal{F} = \{\mathcal{F}_t\}_{t \in \mathbb{N}}$. We say that $\mathcal{F}$ is input partition resistant for degree k if no PPT adversary successfully performs an input partition attack of degree k on $\mathcal{F}$.*

A stronger version of this is for a function to simply not admit any input partitions which would clearly make attacks requiring a partition of the input impossible.

Definition 10 (Input Unpartitionable Function). *Let $v \in \mathbb{N}^t$ be a vector and $f\colon \mathbb{Z}_v \to \{0, \perp\}$ be a function. We say that f is input unpartitionable for degree k if it does not admit an input partition of degree k. If f is input unpartitionable for degree 2, we simply say that it is input unpartitionable.*

4.2 Securing Functions

Now that we have defined the type of attack we aim to defend against, we introduce the input "stamping" function h and define what it means for h to secure a function f.

Definition 11 (Securing a Function). *Let $v_1 \in \mathbb{N}^{t_1}, v_2 \in \mathbb{N}^{t_2}$ be vectors and write $v = v_1 \| v_2$. Let $f\colon \mathbb{Z}_{v_1} \to \{0, 1\}$ and $h\colon \mathbb{Z}_{v_1} \to \mathbb{Z}_{v_2}$ be functions and construct a function $g\colon \mathbb{Z}_v \to \{0, \perp\}$ as follows:*

$$g(a\|b) = \begin{cases} f(a), & h(a) = b \\ \perp, & h(a) \neq b. \end{cases}$$

We say that h completely secures f if g is input unpartitionable.

A slightly less strict definition is the following which defines what it means for a function family to statistically secure a function.

Definition 12 (Statistically Securing a Function). *Let $v \in \mathbb{N}^t$ be a vector, f be a function $f\colon \mathbb{Z}_v \to \{0, 1\}$, and $\mathcal{H}$ be a collection $\mathcal{H} = \{\mathcal{H}_\lambda\}_{\lambda \in \mathbb{N}}$ of function families such that $\mathcal{H}_\lambda$ is a family of functions $h\colon \mathbb{Z}_v \to \mathbb{Z}_{u_\lambda}$ for some $u_\lambda \in \mathbb{N}^{k_\lambda}$. We say that $\mathcal{H}$ statistically secures f if for some negligible function ϵ and for all $\lambda \in \mathbb{N}$,*

$$\Pr_{h \xleftarrow{\$} \mathcal{H}_\lambda} [h \text{ completely secures } f] > 1 - \epsilon(\lambda),$$

where h is sampled uniformly from $\mathcal{H}_\lambda$.

4.3   Necessary and Sufficient Conditions

In this section, we present and prove the necessary and sufficient condition on a function h in order for it to secure every function f. First, we give some useful definitions.

Definition 13. *Let $k \in \mathbb{N}$ be a positive integer and define the equivalence relation $\sim$ on $\mathbb{Z}_k \times \mathbb{Z}_k$ as follows. Two elements $(a, b), (c, d) \in \mathbb{Z}_k \times \mathbb{Z}_k$ are equivalent under $\sim$ if and only if either $(a, b) = (c, d)$ or $a = b$ and $c = d$. We denote by $\mathcal{Z}_k$ the set*

$$\mathcal{Z}_k = \mathbb{Z}_k \times \mathbb{Z}_k / \sim.$$

For a vector $\boldsymbol{v} \in \mathbb{N}^t$ we write $\mathcal{Z}_{\boldsymbol{v}}$ for the set $\mathcal{Z}_{v_1} \times \mathcal{Z}_{v_2} \times \cdots \times \mathcal{Z}_{v_t}$.

Definition 14. *Let $\boldsymbol{v} \in \mathbb{N}^t$ be a vector. Define an operation $* \colon \mathbb{Z}_{\boldsymbol{v}} \times \mathbb{Z}_{\boldsymbol{v}} \to \mathcal{Z}_{\boldsymbol{v}}$ as follows. For two elements $(a_1, \ldots, a_t), (b_1, \ldots, b_t) \in \mathbb{Z}_{\boldsymbol{v}}$,*

$$(a_1, \ldots, a_t) * (b_1, \ldots, b_t) = ((a_1, b_1), \ldots, (a_t, b_t)) \in \mathcal{Z}_{\boldsymbol{v}}.$$

The operation $*$ is essentially a projection of two vectors $\boldsymbol{a}, \boldsymbol{b} \in \mathbb{Z}_{\boldsymbol{v}}$ into $\mathcal{Z}_{\boldsymbol{v}}$.
We now give the characterization.

Definition 15 (Safe Function). *Let $\boldsymbol{v}_1 \in \mathbb{N}^{t_1}, \boldsymbol{v}_2 \in \mathbb{N}^{t_2}$ be vectors. A function $h \colon \mathbb{Z}_{\boldsymbol{v}_1} \to \mathbb{Z}_{\boldsymbol{v}_2}$ is safe if for every $x_{1,1}, x_{1,2}, x_{2,1}, x_{2,2} \in \mathbb{Z}_{\boldsymbol{v}_1}$ it is the case that if both of the following hold:*

$$x_{1,1} * x_{1,2} = x_{2,1} * x_{2,2}$$
$$h(x_{1,1}) * h(x_{1,2}) = h(x_{2,1}) * h(x_{2,2}),$$

then $x_{1,1} = x_{1,2}$ or $x_{1,1} = x_{2,1}$.

Theorem 1. *Let $\boldsymbol{v}_1 \in \mathbb{N}^{t_1}, \boldsymbol{v}_2 \in \mathbb{N}^{t_2}$ be vectors. The function $h \colon \mathbb{Z}_{\boldsymbol{v}_1} \to \mathbb{Z}_{\boldsymbol{v}_2}$ completely secures every function $f \colon \mathbb{Z}_{\boldsymbol{v}_1} \to \{0, 1\}$ if and only if it is safe.*

In order to prove Theorem 1 we first state and prove two lemmas.

Lemma 1. *Let $\boldsymbol{v}_1 \in \mathbb{N}^{t_1}$ and $\boldsymbol{v}_2 \in \mathbb{N}^{t_2}$ be vectors and $\sigma \in S_{t_1+t_2}$. Let $a_1, a_2 \in \mathbb{Z}_{\boldsymbol{v}_1}, c_1, c_2 \in \mathbb{Z}_{\boldsymbol{v}_2}$, and*

$$\{r_1, \ldots, r_k\} = T \subseteq [t_1 + t_2], r_1 < r_2 < \cdots < r_k.$$

Finally, define a function p_T such that for $x \in \mathbb{Z}_{\boldsymbol{v}_1 || \boldsymbol{v}_2}$, $p_T(x) = x_{r_1} x_{r_2} \ldots x_{r_k}$. Then

$$p_T(\sigma(a_1 || c_1)) * p_T(\sigma(a_2 || c_1)) = p_T(\sigma(a_1 || c_2)) * p_T(\sigma(a_2 || c_2))$$

Proof. First, note that this lemma holds in general if and only if it holds for $T = [t_1 + t_2]$. So we will simply show that

$$\sigma(a_1||c_1) * \sigma(a_2||c_1) = \sigma(a_1||c_2) * \sigma(a_2||c_2)$$

which is equivalent to showing that

$$(a_1||c_1) * (a_2||c_1) = (a_1||c_2) * (a_2||c_2).$$

However, this is trivial from the definition of the operation $*$ and the conclusion follows. □

Lemma 2. *Let $v \in \mathbb{N}^t$ be a vector and let $x_{1,1}, x_{1,2}, x_{2,1}, x_{2,2} \in \mathbb{Z}_v$ be given satisfying $x_{1,1} \neq x_{1,2}$ and $x_{1,1} \neq x_{2,1}$ and*

$$x_{1,1} * x_{1,2} = x_{2,1} * x_{2,2}.$$

Then there exist

$$\sigma \in S_t, \ a_1, a_2 \in \mathbb{Z}_{v_1}, c_1, c_2 \in \mathbb{Z}_{v_2}$$

with $a_1 \neq a_2$, $c_1 \neq c_2$, and $\sigma(v_1||v_2) = v$ such that for every $i, j \in \{1, 2\}$,

$$\sigma(a_i||c_j) = x_{i,j}$$

Proof. Let S be the set of indices $j \in [t]$ such that $x_{1,1}^j = x_{1,2}^j$ and let D be the set of indices $j \in [t]$ such that $x_{1,1}^j \neq x_{1,2}^j$. It is clear that S and D partition the set of indices $[t]$. Since $x_{1,1} * x_{1,2} = x_{2,1} * x_{2,2}$, we must have the following relations:

$$\forall j \in S: x_{1,1}^j = x_{1,2}^j \text{ and } x_{2,1}^j = x_{2,2}^j \tag{1}$$

$$\forall j \in D: x_{1,1}^j = x_{2,1}^j \neq x_{1,2}^j = x_{2,2}^j \tag{2}$$

Note that because $x_{1,1} \neq x_{1,2}$, D must be non-empty. S must also be empty, otherwise $x_{1,1} * x_{1,2} = x_{2,1} * x_{2,2}$ would imply that $x_{1,1} = x_{2,1}$. So there must also exist an index $r \in S$ such that $x_{1,1}^r = x_{1,2}^r \neq x_{2,1}^r = x_{2,2}^r$.

Now, enumerating S and D as $S = \{m_1, \ldots, m_k\}$ and $D = \{n_1, \ldots, n_l\}$ with $k + l = t$, we set

$$a_i = x_{i,1}^{m_1} x_{i,1}^{m_2} \ldots x_{i,1}^{m_k} = x_{i,2}^{m_1} x_{i,2}^{m_2} \ldots x_{i,2}^{m_k}$$
$$c_j = x_{1,j}^{n_1} x_{1,j}^{n_2} \ldots x_{1,j}^{n_k} = x_{2,j}^{n_1} x_{2,j}^{n_2} \ldots x_{2,j}^{n_k},$$

for $i, j \in \{1, 2\}$ where the equalities to the right follow from (1) and (2), $a_1 \neq a_2$ because of the existence of r as above, and $c_1 \neq c_2$ from the definition of D.

Letting $\sigma \in S_t$ be the permutation such that

$$\sigma(m_1 \, m_2 \, \ldots \, m_k \, n_1 \, n_2 \, \ldots \, n_l) = 1 \, 2 \, \ldots \, t,$$

we find that $\sigma(a_i||c_j) = x_{i,j}$ for every $i, j \in \{1, 2\}$ and we are done. □

Proof (Proof of Theorem 1). Suppose h completely secures every function f and assume for contradiction that there exist $x_{1,1}, x_{1,2}, x_{2,1}, x_{2,2} \in \mathbb{Z}_{v_1}$ with $x_{1,1} \neq x_{1,2}$ and $x_{1,1} \neq x_{2,1}$ such that

$$x_{1,1} * x_{1,2} = x_{2,1} * x_{2,2}$$
$$h(x_{1,1}) * h(x_{1,2}) = h(x_{2,1}) * h(x_{2,2}).$$

Let f be the function satisfying $f(x) = 0$ for every $x \in \mathbb{Z}_{v_1}$ and consider the function

$$g(a||b) = \begin{cases} f(a), & h(a) = b \\ \bot, & h(a) \neq b. \end{cases}$$

We clearly have

$$(x_{1,1} \ || \ h(x_{1,1})) * (x_{1,2} \ || \ h(x_{1,2})) = (x_{2,1} \ || \ h(x_{2,1})) * (x_{2,2} \ || \ h(x_{2,2}))$$

and thus, by Lemma 2 there exist

$$\sigma \in S_{t_1+t_2}, \ a_1, a_2 \in \mathbb{Z}_{u_1} \ c_1, c_2 \in \mathbb{Z}_{u_2}$$

with $a_1 \neq a_2$, $c_1 \neq c_2$, and $\sigma(u_1 || u_2) = v_1 || v_2$ such that for every $i, j \in \{1, 2\}$,

$$\sigma(a_i || c_j) = x_{i,j} || h(x_{i,j}).$$

However, then $g(\sigma(a_i || c_j)) = 0$ for every $i, j \in \{1, 2\}$ which is a contradiction since g would be input unpartitionable if h completely secured the function f.

Conversely, suppose h is safe and let $f \colon \mathbb{Z}_{v_1} \to \{0, 1\}$ be arbitrary. Define

$$g(a||b) = \begin{cases} f(a), & h(a) = b \\ \bot, & h(a) \neq b. \end{cases}$$

and assume for contradiction that there exists an input partition for g of degree two,

$$\mathcal{I}_g^2 = \left(\sigma \in S_{t_1+t_2}, \ \{a_i\}_{i \in [k]} \subseteq \mathbb{Z}_{u_1}, \ \{c_j\}_{j \in [k]} \subseteq \mathbb{Z}_{u_2} \right).$$

For each $i, j \in \{1, 2\}$, write $\sigma(a_i || c_j) = x_{i,j} y_{i,j}$ with $x_{i,j} \in \mathbb{Z}_{v_1}$ and $y_{i,j} \in \mathbb{Z}_{v_2}$ and observe that then $h(x_{i,j}) = y_{i,j}$. Furthermore, we have

$$x_{1,1} * x_{1,2} = x_{2,1} * x_{2,2}$$
$$y_{1,1} * y_{1,2} = y_{2,1} * y_{2,2}.$$

by Lemma 1. Since $h(x_{i,j}) = y_{i,j}$ it follows directly from the condition on h that either $x_{1,1} = x_{1,2}$ or $x_{1,1} = x_{2,1}$. The two cases are symmetric, so assume without loss of generality that $x_{1,1} = x_{1,2}$. Then $y_{1,1} = y_{1,2}$ and we get $\sigma(a_1 || c_1) = \sigma(a_1 || c_2)$. A contradiction as we required $c_1 \neq c_2$. $\qquad \square$

4.4 Lower Bound on the Output Size of Safe Functions

An implementation of a safe function h to secure a function f by constructing the function g of Definition 11 results in an increase in the input size of f and further this extra input must be checked against the output of h. In the context of matrix branching programs, the check of the extra input requires a pass over the input, adding more matrices, which when initialized over multilinear maps is rather costly. Thus, knowledge about the minimal output size of a safe function is helpful in determining the costs of securing a function. In this section, we show that this output size is at least linear in the input size of f.

Theorem 3. *Let* $v_1 = (v_1^1, v_1^2, \ldots, v_1^{t_1}) \in \mathbb{N}^{t_1}$ *and* $v_2 = (v_2^1, v_2^2, \ldots, v_2^{t_2}) \in \mathbb{N}^{t_2}$ *be vectors and let* $h \colon \mathbb{Z}_{v_1} \to \mathbb{Z}_{v_2}$ *be a safe function. If* k *is such that* $v_1^k = \min_{1 \le i \le t_1}(v_1^i)$ *then*

$$\prod_{\substack{1 \le i \le t_1 \\ i \ne k}} v_1^i \le \prod_{i=1}^{t_2}(v_2^i(v_2^i - 1) + 1).$$

Proof. Assume without loss of generality that $k = 1$. Recall that the elements of $\mathbb{Z}_{v_1^i}$ are $(a, b), a \ne b$ for $a, b \in \mathbb{Z}_{v_1^i}$ together with the single element consisting of the $\sim$-equivalence class $\{(b, b) \mid b \in \mathbb{Z}_{v_1^i}\}$ that we denote by (a, a). Now, consider the vector

$$y = ((b, c), (a, a), (a, a), \ldots, (a, a)) \in \mathbb{Z}_{v_1}$$

with $b \ne c$ and let $T = \{(x_1, x_2) \in (\mathbb{Z}_{v_1})^2 \mid x_1 * x_2 = y\}$. We have $|T| = \prod_{i=2}^{t_1} v_1^i$ since for the first index there is only the choice $x_1^1 = b$ and $x_2^2 = c$ and for index $i > 1$ there are v_1^i choices for $x_1^i = x_2^i$. Now, define the function $t \colon T \to \mathbb{Z}_{v_2}$ by $t(x_1, x_2) = h(x_1) * h(x_2)$. By the definition of a safe function, t must be injective. It follows that

$$\prod_{i=2}^{t_1} v_1^i = |T| \le |\mathbb{Z}_{v_2}| = \prod_{i=1}^{t_2}(v_2^i(v_2^i - 1) + 1).$$

$\square$

Corollary 1. *Let* $h \colon \{0, 1\}^{t_1} \to [k]^{t_2}$ *be a safe function. Then*

$$\frac{t_1 - 1}{\log_2(k(k-1) + 1)} \le t_2.$$

Proof. By Theorem 3, we have

$$2^{t_1 - 1} \le (k(k-1) + 1)^{t_2}.$$

Taking the logarithm on both sides of the equation yields the conclusion. $\square$

We will see in the instantiations of safe functions that while we do not achieve an optimal construction, all our constructions have output size linear in the input size of the original function and with fairly small coefficients, so they are asymptotically optimal.

5   Instantiations

We now present three instantations for h. We first give two number theoretic functions which secure any function f, and then we give a combinatorial function that statistically secures any function f.

5.1   Number Theoretical Functions

By the necessary and sufficient condition of Theorem 1 and the definition of a safe function, it seems that a function will secure every other function if it is somewhat non-linear everywhere. This is captured in the following corollary, letting us work with functions over the integers.

Corollary 2. *Suppose the function $h\colon \{0,1\}^n \to \{0,1\}^m$ satisfies that whenever $x_{1,1}, x_{1,2}, x_{2,1}, x_{2,2} \in \{0,1\}^n$ satisfy the relations*

$$x_{1,1} - x_{1,2} = x_{2,1} - x_{2,2}$$
$$h(x_{1,1}) - h(x_{1,2}) = h(x_{2,1}) - h(x_{2,2}).$$

where we consider each term as an integer $x_{i,j} \in [2^n - 1]$ or $h(x_{i,j}) \in [2^m - 1]$ then either $x_{1,1} = x_{1,2}$ or $x_{1,1} = x_{2,1}$. Then h completely secures every function $f\colon \{0,1\}^n \to \{0, \bot\}$.

Proof. This follows immediately from Theorem 1 since $x_{1,1} * x_{1,2} = x_{2,1} * x_{2,2}$ implies $x_{1,1} - x_{1,2} = x_{2,1} - x_{2,2}$ as integers.   □

Intuitively, many functions we know and love would satisfy this as long as they have sufficient non-linearity. Here we list two examples. In terms of output size, note that these functions both produce $n+1$ bits of output, where the minimum possible by Corollary 1 is $\frac{n-1}{\log_2 3}$. So the output size of these two functions is close to optimal.

Proposition 1. *Let p be a prime satisfying $2^n < p < 2^{n+1}$. The function $h\colon \{0,1\}^n \to \{0,1\}^{n+1}$ given by $h(x) = [x^2]_p$ completely secures every function $f\colon \{0,1\}^n \to \{0, \bot\}$.*

Proof. Let $x_{1,1}, x_{1,2}, x_{2,1}, x_{2,2} \in \{0,1\}^n$ be given and suppose $x_{1,1} - x_{2,1} = x_{1,2} - x_{2,2}$ and $h(x_{1,1}) - h(x_{2,1}) = h(x_{1,2}) - h(x_{2,2})$. We will show that $x_{1,1} = x_{1,2}$ or $x_{1,1} = x_{2,1}$, concluding the proof by Corollary 2.

Directly from the conditions on the $x_{i,j}$, we get

$$\begin{aligned}(x_{1,1} + x_{2,1})(x_{1,1} - x_{2,1}) &\equiv h(x_{1,1}) - h(x_{2,1}) \\ &= h(x_{1,2}) - h(x_{2,2}) \\ &\equiv (x_{1,2} + x_{2,2})(x_{1,2} - x_{2,2}) \pmod{p}.\end{aligned}$$

This yields two cases. If $x_{1,1} - x_{2,1} = x_{1,2} - x_{2,2} = 0$ then $x_{1,1} = x_{2,1}$ and we are done. Otherwise $x_{1,1} - x_{2,1} = x_{1,2} - x_{2,2}$ is invertible modulo p since $p > 2^n$ and we get

$$x_{1,1} + x_{2,1} \equiv x_{1,2} + x_{2,2} \pmod{p}.$$

Adding $x_{1,1} - x_{2,1} = x_{1,2} - x_{2,2}$ to both sides yields $2x_{1,1} \equiv 2x_{1,2} \pmod{p}$ which is equivalent to $x_{1,1} \equiv x_{1,2} \pmod{p}$. Hence, $x_{1,1} = x_{1,2}$ since $p > 2^n$ and we are done.   □

Proposition 2. *Let p be a prime satisfying $2^n < p < 2^{n+1}$ with primitive root r. The function $h\colon \{0,1\}^n \to \{0,1\}^{n+1}$ given by $h(x) = [r^x]_p$ completely secures every function $f\colon \{0,1\}^n \to \{0,1\}$.*

Proof. Let $x_{1,1}, x_{1,2}, x_{2,1}, x_{2,2} \in \{0,1\}^n$ be given and suppose $x_{1,1} - x_{2,1} = x_{1,2} - x_{2,2}$ and $h(x_{1,1}) - h(x_{2,1}) = h(x_{1,2}) - h(x_{2,2})$. We will show that $x_{1,1} = x_{1,2}$ or $x_{1,1} = x_{2,1}$, concluding the proof by Corollary 2.

Directly from the conditions on the $x_{i,j}$, we get

$$r^{x_{1,1}}(1 - r^{x_{2,1} - x_{1,1}}) \equiv h(x_{1,1}) - h(x_{2,1})$$
$$= h(x_{1,2}) - h(x_{2,2})$$
$$\equiv r^{x_{1,2}}(1 - r^{x_{2,2} - x_{1,2}}) \pmod{p}.$$

Now we have two cases. First, if $1 - r^{x_{2,1} - x_{1,1}} = 1 - r^{x_{2,2} - x_{1,2}}$ is invertible modulo p then $r^{x_{1,1}} \equiv r^{x_{1,2}} \pmod{p}$, yielding $x_{1,1} = x_{1,2}$ since r has order $p - 1 \geq 2^n$ modulo p. Second, if $1 - r^{x_{2,1} - x_{1,1}} = 1 - r^{x_{2,2} - x_{1,2}}$ is not invertible modulo p then clearly $1 - r^{x_{2,1} - x_{1,1}} \equiv 0 \pmod{p}$ and thus, $x_{2,1} - x_{1,1} = 0$ since the order of r is $\geq 2^n$. It follows that either $x_{1,1} = x_{1,2}$ or $x_{1,1} = x_{2,1}$.   □

5.2   Permutation Hash Functions

We now discuss an instantiation which statistically secures functions f, as opposed to the functions in the previous section which completely secure f. Number theoretical functions like the ones in the previous section are difficult and expensive to implement in the setting of matrix branching programs since these do not generally support operations over a fixed field $\mathbb{Z}_p$. However, matrix branching programs naturally implement operations on the group of permutations, S_k. The functions we define in this section are defined in terms of randomly chosen permutations, and turns out to be a much more practical alternative. This section explains the instantiation and proves statistical security, and Sect. 5.3 describes how to implement it efficiently over CLT13.

Definition 16. *A k-Permutation Hash Function of input size n is a function $h\colon \{0,1\}^n \to [k]$ randomly drawn as follows. Select permutations $\{\pi_{i,b}\}_{i \in [n], b \in \{0,1\}} \xleftarrow{\$} S_k^{2n}$ uniformly at random. For an input $x \in \{0,1\}^n$ let*

$$\sigma_x = \prod_{i=1}^n \pi_{i,x_i}.$$

Then $h(x) = \sigma_x(1)$.

Lemma 3. *Let* $x_{1,1}, x_{1,2}, x_{2,1}, x_{2,2} \in \{0,1\}^n$ *be given such that*

$$x_{1,1} * x_{1,2} = x_{2,1} * x_{2,2},$$

$x_{1,1} \neq x_{1,2}$, *and* $x_{1,1} \neq x_{2,1}$. *Then*

$$\Pr_{h \overset{\$}{\leftarrow} S_k^{2n}} [h(x_{1,1}) * h(x_{1,2}) = h(x_{2,1}) * h(x_{2,2})] \leq \frac{k}{(k-1)^2}.$$

Proof. Write $u = x_{1,1} * x_{1,2} = x_{2,1} * x_{2,2}$ and denote by $x_{a,b}^i$ the ith bit of $x_{a,b}$ and by u_i the ith entry of u. Let $S_d \subset [n], d \in \mathcal{Z}_2$ be the set of indices $i \in [n]$ such that $u_i = d$, where we recall that the elements of $\mathcal{Z}_2$ are the equivalence classes $(0,1), (1,0), (0,0) \sim (1,1)$. We will denote the equivalence class containing $(0,0)$ and $(1,1)$ by (a,a). Thus, the set of indices $[n]$ is partitioned into $[n] = S_{(0,1)} \uplus S_{(1,0)} \uplus S_{(a,a)}$.

Now, it must be the case that there is a $j \in S_{(a,a)}$ such that $x_{1,1}^j = x_{1,2}^j \neq x_{2,1}^j = x_{2,2}^j$. To see this, note that $x_{1,1}$ and $x_{2,1}$ are identical at the indices of $S_{(0,1)}$ and $S_{(1,0)}$ and if $x_{1,1}$ and $x_{2,1}$ also are identical at the indices of $S_{(a,a)}$ then $x_{1,1} = x_{2,1}$ contrary to assumption. Choose j to be maximal. We can assume without loss of generality that there is an $l \in S_{(0,1)} \cup S_{(1,0)}$ such that $l > j$ as follows. Suppose that this is not the case. Then $x_{1,1}^i = x_{1,2}^i = x_{2,1}^i = x_{2,2}^i$ for every $i > j$ since j was maximal and we must have $i \in S_{(a,a)}$ whenever $i > j$. Now, consider the equation $u' = x_{1,1} * x_{2,1} = x_{1,2} * x_{2,2}$ where we simply swap $x_{1,2}$ and $x_{2,1}$ from our original expression. Let $S_d' \subset [n], d \in \mathcal{Z}_2$ be the set of indices $i \in [n]$ such that $u_i' = d$. In this dual situation, $j \in [n] \setminus S_{(a,a)}'$ and still $x_{1,1}^i = x_{1,2}^i = x_{2,1}^i = x_{2,2}^i$ for every $i > j$. Further, we can find a new maximal $j' \in S_{(a,a)}'$ with $j' < j$ such that $x_{1,1}^{j'} \neq x_{1,2}^{j'}$. Thus, in the dual situation j is the l that we were seeking in the original case. From this it follows that we without loss of generality can choose $l > j$ as above.

Define the following permutations, noting that $x_{1,1}^i = x_{2,1}^i$ and $x_{1,2}^i = x_{2,2}^i$ for $i > j$.

$$\tau = \prod_{i=j+1}^{n} \pi_{i, x_{1,1}^i}$$

$$\sigma = \prod_{i=j+1}^{n} \pi_{i, x_{1,2}^i}$$

$$\gamma_{b,c} = \prod_{i=1}^{j-1} \pi_{i, x_{b,c}^i}, b, c \in \{1, 2\}.$$

Then letting $b = x_{1,1}^j = x_{1,2}^j$ we get

$$h(x_{1,1}) = \gamma_{1,1} \circ \pi_{j,b} \circ \tau(1), \quad h(x_{2,1}) = \gamma_{2,1} \circ \pi_{j,1-b} \circ \tau(1)$$
$$h(x_{1,2}) = \gamma_{1,2} \circ \pi_{j,b} \circ \sigma(1), \quad h(x_{2,2}) = \gamma_{2,2} \circ \pi_{j,1-b} \circ \sigma(1).$$

Now, since $x_{1,1}^l = x_{2,1}^l \neq x_{1,2}^l = x_{2,2}^l$, the permutations $\pi_{l,x_{1,1}^l} = \pi_{l,x_{2,1}^l}$ and $\pi_{l,x_{1,2}^l} = \pi_{l,x_{2,2}^l}$ are chosen independently of each other. Thus, it follows from $l > j$ that over the choice of $\{\pi_{i,b}\}$ the permutation τ and the permutation σ are independently and uniformly distributed. Writing $(a,b) = (\tau(1), \sigma(1)) \in \mathbb{Z}_k \times \mathbb{Z}_k$, this means that (a,b) will be uniformly distributed in $\mathbb{Z}_k \times \mathbb{Z}_k$ over the choice of $\{\pi_{i,b}\}$. Thus, with probability $\frac{k-1}{k}$ we have $a \neq b$. We will assume that from now on. The above equations now become

$$h(x_{1,1}) = \gamma_{1,1} \circ \pi_{i,b}(a), \quad h(x_{2,1}) = \gamma_{2,1} \circ \pi_{i,1-b}(a)$$
$$h(x_{1,2}) = \gamma_{1,2} \circ \pi_{i,b}(b), \quad h(x_{2,2}) = \gamma_{2,2} \circ \pi_{i,1-b}(b).$$

Noting that $\pi_{i,b}$ and $\pi_{i,1-b}$ are chosen independently we get that for some $(q,r),(s,t) \in T$ for $T = \mathbb{Z}_k \times \mathbb{Z}_k \setminus \{(a,a) \mid a \in \mathbb{Z}_k\}$ which are independent of each other and each uniformly distributed over T over the choice of $\{\pi_{i,b}\}$, we have

$$h(x_{1,1}) = \gamma_{1,1}(q), \quad h(x_{2,1}) = \gamma_{2,1}(s)$$
$$h(x_{1,2}) = \gamma_{1,2}(r), \quad h(x_{2,2}) = \gamma_{2,2}(t).$$

The γs are chosen independently of the other permutations, so even conditional on the particular choice of the γs the distribution of (q,r) and (s,t) are independent and uniformly distributed across T. This means that the pairs $(h(x_{1,1}), h(x_{1,2}))$ and $(h(x_{2,1}, h(x_{2,2}))$ are independent of each other and are uniformly distributed over subsets $U_1, U_2 \subset [k] \times [k]$, respectively, of size $|T| = k(k-1)$.

Now, taking into account the case we disregarded in the beginning where $a = b$ which has probability $\frac{1}{k}$, we can write

$$\Pr_{h \xleftarrow{\$} S_k^{2n}} [h(x_{1,1}) * h(x_{1,2}) = h(x_{2,1}) * h(x_{2,2})] \leq \frac{1}{k} + \Pr_{\substack{(\alpha,\beta) \xleftarrow{\$} U_1 \\ (\gamma,\delta) \xleftarrow{\$} U_2}} [(\alpha,\beta) \sim (\gamma,\delta)]$$

$$= \frac{1}{k} + \frac{1}{k(k-1)} \sum_{(\alpha,\beta) \in U_1} \Pr_{(\gamma,\delta) \xleftarrow{\$} U_2} [(\alpha,\beta) \sim (\gamma,\delta)]$$

$$= \frac{1}{k} + \frac{R}{k(k-1)}$$

Splitting in the case when $\alpha = \beta$ and $\alpha \neq \beta$, where we know that the former happens in at most k cases, we can calculate

$$R = \sum_{(\alpha,\beta) \in U_1} \Pr_{(\gamma,\delta) \xleftarrow{\$} U_2} [(\alpha,\beta) \sim (\gamma,\delta)]$$

$$\leq \sum_{\substack{(\alpha,\beta) \in U_1 \\ \alpha=\beta}} \Pr_{(\gamma,\delta) \xleftarrow{\$} U_2} [\gamma = \delta] + \sum_{\substack{(\alpha,\beta) \in U_1 \\ \alpha \neq \beta}} \Pr_{(\gamma,\delta) \xleftarrow{\$} U_2} [\alpha = \gamma \text{ and } \beta = \delta)]$$

$$\leq k \cdot \frac{1}{k-1} + k(k-1) \cdot \frac{1}{k(k-1)} = \frac{2k-1}{k-1}.$$

It follows that

$$\Pr_{\{\pi_{i,b}\}} [h(x_{2,1} * h(x_{2,2}) = h(x_{1,1}) * h(x_{1,2})] \leq \frac{1}{k} + \frac{2k-1}{k(k-1)^2} = \frac{k}{(k-1)^2}.$$

□

Lemma 4. *Draw k-permutation hash functions on n input bits, $h_1, \ldots, h_t$ independently at random. Then the probability that there exists $\{x_{b,c}\}_{b,c \in \{1,2\}} \subseteq \{0,1\}^n$ with $x_{1,1} \neq x_{1,2}$, $x_{1,1} \neq x_{2,1}$, $x_{1,1} * x_{1,2} = x_{2,1} * x_{2,2}$, and $h_i(x_{1,1}) * h_i(x_{1,2}) = h_i(x_{2,1}) * h_i(x_{2,2})$ for every $i \in [t]$ is strictly less than $\frac{k^t \cdot 6^n}{(k-1)^{2t}}$.*

Proof. Let $u \in (\mathcal{Z}_2)^n$ be given and let s be the number of entries of u that are (a,a), denoting the equivalence classes of $\mathcal{Z}_2$ as before by $(0,1)$, $(1,0)$, and (a,a). The number of choices for $\{x_{b,c}\}_{b,c \in \{1,2\}} \subseteq \{0,1\}^n$ such that $u = x_{1,1} * x_{1,2} = *x_{2,1} * x_{2,2}$ is 2^{2s} since for all entries $i \in [n]$ such that $u_i = (0,1)$ or $u_i = (1,0)$, $x_{1,1}^i = x_{2,1}^i$ and $x_{1,2}^i = x_{2,2}^i$ are fixed and for all entries $i \in [n]$ such that $u_i = (a,a)$ we have $x_{1,1}^i = x_{2,1}^i$ and $x_{1,2}^i = x_{2,2}^i$, but they are not fixed. Now, note that the number of $u \in (\mathcal{Z}_2)^n$ with exactly s (a,a)-entries is $2^{n-s} \binom{n}{s}$. Summing over all possible u we get the total number of choices $\{x_{b,c}\}_{b,c \in \{1,2\}} \subseteq \{0,1\}^n$ with $x_{1,1} * x_{1,2} = x_{2,1} * x_{2,2}$ to be

$$\sum_{s=0}^{n} 2^{n-s} \binom{n}{s} 2^{2s} = 2^n \sum_{s=0}^{n} 2^s \binom{n}{s} = 6^n.$$

By Lemma 3 the probability that any single choice of $\{x_{b,c}\}_{b,c \in \{1,2\}} \subseteq \{0,1\}^n$ such that $x_{1,1} \neq x_{1,2}$, $x_{1,1} \neq x_{2,1}$ and $x_{1,1} * x_{1,2} = x_{2,1} * x_{2,2}$ satisfies $h_i(x_{1,1}) * h_i(x_{1,2}) = h_i(x_{2,1}) * h_i(x_{2,2})$ for all $i \in [t]$ is strictly less than $\left(\frac{k}{(k-1)^2}\right)^t$. Thus, our conclusion follows immediately by the union bound.

□

For the next theorem we define a function that combines several permutation hash functions into one. Choose k-permutation hash functions $h_1, \ldots, h_t$ as discussed above, and define the main hash function $h(x) = h_1(x) || h_2(x) || \ldots || h_t(x)$.

Theorem 4. *Let $k \geq 3$. If $t \geq \frac{(1+\log_2(3))n+\lambda}{2\log_2(k-1)-\log_2(k)}$ then the function h as defined above statistically secures every function $f \colon \{0,1\}^n \to \{0,1\}$.*

Proof. Fix $k \geq 3$ and n. By Lemma 4, h statistically secures every function if

$$\frac{k^t \cdot 6^n}{(k-1)^{2t}} \leq 2^{-\lambda}.$$

Taking logarithms on both sides, this is equivalent to

$$\log_2(k)t + (1 + \log_2(3))n - 2t\log_2(k-1) \leq -\lambda.$$

Rearranging, this yields that h statistically secures every function for

$$t \geq \frac{(1 + \log_2(3))n + \lambda}{2\log_2(k-1) - \log_2(k)}.$$

□

5.3   Implementing Permutation Hash Functions over CLT13

We now describe how to efficiently secure a branching program over CLT using permutation hash functions. We first describe how to construct a branching program that takes an input $u\|v$ and checks whether $v = h_i(u)$ for a single hash function h_i from the previous section. We then describe a technique that allows evaluating branching programs in parallel as long as they have the same input function. Finally, we use this technique to efficiently add a securing permutation hash function to any matrix branching program over CLT13.

Implementing One Permutation Hash Function Check. Assume we are given a k-permutation hash function $h = h_i$ of input size n as in the previous section, with the corresponding permutations $\{\pi_{i,b}\}_{i,b}$. We construct a branching program BP^h over some $R \cong \mathbb{Z}_p$ that works over inputs in $\mathbb{Z}_v$, where $v = [2,\ldots,2,k] \in \mathbb{Z}^{n+1}$. This branching program will compute h over the first n bits in the input and then check if the result matches the final piece of input.

In the following sections we denote a branching program of length l that works over inputs in $\mathbb{Z}_v$ by the tuple $(\mathbf{mat}, M_0, M_{l+1}, \mathbf{inp})$, where $\mathbf{mat}(i)$ is an indexed family $\{M_{i,c}\}_{c \in \mathbb{Z}_{v_i}}$ for all $i \in [l]$. M_0 and M_{l+1} are "bookend" vectors. This branching program is evaluated over an input $x \in \mathbb{Z}_v$ by computing the following product:

$$M_0 \times \prod_{i=1}^{n} M_{i,x_{\mathbf{inp}(i)}} \times M_{l+1}.$$

For a k-permutation hash function h, let

$$\mathsf{BP}^h = (\mathbf{mat}^h, M_0^h, M_{n+2}^h, \mathbf{inp}^h).$$

The components of BP^h are defined as follows:

1. $\mathbf{mat}^h(1) = \{M_{1,c}^h\}_{c \in \mathbb{Z}_k}$, where $M_{1,c}^h \in M_k(R)$ is the permutation matrix corresponding to the transposition $(1\ c)$.
2. $\mathbf{mat}^h(i) = \{M_{i,b}^h\}_{b \in \{0,1\}}$ for $2 \leq i \leq n+1$, where $M_{i,b}^h \in M_k(R)$ is the permutation matrix corresponding to $\pi_{i-1,b}$.
3. $M_0^h = [1, 0, \ldots, 0] \in R^k$.
4. $M_{n+2}^h = [0, 1, \ldots, 1]^T \in R^k$.
5. $\mathbf{inp}^h(i) = \begin{cases} n+1 & i = 1 \\ i-1 & 2 \leq i \leq n+1. \end{cases}$

Consider an evaluation of the branching program BP^h over an input $u\|v$, where $u \in \{0,1\}^n$ and $v \in \mathbb{Z}_k$. The result is of the form

$$M_0^h \times M_{1,v}^h \times \prod_{i=2}^{n+1} M_{i,u_{i-1}}^h \times M_{n+2}^h. \tag{3}$$

The product $\prod_{i=2}^{n+1} M_{i,u_{i-1}}^h \times M_{n+2}^h$ results in a column vector with a 0 at position $h(u)$ and 1s in every other position. The product of this result with $M_{1,v}^h$ produces $[0, 1, \ldots, 1]^T$ if and only if $h(u) = v$ and otherwise produces a vector with a 0 in a position other than the first and 1s everywhere else. Multiplying by M_0^h thus produces 0 if and only if $h(u) = v$. In conclusion, evaluating BP^h on input $u\|v$ outputs 0 if and only if $h(u) = v$.

Parallel Branching Programs. The CLT13 multilinear map uses a ring of composite order, which allows for a certain type of parallel branching program computation. Namely, we can construct a branching program where each step actuality encodes steps for several branching programs, and the parent branching program evaluates to zero if and only if all of its underlying branching programs do. In this section, we describe how to construct such a parallel computation.

Let n be the dimension chosen by the CLT13 instantiation based on the security parameter. This number is the number of prime factors of the ring order. (We assume it is squarefree.) Let $\mathsf{BP}_1, \mathsf{BP}_2, \ldots, \mathsf{BP}_n$ be the set of branching programs we want to evaluate in parallel. Following [GLW14], we will work over a CLT13 instantiation of dimension n^2. Recall that the plaintext ring for a CLT13 instantiation is of the form $\mathbb{Z}_g \cong \bigoplus_{i=1}^n \mathbb{Z}_{g_i}$ for primes g_i. In our case we think of the ring as being $\mathbb{Z}_g \cong \bigoplus_{i=1}^n \mathbb{Z}_{G_i}$, where each G_i is the product of n primes $g_{i,j}$. We will perform the evaluation of each branching program in a different component $\mathbb{Z}_{G_i}$. This variant of CLT13 is described in Section B6 of [GLW14].

We make several assumptions restricting the types of branching programs that we can execute in parallel. First, assume they are all of the same length l and all take inputs from $\mathbb{Z}_v$. Second, assume the matrices of BP_i are defined over the ring $\mathbb{Z}_{G_i}$ for all i. We also assume the matrices of all the BP_i are of the same size, which is without loss of generality since we can pad them with identity matrices. Finally, assume every BP_i has the same input function inp.

Let $\mathsf{BP}_i = (\mathbf{mat}_i, M_{i,0}, M_{i,l+1}, \mathsf{inp})$, where $\mathbf{mat}_i(j) = \{M_{i,j,c}\}_{c \in \mathbb{Z}_{v_{\mathsf{inp}(j)}}}$. We construct a new branching program $\mathsf{BP}' = (\mathbf{mat}', M_0', M_{l+1}', \mathsf{inp})$ over the ring $\mathbb{Z}_g$, where $\mathbf{mat}'(j) = \{M_{j,c}'\}_{c \in \mathbb{Z}_{v_{\mathsf{inp}(j)}}}$ with $M_{j,c}' \equiv M_{i,j,c} \pmod{G_i}$ for all $i \in [n]$, $j \in [l]$, and $c \in \mathbb{Z}_{v_j}$, and additionally $M_0' \equiv M_{i,0} \pmod{G_i}$ and $M_{l+1}' \equiv M_{i,l+1} \pmod{G_i}$ for all $i \in [n]$. If we evaluate the branching program BP' on $x \in \mathbb{Z}_v$ as the product

$$M_0' \times \prod_{j=1}^{l} M_{j,x_{\mathsf{inp}(j)}}' \times M_{l+1}' \pmod{g},$$

the result, $\mathsf{BP}'(x)$, is zero if and only if

$$M_{i,0} \times \prod_{j=1}^{l} M_{i,j,x_{\mathsf{inp}(j)}} \times M_{i,l+1} \equiv 0 \pmod{G_i}$$

for all $i \in [n]$.

Securing an Arbitrary Branching Program. Assume we have a branching program $\mathsf{BP} = (\mathbf{mat}, M_0, M_{l+1}, \mathbf{inp})$, $\mathbf{mat}(j) = \{M_{j,b}\}_{b\in\{0,1\}}$ over $\{0,1\}^n$ which we would like to secure. We need to construct a new branching program BP' which computes BP but also requires an additional section of input which should be a hash of the first part. BP' must check whether the hash is valid and must always return a nonzero value if it is not.

More formally, let u be an input to BP. Let $h_1, \ldots, h_t$ be k-permutation hash functions on $|u|$ bits. BP' takes input $u\|v$ and checks whether $v_i = h_i(u)$ for all $i \in [t]$. If so BP' returns $\mathsf{BP}(u)$, and if not BP' returns some nonzero value.

Let h_i be implemented by the branching program

$$\mathsf{BP}^{h_i} = (\mathbf{mat}^{h_i}, M_{b_1}^{h_i}, M_{b_2}^{h_i}, \mathbf{inp}^{h_i}),$$

where $\mathbf{mat}^{h_i}(1) = \{M_{1,c}^{h_i}\}_{c\in\mathbb{Z}_k}$ and $\mathbf{mat}^{h_i}(j) = \{M_{j,b}^{h_i}\}_{b\in\{0,1\}}$ for $2 \le i \le n+1$. We need to modify this branching program so that instead of taking an input $u\|v \in \mathbb{Z}_{[2,\ldots,2,k]}$ of length $n + 1$, it takes an input $u\|v \in \mathbb{Z}_{[2,\ldots,2,k,\ldots,k]}$ of length $n + t$ and checks whether $v_i = h_i(u)$. We can do this by altering the input function $\mathbf{inp}^{h_i}$ to set $\mathbf{inp}^{h_i}(1) = n + i$, but this would result in the branching programs BP^{h_i} having different input functions for different values of i, which is not compatible with parallel branching program evaluation. So instead we pad the branching program so that the first t entries are all the identity matrix except for the i'th entry which is $\{M_{1,c}^{h_i}\}_{c\in\mathbb{Z}_k}$. Then the input function can be set to be the same for all i. Specifically, we redefine $\mathbf{mat}^{h_i}$ as follows:

- $\mathbf{mat}^{h_i}(i) = \{M_{1,c}^{h_i}\}_{c\in\mathbb{Z}_k}$
- $\mathbf{mat}^{h_i}(j) = \{I_k\}_{c\in\mathbb{Z}_k}$ for all $1 \le j \le t, j \ne i$
- $\mathbf{mat}^{h_i}(j) = \{M_{j-t+1,b}^{h_i}\}_{b\in\{0,1\}}$ for all $t + 1 \le j \le t + n$

and we redefine $\mathbf{inp}^{h_i}$ as follows:

$$\mathbf{inp}^{h_i}(j) = \mathbf{inp}^h(j) = \begin{cases} n + j & 1 \le j \le t \\ j - t & t + 1 \le j \le t + n. \end{cases}$$

We are now ready to use parallel branching program evaluation to combine the hash function checks with the original branching program functionality. We use $t + 1$ branching programs, one of which is a modified version of BP and the others are modified versions of the BP^{h_i}. Every modified branching program will have length $t + l$ and will share the same input function $\mathbf{inp}'$, so as to facilitate parallel evaluation.

We first define the new input function:

$$\mathbf{inp}'(j) = \begin{cases} n + j & 1 \le j \le t \\ \mathbf{inp}(j - t) & t + 1 \le j \le t + n. \end{cases}$$

The reasoning for this definition will become clear shortly. We modify $\mathsf{BP} = (\mathbf{mat}, M_0, M_{l+1}, \mathbf{inp})$ by padding the branching program with identity matrices at the beginning while leaving the rest of the program unchanged. So $\mathbf{mat}(j) = I$ for $1 \leq i \leq t$, and $\mathbf{mat}(j) = \{M_{j-t,b}\}_{b \in \{0,1\}}$ for $t + 1 \leq j \leq t + l$. Note that BP should now be evaluated using the input function $\mathbf{inp}'$.

Finally, we describe how we modify BP^{h_i} to work with the input function $\mathbf{inp}'$. The problem with the definition of BP^{h_i} given above is that the input function during the latter part of the branching program, where $j > t$, does not match the input function of BP. We could fix this by padding BP with more identity matrices so that the computation of the h_i and the computation of BP would happen sequentially, but this would add to the total length of the resulting parallel branching program. Instead, we make an observation about the computation of h_i which allows for some flexibility in how we define the input function to the program. We will use these observations to rearrange BP^{h_i} so that it matches $\mathbf{inp}'$ exactly.

We observe that changing the order in which we read the input does not affect whether $h = h_1 || \ldots || h_t$ secures a function or not since this is equivalent to for each h_i to permute the order of composition of the permutations of h_i. Since each of the permutations of h_i are chosen uniformly at random, this does not affect the distribution of h_i.

Given this observation, we can redefine $\mathbf{mat}_{h_i}$ as follows without changing its functionality. Let f_j be the smallest r such that $\mathbf{inp}(r) = j$ (we assume that BP reads all of its input at some point such that this is well-defined). Then we set

- $\mathbf{mat}^{h_i}(t + f_j) = \{M^{h_i}_{j+1,b}\}_{b \in \{0,1\}}$ for all $1 \leq j \leq n$.
- $\mathbf{mat}^{h_i}(t + r) = \{I\}_{b \in \{0,1\}}$ for all $r \in [l] \setminus \{f_j\}_{j \in [n]}$.
- $\mathbf{mat}^{h_i}(i) = \{M^{h_i}_{1,c}\}_{c \in \mathbb{Z}_k}$.
- $\mathbf{mat}^{h_i}(j) = \{I_k\}_{c \in \mathbb{Z}_k}$ for all $1 \leq j \leq t, j \neq i$.

Thus we have $t + 1$ branching programs which now share the same input function, and evaluating the branching programs in parallel as described above achieves the functionality of BP' as desired.

Incurred Overhead. There are three ways in which our technique for securing obfuscation against CLT zeroizing attacks can increase the size of the program. We explain and address each of these below.

First, the parallelization requires more primes in the CLT instantiation. As described above, we use a variant of CLT13 from [GLW14] where there is an incurred increase in the dimension. This is to allow secure parallel branching program execution. We note, however, that the number of parallel executions needed, t, is less than the dimension n needed for security in the original CLT13 construction for every interesting branching program (e.g. every branching program that reads every bit of its input).

Second, checking the result of the permutation hash functions requires making the branching program longer by t matrices. This increases the degrees of multilinearity by t.

Third, if the original branching program had breadth q, i.e. each matrix of the branching program was a q by q matrix, then our procedure yields a branching program of breadth $\max\{q, k\}$. So having a large choice of k might increase the size of the branching program. Note that Theorem 4 implies a tradeoff between t and k.

We explore this further with a concrete example. Let us assume that the BP takes 5 passes over its input and that BP has breadth 5. Setting the parameter k, there is a trade-off between the number of encodings needed and the levels of multilinearity. The latter decides the size of each encoding. If we simply set $k = 5$ then the breadth of the branching program stays the same and we get roughly $2n$ extra levels of multilinearity. In our concrete example, this leads to a 40% increase in multilinearity and a 7 factor increase in the number of encodings. If instead we let $k = 2^{10}$, we see a mere 7% increase in the levels of multilinearity. The number of encodings increases drastically since the breadth of the branching program becomes about 2^{10}. However, the most efficient current obfuscation implementation [BISW17] Boneh et al. uses branching programs with breadth in excess of 2^{10}. Thus, for practical implementation this is not unreasonable.

Acknowledgements. We thank Jean-Sebastien Coron for pointing out a technical oversight in an earlier version of this work. We also thank the anonymous referees for their comments.

References

[AGIS14] Ananth, P., Gupta, D., Ishai, Y., Sahai, A.: Optimizing obfuscation: avoiding barrington's theorem. In: Proceedings of the 2014 ACM SIGSAC Conference on Computer and Communications Security, CCS 2014, pp. 646–658. ACM, New York (2014)

[AS17] Ananth, P., Sahai, A.: Projective arithmetic functional encryption and indistinguishability obfuscation from degree-5 multilinear maps. In: Coron, J.-S., Nielsen, J.B. (eds.) EUROCRYPT 2017 Part I. LNCS, vol. 10210, pp. 152–181. Springer, Cham (2017). doi:10.1007/978-3-319-56620-7_6

[BGK+14] Barak, B., Garg, S., Kalai, Y.T., Paneth, O., Sahai, A.: Protecting obfuscation against algebraic attacks. In: Nguyen, P.Q., Oswald, E. (eds.) EUROCRYPT 2014. LNCS, vol. 8441, pp. 221–238. Springer, Heidelberg (2014). doi:10.1007/978-3-642-55220-5_13

[BISW17] Boneh, D., Ishai, Y., Sahai, A., Wu, D.J.: Lattice-based SNARGs and their application to more efficient obfuscation. In: Coron, J.-S., Nielsen, J.B. (eds.) EUROCRYPT 2017 Part III. LNCS, vol. 10212, pp. 247–277. Springer, Cham (2017). doi:10.1007/978-3-319-56617-7_9

[BMSZ16] Badrinarayanan, S., Miles, E., Sahai, A., Zhandry, M.: Post-zeroizing obfuscation: new mathematical tools, and the case of evasive circuits. In: Fischlin, M., Coron, J.-S. (eds.) EUROCRYPT 2016 Part II. LNCS, vol. 9666, pp. 764–791. Springer, Heidelberg (2016). doi:10.1007/978-3-662-49896-5_27

[CGH+15] Coron, J.-S., et al.: Zeroizing without low-level zeroes: new MMAP attacks and their limitations. In: Gennaro, R., Robshaw, M. (eds.) CRYPTO 2015 Part I. LNCS, vol. 9215, pp. 247–266. Springer, Heidelberg (2015). doi:10.1007/978-3-662-47989-6_12

[CGH17] Chen, Y., Gentry, C., Halevi, S.: Cryptanalyses of candidate branching program obfuscators. In: Coron, J.-S., Nielsen, J.B. (eds.) EUROCRYPT 2017 Part III. LNCS, vol. 10212, pp. 278–307. Springer, Cham (2017). doi:10.1007/978-3-319-56617-7_10

[CHL+15] Cheon, J.H., Han, K., Lee, C., Ryu, H., Stehlé, D.: Cryptanalysis of the multilinear map over the integers. In: Oswald, E., Fischlin, M. (eds.) EUROCRYPT 2015 Part I. LNCS, vol. 9056, pp. 3–12. Springer, Heidelberg (2015). doi:10.1007/978-3-662-46800-5_1

[CLLT17] Coron, J.-S., Lee, M.S., Lepoint, T., Tibouchi, M.: Zeroizing attacks on indistinguishability obfuscation over CLT13. In: Fehr, S. (ed.) PKC 2017 Part I. LNCS, vol. 10174, pp. 41–58. Springer, Heidelberg (2017). doi:10.1007/978-3-662-54365-8_3

[CLT13] Coron, J.-S., Lepoint, T., Tibouchi, M.: Practical multilinear maps over the integers. In: Canetti, R., Garay, J.A. (eds.) CRYPTO 2013 Part I. LNCS, vol. 8042, pp. 476–493. Springer, Heidelberg (2013). doi:10.1007/978-3-642-40041-4_26

[GGH13a] Garg, S., Gentry, C., Halevi, S.: Candidate multilinear maps from ideal lattices. In: Johansson, T., Nguyen, P.Q. (eds.) EUROCRYPT 2013. LNCS, vol. 7881, pp. 1–17. Springer, Heidelberg (2013). doi:10.1007/978-3-642-38348-9_1

[GGH+13b] Garg, S., Gentry, C., Halevi, S., Raykova, M., Sahai, A., Waters, B.: Candidate indistinguishability obfuscation and functional encryption for all circuits. In: 2013 IEEE 54th Annual Symposium on Foundations of Computer Science (FOCS), pp. 40–49. IEEE (2013)

[GGH15] Gentry, C., Gorbunov, S., Halevi, S.: Graph-induced multilinear maps from lattices. In: Dodis, Y., Nielsen, J.B. (eds.) TCC 2015 Part II. LNCS, vol. 9015, pp. 498–527. Springer, Heidelberg (2015). doi:10.1007/978-3-662-46497-7_20

[GLSW15] Gentry, C., Lewko, A.B., Sahai, A., Waters, B.: Indistinguishability obfuscation from the multilinear subgroup elimination assumption. In: IEEE 56th Annual Symposium on Foundations of Computer Science, FOCS 2015, Berkeley, CA, USA, 17–20 October 2015, pp. 151–170 (2015)

[GLW14] Gentry, C., Lewko, A., Waters, B.: Witness encryption from instance independent assumptions. In: Garay, J.A., Gennaro, R. (eds.) CRYPTO 2014 Part I. LNCS, vol. 8616, pp. 426–443. Springer, Heidelberg (2014). doi:10.1007/978-3-662-44371-2_24

[GMM+16] Garg, S., Miles, E., Mukherjee, P., Sahai, A., Srinivasan, A., Zhandry, M.: Secure obfuscation in a weak multilinear map model. In: Hirt, M., Smith, A. (eds.) TCC 2016 Part II. LNCS, vol. 9986, pp. 241–268. Springer, Heidelberg (2016). doi:10.1007/978-3-662-53644-5_10

[MSZ16] Miles, E., Sahai, A., Zhandry, M.: Annihilation attacks for multilinear maps: cryptanalysis of indistinguishability obfuscation over GGH13. In: Robshaw, M., Katz, J. (eds.) CRYPTO 2016 Part II. LNCS, vol. 9815, pp. 629–658. Springer, Heidelberg (2016). doi:10.1007/978-3-662-53008-5_22

Zero-Knowledge Proofs

Two-Message Witness Indistinguishability and Secure Computation in the Plain Model from New Assumptions

Saikrishna Badrinarayanan[1(⊠)], Sanjam Garg[2], Yuval Ishai[1,3], Amit Sahai[1], and Akshay Wadia[1]

[1] UCLA, Los Angeles, USA
{saikrishna,sahai}@cs.ucla.edu, akshaywadia@gmail.com
[2] UC Berkeley, Berkeley, USA
sanjamg@berkeley.edu
[3] Technion, Haifa, Israel
yuvali@cs.technion.ac.il

Abstract. We study the feasibility of two-message protocols for secure two-party computation in the plain model, for functionalities that deliver output to one party, with security against malicious parties. Since known impossibility results rule out polynomial-time simulation in this setting, we consider the common relaxation of allowing super-polynomial simulation.

We first address the case of zero-knowledge functionalities. We present a new construction of two-message zero-knowledge protocols with super-polynomial simulation from any (sub-exponentially hard) game-based two-message oblivious transfer protocol, which we call Weak OT. As a corollary, we get the first two-message WI arguments for NP from (sub-exponential) DDH. Prior to our work, such protocols could only be constructed from assumptions that are known to imply non-interactive zero-knowledge protocols (NIZK), which do not include DDH.

We then extend the above result to the case of general single-output functionalities, showing how to construct two-message secure computation protocols with quasi-polynomial simulation from Weak OT. This implies protocols based on sub-exponential variants of several standard assumptions, including Decisional Diffie Hellman (DDH), Quadratic Residuosity Assumption, and N^{th} Residuosity Assumption. Prior works on two-message protocols either relied on some trusted setup (such as a common reference string) or were restricted to special functionalities such as blind signatures. As a corollary, we get three-message protocols for two-output functionalities, which include coin-tossing as an interesting special case. For both types of functionalities, the number of messages (two or three) is optimal.

Finally, motivated by the above, we further study the Weak OT primitive. On the positive side, we show that Weak OT can be based on any semi-honest 2-message OT with a short second message. This simplifies a previous construction of Weak OT from the Nth Residuosity Assumption. We also present a construction of Weak OT from Witness Encryption (WE) and injective one-way functions, implying the first

T. Takagi and T. Peyrin (Eds.): ASIACRYPT 2017, Part III, LNCS 10626, pp. 275–303, 2017.
https://doi.org/10.1007/978-3-319-70700-6_10

construction of two-message WI arguments from WE. On the negative side, we show that previous constructions of Weak OT do not satisfy simulation-based security even if the simulator can be computationally unbounded.

1   Introduction

There has been a long line of work on minimizing the round complexity of protocols for secure two-party computation (see, e.g., [9,19,28,29,35] and references therein). In the present work we continue the study of this question, focusing on protocols in the "plain model," which do not rely on any form of set-up, and where security is based on standard cryptographic assumptions.

We will start by addressing the case of computing functions that depend on the inputs of the two parties and deliver an output to one party. (The general case will be discussed later.) For such single-output functions, it is clear that two messages are necessary: the first by the "receiver" who receives the output and the second by the "sender." The main question we ask is under what assumptions two messages are also sufficient. Two-message protocols, also referred to as "non-interactive secure computation" (NISC) protocols [1,31], have the qualitative advantage of allowing one party to go offline (after sending its message) while waiting for the other party to respond.

For security against semi-honest parties, the situation is well understood: such general two-message protocols exist if a two-message oblivious transfer (OT) protocol with security against semi-honest parties exists [9,41]. This assumption is also necessary, since OT is a simple special case of general secure computation.

The situation is far more complex when considering security against malicious parties. For protocols with black-box simulation, four messages are necessary and are also sufficient under standard assumptions [21,35]. This can be improved to two messages by using standard setup assumptions such as a common reference string [9,29,31]. In the plain model, however, two-message protocols that satisfy the standard notion of security are known not to exist, even when allowing non-black-box simulation and even for the special of zero-knowledge [4,23]. To get around this impossibility, Pass [39] suggested considering simulation whose running time is super-polynomial, but not necessarily unbounded, and realized two-message zero-knowledge in this model. General secure computation with super-polynomial simulation was first studied by Prabhakaran and Sahai [40] and by Barak and Sahai [6] in the context of concurrent security (with protocols requiring multiple rounds of interaction).

Secure computation with super-polynomial simulation is motivated by the fact that it captures the desirable security goals for the typical case of computing "non-cryptographic" functions, where even an unbounded simulator does not get a meaningful advantage. Moreover, using complexity leveraging, such protocols can be "as good" as standard protocols even for computing cryptographic functions, provided that the level of security of the primitives or other protocols with which they interact is sufficient to protect against an adversary with the same running time as the simulator. See Sect. 1.2 for further details.

The above discussion motivates the following question:

Under what assumptions can we construct two-message secure computation protocols with super-polynomial simulation in the plain model?

A natural first step is to study the above question for the special case of zero-knowledge, which captures functions that take input from only one party. Zero-knowledge protocols with unbounded simulation are equivalent to *witness indistinguishable* (WI) protocols. Two-message WI protocols for NP (also called private coin ZAPs) can be constructed from non-interactive zero-knowledge (NIZK) protocols [?]. These were used in [39] to obtain 2-message zero-knowledge arguments with quasi-polynomial simulation. They were further used in [20] to obtain two-message blind signatures in the plain model, which can be viewed as another instance of general secure two-party computation.

While it is known that NIZK can be based on standard assumptions such as trapdoor permutations and bilinear maps [8,10,16,26], there are several other well studied assumptions, such as the DDH assumption or even a strong assumption such as Witness Encryption [17], that are not known to imply NIZK or even 2-message WI arguments for NP. As far as we know, all non-trivial instances of 2-message protocols in the plain model appearing in the literature (even ones with unbounded simulation) require either NIZK or bilinear maps [25].

1.1 Our Contribution

We essentially settle the above question, showing that general two-message secure computation in the plain model with super-polynomial simulation is implied by any (sub-exponentially secure) "game-based" two-message OT protocol. Such a protocol is required to be secure with super-polynomial simulation against a malicious receiver, and is only required to satisfy indistinguishability-based security against the sender (i.e., the sender cannot distinguish between the two possible selection bits of the receiver). From here on, we refer to such an OT protocol as *Weak OT*. Weak OT protocols can be easily constructed from the DDH assumption [2,38] (which is not known to imply NIZK) and are also known to follow from the Quadratic Residuosity Assumption and the Nth Residuosity Assumption (i.e., the security of the Paillier cryptosystem) [27].

The above result essentially settles our main question, since Weak OT can be viewed as the simplest analogue of two-message semi-honest OT for the case of security against malicious parties. As a corollary of our main result, Weak OT implies 3-message protocols with super-polynomial simulation in the plain model for functions that deliver outputs to both parties. This includes (multi-output) coin-tossing as an important special case. Motivated by the usefulness of Weak OT, we further study this primitive, obtaining several new positive and negative results.

We now give a more detailed account of our results.

1. We start by studying the Weak OT primitive described above (and formally defined in Sect. 2) and explore the feasibility of using it for secure computation with super-polynomial simulation. We show that Weak OT protocols

may not even be secure with unbounded simulation. We demonstrate this by constructing a protocol (only a slight modification of the protocol in [2,38]) that achieves the game based notion but suffers from a real attack. Concretely, we show a malicious sender strategy for this protocol such that even a single instance of execution of the protocol with the malicious sender would suffer from the attack. This is counter-intuitive because in a single instance of OT, any probabilistic mapping from the receiver's input to its output can be realized by a malicious sender in the ideal model, and so simulation seems easy. However, in our attack, the receiver's output becomes a value that *cannot be known* to the sender. This attack not only violates the intuitive notion of correctness and security, but it provably cannot be simulated even by an unbounded simulator. This impossibility result shows that proving security using a super-polynomial simulator, which is the setting in the rest of our work, is non-trivial and interesting.

2. Based on any (sub-exponentially secure) Weak OT, we construct a secure protocol for two-message zero knowledge argument of knowledge with quasi-polynomial simulation in the plain model. This implies the first such protocols, and even the first 2-message WI protocols, under assumptions that are not known to imply NIZK. More precisely, we prove the following:

Theorem 1. *Assuming the existence of sub-exponentially secure Weak OT, there exist two-message zero knowledge arguments (with argument of knowledge) for NP in the plain model with quasi-polynomial simulation.*

In particular, we get the following new corollary:

Theorem 2. *Two-message witness indistinguishable arguments for NP can be based on the sub-exponentially hard Decisional Diffie-Hellman Assumption.*

3. Using a variant of the "GMW paradigm" [22], we extend the above result to the case of general secure computation. Concretely, we prove the following theorem:

Theorem 3. *Two Message Secure Computation protocols with quasi-polynomial simulation in the plain model for general single-output functionalities can be based on any sub-exponentially secure Weak OT.*

As a corollary, we get the first general 2-message protocols in the plain model.

Corollary 1. *Two Message Secure Computation protocols with quasi-polynomial simulation for general single-output functionalities can be based on any of the following sub-exponentially hard assumptions: (1) Decisional Diffie-Hellman Assumption; (2) Quadratic Residuosity Assumption; (3) N^{th} Residuosity Assumption.*

While such protocols are not very hard to obtain from previous two-message zero-knowledge protocols with super-polynomial simulation, we are not aware of such a result in the literature. Moreover, the DDH-based construction crucially depends on our new construction of 2-message zero-knowledge protocols.

Secure two-message protocols for single-output functionalities imply secure three-message protocols for two-output functionalities. Concretely, we get the following corollary.

Corollary 2. *Three-message secure protocols with quasi-polynomial simulation for general two-output functionalities (satisfying "security with abort") can be based on sub-exponentially secure Weak OT, and in particular on sub-exponential DDH.*

A particularly useful special case is that of (multi-bit) coin-tossing where neither party has any input and both parties get a uniformly random string as output. Here quasi-polynomial simulation seems enough for all natural applications. Despite the large body of work on coin-tossing, we are not aware of any previous multi-bit coin-tossing protocol in the plain model that provides a meaningful notion of security (even a game-based notion) with only 3 messages under standard assumptions. Our coin-tossing result should be compared with the 5-message protocol from [35] (which is optimal for standard black-box simulation) and a recent 4-message protocol from [28] which is secure with inverse-polynomial simulation error.

4. To further expand the class of assumptions on which we can base our general protocols, we provide new constructions of Weak OT satisfying the game based notion.

   The first construction is based on any high rate semi-honest secure OT which in turn can be reduced to any high rate additively homomorphic encryption. Concretely, we need a semi-honest one-out-of-two string-OT protocol in which the output length on a pair of strings of length ℓ is smaller than $c\ell$ for some constant $c < 2$. As a corollary, by instantiating the high rate homomorphic encryption scheme using a construction of Damgård and Jurik [14], we simplify the construction and analysis of Weak OT from the Nth Residuosity Assumption of Halevi and Kalai [27]. In particular, our construction only relies on the semantic security of the DJ cryptosystem and simple "syntactic" properties (homomorphism and ciphertext length) and does not involve smooth projective hash functions. This general new construction of Weak OT could potentially lead to basing our general protocols on other assumptions, such as lattice-based assumptions. The construction is presented in Sect. 6.

   Our second construction of Weak OT builds on Witness Encryption [17] and any injective one way function. This is described in Sect. 7. As a corollary, all of the results discussed above can also be based on WE (and injective one-way functions).

At the heart of our two-message secure computation protocols is a two-message protocol for zero-knowledge from sub-exponential security of game based OT. Note that this is contrast to the construction of Pass [39] who gave a construction based on NIZKs. Our alternative new construction avoids the use of NIZKs and is what enables our new results that provide constructions under alternative assumptions. This construction of zero-knowledge is provided in Sect. 4. The construction of two-message secure computation using this zero-knowledge protocol is provided in Sect. 5.

1.2   Discussion and Related Work

In this section we discuss the two key features of our protocols: super-polynomial simulation and security in the plain model, and survey some related work.

What good is super-polynomial simulation? Intuitively speaking, the notion of super-polynomial simulation (SPS) guarantees that the real world adversary does not learn anything more than an ideal world adversary running in super-polynomial time. So, what does the SPS ideal world adversary learn? For information theoretic functionalities (example, Yao's millionaire problem), the running time of the ideal-world simulator does not affect security in any sense. In particular the computational power awarded to the ideal world adversary is useless for learning anything about the input of the honest party. It does not rule out the possibility that the adversary learns some super-polynomial function of its view but this is irrelevant for the task at hand. On the other hand, for cryptographic functionalities, the adversary's ability to run in super-polynomial time is indeed problematic as it could potentially harm the security of the functionality itself. However, at an often small cost to efficiency, it is almost always possible to choose a higher security parameter for the cryptographic operations performed by the functionality such that meaningful security can be obtained (see e.g. [20] for the example of blind signatures). SPS is commonly used in cryptography. In fact, any zero knowledge protocol with super polynomial simulation is a witness indistinguishable protocol.

Relation to concurrently secure computation. The notion of concurrently secure super-polynomial simulation [6,18,40] and its variants [11] have been extensively studied in the literature. This notion is known to be impossible [5, 24,37] to achieve with polynomial-time simulation. The notion of two-message secure computation that we study implies the notion of concurrently secure computation, in the restricted setting where the adversary is allowed to play as a sender or as a receiver across all concurrent sessions (the so-called "fixed-roles" setting). This improves on the round complexity of known solutions.

Recently, Dottling et al. [15] constructed two round two-party computation protocols for certain functionalities that is secure against semi-honest senders and malicious receivers. However, they consider a game-based security notion against a malicious receiver and this is incomparable to our setting.

Concurrent and subsequent work. Concurrent to our work, Jain et al. [34] construct protocols that are similar to our two-round protocols. While their focus is on polynomial time distinguisher-dependent simulation, we focus on super-polynomial simulation. Therefore, the only result in common between the two papers is two-round witness indistinguishability for NP from Weak OT. Our proof of WI is significantly simpler than theirs, because our analysis is via super-polynomial simulation. Our paper also contains additional results on Weak OT (both negative and positive) that simplify previous constructions and extend the set of assumptions on which both our and their round-optimal protocols can be based.

Subsequent to our work, Khurana and Sahai [36] use our two-message secure computation protocol crucially to build two-message non-malleable commitments with respect to commitment from sub-exponentially hard DDH.

Even though we have a straight line simulation, our protocol doesnt extend to the UC/concurrent setting because that requires non-malleability. A very recent follow-up work by Badrinarayanan et al. [3] achieves concurrent security in the MPC setting by building on our techniques (along with other techniques), using 3 rounds of simultaneous message exchange. Note that, in contrast, our protocols use only 2 rounds of unidirectional message exchange.

1.3 Technical Overview

The new 2-round SPS-Zero Knowledge protocol from 2-round Weak OT. The technical heart of our result is a new 2-round super-polynomial simulation secure zero knowledge protocol (SPS-ZK) from a 2-round weak OT. The weak OT protocol we use has statistical sender's security but only T-chooser's security. That is, the receiver's choice bit is secure against all adversaries running in time $T.\text{poly}(\lambda)$. Additionally, we will also use a T-time extractable commitment protocol. To ease the exposition, let's allow the simulator to run in exponential time. Then, by running the protocol with an appropriately smaller security parameter, we can rely on just quasi-polynomial simulation.

The main idea behind the new zero knowledge protocol is to "squash" a parallelized version of Blum's 3-round zero knowledge protocol for Hamiltonicity, by making use of the 2-round Weak OT protocol. Our technique applies more generally to parallelized Σ-protocols with "special soundness", but here we will focus on Blum's protocol for clarity. Recall that in Blum's protocol, the prover generates an initial message α, and prepares two responses γ_0, γ_1. The verifier then sends a random bit $\beta \in \{0, 1\}$, and the prover responds with γ_β.

To squash this protocol to two rounds, we first have the verifier choose β at the start, and then use β as its input in the role of receiver in the Weak OT protocol. Note that this intuitively keeps β hidden from the prover. Then, the prover sends α separately as part of its message, but also uses γ_0 and γ_1 as its inputs in the role of sender in the Weak OT protocol. Thus, the verifier learns only α and γ_β and can then verify the original Blum proof. This protocol can be repeated in parallel to boost soundness. We will now discuss how to establish SPS zero knowledge and computational soundness separately.

Zero Knowledge: No rewinding allowed. First, observe that we can't directly use the same proof strategy as in Blum's protocol as we can not rewind the adversary here. In our protocol, since the verifier sends just one message, if we try rewinding the malicious verifier, it could just keep sending the same message over and over. Thus, there is nothing to be gained from rewinding.

To establish zero-knowledge, we will use complexity leveraging to construct a super-polynomial simulator running in time $T_1 \cdot \text{poly}(\lambda)$, where $T_1 > T$, that can extract β from the verifier's first message in the Weak OT protocol. Now that the simulator knows β, simulation at first glance appears to be straightforward, since it needs to place a correct value only in γ_β. This can be done by just invoking the zero knowledge simulator of Blum's protocol. However, there is a subtle flaw in this argument due to the Weak OT protocol, as we discuss now in further detail.

Before we can see the flaw, we have to briefly discuss soundness. In order for soundness to hold, we need that the prover cannot somehow "malleate" the verifier's first OT message into a successful second OT response by the prover. The way we will achieve such non-malleability is by adding a weak commitment that can be extracted in time $T < T_1$. Recall that it is impossible for an adversary to take as input a T_1-strong commitment $C_{T_1}(m)$, and produce a T-weak commitment $C_T(m')$ in a way that causes m' to depend on m. This is easy to see: An adversary running in time T can anyway break $C_T(m')$ and recover m'. It could then use m' to predict m, thereby breaking the T_1-secure commitment $C_{T_1}(m)$ – a contradiction to the stronger security of C_{T_1} since $T_1 > T$.

In the case of zero knowledge, recall that our simulator runs in time $T_1 \cdot \text{poly}(\lambda)$, where $T_1 > T$. Note that the OT protocol does not have receiver's security against adversaries running in time $T_1 \cdot \text{poly}(\lambda)$, since it needs to extract β from the first message of the weak OT. But then, it can anyway break the commitment scheme since $T_1 > T$. Therefore, now, in order for the commitment to be stronger than the OT, we need $T_1 < T$, whereas for proving soundness, we require that $T_1 > T$. (Since we require that the time taken to break the commitment is lesser than the time taken to break the chooser's security in the OT protocol.) This a fundamental contradiction that suggests that perhaps our goal is impossible to achieve!

We fix this by exploiting the special structure of our protocol: Recall our observation that a cheating verifier, which is without loss of generality deterministic, if rewound, would just keep sending the same first message. Now, we want to exploit this fact to keep $T_1 > T$ as needed by soundness, and argue zero knowledge in a different way: The simulation strategy itself is the same as before. That is, the simulator runs in time $T_1 \cdot \text{poly}(\lambda)$ and extracts β from the first message of the weak OT. It then invokes the simulator of Blum's protocol and produces the prover's message. This second phase runs in polynomial time. Now, let's consider the reduction that breaks the commitment scheme by interacting with the malicious PPT verifier. The reduction, given the commitment to be broken as an external challenge, includes it as part of the prover's message (more specifically, includes it as a commitment to $\gamma_{1-\beta}$ in the string α). Now, based on the PPT verifier's guess it breaks the commitment. The only stage in the reduction that runs in super-polynomial time is when it breaks the initial message of the verifier to extract β. Therefore, let's consider the malicious verifier with the "best possible" initial message and fix this message. The value β^* extracted from this can just be given as auxiliary input (non-uniform advice) to the reduction! So, now, the reduction is a non-uniform PPT machine. Therefore, if the PPT reduction can now break the commitment scheme, we will achieve a contradiction. Note that the auxiliary input is also given to the external challenger of the commitment scheme.

Soundness. To establish soundness of the protocol, we in fact prove a stronger property: that our protocol is an argument of knowledge. (We will anyway need this later when we construct the two message secure computation protocol for any general functionality). We will construct an extractor, that, running in super-polynomial time $T \cdot \mathrm{poly}(\lambda)$, can extract out γ_0 and γ_1 from the prover's initial message by running the commitment extractor. Blum's protocol is designed so that α, γ_0, and γ_1 together yield knowledge of the Hamiltonian cycle in the original graph and hence the extractor learns the witness. We will then show that if the extractor fails, but the malicious prover succeeds in giving a correct proof, we can use this prover to break the T-chooser's security of the OT protocol by using the external OT challenge in the verifier's first message against this malicious prover. Several challenges arise when trying to establish soundness. We discuss them now.

Recall that the aim is to show that if the malicious prover succeeds in giving a valid proof but the extractor fails, then the reduction will break the T-chooser's security of the OT protocol. Note that the reduction can run in time $T.\mathrm{poly}(\lambda)$. The idea here was that the reduction interacts with the malicious prover and embeds the external OT challenge (of the OT receiver) in one of the indices as part of the verifier's first message. After checking that the proof is valid, the reduction can extract both γ_0 and γ_1 from α by running the T-time commitment extractor and then run the BlumExt to obtain the choice bit of the OT challenge. However, there is a subtle issue here that in order to check that the proof is valid, the reduction needs to run the third stage OT algorithm to recover γ_β. But, since it did not generate the first OT message, the reduction does not have the associated state that was used in that generation and hence cannot validate the proof (the state output by the first OT algorithm will be needed as input for the third stage).

We fix this by using a simple combinatorial argument. We consider a new verifier strategy where the verifier checks the proof at all indices except one and this choice is not revealed to the prover. It can be easily seen that the success probability of the malicious prover is as much, if not more, against this new verifier as well. Now, the reduction no longer needs to verify the proof at the index where the OT challenge was embedded. Also, if the malicious prover has to produce a valid proof, with probability close to 1, it still needs to produce a valid proof at every index since it can guess the missed out index with very small probability. Therefore, the Blum extraction would still work correctly on the embedded index and the reduction can break the OT receiver's security.

Two message secure computation. Given any weak OT protocol and the two message secure zero knowledge protocol from above, we compile them together using Yao's garbled circuits construction to produce a two message secure computation protocol for any general functionality. In fact, we don't need the full power of the zero knowledge protocol from above. In this construction, we will only need the weaker notion of witness indistinguishability(WI) which is anyway implied by SPS zero knowledge.

Consider a sender with input x and a receiver with input y and let the function they're computing be f. In the first round, the receiver, using each bit of his input, computes the first message of the weak OT protocol and sends this across. In addition, he also initiates a WI protocol with the sender and sends the first message of the verifier. Finally, he also sends the output of a one way function OWF that is not invertible in time $T.\mathrm{poly}(\lambda)$ (but is invertible in time $T_1.\mathrm{poly}(\lambda)$ where $T_1 > T$). Looking ahead, this value will help the simulator against a cheating receiver to generate a proof using the trapdoor statement. In response, the sender computes a garbled circuit that has his input hardwired into it and then runs the OT algorithm using the garbled keys as his input to the OT. Also, he computes a commitment c_1 to his input and another commitment c_2 to 0 which will prove to be useful for the simulator. He then computes a WI proof that he computed the commitment c_1 correctly, ran the OT algorithm correctly and computed the garbled circuit correctly. It is easy to see that the receiver, after checking the validity of the proof, can recover the garbled keys corresponding to his input using the OT and evaluate the garbled circuit to obtain the output of the function. The trapdoor statement in the WI proof will basically say that the prover knows the pre-image to the output of the OWF and the commitment c_2 is a commitment to this pre-image. Notice that we don't need the full expressiveness of the zero knowledge property. It is enough to have just witness indistinguishability and the simulator against a malicious receiver, just extracts the pre-image of the one-way function OWF and uses the trapdoor statement to prove that the pre-image is correct.

Similar to the proof of the zero knowledge protocol, the key tool in order to prove security is complexity leveraging. The main obstacle we face is very similar to the one faced in the case of the zero knowledge protocol. In particular, for proving security against a malicious receiver, we will need to break the chooser's security of the OT protocol and then reduce the security of our protocol to the hiding of the commitment scheme. Therefore, we will need $T_1 < T$. However, to prove security against a malicious sender, we will require that $T < T_1$, following a similar argument as in the case of the soundness of the zero knowledge protocol. As in the case of our zero knowledge protocol, we fix this issue by considering an intermediate hybrid where non-uniform advice can provide key information embedded in the malicious receiver's fixed first message. This advice allows us to consider experiments that do not incur the running time needed to actually extract the information that was present in the first message of the receiver.

2   Preliminaries

Let λ denote the security parameter. We say that a function is *negligible* in the security parameter λ, if it is asymptotically smaller than the inverse of any fixed polynomial. Otherwise, the function is said to be *non-negligible* in λ. We say that an event happens with *overwhelming* probability if it happens with a probability $p(\lambda) = 1 - \nu(\lambda)$ where $\nu(\lambda)$ is a negligible function of λ. In this section, we define the primitives studied in this paper. We will start by defining a weaker indistinguishability based notion for oblivious transfer and then subsequently describe the simulation based notion for general functionalities.

We write $y = A(x; r)$ when the algorithm A on input x and randomness r, outputs y. We write $y \leftarrow A(x)$ for the process of picking r at random and setting $y = A(x; r)$. We also write $y \leftarrow S$ for sampling y uniformly at random from the set S. Some more primitives are defined in the full version.

Weak OT. In this paper, we consider a 1-out-of-2 *Oblivious Transfer* protocol (similar to [2,27,38]) where one party, the *sender*, has input composed of two strings (M_0, M_1) and the input of the second party, the *chooser*, is a bit c. The chooser should learn M_c and nothing regarding M_{1-c} while the sender should gain no information about c. We give a definition for the setting where the sender is protected information theoretically while the chooser is protected only computationally.

Definition 1 (Weak OT). *The chooser runs the algorithm* OT_1 *which takes* 1^λ *and a choice bit* $c \in \{0, 1\}$ *as input and outputs* $(\mathsf{ot}_1, \mathsf{state})$. *Chooser then sends* ot_1 *to the sender, who obtains* ot_2 *by evaluating* $\mathsf{OT}_2(1^\lambda, \mathsf{ot}_1, M_0, M_1)$, *where* M_0 *and* M_1 *(such that* $M_0, M_1 \in \{0, 1\}^\lambda$*) are its inputs. The sender then sends* ot_2 *to the chooser who obtains* M_c *by evaluating* $\mathsf{OT}_3(1^\lambda, \mathsf{ot}_2, \mathsf{state})$.

- **Perfect correctness.** *For every choice bit* $c \in \{0, 1\}$ *of the chooser and input messages* M_0 *and* M_1 *of the sender we require that, if* $(\mathsf{ot}_1, \mathsf{state}) \leftarrow \mathsf{OT}_1(1^\lambda, c)$, $\mathsf{ot}_2 \leftarrow \mathsf{OT}_2(1^\lambda, \mathsf{ot}_1, M_0, M_1)$, *then* $\mathsf{OT}_3(1^\lambda, \mathsf{ot}_2, \mathsf{state}) = M_c$ *with probability 1. We speak of statistical correctness if this probability is overwhelming in* λ.
- **Chooser's security.** *We require that for every non-uniform polynomial-time adversary* $\mathcal{A}$, $|\Pr[\mathcal{A}(\mathsf{OT}_1(1^\lambda, 0)) = 1] - \Pr[\mathcal{A}(\mathsf{OT}_1(1^\lambda, 1)) = 1]|$ *is negligible in* λ.
  We speak of T-*chooser's security if the above condition holds against all non-uniform adversaries* $\mathcal{A}$ *running in time* $T \cdot \mathrm{poly}(\lambda)$.
- **Statistical sender's security.** *We define an unbounded*[1] *time extractor* OTExt *such that* OTExt *on any input* ot_1 *outputs 0 if there exists some random coins such that* $\mathsf{OT}_1(1^\lambda, 0)$ *outputs* ot_1, *and 1 otherwise.*
  Then for any value of ot_1, *and any* K_0, K_1, L_0, L_1 *with* $K_{\mathsf{OTExt}(\mathsf{ot}_1)} = L_{\mathsf{OTExt}(\mathsf{ot}_1)}$, *we have that* $\mathsf{OT}_2(1^\lambda, \mathsf{ot}_1, K_0, K_1)$ *and* $\mathsf{OT}_2(1^\lambda, \mathsf{ot}_1, L_0, L_1)$ *are statistically indistinguishable. We speak of computational sender's security if for all non-uniform polynomial time adversaries* $\mathcal{A}$ *we have that* $|\Pr[\mathcal{A}(\mathsf{OT}_2(1^\lambda, \mathsf{ot}_1, K_0, K_1)) = 1] - \Pr[\mathcal{A}(\mathsf{OT}_2(1^\lambda, \mathsf{ot}_1, L_0, L_1)) = 1]|$ *is negligible in* λ.

T-secure Weak OT. Finally, we define T-secure Weak OT to be a Weak OT protocol with T-chooser's security. Note that we can claim that any Weak

---

[1] Note that fixing the parameters of the scheme, we can bound the running time of the extractor by some sub-exponential function but we avoid it to keep notation simple and avoid unnecessary parameters.

OT protocol with chooser's security based on a set of assumptions Υ, is also a T-secure Weak OT protocol if each assumption in Υ is additionally assumed to be secure against all non-uniform adversaries running in time $T \cdot \mathsf{poly}(\lambda)$. Note that this additionally relies on the fact that the security reduction for proving chooser's security of the underlying protocol is tight up to a multiplicative polynomial factor, in the security parameter.

Naor-Pinkas and Aiello et al. [2,38] provided a construction of a Weak OT protocol based on the Decisional Diffie-Hellman assumption. Subsequently, Halevi and Kalai [27] provided an instantiation based on any smooth projective hash function. Further, note that the above definition is not a simulation-based definition but rather an indistinguishability-based one. Although it is a meaningful notion and is sufficient for some applications, it is still weaker than the simulation-based (described next) notion.

Two Message Secure Computation via super-polynomial simulation. The simulation-based definition compares the "real world," where the parties (the sender and the receiver) execute the protocol, to an "ideal world," where no message is exchanged between the parties; rather, there is a trusted party that takes an input from both parties, computes the output of the functionality on these inputs, and sends the corresponding output to each party. Loosely speaking, the simulation (resp., super-polynomial simulation)-based definition asserts that for every efficient adversary $\mathcal{A}$ (controlling either the sender or the receiver) in the real world there exists an efficient (resp., super-polynomial) simulator $\mathcal{S}$, controlling the same party in the "ideal world," so that the outputs of the parties in the ideal world are computationally indistinguishable from their outputs in the real world. In particular, the simulator $\mathcal{S}$ needs to simulate the view of the adversary $\mathcal{A}$ in a computationally indistinguishable manner.

Next, we formally define a Two Message Secure Computation protocol $\langle S, R \rangle$, between a *sender* S with input x and a *receiver* R with input y. The receiver should learn $f(x, y)$ and nothing else[2] while the sender should gain no information about y. More formally we will define this notion by comparing a two-round realization in the real-world with an ideal world scenario.

Real World. A Two Message Secure Computation protocol $\langle S, R \rangle$ is defined by three probabilistic algorithms $(\mathsf{NISC}_1, \mathsf{NISC}_2, \mathsf{NISC}_3)$ as follows. The receiver runs the algorithm NISC_1 which takes the receiver's input $y \in \{0,1\}^\lambda$ as input and outputs $(\mathsf{nisc}_1, state)$. The receiver then sends nisc_1 to the sender, who obtains nisc_2 by evaluating $\mathsf{NISC}_2(\mathsf{nisc}_1, x)$, where $x \in \{0,1\}^\lambda$ is the sender's input.[3] The sender then sends nisc_2 to the receiver who obtains $f(x, y)$ by evaluating $\mathsf{NISC}_3(\mathsf{nisc}_2, state)$.

At the onset of the computation the *real world adversary* $\mathcal{A}$ corrupting either the sender S or the receiver R, receives some auxiliary information z. Next, the

---

[2] Unlike Weak OT in which the sender is protected information theoretically this notion will provide only computational security for the sender.

[3] For simplicity of notation we denote the lengths of the inputs of the sender and the receiver by λ. In general they could be arbitrary polynomials in the security parameter λ.

computation proceeds as described above where the honest party sends messages as prescribed by the protocol and the adversary $\mathcal{A}$ sends arbitrary messages on behalf on the corrupted party. At the end of the computation the uncorrupted party outputs whatever is specified in the protocol. The corrupted party outputs any arbitrary PPT function of the view of $\mathcal{A}$. The overall output of the real-world experiment consists of all the values output by all parties at the end of the protocol, and this random variable is denoted by $\mathsf{REAL}_{\mathcal{A}}^{\langle S,R \rangle}(1^k, x, y, z)$. Let $\mathsf{REAL}_{\mathcal{A}}^{\langle S,R \rangle}$ denote the ensemble $\{\mathsf{REAL}_{\mathcal{A}}^{\langle S,R \rangle}(1^k, x, y, z)\}_{k \in \mathbb{N}, x, y \in \{0,1\}^\lambda, z \in \{0,1\}^*}$.

Ideal World. In the ideal world experiment, the sender S and the receiver R interact with a *trusted party* for computing a function $f : \{0,1\}^\lambda \times \{0,1\}^\lambda \to \{0,1\}^\lambda$. The ideal world computation in presence of the *ideal world adversary* $\mathcal{S}$ corrupting either the sender S or the receiver R, and an (incorruptible) *trusted party* $\mathcal{F}$, proceeds as follows. First, as in the real-life model, $\mathcal{S}$ gets auxiliary information z. Next, the ideal world adversary $\mathcal{S}$ generates any arbitrary input on behalf of the corrupted party, which it sends to the trusted party $\mathcal{F}$. The honest party sends its input to the trusted party $\mathcal{F}$. At this point the ideal functionality evaluates the output and sends it to the receiver. The honest receiver outputs this value. The adversarial receiver $\mathcal{S}$ outputs an arbitrary value. Note that $\mathcal{S}$ is allowed to run in super-polynomial time. In this work, we will focus by default on simulators running in quasi-polynomial time - i.e. $n^{\mathsf{poly}(\log(n))}$ where n is the security parameter. (See Definition 6 in [39] for a definition of quasi-polynomial simulation in the context of zero-knowledge protocols.)

The ideal world output consists of all the values output by all parties at the end of the protocol. We denote this random variable by $\mathsf{IDEAL}_{\mathcal{S}}^{\mathcal{F}}(1^k, x, y, z)$ and $\mathsf{IDEAL}_{\mathcal{S}}^{\mathcal{F}}$ denotes the ensemble $\{\mathsf{IDEAL}_{\mathcal{S}}^{\mathcal{F}}(1^k, x, y, z)\}_{k \in \mathbb{N}, x, y \in \{0,1\}^\lambda, z \in \{0,1\}^*}$.

Equivalence of Computations. Informally, we require that executing a protocol $\langle S, R \rangle$ in the real world roughly emulates the ideal process for evaluating f.

Definition 2. *Let f be any polynomial time computable function on two inputs and let $\langle S, R \rangle$ be a protocol between a sender S and a receiver R. We say that $\langle S, R \rangle$ two-message securely evaluates f if for every PPT real world adversary $\mathcal{A}$ there exists an ideal world adversary $\mathcal{S}$, such that $\mathsf{REAL}_{\mathcal{A}}^{\langle S,R \rangle} \overset{c}{\approx} \mathsf{IDEAL}_{\mathcal{S}}^{\mathcal{F}}$.*

Stricter Simulation. As described above the ideal world adversary is allowed to execute in super-polynomial time. We will consider a stricter notion of simulation under which a simulator is allowed to execute in super-polynomial time prior to its interaction with the ideal functionality. The simulator is subsequently restricted to be polynomial time. We discuss this more formally in the full version.

In the full version, we define the notion of Two Message Secure Computation for two specific functionalities, namely, zero-knowledge and Parallel OT.

3 Difficulties in Constructing Two Message Secure Computation Protocols

Goldreich and Oren [23] showed that it is impossible to construct 2-round zero-knowledge arguments for languages outside BPP. As explained in [13], this result extends in a straightforward manner to show the impossibility of constructing 2-round T-zero-knowledge[4] arguments for T-hard languages, that are sound against cheating provers running in time $T \cdot \mathsf{poly}(\lambda)$. More recent works [12,13] gave a black-box impossibility result ruling out 2-round zero-knowledge sound against polynomial-time cheating provers (based on T-hard falsifiable assumptions). Note that since Two Message Secure Computation for OT implies 2-round zero-knowledge arguments we can obtain analogous impossibility results for Two Message Secure Computation for OT. It is also interesting to note that the positive result for Two Message Secure Computation for OT obtained in this paper assume that the underlying assumption is T'-hard for T' that is strictly more than the running time of the distinguisher. Thus, these results are essentially tight.

The aforementioned impossibility result only rules out black-box reductions to falsifiable assumptions. As a starting point, based on the premise that known instances of Weak OT protocols such as [2,27,38] are not known to be susceptible to any attacks in the simulation setting, one may conjecture that that in fact all Weak OT protocols can be proven secure under a simulation based definition when unbounded simulation is allowed and we are willing to make strong (possibly non-falsifiable) assumptions.

In fact, it is argued in [27] (Sect. 3) that any Weak OT protocol provides simulation-based security in the standard sense for the case of a malicious sender, assuming that the simulator is allowed to reset the sender. This is argued as follows. The simulator (who does not know the choice bit of the actual receiver) simulates the (honest) receiver first with choice bit $b = 0$, and then it resets the sender and simulates the honest receiver with choice bit $b = 1$. This way the simulator extracts both messages M_0 and M_1 from the corrupted sender. It then gives (M_0, M_1) to the trusted party. Then the simulator uses the view of the cheating sender in the first execution (with the choice-bit $b = 0$). They argue that this view is indistinguishable from the "real world" view based on the receiver's security of the Weak OT protocol, and from the fact that the sender does not receive any output from the trusted party. On the other hand, they argue that Weak OT does not give standard simulation-based guarantee in the case that the receiver is corrupted, because a malicious receiver is not guaranteed to "know its own choice bit b," and therefore the simulator does not know which input bit to send to the trusted party. However, they point out that this does guarantee an exponential time simulation of the receiver's view of the interaction.

---

[4] Recall that in T-zero-knowledge protocols the simulator and the distinguisher run in time $T \cdot \mathsf{poly}(\lambda)$.

Note that the argument from [27] is only applicable when the view of a cheating sender or the view of a cheating receiver is considered by itself. We show that if (as per standard definitions) joint distributions of the outputs of both parties are considered, then proving simulation based security for a Weak OT protocol even when unbounded simulation is allowed is very problematic. We demonstrate this by constructing a protocol that can be proved to be Weak OT under reasonable assumptions but suffers for a real attack under a simulation based definition. In particular, we show a malicious sender strategy such that even a single instance of execution of the protocol with a malicious sender can not be simulated by any unbounded simulator.

The protocol that we construct is only a slight modification of known protocols [2,38] and highlights at the very least, the obstacles that we face even in proving security of specific protocols. We start by recalling the ElGamal encryption scheme abstractly. Let G be a multiplicative subgroup of Z_q^* of order p, where p and q are primes and p is of length λ that divides $q-1$. Let g be the generator of this group G. The public key for ElGamal encryption is generated by sampling $x \leftarrow Z_p^*$ and setting the public key to be (p,q,g,h) where $h = g^x$. The encryption procedure $Enc((p,q,g,h),m)$ is defined follows: Choose $r \in Z_p^*$ and output $(g^r, m \cdot h^r)$. The decryption procedure $Dec((u,v),x)$ outputs $\frac{v}{u^x}$. Let $e(\cdot)$ be some invertible encoding function mapping Z_p to G. Then *circular security* of ElGamal implies that the encryption scheme remains semantically secure even when an encryption of $e(x)$ is given to the adversary. In particular semantic security is preserved when $Enc((p,q,g,h),e(x))$ is included in the public key. As pointed out in [?], it is unlikely that it would follow from the DDH assumption.

Lemma 1. *Assuming that ElGamal is circularly secure, there exists a Weak OT protocol and a real world cheating sender S^* strategy for this protocol such that it can not be simulated by any (unbounded) ideal world simulator.*

Proof. We will start by giving the protocol. The protocol used in our counter example is very similar to the DDH based Weak OT protocols from [2,38]. The only difference being that our protocol includes an encryption E of $e(a)$ along with its first message. This value is not used by the protocol itself but however will be useful for the malicious sender that we will construct.

1. $(\mathsf{ot}_1, state) \leftarrow \mathsf{OT}_1(c)$: Sample $a, b \leftarrow \mathbb{Z}_p$. Compute $x := g^a$, $y := g^b$, $z := g^{ab+c}$ and $E = (g^r, e(a) \cdot g^{br})$. The output ot_1 is then the tuple (x, y, z, E).
2. $\mathsf{ot}_2 \leftarrow \mathsf{OT}_2(\mathsf{ot}_1, M_0, M_1)$: sample $t_0, s_0, t_1, s_1 \leftarrow \mathbb{Z}_p$. For each $i \in \{0,1\}$, compute $w_i := x^{s_i} g^{t_i}$ and $u_i := \left(z \cdot g^{-i}\right)^{s_i} y^{t_i} M_i$. The output ot_2 is then the tuple (w_0, u_0, w_1, u_1).
3. $\mathsf{OT}_3(\mathsf{ot}_2, state)$: Compute M_c as $u_c \cdot w_c^{-b}$.

The above protocol is a Weak OT protocol. The argument follows directly from the proof of [2,38] except that in our case the chooser's security will be based on the circular security of ElGamal.

We will now provide an attack that specifies a particular cheating strategy for a malicious sender (in a single instance of execution of the protocol) that can not be simulated by any unbounded simulator. In particular we will provide an efficient malicious sender strategy such that for every unbounded simulator we have that the joint distributions of the sender's view and the receiver's output in the real world and the ideal world are efficiently distinguishable.

Our cheating sender proceeds as follows: On receiving the message (x, y, z, E) it proceeds by setting ot_2 to be the tuple (E, E). On receiving this message R, regardless of the value of b, outputs $e(a)$. Note that in the real world the joint distribution of the view of the sender and the output of the honest receiver $((x, y, z, E), e(a))$ is sufficient for the distinguisher to efficiently compute the input of the honest receiver. The distinguisher computes the honest receiver's input c as follows:

- Given : (x, y, z, E, a)
- Compute g^{ab} as y^a. (since $y = g^b$)
- Then, compute c as $\frac{z}{g^{ab}}$ (since $z = g^{ab+c}$).

However no unbounded simulator can simulate this distribution in the ideal world.

The protocol used in describing the above attack is a simple modification of the Naor-Pinkas/Aiello-Ishai-Reingold protocol where the honest receiver with its first message includes an encryption of the secret key. It is reasonable to assume that ElGamal is indeed circularly secure. Furthermore if it was possible to efficiently obtain an encryption of the secret key given the public key then the counterexample presented above would extend to the Naor-Pinkas/Aiello-Ishai-Reingold protocol. We do not believe that such a procedure exists. But it seems likely that the existence of such a procedure that efficiently concocts an encryption of the secret key can not be ruled out under the DDH assumption alone. Based on this conjecture we can claim that Naor-Pinkas/Aiello-Ishai-Reingold protocol can not be proved secure with an unbounded simulator under the DDH assumption. We stress that we do not make any of these speculative assumptions elsewhere in the paper. We use them here just to possibly explain obstacles in coming up with proofs for known protocols under reasonable assumptions.

4   Zero-Knowledge from Weak OT

In the following two sections, we will prove that a secure realization of T-secure Weak OT protocol (Weak OT with T-chooser's security), where T is an appropriate super-polynomial function in the security parameter, suffices for realizing Two Message Secure Computation for any functionality. We will provide this construction in two steps. First, in this section, we will show that a T-secure Weak OT protocol suffices for constructing a Two Message Secure Computation protocol for the zero-knowledge functionality. In the next section, we will show that zero-knowledge and Weak OT suffice for realizing Two Message Secure Computation for any functionality.

Before we describe the protocol, let's list the primitives used. Let T, T_1 be some super-polynomial functions in the security parameter λ with $T < T_1$.

Parameters:

- $(\mathsf{OT}_1, \mathsf{OT}_2, \mathsf{OT}_3)$ be functions corresponding to a T-secure Weak OT protocol. That is, it is secure against all adversaries running in time $T.\mathrm{poly}(\lambda)$, but can be broken by adversaries running in time $T_1.\mathrm{poly}(\lambda)$.
- $\mathcal{C} = (\mathsf{Com}, \mathsf{Open})$ be a non-interactive T-extractable commitment scheme with non-uniform hiding. (see the full version for definition).

The construction of the protocol appears in Fig. 1.

Notation for (Modified) Blum's Hamiltonicity Protocol.

- The distribution $\mathcal{D}(\cdot, \cdot)$ on input x and witness w generates $(\alpha, \gamma^0, \gamma^1)$ as follows. Sample (a, b^0, b^1) such that a is the first message of Blum's Hamiltonicity protocol (instantiated with commitment $\mathcal{C}$) and b^0 and b^1 are the response on challenges 0 and 1 respectively. Let $\alpha = (a, c^0, c^1)$ where $c^0 = \mathsf{Com}(b^0; r^0)$ and $c^1 = \mathsf{Com}(b^1; r^1)$ and let $\gamma^0 = (b^0, r^0)$ and $\gamma^1 = (b^1, r^1)$.
- Let $\mathcal{V}_{Blum}$ be the (modified) verification algorithm for the Blum's Hamiltonicity protocol. More specifically, $\mathcal{V}_{Blum}$ on input $(x, \alpha, \beta, \gamma)$ outputs 1 if the underlying transcript is an accepting transcript of the Blum's Hamiltonicity protocol.
- $\mathcal{S}_{Blum}$ on input (x, β) generates a simulated accepting transcript $(\alpha, \beta, \gamma^0, \gamma^1)$ such that it is computationally indistinguishable from a real transcript.
- Finally, we will also use the extractor for Blum's Hamiltonicity protocol, denoted by $\mathsf{BlumExt}$. The extractor on input x, α, γ^0 and γ^1 outputs the Hamiltonian cycle in x or $(\perp, \beta)$ such that for no value of γ, $\mathcal{V}_{Blum}(x, \alpha, 1 - \beta, \gamma) = 1$. The extractor $\mathsf{BlumExt}$ runs in time $T.\mathrm{poly}(\lambda)$.

Lemma 2. *Assuming that $(\mathsf{OT}_1, \mathsf{OT}_2, \mathsf{OT}_3)$ is a 2^{λ^ϵ}-secure Weak OT protocol and $\mathcal{C} = (\mathsf{Com}, \mathsf{Open})$ is a 2^{λ^ϵ}-extractable and non-uniformly hiding non-interactive[5] commitment scheme for some constant $0 < \epsilon < 1$, we have that the protocol π_{ZK} described in Fig. 1 with the parameters described above is a two message zero-knowledge argument for NP with quasi-polynomial simulation.*

This lemma immediately implies the following theorem.

Theorem 4. *Two round protocols for the zero-knowledge functionality with quasi-polynomial simulation can be based on sub-exponentially hard Decisional Diffie-Hellman Assumption.*

---

[5] Note that we present our protocol in terms of non-interactive commitments (implied by one-to-one OWFs) just for simplicity of notation. We can instead use Naor's two round commitment scheme [?] that can be based on one-way functions. The only difference being that Naor's commitment is statistically binding instead of being perfectly binding. All claims which rely on this lemma will be stated keeping this simplification in mind.

$$\pi_{ZK}$$

Common Input: A graph x.
Auxiliary Input for P: w such that w is a Hamiltonian cycle in the graph x.

1. $(\mathsf{zk}_1, \mathsf{zkst}) \leftarrow \mathsf{ZK}_1(1^\lambda)$:
   - For each $i \in [\lambda]$, V samples $\beta_i \leftarrow \{0,1\}$ and generates $(\mathsf{ot}_{1,i}, state_i) \leftarrow \mathsf{OT}_1(\beta_i)$.
   - It sets $\mathsf{zk}_1 := \{\mathsf{ot}_{1,i}\}_{i \in [\lambda]}$ and $\mathsf{zkst} := \{state_i\}_{i \in [\lambda]}$.
2. $\mathsf{zk}_2 \leftarrow \mathsf{ZK}_2(1^\lambda, \mathsf{zk}_1, x, w)$:
   - P parses zk_1 as $\{\mathsf{ot}_{1,i}\}_{i \in [\lambda]}$.
   - For each $i \in [\lambda]$, P generates $(\alpha_i, \gamma_i^0, \gamma_i^1) \leftarrow \mathcal{D}(x, w)$, $\mathsf{ot}_{2,i} \leftarrow \mathsf{OT}_2(\mathsf{ot}_{1,i}, \gamma_i^0, \gamma_i^1)$.
   - It then sets $\mathsf{zk}_2 := \{\mathsf{ot}_{2,i}, \alpha_i\}_{i \in [\lambda]}$.
3. $\mathsf{ZK}_3(1^\lambda, \mathsf{zk}_2, x, \mathsf{zkst})$:
   - V parses zkst as $\{state_i\}_{i \in [\lambda]}$ and zk_2 as $\{\mathsf{ot}_{2,i}, \alpha_i\}_{i \in [\lambda]}$.
   - For each $i \in [\lambda]$, V obtains $\gamma_i^{\beta_i}$ as $\mathsf{OT}_3(\mathsf{ot}_{2,i}, state_i)$ and checks to see if $\mathcal{V}_{Blum}(x, \alpha_i, \beta_i, \gamma_i^{\beta_i}) = 1$.
   - Output 1 if all the checks pass and 0 otherwise.

Fig. 1. Two message secure computation for zero-knowledge

4.1 Security Proof

The correctness of the scheme follows from the correctness of Blum's Hamiltonicity protocol. We will now give proofs for the simulation of the prover (argument of knowledge) and of the verifier (zero-knowledge).

Remark: In the security proofs in this section and the next, the simulator will run in time $T_1 \cdot \mathsf{poly}(\lambda)$. Notice that when we instantiate the primitives, $T = 2^{\lambda^\epsilon}$ and $T_1 = 2^\lambda$. This corresponds to an exponential time simulator whereas we require the simulator to only run in quasi-polynomial time. We will use the standard trick of using a smaller security parameter to address this. Let $\lambda = \log^2(k)$. We will now use k as the security parameter in our protocols. Note that the assumptions are still sub-exponentially secure with respect to λ. However, the simulator now runs in time $2^{\log^2(k)} = k^{\log(k)}$ which is quasi-polynomial in the security parameter k.

Argument of Knowledge. First, we note that arguing argument of knowledge also implicitly captures soundness of the protocol. In order to argue the argument of knowledge property, we need to construct an extractor Ext with the following property: we require that for any PPT malicious prover P^* such that $(\mathsf{zk}_2, x^*) \leftarrow P^*(1^\lambda, \mathsf{zk}_1)$ and $\mathsf{ZK}_3(1^\lambda, \mathsf{zk}_2, x^*, \mathsf{zkst}) = 1$ where $(\mathsf{zk}_1, \mathsf{zkst}) \leftarrow \mathsf{ZK}_1(1^\lambda)$ we have that the extractor algorithm Ext running in time $T \cdot \mathsf{poly}(\lambda)$ on input $(\mathsf{zk}_2, x^*, \mathsf{zkst})$ outputs a Hamiltonian cycle in the graph x^*. The extractor is described in Fig. 2.

Now we will argue that this extraction procedure described above successfully extracts a cycle in x^* with overwhelming probability. We will prove this

---

Input: $(\mathsf{zk}_2, x^*, \mathsf{zkst})$.
The extractor does the following:

- For each $i \in [\lambda]$, recall that $\alpha_i = (a_i, c_i^0, c_i^1)$ where c_i^0 and c_i^1 are commitments to strings γ_i^0 and γ_i^1 respectively. Run the T-time extractor ComExt on inputs c_i^0 and c_i^1 to obtain γ_i^0 and γ_i^1.
- For each $i \in [\lambda]$, execute the T-time extractor BlumExt on input $(\alpha, \gamma_i^0, \gamma_i^1)$. If BlumExt outputs a Hamiltonian cycle in graph x^*, then abort everything else and output the extracted cycle. On the other hand, if BlumExt outputs $(\perp, \cdot)$ for every $i \in [\lambda]$, then output $\perp$.

Fig. 2. Extraction strategy against a malicious prover

by reaching a contradiction. Lets assume that there exists a PPT cheating prover P* such that it succeeds in generating accepting proofs even though the extraction of the witness fails. More formally, lets P* be a PPT adversary such that, $\epsilon = \Pr[\mathsf{ZK}_3(1^\lambda, \mathsf{zk}_2, x^*, \mathsf{zkst}) = 1 \bigwedge \mathsf{Ext}(\mathsf{zk}_2, x^*, \mathsf{zkst}) = \perp : (\mathsf{zk}_1, \mathsf{zkst}) \leftarrow \mathsf{ZK}_1(1^\lambda), (\mathsf{zk}_2, x^*) \leftarrow \mathsf{P}^*(1^\lambda, \mathsf{zk}_1)]$ is non-negligible.

Then we will use such an adversarial prover P^* and construct an adversary contradicting T-chooser's security of the Weak OT protocol. We proceed with the following hybrids:

- H_0: This is the real game with the guarantee that ϵ is non-negligible.
- H_1: Recall than in H_0 ZK_3 outputs 1 only if $\mathcal{V}_{Blum}(x, \alpha_i, \beta_i, \gamma_i^{\beta_i}) = 1$ for every $i \in [\lambda]$. In H_1 we modify ZK_3 and denote it by ZK_3'. ZK_3' samples a random subset $S \subset [\lambda]$ such that $|S| = (\lambda - 1)$ and check $\mathcal{V}_{Blum}(x, \alpha_i, \beta_i, \gamma_i^{\beta_i}) = 1$ for all $i \in S$ (as opposed to all $i \in [\lambda]$). We have that, $\Pr[\mathsf{ZK}_3'(1^\lambda, \mathsf{zk}_2, x^*, \mathsf{zkst}) = 1 \bigwedge \mathsf{Ext}(\mathsf{zk}_2, x^*, \mathsf{zkst}) = \perp : (\mathsf{zk}_1, \mathsf{zkst}) \leftarrow \mathsf{ZK}_1(1^\lambda), (\mathsf{zk}_2, x^*) \leftarrow \mathsf{P}^*(1^\lambda, \mathsf{zk}_1)]$ is at least ϵ.

Let $R \subseteq [\lambda] \backslash S$, be a set such that $j \in R$ if $\mathcal{V}_{Blum}(x, \alpha_j, \beta_j, \gamma_j^{\beta_j}) = 1$. Clearly, $0 \leq |R| \leq 1$ since there is only one index not in S. Further, let E be the event such that $|R| = 1$. Now, its easy to see that the only way $|R|$ could be 0 is if the malicious prover P* was able to guess S correctly. This can happen with probability at most $\frac{1}{\lambda}$ (i.e. probability that P* correctly guesses which random index was not part of S). Therefore, $\Pr[\neg E] = \frac{1}{\lambda}$ and so, $\Pr[E] = (1 - \frac{1}{\lambda})$. That is, probability that the other index not part of set S belongs to set R is at least $(1 - \frac{1}{\lambda})$.

Using this malicious prover P*, we will construct an adversary $\mathcal{A}$ that contradicts the T-chooser's security of the Weak OT protocol. Note that commitments can be broken in time $T.\mathsf{poly}(\lambda)$ but the chooser's bit in the OT protocol is assumed to be secure against adversaries running in time $T.\mathsf{poly}(\lambda)$. The adversary $\mathcal{A}$ obtains an external challenge $\mathsf{OT}_1(b)$ for a random $b \in \{0, 1\}$, and it needs to guess b. It does the following:

- Invoke P* and embed the challenge in one of the random locations $i^* \leftarrow [\lambda]$. That is, as part of the first message of the verifier, for index i^*, set the external challenge $OT_1(b)$ as ot_{1,i^*} and $state_{i^*} = i^*$. For all other indices $i \in [\lambda]$, choose a random bit β_i and compute $(ot_{1,i}, state_i) \leftarrow OT_1(\beta_i)$.

- Set $zkst := \{state_i\}_{i \in [\lambda]}$ and send $zk_1 = \{ot_{1,i}\}_{i \in [\lambda]}$.
- Obtain the message zk_2 from P^* and run the algorithm ZK'_3 on input $(1^\lambda, x^*, zk_2, zkst)$ using the random set S to be $[\lambda] \setminus \{i^*\}$.
- If the algorithm ZK'_3 outputs 0 then $\mathcal{A}$ outputs a random bit.
- On the other hand if ZK'_3 outputs 1, similar to the extractor described above, $\mathcal{A}$ first runs ComExt to extract $\gamma_{i^*}^0$ and $\gamma_{i^*}^1$. Then, it outputs b' where $(\perp, b')$ is the output of BlumExt on input $(\alpha_i^*, \gamma_{i^*}^0, \gamma_{i^*}^1)$.

Analysis:

When ZK'_3 outputs 0 (which happens with probability $1 - \epsilon$) then $\mathcal{A}$'s guess about b will be correct with probability at least $\frac{1}{2}$. On the other hand when ZK'_3 outputs 1, we will have that with probability at least $(1 - \frac{1}{\lambda})$ $i^* \in R$ and hence b' where $(\perp, b')$ is the output of BlumExt on input $(\alpha, \gamma_{i^*}^0, \gamma_{i^*}^1)$ will be the correct guess for b. Compiling together the two cases we have that $\mathcal{A}$ guesses the bit b correctly with probability at least $(1 - \epsilon).1/2 + \epsilon(1 - \frac{1}{\lambda}) = \frac{1}{2} + \frac{\epsilon}{2} - \frac{\epsilon}{\lambda}$ which is non-negligible if ϵ is non-negligible. This contradicts the T-chooser's security of the OT protocol.

This completes the proof of argument of knowledge.

Zero-Knowledge. In order to show zero-knowledge (or simulating a malicious verifier V^*) we need to construct a simulator $\mathcal{S}$ satisfying Definition 2. Let's consider a malicious verifier V^* described using a pair of algorithms (V_1^*, V_2^*). The simulation strategy is described in Fig. 3. Note that the simulator runs in time $T_1.poly(\lambda)$.

Common Input: A graph x.

1. $(zk_1, zkst) \leftarrow V_1^*(x, 1^\lambda)$: The malicious verifier runs V_1^* computes its first message zk_1 to be sent to the prover and some associated state $zkst$.
2. The simulator $\mathcal{S}$ does the following:
   - Parse zk_1 as $\{ot_{1,i}\}_{i \in [\lambda]}$.
   - For each $i \in \lambda$, run $OTExt(ot_{1,i})$ to extract the challenge bit β_i of the verifier. By the sender's security in the OT protocol, we know that the extraction succeeds with non negligible probability. Note that this requires time $T_1.poly(\lambda)$.
   - For each $i \in \lambda$, run $\mathcal{S}_{Blum}$ on input (x, β_i) to produce $(\alpha_i, \gamma_i^0, \gamma_i^1)$. (where γ_i^{1-b} is in fact a dummy message)
   - Compute $ot_{2,i} \leftarrow OT_2(ot_{1,i}, \gamma_i^0, \gamma_i^1)$.
   - Set $zk_2 := \{ot_{2,i}, \alpha_i\}_{i \in [\lambda]}$ and send it to the verifier.
3. $V_2^*(x, \lambda, zkst, zk_2)$: The malicious verifier runs V_2^* outputs either 0 or 1.

Fig. 3. Simulation strategy against a malicious verifier

Claim. The simulation strategy described in Fig. 3 is secure against a malicious verifier.

Proof. Using a series of hybrid arguments, we will show that the view of the malicious verifier in the ideal world is computationally indistinguishable from its view in the real world.

Let's assume to the contrary that there exists a PPT malicious verifier $V^* = (V_1^*, V_2^*)$ that has a non negligible probability ϵ of distinguishing its view in the real world from the ideal world. Let's consider the "best possible" initial message of the verifier - i.e. the output of the algorithm V_1^* that produces the highest distinguishing probability between the views in the real and ideal worlds. Let's fix this message as the initial message zk_1^* of the verifier. That is, consider V_1^* to be a deterministic algorithm that takes as input the randomness used to output this best possible message.

Essentially, given any PPT malicious adversary $\widehat{R^*}$ that can distinguish the two views with non-negligible probability ϵ, we are transforming it into a new deterministic adversary R^* such that the randomness used to produce this best possible initial message is hardwired inside it. Therefore, even R^* can distinguish the two views with probability at least ϵ.

Using this malicious verifier, we can construct a non-uniform PPT adversary $\mathcal{A}$ that breaks either the hiding property of the commitment scheme or the sender's security of the OT protocol. Note that the commitment scheme is secure against all PPT adversaries (it is only assumed to be broken by an adversary running in time $T.\mathrm{poly}(\lambda)$) and the OT protocol in fact has statistical security and hence is secure against all PPT adversaries. Thus, this would lead to a contradiction. In our reduction, the non-uniform advice (or auxiliary input) given to the adversary $\mathcal{A}$ is the set of challenge bits $\{\beta_i^*\}_{i \in [\lambda]}$ of the verifier V_1^* that was used to generate the fixed first message (observe that this is exactly what the simulator in the ideal world extracts in the first step by running the OTExt algorithm). These challenge bits are also accessible to the second stage verifier V_2^* as part of the state - zkst that is output by V_1^*. We will now describe the reduction. $\mathcal{A}$ acts as the prover in its interaction with the malicious verifier V^*.

1. Hybrid 0: This is the real experiment where the message sent to the verifier $zk_2 = \{ot_{2,i}, \alpha_i\}_{i \in [\lambda]}$ is computed using the algorithm $\mathsf{ZK}_2(1^\lambda, zk_1^*, x, w)$. Here, $\alpha_i = (a_i, c_i^0, c_i^1)$ where $c_i^0 = \mathsf{Com}(b_i^0; r_i^0)$, $c_i^1 = \mathsf{Com}(b_i^1; r_i^1)$, $\gamma_i^0 = (b_i^0, r_i^0)$ and $\gamma_i^1 = (b_i^1, r_i^1)$. Also, $ot_{2,i} \leftarrow \mathsf{OT}_2(ot_{1,i}, \gamma_i^0, \gamma_i^1)$.
2. Hybrid 1: For each $i \in \lambda$, compute $\gamma_i^{(1-\beta_i^*)} = (\bot, \bot)$.
3. Hybrid 2: For each $i \in \lambda$, compute $c_i^{(1-\beta_i^*)} = \mathsf{Com}(\bot; r_i^{(1-\beta_i^*)})$. Observe that this is same as the ideal world experiment since the simulator would do exactly this: replace the entries corresponding to the positions not challenged by the verifier (i.e. positions $(1 - \beta_i^*)$) using $\bot$.

We defer the argument for the indistinguishability of the hybrids to the full version.

5   Two Message Secure Computation from Weak OT

In this section, we show that Weak OT together with two message witness indistinguishability gives an immediate construction of Two Message Secure Computation for the general functionality. This construction is obtained by compiling the Yao's garbled circuit construction (see the full version for the definition) and Weak OT protocol with our zero-knowledge protocol. Note that the zero knowledge protocol from Sect. 4 already satisfies this requirement. (and gives a much stronger functionality).

Let's consider two parties : the sender S with input x and the receiver R with input y that wish to securely compute any general function f. Before we describe the protocol, let's list the primitives used. Let T, T_1 be some super-polynomial functions in the security parameter λ with $T < T_1$.

Parameters:

- $(\mathsf{OT}_1, \mathsf{OT}_2, \mathsf{OT}_3)$ be functions corresponding to a T-secure Weak OT protocol. That is, its secure against all adversaries running in time $T.\mathrm{poly}(\lambda)$, but can be broken by adversaries running in time $T_1.\mathrm{poly}(\lambda)$.
- $(\mathsf{WI}_1, \mathsf{WI}_2, \mathsf{WI}_3)$ be a two message secure computation protocol for the witness indistinguishability functionality. This protocol is secure against all adversaries running in time $T.\mathrm{poly}(\lambda)$, but can be broken by adversaries running in time $T_1.\mathrm{poly}(\lambda)$.
- $(\mathsf{Garble}, \mathsf{GCEval})$ be the algorithms corresponding to Yao's garbled circuit construction that is secure against all adversaries running in time $T.\mathrm{poly}(\lambda)$, but can be broken by adversaries running in time $T_1.\mathrm{poly}(\lambda)$.
- com be a commitment scheme that is extractable in time $T.\mathrm{poly}(\lambda)$.
- let OWF be a one-way function that is not invertible in time $T.\mathrm{poly}(\lambda)$ but can be inverted by an attacker $\mathcal{A}_{\mathsf{OWF}}$ running in time $T_1.\mathrm{poly}(\lambda)$.

In Fig. 4 we describe the construction of our Two Message Secure Computation protocol. We will next prove its security.

Lemma 3. *If* $(\mathsf{OT}_1, \mathsf{OT}_2, \mathsf{OT}_3)$ *is a sub-exponentially secure Weak OT protocol,* $(\mathsf{WI}_1, \mathsf{WI}_2, \mathsf{WI}_3)$ *is a sub-exponentially secure two message witness indistinguishable argument for NP and sub-exponentially secure one-way functions exist, the protocol presented in Fig. 4 with the parameters described above is a Two Message Secure Computation protocol with quasi-polynomial simulation for any general function.*

This lemma immediately implies the following theorem.

Theorem 5. *Two Message Secure Computation protocols with quasi-polynomial simulation for general functionalities can be based on any of the following sub-exponential assumptions: (1) Decisional Diffie-Hellman Assumption; (2) Quadratic Residuosity Assumption; (3) N^{th} Residuosity Assumption; or (4) Witness Encryption (together with one-to-one one-way functions).*

$\langle S, R \rangle$

Inputs: The sender S gets input x and the receiver R gets input y. Both S and R get the function f they want to evaluate as input.

Output: R expects to receive $f(x, y)$ as output.

1. $(\mathsf{nisc}_1, state) \leftarrow \mathsf{NISC}_1(1^\lambda, y)$:
   - R generates $(\mathsf{ot}_{1,i}, state_i) \leftarrow \mathsf{OT}_1(y_i)$ for each $i \in [\lambda]$ and $(\mathsf{wi}_1, \mathsf{wist}) \leftarrow \mathsf{WI}_1(1^\lambda)$.
   - R chooses a random string $z \leftarrow \{0,1\}^\lambda$ and computes $Z = \mathsf{OWF}(z)$.
   - It sets $\mathsf{nisc}_1 := (\mathsf{ot}_{1,1}, \ldots, \mathsf{ot}_{1,\lambda}, \mathsf{wi}_1, Z)$,
     $state = (state_1, \ldots, state_\lambda, \mathsf{wist}, z, Z)$ and sends nisc_1 to S.
2. $\mathsf{nisc}_2 \leftarrow \mathsf{NISC}_2(\mathsf{nisc}_1, x)$:
   - S parses nisc_1 as $(\mathsf{ot}_{1,1}, \ldots, \mathsf{ot}_{1,\lambda}, \mathsf{wi}_1, Z)$.
   - It computes $c_1 = \mathsf{com}(x; r_1)$ and $c_2 = \mathsf{com}(0; r_2)$ using randomness r_1, r_2 respectively.
   - S samples a 2λ key tuple $\overline{K} = \{K_{i,b}\}_{i \in [\lambda], b \in \{0,1\}}$, where $K_{i,b} \in \{0,1\}^\lambda$ and generates a garbled circuit $GC := \mathsf{Garble}(\overline{K}, C; \omega)$ where $C_x(y)$ is a circuit that evaluates $f(x, y)$ on input y.
   - Then, S generates $\mathsf{ot}_{2,i} := \mathsf{OT}_2(\mathsf{ot}_{1,i}, K_{i,0}, K_{i,1}; \omega_i)$ for each $i \in [\lambda]$.
   - After that, S computes $\mathsf{wi}_2 \leftarrow \mathsf{WI}_2(1^\lambda, \mathsf{wi}_1, (\mathsf{ot}_{2,1}, \ldots, \mathsf{ot}_{2,\lambda}, GC, c_1, c_2, Z),$
     $(\omega, \{\omega_i\}_{i \in [\lambda]}, x, r_1, \perp, \perp))$ for the statement $(\mathsf{ot}_{2,1}, \ldots, \mathsf{ot}_{2,\lambda}, GC, c_1, c_2, Z)$
     $\in L$, where L contains tuples for which there exists
     either a witness $\Omega = (\overline{K}, C, \omega, \omega_1, \ldots, \omega_\lambda, x, r_1)$ such that:

     $$(1) \quad GC := \mathsf{Garble}(\overline{K}, C; \omega) \bigwedge c_1 = \mathsf{com}(x; r_1) \bigwedge$$
     $$\forall i \in [\lambda], \mathsf{ot}_{2,i} = \mathsf{OT}_2(\mathsf{ot}_{1,i}, K_{i,0}, K_{i,1}; \omega_i)$$

     (OR) there exists a witness $\Omega_2 = (z, r_2)$ such that:

     $$(2) \quad \mathsf{OWF}(z) = Z \bigwedge c_2 = \mathsf{com}(z; r_2)$$

   - Finally, S sets $\mathsf{nisc}_2 := (\mathsf{ot}_{2,1}, \ldots, \mathsf{ot}_{2,\lambda}, GC, \mathsf{wi}_2, c_1, c_2)$ and sends it to R.
3. $\mathsf{NISC}_3(\mathsf{nisc}_2, state)$:
   - R parses nisc_2 as $(\mathsf{ot}_{2,1}, \ldots, \mathsf{ot}_{2,\lambda}, GC, \mathsf{wi}_2, c_1, c_2)$ and $state$ as $(state_1, \ldots, state_\lambda, \mathsf{wist}, z, Z)$.
   - If $\mathsf{WI}_3(1^\lambda, \mathsf{wi}_2, (\mathsf{ot}_{2,1}, \ldots, \mathsf{ot}_{2,\lambda}, GC, c_1, c_2), \mathsf{wist}) = 0$ then R outputs $\perp$.
   - Otherwise, for each $i \in [\lambda]$ it obtains $K_{i,y_i} = \mathsf{OT}_3(\mathsf{ot}_{2,i}, state_i)$ and outputs $\mathsf{GCEval}(\overline{K}_y, GC)$.

Fig. 4. Two message secure computation for a general function f

We can easily transform the above protocol to a setting where both parties are required to receive outputs by adding an extra round. Now, for the special case of coin tossing, we get the following corollary:

Corollary 3. *Three round secure coin tossing protocols with quasi-polynomial simulation can be based on any of the following sub-exponential assumptions: (1)*

Decisional Diffie-Hellman Assumption; (2) Quadratic Residuosity Assumption; (3) N^{th} Residuosity Assumption; or (4) Witness Encryption (together with one-to-one one-way functions).

In order to get better efficiency we can use the Two Message Secure Computation protocol in Fig. 4 and obtain a protocol (Parallel OT) for realizing the functionality that allows for $\text{poly}(\lambda)$-parallel oblivious transfer invocations. We can then use this protocol in order to instantiate the protocols of Ishai et al. [31,33].

We defer the proof to the full version.

6   Weak OT from High Rate Semi-honest OT

In this section, we first give a generic construction of two message Weak OT from any high rate two message semi-honest OT.

Parameters:
As defined earlier, let λ be the security parameter. Consider a sender S with inputs $m_0, m_1 \in \{0,1\}^n$ where $n = \text{poly}(\lambda)$ and a receiver R with choice bit b who wish to run a Weak OT protocol. Let $\text{OT}^{sh} = (\text{OT}_1^{sh}, \text{OT}_2^{sh}, \text{OT}_3^{sh})$ be a two message semi-honest secure OT protocol with high rate c (>0.5). The rate of the OT protocol is defined as the ratio of the size of one of the sender's input strings to the size of the sender's message. That is, $\text{rate} = \frac{|m_0|}{|\text{OT}_2^{sh}(m_0,m_1,\text{OT}_1^{sh}(b))|}$. Let $\text{Ext} : \{0,1\}^s \times \{0,1\}^d \to \{0,1\}^n$ be a (k,ϵ) strong seeded randomness extractor (defined in the full version), where $s = \frac{c}{2c-1} \cdot (n + 2 \log(1/\epsilon))$, $d = s$ and $k = (n + 2 \log(1/\epsilon))$ for any $\epsilon = 2^{-\lambda}$. Recall that we know how to construct such a strong seeded extractor using the Leftover Hash Lemma [30]. We prove the following theorem.

Theorem 6. *Assuming $\text{OT}^{sh} = (\text{OT}_1^{sh}, \text{OT}_2^{sh}, \text{OT}_3^{sh})$ is a high rate (>0.5) two message semi-honest OT protocol, the protocol in Fig. 5 with the parameters described above is a Weak OT protocol.*

We defer the proof to the full version.

6.1   Weak OT from High Rate Linear Homomorphic Encryption

In this section, we describe the construction of two message semi-honest OT from any linear homomorphic encryption as in [32]. Let $\text{LHE} = (\text{Setup}, \text{Enc}, \text{Dec}, \text{Add}, \text{Const.Mul})$ be any linear homomorphic encryption scheme (defined in the full version). We prove the following theorem.

Theorem 7. *Assuming LHE is a high rate (>0.5) linear homomorphic encryption scheme, the protocol in Fig. 6 is a high rate (>0.5) two message semi-honest OT protocol.*

We defer the proof to the full version.

Finally, as a corollary of the Theorems 6 and 7, we get a construction of Weak OT from any high rate linear homomorphic encryption scheme. Formally:

$$\pi_{WeakOT}$$

1. $(\mathsf{ot}_1, state) \leftarrow \mathsf{OT}_1(\mathsf{b})$:
   - Compute $(\mathsf{ot}_1^{sh}, state) = \mathsf{OT}_1^{sh}(\mathsf{b})$.
   - Set the output $\mathsf{ot}_1 = \mathsf{ot}_1^{sh}$.

2. $\mathsf{ot}_2 \leftarrow \mathsf{OT}_2(\mathsf{ot}_1, \mathsf{m}_0, \mathsf{m}_1)$:
   - Pick two random strings $\mathsf{r}_0, \mathsf{r}_1 \in \{0,1\}^{s(n)}$ and two random seeds $\mathsf{s}_0, \mathsf{s}_1 \in \{0,1\}^d$.
   - Compute $\mathsf{ot}_2^{sh} = \mathsf{OT}_2^{sh}(\mathsf{ot}_1, \mathsf{r}_0, \mathsf{r}_1)$.
   - Compute $\mathsf{S}_0 = \mathsf{m}_0 \oplus \mathsf{Ext}(\mathsf{r}_0, \mathsf{s}_0)$ and $\mathsf{S}_1 = \mathsf{m}_1 \oplus \mathsf{Ext}(\mathsf{r}_1, \mathsf{s}_1)$.
   - Output $\mathsf{ot}_2 = (\mathsf{ot}_2^{sh}, \mathsf{S}_0, \mathsf{s}_0, \mathsf{S}_1, \mathsf{s}_1)$.

3. $\mathsf{OT}_3(\mathsf{ot}_2, state)$:
   - Parse ot_2 as $(\mathsf{ot}_2^{sh}, \mathsf{S}_0, \mathsf{s}_0, \mathsf{S}_1, \mathsf{s}_1)$
   - Compute $\mathsf{r}_\mathsf{b} = \mathsf{OT}_3^{sh}(\mathsf{ot}_2^{sh}, state)$.
   - Compute the output $\mathsf{m}_\mathsf{b} = (\mathsf{S}_\mathsf{b} \oplus \mathsf{Ext}(\mathsf{r}_\mathsf{b}, \mathsf{s}_\mathsf{b}))$.

Fig. 5. Weak OT from semi-honest OT

$$\pi_{\mathsf{OT}^{sh}}$$

1. $(\mathsf{ot}_1^{sh}, state) \leftarrow \mathsf{OT}_1^{sh}(\mathsf{b})$:
   - Generate $(\mathsf{sk}, \mathsf{pk}) \leftarrow \mathsf{Setup}(1^\lambda)$ for the encryption scheme.
   - Compute the output $\mathsf{ot}_1^{sh} := (\mathsf{pk}, \mathsf{Enc}_{\mathsf{pk}}(\mathsf{b}; \mathsf{r}))$ usign randomness r. Let $state$ be $(\mathsf{b}, \mathsf{pk}, \mathsf{sk})$.

2. $\mathsf{ot}_2^{sh} \leftarrow \mathsf{OT}_2^{sh}(\mathsf{ot}_1, \mathsf{m}_0, \mathsf{m}_1)$:
   - Compute $\mathsf{ct}_0 = \mathsf{Const.Mul}(\mathsf{ot}_1, \mathsf{m}_1)$.
   - Let $\mathsf{ct}_{1,1} = \mathsf{Const.Mul}(\mathsf{Enc}_{\mathsf{pk}}(1; \mathsf{r}'), \mathsf{m}_0)$ where r' is a random string. Let $\mathsf{ct}_{1,2} = \mathsf{Const.Mul}(\mathsf{ot}_1, -1)$ and $\mathsf{ct}_{1,3} = \mathsf{Const.Mul}(\mathsf{ct}_{1,2}, \mathsf{m}_1)$.
   - Compute $\mathsf{ct}_1 = \mathsf{Add}(\mathsf{ct}_{1,1}, \mathsf{ct}_{1,3})$.
   - Compute $\mathsf{ct} = \mathsf{Add}(\mathsf{ct}_0, \mathsf{ct}_1)$. That is, $\mathsf{ct} = \mathsf{Enc}_{\mathsf{pk}}(\mathsf{b} \cdot \mathsf{m}_1 + (1 - \mathsf{b}) \cdot \mathsf{m}_0)$.
   - Output $\mathsf{ot}_2 = \mathsf{ct}$.

3. $\mathsf{OT}_3(\mathsf{ot}_2, state)$:
   - Parse $state$ as $(\mathsf{b}, \mathsf{pk}, \mathsf{sk})$ and ot_2 as (ct)
   - Compute the output $\mathsf{m}_\mathsf{b} = \mathsf{Dec}_{\mathsf{sk}}(\mathsf{ct})$.

Fig. 6. OT^{sh} from linear homomorphic encryption

Corollary 4. *High rate (>0.5) linear homomorphic encryption implies Weak OT.*

6.2   Weak OT from N^{th} Residuosity Assumption

Finally, we instantiate the high rate linear homomorphic encryption scheme using a construction where the size of the ciphertext is λ more than the size of the plaintext. Such an encryption scheme can be built based on the Nth Residuosity Assumption [14,32]. As a result, we get the following corollary:

Corollary 5. *The* N^{th} *Residuosity Assumption implies Weak OT.*

An earlier construction of Weak OTbased on the Nth Residuosity Assumption appeared in [27]. In that construction, they first construct Weak OT from any smooth projective hash function which is then instantiated based on the Nth Residuosity Assumption using a complex transformation. Our construction and analysis are arguably simpler.

7 Weak OT from Witness Encryption

We defer the details of this section to the full version. Formally, we show the following lemma:

Lemma 4. *Assuming injective one-way functions exist and a non-uniform witness encryption scheme exists, there exists a secure Weak OT protocol.*

Acknowledgements. The second author's research is supported in part from 2017 AFOSR YIP Award, DARPA/ARL SAFEWARE Award W911NF15C0210, DARPA Brandeis program under Contract N66001-15-C-4065, AFOSR Award FA9550-15-1-0274, NSF CRII Award 1464397, and research grants by the Okawa Foundation, Visa Inc., and Center for Long-Term Cybersecurity (CLTC, UC Berkeley). The views expressed are those of the author and do not reflect the official policy or position of the funding agencies.

The third author's research is supported in part by a DARPA/ARL SAFEWARE award, NSF Frontier Award 1413955, NSF grants 1619348, 1228984, 1136174, and 1065276, NSF-BSF grant 2015782, ISF grant 1709/14, BSF grant 2012378, a Xerox Faculty Research Award, a Google Faculty Research Award, an equipment grant from Intel, and an Okawa Foundation Research Grant. This material is based upon work supported by the Defense Advanced Research Projects Agency through the ARL under Contract W911NF-15-C-0205. The views expressed are those of the authors and do not reflect the official policy or position of the Department of Defense, the National Science Foundation, or the U.S. Government.

The fourth author's research is supported in part from a DARPA/ARL SAFE-WARE award, NSF Frontier Award 1413955, NSF grants 1619348, 1228984, 1136174, and 1065276, BSF grant 2012378, a Xerox Faculty Research Award, a Google Faculty Research Award, an equipment grant from Intel, and an Okawa Foundation Research Grant. This material is based upon work supported by the Defense Advanced Research Projects Agency through the ARL under Contract W911NF-15-C-0205. The views expressed are those of the authors and do not reflect the official policy or position of the Department of Defense, the National Science Foundation, or the U.S. Government.

References

1. Afshar, A., Mohassel, P., Pinkas, B., Riva, B.: Non-interactive secure computation based on cut-and-choose. In: Nguyen, P.Q., Oswald, E. (eds.) EUROCRYPT 2014. LNCS, vol. 8441, pp. 387–404. Springer, Heidelberg (2014). doi:10.1007/978-3-642-55220-5_22

2. Aiello, B., Ishai, Y., Reingold, O.: Priced oblivious transfer: how to sell digital goods. In: Pfitzmann, B. (ed.) EUROCRYPT 2001. LNCS, vol. 2045, pp. 119–135. Springer, Heidelberg (2001). doi:10.1007/3-540-44987-6_8
3. Badrinarayanan, S., Goyal, V., Jain, A., Khurana, D., Sahai, A.: Round optimal concurrent MPC via strong simulation. In: TCC (2017)
4. Barak, B., Lindell, Y., Vadhan, S.P.: Lower bounds for non-black-box zero knowledge. In: Proceedings of 44th Symposium on Foundations of Computer Science, FOCS 2003, 11–14 October 2003, Cambridge, MA, USA, pp. 384–393. IEEE Computer Society (2003). https://doi.org/10.1109/SFCS.2003.1238212
5. Barak, B., Prabhakaran, M., Sahai, A.: Concurrent non-malleable zero knowledge. In: Proceedings of 47th Annual IEEE Symposium on Foundations of Computer Science, FOCS 2006, 21–24 October 2006, Berkeley, California, USA, pp. 345–354. IEEE Computer Society (2006). https://doi.org/10.1109/FOCS.2006.21
6. Barak, B., Sahai, A.: How to play almost any mental game over the net - concurrent composition via super-polynomial simulation. In: Proceedings of 46th Annual IEEE Symposium on Foundations of Computer Science, FOCS 2005, 23–25 October 2005, Pittsburgh, PA, USA, pp. 543–552. IEEE Computer Society (2005). https://doi.org/10.1109/SFCS.2005.43
7. Biham, E. (ed.): Advances in Cryptology - EUROCRYPT 2003. LNCS, vol. 2656. Springer, Heidelberg (2003). doi:10.1007/3-540-39200-9
8. Blum, M., Feldman, P., Micali, S.: Non-interactive zero-knowledge and its applications (extended abstract). In: STOC (1988)
9. Cachin, C., Camenisch, J., Kilian, J., Müller, J.: One-round secure computation and secure autonomous mobile agents. In: Montanari, U., Rolim, J.D.P., Welzl, E. (eds.) ICALP 2000. LNCS, vol. 1853, pp. 512–523. Springer, Heidelberg (2000). doi:10.1007/3-540-45022-X_43
10. Canetti, R., Halevi, S., Katz, J.: A forward-secure public-key encryption scheme. In: Biham [7], pp. 255–271. doi:10.1007/3-540-39200-9_16
11. Canetti, R., Lin, H., Pass, R.: Adaptive hardness and composable security in the plain model from standard assumptions. In: 51th Annual IEEE Symposium on Foundations of Computer Science, FOCS 23–26 October 2010, Las Vegas, Nevada, USA, pp. 541–550. IEEE Computer Society (2010). https://doi.org/10.1109/FOCS.2010.86
12. Chung, K.M., Lui, E., Mahmoody, M., Pass, R.: Unprovable security of two-message zero knowledge. Cryptology ePrint Archive, Report 2012/711 (2012)
13. Dachman-Soled, D., Jain, A., Kalai, Y.T., Lopez-Alt, A.: On the (in)security of the fiat-shamir paradigm, revisited. Cryptology ePrint Archive, Report 2012/706 (2012)
14. Damgård, I., Jurik, M.: A generalisation, a simpli.cation and some applications of paillier's probabilistic public-key system. In: Kim, K. (ed.) PKC 2001. LNCS, vol. 1992, pp. 119–136. Springer, Heidelberg (2001). doi:10.1007/3-540-44586-2_9
15. Döttling, N., Fleischhacker, N., Krupp, J., Schröder, D.: Two-message, oblivious evaluation of cryptographic functionalities. In: Robshaw, M., Katz, J. (eds.) CRYPTO 2016. LNCS, vol. 9816, pp. 619–648. Springer, Heidelberg (2016). doi:10.1007/978-3-662-53015-3_22
16. Feige, U., Lapidot, D., Shamir, A.: Multiple noninteractive zero knowledge proofs under general assumptions. SIAM J. Comput. 29, 1–28 (1999)
17. Garg, S., Gentry, C., Sahai, A., Waters, B.: Witness encryption and its applications. In: STOC (2013)

18. Garg, S., Goyal, V., Jain, A., Sahai, A.: Concurrently secure computation in constant rounds. In: Pointcheval, D., Johansson, T. (eds.) EUROCRYPT 2012. LNCS, vol. 7237, pp. 99–116. Springer, Heidelberg (2012). doi:10.1007/978-3-642-29011-4_8

19. Garg, S., Mukherjee, P., Pandey, O., Polychroniadou, A.: The exact round complexity of secure computation. In: Fischlin, M., Coron, J.-S. (eds.) EUROCRYPT 2016. LNCS, vol. 9666, pp. 448–476. Springer, Heidelberg (2016). doi:10.1007/978-3-662-49896-5_16

20. Garg, S., Rao, V., Sahai, A., Schröder, D., Unruh, D.: Round optimal blind signatures. In: Rogaway, P. (ed.) CRYPTO 2011. LNCS, vol. 6841, pp. 630–648. Springer, Heidelberg (2011). doi:10.1007/978-3-642-22792-9_36

21. Goldreich, O., Krawczyk, H.: On the composition of zero-knowledge proof systems. In: Paterson, M.S. (ed.) ICALP 1990. LNCS, vol. 443, pp. 268–282. Springer, Heidelberg (1990). doi:10.1007/BFb0032038

22. Goldreich, O., Micali, S., Wigderson, A.: How to play any mental game or a completeness theorem for protocols with honest majority. In: Proceedings of the 19th Annual ACM Symposium on Theory of Computing, New York, New York, USA. pp. 218–229 (1987). http://doi.acm.org/10.1145/28395.28420

23. Goldreich, O., Oren, Y.: Definitions and properties of zero-knowledge proof systems. J. Cryptol. 7(1), 1–32 (1994). https://doi.org/10.1007/BF00195207

24. Goyal, V.: Positive results for concurrently secure computation in the plain model. In: FOCS (2012)

25. Groth, J., Ostrovsky, R., Sahai, A.: Non-interactive zaps and new techniques for NIZK. In: Dwork, C. (ed.) CRYPTO 2006. LNCS, vol. 4117, pp. 97–111. Springer, Heidelberg (2006). doi:10.1007/11818175_6

26. Groth, J., Ostrovsky, R., Sahai, A.: Perfect non-interactive zero knowledge for NP. In: Vaudenay, S. (ed.) EUROCRYPT 2006. LNCS, vol. 4004, pp. 339–358. Springer, Heidelberg (2006). doi:10.1007/11761679_21

27. Halevi, S., Kalai, Y.T.: Smooth projective hashing and two-message oblivious transfer. J. Cryptology 25(1), 158–193 (2012)

28. Hazay, C., Venkitasubramaniam, M.: What security can we achieve within 4 rounds? In: Zikas, V., Prisco, R. (eds.) SCN 2016. LNCS, vol. 9841, pp. 486–505. Springer, Cham (2016). doi:10.1007/978-3-319-44618-9_26

29. Horvitz, O., Katz, J.: Universally-composable two-party computation in two rounds. In: Menezes, A. (ed.) CRYPTO 2007. LNCS, vol. 4622, pp. 111–129. Springer, Heidelberg (2007). doi:10.1007/978-3-540-74143-5_7

30. Impagliazzo, R., Levin, L.A., Luby, M.: Pseudo-random generation from one-way functions (extended abstracts). In: Johnson, D.S. (ed.) Proceedings of the 21st Annual ACM Symposium on Theory of Computing, May 14–17, 1989, Seattle, Washington, USA, pp. 12–24. ACM (1989). http://doi.acm.org/10.1145/73007.73009

31. Ishai, Y., Kushilevitz, E., Ostrovsky, R., Prabhakaran, M., Sahai, A.: Efficient non-interactive secure computation. In: Paterson, K.G. (ed.) EUROCRYPT 2011. LNCS, vol. 6632, pp. 406–425. Springer, Heidelberg (2011). doi:10.1007/978-3-642-20465-4_23

32. Ishai, Y., Paskin, A.: Evaluating branching programs on encrypted data. In: Vadhan, S.P. (ed.) TCC 2007. LNCS, vol. 4392, pp. 575–594. Springer, Heidelberg (2007). doi:10.1007/978-3-540-70936-7_31

33. Ishai, Y., Prabhakaran, M., Sahai, A.: Founding cryptography on oblivious transfer – efficiently. In: Wagner, D. (ed.) CRYPTO 2008. LNCS, vol. 5157, pp. 572–591. Springer, Heidelberg (2008). doi:10.1007/978-3-540-85174-5_32

34. Jain, A., Kalai, Y.T., Khurana, D., Rothblum, R.: Distinguisher-dependent simulation in two rounds and its applications. In: Katz, J., Shacham, H. (eds.) CRYPTO 2017 Part II. LNCS, vol. 10402, pp. 158–189. Springer, Cham (2017). doi:10.1007/978-3-319-63715-0_6

35. Katz, J., Ostrovsky, R.: Round-optimal secure two-party computation. In: Franklin, M. (ed.) CRYPTO 2004. LNCS, vol. 3152, pp. 335–354. Springer, Heidelberg (2004). doi:10.1007/978-3-540-28628-8_21

36. Khurana, D., Sahai, A.: Two-message non-malleable commitments from standard sub-exponential assumptions. IACR Cryptol. ePrint Arch. **2017**, 291 (2017)

37. Lindell, Y.: Lower bounds for concurrent self composition. In: Naor, M. (ed.) TCC 2004. LNCS, vol. 2951, pp. 203–222. Springer, Heidelberg (2004). doi:10.1007/978-3-540-24638-1_12

38. Naor, M., Pinkas, B.: Efficient oblivious transfer protocols. In: Kosaraju, S.R. (ed.) Proceedings of the Twelfth Annual Symposium on Discrete Algorithms, 7–9 January 2001, Washington, DC, USA, pp. 448–457. ACM/SIAM (2001). http://dl.acm.org/citation.cfm?id=365411.365502

39. Pass, R.: Simulation in quasi-polynomial time, and its application to protocol composition. In: Biham [7], pp. 160–176. doi:10.1007/3-540-39200-9_10

40. Prabhakaran, M., Sahai, A.: New notions of security: achieving universal composability without trusted setup. In: Babai, L. (ed.) Proceedings of the 36th Annual ACM Symposium on Theory of Computing, Chicago, IL, USA, 13–16 June 2004, pp. 242–251. ACM (2004). http://doi.acm.org/10.1145/1007352.1007394

41. Yao, A.C.: How to generate and exchange secrets (extended abstract). In: 27th Annual Symposium on Foundations of Computer Science, Toronto, Canada, 27–29, pp. 162–167 (1986). http://dx.doi.org/10.1109/SFCS.1986.25

Zero-Knowledge Arguments for Lattice-Based PRFs and Applications to E-Cash

Benoît Libert[1,2]([⊠]), San Ling[3], Khoa Nguyen[3], and Huaxiong Wang[3]

[1] CNRS, Laboratoire LIP, Lyon, France
benoit.libert@ens-lyon.fr
[2] ENS de Lyon, Laboratoire LIP (U. Lyon, CNRS, ENSL, INRIA, UCBL),
Lyon, France
[3] School of Physical and Mathematical Sciences, Nanyang Technological University,
Singapore, Singapore
{lingsan,khoantt,hxwang}@ntu.edu.sg

Abstract. Beyond their security guarantees under well-studied assumptions, algebraic pseudo-random functions are motivated by their compatibility with efficient zero-knowledge proof systems, which is useful in a number of privacy applications like digital cash. We consider the problem of proving the correct evaluation of lattice-based PRFs based on the Learning-With-Rounding (LWR) problem introduced by Banerjee *et al.* (Eurocrypt'12). Namely, we are interested zero-knowledge arguments of knowledge of triples (y, k, x) such that $y = F_k(x)$ is the correct evaluation of a PRF for a secret input x and a committed key k. While analogous statements admit efficient zero-knowledge protocols in the discrete logarithm setting, they have never been addressed in lattices so far. We provide such arguments for the key homomorphic PRF of Boneh *et al.* (Crypto'13) and the generic PRF implied by the LWR-based pseudo-random generator. As an application of our ZK arguments, we design the first compact e-cash system based on lattice assumptions. By "compact", we mean that the complexity is at most logarithmic in the value of withdrawn wallets. Our system can be seen as a lattice-based analogue of the first compact e-cash construction due to Camenisch, Hohenberger and Lysyanskaya (Eurocrypt'05).

Keywords: Lattices · Pseudo-random functions · Zero-knowledge arguments · E-cash systems · Anonymity

1 Introduction

Since the seminal results of Ajtai [2] and Regev [85], lattice-based cryptography has been a very active area which undergone quite rapid development, notably with the advent of lattice trapdoors [52,76] and homomorphic encryption [51]. Not only does it enable powerful functionalities, it also offers many advantages over conventional number-theoretic techniques, like simpler arithmetic operations, its conjectured resistance to quantum attacks or a better asymptotic efficiency.

© International Association for Cryptologic Research 2017
T. Takagi and T. Peyrin (Eds.): ASIACRYPT 2017, Part III, LNCS 10626, pp. 304–335, 2017.
https://doi.org/10.1007/978-3-319-70700-6_11

The design of numerous cryptographic protocols appeals to zero-knowledge proofs [55] to prove properties about encrypted or committed values so as to enforce honest behavior on behalf of participants or protect the privacy of users. In the lattice settings, efficient zero-knowledge proofs are non-trivial to construct. While natural solutions exist for proving knowledge of secret keys [64,70,73,77], they are only known to work for very specific languages. When it comes to proving general circuit satisfiability, the best known methods rely on the ring variants [14,89] of the Learning-With-Errors (LWE) and Short Integer Solution (SIS) problems and are not known to readily carry over to standard lattices. In the standard model, the problem is even trickier as we do not have a lattice-based counterpart of Groth-Sahai proofs [58] and efficient non-interactive proof systems are only available for specific problems [84].

In this paper, we consider the natural problem of proving the correct evaluation of lattice-based pseudo-random functions (PRFs) w.r.t. committed keys and inputs. This problem arises in numerous protocols where a user has to deterministically generate a random-looking value without betraying his identity.

We provide zero-knowledge arguments of correct evaluation for the LWE-based PRF of Boneh, Lewi, Montgomery and Raghunathan (BLMR) [17] as well as the construction generically obtained from pseudo-random generators via the Goldreich-Goldwasser-Micali (GGM) methodology [54]. As an application of our arguments, we provide the first lattice-based realization of the compact e-cash primitive of Camenisch, Hohenberger and Lysyanskaya [22].

Introduced by Chaum [33,34], electronic cash is the digital counterpart of regular money. As envisioned in [33], digital cash involve a bank and several users and merchants. It allows users to withdraw digital coins from the bank in such a way that e-coins can later be spent at merchants. In the on-line setting [33,35,36], merchants contact the bank before accepting any payment so that the bank is involved in all transactions to prevent double-spendings. In the (usually preferred) off-line model [37], the merchant accepts payments without any interaction with the bank: the deposit phase is postponed to a later moment where the merchant can return many coins at once. In all cases, when a merchant returns coins back to the bank, the latter should infer no information as to when and by whom the returned coins were withdrawn. Should the bank collude with the merchant, it remains unable to link a received coin to a particular execution of the withdrawal protocol. Of course, dishonest users should not be able to spend more coins than they withdrew without being identified. While fair e-cash systems [88] resort to an off-line trusted authority to call out cheaters, classical e-cash [37] allows identifying double-spenders without any TTP. In 2005, Camenisch, Hohenberger and Lysyanskaya [22] advocated e-cash solutions with *compactness* property: namely, a compact e-cash scheme allows a user to withdraw a wallet of 2^L coins in such a way that the complexity of spending and withdrawal protocols does not exceed $\mathcal{O}(L + \lambda)$, where λ is the security parameter. The constructions of [22] elegantly combine signature schemes with efficient protocols [25,26], number theoretic pseudo-random functions [44] and zero-knowledge proofs, making it possible to store a wallet using only $\mathcal{O}(L + \lambda)$ bits.

1.1   Our Contributions

OUR RESULTS. We describe the first compact e-cash system [22] based on lattice assumptions. Here, consistently with the literature on e-cash, "compactness" refers to schemes where the withdrawal, spending and deposit phases have at most logarithmic complexities in the maximal value of withdrawn wallets (analogously to the solutions of [22] where the term "compact" was introduced). The security of our scheme is proved in the random oracle model [11] under the Short Integer Solution (SIS) and LWE assumptions.

As a crucial ingredient of our solution, we provide zero-knowledge arguments vouching for the correct evaluation of lattice-based pseudo-random functions. More precisely, we construct arguments of knowledge of a committed seed $\mathbf{k}$, a secret input J and an output $\mathbf{y}$ satisfying $\mathbf{y} = F_{\mathbf{k}}(J)$. We describe such arguments for the key-homomorphic PRF of Boneh *et al.* [17] and the PRF obtained by applying the Goldreich-Goldwasser-Micali (GGM) [54] paradigm. As a building block, we provide zero-knowledge arguments for statements related to the Learning-With-Rounding (LWR) problem of Banerjee, Peikert and Rosen [8]. Given a public random matrix $\mathbf{A} \in \mathbb{Z}_q^{n \times m}$, it requires to tell apart vectors $\lfloor \mathbf{A}^T \cdot \mathbf{s} \rfloor_p = \lfloor (p/q) \cdot \mathbf{A}^T \cdot \mathbf{s} \rfloor \in \mathbb{Z}_p^m$ from the uniform distribution $U(\mathbb{Z}_p^m)$ over $\mathbb{Z}_p^m$, where $q > p \geq 2$. A crucial step of our argument system consists in demonstrating the correct computation of the rounding step: i.e., proving that $\mathbf{y} = \lfloor \mathbf{x} \rfloor_p$, for $\mathbf{x} \in \mathbb{Z}_q^m$ satisfying some additional context-dependent constraints.

We believe that our zero-knowledge arguments can find use cases in many other applications involving PRFs, and where zero-knowledge proofs constrain participants not to deviate from the protocol. Examples include privacy-preserving de-centralized e-cash systems [12,39,57], stateful anonymous credentials [40], n-times periodic anonymous authentication [21], traceable ring signatures [50], anonymous survey systems [59], password-protected secret sharing [61] or unlinkable pseudonyms for privacy-preserving distributed databases [24]. We also think of distributed PRFs [75,80], where servers holding a polynomial share $\mathbf{k}_i$ of the seed $\mathbf{k}$ can prove the correctness of their contribution w.r.t. to their committed share $\mathbf{k}_i$. Our arguments may also prove useful in the context of oblivious PRF evaluations [49,62], where one party holds a PRF key $\mathbf{k}$ and must convince the other party that $\mathbf{k}$ was correctly used in oblivious computations.

OUR TECHNIQUES. In order to convince a verifier of the correct evaluation of LWR-based PRFs, the first step is to provide evidence that the underlying rounding operation is properly carried out. For dimension $m > 1$ and moduli $q > p \geq 2$, identify $\mathbb{Z}_q, \mathbb{Z}_p$ as the set $[0, q-1]$ and $[0, p-1]$, respectively, and consider the function $\lfloor \cdot \rfloor_p : \mathbb{Z}_q^m \rightarrow \mathbb{Z}_p^m : \mathbf{x} \mapsto \mathbf{y} = \lfloor (p/q) \cdot \mathbf{x} \rfloor \mod p$. We observe that, one knows secret vector $\mathbf{x} \in [0, q-1]^m$ such that $\lfloor \mathbf{x} \rfloor_p = \mathbf{y}$ for a given $\mathbf{y} \in [0, p-1]^m$, if and only if one knows $\mathbf{x}, \mathbf{z} \in [0, q-1]^m$ such that

$$p \cdot \mathbf{x} = q \cdot \mathbf{y} + \mathbf{z} \mod pq. \tag{1}$$

This crucial observation gives us a modular equation where the secret vectors $\mathbf{x}, \mathbf{z}$ are "small" relatively to the modulus pq. To prove that we know such secret vectors (where $\mathbf{x}$ may satisfy additional statements, e.g., it is committed, or certified, or is the output of other algorithms), we exploit Ling *et al.*'s decomposition-extension framework [70], which interacts well with Stern's permuting technique [86]. Specifically, we employ a matrix $\mathbf{H}_{m,q-1} \in \mathbb{Z}_q^{m \times \bar{m}}$, where $\bar{m} = m\lceil \log q \rceil$, that allows to compute $\tilde{\mathbf{x}}, \tilde{\mathbf{z}} \in \{0,1\}^{\bar{m}}$ such that $\mathbf{H}_{m,q-1} \cdot \tilde{\mathbf{x}} = \mathbf{x}$ and $\mathbf{H}_{m,q-1} \cdot \tilde{\mathbf{z}} = \mathbf{z}$. Then, we let $\mathsf{B}_{\bar{m}}^2$ be the set of all vectors in $\{0,1\}^{2\bar{m}}$ that have fixed Hamming weight $\bar{m}$, and append $\bar{m}$ suitable entries to $\tilde{\mathbf{x}}, \tilde{\mathbf{z}}$ to obtain $\hat{\mathbf{x}}, \hat{\mathbf{z}} \in \mathsf{B}_{\bar{m}}^2$. Now, Eq. (1) is rewritten as:

$$\left(p \cdot [\mathbf{H}_{m,q-1} \mid \mathbf{0}^{m \times \bar{m}}]\right) \cdot \hat{\mathbf{x}} - [\mathbf{H}_{m,q-1} \mid \mathbf{0}^{m \times \bar{m}}] \cdot \hat{\mathbf{z}} = q \cdot \mathbf{y} \bmod pq. \qquad (2)$$

Note that, one knows $\mathbf{x}, \mathbf{z} \in [0, q-1]^m$ satisfying (1) if and only if one can compute $\hat{\mathbf{x}}, \hat{\mathbf{z}} \in \mathsf{B}_{\bar{m}}^2$ satisfying (2). Moreover, as the constraint of $\hat{\mathbf{x}}, \hat{\mathbf{z}}$ is invariant under permutation (namely, $\hat{\mathbf{x}}, \hat{\mathbf{z}} \in \mathsf{B}_{\bar{m}}^2$ if and only if $\pi_x(\hat{\mathbf{x}}), \pi_z(\hat{\mathbf{z}}) \in \mathsf{B}_{\bar{m}}^2$, where π_x, π_z are permutations of $2\bar{m}$ elements), the latter statement can be handled via Stern's technique. Our method is readily extended to prove that the underlying vector $\mathbf{x}$ satisfies additional statements.

Let us now consider the problem of proving a correct evaluation of the Boneh *et al.* PRF [17]. The function uses public binary matrices $\mathbf{P}_0, \mathbf{P}_1 \in \{0,1\}^{m \times m}$ and a secret seed $\mathbf{k} \in \mathbb{Z}_q^m$ which allows mapping an input $J \in \{0,1\}^L$ to

$$F_{\mathbf{k}}(J) = \left\lfloor \mathbf{P}_{J[L]} \cdot \mathbf{P}_{J[L-1]} \cdots \mathbf{P}_{J[1]} \cdot \mathbf{k} \right\rfloor_p.$$

We consider the evaluation process iteratively and transform intermediate witnesses using the decomposition-extension framework [70], so that they nicely interact with Stern's permuting technique [86]. Namely, we define a sequence $\{\mathbf{x}_i\}_{i=0}^L$ which is initialized with $\mathbf{x}_0 = \mathbf{k} \in \mathbb{Z}_q^m$, iteratively computed as $\mathbf{x}_i = \mathbf{P}_{J[i]} \cdot \mathbf{x}_{i-1} \in \mathbb{Z}_q^m$, for each $i \in [1, L]$, and eventually yields the output $\mathbf{y} = \lfloor \mathbf{x}_L \rfloor_p$. For each $i \in [1, L]$, we translate the equation $\mathbf{x}_i = \mathbf{P}_{J[i]} \cdot \mathbf{x}_{i-1} \bmod q$ into

$$\mathbf{x}_i = [\mathbf{P}_0 \mid \mathbf{P}_1] \cdot \mathbf{t}_{i-1} \bmod q, \qquad \text{with} \qquad \mathbf{t}_{i-1} = \begin{pmatrix} \overline{J[i]} \cdot \mathbf{x}_{i-1} \\ J[i] \cdot \mathbf{x}_{i-1} \end{pmatrix}$$

and where $J[i]$ and $\overline{J[i]} = 1 - J[i]$ are part of the witnesses. Using suitable decomposition-extension techniques [69,70] on vectors $\{\mathbf{x}_i\}_{i=0}^L, \{\mathbf{t}_i\}_{i=1}^L$, we manage to express all the L iterative equations by just one equation of the form $\mathbf{M}_1 \cdot \mathbf{w}_1 = \mathbf{u}_1 \bmod q$, for some public matrix $\mathbf{M}_1$ and vector $\mathbf{u}_1$ over $\mathbb{Z}_q$, while $\mathbf{w}_1$ is a binary vector containing secret bits of all the witnesses and fitting a certain pattern. Meanwhile, the rounding step $\mathbf{y} = \lfloor \mathbf{x}_L \rfloor_p$, as discussed above, would yield an equation of the form $\mathbf{M}_2 \cdot \mathbf{w}_2 = \mathbf{u}_2 \bmod pq$, where $\mathbf{w}_2$ is correlated to $\mathbf{w}_1$. Furthermore, our applications require to additionally prove that a binary representation of the seed $\mathbf{x}_0 = \mathbf{k}$ is properly committed or certified, while the commitment or signature scheme may use a different modulus. Thus, we eventually have to handle relations of the form $\mathbf{M}_i \cdot \mathbf{w}_i = \mathbf{u}_i \bmod q_i$ for several moduli $q_1, \ldots, q_N$ when, for distinct $i, j \in [N]$, witnesses $\mathbf{w}_i, \mathbf{w}_j$ may have entries in common. An abstraction of Stern's protocol was recently suggested by

Libert *et al.* [68] to address a similar setting when one has to prove a number of linear relations. Unfortunately, their framework, which deals with a unique modulus, does not directly cover our setting. To overcome this problem, we thus put forward a generalization of Libert *et al.*'s framework, so as to handle correlated witnesses across relations modulo distinct integers.

The above techniques thus smoothly interact with the pseudo-random functions of Boneh *et al.* [17] and the PRG of [8]. Unfortunately, we did not manage to extend them to other existing PRFs [7,8,45] based on the hardness of LWR. In the synthesizer-based construction of Banerjee *et al.* [8], the difficulty is the lack of public matrices which would help us reduce the statement to an assertion of the form $\mathbf{M} \cdot \mathbf{w} = \mathbf{u}$, for some witness $\mathbf{w} \in \mathbb{Z}^m$ and public $\mathbf{M} \in \mathbb{Z}_q^{n \times m}$, $\mathbf{u} \in \mathbb{Z}_q^n$. Our zero-knowledge arguments do not appear to carry over to the key homomorphic functions of Banerjee and Peikert [7] either as they rely on a more complex tree-like structure. The fact that our techniques do *not* apply to all known lattice-based PRFs emphasizes that they are far more innovative than just an application of generic zero-knowledge paradigms, which would resort to a circuit decomposition of the evaluation algorithm and proceed in a gate-by-gate manner. Indeed, we process all statements without decomposing the arithmetic operations into a circuit.

Our compact e-cash construction builds on the design principle of Camenisch *et al.* [22] which combines signatures with efficient protocols [25,26], algebraic pseudo-random functions [44] and zero-knowledge proofs. In the lattice setting, we take the same approach by combining (a variant of) the signature scheme with efficient protocols of [68] and the PRF of [17]. While the GGM-based PRF of [8] would allow a more efficient choice of parameters, we chose to instantiate our system with the realization of Boneh *et al.* [17] since it simplifies the description and the security proof (specifically, we do not have to rely on the pseudo-randomness of the function in one of the security properties). However, our scheme can be modified to rely on the PRF built on the LWR-based PRG.

As in [22], the withdrawal phase allows the user to obtain a wallet of value $2^L - 1$ which consists of two PRF seeds, a counter and a signature generated by the bank on committed values. In the withdrawal protocol, the PRF seeds are obliviously signed (and bound to the user's secret key) by the bank using a signature scheme with efficient protocols [25,26]. The first seed $\mathbf{k}$ is used to derive each coin's serial number $\mathbf{y}_S = F_{\mathbf{k}}(J) \in \mathbb{Z}_p^m$ as a pseudo-random function of an L-bit counter $J \in \{0,1\}^L$ which denotes the number of previously spent coins. By spending the same coin twice, the user is thus forced to use the same serial number in two distinct transactions, making the cheating attempt detectable.

The second PRF seed $\mathbf{t}$ is used to compute a security tag $\mathbf{y}_T$ that allows identifying double-spenders. This tag is a vector $\mathbf{y}_T = PK_{\mathcal{U}} + H(\mathtt{info}) \cdot F_{\mathbf{t}}(J) \in \mathbb{Z}_p^m$, where $PK_{\mathcal{U}}$ is the user's public key and $H(\mathtt{info}) \in \mathbb{Z}_p^{m \times m}$ is a matrix generated by hashing some transaction-specific information supplied by the merchant. From two coins that share the same serial number $\mathbf{y}_S$ and distinct security tags $\mathbf{y}_{T,1} = PK_{\mathcal{U}} + H(\mathtt{info}_1) \cdot F_{\mathbf{t}}(J)$ and $\mathbf{y}_{T,2} = PK_{\mathcal{U}} + H(\mathtt{info}_2) \cdot F_{\mathbf{t}}(J)$, the difference $\mathbf{y}_{T,1} - \mathbf{y}_{T,2} = (H(\mathtt{info}_1) - H(\mathtt{info}_2)) \cdot F_{\mathbf{t}}(J)$ allows computing the PRF value $F_t(J) = (H(\text{info}_1) - H(\text{info}_2))^{-1} \cdot (\mathbf{y}_{T,1} - \mathbf{y}_{T,2}) \in \mathbb{Z}_p^m$ (and then $PK_{\mathcal{U}}$) whenever $H(\text{info}_1) - H(\text{info}_2)$ is invertible over $\mathbb{Z}_p$. This property is precisely ensured by the Full-Rank Difference function of Agrawal et $al.$ [1], which comes in handy to instantiate $H : \{0,1\}^* \to \mathbb{Z}_p^{m \times m}$. In contrast with [1], the Full-Rank Difference function is utilized in the scheme while [1] uses it in security proofs.

1.2   Related Work

E-CASH. Chaum's pioneering work [33,34] inspired a large body of research towards efficient e-cash systems [37,38,47,82,83,87] with better properties during two decades. The first compact realization was given by Camenisch et $al.$ [22] whose techniques served as a blueprint for many subsequent e-cash systems with additional features such as refined accountability-anonymity tradeoffs [23], coin endorsement [27], or security proofs in the standard model [10]. The authors of [22] extended their schemes with a coin tracing mechanism whereby all the coins of a double-spender can be traced once this user has been identified.

Divisible e-cash [82] allow users to withdraw a wallet of value 2^L in such a way that each spending may involve transactions of variable amounts. While the early constructions [82,83] only provided weaker anonymity properties, Canard and Gouget gave truly anonymous realizations [28] using tree-based techniques which were subsequently made scalable [29,30]. The recent adoption of de-centralized payment systems [79] has triggered a new line of research towards strengthening the privacy of Bitcoin (see [12,57,78] and references therein).

To our knowledge, all truly private compact e-cash systems rely on discrete-logarithm-based techniques, either because of the underlying pseudo-random function [10,22] or via accumulators [5] (or both). In the lattice setting, we are not aware of any compact e-cash realization and neither do we know of any proofs of correct PRF evaluation with or without random oracles. In particular, it remains an open problem to build verifiable random functions [74] from lattices.

LATTICES AND ZERO-KNOWLEDGE PROOFS. Existing methods of proving relations appearing in lattice-based cryptosystems belong to two main families. The first family, introduced by Lyubashevsky [73], uses "rejection sampling" techniques, and recently lead to relatively efficient proofs of knowledge of small secret vectors [9,13,14,42,43]. However, due to the nature of "rejection sampling" mechanisms, even the honest prover may fail to convince the verifier with a tiny probability: i.e., protocols in this family do not have perfect completeness. Furthermore, when proving knowledge of vectors having norm bound β, the knowledge extractor of these protocols is only guaranteed to produce witnesses of norm bound $g \cdot \beta$, for some factor $g > 1$. This factor, called the "soundness slack" in [9,42], may have an undesirable consequence: if an extracted witness has to be used in the security proof to solve a challenge SIS instance, we have to rely on the $\text{SIS}_{g \cdot \beta}$ assumption, which is stronger than the SIS_β assumption required by the protocol itself. Moreover, in some advanced protocols such as those considered in this work, the coordinates of extracted vectors are expected to be in $\{0,1\}$ and/or satisfy a specific pattern. Such issues seem hard to tackle using this family of protocols.

The second family, initiated by Ling *et al.* [70], rely on "decomposition-extension" techniques in lattice-based analogues [64] of Stern's protocol [86]. Stern-like systems are less efficient than those of the first family because each protocol execution admits a constant soundness error, requiring the protocols to be repeated $\omega(\log \lambda)$ times in order to achieve a negligible soundness error. On the upside, Stern-like protocols do have perfect completeness and are capable of handling a wide range of lattice-based relations [66–69,71], especially when the witnesses are not only required to be small or binary, but should also have prescribed arrangements of coordinates. Moreover, unlike protocols of the first family, the extractor of Stern-like protocols are able to output witness vectors having exactly the same properties as those expected from valid witnesses. This feature is often crucial in the design of advanced cryptographic constructions involving zero-knowledge proofs. Additionally, the "soundness slack" issue is completely avoided, so that the hardness assumptions are kept "in place".

When it comes to proving the correct evaluation of AES-like secret key primitives, several works [32,48,63] built zero-knowledge proofs upon garbled circuits or multi-party computation [53,60], which may lead to truly practical proofs [53] even for non-algebraic statements. However, the garbled circuit paradigm [63] inherently requires interactive proofs (and cannot be made non-interactive via Fiat-Shamir [46]), making it unsuitable to our applications where coins must carry a non-interactive proof. While Giacomelli *et al.* [53] successfully designed efficient non-interactive proofs for SHA-256 evaluations, these remain of linear length in the circuit size and efficiently combining them with proofs of algebraic statements is non-trivial here: in the e-cash setting, our goal is to prove the correct evaluation of LWE-based symmetric primitives for committed inputs and keys. To our knowledge, known results on the smooth integration of algebraic and non-algebraic statements [32] are obtained by tweaking the approach of Jawurek *et al.* [63], which requires interaction.

Despite the scarcity of truly efficient zero-knowledge proofs in the lattice-setting, a recent body of work successfully designed proof systems in privacy-preserving protocols [13,56,64,65,71,81]. These results, however, only considered ring signatures [19,64], group signatures [13,56,65,66,71,81], group encryption [67] or building blocks [68] for anonymous credentials [35]. As of the time of writing, lattice-based realizations of anonymous e-cash still remain lacking.

2   Background and Definitions

Vectors are denoted in bold lower-case letters and bold upper-case letters will denote matrices. The Euclidean and infinity norm of any vector $\mathbf{b} \in \mathbb{R}^n$ will be denoted by $\|\mathbf{b}\|$ and $\|\mathbf{b}\|_\infty$, respectively. The Euclidean norm of matrix $\mathbf{B} \in \mathbb{R}^{m \times n}$ with columns $(\mathbf{b}_i)_{i \leq n}$ is $\|\mathbf{B}\| = \max_{i \leq n} \|\mathbf{b}_i\|$. When $\mathbf{B}$ has full column-rank, we let $\widetilde{\mathbf{B}}$ denote its Gram-Schmidt orthogonalization.

When S is a finite set, we denote by $U(S)$ the uniform distribution over S and by $x \hookleftarrow U(S)$ the action of sampling x according to this distribution.

For any $\mathbf{x} \in \mathbb{Z}_q^m$, the notation $\lfloor \mathbf{x} \rceil_p$ stands for the result of the rounding operation $\lfloor \mathbf{x} \rceil_p = \lfloor (p/q) \cdot \mathbf{x} \rceil \bmod p$. Intuitively, the mapping $\lfloor \cdot \rceil_p : \mathbb{Z}_q^m \to \mathbb{Z}_p^m$ can be seen as dividing $\mathbb{Z}_q$ into p intervals of size (q/p) and sending each coordinate of $\mathbf{x} \in \mathbb{Z}_q^m$ to the interval it belongs to.

The column concatenation of matrices $\mathbf{A} \in \mathbb{R}^{n \times k}$ and $\mathbf{B} \in \mathbb{R}^{n \times m}$ is denoted by $[\mathbf{A} \mid \mathbf{B}] \in \mathbb{R}^{n \times (k+m)}$. When concatenating column vectors $\mathbf{x} \in \mathbb{R}^k$ and $\mathbf{y} \in \mathbb{R}^m$, for simplicity, we often use the notation $(\mathbf{x} \| \mathbf{y}) \in \mathbb{R}^{k+m}$ (instead of $(\mathbf{x}^\top \| \mathbf{y}^\top)^\top$).

2.1   Lattices

A lattice L is the set of integer linear combinations of linearly independent basis vectors $(\mathbf{b}_i)_{i \le n}$ living in $\mathbb{R}^m$. We work with q-ary lattices, for some prime q.

Definition 1. *Let $m \ge n \ge 1$, a prime $q \ge 2$ and $\mathbf{A} \in \mathbb{Z}_q^{n \times m}$ and $\mathbf{u} \in \mathbb{Z}_q^n$, define $\Lambda_q(\mathbf{A}) := \{\mathbf{e} \in \mathbb{Z}^m \mid \exists \mathbf{s} \in \mathbb{Z}_q^n \ s.t. \ \mathbf{A}^T \cdot \mathbf{s} = \mathbf{e} \bmod q\}$ as well as*

$$\Lambda_q^\perp(\mathbf{A}) := \{\mathbf{e} \in \mathbb{Z}^m \mid \mathbf{A} \cdot \mathbf{e} = \mathbf{0}^n \bmod q\}, \quad \Lambda_q^\mathbf{u}(\mathbf{A}) := \{\mathbf{e} \in \mathbb{Z}^m \mid \mathbf{A} \cdot \mathbf{e} = \mathbf{u} \bmod q\}$$

For any arbitrary $\mathbf{t} \in \Lambda_q^\mathbf{u}(\mathbf{A})$, we also define the shifted lattice $\Lambda_q^\mathbf{u}(\mathbf{A}) = \Lambda_q^\perp(\mathbf{A}) + \mathbf{t}$.

For a lattice L, a vector $\mathbf{c} \in \mathbb{R}^m$ and a real number $\sigma > 0$, define the function $\rho_{\sigma,\mathbf{c}}(\mathbf{x}) = \exp(-\pi \|\mathbf{x} - \mathbf{c}\|^2 / \sigma^2)$. The discrete Gaussian distribution of support L, center $\mathbf{c}$ and parameter σ is defined as $D_{L,\sigma,\mathbf{c}}(\mathbf{y}) = \rho_{\sigma,\mathbf{c}}(\mathbf{y}) / \rho_{\sigma,\mathbf{c}}(L)$ for any $\mathbf{y} \in L$, where $\rho_{\sigma,\mathbf{c}}(L) = \sum_{\mathbf{x} \in L} \rho_{\sigma,\mathbf{c}}(\mathbf{x})$. We denote by $D_{L,\sigma}(\mathbf{y})$ the distribution centered in $\mathbf{c} = \mathbf{0}^m$ and exploit the fact that samples from $D_{L,\sigma}$ are short w.h.p.

Lemma 1 [6, Lemma 1.5]. *For any lattice $L \subseteq \mathbb{R}^m$ and positive real number $\sigma > 0$, we have $\mathrm{Pr}_{\mathbf{b} \hookleftarrow D_{L,\sigma}}[\|\mathbf{b}\| \le \sqrt{m}\sigma] \ge 1 - 2^{-\Omega(m)}$.*

It is well-known that Gaussian distributions with lattice support can be efficiently sampled from a sufficiently short basis of the lattice.

Lemma 2 [20, Lemma 2.3]. *There exists a PPT algorithm GPVSample that takes as inputs a basis $\mathbf{B}$ of a lattice $L \subseteq \mathbb{Z}^n$ and a rational $\sigma \ge \|\widetilde{\mathbf{B}}\| \cdot \Omega(\sqrt{\log n})$, and outputs vectors $\mathbf{b} \in L$ with distribution $D_{L,\sigma}$.*

We rely on the trapdoor generation algorithm of Alwen and Peikert [4].

Lemma 3 [4, Theorem 3.2]. *There exists a PPT algorithm TrapGen that takes as inputs 1^n, 1^m and an integer $q \ge 2$ with $m \ge \Omega(n \log q)$, and outputs a matrix $\mathbf{A} \in \mathbb{Z}_q^{n \times m}$ and a basis $\mathbf{T_A}$ of $\Lambda_q^\perp(\mathbf{A})$ such that $\mathbf{A}$ is within statistical distance $2^{-\Omega(n)}$ to $U(\mathbb{Z}_q^{n \times m})$, and $\|\widetilde{\mathbf{T_A}}\| \le \mathcal{O}(\sqrt{n \log q})$.*

We utilize the basis delegation algorithm [31] that inputs a trapdoor for $\mathbf{A} \in \mathbb{Z}_q^{n \times m}$ and produces a trapdoor for any $\mathbf{B} \in \mathbb{Z}_q^{n \times m'}$ containing $\mathbf{A} \in \mathbb{Z}_q^{n \times m}$ as a submatrix.

Lemma 4 [31, Lemma 3.2]. *There exists a PPT algorithm that inputs a matrix* $\mathbf{B} \in \mathbb{Z}_q^{n \times m'}$ *whose first* m *columns span* $\mathbb{Z}_q^n$, *and a basis* $\mathbf{T_A}$ *of* $\Lambda_q^{\perp}(\mathbf{A})$ *where* $\mathbf{A}$ *is an* $n \times m$ *submatrix of* $\mathbf{B}$, *and outputs a basis* $\mathbf{T_B}$ *of* $\Lambda_q^{\perp}(\mathbf{B})$ *with* $\|\widetilde{\mathbf{T_B}}\| \leq \|\widetilde{\mathbf{T_A}}\|$.

Our security proofs use a technique introduced by Agrawal *et al.* [1].

Lemma 5 [1, Theorem 19]. *There exists a PPT algorithm that inputs matrices* $\mathbf{A}, \mathbf{C} \in \mathbb{Z}_q^{n \times m}$, *a small-norm matrix* $\mathbf{R} \in \mathbb{Z}^{m \times m}$, *a short basis* $\mathbf{T_C} \in \mathbb{Z}^{m \times m}$ *of* $\Lambda_q^{\perp}(\mathbf{C})$, *a vector* $\mathbf{u} \in \mathbb{Z}_q^n$ *and a rational* σ *such that* $\sigma \geq \|\widetilde{\mathbf{T_C}}\| \cdot \Omega(\sqrt{\log n})$, *and outputs vectors* $\mathbf{b} \in \mathbb{Z}^{2m}$ *such that* $[\mathbf{A} \mid \mathbf{A} \cdot \mathbf{R} + \mathbf{C}] \cdot \mathbf{b} = \mathbf{u} \bmod q$ *and with distribution statistically close to* $D_{L,\sigma}$ *where* L *denotes the shifted lattice* $\{\mathbf{x} \in \mathbb{Z}^{2m} : [\mathbf{A} \mid \mathbf{A} \cdot \mathbf{R} + \mathbf{C}] \cdot \mathbf{x} = \mathbf{u} \bmod q\}$.

2.2 Hardness Assumptions

Definition 2. *Let* $m, n, q \in \mathbb{N}$ *with* $m > n$ *and* $\beta > 0$. *The Short Integer Solution problem* $\mathsf{SIS}_{m,q,\beta}$ *is, given* $\mathbf{A} \hookleftarrow U(\mathbb{Z}_q^{n \times m})$, *find* $\mathbf{x} \in \Lambda^{\perp}(\mathbf{A})$ *with* $0 < \|\mathbf{x}\| \leq \beta$.

Definition 3. *Let* q, α *be functions of a parameter* n. *For a secret* $\mathbf{s} \in \mathbb{Z}_q^n$, *the distribution* $A_{q,\alpha,\mathbf{s}}$ *over* $\mathbb{Z}_q^n \times \mathbb{Z}_q$ *is obtained by sampling* $\mathbf{a} \hookleftarrow U(\mathbb{Z}_q^n)$ *and a noise* $e \hookleftarrow D_{\mathbb{Z},\alpha q}$, *and returning* $(\mathbf{a}, \langle \mathbf{a}, \mathbf{s} \rangle + e)$. *The Learning-With-Errors problem* $\mathsf{LWE}_{q,\alpha}$ *is, for* $\mathbf{s} \hookleftarrow U(\mathbb{Z}_q^n)$, *to distinguish between arbitrarily many independent samples from* $U(\mathbb{Z}_q^n \times \mathbb{Z}_q)$ *and the same number of samples from* $A_{q,\alpha,\mathbf{s}}$.

If $q \geq \sqrt{n}\beta$ and $m, \beta \leq \mathsf{poly}(n)$, then standard worst-case lattice problems with approximation factors $\gamma = \widetilde{\mathcal{O}}(\beta\sqrt{n})$ reduce to $\mathsf{SIS}_{m,q,\beta}$ (see, e.g., [52, Sec. 9]). Similarly, if $\alpha q = \Omega(\sqrt{n})$, standard worst-case lattice problems with approximation factors $\gamma = \mathcal{O}(\alpha/n)$ reduce [20,85] to $\mathsf{LWE}_{q,\alpha}$. In the design of deterministic primitives like PRFs, the following variant of LWE comes in handy.

Definition 4 [8]. *Let* q, p, m *be functions of a security parameter* n *such that* $q > p \geq 2$ *and* $m > n$. *The Learning-With-Rounding (LWR) problem is to distinguish the distribution* $\{(\mathbf{A}, \lfloor \mathbf{A}^T \cdot \mathbf{s} \rceil_p) \mid \mathbf{A} \hookleftarrow U(\mathbb{Z}_q^{n \times m}), \mathbf{s} \hookleftarrow U(\mathbb{Z}_q^n)\}$ *from the distribution* $\{(\mathbf{A}, \mathbf{y}) \mid \mathbf{A} \hookleftarrow U(\mathbb{Z}_q^{n \times m}), \mathbf{y} \hookleftarrow U(\mathbb{Z}_p^m)\}$.

Banerjee *et al.* [8] proved that LWR is as hard as LWE when q and the modulus-to-error ratio are super-polynomial. Alwen *et al.* [3] showed that, when the number m of samples is fixed in advance, LWR retains its hardness for polynomial moduli. Bogdanov *et al.* [15] generalized the result of [3] to get rid of restrictions on the modulus q. For such parameters, their result implies the security of the LWR-based PRG, which stretches the seed $\mathbf{s} \in \mathbb{Z}_q^n$ into $\lfloor \mathbf{A}^T \cdot \mathbf{s} \rceil_p \in \mathbb{Z}_p^m$.

2.3 Syntactic Definitions for Off-Line Compact E-Cash

An off-line e-cash system involves a bank $\mathcal{B}$, many users $\mathcal{U}$ and merchants $\mathcal{M}$. In the syntax defined by Camenisch, Hohenberger and Lysyanskaya [22], all these parties interact together via the following algorithms and protocols:

ParGen(1^λ): inputs a security parameter 1^λ and outputs public parameters par.

In the following, we assume that par are available to all parties although we sometimes omit them from the inputs of certain algorithms.

BKeygen(par): generates a bank's key pair $(SK_\mathcal{B}, PK_\mathcal{B})$ which allows $\mathcal{B}$ to issue wallets of value $2^L \in \mathsf{poly}(\lambda)$ (we assume that L is part of par).

UKeygen(par): generates a user key pair $(SK_\mathcal{U}, PK_\mathcal{U})$.

Withdraw$(\mathcal{U}(PK_\mathcal{B}, SK_\mathcal{U}), \mathcal{B}(PK_\mathcal{U}, SK_\mathcal{B}))$: is an interactive protocol between a user $\mathcal{U}$ and the bank $\mathcal{B}$. The user $\mathcal{U}$ obtains either a wallet $\mathcal{W}$ of 2^L coins or an error message $\bot$. The bank outputs some state information $T_\mathcal{W}$ which allows identifying $\mathcal{U}$, should he overspend.

Spend$(\mathcal{U}(\mathcal{W}, PK_\mathcal{B}, PK_\mathcal{M}, \mathtt{info}), \mathcal{M}(SK_\mathcal{M}, PK_\mathcal{B}, 2^L))$: is a protocol whereby the user $\mathcal{U}$, on input of public keys $PK_\mathcal{M}, PK_\mathcal{B}$ and some transaction-specific meta data $\mathtt{info}$, spends a coin from his wallet $\mathcal{W}$ to merchant $\mathcal{M}$. The merchant obtains a coin $coin$ comprised of a serial number and a proof of validity. $\mathcal{U}$'s output is an updated wallet $\mathcal{W}'$.

VerifyCoin(par, $PK_\mathcal{M}, PK_\mathcal{B}, coin$): is a non-interactive coin verification algorithm. On input of a purported coin and the public keys $PK_\mathcal{M}$, $PK_\mathcal{B}$ of the bank and the merchant, it outputs 0 or 1.

Deposit$(\mathcal{M}(SK_\mathcal{M}, coin, PK_\mathcal{B}), \mathcal{B}(PK_\mathcal{M}, SK_\mathcal{B}, \mathsf{state}_\mathcal{B})))$: is a protocol allowing the merchant $\mathcal{M}$ to deposit a received coin $coin$ into its account at the bank $\mathcal{B}$. $\mathcal{M}$ outputs $\bot$ if the protocol fails and nothing if it succeeds. The bank $\mathcal{B}$ outputs "accept" and updates its state $\mathsf{state}_\mathcal{B}$ by adding an entry $(PK_\mathcal{M}, coin)$ if VerifyCoin(par, $PK_\mathcal{M}, PK_\mathcal{B}, coin) = 1$ and no double-spending is detected. Otherwise, if VerifyCoin(par, $PK_\mathcal{M}, PK_\mathcal{B}, coin) = 1$ and $\mathsf{state}_\mathcal{B}$ already contains a coin with the same serial number, it outputs "user". If VerifyCoin(par, $PK_\mathcal{M}, PK_\mathcal{B}, coin) = 0$ or $\mathsf{state}_\mathcal{B}$ already contains an entry $(PK_\mathcal{M}, coin)$, it outputs "merchant".

Identify$(\mathsf{par}, PK_\mathcal{B}, coin_1, coin_2)$: is an algorithm that allows the bank $\mathcal{B}$ to identify a double-spender on input of two coins $coin_1$, $coin_2$ with identical serial numbers. The bank outputs the double-spender's public key $PK_\mathcal{U}$ and a proof Π_G that $\mathcal{U}$ indeed overspent.

Like [22], we assume that wallets $\mathcal{W}$ contain a counter J, initialized to 0, which indicates the number of previously spent coins. We also assume that each coin contains a serial number S, a proof of validity π as well as some information on the merchant's public key $PK_\mathcal{M}$ and some meta-data $\mathtt{info}$.

Following [22], we say that an off-line e-cash system is *compact* if the bitlength of the wallet $\mathcal{W}$ and the communication/computational complexities of all protocols is at most logarithmic in the value of the wallet (i.e., linear in L).

We use the security definitions of [22], which formalize security requirements called *anonymity*, *balance*, *double-spender identification* and *exculpability*.

Informally, the *balance* property considers colluding users interacting with a honest bank and attempting to spend more coins than they withdraw. This property is broken if the adversary manages to spend a coin of which the serial number does not match the serial number of any legally withdrawn coin. *Double-spender identification* complements the balance property by requiring that a malicious user be unable to output two coins with the same serial number without being caught by the Identify algorithm. *Anonymity* mandates that, when a merchant returns a received coin to the bank, even if they collude, they cannot infer anything as to when and by whom the coin was withdrawn. The *exculpability* property captures that honest users cannot be falsely accused of being double-spenders: the adversary controls the bank and wins if it outputs two coins with the same serial number and such that Identify points to a well-behaved user. The formal definitions of these properties are recalled in the full version of the paper.

3   Warm-Up: Permutations, Decompositions, Extensions

This section presents various notations and techniques that appeared (in slightly different forms) in earlier works on Stern-like protocols [66,68–71], and that will be used extensively throughout this work.

PERMUTATIONS.   For any positive integer m, we define the following sets.

- $\mathcal{S}_{\mathsf{m}}$: the set of all permutations of m elements.
- $\mathsf{B}_{\mathsf{m}}^2$: the set of binary vectors in $\{0,1\}^{2\mathsf{m}}$ with Hamming weight m. Note that for any $\mathbf{v} \in \mathbb{Z}^{2\mathsf{m}}$ and $\pi \in \mathcal{S}_{2m}$, we have:

$$\mathbf{v} \in \mathsf{B}_{\mathsf{m}}^2 \iff \pi(\mathbf{v}) \in \mathsf{B}_{\mathsf{m}}^2. \tag{3}$$

- $\mathsf{B}_{\mathsf{m}}^3$: the set of vectors in $\{-1,0,1\}^{3\mathsf{m}}$ that have exactly m coordinates equal to j, for every $j \in \{-1,0,1\}$. Note that for any $\mathbf{w} \in \mathbb{Z}^{3\mathsf{m}}$ and $\phi \in \mathcal{S}_{3m}$:

$$\mathbf{w} \in \mathsf{B}_{\mathsf{m}}^3 \iff \phi(\mathbf{w}) \in \mathsf{B}_{\mathsf{m}}^3. \tag{4}$$

For bit $c \in \{0,1\}$ and integer vector $\mathbf{v}$ of any dimension m, we denote by $\mathsf{Expand}(c,\mathbf{v})$ the vector $\begin{pmatrix} \bar{c} \cdot \mathbf{v} \\ c \cdot \mathbf{v} \end{pmatrix} \in \mathbb{Z}^{2\mathsf{m}}$, where $\bar{c}$ denotes the bit $1-c$.

For any positive integer m, bit $b \in \{0,1\}$, and permutation $\pi \in \mathcal{S}_{\mathsf{m}}$, we denote by $T_{b,\pi}$ the permutation that transforms the vector $\mathbf{v} = \begin{pmatrix} \mathbf{v}_0 \\ \mathbf{v}_1 \end{pmatrix} \in \mathbb{Z}^{2\mathsf{m}}$, where $\mathbf{v}_0, \mathbf{v}_1 \in \mathbb{Z}^{\mathsf{m}}$, into the vector $T_{b,\pi}(\mathbf{v}) = \begin{pmatrix} \pi(\mathbf{v}_b) \\ \pi(\mathbf{v}_{\bar{b}}) \end{pmatrix}$. Namely, $T_{b,\pi}$ first rearranges the 2 blocks of $\mathbf{v}$ according to b (it keeps the arrangement of blocks if $b=0$ and swaps them if $b=1$), then it permutes each block according to π.

Observe that the following equivalence holds for all $\mathfrak{m} \in \mathbb{Z}_+$, $b, c \in \{0, 1\}$, $\pi \in \mathcal{S}_\mathfrak{m}$, $\mathbf{v} \in \mathbb{Z}^\mathfrak{m}$:

$$\mathbf{z} = \mathsf{expand}(c, \mathbf{v}) \iff T_{b,\pi}(\mathbf{z}) = \mathsf{expand}(c \oplus b, \pi(\mathbf{v})), \tag{5}$$

where $\oplus$ denotes the addition operation modulo 2.

DECOMPOSITIONS. For any $B \in \mathbb{Z}_+$, define $\delta_B := \lfloor \log_2 B \rfloor + 1 = \lceil \log_2(B+1) \rceil$ and the sequence $B_1, \ldots, B_{\delta_B}$, where $B_j = \lfloor \frac{B + 2^{j-1}}{2^j} \rfloor$, for each $j \in [1, \delta_B]$. As observed in [70,72], it satisfies $\sum_{j=1}^{\delta_B} B_j = B$ and any integer $v \in [0, B]$ can be decomposed to $\mathsf{idec}_B(v) = (v^{(1)}, \ldots, v^{(\delta_B)})^\top \in \{0, 1\}^{\delta_B}$ such that $\sum_{j=1}^{\delta_B} B_j \cdot v^{(j)} = v$. We describe this decomposition procedure in a deterministic manner as follows:

1. $v' := v$
2. For $j = 1$ to δ_B do:
   (i) If $v' \geq B_j$ then $v^{(j)} := 1$, else $v^{(j)} := 0$;
   (ii) $v' := v' - B_j \cdot v^{(j)}$.
3. Output $\mathsf{idec}_B(v) = (v^{(1)}, \ldots, v^{(\delta_B)})^\top$.

Next, for any positive integers $\mathfrak{m}, B$, we define the matrix:

$$\mathbf{H}_{\mathfrak{m}, B} := \begin{bmatrix} B_1 \ldots B_{\delta_B} & & & \\ & B_1 \ldots B_{\delta_B} & & \\ & & \ddots & \\ & & & B_1 \ldots B_{\delta_B} \end{bmatrix} \in \mathbb{Z}^{\mathfrak{m} \times \mathfrak{m}\delta_B}, \tag{6}$$

and the following injective functions:

(i) $\mathsf{vdec}_{\mathfrak{m}, B} : [0, B]^\mathfrak{m} \to \{0, 1\}^{\mathfrak{m}\delta_B}$ that maps the vector $\mathbf{v} = (v_1, \ldots, v_\mathfrak{m})$ to $(\mathsf{idec}_B(v_1)\| \ldots \|\mathsf{idec}_B(v_\mathfrak{m}))$. Note that $\mathbf{H}_{\mathfrak{m}, B} \cdot \mathsf{vdec}_{\mathfrak{m}, B}(\mathbf{v}) = \mathbf{v}$.
(ii) $\mathsf{vdec}'_{\mathfrak{m}, B} : [-B, B]^\mathfrak{m} \to \{-1, 0, 1\}^{\mathfrak{m}\delta_B}$ that decomposes $\mathbf{w} = (w_1, \ldots, w_\mathfrak{m})$ into the vector $(\sigma(w_1) \cdot \mathsf{idec}_B(|w_1|)\| \ldots \|\sigma(w_\mathfrak{m}) \cdot \mathsf{idec}_B(|w_\mathfrak{m}|))$ such that, for each $i \in [\mathfrak{m}]$, we have: $\sigma(w_i) = 0$ if $w_i = 0$; $\sigma(w_i) = -1$ if $w_i < 0$; $\sigma(w_i) = 1$ if $w_i > 0$. Note that $\mathbf{H}_{\mathfrak{m}, B} \cdot \mathsf{vdec}'_{\mathfrak{m}, B}(\mathbf{w}) = \mathbf{w}$.

EXTENSIONS. We define following extensions of matrices and vectors.

- For any $\mathfrak{m}, B \in \mathbb{Z}_+$, define $\widehat{\mathbf{H}}_{\mathfrak{m}, B} \in \mathbb{Z}^{\mathfrak{m} \times 2\mathfrak{m}\delta_B}$, $\check{\mathbf{H}}_{\mathfrak{m}, B} \in \mathbb{Z}^{\mathfrak{m} \times 3\mathfrak{m}\delta_B}$ as follows:

$$\widehat{\mathbf{H}}_{\mathfrak{m}, B} := \begin{bmatrix} \mathbf{H}_{\mathfrak{m}, B} \mid \mathbf{0}^{\mathfrak{m} \times \mathfrak{m}\delta_B} \end{bmatrix}; \qquad \check{\mathbf{H}}_{\mathfrak{m}, B} := \begin{bmatrix} \mathbf{H}_{\mathfrak{m}, B} \mid \mathbf{0}^{\mathfrak{m} \times 2\mathfrak{m}\delta_B} \end{bmatrix}.$$

- Given $\mathbf{v} \in \{0, 1\}^\mathfrak{m}$, define $\mathsf{TwoExt}(\mathbf{v}) := (\mathbf{v}\|\mathbf{0}^{\mathfrak{m}-n_0}\|\mathbf{1}^{\mathfrak{m}-n_1}) \in \mathsf{B}_\mathfrak{m}^2$, where n_0, n_1 are the number of coordinates in $\mathbf{v}$ equal to 0 and 1, respectively.
- Given $\mathbf{v} \in [-1, 0, 1]^\mathfrak{m}$, define

$$\mathsf{ThreeExt}(\mathbf{v}) := (\mathbf{v}\|\mathbf{0}^{\mathfrak{m}-n_0}\|\mathbf{1}^{\mathfrak{m}-n_1}\|-\mathbf{1}^{\mathfrak{m}-n_{-1}}) \in \mathsf{B}_\mathfrak{m}^3,$$

where n_0, n_1, n_{-1} are the number of coordinates in $\mathbf{v}$ equal to 0, 1, and -1, respectively.

Note that, if $\mathbf{x} \in [0, B]^m$ and $\mathbf{y} \in [-B, B]^m$, then we have:

$$\mathsf{TwoExt}\big(\mathsf{vdec}_{m,B}(\mathbf{x})\big) \in \mathsf{B}^2_{m\delta_B} \text{ and } \widehat{\mathbf{H}}_{m,B} \cdot \mathsf{TwoExt}\big(\mathsf{vdec}_{m,B}(\mathbf{x})\big) = \mathbf{x}, \quad (7)$$

$$\mathsf{ThreeExt}\big(\mathsf{vdec}'_{m,B}(\mathbf{y})\big) \in \mathsf{B}^3_{m\delta_B} \text{ and } \breve{\mathbf{H}}_{m,B} \cdot \mathsf{ThreeExt}\big(\mathsf{vdec}'_{m,B}(\mathbf{y})\big) = \mathbf{y}. \quad (8)$$

In the framework of Stern-like protocols [64,68,70,71,86], the above techniques are useful when it comes proving in zero-knowledge the possession of integer vectors satisfying several different constraints:

Case 1: $\mathbf{x} \in [0, B]^m$. We equivalently prove $\hat{\mathbf{x}} = \mathsf{TwoExt}\big(\mathsf{vdec}_{m,B}(\mathbf{x})\big) \in \mathsf{B}^2_{m\delta_B}$. To do this, pick $\pi \hookleftarrow U(\mathcal{S}_{2m\delta_B})$, and convince the verifier that $\pi(\hat{\mathbf{x}}) \in \mathsf{B}^2_{m\delta_B}$.

Case 2: $\mathbf{x} \in [-B, B]^m$. We equivalently prove $\breve{\mathbf{x}} = \mathsf{ThreeExt}\big(\mathsf{vdec}'_{m,B}(\mathbf{y})\big) \in \mathsf{B}^3_{m\delta_B}$. To do this, pick $\pi \hookleftarrow U(\mathcal{S}_{3m\delta_B})$, and convince the verifier that $\pi(\breve{\mathbf{x}}) \in \mathsf{B}^3_{m\delta_B}$.

Case 3: $\mathbf{x} = \mathsf{expand}(c, \mathbf{v})$, where $\mathbf{v}$ satisfies one of the above two constraints. To hide $\mathbf{v}$, we use the respective decomposition-extension-permutation technique. To hide the bit c, we pick a "one-time pad" $b \hookleftarrow U(\{0, 1\})$ and exploit the equivalence observed in (5). Looking ahead, this technique will be used in Sect. 4.3 to hide the bits of the PRF input J and those of a signature component $\tau \in \{0, 1\}^\ell$ in Sect. 5.

4   Zero-Knowledge Arguments for Lattice-Based PRFs

Here, we first give an abstraction of Stern's protocol [86]. With this abstraction in mind, we then present our techniques for achieving zero-knowledge arguments for the BLMR PRF [17].

In the full version of the paper, we adapt these techniques to the PRF generically implied by the GGM [54] paradigm. While slightly more complex to describe, the GGM-based construction allows for a better choice of parameters since, owing to the result of Bogdanov et al. [15], it allows instantiating the LWR-based PRG with polynomial-size moduli.

4.1   An Abstraction of Stern's Protocol

In [86], Stern proposed a zero-knowledge protocol for the Syndrome Decoding problem, in which the main idea is to use a random permutation over coordinates of a secret vector to prove that the latter satisfies a given constraint (e.g., having fixed Hamming weight). Later on, Stern's protocol was adapted to the lattice setting by Kawachi et al. [64] and refined by Ling et al. [70] to handle statements related to the SIS and LWE problems. Subsequently, the protocol was further developed to design several lattice-based systems [66,69,71]. Recently, Libert et al. [68] suggested an abstraction that addresses the setting where one has to prove knowledge of small secret vectors satisfying a number of modular linear equations with respect to one modulus. While their generalization subsumes many relations that naturally appear in privacy-preserving protocols involving

1. **Commitment:** $\mathcal{P}$ samples $\phi \leftarrow U(\mathcal{S})$, $\mathbf{r}_1 \leftarrow U(\mathbb{Z}_{q_1}^{d_1}), \ldots, \mathbf{r}_N \leftarrow U(\mathbb{Z}_{q_N}^{d_N})$, and computes $\mathbf{r} = (\mathbf{r}_1 \| \ldots \| \mathbf{r}_N)$, $\mathbf{z} = \mathbf{w} \boxplus \mathbf{r}$.
   Then $\mathcal{P}$ samples randomness ρ_1, ρ_2, ρ_3 for COM, and sends CMT $= (C_1, C_2, C_3)$ to $\mathcal{V}$, where $C_1 = \mathsf{COM}(\phi, \{\mathbf{M}_i \cdot \mathbf{r}_i \bmod q_i\}_{i \in [N]}; \rho_1)$, and

$$C_2 = \mathsf{COM}(\Gamma_\phi(\mathbf{r}); \rho_2), \quad C_3 = \mathsf{COM}(\Gamma_\phi(\mathbf{z}); \rho_3).$$

2. **Challenge:** $\mathcal{V}$ sends a challenge $Ch \leftarrow U(\{1, 2, 3\})$ to $\mathcal{P}$.
3. **Response:** $\mathcal{P}$ sends RSP computed according to Ch, as follows:
   - $Ch = 1$: RSP $= (\mathbf{t}, \mathbf{s}, \rho_2, \rho_3)$, where $\mathbf{t} = \Gamma_\phi(\mathbf{w})$ and $\mathbf{s} = \Gamma_\phi(\mathbf{r})$.
   - $Ch = 2$: RSP $= (\pi, \mathbf{x}, \rho_1, \rho_3)$, where $\pi = \phi$ and $\mathbf{x} = \mathbf{z}$.
   - $Ch = 3$: RSP $= (\psi, \mathbf{y}, \rho_1, \rho_2)$, where $\psi = \phi$ and $\mathbf{y} = \mathbf{r}$.

Verification: Receiving RSP, $\mathcal{V}$ proceeds as follows:

- $Ch = 1$: Check that $\mathbf{t} \in$ VALID, and $C_2 = \mathsf{COM}(\mathbf{s}; \rho_2)$, $C_3 = \mathsf{COM}(\mathbf{t} \boxplus \mathbf{s}; \rho_3)$.
- $Ch = 2$: Parse $\mathbf{x} = (\mathbf{x}_1 \| \ldots \| \mathbf{x}_N)$, where $\mathbf{x}_i \in \mathbb{Z}_{q_i}^{d_i}$ for all $i \in [N]$, and check that

$$C_1 = \mathsf{COM}(\pi, \{\mathbf{M}_i \cdot \mathbf{x}_i - \mathbf{u}_i \bmod q_i\}_{i \in [N]}; \rho_1), \quad C_3 = \mathsf{COM}(\Gamma_\pi(\mathbf{x}); \rho_3).$$

- $Ch = 3$: Parse $\mathbf{y} = (\mathbf{y}_1 \| \ldots \| \mathbf{y}_N)$, where $\mathbf{y}_i \in \mathbb{Z}_{q_i}^{d_i}$ for all $i \in [N]$, and check that

$$C_1 = \mathsf{COM}(\psi, \{\mathbf{M}_i \cdot \mathbf{y}_i \bmod q_i\}_{i \in [N]}; \rho_1), \quad C_2 = \mathsf{COM}(\Gamma_\psi(\mathbf{y}); \rho_2).$$

In each case, $\mathcal{V}$ outputs 1 if and only if all the conditions hold.

Fig. 1. Our abstract protocol.

lattices, it is not immediately applicable to the statements considered in this paper since we have to work with more than one modulus.

We thus put forward a new abstraction of Stern's protocol [86] that handles modular equations with respect to $N \geq 1$ moduli $q_1, \ldots, q_N$, where secret witnesses may simultaneously appear across multiple equations.

Let n_i and $d_i \geq n_i$ be positive integers, and let $d = d_1 + \cdots + d_N$. Suppose that VALID is a subset of $\{-1, 0, 1\}^d$ and $\mathcal{S}$ is a finite set such that every $\phi \in \mathcal{S}$ can be associated with a permutation Γ_ϕ of d elements satisfying the conditions

$$\begin{cases} \mathbf{w} \in \mathsf{VALID} \iff \Gamma_\phi(\mathbf{w}) \in \mathsf{VALID}; \\ \text{If } \mathbf{w} \in \mathsf{VALID} \text{ and } \phi \text{ is uniform in } \mathcal{S}, \text{ then } \Gamma_\phi(\mathbf{w}) \text{ is uniform in } \mathsf{VALID}. \end{cases} \quad (9)$$

In our abstract protocol, for public matrices $\{\mathbf{M}_i \in \mathbb{Z}_{q_i}^{n_i \times d_i}\}_{i \in [N]}$ and vectors $\mathbf{u}_i \in \mathbb{Z}_{q_i}^{n_i}$, the prover argues in zero-knowledge the possession of integer vectors $\{\mathbf{w}_i \in \{-1, 0, 1\}^{d_i}\}_{i \in [N]}$ such that:

$$\mathbf{w} = (\mathbf{w}_1 \| \ldots \| \mathbf{w}_N) \in \mathsf{VALID}, \quad (10)$$

$$\forall i \in [N] : \mathbf{M}_i \cdot \mathbf{w}_i = \mathbf{u}_i \bmod q_i. \quad (11)$$

Looking ahead, all the statements considered in Sects. 4.3, 5 will be reduced to the above setting, wherein secret vectors $\mathbf{w}_1, \ldots, \mathbf{w}_N$ are mutually related, e.g., some entries of $\mathbf{w}_i$ also appear in $\mathbf{w}_j$.

The main ideas driving our protocol are as follows. To prove (10), the prover samples $\phi \leftarrow U(\mathcal{S})$ and provides evidence that $\Gamma_\phi(\mathbf{w}) \in$ VALID. The verifier should be convinced while learning nothing else, owing to the aforementioned properties of the sets VALID and $\mathcal{S}$. Meanwhile, to prove that Eq. (11) hold, the prover uses masking vectors $\{\mathbf{r}_i \leftarrow U(\mathbb{Z}_{q_i}^{d_i})\}_{i \in [N]}$ and demonstrates instead that $\mathbf{M}_i \cdot (\mathbf{w}_i + \mathbf{r}_i) = \mathbf{u}_i + \mathbf{M}_i \cdot \mathbf{r}_i \bmod q_i$.

The interaction between prover $\mathcal{P}$ and verifier $\mathcal{V}$ is described in Fig. 1. The common input consists of $\{\mathbf{M}_i \in \mathbb{Z}_{q_i}^{n_i \times d_i}\}_{i \in [N]}$ and $\mathbf{u}_i \in \mathbb{Z}_{q_i}^{n_i}$, while $\mathcal{P}$'s secret input is $\mathbf{w} = (\mathbf{w}_1 \| \ldots \| \mathbf{w}_N)$. The protocol makes use of a statistically hiding and computationally binding string commitment scheme COM such as the SIS-based commitment of [64]. For simplicity of presentation, for vectors $\mathbf{w} = (\mathbf{w}_1 \| \ldots \| \mathbf{w}_N) \in \mathbb{Z}^d$ and $\mathbf{r} = (\mathbf{r}_1 \| \ldots \| \mathbf{r}_N) \in \mathbb{Z}^d$, we denote by $\mathbf{w} \boxplus \mathbf{r}$ the operation that computes $\mathbf{z}_i = \mathbf{w}_i + \mathbf{r}_i \bmod q_i$ for all $i \in [N]$, and outputs d-dimensional integer vector $\mathbf{z} = (\mathbf{z}_1 \| \ldots \| \mathbf{z}_N)$. We note that, for all $\phi \in \mathcal{S}$, if $\mathbf{t} = \Gamma_\phi(\mathbf{w})$ and $\mathbf{s} = \Gamma_\phi(\mathbf{r})$, then we have $\Gamma_\phi(\mathbf{w} \boxplus \mathbf{r}) = \mathbf{t} \boxplus \mathbf{s}$.

The properties of our protocol are summarized in the following theorem.

Theorem 1. *Suppose that* COM *is a statistically hiding and computationally binding string commitment. Then, the protocol of Fig. 1 is a zero-knowledge argument of knowledge for the given statement, with perfect completeness, soundness error $2/3$, and communication cost $\mathcal{O}\big(\sum_{i=1}^{N} d_i \cdot \log q_i\big)$. In particular:*

- *There exists an efficient simulator that, on input $\{\mathbf{M}_i, \mathbf{u}_i\}_{i \in [N]}$, outputs an accepted transcript statistically close to that produced by the real prover.*
- *There exists an efficient knowledge extractor that, on input a commitment CMT as well as valid responses $(\mathrm{RSP}_1, \mathrm{RSP}_2, \mathrm{RSP}_3)$ to all three possible values of the challenge Ch, outputs a witness $\mathbf{w}' = (\mathbf{w}'_1 \| \ldots \| \mathbf{w}'_N) \in$ VALID such that $\mathbf{M}_i \cdot \mathbf{w}'_i = \mathbf{u}_i \bmod q_i$, for all $i \in [N]$.*

The proof of Theorem 1 employs standard simulation and extraction techniques of Stern-like protocols [64,68–70], and is deferred to the full version of the paper.

4.2 Transforming the LWR Relation

Let $q \geq p \geq 2$, $m \geq 1$, and let $\mathbb{Z}_q = [0, q-1]$ and $\mathbb{Z}_p = [0, p-1]$. Consider the LWR rounding function: $\lfloor \cdot \rceil_p : \mathbb{Z}_q^m \to \mathbb{Z}_p^m : \mathbf{x} \mapsto \mathbf{y} = \lfloor (p/q) \cdot \mathbf{x} \rceil \bmod p$.

On the road towards zero-knowledge arguments for LWR-based PRFs, we have to build a sub-protocol that allows proving knowledge of a secret vector $\mathbf{x} \in \mathbb{Z}_q^m$ satisfying, among other statements, the property of rounding to a given $\mathbf{y} \in \mathbb{Z}_p^m$: i.e., $\lfloor \mathbf{x} \rceil_p = \mathbf{y}$. To our knowledge, such a sub-protocol is not available in the literature for the time being and must be designed from scratch.

Our crucial observation is that one knows $\mathbf{x} \in [0, q-1]^m$ such that $\lfloor \mathbf{x} \rceil_p = \mathbf{y}$, if and only if one can compute $\mathbf{x}, \mathbf{z} \in [0, q-1]^m$ such that:

$$p \cdot \mathbf{x} = q \cdot \mathbf{y} + \mathbf{z} \bmod pq. \tag{12}$$

This observation allows us to transform the LWR relation into an equivalent form that can be handled using the Stern-like techniques provided in Sect. 3. Let $\widehat{\mathbf{x}} = \mathsf{TwoExt}\big(\mathsf{vdec}_{m,q-1}(\mathbf{x})\big)$ and $\widehat{\mathbf{z}} = \mathsf{TwoExt}\big(\mathsf{vdec}_{m,q-1}(\mathbf{z})\big)$. Then we have $\mathbf{x} = \widehat{\mathbf{H}}_{m,q-1} \cdot \widehat{\mathbf{x}}$ and $\mathbf{z} = \widehat{\mathbf{H}}_{m,q-1} \cdot \widehat{\mathbf{z}}$, so that Eq. (12) can be written as:

$$(p \cdot \widehat{\mathbf{H}}_{m,q-1}) \cdot \widehat{\mathbf{x}} - \widehat{\mathbf{H}}_{m,q-1} \cdot \widehat{\mathbf{z}} = q \cdot \mathbf{y} \bmod pq. \tag{13}$$

Note that one knows $\mathbf{x}, \mathbf{z} \in [0, q-1]^m$ satisfying (12) if and only if one can compute $\widehat{\mathbf{x}}, \widehat{\mathbf{z}} \in \mathsf{B}^2_{m\delta_{q-1}}$ satisfying (13). Furthermore, Stern's framework allows proving the latter in zero-knowledge using random permutations.

4.3 Argument of Correct Evaluation for the BLMR PRF

We now consider the problem of proving the correct evaluation of the BLMR pseudo-random function from [17]. Namely, we would like to prove that a given $\mathbf{y} = \big\lfloor \prod_{i=1}^{L} \mathbf{P}_{J[L+1-i]} \cdot \mathbf{k} \big\rfloor_p \in \mathbb{Z}_p^m$ is the correct evaluation for a committed seed $\mathbf{k} \in \mathbb{Z}_q^m$ and a secret input $J[1] \ldots J[L] \in \{0,1\}^L$, where $\mathbf{P}_0, \mathbf{P}_1 \in \{0,1\}^{m \times m}$ are public binary matrices, while revealing neither $\mathbf{k}$ nor $J[1] \ldots J[L]$. We assume public matrices $\mathbf{D}_0 \in \mathbb{Z}_{q_s}^{n \times m_0}$, $\mathbf{D}_1 \in \mathbb{Z}_{q_s}^{n \times \bar{m}}$, for some modulus q_s and integers m_0 and $\bar{m} = m\delta_{q-1}$, which are used to compute a KTX commitment [64] $\mathbf{c} = \mathbf{D}_0 \cdot \mathbf{r} + \mathbf{D}_1 \cdot \tilde{\mathbf{k}} \in \mathbb{Z}_{q_s}^n$ to the decomposition $\tilde{\mathbf{k}} = \mathsf{vdec}_{m,q-1}(\mathbf{k}) \in \{0,1\}^{\bar{m}}$ of the seed $\mathbf{k}$, where $\mathbf{r} \in [-\beta, \beta]^{m_0}$ is a discrete Gaussian vector (for some small integer β), and $\tilde{\mathbf{k}}$ satisfies $\mathbf{H}_{m,q-1} \cdot \tilde{\mathbf{k}} = \mathbf{k}$.

We first note that, in the evaluation process of $\mathbf{y} = \big\lfloor \prod_{i=1}^{L} \mathbf{P}_{J[L+1-i]} \cdot \mathbf{k} \big\rfloor_p$, one works with vectors $\{\mathbf{x}_i \in \mathbb{Z}_q^m\}_{i=0}^L$ such that $\mathbf{x}_0 = \mathbf{k}$, $\mathbf{x}_i = \mathbf{P}_{J[i]} \cdot \mathbf{x}_{i-1} \bmod q$ for each $i \in \{1, \ldots, L\}$, and $\mathbf{y} = \lfloor \mathbf{x}_L \rfloor_p$. We further observe that the iterative equation $\mathbf{x}_i = \mathbf{P}_{J[i]} \cdot \mathbf{x}_{i-1} \bmod q$ is equivalent to:

$$\mathbf{x}_i = \mathbf{P}_0 \cdot (\overline{J[i]} \cdot \mathbf{x}_{i-1}) + \mathbf{P}_1 \cdot (J[i] \cdot \mathbf{x}_{i-1}) = \big[\mathbf{P}_0 \,|\, \mathbf{P}_1 \big] \cdot \begin{pmatrix} \overline{J[i]} \cdot \mathbf{x}_{i-1} \\ J[i] \cdot \mathbf{x}_{i-1} \end{pmatrix} \bmod q. \tag{14}$$

Intuitively, this observation allows us to move the secret bit $J[i]$ from the "matrix side" to the "vector side" in order to make the equation compatible with Stern-like protocols. Next, for each $i \in \{0, \ldots, L\}$, we form the vector $\widehat{\mathbf{x}}_i = \mathsf{TwoExt}\big(\mathsf{vdec}_{m,q-1}(\mathbf{x}_i)\big) \in \mathsf{B}^2_{\bar{m}}$. Equation (14) can then be written as:

$$\widehat{\mathbf{H}}_{m,q-1} \cdot \widehat{\mathbf{x}}_i = \big[\mathbf{P}_0 \cdot \widehat{\mathbf{H}}_{m,q-1} \,|\, \mathbf{P}_1 \cdot \widehat{\mathbf{H}}_{m,q-1} \big] \cdot \mathsf{expand}(J[i], \widehat{\mathbf{x}}_{i-1}) \bmod q.$$

Let $\mathbf{P} = \big[\mathbf{P}_0 \cdot \widehat{\mathbf{H}}_{m,q-1} \,|\, \mathbf{P}_1 \cdot \widehat{\mathbf{H}}_{m,q-1} \big] \in \mathbb{Z}_q^{m \times 4\bar{m}}$, and $\{\mathbf{s}_{i-1} = \mathsf{expand}(J[i], \widehat{\mathbf{x}}_{i-1})\}_{i=1}^L$, we have the L equations:

$$\begin{cases} \mathbf{P} \cdot \mathbf{s}_0 - \widehat{\mathbf{H}}_{m,q-1} \cdot \widehat{\mathbf{x}}_1 = \mathbf{0} \bmod q, \\ \quad \vdots \\ \mathbf{P} \cdot \mathbf{s}_{L-1} - \widehat{\mathbf{H}}_{m,q-1} \cdot \widehat{\mathbf{x}}_L = \mathbf{0} \bmod q, \end{cases} \tag{15}$$

Regarding the rounding step $\lfloor \mathbf{x}_L \rceil_p = \mathbf{y} \in \mathbb{Z}_p^m$, using the transformations of Sect. 4.2, we obtain the following equation for $\widehat{\mathbf{z}} \in \mathsf{B}_{\bar{m}}^2$:

$$(p \cdot \widehat{\mathbf{H}}_{m,q-1}) \cdot \widehat{\mathbf{x}}_L - \widehat{\mathbf{H}}_{m,q-1} \cdot \widehat{\mathbf{z}} = q \cdot \mathbf{y} \bmod pq, \qquad (16)$$

As for the commitment relation, we have the equation $\mathbf{c} = \mathbf{D}_0 \cdot \mathbf{r} + \mathbf{D}_1 \cdot \tilde{\mathbf{k}} \bmod q_s$, where $\mathbf{H}_{m,q-1} \cdot \tilde{\mathbf{k}} = \mathbf{x}_0$. We let $\check{\mathbf{r}} = \mathsf{ThreeExt}(\mathsf{vdec}'_{m_0,\beta}(\mathbf{r})) \in \mathsf{B}_{m_0 \delta_\beta}^3$ and remark that $\mathsf{TwoExt}(\tilde{\mathbf{k}}) = \widehat{\mathbf{x}}_0$. Then, we have:

$$[\mathbf{D}_0 \cdot \check{\mathbf{H}}_{m_0,\beta}] \cdot \check{\mathbf{r}} + [\mathbf{D}_1 \mid \mathbf{0}^{n \times \bar{m}}] \cdot \widehat{\mathbf{x}}_0 = \mathbf{c} \bmod q_s. \qquad (17)$$

Our goal is now reduced to proving the possession of $J[1] \dots J[L] \in \{0,1\}^L$, $\widehat{\mathbf{x}}_0, \dots, \widehat{\mathbf{x}}_L, \widehat{\mathbf{z}} \in \mathsf{B}_{\bar{m}}^2$ and $\check{\mathbf{r}} \in \mathsf{B}_{m_0 \delta_\beta}^3$, satisfying Eqs. (17), (15) and (16). Next, we let $q_1 = q_s$, $q_2 = q$, $q_3 = pq$, and proceed as follows.

Regarding Eq. (17), letting $\mathbf{M}_1 = [\mathbf{D}_0 \cdot \check{\mathbf{H}}_{m_0,\beta} \mid \mathbf{D}_1 \mid \mathbf{0}^{n \times \bar{m}}]$, $\mathbf{u}_1 = \mathbf{c}$ and $\mathbf{w}_1 = (\check{\mathbf{r}} \| \widehat{\mathbf{x}}_0)$, the equation becomes:

$$\mathbf{M}_1 \cdot \mathbf{w}_1 = \mathbf{u}_1 \bmod q_1.$$

Next, we unify the L equations in (15). To this end, we define

$$\mathbf{M}_2 = \begin{bmatrix} \mathbf{P} & -\widehat{\mathbf{H}}_{m,q-1} & & \\ & \ddots & \ddots & \\ & & \mathbf{P} & -\widehat{\mathbf{H}}_{m,q-1} \end{bmatrix}, \quad \mathbf{u}_2 = \mathbf{0},$$

and $\mathbf{w}_2 = (\mathbf{s}_0 \| \widehat{\mathbf{x}}_1 \| \cdots \| \mathbf{s}_{L-1} \| \widehat{\mathbf{x}}_L)$. Then, (15) can be equivalently written as:

$$\mathbf{M}_2 \cdot \mathbf{w}_2 = \mathbf{u}_2 \bmod q_2.$$

As for Eq. (16), let $\mathbf{M}_3 = [(p \cdot \widehat{\mathbf{H}}_{m,q-1}) \mid -\widehat{\mathbf{H}}_{m,q-1}]$, $\mathbf{u}_3 = q \cdot \mathbf{y}$ and $\mathbf{w}_3 = (\widehat{\mathbf{x}}_L \| \widehat{\mathbf{z}})$. Then, we obtain:

$$\mathbf{M}_3 \cdot \mathbf{w}_3 = \mathbf{u}_3 \bmod q_3.$$

Now, we let $d_1 = 3m_0 \delta_\beta + 2\bar{m}$, $d_2 = 6L\bar{m}$ and $d_3 = 4\bar{m}$ be the dimensions of $\mathbf{w}_1, \mathbf{w}_2$ and $\mathbf{w}_3$, respectively, and $d = d_1 + d_2 + d_3$. We form the vector $\mathbf{w} = (\mathbf{w}_1 \| \mathbf{w}_2 \| \mathbf{w}_3) \in \{-1, 0, 1\}^d$, which has the form:

$$\mathbf{w} = (\check{\mathbf{r}} \| \widehat{\mathbf{x}}_0 \| \mathbf{s}_0 \| \widehat{\mathbf{x}}_1 \| \cdots \| \mathbf{s}_{L-1} \| \widehat{\mathbf{x}}_L \| \widehat{\mathbf{x}}_L \| \widehat{\mathbf{z}}). \qquad (18)$$

At this point, we have come close to reducing our statement to an instance of the one considered in Sect. 4.1. Next, let us specify the set VALID containing $\mathbf{w}$, the set $\mathcal{S}$ and the associated permutation Γ_ϕ satisfying conditions in 9.

Let VALID be the set of all vectors in $\{-1, 0, 1\}^d$ having the form (18), where

- $\check{\mathbf{r}} \in \mathsf{B}_{m_0 \delta_\beta}^3$, and $\widehat{\mathbf{x}}_0, \dots, \widehat{\mathbf{x}}_L, \widehat{\mathbf{z}} \in \mathsf{B}_{\bar{m}}^2$.
- $\{\mathbf{s}_{i-1} = \mathsf{expand}(J[i], \widehat{\mathbf{x}}_{i-1})\}_{i=1}^L$, for some $J[1] \dots J[L] \in \{0,1\}^L$.

It can be seen that our vector $\mathbf{w}$ belongs to this tailored set VALID.

Now, we define $\mathcal{S} := \mathcal{S}_{3m_0\delta_\beta} \times (\mathcal{S}_{2\bar{m}})^{L+2} \times \{0,1\}^L$. Then, for any set element $\phi = (\phi_r, \phi_0, \phi_1, \ldots, \phi_L, \phi_z, b_1 \ldots b_L) \in \mathcal{S}$, let Γ_ϕ be the permutation that transforms vector $\mathbf{w} \in \mathbb{Z}^d$ of the form (18) to vector $\Gamma_\phi(\mathbf{w})$ of the form:

$$\Gamma_\phi(\mathbf{w}) = \big(\phi_r(\breve{\mathbf{r}}) \,\|\, \phi_0(\widehat{\mathbf{x}}_0) \,\|\, T_{b_1,\phi_0}(\mathbf{s}_0) \,\|\, \phi_1(\widehat{\mathbf{x}}_1) \,\|\, \cdots \,\|\, T_{b_L,\phi_{L-1}}(\mathbf{s}_{L-1}) \,\|\, \phi_L(\widehat{\mathbf{x}}_L)$$
$$\|\, \phi_L(\widehat{\mathbf{x}}_L) \,\|\, \phi_z(\widehat{\mathbf{z}})\big).$$

Thanks to the equivalences (3), (4), (5) from Sect. 3, we have $\mathbf{w} \in$ VALID if and only if $\Gamma_\phi(\mathbf{w}) \in$ VALID. Furthermore, if $\phi \leftarrow U(\mathcal{S})$, then $\Gamma_\phi(\mathbf{w})$ is uniform in VALID. Said otherwise, the conditions in (9) are satisfied.

Given the above transformations and specifications, we can now run the abstract protocol of Fig. 1 to prove knowledge of $\mathbf{w} = (\mathbf{w}_1\|\mathbf{w}_2\|\mathbf{w}_3) \in$ VALID satisfying $\{\mathbf{M}_i \cdot \mathbf{w}_i = \mathbf{u}_i \bmod q_i\}_{i=1,2,3}$, where public matrices/vectors $\{\mathbf{M}_i, \mathbf{u}_i\}_{i=1,2,3}$ are as constructed above. As a result, we obtain a statistical zero-knowledge argument of knowledge for the statement described at the beginning of this section. For simulation, we run the simulator of Theorem 1 with public input $\{\mathbf{M}_i, \mathbf{u}_i\}_{i=1,2,3}$. For extraction (see also the full version of the paper), we first run the knowledge extractor of Theorem 1, to obtain $\mathbf{w}' = (\mathbf{w}_1'\|\mathbf{w}_2'\|\mathbf{w}_3') \in$ VALID such that $\{\mathbf{M}_i \cdot \mathbf{w}_i' = \mathbf{u}_i \bmod q_i\}_{i=1,2,3}$ and then reverse the witness transformations to get $\mathbf{k}' \in \mathbb{Z}_q^m$, $J'[1] \ldots J'[L] \in \{0,1\}^L$ and $\mathbf{r}' \in [-\beta,\beta]^{m_0}$, $\tilde{\mathbf{k}}' \in \{0,1\}^{\bar{m}}$ satisfying:

$$\mathbf{y} = \Big\lfloor \prod_{i=1}^{L} \mathbf{P}_{J'[L+1-i]} \cdot \mathbf{k}' \Big\rfloor_p, \quad \mathbf{c} = \mathbf{D}_0 \cdot \mathbf{r}' + \mathbf{D}_1 \cdot \mathbf{k}' \bmod q_s, \quad \mathbf{H}_{m,q-1} \cdot \tilde{\mathbf{k}}' = \mathbf{k}'.$$

The protocol has communication cost $\mathcal{O}(d_1 \cdot \log q_1 + d_2 \cdot \log q_2 + d_3 \cdot \log q_3)$. For a typical setting of parameters (as in Sect. 5), this cost is of order $\widetilde{\mathcal{O}}(\lambda \cdot L)$, where λ is the security parameter (and L is the input length of the PRF).

5   Description of Our Compact E-Cash System

This section describes our e-cash system. We do not present a general construction from lower level primitives because such a construction is already implicit in the work of Camenisch et al. [22] of which we follow the blueprint. To avoid repeating it, we directly show how to apply the same design principle in lattices using carefully chosen primitives that interact with our zero-knowledge proofs.

Like [22], our scheme combines signatures with efficient protocols and pseudorandom functions which support proofs of correct evaluation. Our e-cash system builds on the signature scheme with efficient protocols of Libert et al. [68]. The latter is a variant of the SIS-based signatures described by Boyen [18] and Böhl et al. [16]. We actually use a simplified version of their scheme which is recalled in the full version of the paper and dispenses with the need to encode messages in a special way.

As in [22], our withdrawal protocol involves a step where the bank and the user jointly compute a seed $\mathbf{k} = \mathbf{k}_0 + \mathbf{k}_1 \in \mathbb{Z}_q^m$, which will be uniform over $\mathbb{Z}_q^m$ as long as one of the two parties is honest. The reason is that the identification of double-spenders can only be guaranteed if two distinct small-domain PRFs with independent random keys never collide, except with negligible probability. To jointly generate the PRF seed $\mathbf{k}$, the protocol of [22] relies on the homomorphic property of the commitment scheme used in their oblivious signing protocol. In our setting, one difficulty is that the underlying KTX commitment [64] has message space $\{0,1\}^{m\lceil \log q \rceil}$ and is not homomorphic over $\mathbb{Z}_q^m$. To solve this problem, our withdrawal protocol lets the user obtain the bank's signature on a message containing the binary decompositions of $\mathbf{k}_0$ and $\mathbf{k}_1$, so that the sum $\mathbf{k} = \mathbf{k}_0 + \mathbf{k}_1$ is only reconstructed during the spending phase.

At the withdrawal step, the user also chooses a second PRF seed $\mathbf{t} \in \mathbb{Z}_q^m$ of its own. The withdrawal protocol ends with the user obtaining a signature on the committed messages $(\mathbf{e}_u, \tilde{\mathbf{k}}_0, \tilde{\mathbf{k}}_1, \tilde{\mathbf{t}})$, where $(\tilde{\mathbf{k}}_0, \tilde{\mathbf{k}}_1, \tilde{\mathbf{t}})$ are bitwise decompositions of PRF seeds and $\mathbf{e}_u$ is the user's private key for which the corresponding public key is a GPV syndrome $PK_{\mathcal{U}} = \mathbf{F} \cdot \mathbf{e}_u \in \mathbb{Z}_p^m$, for a random matrix $\mathbf{F} \in \mathbb{Z}_p^{m \times m\lceil \log q \rceil}$.

In each spent coin, the user computes a serial number $\mathbf{y}_S = F_{\mathbf{k}}(J) \in \mathbb{Z}_p^m$ consisting of a PRF evaluation under $\mathbf{k} \in \mathbb{Z}_p^m$ and generates a NIZK argument that $\mathbf{y}_S$ is the correct evaluation for the secret index J and the key $\mathbf{k} = \mathbf{k}_0 + \mathbf{k}_1$ contained in the certified wallet. Note that the argument does not require a separate commitment to $\mathbf{k}$ since the bank's signature $sig_{\mathcal{B}}$ on the message $(\mathbf{e}_u, \tilde{\mathbf{k}}_0, \tilde{\mathbf{k}}_1, \tilde{\mathbf{t}})$ already contains a commitment to the bits of $(\mathbf{k}_0, \mathbf{k}_1)$. Since $sig_{\mathcal{B}}$ and $(\mathbf{e}_u, \tilde{\mathbf{k}}_0, \tilde{\mathbf{k}}_1, \tilde{\mathbf{t}})$ are part of the witnesses that the user argues knowledge of, it is eventually the bank's public key that commits the user to the seed $\mathbf{k}$.

In each coin, the identification of double-spenders is enabled by a security tag $\mathbf{y}_T = PK_{\mathcal{U}} + H_{\mathrm{FRD}}(R) \cdot F_{\mathbf{t}}(J) \in \mathbb{Z}_p^m$, where $H_{\mathrm{FRD}}(R)$ is a Full-Rank Difference function [1,41] of some transaction-specific information. If two coins share the same serial number $\mathbf{y}_S$, the soundness of the argument system implies that the two security tags $\mathbf{y}_{T,1}, \mathbf{y}_{T,2}$ hide the same $PK_{\mathcal{U}}$. By the Full Rank Difference property, subtracting $\mathbf{y}_{T,1} - \mathbf{y}_{T,2}$ exposes $F_{\mathbf{t}}(J) \in \mathbb{Z}_p^m$ and, in turn, $PK_{\mathcal{U}} \in \mathbb{Z}_p^m$.

The details of the underlying argument system are given in Sect. 5.2, where we show that the considered statement reduces to an instance of the abstraction given in Sect. 4.1. On the way, we use a combination our transformation techniques for the BLMR PRF from Sect. 4.3 and the Stern-like techniques for the signature scheme of [68].

5.1   Description

In the description below, we use the injective function $\mathsf{vdec}_{n,q-1}(\cdot)$ defined in Sect. 3, which maps a vector $\mathbf{v} \in \mathbb{Z}_q^n$ to the vector $\mathsf{vdec}_{n,q-1}(\mathbf{v}) \in \{0,1\}^{n\lceil \log_2 q \rceil}$.

ParGen$(1^\lambda, 1^L)$: Given a security parameter $\lambda > 0$ and an integer $L > 0$ such that 2^L is the desired value of wallets, public parameters are chosen as follows.

1. Choose a lattice parameter $n = \mathcal{O}(\lambda)$. Choose parameters that will be used by the BLMR pseudo-random function [17]: an LWE parameter $\alpha = 2^{-\omega(\log^{1+c}(n))}$ for some constant $c > 0$; moduli $p = 2^{\log^{1+c}(n)}$ and $q = \mathcal{O}(\sqrt{n}/\alpha)$ such that p divides q; and dimension $m = \lceil n \log q \rceil$. Pick another prime modulus $q_s = \widetilde{\mathcal{O}}(n^4)$ to be used by the signature scheme. Pick an integer $\ell = \Theta(\lambda)$, a Gaussian parameter $\sigma = \Omega(\sqrt{n \log q_s} \log n)$, and an infinity norm bound $\beta = \sigma \cdot \omega(\log n)$. Let $\delta_{q_s-1} = \lceil \log_2(q_s) \rceil$, $\delta_{q-1} = \lceil \log_2(q) \rceil$, $\delta_{p-1} = \lceil \log_2(p) \rceil$.
   We will use an instance of the signature scheme with efficient protocols from [68], where matrices $(\mathbf{A}, \{\mathbf{A}_j\}_{j=0}^{\ell}, \mathbf{D}), \{\mathbf{D}_k\}_{k=0}^{4}$ do not all have the same number of columns. Specifically, let $m_s = m_0 = 2n\delta_{q_s-1}$ and define the length of message blocks to be $m_1 = m_2 = m_3 = m_4 = \bar{m} = m\delta_{q-1}$. We also use an additional matrix $\mathbf{F} \in \mathbb{Z}_p^{m \times m_f}$, where $m_f = \bar{m} = m\delta_{q-1}$.

2. Choose a commitment key CK for a statistically hiding commitment where the message space is $\{0,1\}^{m_1} \times \{0,1\}^{m_2} \times \{0,1\}^{m_3}$. This commitment key $CK = ([\mathbf{D}'_0 \mid \mathbf{D}''_0], \mathbf{D}_1, \mathbf{D}_2, \mathbf{D}_3, \mathbf{D}_4)$ consists of random matrices $\mathbf{D}'_0, \mathbf{D}''_0 \hookleftarrow U(\mathbb{Z}_{q_s}^{n \times m_0}), \mathbf{D}_1 \hookleftarrow U(\mathbb{Z}_{q_s}^{n \times m_1}), \mathbf{D}_2 \hookleftarrow U(\mathbb{Z}_{q_s}^{n \times m_2}), \mathbf{D}_3 \hookleftarrow U(\mathbb{Z}_{q_s}^{n \times m_3}), \mathbf{D}_4 \hookleftarrow U(\mathbb{Z}_{q_s}^{n \times m_4})$.

3. Select two binary matrices $\mathbf{P}_0, \mathbf{P}_1 \in \{0,1\}^{m \times m}$ uniformly among $\mathbb{Z}_q$-invertible matrices.

4. Finally, choose a full-rank difference function $H_{\mathrm{FRD}} : \mathbb{Z}_p^m \rightarrow \mathbb{Z}_p^{m \times m}$ such as [1], a collision-resistant hash function $H_0 : \{0,1\}^* \rightarrow \mathbb{Z}_p^m$ and another hash function $H : \{0,1\}^* \rightarrow \{1,2,3\}^\kappa$, for some $\kappa = \omega(\log \lambda)$, which will be modeled as a random oracle in the security analysis.

We define

$$\mathsf{par} := \Big(\mathbf{F}, \ \{\mathbf{P}_0, \ \mathbf{P}_1\}, \ H_{\mathrm{FRD}}, \ H_0, \ H, \ CK \Big).$$

where $CK = (\mathbf{D}_0 = [\mathbf{D}'_0 \mid \mathbf{D}''_0], \mathbf{D}_1, \mathbf{D}_2, \mathbf{D}_3, \mathbf{D}_4)$.

BKeygen$(1^\lambda, \mathsf{par})$: The bank $\mathcal{B}$ generates a key pair for the signature scheme with efficient protocols. This is done as follows.

1. Run $\mathsf{TrapGen}(1^n, 1^{m_s}, q_s)$ to get $\mathbf{A} \in \mathbb{Z}_{q_s}^{n \times m_s}$ and a short basis $\mathbf{T_A}$ of $\Lambda_{q_s}^{\perp}(\mathbf{A})$. This basis allows computing short vectors in $\Lambda_{q_s}^{\perp}(\mathbf{A})$ with a Gaussian parameter σ. Next, choose matrices $\mathbf{A}_0, \mathbf{A}_1, \ldots, \mathbf{A}_\ell \hookleftarrow U(\mathbb{Z}_{q_s}^{n \times m_s})$.

2. Choose $\mathbf{D} \hookleftarrow U\big(\mathbb{Z}_{q_s}^{n \times (m_s/2)}\big)$ and a random vector $\mathbf{u} \hookleftarrow U(\mathbb{Z}_{q_s}^n)$.
   The private key consists of $SK_\mathcal{B} := \mathbf{T_A}$ while the public key is

$$PK_\mathcal{B} := \big(\mathbf{A}, \ \{\mathbf{A}_j\}_{j=0}^{\ell}, \ \mathbf{D}, \ \mathbf{u}\big).$$

UKeygen$(1^\lambda, \mathsf{par})$: As a secret key, the user picks $SK_\mathcal{U} := \mathbf{e}_u \hookleftarrow U(\{0,1\}^{m_f})$ at random and computes his public key $PK_\mathcal{U}$ as a syndrome $PK_\mathcal{U} = \mathbf{F} \cdot \mathbf{e}_u \in \mathbb{Z}_p^m$.

Withdraw$\big(\mathcal{U}(PK_\mathcal{B}, SK_\mathcal{U}, 2^L), \mathcal{B}(PK_\mathcal{U}, SK_\mathcal{B}, 2^L)\big)$: The bank $\mathcal{B}$, which has a key pair $(SK_\mathcal{B}, PK_\mathcal{B})$, interacts with $\mathcal{U}$, who has $SK_\mathcal{U} = \mathbf{e}_u$, as follows.

1. $\mathcal{U}$ picks $\mathbf{t}, \mathbf{k}_0 \hookleftarrow U(\mathbb{Z}_q^m)$ and computes $\tilde{\mathbf{t}} = \mathsf{vdec}_{m,q-1}(\mathbf{t}) \in \{0,1\}^{\bar{m}}$, $\tilde{\mathbf{k}}_0 = \mathsf{vdec}_{m,q-1}(\mathbf{k}_0) \in \{0,1\}^{\bar{m}}$. Then, he generates a commitment to the 3-block message $(\mathbf{e}_u, \tilde{\mathbf{t}}, \tilde{\mathbf{k}}_0) \in \{0,1\}^{m_f} \times \{0,1\}^{\bar{m}} \times \{0,1\}^{\bar{m}}$. To this end, $\mathcal{U}$ samples $\mathbf{r}_0 \hookleftarrow D_{\mathbb{Z}^{m_s},\sigma}$ and computes

$$\mathbf{c}_\mathcal{U} = \mathbf{D}_0' \cdot \mathbf{r}_0 + \mathbf{D}_1 \cdot \mathbf{e}_u + \mathbf{D}_2 \cdot \tilde{\mathbf{t}} + \mathbf{D}_3 \cdot \tilde{\mathbf{k}}_0 \in \mathbb{Z}_{q_s}^n, \tag{19}$$

which is sent to $\mathcal{B}$. In addition, $\mathcal{U}$ generates an interactive zero-knowledge argument of knowledge of an opening

$$(\mathbf{r}_0, \mathbf{e}_u, \tilde{\mathbf{t}}, \tilde{\mathbf{k}}_0) \in D_{\mathbb{Z}^{m_s},\sigma} \times \{0,1\}^{m_f} \times \{0,1\}^{\bar{m}} \times \{0,1\}^{\bar{m}}$$

of $\mathbf{c}_\mathcal{U} \in \mathbb{Z}_{q_s}^n$ satisfying (19) and such that $PK_\mathcal{U} = \mathbf{F} \cdot \mathbf{e}_u \in \mathbb{Z}_p^m$. We note that this argument system is obtained via a straightforward adaptation of the Stern-like protocol from [68].

2. If the argument of step 1 verifies, $\mathcal{B}$ samples $\mathbf{r}_1 \hookleftarrow D_{\mathbb{Z}^{m_s},\sigma}$, $\mathbf{k}_1 \hookleftarrow U(\mathbb{Z}_q^m)$ and computes $\tilde{\mathbf{k}}_1 = \mathsf{vdec}_{m,q-1}(\mathbf{k}_1) \in \{0,1\}^{\bar{m}}$ and a re-randomized version of $\mathbf{c}_\mathcal{U}$ which is obtained as $\mathbf{c}_\mathcal{U}' = \mathbf{c}_\mathcal{U} + \mathbf{D}_0'' \cdot \mathbf{r}_1 + \mathbf{D}_4 \cdot \tilde{\mathbf{k}}_1 \in \mathbb{Z}_{q_s}^n$. It defines $\mathbf{u}_\mathcal{U} = \mathbf{u} + \mathbf{D} \cdot \mathsf{vdec}_{n,q_s-1}(\mathbf{c}_\mathcal{U}') \in \mathbb{Z}_{q_s}^n$. Next, $\mathcal{B}$ randomly picks $\tau \hookleftarrow \{0,1\}^\ell$ and uses $\mathbf{T_A}$ to compute a delegated basis $\mathbf{T}_\tau \in \mathbb{Z}^{2m_s \times 2m_s}$ for the matrix $\mathbf{A}_\tau \in \mathbb{Z}_{q_s}^{n \times 2m_s}$ defined as

$$\mathbf{A}_\tau = [\mathbf{A} \mid \mathbf{A}_0 + \sum_{j=1}^{\ell} \tau[j] \cdot \mathbf{A}_j] \in \mathbb{Z}_{q_s}^{n \times 2m_s}. \tag{20}$$

Using $\mathbf{T}_\tau \in \mathbb{Z}^{2m_s \times 2m_s}$, $\mathcal{B}$ samples a short vector $\mathbf{v} \in \mathbb{Z}^{2m_s}$ in $D_{\Lambda_{q_s}^{\mathbf{u}_\mathcal{U}}(\mathbf{A}_\tau),\sigma}$. It returns $\mathbf{k}_1 \in \mathbb{Z}_q^m$ and the vector $(\tau, \mathbf{v}, \mathbf{r}_1) \in \{0,1\}^\ell \times \mathbb{Z}^{2m_s} \times \mathbb{Z}^{m_s}$ to $\mathcal{U}$.

3. $\mathcal{U}$ computes $\mathbf{r} = \begin{bmatrix} \mathbf{r}_0 \\ \mathbf{r}_1 \end{bmatrix} \in \mathbb{Z}^{2m_s}$ and verifies that

$$\mathbf{A}_\tau \cdot \mathbf{v} = \mathbf{u} + \mathbf{D} \cdot \mathsf{vdec}_{n,q_s-1}\Big(\mathbf{D}_0 \cdot \mathbf{r} + \mathbf{D}_1 \cdot \mathbf{e}_u + \mathbf{D}_2 \cdot \mathsf{vdec}_{m,q-1}(\mathbf{t})$$

$$+\mathbf{D}_3 \cdot \mathsf{vdec}_{m,q-1}(\mathbf{k}_0) + \mathbf{D}_4 \cdot \mathsf{vdec}_{m,q-1}(\mathbf{k}_1)\Big) \bmod q_s$$

and $\|\mathbf{v}\| \leq \sigma\sqrt{2m_s}$, $\|\mathbf{r}_1\| \leq \sigma\sqrt{m_s}$. If so, $\mathcal{U}$ saves the wallet

$$\mathcal{W} := \Big(\mathbf{e}_u, \mathbf{t}, \mathbf{k}_0, \mathbf{k}_1, sig_\mathcal{B} = (\tau, \mathbf{v}, \mathbf{r}), J = 0\Big),$$

where $J \in \{0, \dots, 2^L-1\}$ is a counter initialized to 0 (otherwise, it outputs $\perp$). The bank $\mathcal{B}$ records a debit of 2^L for the account $PK_\mathcal{U}$.

Spend$\big(\mathcal{U}(\mathcal{W}, PK_\mathcal{B}, PK_\mathcal{M}, \mathsf{info}), \mathcal{M}(SK_\mathcal{M}, PK_\mathcal{B}, 2^L)\big)$: The user $\mathcal{U}$, on input of a wallet $\mathcal{W} = \big(\mathbf{e}_u, \mathbf{t}, \mathbf{k}_0, \mathbf{k}_1, sig_\mathcal{B} = (\tau, \mathbf{v}, \mathbf{r}), J\big)$, outputs $\perp$ if $J > 2^L - 1$. Otherwise, it runs the following protocol with $\mathcal{M}$.

1. Hash info $\in \{0,1\}^*$ and $PK_{\mathcal{M}}$ to obtain $R = H_0(PK_{\mathcal{M}}, \text{info}) \in \mathbb{Z}_p^m$.
2. Compute $\mathbf{k} = \mathbf{k}_0 + \mathbf{k}_1 \bmod q$, which will serve a PRF seed $\mathbf{k} \in \mathbb{Z}_q^m$. Using $\mathbf{k}$, compute the serial number

$$\mathbf{y}_S = \lfloor \prod_{i=1}^{L} \mathbf{P}_{J[L+1-i]} \cdot \mathbf{k} \rfloor_p \in \mathbb{Z}_p^m, \tag{21}$$

where $J[1] \ldots J[L] \in \{0,1\}^L$ is the representation of $J \in \{0, \ldots, 2^L - 1\}$.
3. Using the PRF seed $\mathbf{t} \in \mathbb{Z}_q^m$, compute the security tag

$$\mathbf{y}_T = PK_{\mathcal{U}} + H_{\mathrm{FRD}}(R) \cdot \lfloor \prod_{i=1}^{L} \mathbf{P}_{J[L+1-i]} \cdot \mathbf{t} \rfloor_p \in \mathbb{Z}_p^m. \tag{22}$$

4. Generate a non-interactive argument of knowledge π_K to prove that:
   (i) The given serial number $\mathbf{y}_S$ is the correct output of the PRF with key $\mathbf{k} = \mathbf{k}_0 + \mathbf{k}_1 \bmod q$ and input $J[1] \ldots J[L]$ (Eq. (21));
   (ii) The same input $J[1] \ldots J[L]$ and another key $\mathbf{t}$ involve in the generation of the security tag $\mathbf{y}_T$ of the form (22);
   (iii) The PRF keys $\mathbf{k}_0, \mathbf{k}_1, \mathbf{t}$ and the secret key $\mathbf{e}_u$ that corresponds to $PK_{\mathcal{U}}$ in (22) were certified by the bank.
   This is done by running the interactive zero-knowledge argument presented in Sect. 5.2, which can be seen as a combination of 2 instances of the protocol for the PRF layer from Sect. 4.3 and one instance of the protocol for the signature layer from [68]. The argument is repeated $\kappa = \omega(\log \lambda)$ times to achieve negligible soundness error, and then made non-interactive using the Fiat-Shamir heuristic [46] as a triple $\pi_K = (\{\mathsf{Comm}_{K,j}\}_{j=1}^\kappa, \mathsf{Chall}_K, \{\mathsf{Resp}_{K,j}\}_{j=1}^\kappa)$ where

$$\mathsf{Chall}_K = H(R, \mathbf{y}_S, \mathbf{y}_T, \{\mathsf{Comm}_{K,j}\}_{j=1}^t) \in \{1, 2, 3\}^\kappa.$$

$\mathcal{U}$ sends $coin = (R, \mathbf{y}_S, \mathbf{y}_T, \pi_K)$ to $\mathcal{M}$ who outputs $coin$ if VerifyCoin accepts it and $\bot$ otherwise. $\mathcal{U}$ outputs an updated wallet $\mathcal{W}'$, where J is incremented. We note that $coin$ has bit-size $\widetilde{\mathcal{O}}(L \cdot \lambda + \lambda^2)$, which is inherited from that of the underlying zero-knowledge argument system of Sect. 5.2.

VerifyCoin$(par, PK_{\mathcal{M}}, PK_{\mathcal{B}}, coin)$: Parse the coin as $coin = (R, \mathbf{y}_S, \mathbf{y}_T, \pi_K)$ and output 1 if and only if π_K properly verifies.

Deposit$(\mathcal{M}(SK_{\mathcal{M}}, coin, PK_{\mathcal{B}})), \mathcal{B}(PK_{\mathcal{M}}, SK_{\mathcal{B}}, \text{state}_{\mathcal{B}}))$: $coin = (R, \mathbf{y}_S, \mathbf{y}_T, \pi_K)$ is sent by $\mathcal{M}$ to the bank $\mathcal{B}$. If VerifyCoin$(par, PK_{\mathcal{M}}, PK_{\mathcal{B}}, coin) = 1$ and if serial number $\mathbf{y}_S$ does not already appear in any coin of the list state$_{\mathcal{B}}$, $\mathcal{B}$ accepts $coin$, adds $(R, \mathbf{y}_S)$ in state$_{\mathcal{B}}$ and credits $PK_{\mathcal{M}}$'s account. Otherwise, $\mathcal{B}$ returns "user" or "merchant" depending on which party is declared faulty.

Identify$(par, PK_{\mathcal{B}}, coin_1, coin_2))$: Given two coins $coin_1 = (R_1, \mathbf{y}_S, \mathbf{y}_{T,1}, \pi_{K,1})$, $coin_2 = (R_2, \mathbf{y}_S, \mathbf{y}_{T,2}, \pi_{K,2})$ with verifying proofs $\pi_{K,1}, \pi_{K,2}$ and the same serial number $\mathbf{y}_S \in \mathbb{Z}_p^m$ in distinct transactions $R_1 \neq R_2$, output $\perp$ if $\mathbf{y}_{T,1} = \mathbf{y}_{T,2}$. Otherwise, compute

$$\mathbf{y}_T' = \big(H_{\text{FRD}}(R_1) - H_{\text{FRD}}(R_2)\big)^{-1} \cdot (\mathbf{y}_{T,1} - \mathbf{y}_{T,2}) \in \mathbb{Z}_p^m$$

and then $PK_{\mathcal{U}} = \mathbf{y}_{T,1} - H_{\text{FRD}}(R_1) \cdot \mathbf{y}_T' \in \mathbb{Z}_p^m$. The proof Π_G that $\mathcal{U}$ is guilty simply consists of the two coins $coin_1$, $coin_2$ and the public key $PK_{\mathcal{U}}$.

In the full version of the paper, we show how to extend the scheme with a mechanism allowing to trace all the coins of an identified double-spender. Like Camenisch *et al.* [22], we can add this feature via a verifiable encryption step during the withdrawal phase. For this purpose, however, [22] crucially relies on properties of groups with a bilinear map that are not available here. To overcome this difficulty, we slightly depart from the design principle of [22] in that we rather use a secret-key verifiable encryption based on the hardness of LWE.

5.2   The Underlying Argument System of Our E-Cash Scheme

We now present the argument system employed by the Spend algorithm of the e-cash scheme in Sect. 5.1. This protocol is summarized as follows.

Let parameters λ, n, p, q, q_s, m, β, L, ℓ, $\bar{m} = m\delta_{q-1}$, $m_s = 2n\delta_{q_s-1}$ be as specified in Sect. 5.1. The public input consists of:

$$\begin{cases} \mathbf{D} \in \mathbb{Z}_{q_s}^{n \times (m_s/2)}; \ \mathbf{D}_0 \in \mathbb{Z}_{q_s}^{n \times 2m_s}; \ \{\mathbf{D}_k \in \mathbb{Z}_{q_s}^{n \times \bar{m}}\}_{k=1}^4; \ \mathbf{A}, \{\mathbf{A}_j\}_{j=0}^\ell \in \mathbb{Z}_{q_s}^{n \times m_s}; \\ \mathbf{F} \in \mathbb{Z}_p^{m \times \bar{m}}; \ \mathbf{u} \in \mathbb{Z}_{q_s}^n; \ \mathbf{P}_0, \mathbf{P}_1 \in \{0,1\}^{m \times m}; \ H_{\text{FRD}}(R) \in \mathbb{Z}_p^{m \times m}; \ \mathbf{y}_S, \mathbf{y}_T \in \mathbb{Z}_p^m. \end{cases}$$

The prover's goal is to prove in zero-knowledge the possession of

$$\begin{cases} \mathbf{v}_1, \mathbf{v}_2 \in [-\beta, \beta]^{m_s}; \ \mathbf{r} \in [-\beta, \beta]^{2m_s}; \ \tilde{\mathbf{w}} \in \{0,1\}^{m_s/2}; \\ \mathbf{e}_u, \tilde{\mathbf{k}}_0, \tilde{\mathbf{k}}_1, \tilde{\mathbf{t}} \in \{0,1\}^{\bar{m}}; \ \mathbf{y}_T' \in \mathbb{Z}_p^m; \ \mathbf{k} \in \mathbb{Z}_q^m; \\ \mathbf{k}_0 = \mathbf{H}_{m,q-1} \cdot \tilde{\mathbf{k}}_0 \in \mathbb{Z}_q^m; \ \mathbf{k}_1 = \mathbf{H}_{m,q-1} \cdot \tilde{\mathbf{k}}_1 \in \mathbb{Z}_q^m; \ \mathbf{t} = \mathbf{H}_{m,q-1} \cdot \tilde{\mathbf{t}} \in \mathbb{Z}_q^m; \\ \tau[1] \ldots \tau[\ell] \in \{0,1\}^\ell; \ J[1] \ldots J[L] \in \{0,1\}^L, \end{cases}$$

such that the following equations hold:

$$\mathbf{A} \cdot \mathbf{v}_1 + \mathbf{A}_0 \cdot \mathbf{v}_2 + \sum_{j=1}^\ell \mathbf{A}_j \cdot (\tau[j] \cdot \mathbf{v}_2) - \mathbf{D} \cdot \tilde{\mathbf{w}} = \mathbf{u} \in \mathbb{Z}_{q_s}^n, \tag{23}$$

$$\mathbf{D}_0 \cdot \mathbf{r} + \mathbf{D}_1 \cdot \mathbf{e}_u + \mathbf{D}_2 \cdot \tilde{\mathbf{t}} + \mathbf{D}_3 \cdot \tilde{\mathbf{k}}_0 + \mathbf{D}_4 \cdot \tilde{\mathbf{k}}_1 - \mathbf{H}_{n,q_s-1} \cdot \tilde{\mathbf{w}} = \mathbf{0} \in \mathbb{Z}_{q_s}^n, \tag{24}$$

$$\mathbf{k} = \mathbf{k}_0 + \mathbf{k}_1 \in \mathbb{Z}_q^m; \quad \mathbf{y}_S = \left\lfloor \prod_{i=1}^L \mathbf{P}_{J[L+1-i]} \cdot \mathbf{k} \right\rfloor_p \in \mathbb{Z}_p^m, \tag{25}$$

$$\mathbf{y}_T' = \left\lfloor \prod_{i=1}^L \mathbf{P}_{J[L+1-i]} \cdot \mathbf{t} \right\rfloor_p \in \mathbb{Z}_p^m, \quad \mathbf{y}_T = \mathbf{F} \cdot \mathbf{e}_u + H_{\text{FRD}}(R) \cdot \mathbf{y}_T' \in \mathbb{Z}_p^m. \tag{26}$$

Our strategy is to reduce the above statement to an instance of the abstraction in Sect. 4.1. To this end, we will combine the zero-knowledge proofs of signatures from the Stern-like techniques of [68] and our techniques for the PRF layer from Sect. 4.3. Specifically, we let $q_1 = q_s$, $q_2 = q$, $q_3 = pq$, $q_4 = p$, and perform the following transformations.

Regarding the two equations of the signature relation in (23)–(24), we apply the following decompositions and/or extensions to the underlying secret vectors:

$$\begin{cases} \{\check{\mathbf{v}}_i = \mathsf{ThreeExt}\big(\mathsf{vdec}'_{m_s,\beta}(\mathbf{v}_i)\big) \in \mathsf{B}^3_{m_s\delta_\beta}\}_{i=1}^2; \quad \{\mathbf{c}_j = \mathsf{expand}(\tau[j], \check{\mathbf{v}}_2)\}_{j=1}^\ell; \\ \check{\mathbf{r}} = \mathsf{ThreeExt}\big(\mathsf{vdec}'_{2m_s,\beta}(\mathbf{r})\big) \in \mathsf{B}^3_{2m_s\delta_\beta}; \quad \widehat{\mathbf{w}} = \mathsf{TwoExt}(\tilde{\mathbf{w}}) \in \mathsf{B}^2_{m_s/2}; \\ \widehat{\mathbf{e}} = \mathsf{TwoExt}(\mathbf{e}_u) \in \mathsf{B}^2_{\bar{m}}; \quad \forall \alpha \in \{\mathbf{t}, \mathbf{k}_0, \mathbf{k}_1\} : \widehat{\alpha} = \mathsf{TwoExt}(\tilde{\alpha}) \in \mathsf{B}^2_{\bar{m}}. \end{cases} \quad (27)$$

At the same time, we also transform the associated public matrices $\mathbf{A}$, $\{\mathbf{A}_j\}_{j=0}^\ell$, $\mathbf{D}$, $\{\mathbf{D}_j\}_{j=0}^4$, $\mathbf{H}_{n,q_s-1}$ accordingly, so that the equations are preserved. Next, we combine the vectors obtained in (27) into:

$$\mathbf{w}_1 = \big(\check{\mathbf{v}}_1 \,\|\, \check{\mathbf{v}}_2 \,\|\, \mathbf{c}_1 \,\|\, \ldots \,\|\, \mathbf{c}_\ell \,\|\, \check{\mathbf{r}} \,\|\, \widehat{\mathbf{w}} \,\|\, \widehat{\mathbf{e}} \,\|\, \widehat{\mathbf{t}} \,\|\, \widehat{\mathbf{k}}_0 \,\|\, \widehat{\mathbf{k}}_1\big) \in \{-1, 0, 1\}^{d_1}, \quad (28)$$

where $d_1 = 6(\ell + 2)m_s\delta_\beta + m_s + 8\bar{m}$. We observe that the two equations can be unified into just one equation of the form $\mathbf{M}_1 \cdot \mathbf{w}_1 = \mathbf{u}_1 \bmod q_1$, where $\mathbf{M}_1 \in \mathbb{Z}_{q_1}^{2n \times d_1}$ is built from public matrices, and $\mathbf{u}_1 = (\mathbf{u} \,\|\, \mathbf{0}) \in \mathbb{Z}_{q_1}^{2n}$.

We now consider equations in (25) and (26), which involve PRF evaluations. We note that, for all $\alpha \in \{\mathbf{t}, \mathbf{k}_0, \mathbf{k}_1\}$ appearing in this layer, we have the connection

$$\alpha = \mathbf{H}_{m,q-1} \cdot \tilde{\alpha} = \widehat{\mathbf{H}}_{m,q-1} \cdot \widehat{\alpha},$$

where $\widehat{\alpha}$ is constructed in 27. To transform the equation $\mathbf{k} = \mathbf{k}_0 + \mathbf{k}_1 \in \mathbb{Z}_q^m$ in (25), we let $\widehat{\mathbf{k}} = \mathsf{TwoExt}(\mathsf{vdec}_{m,q-1}(\mathbf{k})) \in \mathsf{B}^2_{\bar{m}}$, and rewrite the equation as

$$\widehat{\mathbf{H}}_{m,q-1} \cdot \widehat{\mathbf{k}}_0 + \widehat{\mathbf{H}}_{m,q-1} \cdot \widehat{\mathbf{k}}_1 - \widehat{\mathbf{H}}_{m,q-1} \cdot \widehat{\mathbf{k}} = \mathbf{0} \bmod q.$$

Letting $\mathbf{M}_{k,2} = [\widehat{\mathbf{H}}_{m,q-1} \mid \widehat{\mathbf{H}}_{m,q-1} \mid -\widehat{\mathbf{H}}_{m,q-1}]$ and $\mathbf{u}_{k,2} = \mathbf{0}$, we have the equation $\mathbf{M}_{k,2} \cdot \mathbf{w}_{k,2} = \mathbf{u}_{k,2} \bmod q_2$, where:

$$\mathbf{w}_{k,2} = (\widehat{\mathbf{k}}_0 \,\|\, \widehat{\mathbf{k}}_1 \,\|\, \widehat{\mathbf{k}}). \quad (29)$$

The evaluation process of $\mathbf{y}_S$ in (25) is handled as in Sect. 4.3, resulting in equations $\mathbf{M}_{S,2} \cdot \mathbf{w}_{S,2} = \mathbf{u}_{S,2} \bmod q_2$, and $\mathbf{M}_{S,3} \cdot \mathbf{w}_{S,3} = \mathbf{u}_{S,3} \bmod q_3$, where

$$\mathbf{w}_{S,2} = \big(\mathbf{s}_{S,0} \,\|\, \widehat{\mathbf{x}}_{S,1} \,\|\, \cdots \,\|\, \mathbf{s}_{S,L-1} \,\|\, \widehat{\mathbf{x}}_{S,L}\big); \quad \mathbf{w}_{S,3} = \big(\widehat{\mathbf{x}}_{S,L} \,\|\, \widehat{\mathbf{z}}_S\big), \quad (30)$$

satisfy $\{\widehat{\mathbf{x}}_{S,i}\}_{i=1}^L$, $\widehat{\mathbf{z}}_S \in \mathsf{B}^2_{\bar{m}}$ and

$$\mathbf{s}_{S,0} = \mathsf{expand}(J[1], \widehat{\mathbf{k}}); \qquad \{\mathbf{s}_{S,i-1} = \mathsf{expand}(J[i], \widehat{\mathbf{x}}_{S,i-1})\}_{i=2}^L.$$

Regarding the evaluation of $\mathbf{y}'_T$ in (26), equations appearing in the iteration step can also be unified into one of the form $\mathbf{M}_{T,2} \cdot \mathbf{w}_{T,2} = \mathbf{u}_{T,2} \bmod q_2$, where

$$\mathbf{w}_{T,2} = \big(\mathbf{s}_{T,0} \,\|\, \widehat{\mathbf{x}}_{T,1} \,\|\, \cdots \,\|\, \mathbf{s}_{T,L-1} \,\|\, \widehat{\mathbf{x}}_{T,L}\big), \quad (31)$$

satisfy

$\{\widehat{\mathbf{x}}_{T,i} \in \mathsf{B}_{\bar{m}}^2\}_{i=1}^L$; $\mathbf{s}_{T,0} = \mathsf{expand}(J[1], \widehat{\mathbf{t}})$; $\{\mathbf{s}_{T,i-1} = \mathsf{expand}(J[i], \widehat{\mathbf{x}}_{T,i-1})\}_{i=2}^L$.

Meanwhile, the rounding step is handled in a slightly different manner as the output $\mathbf{y}_T' \in \mathbb{Z}_p^m$ is hidden. Letting $\widehat{\mathbf{y}}_T' = \mathsf{TwoExt}\bigl(\mathsf{vdec}_{m,p-1}(\mathbf{y}_T')\bigr) \in \mathsf{B}_{m\delta_{p-1}}^2$, we are presented with the equation

$$(p \cdot \widehat{\mathbf{H}}_{m,q-1}) \cdot \widehat{\mathbf{x}}_{T,L} - \widehat{\mathbf{H}}_{m,q-1} \cdot \widehat{\mathbf{z}}_T - (q \cdot \widehat{\mathbf{H}}_{m,p-1}) \cdot \widehat{\mathbf{y}}_T' = \mathbf{0} \bmod pq,$$

where $\widehat{\mathbf{z}}_T \in \mathsf{B}_{\bar{m}}^2$. This equation can be written as $\mathbf{M}_{T,3} \cdot \mathbf{w}_{T,3} = \mathbf{u}_{T,3} \bmod q_3$, where $\mathbf{M}_{T,3} = \bigl[p \cdot \widehat{\mathbf{H}}_{m,q-1} \mid -\widehat{\mathbf{H}}_{m,q-1} \mid -q \cdot \widehat{\mathbf{H}}_{m,p-1}\bigr]$, $\mathbf{u}_{T,3} = \mathbf{0}$, and

$$\mathbf{w}_{T,3} = (\widehat{\mathbf{x}}_{T,L} \parallel \widehat{\mathbf{z}}_T \parallel \widehat{\mathbf{y}}_T'). \tag{32}$$

Furthermore, we observe that, the three equations modulo q_2, as well as the two equations modulo q_3 we have obtained above can be unified as follows. Let

$$\mathbf{M}_2 = \begin{bmatrix} \mathbf{M}_{k,2} & & \\ & \mathbf{M}_{S,2} & \\ & & \mathbf{M}_{T,2} \end{bmatrix}; \; \mathbf{u}_2 = \begin{pmatrix} \mathbf{u}_{k,2} \\ \mathbf{u}_{S,2} \\ \mathbf{u}_{T,2} \end{pmatrix}; \; \mathbf{M}_3 = \begin{bmatrix} \mathbf{M}_{S,3} & \\ & \mathbf{M}_{T,3} \end{bmatrix}; \; \mathbf{u}_3 = \begin{pmatrix} \mathbf{u}_{S,3} \\ \mathbf{u}_{T,3} \end{pmatrix},$$

then we have $\mathbf{M}_2 \cdot \mathbf{w}_2 = \mathbf{u}_2 \bmod q_2$ and $\mathbf{M}_3 \cdot \mathbf{w}_3 = \mathbf{u}_3 \bmod q_3$, where

$$\mathbf{w}_2 = (\mathbf{w}_{k,2} \parallel \mathbf{w}_{S,2} \parallel \mathbf{w}_{T,2}) \in \{-1, 0, 1\}^{d_2}; \tag{33}$$

$$\mathbf{w}_3 = (\mathbf{w}_{S,3} \parallel \mathbf{w}_{T,3}) \in \{-1, 0, 1\}^{d_3}, \tag{34}$$

for $\mathbf{w}_{k,2}, \mathbf{w}_{S,2}, \mathbf{w}_{T,2}, \mathbf{w}_{S,3}, \mathbf{w}_{T,3}$ defined by (29)–(32), and for $d_2 = 6\bar{m}(2L+1)$, $d_3 = 8\bar{m} + 2m\delta_{p-1}$.

Now, the remaining equation in (26) can be written as:

$$\bigl[\mathbf{F} \mid \mathbf{0}^{m \times \bar{m}}\bigr] \cdot \widehat{\mathbf{e}} + \bigl(H_{\mathsf{FRD}}(R) \cdot \widehat{\mathbf{H}}_{m,p-1}\bigr) \cdot \widehat{\mathbf{y}}_T' = \mathbf{y}_T \bmod p,$$

where $\widehat{\mathbf{e}}$ and $\widehat{\mathbf{y}}_T'$ are as constructed earlier. We therefore obtain the equation $\mathbf{M}_4 \cdot \mathbf{w}_4 = \mathbf{u}_4 \bmod q_4$, where $\mathbf{M}_4 = \bigl[H_{\mathsf{FRD}}(R) \cdot \widehat{\mathbf{H}}_{m,p-1} \mid \mathbf{F} \mid \mathbf{0}^{m \times \bar{m}}\bigr]$, $\mathbf{u}_4 = \mathbf{y}_T$ and, for $d_4 = 2\bar{m} + 2m\delta_{p-1}$,

$$\mathbf{w}_4 = (\widehat{\mathbf{y}}_T' \parallel \widehat{\mathbf{e}}) \in \{-1, 0, 1\}^{d_4}. \tag{35}$$

At this point, we have transformed all the considered equations into four equations $\{\mathbf{M}_i \cdot \mathbf{w}_i = \mathbf{u}_i \bmod q_i\}_{i=1}^4$. We then let $d = \sum_{i=1}^4 d_i$ and, for $\mathbf{w}_1, \mathbf{w}_2, \mathbf{w}_3, \mathbf{w}_4$ defined by (28), (33), (34), (35), respectively, let

$$\mathbf{w} = (\mathbf{w}_1 \parallel \mathbf{w}_2 \parallel \mathbf{w}_3 \parallel \mathbf{w}_4) \in \{-1, 0, 1\}^d. \tag{36}$$

Let us now specify the set VALID containing $\mathbf{w}$, the set $\mathcal{S}$ and the associated permutation Γ_ϕ, satisfying conditions in 9.

Let VALID be the set of all vectors in $\{-1, 0, 1\}^d$ having the form (36) (which follows from (28)–(35)), whose block-vectors satisfy the following conditions:

$$\begin{cases} \check{\mathbf{v}}_1, \check{\mathbf{v}}_2 \in \mathsf{B}_{m_s \delta_\beta}^3; \quad \check{\mathbf{r}} \in \mathsf{B}_{2m_s \delta_\beta}^3; \quad \widehat{\mathbf{w}} \in \mathsf{B}_{m_s/2}^2; \quad \mathbf{y}_T' \in \mathsf{B}_{m\delta_{p-1}}^2; \\ \widehat{\mathbf{e}}, \widehat{\mathbf{t}}, \widehat{\mathbf{k}}_0, \widehat{\mathbf{k}}_1, \widehat{\mathbf{k}}, \widehat{\mathbf{x}}_{S,1}, \dots, \widehat{\mathbf{x}}_{S,L}, \widehat{\mathbf{x}}_{T,1}, \dots, \widehat{\mathbf{x}}_{T,L}, \widehat{\mathbf{z}}_S, \widehat{\mathbf{z}}_T \in \mathsf{B}_{\bar{m}}^2; \\ \{\mathbf{c}_j = \mathsf{expand}(\tau[j], \check{\mathbf{v}}_2)\}_{j=1}^\ell; \quad \mathbf{s}_{S,0} = \mathsf{expand}(J[1], \widehat{\mathbf{k}}); \quad \mathbf{s}_{T,0} = \mathsf{expand}(J[1], \widehat{\mathbf{t}}); \\ \{\mathbf{s}_{S,i-1} = \mathsf{expand}(J[i], \widehat{\mathbf{x}}_{S,i-1}), \; \mathbf{s}_{T,i-1} = \mathsf{expand}(J[i], \widehat{\mathbf{x}}_{T,i-1})\}_{i=2}^L, \end{cases}$$

for some $\tau[1] \dots \tau[\ell] \in \{0,1\}^\ell$ and some $J[1] \dots J[L] \in \{0,1\}^L$. By construction, our vector $\mathbf{w}$ belongs to this tailored set VALID.

Now, we define

$$\mathcal{S} := (\mathcal{S}_{3m_s\delta_\beta})^2 \times \mathcal{S}_{6m_s\delta_\beta} \times \mathcal{S}_{m_s} \times \mathcal{S}_{2m\delta_{p-1}} \times (\mathcal{S}_{2\bar{m}})^{2L+7} \times \{0,1\}^\ell \times \{0,1\}^L.$$

Then, for any element $\phi \in \mathcal{S}$ of the form

$$\phi = \big(\phi_{\check{\mathbf{v}}_1}, \ \phi_{\check{\mathbf{v}}_2}, \ \phi_{\check{\mathbf{r}}}, \ \phi_{\widehat{\mathbf{w}}}, \ \phi_{\mathbf{y}'_T}, \ \phi_{\widehat{\mathbf{e}}}, \ \phi_{\widehat{\mathbf{t}}}, \ \phi_{\widehat{\mathbf{k}}_0}, \ \phi_{\widehat{\mathbf{k}}_1}, \ \phi_{\widehat{\mathbf{k}}}, \ \phi_{\widehat{\mathbf{x}}_{S,1}}, \ \dots, \ \phi_{\widehat{\mathbf{x}}_{S,L}},$$
$$\phi_{\widehat{\mathbf{x}}_{T,1}}, \ \dots, \ \phi_{\widehat{\mathbf{x}}_{T,L}}, \ \phi_{\widehat{\mathbf{z}}_S}, \ \phi_{\widehat{\mathbf{z}}_T}, \ a[1] \dots a[\ell], \ b[1] \dots b[L]\big),$$

let Γ_ϕ be the permutation that, on input vector $\mathbf{w} \in \mathbb{Z}^d$ of the form (18) (which is implied by (28)–(35)), it transforms the block-vectors of $\mathbf{w}$ as follows:

- Apply permutation ϕ_α to block α, for all

$$\alpha \in \big\{\check{\mathbf{v}}_1, \ \check{\mathbf{v}}_2, \ \check{\mathbf{r}}, \ \widehat{\mathbf{w}}, \ \mathbf{y}'_T, \ \widehat{\mathbf{e}}, \ \widehat{\mathbf{t}}, \ \widehat{\mathbf{k}}_0, \ \widehat{\mathbf{k}}_1, \ \widehat{\mathbf{k}}, \ \widehat{\mathbf{x}}_{S,1}, \dots, \widehat{\mathbf{x}}_{S,L}, \ \widehat{\mathbf{x}}_{T,1}, \dots, \widehat{\mathbf{x}}_{T,L}, \ \widehat{\mathbf{z}}_S, \ \widehat{\mathbf{z}}_T\big\}.$$

- For $j \in [\ell]$, apply permutation $T_{a[j],\phi_{\check{\mathbf{v}}_2}}$ to block $\mathbf{c}_j$.
- Apply permutation $T_{b[1],\phi_{\widehat{\mathbf{k}}}}$ to block $\mathbf{s}_{S,0}$, and $T_{b[1],\phi_{\widehat{\mathbf{t}}}}$ to block $\mathbf{s}_{T,0}$.
- For $i \in [2,L]$, apply permutation $T_{b[i],\phi_{\widehat{\mathbf{x}}_{S,i-1}}}$ to block $\mathbf{s}_{S,i-1}$, and permutation $T_{b[i],\phi_{\widehat{\mathbf{x}}_{T,i-1}}}$ to block $\mathbf{s}_{T,i-1}$.

It can be checked that, we have $\mathbf{w} \in$ VALID if and only if $\Gamma_\phi(\mathbf{w}) \in$ VALID, thanks to the equivalences (3), (4), (5) from Sect. 3. Furthermore, if $\phi \leftarrow U(\mathcal{S})$, then $\Gamma_\phi(\mathbf{w})$ is uniform in VALID. In other words, the conditions in 9 are satisfied.

Given the above transformations and specifications, we can run the abstract protocol of Fig. 1 to prove knowledge of $\mathbf{w} = (\mathbf{w}_1 \| \mathbf{w}_2 \| \mathbf{w}_3\mathbf{w}_4) \in$ VALID satisfying $\{\mathbf{M}_i \cdot \mathbf{w}_i = \mathbf{u}_i \bmod q_i\}_{i \in [4]}$, where public matrices/vectors $\{\mathbf{M}_i, \mathbf{u}_i\}_{i \in [4]}$ are as constructed above. As a result, we obtain a statistical zero-knowledge argument of knowledge for the considered statement.

Each round of the protocol has communication cost $\mathcal{O}(\sum_{i=1}^4 d_i \cdot \log q_i)$. For the parameters in Sect. 5.1, this cost is of order $\widetilde{\mathcal{O}}(L \cdot \lambda + \lambda^2)$. In the Spend algorithm, the protocol is repeated $\kappa = \omega(\log \lambda)$ to achieve negligible soundness error. The global communication cost is $\widetilde{\mathcal{O}}(L \cdot \lambda + \lambda^2) \cdot \omega(\log \lambda) = \widetilde{\mathcal{O}}(L \cdot \lambda + \lambda^2)$.

6   Security

We now state the security results for which proofs are available in the full version of the paper.

Theorem 2. *The scheme guarantees balance under the* SIS *assumption in the random oracle model.*

Theorem 3 shows that, under the SIS assumption and assuming the collision-resistance of H_0, double-spenders can always be identified by the bank. Analogously to the security proof of Camenisch *et al.* [22] which relies on a similar feature of the Dodis-Yampolskiy PRF [44], the proof uses some range-disjointness property of the underlying small-domain PRF: namely, two functions keyed by independent keys should have disjoint ranges with high probability. In the full paper, we prove this property unconditionally for the BLMR PRF [17].

Theorem 3. *If H_0 is a collision-resistant hash function and H is modeled as a random oracle, the scheme guarantees the identification of double spenders under the SIS assumption.*

Theorem 4. *The scheme provides strong exculpability under the SIS assumption in the random oracle model.*

Theorem 5. *The scheme provides anonymity under the LWE assumption in the random oracle model.*

Our scheme can be modified so as to use the more efficient LWR-based PRF based on the GGM technique. This allows significantly improving the choice of parameters at the expense of a longer description and a more complex proof for the identification property. The reason is that, in the GGM-based PRF, the range disjointness property (for small domains) does not appear to be provable in the statistical sense. This can be addressed by relying on the pseudo-randomness of the function, as in the security proof of Belenkiy *et al.* [10]. Relying on the pseudo-randomness is perhaps counter-intuitive since the adversary knows the PRF seed in the proof of the identification property. Nevertheless, the reduction still works as in [10, Appendix F] when the domain has polynomial size.

Acknowledgements. Part of this research was funded by Singapore Ministry of Education under Research Grant MOE2016-T2-2-014(S), by the French ANR ALAMBIC project (ANR-16-CE39-0006) and by BPI-France in the context of the national project RISQ (P141580).

References

1. Agrawal, S., Boneh, D., Boyen, X.: Efficient lattice (H)IBE in the standard model. In: Gilbert, H. (ed.) EUROCRYPT 2010. LNCS, vol. 6110, pp. 553–572. Springer, Heidelberg (2010). doi:10.1007/978-3-642-13190-5_28
2. Ajtai, M.: Generating hard instances of the short basis problem. In: Wiedermann, J., Emde Boas, P., Nielsen, M. (eds.) ICALP 1999. LNCS, vol. 1644, pp. 1–9. Springer, Heidelberg (1999). doi:10.1007/3-540-48523-6_1
3. Alwen, J., Krenn, S., Pietrzak, K., Wichs, D.: Learning with rounding, revisited. In: Canetti, R., Garay, J.A. (eds.) CRYPTO 2013. LNCS, vol. 8042, pp. 57–74. Springer, Heidelberg (2013). doi:10.1007/978-3-642-40041-4_4
4. Alwen, J., Peikert, C.: Generating shorter bases for hard random lattices. In: STACS 2009 (2009)

5. Au, M.H., Wu, Q., Susilo, W., Mu, Y.: Compact e-cash from bounded accumulator. In: Abe, M. (ed.) CT-RSA 2007. LNCS, vol. 4377, pp. 178–195. Springer, Heidelberg (2006). doi:10.1007/11967668_12
6. Banaszczyk, W.: New bounds in some transference theorems in the geometry of number **296** (1993)
7. Banerjee, A., Peikert, C.: New and improved key-homomorphic pseudorandom functions. In: Garay, J.A., Gennaro, R. (eds.) CRYPTO 2014. LNCS, vol. 8616, pp. 353–370. Springer, Heidelberg (2014). doi:10.1007/978-3-662-44371-2_20
8. Banerjee, A., Peikert, C., Rosen, A.: Pseudorandom functions and lattices. In: Pointcheval, D., Johansson, T. (eds.) EUROCRYPT 2012. LNCS, vol. 7237, pp. 719–737. Springer, Heidelberg (2012). doi:10.1007/978-3-642-29011-4_42
9. Baum, C., Damgård, I., Larsen, K.G., Nielsen, M.: How to prove knowledge of small secrets. In: Robshaw, M., Katz, J. (eds.) CRYPTO 2016. LNCS, vol. 9816, pp. 478–498. Springer, Heidelberg (2016). doi:10.1007/978-3-662-53015-3_17
10. Belenkiy, M., Chase, M., Kohlweiss, M., Lysyanskaya, A.: Compact e-cash and simulatable VRFs revisited. In: Shacham, H., Waters, B. (eds.) Pairing 2009. LNCS, vol. 5671, pp. 114–131. Springer, Heidelberg (2009). doi:10.1007/978-3-642-03298-1_9
11. Bellare, M., Rogaway, P.: Random oracles are practical: a paradigm for designing efficient protocols. In: ACM-CCS 1993 (1993)
12. Ben-Sasson, E., Chiesa, A., Garman, C., Green, M., Miers, I., Tromer, E., Virza, M.: Zerocash: decentralized anonymous payments from bitcoin. In: IEEE S&P 2014 (2014)
13. Benhamouda, F., Camenisch, J., Krenn, S., Lyubashevsky, V., Neven, G.: Better zero-knowledge proofs for lattice encryption and their application to group signatures. In: Sarkar, P., Iwata, T. (eds.) ASIACRYPT 2014. LNCS, vol. 8873, pp. 551–572. Springer, Heidelberg (2014). doi:10.1007/978-3-662-45611-8_29
14. Benhamouda, F., Krenn, S., Lyubashevsky, V., Pietrzak, K.: Efficient zero-knowledge proofs for commitments from learning with errors over rings. In: Pernul, G., Ryan, P.Y.A., Weippl, E. (eds.) ESORICS 2015. LNCS, vol. 9326, pp. 305–325. Springer, Cham (2015). doi:10.1007/978-3-319-24174-6_16
15. Bogdanov, A., Guo, S., Masny, D., Richelson, S., Rosen, A.: On the hardness of learning with rounding over small modulus. In: Kushilevitz, E., Malkin, T. (eds.) TCC 2016. LNCS, vol. 9562, pp. 209–224. Springer, Heidelberg (2016). doi:10.1007/978-3-662-49096-9_9
16. Böhl, F., Hofheinz, D., Jager, T., Koch, J., Striecks, C.: Confined guessing: new signatures from standard assumptions. J. Cryptol. **28**(1) (2015)
17. Boneh, D., Lewi, K., Montgomery, H., Raghunathan, A.: Key homomorphic PRFs and their applications. In: Canetti, R., Garay, J.A. (eds.) CRYPTO 2013. LNCS, vol. 8042, pp. 410–428. Springer, Heidelberg (2013). doi:10.1007/978-3-642-40041-4_23
18. Boyen, X.: Lattice mixing and vanishing trapdoors: a framework for fully secure short signatures and more. In: Nguyen, P.Q., Pointcheval, D. (eds.) PKC 2010. LNCS, vol. 6056, pp. 499–517. Springer, Heidelberg (2010). doi:10.1007/978-3-642-13013-7_29
19. Brakerski, Z., Kalai, Y.T.: A framework for efficient signatures, ring signatures and identity based encryption in the standard model. IACR Cryptol. ePrint Arch. **2010**, 86 (2010)
20. Brakerski, Z., Langlois, A., Peikert, C., Regev, O., Stehlé, D.: On the classical hardness of learning with errors. In: STOC 2013 (2013)

21. Camenisch, J., Hohenberger, S., Kohlweiss, M., Lysyanskaya, A., Meyerovich, M.: How to win the clone wars: efficient periodic n-times anonymous authentication. In: ACM-CCS (2006)
22. Camenisch, J., Hohenberger, S., Lysyanskaya, A.: Compact e-cash. In: Cramer, R. (ed.) EUROCRYPT 2005. LNCS, vol. 3494, pp. 302–321. Springer, Heidelberg (2005). doi:10.1007/11426639_18
23. Camenisch, J., Hohenberger, S., Lysyanskaya, A.: Balancing accountability and privacy using e-cash (extended abstract). In: De Prisco, R., Yung, M. (eds.) SCN 2006. LNCS, vol. 4116, pp. 141–155. Springer, Heidelberg (2006). doi:10.1007/11832072_10
24. Camenisch, J., Lehmann, A.: (un)linkable pseudonyms for governmental databases. In: ACM-CCS 2015 (2015)
25. Camenisch, J., Lysyanskaya, A.: A signature scheme with efficient protocols. In: Cimato, S., Persiano, G., Galdi, C. (eds.) SCN 2002. LNCS, vol. 2576, pp. 268–289. Springer, Heidelberg (2003). doi:10.1007/3-540-36413-7_20
26. Camenisch, J., Lysyanskaya, A.: Signature schemes and anonymous credentials from bilinear maps. In: Franklin, M. (ed.) CRYPTO 2004. LNCS, vol. 3152, pp. 56–72. Springer, Heidelberg (2004). doi:10.1007/978-3-540-28628-8_4
27. Camenisch, J., Lysyanskaya, A., Meyerovich, M.: Endorsed e-cash. In: IEEE S&P 2007 (2007)
28. Canard, S., Gouget, A.: Divisible e-cash systems can be truly anonymous. In: Naor, M. (ed.) EUROCRYPT 2007. LNCS, vol. 4515, pp. 482–497. Springer, Heidelberg (2007). doi:10.1007/978-3-540-72540-4_28
29. Canard, S., Pointcheval, D., Sanders, O., Traoré, J.: Divisible e-cash made practical. In: Katz, J. (ed.) PKC 2015. LNCS, vol. 9020, pp. 77–100. Springer, Heidelberg (2015). doi:10.1007/978-3-662-46447-2_4
30. Canard, S., Pointcheval, D., Sanders, O., Traoré, J.: Scalable divisible e-cash. In: Malkin, T., Kolesnikov, V., Lewko, A.B., Polychronakis, M. (eds.) ACNS 2015. LNCS, vol. 9092, pp. 287–306. Springer, Cham (2015). doi:10.1007/978-3-319-28166-7_14
31. Cash, D., Hofheinz, D., Kiltz, E., Peikert, C.: Bonsai trees, or how to delegate a lattice basis. In: Gilbert, H. (ed.) EUROCRYPT 2010. LNCS, vol. 6110, pp. 523–552. Springer, Heidelberg (2010). doi:10.1007/978-3-642-13190-5_27
32. Chase, M., Ganesh, C., Mohassel, P.: Efficient zero-knowledge proof of algebraic and non-algebraic statements with applications to privacy preserving credentials. In: Robshaw, M., Katz, J. (eds.) CRYPTO 2016. LNCS, vol. 9816, pp. 499–530. Springer, Heidelberg (2016). doi:10.1007/978-3-662-53015-3_18
33. Chaum, D.: Blind signatures for untraceable payments. In: Chaum, D., Rivest, R.L., Sherman, A.T. (eds.) Crypto 1982, pp. 199–203. Springer, Heidelberg (1982). doi:10.1007/978-1-4757-0602-4_18
34. Chaum, D.: Blind signature system. In: Chaum, D. (ed.) Crypto 1983, p. 153. Springer, Heidelberg (1983). doi:10.1007/978-1-4684-4730-9_14
35. Chaum, D.: Security without identification: transactions system to make big brother obsolete. Commun. ACM 28(10) (1985)
36. Chaum, D.: On-line cash checks. In: Eurocrypt 1989 (1989)
37. Chaum, D., Fiat, A., Naor, M.: Untraceable electronic cash. In: Goldwasser, S. (ed.) CRYPTO 1988. LNCS, vol. 403, pp. 319–327. Springer, New York (1990). doi:10.1007/0-387-34799-2_25
38. Chaum, D., Pedersen, T.P.: Transferred cash grows in size. In: Rueppel, R.A. (ed.) EUROCRYPT 1992. LNCS, vol. 658, pp. 390–407. Springer, Heidelberg (1993). doi:10.1007/3-540-47555-9_32

39. Chiesa, A., Green, M., Liu, J., Miao, P., Miers, I., Mishra, P.: Decentralized anonymous micropayments. In: Coron, J.-S., Nielsen, J.B. (eds.) EUROCRYPT 2017. LNCS, vol. 10211, pp. 609–642. Springer, Cham (2017). doi:10.1007/978-3-319-56614-6_21

40. Coull, S., Green, M., Hohenberger, S.: Controlling access to an oblivious database using stateful anonymous credentials. In: Jarecki, S., Tsudik, G. (eds.) PKC 2009. LNCS, vol. 5443, pp. 501–520. Springer, Heidelberg (2009). doi:10.1007/978-3-642-00468-1_28

41. Cramer, R., Damgård, I.: On the amortized complexity of zero-knowledge protocols. In: Halevi, S. (ed.) CRYPTO 2009. LNCS, vol. 5677, pp. 177–191. Springer, Heidelberg (2009). doi:10.1007/978-3-642-03356-8_11

42. Cramer, R., Damgård, I., Xing, C., Yuan, C.: Amortized complexity of zero-knowledge proofs revisited: achieving linear soundness slack. In: Coron, J.-S., Nielsen, J.B. (eds.) EUROCRYPT 2017. LNCS, vol. 10210, pp. 479–500. Springer, Cham (2017). doi:10.1007/978-3-319-56620-7_17

43. del Pino, R., Lyubashevsky, V.: Amortization with fewer equations for proving knowledge of small secrets. Cryptology ePrint Archive: Report 2017/280 (2017)

44. Dodis, Y., Yampolskiy, A.: A verifiable random function with short proofs and keys. In: Vaudenay, S. (ed.) PKC 2005. LNCS, vol. 3386, pp. 416–431. Springer, Heidelberg (2005). doi:10.1007/978-3-540-30580-4_28

45. Döttling, N., Schröder, D.: Efficient pseudorandom functions via on-the-fly adaptation. In: Gennaro, R., Robshaw, M. (eds.) CRYPTO 2015. LNCS, vol. 9215, pp. 329–350. Springer, Heidelberg (2015). doi:10.1007/978-3-662-47989-6_16

46. Fiat, A., Shamir, A.: How to prove yourself: practical solutions to identification and signature problems. In: Odlyzko, A.M. (ed.) CRYPTO 1986. LNCS, vol. 263, pp. 186–194. Springer, Heidelberg (1987). doi:10.1007/3-540-47721-7_12

47. Franklin, M., Yung, M.: Secure and efficient off-line digital money (extended abstract). In: Lingas, A., Karlsson, R., Carlsson, S. (eds.) ICALP 1993. LNCS, vol. 700, pp. 265–276. Springer, Heidelberg (1993). doi:10.1007/3-540-56939-1_78

48. Frederiksen, T.K., Nielsen, J.B., Orlandi, C.: Privacy-free garbled circuits with applications to efficient zero-knowledge. In: Oswald, E., Fischlin, M. (eds.) EUROCRYPT 2015. LNCS, vol. 9057, pp. 191–219. Springer, Heidelberg (2015). doi:10.1007/978-3-662-46803-6_7

49. Freedman, M.J., Ishai, Y., Pinkas, B., Reingold, O.: Keyword search and oblivious pseudorandom functions. In: Kilian, J. (ed.) TCC 2005. LNCS, vol. 3378, pp. 303–324. Springer, Heidelberg (2005). doi:10.1007/978-3-540-30576-7_17

50. Fujisaki, E., Suzuki, K.: Traceable ring signature. In: Okamoto, T., Wang, X. (eds.) PKC 2007. LNCS, vol. 4450, pp. 181–200. Springer, Heidelberg (2007). doi:10.1007/978-3-540-71677-8_13

51. Gentry, C.: Fully homomorphic encryption using ideal lattices. In: STOC 2009 (2009)

52. Gentry, C., Peikert, C., Vaikuntanathan, V.: Trapdoors for hard lattices and new cryptographic constructions. In: STOC 2008 (2008)

53. Giacomelli, I., Madsen, J., Orlandi, C.: Zkboo: faster zero-knowledge for Boolean circuits. In: USENIX Security Symposium (2016)

54. Goldreich, O., Goldwasser, S., Micali, S.: How to construct random functions. J. ACM **33** (1986)

55. Goldwasser, S., Micali, S., Rackoff, C.: The knowledge complexity of interactive proof-systems. In: STOC (1985)

56. Gordon, S.D., Katz, J., Vaikuntanathan, V.: A group signature scheme from lattice assumptions. In: Abe, M. (ed.) ASIACRYPT 2010. LNCS, vol. 6477, pp. 395–412. Springer, Heidelberg (2010). doi:10.1007/978-3-642-17373-8_23

57. Green, M., Miers, I.: Bolt: anonymous payment channels for decentralized currencies. Cryptology ePrint Archive: Report 2016/701 (2016)

58. Groth, J., Sahai, A.: Efficient non-interactive proof systems for bilinear groups. In: Smart, N. (ed.) EUROCRYPT 2008. LNCS, vol. 4965, pp. 415–432. Springer, Heidelberg (2008). doi:10.1007/978-3-540-78967-3_24

59. Hohenberger, S., Myers, S., Pass, R., Shelat, A.: ANONIZE: a large-scale anonymous survey system. In: IEEE S&P 2014 (2014)

60. Ishai, Y., Kushilevitz, E., Ostrovksy, R., Sahai, A.: Zero-knowledge from secure multiparty computation. In: STOC 2007 (2007)

61. Jarecki, S., Kiayias, A., Krawczyk, H.: Round-optimal password-protected secret sharing and T-PAKE in the password-only model. In: Sarkar, P., Iwata, T. (eds.) ASIACRYPT 2014. LNCS, vol. 8874, pp. 233–253. Springer, Heidelberg (2014). doi:10.1007/978-3-662-45608-8_13

62. Jarecki, S., Liu, X.: Efficient oblivious pseudorandom function with applications to adaptive OT and secure computation of set intersection. In: Reingold, O. (ed.) TCC 2009. LNCS, vol. 5444, pp. 577–594. Springer, Heidelberg (2009). doi:10.1007/978-3-642-00457-5_34

63. Jawurek, M., Kerschbaum, F., Orlandi, C.: Zero-knowledge using garbled circuits: how to prove non-algebraic statements efficiently. In: ACM-CCS (2013)

64. Kawachi, A., Tanaka, K., Xagawa, K.: Concurrently secure identification schemes based on the worst-case hardness of lattice problems. In: Pieprzyk, J. (ed.) ASIACRYPT 2008. LNCS, vol. 5350, pp. 372–389. Springer, Heidelberg (2008). doi:10.1007/978-3-540-89255-7_23

65. Laguillaumie, F., Langlois, A., Libert, B., Stehlé, D.: Lattice-based group signatures with logarithmic signature size. In: Sako, K., Sarkar, P. (eds.) ASIACRYPT 2013. LNCS, vol. 8270, pp. 41–61. Springer, Heidelberg (2013). doi:10.1007/978-3-642-42045-0_3

66. Langlois, A., Ling, S., Nguyen, K., Wang, H.: Lattice-based group signature scheme with verifier-local revocation. In: Krawczyk, H. (ed.) PKC 2014. LNCS, vol. 8383, pp. 345–361. Springer, Heidelberg (2014). doi:10.1007/978-3-642-54631-0_20

67. Libert, B., Ling, S., Mouhartem, F., Nguyen, K., Wang, H.: Zero-knowledge arguments for matrix-vector relations and lattice-based group encryption. In: Cheon, J.H., Takagi, T. (eds.) ASIACRYPT 2016. LNCS, vol. 10032, pp. 101–131. Springer, Heidelberg (2016). doi:10.1007/978-3-662-53890-6_4

68. Libert, B., Ling, S., Mouhartem, F., Nguyen, K., Wang, H.: Signature schemes with efficient protocols and dynamic group signatures from lattice assumptions. In: Cheon, J.H., Takagi, T. (eds.) ASIACRYPT 2016. LNCS, vol. 10032, pp. 373–403. Springer, Heidelberg (2016). doi:10.1007/978-3-662-53890-6_13

69. Libert, B., Ling, S., Nguyen, K., Wang, H.: Zero-knowledge arguments for lattice-based accumulators: logarithmic-size ring signatures and group signatures without trapdoors. In: Fischlin, M., Coron, J.-S. (eds.) EUROCRYPT 2016. LNCS, vol. 9666, pp. 1–31. Springer, Heidelberg (2016). doi:10.1007/978-3-662-49896-5_1

70. Ling, S., Nguyen, K., Stehlé, D., Wang, H.: Improved zero-knowledge proofs of knowledge for the ISIS problem, and applications. In: Kurosawa, K., Hanaoka, G. (eds.) PKC 2013. LNCS, vol. 7778, pp. 107–124. Springer, Heidelberg (2013). doi:10.1007/978-3-642-36362-7_8

71. Ling, S., Nguyen, K., Wang, H.: Group signatures from lattices: simpler, tighter, shorter, ring-based. In: Katz, J. (ed.) PKC 2015. LNCS, vol. 9020, pp. 427–449. Springer, Heidelberg (2015). doi:10.1007/978-3-662-46447-2_19

72. Lipmaa, H., Asokan, N., Niemi, V.: Secure vickrey auctions without threshold trust. In: Blaze, M. (ed.) FC 2002. LNCS, vol. 2357, pp. 87–101. Springer, Heidelberg (2003). doi:10.1007/3-540-36504-4_7

73. Lyubashevsky, V.: Lattice-based identification schemes secure under active attacks. In: Cramer, R. (ed.) PKC 2008. LNCS, vol. 4939, pp. 162–179. Springer, Heidelberg (2008). doi:10.1007/978-3-540-78440-1_10

74. Micali, S., Rabin, M., Vadhan, S.: Verifiable random functions. In: FOCS (1999)

75. Micali, S., Sidney, R.: A simple method for generating and sharing pseudo-random functions, with applications to clipper-like key escrow systems. In: Coppersmith, D. (ed.) CRYPTO 1995. LNCS, vol. 963, pp. 185–196. Springer, Heidelberg (1995). doi:10.1007/3-540-44750-4_15

76. Micciancio, D., Peikert, C.: Trapdoors for lattices: simpler, tighter, faster, smaller. In: Pointcheval, D., Johansson, T. (eds.) EUROCRYPT 2012. LNCS, vol. 7237, pp. 700–718. Springer, Heidelberg (2012). doi:10.1007/978-3-642-29011-4_41

77. Micciancio, D., Vadhan, S.P.: Statistical zero-knowledge proofs with efficient provers: lattice problems and more. In: Boneh, D. (ed.) CRYPTO 2003. LNCS, vol. 2729, pp. 282–298. Springer, Heidelberg (2003). doi:10.1007/978-3-540-45146-4_17

78. Miers, I., Garman, C., Green, M., Rubin, A.: Zerocoin: anonymous distributed e-cash from bitcoin. In: IEEE S&P (2013)

79. Nakamoto, S.: Bitcoin: a peer-to-peer electronic cash system. www.bitcoin.org

80. Naor, M., Pinkas, B., Reingold, O.: Distributed pseudo-random functions and KDCs. In: Stern, J. (ed.) EUROCRYPT 1999. LNCS, vol. 1592, pp. 327–346. Springer, Heidelberg (1999). doi:10.1007/3-540-48910-X_23

81. Nguyen, P.Q., Zhang, J., Zhang, Z.: Simpler efficient group signatures from lattices. In: Katz, J. (ed.) PKC 2015. LNCS, vol. 9020, pp. 401–426. Springer, Heidelberg (2015). doi:10.1007/978-3-662-46447-2_18

82. Okamoto, T., Ohta, K.: Universal electronic cash. In: Feigenbaum, J. (ed.) CRYPTO 1991. LNCS, vol. 576, pp. 324–337. Springer, Heidelberg (1992). doi:10.1007/3-540-46766-1_27

83. Okamoto, T.: An efficient divisible electronic cash scheme. In: Coppersmith, D. (ed.) CRYPTO 1995. LNCS, vol. 963, pp. 438–451. Springer, Heidelberg (1995). doi:10.1007/3-540-44750-4_35

84. Peikert, C., Vaikuntanathan, V.: Noninteractive statistical zero-knowledge proofs for lattice problems. In: Wagner, D. (ed.) CRYPTO 2008. LNCS, vol. 5157, pp. 536–553. Springer, Heidelberg (2008). doi:10.1007/978-3-540-85174-5_30

85. Regev, O.: On lattices, learning with errors, random linear codes, and cryptography. In: STOC (2005)

86. Stern, J.: A new paradigm for public key identification. IEEE Trans. Inf. Theory 42(6) (1996)

87. Tsiounis, Y.: Efficient electronic cash: new notions and techniques. Ph.D. thesis, Northeastern University (1997)

88. von Solms, S., Naccache, D.: On blind signatures and perfect crimes. Comput. Secur. 11 (1992)

89. Xie, X., Xue, R., Wang, M.: Zero knowledge proofs from ring-LWE. In: Abdalla, M., Nita-Rotaru, C., Dahab, R. (eds.) CANS 2013. LNCS, vol. 8257, pp. 57–73. Springer, Cham (2013). doi:10.1007/978-3-319-02937-5_4

Linear-Time Zero-Knowledge Proofs for Arithmetic Circuit Satisfiability

Jonathan Bootle[1]([✉]), Andrea Cerulli[1], Essam Ghadafi[2], Jens Groth[1], Mohammad Hajiabadi[3], and Sune K. Jakobsen[1]

[1] University College London, London, UK
{jonathan.bootle.14,andrea.cerulli.13, j.groth,s.jakobsen}@ucl.ac.uk
[2] University of the West of England, Bristol, UK
essam.ghadafi@uwe.ac.uk
[3] University of California, Berkeley, CA, USA
mdhajiabadi@berkeley.edu

Abstract. We give computationally efficient zero-knowledge proofs of knowledge for arithmetic circuit satisfiability over a large field. For a circuit with N addition and multiplication gates, the prover only uses $\mathcal{O}(N)$ multiplications and the verifier only uses $\mathcal{O}(N)$ additions in the field. If the commitments we use are statistically binding, our zero-knowledge proofs have unconditional soundness, while if the commitments are statistically hiding we get computational soundness. Our zero-knowledge proofs also have sub-linear communication if the commitment scheme is compact.

Our construction proceeds in three steps. First, we give a zero-knowledge proof for arithmetic circuit satisfiability in an ideal linear commitment model where the prover may commit to secret vectors of field elements, and the verifier can receive certified linear combinations of those vectors. Second, we show that the ideal linear commitment proof can be instantiated using error-correcting codes and non-interactive commitments. Finally, by choosing efficient instantiations of the primitives we obtain linear-time zero-knowledge proofs.

Keywords: Zero-knowledge · Arithmetic circuit · Ideal linear commitments

1 Introduction

A zero-knowledge proof [GMR85] is a protocol between two parties: a prover and a verifier. The prover wants to convince the verifier that an instance u belongs to a specific language $\mathcal{L}_\mathcal{R}$ in NP. She has a witness w such that (u, w) belongs

The research leading to these results has received funding from the European Research Council under the European Union's Seventh Framework Programme (FP/2007–2013)/ERC Grant Agreement no. 307937.

E. Ghadafi and M. Hajiabadi—Part of the work was done while at University College London.

T. Takagi and T. Peyrin (Eds.): ASIACRYPT 2017, Part III, LNCS 10626, pp. 336–365, 2017.
https://doi.org/10.1007/978-3-319-70700-6_12

to the NP relation $\mathcal{R}$ defining the language, but wants to convince the verifier that the statement $u \in \mathcal{L}_{\mathcal{R}}$ is true without revealing the witness or any other confidential information.

Zero-knowledge proofs are widely used in cryptography since it is often useful to verify that a party is following a protocol without requiring her to divulge secret keys or other private information. Applications range from digital signatures and public-key encryption to secure multi-party computation and verifiable cloud computing.

Efficiency is crucial for large and complex statements such as those that may arise in the latter applications. Important efficiency parameters include the time complexity of the prover, the time complexity of the verifier, the amount of communication measured in bits, and the number of rounds the prover and verifier need to interact. Three decades of research on zero-knowledge proofs have gone into optimizing these efficiency parameters and many insights have been learned.

For zero-knowledge proofs with unconditional soundness where it impossible for any cheating prover to convince the verifier of a false statement, it is possible to reduce communication to the witness size [IKOS09, KR08, GGI+14]. For zero-knowledge arguments where it is just computationally intractable for the prover to cheat the verifier we can do even better and get sub-linear communication complexity [Kil92].

There are many constant-round zero-knowledge proofs and arguments, for instance Bellare et al. [BJY97] construct four round arguments based on one-way functions. In the common reference string model, it is even possible to give non-interactive proofs where the prover computes a convincing zero-knowledge proof directly without receiving any messages from the verifier [BFM88].

The verifier computation is in general at least proportional to the instance size because the verifier must read the entire instance in order to verify it. However, the verifier computation can be sub-linear in the time it takes for the relation to verify a witness for the statement being true [GKR08], which is useful in both zero-knowledge proofs and verifiable computation.

Having reduced the cost of many other efficiency parameters, today the major bottleneck is the prover's computation. Classical number-theoretic constructions for circuit satisfiability such as [CD98] require a linear number of exponentiations, i.e., the cost is $\mathcal{O}(\lambda N)$ group multiplications where N is the number of gates and λ is a security parameter. Relying on different techniques and underlying cryptography [DIK10] has reduced the computational overhead further to being $\mathcal{O}(\log(\lambda))$. This leaves a tantalizing open question of whether we can come all the way down to constant overhead $\mathcal{O}(1)$, i.e., make the prover's cost within a constant factor of the time it takes to verify $(u, w) \in R$ directly.

1.1   Our Contributions

We construct zero-knowledge proofs of knowledge for the satisfiability of arithmetic circuits. An instance is an arithmetic circuits with N fan-in 2 addition and multiplication gates over a finite field $\mathbb{F}$ and a specification of the values of some of the wires. A witness consists of the remaining wires such that the values are consistent with the gates and the wire values specified in the instance.

Our zero-knowledge proofs are highly efficient asymptotically:

- Prover time is $\mathcal{O}(N)$ field additions and multiplications.
- Verifier time is $\mathcal{O}(N)$ field *additions*.

This is optimal up to a constant factor for both the prover and verifier. The prover only incurs a constant overhead compared to the time needed to evaluate the circuit from scratch given an instance and a witness, and for instances of size equivalent to $\Omega(N)$ field elements the verifier only incurs a constant overhead compared to the time it takes to read the instance. The constants are large, so we do not recommend implementing the zero-knowledge proof as it is, but from a theoretical perspective we consider it a big step forward to get constant overhead for both prover and verifier.

Our zero-knowledge proofs have perfect completeness, i.e., when the prover knows a satisfactory witness she is always able to convince the verifier. Our constructions are proofs of knowledge, that is, not only does the prover demonstrate the statement is true but also that she knows a witness. The proofs have special honest-verifier zero-knowledge, which means that given a set of verifier challenges it is possible to simulate a proof answering the challenges without knowing a witness. The flavour of knowledge soundness and special honest-verifier zero-knowledge depends on the underlying commitment scheme we use. When instantiated with statistically binding commitment schemes, we obtain proofs (statistically knowledge sound) with computational zero-knowledge. When we use statistically hiding commitments we obtain arguments of knowledge with statistical special honest verifier zero-knowledge. The communication complexity of our proofs with unconditional soundness is only $\mathcal{O}(N)$ field elements, while our arguments with computational soundness have sub-linear communication of $\mathrm{poly}(\lambda)\sqrt{N}$ field elements when the commitments are compact. Round complexity is also low, when we optimize for computational efficiency for prover and verifier we only use $\mathcal{O}(\log\log N)$ rounds.

1.2   Construction and Techniques

Our construction is modular and consists of three steps. First, we construct a proof in a communication model we call the Ideal Linear Commitment (ILC) channel. In the ILC model, the prover can commit vectors of secret field elements to the channel. The verifier may later query openings to linear combinations of the committed vectors, which the channel will answer directly. We show that idealizing the techniques by Groth et al. [Gro09, BCC+16] gives us efficient proofs in the ideal linear commitment model. By optimizing primarily for prover computation and secondarily for round efficiency, we get a round complexity of $\mathcal{O}(\log\log N)$ rounds, which is better than the $\mathcal{O}(\log N)$ rounds of Bootle et al. [BCC+16] that optimized for communication complexity.

Next, we compile proofs in the ILC model into proof and argument systems using non-interactive commitment schemes; however, unlike previous works we do not commit directly to the vectors. Instead, we encode the vectors as randomized codewords using a linear error-correcting code. We now consider the codewords as rows of a matrix and commit to the columns of that matrix. When the verifier asks for a linear combination of the vectors, the prover simply tells the verifier what the linear combination is. However, the verifier does not have to have blind confidence in the prover because she can ask for openings of some of the committed columns and use them to spot check that the resulting codeword is correct.

Finally, we instantiate the scheme with concrete error-correcting codes and non-interactive commitment schemes. We use the error-correcting codes of Druk and Ishai [DI14], which allow the encoding of k field elements using $\mathcal{O}(k)$ additions in the field. Statistically hiding commitment schemes can be constructed from collision-resistant hash functions, and using the recent hash functions of Applebaum et al. [AHI+17] we can hash t field elements at a cost equivalent to $\mathcal{O}(t)$ field additions. Statistically binding commitment schemes on the other hand can be built from pseudorandom number generators. Using the linear-time computable pseudorandom number generators of Ishai et al. [IKOS08] we get linear-time computable statistically binding commitments. Plugging either of the commitment schemes into our construction yields zero-knowledge proofs with linear-time computation for both prover and verifier.

1.3 Related Work

There is a rich body of research on zero-knowledge proofs. Early practical zero-knowledge proofs such as Schnorr [Sch91] and Guillou-Quisquater [GQ88] used number-theoretic assumptions. There have been several works extending these results to prove more general statements [CDS94, CD98, Gro09, BCC+16] with the latter giving discrete-logarithm based arguments for arithmetic circuit satisfiability with logarithmic communication complexity and a linear number of exponentiations for the prover, i.e., a computation cost of $\mathcal{O}(\lambda N)$ group multiplications for λ-bit exponents and a circuit with N multiplication gates.

Ishai et al. [IKOS09] showed how to use secure multi-party computation (MPC) protocols to construct zero-knowledge proofs. The intuition behind this generic construction is that the prover first executes in *her head* an MPC protocol for computing a circuit verifying some relation R and then commits to the views of all the virtual parties. The verifier asks the prover to open a subset of those views and then verifies their correctness and consistency with each other. Soundness and zero-knowledge follow from robustness and privacy of the MPC protocol. Applying this framework to efficient MPCs gives asymptotically efficient zero-knowledge proofs. For example, the perfectly secure MPC of [DI06] is used in [IKOS09] to obtain zero-knowledge proofs for the satisfiability of Boolean circuits with communication linear in the circuit size, $\mathcal{O}(N)$, and a computational cost of $\Omega(\lambda N)$, for circuits of size N and security parameter λ. Damgård et al. [DIK10] used the MPC framework to construct zero-knowledge proofs for the satisfiability of arithmetic circuits. Their construction has more balanced efficiency and achieves $\mathcal{O}(\text{polylog}(\lambda)N)$ complexity for both computation and communication.

Jawurek et al. [JKO13] gave a very different approach to building zero-knowledge proofs based on garbled circuits. Their approach proved [FNO15, CGM16] to be very efficient in practice for constructing proofs for languages represented as Boolean circuits. These techniques are appealing for proving small statements as they require only a constant number of symmetric-key operations per gate, while the main bottleneck is in their communication complexity. Asymptotically, this approach yields computational and communication complexity of $\mathcal{O}(\lambda N)$ bit operations and bits, respectively, when λ is the cost of a single symmetric-key operation. Recently, these techniques found applications in zero-knowledge proofs for checking the execution of RAM programs [HMR15, MRS17]. For instances that can be represented as RAM programs terminating in T steps and using memory of size M, these zero-knowledge proofs yield communication and computation with $\text{polylog}(M)$ overhead compared to the running time T of the RAM program.

Cramer et al. [CDP12] introduce zero-knowledge proofs for verifying multiplicative relations of committed values using techniques related to ours. When applied to zero-knowledge proofs for the satisfiability of Boolean circuits, the asymptotic communication and computation complexities of [CDP12] are close to [IKOS09], although the constants are smaller. Unlike [CDP12], we do not require any homomorphic property from the commitment scheme, and instead of relying on linear secret sharing schemes with product reconstruction, we use linear error-correcting codes.

In past years, a lot of attention has been dedicated to the study of succinct non-interactive arguments of knowledge (SNARKs) [Gro10, BCCT12, GGPR13, BCCT13, PHGR13, BCG+13, Gro16]. These are very compact arguments offering very efficient verification time. In the most efficient cases, the arguments consist of only a constant number of group elements and verification consists of a constant number of pairings and a number of group exponentiations that is linear in the instance size but independent of the witness size. The main bottleneck of these arguments is the computational complexity of the prover which requires $\mathcal{O}(N)$ group exponentiations.

Recently, Ben-Sasson et al. [BSCS16] proposed the notion of interactive oracle proofs (IOPs), which are interactive protocols where the prover may send a probabilisticaly checkable proof (PCP) in each round. Ben-Sasson et al. [BSCG+16] construct a 3-round public-coin IOP (with soundness error 1/2) for Boolean circuit satisfiability with linear proof length and quasi-linear running times for both the prover and the verifier. Moreover, the constructed IOP has constant query complexity (the number of opening queries requested by the verifier), while prior PCP constructions require sub-linear query complexity. Another follow-up work by Ben-Sasson et al. [BSCGV16] gives 2-round zero-knowledge IOPs (duplex PCPs) for any language in NTIME$(T(n))$ with quasi-linear prover computation in $n + T(n)$.

Efficiency Comparison. All the proofs we list above have super-linear cost for the prover. This means our zero-knowledge proofs are the most efficient zero-knowledge proofs for arithmetic circuits for the prover. We also know that our verification time is optimal for an instance of size $\Omega(N)$ field elements since the verification time is comparable to the time it takes just to read the instance.

Another well-studied class of languages is Boolean circuit satisfiability but here our techniques do not fare as well since there would be an overhead in representing bits as field elements. We therefore want to make clear that our claim of high efficiency and a significant performance improvement over the state of the art relates only to arithmetic circuits. Nevertheless, we find the linear cost for arithmetic circuits a significant result in itself. This is the first time for any general class of NP-complete language that true linear cost is achieved for the prover when compared to the time it takes to evaluate the statement directly given the prover's witness.

2 Preliminaries

2.1 Notation and Computational Model

We write $y \leftarrow A(x)$ for an algorithm outputting y on input x. When the algorithm is randomized, and we wish to explicitly refer to a particular choice of random coins r chosen by the algorithm, we write $y \leftarrow A(x; r)$. We write PPT/DPT for algorithms running in probabilistic polynomial time and deterministic polynomial time in the size of their inputs. Typically, the size of inputs and output will be polynomial in a *security parameter* λ, with the intention that larger λ means better security. For functions $f, g : \mathbb{N} \rightarrow [0, 1]$, we write $f(\lambda) \approx g(\lambda)$ if $|f(\lambda) - g(\lambda)| = \frac{1}{\lambda^{\omega(1)}}$. We say a function f is *overwhelming* if $f(\lambda) \approx 1$ and f is *negligible* if $f(\lambda) \approx 0$.

Throughout the paper, we will be working over a finite field $\mathbb{F}$. To get negligible risk of an adversary breaking our zero-knowledge proofs, we need the field to be large enough such that $\log |\mathbb{F}| = \omega(\lambda)$. When considering efficiency of our zero-knowledge proofs, we will assume the prover and verifier are RAM machines where operations on W-bit words have unit cost. We assume a field element is represented by $\mathcal{O}(\frac{\log |\mathbb{F}|}{W})$ words and that additions in $\mathbb{F}$ carry a cost of $\mathcal{O}\left(\frac{\log |\mathbb{F}|}{W}\right)$ machine operations. We expect multiplications to be efficiently computable as well but at a higher cost of $\omega\left(\frac{\log |\mathbb{F}|}{W}\right)$ machine operations.

For a positive integer n, $[n]$ denotes the set $\{1, \ldots, n\}$. We use bold letters such as $\boldsymbol{v}$ for row vectors. For $\boldsymbol{v} \in \mathbb{F}^n$ and a set $J = \{j_1, \ldots, j_k\} \subset [n]$ with $j_1 < \cdots < j_k$ we define the vector $\boldsymbol{v}|_J$ to be $(\boldsymbol{v}_{j_1}, \ldots, \boldsymbol{v}_{j_k})$. Similarly, for a matrix $V \in \mathbb{F}^{m \times n}$ we let $V|_J \in \mathbb{F}^{m \times k}$ be the submatrix of V restricted to the columns indicated in J.

2.2 Proofs of Knowledge

A *proof system* is defined by a triple of stateful PPT algorithms $(\mathcal{K}, \mathcal{P}, \mathcal{V})$, which we call the setup *generator*, the *prover* and *verifier*, respectively. The setup

generator $\mathcal{K}$ creates public parameters pp that will be used by the prover and
the verifier. We think of pp as being honestly generated, however, in the proofs
we construct it consists of parts that are either publicly verifiable or could be
generated by the verifier, so we use the public parameter model purely for simplicity and efficiency of our proofs, not for security.

The prover and verifier communicate with each other through a *commu-
nication channel* $\stackrel{\text{chan}}{\longleftrightarrow}$. When $\mathcal{P}$ and $\mathcal{V}$ interact on inputs s and t through a
communication channel $\stackrel{\text{chan}}{\longleftrightarrow}$ we let $\text{view}_{\mathcal{V}} \leftarrow \langle \mathcal{P}(s) \stackrel{\text{chan}}{\longleftrightarrow} \mathcal{V}(t) \rangle$ be the view of the
verifier in the execution, i.e., all inputs he gets including random coins and let
$\text{trans}_{\mathcal{P}} \leftarrow \langle \mathcal{P}(s) \stackrel{\text{chan}}{\longleftrightarrow} \mathcal{V}(t) \rangle$ denote the transcript of the communication between
prover and channel. This overloads the notation $\leftarrow \langle \mathcal{P}(s) \stackrel{\text{chan}}{\longleftrightarrow} \mathcal{V}(t) \rangle$ but it will
always be clear from the variable name if we get the verifier's view or the prover's
transcript. At the end of the interaction the verifier accepts or rejects. We write
$\langle \mathcal{P}(s) \stackrel{\text{chan}}{\longleftrightarrow} \mathcal{V}(t) \rangle = b$ depending on whether the verifier rejects ($b = 0$) or accepts
($b = 1$).

In the *standard channel* $\longleftrightarrow$, all messages are forwarded between prover and
verifier. We also consider an *ideal linear commitment* channel, $\stackrel{\text{ILC}}{\longleftrightarrow}$, or simply
ILC, described in Fig. 1. When using the ILC channel, the prover can submit a
`commit` command to commit to vectors of field elements of some fixed length k,
specified in pp_{ILC}. The vectors remain secretly stored in the channel, and will not
be forwarded to the verifier. Instead, the verifier only learns how many vectors
the prover has committed to. The verifier can submit a `send` command to the
ILC to send field elements to the prover. In addition, the verifier can also submit
`open` queries to the ILC for obtaining the opening of any linear combinations of

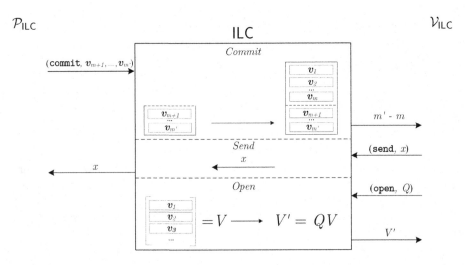

Fig. 1. Description of the ILC channel.

the vectors sent by the prover. We stress that the verifier can request several linear combinations within a single open query, as depicted in Fig. 1.

In a proof system over the ILC channel, sequences of commit, send and open queries could alternate in an arbitrary order. We call a proof system over the ILC channel *non-adaptive* if the verifier only makes one *open* query to the ILC channel before terminating his interaction with the channel, otherwise we call it *adaptive*. Although adaptive proof systems are allowed by the model, in this paper we will only consider non-adaptive ILC proof systems to simplify the exposition.

We remark that ILC proof systems are different from linear interactive proofs considered in [BCI+13]. In linear interactive proofs both the prover and verifier send vectors of field elements, but the prover can only send linear (or affine) transformations of the verifier's previously sent vectors. However, for our constructions it is important that the prover can compute on field elements received by the verifier and for instance evaluate polynomials.

We say a proof system is *public coin* if the verifier's messages to the communication channel are chosen uniformly at random and independently of the actions of the prover, i.e., the verifier's messages to the prover correspond to the verifier's randomness ρ.

We will consider relations $\mathcal{R}$ consisting of tuples (pp, u, w), and define $\mathcal{L}_{\mathcal{R}} = \{(pp, u) | \exists w : (pp, u, w) \in \mathcal{R}\}$. We refer to u as the *instance* and w as the *witness* that $(pp, u) \in \mathcal{L}_{\mathcal{R}}$. The *public parameter* pp will specify the security parameter λ, perhaps implicitly through its length, and may also contain other parameters used for specifying the specific relation, e.g. a description of a field. Typically, pp will also contain parameters that do not influence membership of $\mathcal{R}$ but may aid the prover and verifier, for instance, a description of an encoding function that they will use.

We will construct SHVZK proofs of knowledge for the relation $\mathcal{R}_{AC}$, where the instances are arithmetic circuits over a field $\mathbb{F}$ specified by pp. An instance consists of many fan-in 2 addition and multiplication gates over $\mathbb{F}$, a description of how wires in the circuit connect to the gates, and values assigned to some of the input wires. Witnesses w are the remaining inputs such that the output of the circuit is 0. For an exact definition of how we represent an arithmetic circuit, see Sect. 3. We would like to stress the fact that the wiring of the circuit is part of the instance and we allow a fully adaptive choice of the arithmetic circuit. This stands in contrast to pairing-based SNARKs that usually only consider circuits with fixed wires, i.e., the arithmetic circuit is partially non-adaptive, and getting full adaptivity through a universal circuit incurs an extra efficiency overhead.

The protocol $(\mathcal{K}, \mathcal{P}, \mathcal{V})$ is called a *proof of knowledge* over communication channel $\overset{chan}{\longleftrightarrow}$ for relation $\mathcal{R}$ if it has perfect completeness and computational knowledge soundness as defined below.

Definition 1 (Perfect Completeness). *The proof is* perfectly complete *if for all PPT adversaries* $\mathcal{A}$

$$\Pr\left[\begin{array}{c} pp \leftarrow \mathcal{K}(1^\lambda); (u, w) \leftarrow \mathcal{A}(pp) : \\ (pp, u, w) \notin \mathcal{R} \ \lor \ \langle \mathcal{P}(pp, u, w) \overset{chan}{\longleftrightarrow} \mathcal{V}(pp, u) \rangle = 1 \end{array} \right] = 1.$$

Definition 2 (Knowledge soundness). *A public-coin proof system has com-putational (strong black-box) knowledge soundness if for all DPT $\mathcal{P}^*$ there exists an expected PPT extractor $\mathcal{E}$ such that for all PPT adversaries $\mathcal{A}$*

$$\Pr\left[\begin{array}{c} pp \leftarrow \mathcal{K}(1^\lambda); (u, s) \leftarrow \mathcal{A}(pp); w \leftarrow \mathcal{E}^{\langle \mathcal{P}^*(s) \overset{chan}{\longleftrightarrow} \mathcal{V}(pp,u)\rangle}(pp, u): \\ b = 1 \ \wedge \ (pp, u, w) \notin \mathcal{R} \end{array}\right] \approx 0.$$

Here the oracle $\langle \mathcal{P}^*(s) \overset{chan}{\longleftrightarrow} \mathcal{V}(pp, u)\rangle$ runs a full protocol execution and if the proof is successful it returns a transcript of the prover's communication with the channel. The extractor $\mathcal{E}$ can ask the oracle to rewind the proof to any point in a previous transcript and execute the proof again from this point on with fresh public-coin challenges from the verifier. We define $b \in \{0, 1\}$ to be the verifier's output in the first oracle execution, i.e., whether it accepts or not, and we think of s as the state of the prover. The definition can then be paraphrased as saying that if the prover in state s makes a convincing proof, then we can extract a witness.

If the definition holds also for unbounded $\mathcal{P}^*$ and $\mathcal{A}$ we say the proof has statistical knowledge soundness.

If the definition of knowledge soundness holds for a non-rewinding extractor, i.e., a single transcript of the prover's communication with the communication channel suffices, we say the proof system has knowledge soundness with straight-line extraction.

We will construct public-coin proofs that have special honest-verifier zero-knowledge. This means that if the verifier's challenges are known, or even adver-sarially chosen, then it is possible to simulate the verifier's view without the witness. In other words, the simulator works for verifiers who may use adver-sarial coins in choosing their challenges but they follow the specification of the protocol as an honest verifier would.

Definition 3 (Special Honest-Verifier Zero-Knowledge). *The proof of knowledge is computationally special honest-verifier zero-knowledge (SHVZK) if there exists a PPT simulator $\mathcal{S}$ such that for all stateful interactive PPT adversaries $\mathcal{A}$ that output (u, w) such that $(pp, u, w) \in R$ and randomness ρ for the verifier*

$$\Pr\left[\begin{array}{c} pp \leftarrow \mathcal{K}(1^\lambda); (u, w, \rho) \leftarrow \mathcal{A}(pp); \\ \mathsf{view}_\mathcal{V} \leftarrow \langle \mathcal{P}(pp, u, w) \overset{chan}{\longleftrightarrow} \mathcal{V}(pp, u; \rho)\rangle : \mathcal{A}(\mathsf{view}_\mathcal{V}) = 1 \end{array}\right]$$
$$\approx \Pr\left[pp \leftarrow \mathcal{K}(1^\lambda); (u, w, \rho) \leftarrow \mathcal{A}(pp); \mathsf{view}_\mathcal{V} \leftarrow \mathcal{S}(pp, u, \rho) : \mathcal{A}(\mathsf{view}_\mathcal{V}) = 1\right].$$

We say the proof is statistically SHVZK *if the definition holds also against unbounded adversaries, and we say the proof is* perfect SHVZK *if the probabilities are exactly equal.*

From Honest-Verifier to General Zero-Knowledge. Honest-verifier zero-knowledge only guarantees the simulator works for verifiers following the proof system specifications. It might be desirable to consider general zero-knowledge where the simulator works for arbitrary malicious verifiers that may deviate from the specification of the proof. However, honest-verifier zero-knowledge is a first natural stepping stone to get efficient zero-knowledge proofs. We recall that our proofs are public coin, which means that the verifier's messages are chosen uniformly at random and independently from the messages received from the verifier. Below we recall few options to obtain general zero-knowledge proofs from a public-coin SHVZK proof. All these transformations are very efficient in terms of computation and communication such that the efficiency properties of our special honest-verifier zero-knowledge protocols are preserved.

In the Fiat-Shamir transform [FS86] the verifier's challenges are computed using a cryptographic hash function applied to the transcript up to the challenge. The Fiat-Shamir transform is more generally used to turn a public-coin proof into a non-interactive one. Since interaction with the verifier is no longer needed, general zero-knowledge is immediately achieved. If the hash function can be computed in linear time in the input size, then the Fiat-Shamir transform only incurs an additive linear overhead in computation for the prover and verifier. The drawback of the Fiat-Shamir transform is that security is usually proved in the random oracle model [BR93] where the hash function is modelled as an ideal random function.

Assuming a common reference string and relying on trapdoor commitments, Damgård [Dam00] gave a transformation yielding concurrently secure protocols for Σ-Protocols. The transformation can be optimized [Gro04] using the idea that for each public-coin challenge x, the prover first commits to a value x', then the verifier sends a value x'', after which the prover opens the commitment and uses the challenge $x = x' + x''$. The coin-flipping can be interleaved with the rest of the proof, which means the transformation preserves the number of rounds and only incurs a very small efficiency cost to do the coin-flipping for the challenges.

If one does not wish to rely on a common reference string for security, one can use a private-coin transformation where the verifier does not reveal the random coins used to generate the challenges sent to the prover (hence the final protocol is no longer public coin). One example is the Micciancio and Petrank [MP03] transformation (yielding concurrently secure protocols) while incurring a small overhead of $\omega(\log \lambda)$ with respect to the number of rounds as well as the computational and communication cost in each round. The transformation preserves the soundness and completeness errors of the original protocol; however, it does not preserve statistical zero-knowledge as the obtained protocol only has computational zero-knowledge.

There are other public-coin transformations to general zero-knowledge e.g. Goldreich et al. [GSV98]. The transformation relies on a random-selection protocol between the prover and verifier to specify a set of messages and restricting the verifier to choose challenges from this set. This means to get negligible soundness error these transformations require $\omega(1)$ sequential repetitions so the round complexity goes up.

2.3   Linear-Time Linear Error-Correcting Codes

A *code* over an alphabet Σ is a subset $\mathcal{C} \subseteq \Sigma^n$. A code $\mathcal{C}$ is associated with an encoding function $E_{\mathcal{C}} : \Sigma^k \to \Sigma^n$ mapping messages of length k into *codewords* of length n. We assume there is a setup algorithm $\mathsf{Gen}_{E_{\mathcal{C}}}$ which takes as input a finite field $\mathbb{F}$ and the parameter $k \in \mathbb{N}$, and outputs an encoding function $E_{\mathcal{C}}$.

In what follows, we restrict our attention to $\mathbb{F}$-*linear codes* for which the alphabet is a finite field $\mathbb{F}$, the code $\mathcal{C}$ is a k-dimensional linear subspace of $\mathbb{F}^n$, and $E_{\mathcal{C}}$ is an $\mathbb{F}$-linear map. The *rate* of the code is defined to be $\frac{k}{n}$. The *Hamming distance* between two vectors $\boldsymbol{x}, \boldsymbol{y} \in \mathbb{F}^n$ is denoted by $\mathsf{hd}(\boldsymbol{x}, \boldsymbol{y})$ and corresponds to the number of coordinates in which $\boldsymbol{x}, \boldsymbol{y}$ differ. The *(minimum) distance* of a code is defined to be the minimum Hamming distance $\mathsf{hd}_{\min}$ between distinct codewords in $\mathcal{C}$. We denote by $[n, k, \mathsf{hd}_{\min}]_{\mathbb{F}}$ a linear code over $\mathbb{F}$ with length n, dimension k and minimum distance $\mathsf{hd}_{\min}$. The *Hamming weight* of a vector $\boldsymbol{x}$ is $\mathsf{wt}(\boldsymbol{x}) = |\{i \in [n] : \boldsymbol{x}_i \neq 0\}|$.

In the next sections, we will use families of linear codes achieving asymptotically good parameters. More precisely, we require codes with linear length, $n = \Theta(k)$, and linear distance, $\mathsf{hd}_{\min} = \Theta(k)$, in the *dimension* k of the code. We recall that random linear codes achieve with high probability the best trade-off between distance and rate. However, in this work we are particularly concerned with computational efficiency of the encoding procedure and random codes are not known to be very efficient.

To obtain zero-knowledge proofs and arguments with linear cost for prover and verifier, we need to use codes that can be encoded in linear time. Starting from the seminal work of Spielman [Spi95], there has been a rich stream of research [GI01, GI02, GI03, GI05, DI14, CDD+16] regarding linear codes with linear-time encoding. Our construction can be instantiated, for example, with one of the families of codes presented by Druk and Ishai [DI14]. These are defined over a generic finite field $\mathbb{F}$ and meets all the above requirements.

Theorem 1 ([DI14]). *There exist constants $c_1 > 1$ and $c_2 > 0$ such that for every finite field $\mathbb{F}$ there exists a family of $[\lceil c_1 k \rceil, k, \lfloor c_2 k \rfloor]_{\mathbb{F}}$ linear codes, which can be encoded by a uniform family of linear-size arithmetic circuit of addition gates.*

2.4   Commitment Schemes

A non-interactive commitment scheme allows a sender to commit to a secret message and later reveal the message in a verifiable way. Here we are interested in commitment schemes that take as input an arbitrary length message so the message space is $\{0, 1\}^*$. A commitment scheme is defined by a pair of PPT algorithms (Setup, Commit).

$\mathsf{Setup}(1^\lambda) \to ck$: Given a security parameter, this returns a commitment key ck.
$\mathsf{Commit}_{ck}(m) \to c$: Given a message m, this picks a randomness $r \leftarrow \{0, 1\}^{\mathsf{poly}(\lambda)}$
  and computes a commitment $c = \mathsf{Commit}_{ck}(m; r)$.

A commitment scheme must be *binding* and *hiding*. The binding property means that it is infeasible to open a commitment to two different messages, whereas the hiding property means that the commitment does not reveal anything about the committed message.

Definition 4 (Binding). *A commitment scheme is* computationally binding *if for all PPT adversaries* $\mathcal{A}$

$$\Pr\left[\begin{array}{c} ck \leftarrow \mathsf{Setup}(1^\lambda);\ (m_0, r_0, m_1, r_1) \leftarrow \mathcal{A}(ck): \\ m_0 \neq m_1 \ \wedge\ \mathsf{Commit}_{ck}(m_0; r_0) = \mathsf{Commit}_{ck}(m_1; r_1) \end{array}\right] \approx 0.$$

If this holds also for unbounded adversaries, we say the commitment scheme is statistically binding.

Definition 5 (Hiding). *A commitment scheme is* computationally hiding *if for all PPT stateful adversaries* $\mathcal{A}$

$$\Pr\left[\begin{array}{c} ck \leftarrow \mathsf{Setup}(1^\lambda);\ (m_0, m_1) \leftarrow \mathcal{A}(ck);\ b \leftarrow \{0,1\}; \\ c \leftarrow \mathsf{Commit}_{ck}(m_b):\ \mathcal{A}(c) = b \end{array}\right] \approx \frac{1}{2},$$

where $\mathcal{A}$ *outputs messages of equal length* $|m_0| = |m_1|$. *If the definition holds also for unbounded adversaries, we say the commitment scheme is* statistically hiding.

We will be interested in using highly efficient commitment schemes. We say a commitment scheme is *linear-time* if the time to compute $\mathsf{Commit}_{ck}(m)$ is $\mathrm{poly}(\lambda) + \mathcal{O}(|m|)$ bit operations, which we assume corresponds to $\mathrm{poly}(\lambda) + \mathcal{O}(\frac{|m|}{W})$ machine operations on our W-bit RAM machine. We will also be interested in having small size commitments. We say a commitment scheme is compact if there is a polynomial $\ell(\lambda)$ such that commitments have size at most $\ell(\lambda)$ regardless of how long the message is. We say a commitment scheme is *public coin* if there is a polynomial $\ell(\lambda)$ such that $\mathsf{Setup}(1^\lambda)$ picks the commitment key uniformly at random as $ck \leftarrow \{0,1\}^{\ell(\lambda)}$. We will now discuss some candidate linear-time commitment schemes. Applebaum et al. [AHI+17] gave constructions of low-complexity families of collision-resistant hash functions, where it is possible to evaluate the hash function in linear time in the message size. Their construction is based on the binary shortest vector problem assumption, which is related to finding non-trivial low-weight vectors in the null space of a matrix over $\mathbb{F}_2$. To get down to linear-time complexity, they conjecture the binary shortest vector problem is hard when the matrix is sparse, e.g., an LDPC parity check matrix [Gal62]. Halevi and Micali [HM96] show that a collision-resistant hash function gives rise to a compact statistically hiding commitment scheme. Their transformation is very efficient, so starting with a linear-time hash function, one obtains a linear-time statistically hiding compact commitment scheme. Moreover, if we instantiate the hash function with the one by Applebaum et al. [AHI+17], which is public coin, we obtain a linear-time public-coin statistically hiding commitment scheme. Ishai et al. [IKOS08] propose linear-time computable pseudorandom generators. If we have statistically binding commitment scheme this means we can commit to an arbitrary length message m by picking a seed s for the pseudorandom generator, stretch it to $t = \mathrm{PRG}(s)$ of length $|m|$ and let $(\mathsf{Commit}_{ck}(s), t \oplus m)$ be the commitment to m. Assuming the commitment scheme is statistically binding, this gives us a linear-time statistically binding commitment scheme for arbitrary length messages. It can also easily be seen that commitments have the same length as the messages plus an additive polynomial overhead that depends only on the security parameter. The construction also preserves the public-coin property of the seed commitment scheme.

3   Zero-Knowledge Proofs for Arithmetic Circuit Satisfiability in the Ideal Linear Commitment Model

In this section, we construct a SHVZK proof of knowledge for arithmetic circuit satisfiability relations $\mathcal{R}_{AC}$ in the ILC model. Our proof can be seen as an abstraction of the zero-knowledge argument of Groth [Gro09] in an idealized vector commitment setting. In the ILC model, the prover can **commit** to vectors in $\mathbb{F}^k$ by sending them to the channel. The ILC channel stores the received vectors and communicates to verifier the number of vectors it received. The verifier can **send** messages to the prover via the ILC channel, which in the case of Groth's and our proof system consist of field elements in $\mathbb{F}$. Finally, the verifier can query the channel to **open** arbitrary linear combinations of the committed vectors sent by the prover. The field $\mathbb{F}$ and the vector length k is specified by the public parameter pp_{ILC}. It will later emerge that to get the best communication and computation complexity for arithmetic circuits with N gates, k should be approximately $\sqrt{N}$.

Consider a circuit with a total of N fan-in 2 gates, which can be either addition gates or multiplication gates over a field $\mathbb{F}$. Each gate has two inputs (left and right) and one output wire, and each output wire can potentially be attached as input to several other gates. In total, we have $3N$ inputs and outputs to gates. Informally, the description of an arithmetic circuit consists of a set of gates, the connection of wires between gates and known values assigned to some of the inputs and outputs. A circuit is said to be satisfiable if there exists an assignment complying with all the gates, the wiring, and the known values specified in the instance.

At a high level, the idea of the proof is for the prover to commit to the $3N$ inputs and outputs of all the gates in the circuit, and then prove that these assignments are consistent with the circuit description. This amounts to performing the following tasks:

- Prove for each value specified in the instance that this is indeed the value the prover has committed to.
- Prove for each addition gate that the committed output is the sum of the committed inputs.

– Prove for each multiplication gate that the committed output is the product of the committed inputs.
– Prove for each wire that all committed values corresponding to this wire are the same.

To facilitate these proofs, we arrange the committed values into row vectors $v_i \in \mathbb{F}^k$ similarly to [Gro09]. Without loss of generality we assume both the number of addition gates and the number of multiplication gates are divisible by k, which can always be satisfied by adding few dummy gates to the circuit. We can then number addition gates from $(1,1)$ to (m_A, k) and multiplication gates $(m_A + 1, 1)$ to $(m_A + m_M, k)$. We insert assignments to left inputs, right inputs and outputs of addition gates into entries of three matrices $A, B, C \in \mathbb{F}^{m_A \times k}$, respectively. We sort entries to the matrices so that wires attached to the same gate correspond to the same entry of the three matrices, as shown in Fig. 2. A valid assignment to the wires then satisfies $A + B = C$. We proceed in a similar way for the $m_M \cdot k$ multiplication gates to obtain three matrices $D, E, F \in \mathbb{F}^{m_M \times k}$ such that $D \circ E = F$, where $\circ$ denotes the Hadamard (i.e. entry-wise) product of matrices. All the committed wires then constitute a matrix

$$V = \begin{pmatrix} A \\ B \\ C \\ D \\ E \\ F \end{pmatrix} \in \mathbb{F}^{(3m_A + 3m_M) \times k}$$

Without loss of generality, we also assume the gates to be sorted so that the wire values specified in the instance correspond to *full* rows in V. Again, this is without loss of generality because we can always add a few dummy gates to the circuit and to the instance to complete a row.

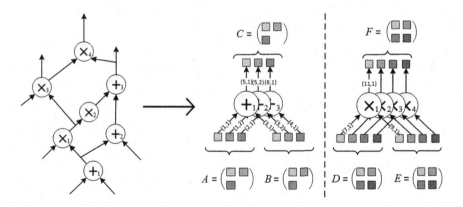

Fig. 2. Representation of an arithmetic circuit and arrangements of the wires into 6 matrices.

With these transformations in mind, let us write the arithmetic circuit relation as follows

$$
\mathcal{R}_{\mathsf{AC}} = \left\{ \begin{array}{l}
(pp, u, w) = \Big((\mathbb{F}, k, *),\ (m_A, m_M, \pi, \{\boldsymbol{v}_i\}_{i \in S}),\ (\{\boldsymbol{v}_i\}_{i \in \bar{S}}) \Big) : \\
\quad m = 3m_A + 3m_M \ \wedge\ \pi \in \Sigma_{[m] \times [k]} \\
\quad \wedge\ S \subseteq [m] \qquad\qquad\quad \wedge\ \bar{S} = [m] \setminus S \\
\quad \wedge\ A = (\boldsymbol{v}_i)_{i=1}^{m_A} \qquad\quad \wedge\ D = (\boldsymbol{v}_i)_{i=3m_A+1}^{3m_A+m_M} \\
\quad \wedge\ B = (\boldsymbol{v}_i)_{i=m_A+1}^{2m_A} \qquad \wedge\ E = (\boldsymbol{v}_i)_{i=3m_A+m_M+1}^{3m_A+2m_M} \\
\quad \wedge\ C = (\boldsymbol{v}_i)_{i=2m_A+1}^{3m_A} \qquad \wedge\ F = (\boldsymbol{v}_i)_{i=3m_A+2m_M+1}^{3m_A+3m_M} \\
\quad \wedge\ A + B = C \qquad\qquad \wedge\ D \circ E = F \\
\quad \wedge\ V = (\boldsymbol{v}_i)_{i=1}^{m} \qquad\qquad \wedge\ V_{i,j} = V_{\pi(i,j)} \ \forall\ (i,j) \in [m] \times [k]
\end{array} \right\}
$$

The role of the permutation π is to specify the wiring of the arithmetic circuit. For each wire, we can write a cycle $((i_1, j_1), \ldots, (i_t, j_t))$ that lists the location of the committed values corresponding to this wire. Then we let $\pi \in \Sigma_{[m] \times [k]}$ be the product of all these cycles, which unambiguously defines the wiring of the circuit. To give an example using the circuit in Fig. 2, the output wire of the first addition gate also appears as input of the first multiplication gate and the second addition gate. Therefore, if they appear as entries $(5, 1), (9, 1), (1, 2)$ in the matrix V defined by the rows $\boldsymbol{v}_i$, then we would have the cycle $((5, 1), (9, 1), (1, 2))$ indicating entries that must to be identical. The output of the second addition gate feeds into the third addition gate, so this might give us a cycle $((5, 2), (4, 1))$ of entries that should have the same value. The permutation π is the product of all these cycles that define which entries should have the same value.

In the proof for arithmetic circuit satisfiability, the prover starts by committing to all values $\{\boldsymbol{v}_i\}_{i=1}^{m}$. She will then call suitable sub-proofs to handle the four constraints these committed values should specify. We describe all the sub-proofs after the main proof given in Fig. 3 and refer to the full version of the paper for the detailed constructions.

$\mathcal{P}_{\mathsf{ILC}}(pp_{\mathsf{ILC}}, u, w)$	$\mathcal{V}_{\mathsf{ILC}}(pp_{\mathsf{ILC}}, u)$
– Parse $u = (m_A, m_M, \pi, \{\boldsymbol{v}_i\}_{i \in S})$	– Parse $u = (m_A, m_M, \pi, \{\boldsymbol{u}_i\}_{i \in S})$
– Parse $w = (\{\boldsymbol{v}_i\}_{i \in \bar{S}})$	– Run $\mathcal{V}_{\mathsf{eq}}(pp_{\mathsf{ILC}}, (\{\boldsymbol{u}_i\}_{i \in S}, [U]))$
– Send $(\mathtt{commit}, \{\boldsymbol{v}_i\}_{i=1}^{m})$ to the ILC channel	– Run $\mathcal{V}_{\mathsf{sum}}(pp_{\mathsf{ILC}}, ([A], [B], [C]))$
– The vectors define $V \in \mathbb{F}^{m \times k}$ and sub-	– Run $\mathcal{V}_{\mathsf{prod}}(pp_{\mathsf{ILC}}, ([D], [E], [F]))$
matrices A, B, C, D, E, F as described earlier	– Run $\mathcal{V}_{\mathsf{perm}}(pp_{\mathsf{ILC}}, (\pi, [V], [V]))$
– Let $U = (\boldsymbol{v}_i)_{i \in S}$	– Return 1 if all the sub-proofs are
– Run $\mathcal{P}_{\mathsf{eq}}(pp_{\mathsf{ILC}}, (\{\boldsymbol{v}_i\}_{i \in S}, [U]))$	accepted and 0 otherwise
– Run $\mathcal{P}_{\mathsf{sum}}(pp_{\mathsf{ILC}}, ([A], [B], [C]))$	
– Run $\mathcal{P}_{\mathsf{prod}}(pp_{\mathsf{ILC}}, ([D], [E], [F]))$	
– Run $\mathcal{P}_{\mathsf{perm}}(pp_{\mathsf{ILC}}, (\pi, [V], [V]))$	

Fig. 3. Arithmetic circuit satisfiability proof in the ILC model.

Here we use the convention that when vectors or matrices are written in square brackets, i.e., when we write $[A]$ in the instance, it means that these are values that have already been committed to the ILC channel. The prover knows these values, but the verifier may not know them. The first sub-proof $\left\langle \mathcal{P}_{eq}\left(pp_{\mathsf{ILC}}, (\{v_i\}_{i\in S}, [U])\right) \xleftrightarrow{\mathsf{ILC}} \mathcal{V}_{eq}\left(pp_{\mathsf{ILC}}, (\{u_i\}_{i\in S}, [U])\right)\right\rangle$ allows the verifier to check that values included in the instance are contained in the corresponding commitments the prover previously sent to the ILC channel. The second sub-proof $\left\langle \mathcal{P}_{sum}\left(pp_{\mathsf{ILC}}, ([A], [B], [C])\right) \xleftrightarrow{\mathsf{ILC}} \mathcal{V}_{sum}\left(pp_{\mathsf{ILC}}, ([A], [B], [C])\right)\right\rangle$ is used to prove the committed matrices A, B and C satisfy $A + B = C$. The sub-proof $\left\langle \mathcal{P}_{prod}\left(pp_{\mathsf{ILC}}, ([D], [E], [F])\right) \xleftrightarrow{\mathsf{ILC}} \mathcal{V}_{prod}\left(pp_{\mathsf{ILC}}, ([D], [E], [F])\right)\right\rangle$ is used to prove that the committed matrices D, E and F satisfy $D \circ E = F$. The last sub-proof $\left\langle \mathcal{P}_{perm}\left(pp_{\mathsf{ILC}}, (\pi, [A], [B])\right) \xleftrightarrow{\mathsf{ILC}} \mathcal{V}_{perm}\left(pp_{\mathsf{ILC}}, (\pi, [A], [B])\right)\right\rangle$ is used to prove that A has the same entries as B except they have been permuted according to the permutation π. Note that when we call the permutation sub-proof with $B = A$, then the statement is that A remains unchanged when we permute the entries according to π. This in turn means that all committed values that lie on the same cycle in the permutation must be identical, i.e., the matrix A respects the wiring of the circuit.

Theorem 2. $(\mathcal{K}_{\mathsf{ILC}}, \mathcal{P}_{\mathsf{ILC}}, \mathcal{V}_{\mathsf{ILC}})$ *is a proof system for* $\mathcal{R}_{\mathsf{AC}}$ *in the* ILC *model with perfect completeness, statistical knowledge soundness with straight-line extraction, and perfect special honest-verifier zero-knowledge.*

Proof. Perfect completeness follows from the perfect completeness of the sub-proofs.

Perfect SHVZK follows from the perfect SHVZK of the sub-proofs. A simulated transcript is obtained by combining the outputs of the simulators of all the sub-proofs.

Also statistical knowledge soundness follows from the knowledge soundness of the sub-proofs. The statistical knowledge soundness of the equality sub-proof guarantees that commitments to values included in the instance indeed contain the publicly known values. The correctness of the addition gates and multiplication gates follows from the statistical knowledge soundness of the respective sub-proofs. Finally, as we have argued above, the permutation sub-proof guarantees the committed values respect the wiring of the circuit.

Since all sub-proofs have knowledge soundness with straight line extraction, so does the main proof. □

The efficiency of our arithmetic circuit satisfiability proof in the ILC model is given in Fig. 4. A detailed breakdown of the costs of each sub-protocol can be found in the full version of the paper. The asymptotic results displayed below are obtained when the parameter k specified by pp_{ILC} is approximately $\sqrt{N}$. The query complexity qc is the number of linear combinations the verifier queries from the ILC channel in the opening query. The verifier communication C_{ILC} is the number of messages sent from the verifier to the prover via the ILC channel

Prover computation	$T_{\mathcal{P}_{ILC}} = \mathcal{O}(N)$ multiplications in $\mathbb{F}$
Verifier computation	$T_{\mathcal{V}_{ILC}} = \mathcal{O}(N)$ additions in $\mathbb{F}$
Query complexity	qc $= 20$
Verifier communication	$C_{ILC} = \mathcal{O}(\log\log(N))$ field elements
Round complexity	$\mu = \mathcal{O}(\log\log(N))$
Total number of committed vectors	$t = \mathcal{O}\left(\sqrt{N}\right)$ vectors in $\mathbb{F}^k$

Fig. 4. Efficiency of our arithmetic circuit satisfiability proof in the ILC model $(\mathcal{K}_{ILC}, \mathcal{P}_{ILC}, \mathcal{V}_{ILC})$ for $(pp, u, w) \in \mathcal{R}_{AC}$.

and in our proof system it is proportional to the number of rounds. Let μ be the number of rounds in the ILC proof and $t_1, \ldots, t_\mu$ be the numbers of vectors that the prover sends to the ILC channel in each round, and let $t = \sum_{i=1}^{\mu} t_i$.

4 Compiling Ideal Linear Commitment Proofs into Standard Proofs

In this section, we show how to compile a proof of knowledge with straight-line extraction for relation $\mathcal{R}$ over the communication channel ILC into a proof of knowledge without straight-line extraction for the same relation over the standard channel. Recall that the ILC channel allows the prover to submit vectors of length k to the channel and the verifier can then query linear combinations of those vectors.

The idea behind the compilation of an ILC proof is that instead of committing to vectors $\boldsymbol{v}_\tau$ using the channel ILC, the prover encodes each vector $\boldsymbol{v}_\tau$ as $\mathsf{E}_C(\boldsymbol{v}_\tau)$ using a linear error-correcting code E_C. In any given round, we can think of the codewords as rows $\mathsf{E}_C(\boldsymbol{v}_\tau)$ in a matrix $\mathsf{E}_C(V)$. However, instead of committing to the rows of the matrix, the prover commits to the columns of the matrix. When the verifier wants to open a linear combination of the original vectors, he sends the coefficients $\boldsymbol{q} = (q_1, \ldots, q_t)$ of the linear combination to the prover, and the prover responds with the linear combination $\boldsymbol{v}_{(\boldsymbol{q})} \leftarrow \boldsymbol{q}V$. Notice that we will use the notation $\boldsymbol{v}_{(\boldsymbol{q})}$, and later on $\boldsymbol{v}_{(\gamma)}$, to denote vectors that depend on $\boldsymbol{q}$ and γ: the $\boldsymbol{q}$ and γ are not indices. Now, to spot check that the prover is not giving a wrong $\boldsymbol{v}_{(\boldsymbol{q})}$, the verifier may request the jth element of each committed codeword $\boldsymbol{e}_\tau$. This corresponds to revealing the jth column of error-corrected matrix $\mathsf{E}_C(V)$. Since the code E_C is linear, the revealed elements should satisfy $\mathsf{E}_C(\boldsymbol{v}_{(\boldsymbol{q})})_j = \sum_{\tau=1}^{t} q_\tau \mathsf{E}_C(\boldsymbol{v}_\tau)_j = \boldsymbol{q}(\mathsf{E}_C(V)|_j)$. The verifier will spot check on multiple columns, so that if the code has sufficiently large minimum distance and the prover gives a wrong $\boldsymbol{v}_{(\boldsymbol{q})}$, then with overwhelming probability, the verifier will open at least one column j where the above equality does not hold.

Revealing entries in a codeword may leak information about the encoded vector. To get SHVZK, instead of using E_C, we use a randomized encoding $\tilde{\mathsf{E}}_C$ defined by $\tilde{\mathsf{E}}_C(\boldsymbol{v}; \boldsymbol{r}) = (\mathsf{E}_C(\boldsymbol{v}) + \boldsymbol{r}, \boldsymbol{r})$. This doubles the code-length to $2n$ but ensures that when you reveal entry j, but not entry $j + n$, then the verifier only learns a random field element. The spot checking technique using $\tilde{\mathsf{E}}_C$ is illustrated in Fig. 5. In the following, we use the notation $\boldsymbol{e}_\tau = (\mathsf{E}_C(\boldsymbol{v}_\tau) + \boldsymbol{r}_\tau, \boldsymbol{r}_\tau)$ and $E = (\mathsf{E}_C(V) + R, R)$. We also add a check, where the verifier sends an extra

$$
\begin{pmatrix} \boldsymbol{v}_0 \\ \vdots \\ \boldsymbol{v}_t \end{pmatrix} \xrightarrow{\tilde{\mathsf{E}}_C} \left(\begin{array}{c|c} \mathsf{E}_C(\boldsymbol{v}_0) + \boldsymbol{r}_0 & \boldsymbol{r}_0 \\ \vdots & \vdots \\ \mathsf{E}_C(\boldsymbol{v}_t) + \boldsymbol{r}_t & \boldsymbol{r}_t \end{array} \right)
$$

$$
\boldsymbol{q} \downarrow \qquad\qquad \boldsymbol{q} \downarrow_{j_1} \cdots \boldsymbol{q} \downarrow_{j_\lambda} \quad \boldsymbol{q} \downarrow_{j_{\lambda+1}} \cdots \boldsymbol{q} \downarrow_{j_{2\lambda}}
$$

$$
\begin{pmatrix} \boldsymbol{v}_{(\boldsymbol{q})} \end{pmatrix} \xrightarrow{\tilde{\mathsf{E}}_C} \left(\begin{array}{c|c} \mathsf{E}_C(\boldsymbol{v}_{(\boldsymbol{q})}) + \boldsymbol{r}_{(\boldsymbol{q})} & \boldsymbol{r}_{(\boldsymbol{q})} \end{array} \right)
$$

Fig. 5. Vectors $\boldsymbol{v}_\tau$ organized in matrix V are encoded row-wise as matrix $E = \tilde{\mathsf{E}}_C(V; R)$. The vertical line in the right matrix and vector denotes concatenation of matrices respectively vectors. The prover commits to each column of E. When the prover given $\boldsymbol{q}$ wants to reveal the linear combination $\boldsymbol{v}_{(\boldsymbol{q})} = \boldsymbol{q}V$ she also reveals $\boldsymbol{r}_{(\boldsymbol{q})} = \boldsymbol{q}R$. The verifier now asks for openings of 2λ columns $J = \{j_1, \ldots, j_{2\lambda}\}$ in E and verifies for these columns that $\boldsymbol{q}E|_J = \tilde{\mathsf{E}}_C(\boldsymbol{v}_{(\boldsymbol{q})}; \boldsymbol{r}_{(\boldsymbol{q})})|_J$. To avoid revealing any information about $\mathsf{E}_C(V)$, we must ensure that $\forall j \in [n] : j \in J \Rightarrow j + n \notin J$. If the spot checks pass, the verifier believes that $\boldsymbol{v}_{(\boldsymbol{q})} = \boldsymbol{q}V$.

random linear combination $\boldsymbol{\gamma} \in \mathbb{F}^t$ to ensure that if a malicious prover commits to values of $\boldsymbol{e}_\tau$ that are far from being codewords, the verifier will most likely reject. The reason the challenges $\boldsymbol{q}$ from the ILC proof are not enough to ensure this is that they are not chosen uniformly at random. One could, for instance, imagine that there was a vector $\boldsymbol{v}_\tau$ that was never queried in a non-trivial way, and hence the prover could choose it to be far from a codeword. To make sure this extra challenge $\boldsymbol{\gamma}$ does not reveal information to the verifier, the prover picks a random blinding vector $\boldsymbol{v}_0$, which is added as the first row of V and will be added to the linear combination of the challenge $\boldsymbol{\gamma}$.

4.1   Construction

Let $(\mathcal{K}_{\mathsf{ILC}}, \mathcal{P}_{\mathsf{ILC}}, \mathcal{V}_{\mathsf{ILC}})$ be a *non-adaptive* μ-round SHVZK proof of knowledge with straight-line extraction over ILC for a relation $\mathcal{R}$. Here, non-adaptive means that the verifier waits until the last round before querying linear combinations of vectors and they are queried all at once instead of the queries depending on each other.[1] Let $\mathsf{Gen}_{\mathsf{E}_C}$ be a generator that given field $\mathbb{F}$ and length parameter

---

[1] The construction can be easily modified to an adaptive ILC proof. For each round of queries in the ILC proof, there will one extra round in the compiled proof.

k outputs a constant rate linear code E_C that is linear-time computable given its description and has linear minimum distance. Define the $\tilde{\mathsf{E}}_C$ with code length $2n$ as above: $\tilde{\mathsf{E}}_C(\boldsymbol{v}; \boldsymbol{r}) = (\mathsf{E}_C(\boldsymbol{v}) + \boldsymbol{r}, \boldsymbol{r})$. Finally, let $(\mathsf{Setup}, \mathsf{Commit})$ be a non-interactive commitment scheme.

We now define a proof of knowledge $(\mathcal{K}, \mathcal{P}, \mathcal{V})$ in Fig. 6, where we use the following notation: given matrices $V_1, \ldots, V_\mu, R_1, \ldots, R_\mu$ and $E_1, \ldots, E_\mu$ we define

---

$\mathcal{P}(pp, u, w)$

- **Parse input**:
  - Parse $pp = (pp_{\mathsf{ILC}}, \mathsf{E}_C, ck)$
  - Parse $pp_{\mathsf{ILC}} = (\mathbb{F}, k)$
  - Get n from E_C
- **Round 1**:
  - $\boldsymbol{v}_0 \leftarrow \mathbb{F}^k$
  - $\boldsymbol{e}_0 \leftarrow \tilde{\mathsf{E}}_C(\boldsymbol{v}_0; \boldsymbol{r}_0)$
  - $(\mathtt{commit}, V_1) \leftarrow \mathcal{P}_{\mathsf{ILC}}(pp_{\mathsf{ILC}}, u, w)$
  - $E_1 \leftarrow \tilde{\mathsf{E}}_C(V_1; R_1)$
  - Let $E_{01} = \begin{pmatrix} \boldsymbol{e}_0 \\ E_1 \end{pmatrix}$
  - $\boldsymbol{c}_1 = \mathsf{Commit}(E_{01}; \boldsymbol{s}_1)$
  - Send $(\boldsymbol{c}_1, t_1)$ to $\mathcal{V}$
- **Rounds $2 \le i \le \mu$**:
  - Get challenge x_{i-1} from $\mathcal{V}$
  - $(\mathtt{commit}, V_i) \leftarrow \mathcal{P}_{\mathsf{ILC}}(x_{i-1})$
  - $E_i \leftarrow \tilde{\mathsf{E}}_C(V_i; R_i)$
  - $\boldsymbol{c}_i = \mathsf{Commit}(E_i; \boldsymbol{s}_i)$
  - Send $(\boldsymbol{c}_i, t_i)$ to $\mathcal{V}$
- **Round $\mu + 1$**:
  - Get $(\boldsymbol{\gamma}, Q)$ from $\mathcal{V}$
  - $\boldsymbol{v}_{(\gamma)} \leftarrow \boldsymbol{v}_0 + \gamma V$
  - $\boldsymbol{r}_{(\gamma)} \leftarrow \boldsymbol{r}_0 + \gamma R$
  - $V_{(Q)} \leftarrow QV$
  - $R_{(Q)} \leftarrow QR$
  - Send $(\boldsymbol{v}_{(\gamma)}, \boldsymbol{r}_{(\gamma)}, V_{(Q)}, R_{(Q)})$ to $\mathcal{V}$
- **Round $\mu + 2$**:
  - Get $J \subset [2n]$ from $\mathcal{V}$
  - Send $(E_{01}|_J, \boldsymbol{s}_1|_J, \ldots, E_\mu, \boldsymbol{s}_\mu|_J)$ to $\mathcal{V}$

$\mathcal{K}(1^\lambda)$

- $pp_{\mathsf{ILC}} \leftarrow \mathcal{K}_{\mathsf{ILC}}(1^\lambda)$
- Parse $pp_{\mathsf{ILC}} = (\mathbb{F}, k)$
- $\mathsf{E}_C \leftarrow \mathsf{Gen}_{\mathsf{E}_C}(\mathbb{F}, k)$
- $ck \leftarrow \mathsf{Setup}(1^\lambda)$
- Return $pp = (pp_{\mathsf{ILC}}, \mathsf{E}_C, ck)$

$\mathcal{V}(pp, u)$

- **Parse input**
  - Parse $pp = (pp_{\mathsf{ILC}}, \mathsf{E}_C, ck)$
  - Parse $pp_{\mathsf{ILC}} = (\mathbb{F}, k)$
  - Get n from E_C
  - Give input (pp_{ILC}, u) to $\mathcal{V}_{\mathsf{ILC}}$
- **Rounds $1 \le i < \mu$**:
  - Receive $(\boldsymbol{c}_i, t_i)$
  - $(\mathtt{send}, x_i) \leftarrow \mathcal{V}_{\mathsf{ILC}}(t_i)$
  - Send x_i to $\mathcal{P}$
- **Round μ**:
  - Receive $(\boldsymbol{c}_\mu, t_\mu)$
  - $\boldsymbol{\gamma} \leftarrow \mathbb{F}^{\sum_{i=1}^\mu t_i}$
  - $(\mathtt{open}, Q) \leftarrow \mathcal{V}_{\mathsf{ILC}}(t_\mu)$
  - Send $(\boldsymbol{\gamma}, Q)$ to $\mathcal{P}$
- **Round $\mu + 1$**:
  - Receive $(\boldsymbol{v}_{(\gamma)}, \boldsymbol{r}_{(\gamma)}, V_{(Q)}, R_{(Q)})$
  - Choose random allowed $J \subset [2n]$
  - Send J to $\mathcal{P}$
- **Round $\mu + 2$**:
  - Receive $(E_{01}|_J, \boldsymbol{s}_1|_J, \ldots, E_\mu, \boldsymbol{s}_\mu|_J)$
  - Check $\boldsymbol{c}_1|_J = \mathsf{Commit}(E_{01}|_J; \boldsymbol{s}_1|_J)$,
    $\ldots, \boldsymbol{c}_\mu|_J = \mathsf{Commit}(E_\mu|_J; \boldsymbol{s}_\mu|_J)$
  - Check $\tilde{\mathsf{E}}_C(\boldsymbol{v}_{(\gamma)}, \boldsymbol{r}_{(\gamma)})|_J = \boldsymbol{e}_0|_J + \gamma E|_J$
  - Check $\tilde{\mathsf{E}}_C(V_{(Q)}, R_{(Q)})|_J = QE|_J$
  - If all checks pass, return decision of $\mathcal{V}_{\mathsf{ILC}}(V_{(Q)})$, else return 0

Fig. 6. Construction of $(\mathcal{K}, \mathcal{P}, \mathcal{V})$ from $(\mathcal{K}_{\mathsf{ILC}}, \mathcal{P}_{\mathsf{ILC}}, \mathcal{V}_{\mathsf{ILC}})$, commitment scheme $(\mathsf{Setup}, \mathsf{Commit})$ and error-correcting code $\mathcal{C}$.

$$V = \begin{pmatrix} V_1 \\ \vdots \\ V_\mu \end{pmatrix} \qquad R = \begin{pmatrix} R_1 \\ \vdots \\ R_\mu \end{pmatrix} \qquad E = \begin{pmatrix} E_1 \\ \vdots \\ E_\mu \end{pmatrix}.$$

The matrices $V_1, \ldots, V_\mu$ are formed by the row vectors $\mathcal{P}_{\mathsf{ILC}}$ commits to, and we let $t_1, \ldots, t_\mu$ be the numbers of vectors in each round, i.e., for all i we have $V_i \in \mathbb{F}^{t_i \times k}$.

We say that a set $J \subset [2n]$ is *allowed* if $|J \cap [n]| = \lambda$ and $|J \setminus [n]| = \lambda$ and there is no $j \in J$ such that $j + n \in J$. In the following we will always assume codewords have length $n \geq 2\lambda$. We use $\tilde{\mathsf{E}}_\mathcal{C}(V; R)$ to denote the function that applies $\tilde{\mathsf{E}}_\mathcal{C}$ *row-wise*. In the protocol for $\mathcal{V}$, we are using that $\tilde{\mathsf{E}}_\mathcal{C}(\boldsymbol{v}; \boldsymbol{r})|_J$ can be computed from just $\boldsymbol{v}$ and $\boldsymbol{r}|_{\{j \in [n] : j \in J \vee j+n \in J\}}$. We use $\mathsf{Commit}(E; \boldsymbol{s})$ to denote the function that applies Commit *column-wise* on E and returns a vector $\boldsymbol{c}$ of $2n$ commitments. We group all $\mathcal{V}_{\mathsf{ILC}}$'s queries in one matrix $Q \in \mathbb{F}^{\mathsf{qc} \times t}$, where t is the total number of vectors committed to by $\mathcal{P}$ and qc is the query complexity of $\mathcal{V}_{\mathsf{ILC}}$, i.e., the total number of linear combinations $\boldsymbol{q}$ that $\mathcal{V}_{\mathsf{ILC}}$ requests to be opened.

4.2   Security Analysis

Theorem 3 (Completeness). *If* $(\mathcal{K}_{\mathsf{ILC}}, \mathcal{P}_{\mathsf{ILC}}, \mathcal{V}_{\mathsf{ILC}})$ *is complete for relation* $\mathcal{R}$ *over* ILC, *then* $(\mathcal{K}, \mathcal{P}, \mathcal{V})$ *in Fig. 6 is complete for relation* $\mathcal{R}$.

Proof. All the commitment openings are correct, so they will be accepted by the verifier. In the execution of $\langle \mathcal{P}(pp, u, w) \longleftrightarrow \mathcal{V}(pp, u) \rangle$, the fact that $\mathsf{E}_\mathcal{C}$ is linear implies $\tilde{\mathsf{E}}_\mathcal{C}$ is linear and hence all the linear checks will be true. If $(pp, u, w) \in \mathcal{R}$ then $(pp_{\mathsf{ILC}}, u, w) \in \mathcal{R}$ and being complete $\langle \mathcal{P}_{\mathsf{ILC}}(pp_{\mathsf{ILC}}, u, w) \overset{\mathsf{ILC}}{\longleftrightarrow} \mathcal{V}_{\mathsf{ILC}}(pp_{\mathsf{ILC}}, stm) \rangle = 1$ so $\mathcal{V}$'s internal copy of $\mathcal{V}_{\mathsf{ILC}}$ will accept. Thus, in this case, $\langle \mathcal{P}(pp, u, w) \longleftrightarrow \mathcal{V}(pp, u) \rangle = 1$, which proves completeness.   □

Theorem 4 (Knowledge Soundness). *If* $(\mathcal{K}_{\mathsf{ILC}}, \mathcal{P}_{\mathsf{ILC}}, \mathcal{V}_{\mathsf{ILC}})$ *is statistically knowledge sound with a straight-line extractor for relation* $\mathcal{R}$ *over* ILC *and* $(\mathsf{Setup}, \mathsf{Commit})$ *is computationally (statistically) binding, then* $(\mathcal{K}, \mathcal{P}, \mathcal{V})$ *as constructed above is computationally (statistically) knowledge sound for relation* $\mathcal{R}$.

Proof. We prove the computational case. The statistical case is similar.

In order to argue that $(\mathcal{K}, \mathcal{P}, \mathcal{V})$ is computationally knowledge sound, we will first show that for every DPT $\mathcal{P}^*$ there exists a deterministic (but not necessarily efficient) $\mathcal{P}^*_{\mathsf{ILC}}$ such that for all PPT $\mathcal{A}$ we have

$$\Pr \left[\begin{array}{c} pp \leftarrow \mathcal{K}(1^\lambda); (pp_{\mathsf{ILC}}, \cdot) = pp; (u, s) \leftarrow \mathcal{A}(pp) : \\ \langle \mathcal{P}^*(s) \longleftrightarrow \mathcal{V}(pp, u; (\rho_{\mathsf{ILC}}, \rho)) \rangle = 1 \\ \wedge \langle \mathcal{P}^*_{\mathsf{ILC}}(s, pp, u) \overset{\mathsf{ILC}}{\longleftrightarrow} \mathcal{V}_{\mathsf{ILC}}(pp_{\mathsf{ILC}}, u; \rho_{\mathsf{ILC}}) \rangle = 0 \end{array} \right] \approx 0. \qquad (1)$$

Note that the randomness ρ_{ILC} in $\mathcal{V}$ which comes from the internal $\mathcal{V}_{\mathsf{ILC}}$ in line two is the same as the randomness used by $\mathcal{V}_{\mathsf{ILC}}$ in line three.

Our constructed $\mathcal{P}^*_{\text{ILC}}$ will run an internal copy of $\mathcal{P}^*$. When the internal $\mathcal{P}^*$ in round i sends a message (c_i, t_i), $\mathcal{P}^*_{\text{ILC}}$ will simulate $\mathcal{P}^*$ on every possible continuation of the transcript, and for each $j = 1, \ldots, 2n$ find the most frequently occurring correct opening $((E_i)_j, (s_i)_j)$ of $(c_i)_j$. $\mathcal{P}^*_{\text{ILC}}$ will then use this to get matrices E^*_i. For each row e^*_τ of these matrices, $\mathcal{P}^*_{\text{ILC}}$ finds a vector $\boldsymbol{v}_\tau$ and randomness $\boldsymbol{r}_\tau$ such that $\mathsf{hd}(\tilde{\mathsf{E}}_\mathcal{C}(\boldsymbol{v}_\tau, \boldsymbol{r}_\tau), e^*_\tau) < \frac{\mathsf{hd}_{\min}}{3}$ if such a vector exists. If for some τ no such vector $\boldsymbol{v}_\tau$ exists, then $\mathcal{P}^*_{\text{ILC}}$ aborts. Otherwise we let V_i and R_i denote the matrices formed by the row vectors $\boldsymbol{v}_\tau$ and $\boldsymbol{r}_\tau$ in round i and $\mathcal{P}^*_{\text{ILC}}$ sends V_i to the ILC. Notice that since the minimum distance of $\tilde{\mathsf{E}}_\mathcal{C}$ is at least $\mathsf{hd}_{\min}$, there is at most one such vector $\boldsymbol{v}_\tau$ for each e^*_τ.

The internal copy of $\mathcal{P}^*$ will expect to get two extra rounds, where in the first it should receive $\boldsymbol{\gamma}$ and Q and should respond with $\boldsymbol{v}^*_{(\gamma)}, \boldsymbol{r}^*_{(\gamma)}, V_{(Q)}$ and $R_{(Q)}$, and in the second it should receive J and send $E_{01}|_J, \boldsymbol{s}_1|_J, \ldots, E_\mu, \boldsymbol{s}_\mu|_J$. Since $\mathcal{P}^*_{\text{ILC}}$ does not send and receive corresponding messages, $\mathcal{P}^*_{\text{ILC}}$ does not have to run this part of $\mathcal{P}^*$. Of course, for each commitment sent by $\mathcal{P}^*$, these rounds are internally simulated many times to get the most frequent opening. Notice that a $\mathcal{V}_{\text{ILC}}$ communicating over ILC with our constructed $\mathcal{P}^*_{\text{ILC}}$ will, on challenge Q receive $V_{(Q)} = QV$ from the ILC.

The verifier $\mathcal{V}$ accepts only if its internal copy of $\mathcal{V}_{\text{ILC}}$ accepts. Hence, the only three ways $\langle \mathcal{P}^*(s) \longleftrightarrow \mathcal{V}(pp, u; (\rho_{\text{ILC}}, \rho)) \rangle$ can accept without $\langle \mathcal{P}^*_{\text{ILC}}(s, pp, u) \xleftrightarrow{\text{ILC}} \mathcal{V}_{\text{ILC}}(pp_{\text{ILC}}, u; \rho_{\text{ILC}}) \rangle$ being accepting are

1. if $\mathcal{P}^*$ makes an opening of a commitment that is not its most frequent opening of that commitment, or
2. if $\mathcal{P}^*_{\text{ILC}}$ has an error because for some τ no $\boldsymbol{v}_\tau, \boldsymbol{r}_\tau$ with $\mathsf{hd}(\tilde{\mathsf{E}}_\mathcal{C}(\boldsymbol{v}_\tau, \boldsymbol{r}_\tau), e^*_\tau) < \frac{\mathsf{hd}_{\min}}{3}$ exists, or
3. if $\mathcal{P}^*$ sends some $V^*_{(Q)} \neq V_{(Q)}$.

We will now argue that for each of these three cases, the probability that they happen and $\mathcal{V}$ accepts is negligible.

Since $\mathcal{P}^*$ runs in polynomial time and the commitment scheme is computationally binding, there is only negligible probability that $\mathcal{P}^*$ sends a valid opening that is not the most frequent. Since $\mathcal{V}$ will reject any opening that is not valid, the probability of $\mathcal{V}$ accepting in case 1 is negligible.

Next, we consider the second case. To do so, define the event Err that E^* is such that for some $\boldsymbol{\gamma}^* \in \mathbb{F}^t$ we have $\mathsf{hd}(\tilde{\mathcal{C}}, \boldsymbol{\gamma}^* E^*) \geq \frac{\mathsf{hd}_{\min}}{3}$. Here $\tilde{\mathcal{C}}$ denotes the image of $\tilde{\mathsf{E}}_\mathcal{C}$, i.e. $\tilde{\mathcal{C}} = \{(c + r, r) : c \in \mathcal{C}, r \in \mathbb{F}^n\}$. Clearly, if $\mathcal{P}^*_{\text{ILC}}$ returns an error because no $\boldsymbol{v}_i, \boldsymbol{r}_i$ with $\mathsf{hd}(\tilde{\mathsf{E}}_\mathcal{C}(\boldsymbol{v}_i, \boldsymbol{r}_i), e^*_i) < \frac{\mathsf{hd}_{\min}}{3}$ exist then we have Err.

The proof of the following claim can be found in the full version of the paper.

Claim. Let $e^*_0, \ldots, e^*_t \in \mathbb{F}^{2n}$. If Err occurs, then for uniformly chosen $\boldsymbol{\gamma} \in \mathbb{F}^t$, there is probability at most $\frac{1}{|\mathbb{F}|}$ that $\mathsf{hd}(\tilde{\mathcal{C}}, e^*_0 + \boldsymbol{\gamma} E^*) < \frac{\mathsf{hd}_{\min}}{6}$.

Thus, if Err then with probability at least $1 - \frac{1}{|\mathbb{F}|}$ the vector $\boldsymbol{\gamma}$ is going to be such that $\mathsf{hd}(\tilde{\mathcal{C}}, e^*_0 + \boldsymbol{\gamma} E^*) \geq \frac{\mathsf{hd}_{\min}}{6}$. If this happens, then for the vectors $(\boldsymbol{v}^*_{(\gamma)}, \boldsymbol{r}^*_{(\gamma)})$ sent by $\mathcal{P}^*$, we must have $\mathsf{hd}(\tilde{\mathsf{E}}_\mathcal{C}(\boldsymbol{v}^*_{(\gamma)}, \boldsymbol{r}^*_{(\gamma)}), e^*_0 + \boldsymbol{\gamma} E^*) \geq \frac{\mathsf{hd}_{\min}}{6}$. This means that either in the first half of the codeword $\tilde{E}_C(\boldsymbol{v}^*_{(\gamma)}, \boldsymbol{r}^*_{(\gamma)})$ or in the second half, there will be at least $\frac{\mathsf{hd}_{\min}}{12}$ values of j where it differs from $\boldsymbol{e}^*_0 + \gamma E^*$. It is easy to see that the λ values of j in one half of $[2n]$ are chosen uniformly and independently at random conditioned on being different.

For each of these j, there is a probability at most $1 - \frac{\mathsf{hd}_{\min}}{12n}$ that $\tilde{E}_C(\boldsymbol{v}_{(\gamma)}, \boldsymbol{r}_{(\gamma)})_j = \boldsymbol{e}^*_{0,j} + \gamma E^*|_j$, and since the j's are chosen uniformly under the condition that they are distinct, given that this holds for the first i values, the probability is even smaller for the $i + 1$'th. Hence, the probability that it holds for all j in this half is negligible. This shows that the probability that Err happens and V accepts is negligible.

Now we turn to case 3, where Err does not happen but $\mathcal{P}^*$ sends a $V^*_{(Q)} \neq V_{(Q)}$. In this case, for all $\boldsymbol{\gamma}^* \in \mathbb{F}^t$, we have $\mathsf{hd}(\tilde{C}, \sum_{\tau=1}^t \gamma^*_\tau \boldsymbol{e}^*_\tau) < \frac{\mathsf{hd}_{\min}}{3}$. In particular, this holds for the vector $\boldsymbol{\gamma}$ given by $\gamma_\tau = 1$ and $\gamma_{\tau'} = 0$ for $\tau' \neq \tau$, so the $\boldsymbol{v}_\tau$'s are well-defined.

For two matrices A and B of the same dimensions, we define their Hamming distance $\mathsf{hd}_2(A, B)$ to be the number of j's such that the jth column of A and jth column of B are different. This agrees with the standard definition of Hamming distance, if we consider each matrix to be a string of column vectors. The proof of the following claim can be found in the full version of the paper.

Claim. Assume $\neg Err$ and let V and R be defined as above. Then for any $\boldsymbol{q} \in \mathbb{F}^t$ there exists an $\boldsymbol{r}_{(q)}$ with $\mathsf{hd}(\tilde{E}_C(\boldsymbol{q}V, \boldsymbol{r}_{(q)}), \boldsymbol{q}E^*) < \frac{\mathsf{hd}_{\min}}{3}$.

In particular, for any $V^*_{(Q)} \neq QV$, and any $R^*_{(Q)}$ we have

$$\mathsf{hd}_2\left(\tilde{E}_C\left(V^*_{(Q)}, R^*_{(Q)}\right), QE^*\right) \geq 2\frac{\mathsf{hd}_{\min}}{3}.$$

This means that if $\neg Err$ occurs and $\mathcal{P}^*$ attempts to open a $V^*_{(Q)} \neq V_{(Q)} = QV$ then

$$\mathsf{hd}_2\left(\tilde{E}_C\left(V^*_{(Q)}, R^*_{(Q)}\right), QE^*\right) \geq 2\frac{\mathsf{hd}_{\min}}{3}.$$

As argued above, if the distance between two strings of length $2n$ is at least $\frac{\mathsf{hd}_{\min}}{3}$, the probability that J will not contain a j such that the two strings differ in position j is negligible. Hence, the probability that $\tilde{E}_C\left(V^*_{(Q)}, R^*_{(Q)}\right)|_J = QE^*|_J$ is negligible. Thus, the probability that $\neg Err$ and V accepts while $\mathcal{V}_{\mathsf{ILC}}$ does not is negligible. This proves (1).

Next, we want to define a *transcript extractor* $\mathcal{T}$ that given rewindable access to $\langle \mathcal{P}^*(s) \longleftrightarrow \mathcal{V}(pp, u)\rangle$ outputs $\widetilde{\mathsf{trans}}_{\mathcal{P}_{\mathsf{ILC}}}$, which we would like to correspond to all messages sent between $\mathcal{P}^*_{\mathsf{ILC}}$ and the channel in $\langle \mathcal{P}^*_{\mathsf{ILC}}(s, pp, u) \xleftrightarrow{\mathsf{ILC}} \mathcal{V}_{\mathsf{ILC}}(pp_{\mathsf{ILC}}, u; \rho_{\mathsf{ILC}})\rangle$. Here ρ_{ILC} is the randomness used by the $\mathcal{V}_{\mathsf{ILC}}$ inside $\mathcal{V}$ in the first execution of $\mathcal{T}$'s oracle $\langle \mathcal{P}^*(s) \longleftrightarrow \mathcal{V}(pp, u)\rangle$. However, we allow $\mathcal{T}$ to fail if $\mathcal{V}$ does not accept in this first transcript and further to fail with negligible probability. Formally, we want $\mathcal{T}$ to run in expected PPT such that for all PPT $\mathcal{A}$:

$$\Pr\left[\begin{array}{c} pp \leftarrow \mathcal{K}(1^\lambda); (pp_{\mathsf{ILC}}, \cdot) = pp; (u,s) \leftarrow \mathcal{A}(pp); \\ \underbrace{\mathsf{trans}_{\mathcal{P}_{\mathsf{ILC}}}}_{} \leftarrow \mathcal{T}^{\langle \mathcal{P}^*(s) \longleftrightarrow \mathcal{V}(pp,u) \rangle}(pp,u); \\ \mathsf{trans}_{\mathcal{P}_{\mathsf{ILC}}} \leftarrow \langle \mathcal{P}^*_{\mathsf{ILC}}(s,pp,u) \xleftrightarrow{\mathsf{ILC}} \mathcal{V}_{\mathsf{ILC}}(pp_{\mathsf{ILC}}, u; \rho_{\mathsf{ILC}}) \rangle : \\ b = 1 \;\wedge\; \mathsf{trans}_{\mathcal{P}_{\mathsf{ILC}}} \neq \underbrace{\mathsf{trans}_{\mathcal{P}_{\mathsf{ILC}}}}_{} \end{array}\right] \approx 0. \qquad (2)$$

Here b is the value output by $\mathcal{V}$ the first time $\mathcal{T}$'s oracle runs $\langle \mathcal{P}^*(s) \longleftrightarrow \mathcal{V}(pp,u) \rangle$, and the randomness ρ_{ILC} used by $\mathcal{V}_{\mathsf{ILC}}$ in the third line is identical to the random value used by the $\mathcal{V}_{\mathsf{ILC}}$ inside $\mathcal{V}$ in the first transcript. On input (pp,u), the transcript extractor $\mathcal{T}$ will first use its oracle to get a transcript of $\langle \mathcal{P}^*(s) \longleftrightarrow \mathcal{V}(pp,u;(\rho_{\mathsf{ILC}},\rho)) \rangle$. If $\mathcal{V}$ rejects, $\mathcal{T}$ will just abort. If $\mathcal{V}$ accepts, $\mathcal{T}$ will rewind the last message of $\mathcal{P}^*$ to get a transcript for a new random challenge J. $\mathcal{T}$ continues this way, until it has an accepting transcript for $2n$ independently chosen sets J. Notice that if there is only one choice of J that results in $\mathcal{V}$ accepting, $\mathcal{P}^*$ will likely have received each allowed challenge around $2n$ times and $\mathcal{T}$ will get the exact same transcript $2n$ times before it is done rewinding. Still, $\mathcal{T}$ runs in expected polynomial time: if a fraction p of all allowed set J results in accept, the expected number of rewindings *given* that the first transcripts accepts is $\frac{2n-1}{p}$. However, the probability that the first run accepts is p, and if it does not accept, $\mathcal{T}$ does not do any rewindings. In total, that gives $\frac{(2n-1)p}{p} = 2n-1$ rewindings in expectation.

We let $J_1, \ldots, J_{2n}$ denote the set of challenges J in the accepting transcripts obtained by $\mathcal{T}$. If $\bigcup_{i=1}^{2n} J_i$ has less than $2n - \frac{\mathsf{hd}_{\min}}{3}$ elements, $\mathcal{T}$ terminates. Otherwise, $\mathcal{T}$ is defined similarly to $\mathcal{P}^*_{\mathsf{ILC}}$: it uses the values of the openings to get at least $2n - \frac{\mathsf{hd}_{\min}}{3}$ columns of each E_i. For each of the row vectors, e_τ, it computes v_τ and r_τ such that $\tilde{\mathsf{E}}_C(v_\tau, r_\tau)$ agrees with e_τ in all entries $(e_\tau)_j$ for which the j'th column have been revealed, if such v exists. Since $\mathcal{T}$ will not correct any errors, finding such v_τ and r_τ corresponds to solving a linear set of equations. Notice that since the minimum distance is more than $2\frac{\mathsf{hd}_{\min}}{3}$ there is at most one such v_τ for each $\tau \in [t]$. If for some τ there is no such v_τ, then $\mathcal{T}$ aborts, otherwise $\mathcal{T}$ use the resulting vectors v_τ as the prover messages to define $\underbrace{\mathsf{trans}_{\mathcal{P}_{\mathsf{ILC}}}}_{}$.

If $|\bigcup_{i=1}^\kappa J_i| < 2n - \frac{\mathsf{hd}_{\min}}{3}$, there are at least $\frac{\mathsf{hd}_{\min}}{6}$ numbers in $[n] \setminus \bigcup_{i=1}^\kappa J_i$ or in $\{n+1, \ldots, 2n\} \setminus \bigcup_{i=1}^\kappa J_i$. In either case, a random allowed J has negligible probability of being contained in $\bigcup_{i=1}^\kappa J_i$. Since $\mathcal{T}$ runs in expected polynomial time, this implies by induction that there is only negligible probability that $|\bigcup_{i=1}^\kappa J_i| < \min(\kappa, 2n - \frac{\mathsf{hd}_{\min}}{3})$ and therefore $|\bigcup_{i=1}^{2n} J_i| < 2n - \frac{\mathsf{hd}_{\min}}{3}$.

Finally, we need to show

Claim. The probability that for some τ there are no v_τ and r_τ such that $\tilde{\mathsf{E}}_C(v_\tau, r_\tau)$ agrees with e_τ on the opened $j \in \bigcup_{i=1}^{2n} J_i$ and $b = 1$ is negligible.

In particular, the probability that $b = 1$ but $\mathcal{T}$ does not extract the transcript of $\mathcal{P}^*_{\mathsf{ILC}}$ is negligible.

Proof. Since we can ignore events that happen with negligible probability, and the expected number of rewindings is polynomial, we can assume that in all the rewindings, $\mathcal{P}^*$ only makes openings to the most common openings. We showed that the probability that $b = 1$ but $\mathcal{P}^*$ sends a $V^*_{(Q)} \neq V$ is negligible and by the same argument the probability that $b = 1$ but $\mathcal{P}^*$ sends $v^*_{(\gamma)} \neq v_{(\gamma)}$ is negligible. Therefore, in the following, we will assume $v^*_{(\gamma)} = v_{(\gamma)}$.

Now suppose that there is some e_τ such that the opened values are inconsistent with being $\tilde{\mathsf{E}}_\mathcal{C}(v_\tau, r_\tau)$ for any r_τ. That is, there is some j such that $j, n + j \in \bigcup_{i=1}^{2n} J_i$ and $(e_\tau)_j - (e_\tau)_{n+j} \neq \mathsf{E}_\mathcal{C}(v)_j$. For uniformly chosen $\gamma_\tau \in \mathbb{F}$, we get that $\gamma_\tau((e_\tau)_j - (e_\tau)_{n+j} - \mathsf{E}_\mathcal{C}(v)_j)$ is uniformly distributed in $\mathbb{F}$. Hence for a random $\gamma \in \mathbb{F}^t$, we have that $\gamma \cdot ((e)_j - (e)_{n+j} - \mathsf{E}_\mathcal{C}(v)_j)$ is uniformly distributed. When $\mathcal{V}$ sends γ, $\mathcal{P}^*$ will respond with $v^*_{(\gamma)} = v_{(\gamma)}$ and some $r^*_{(\gamma)}$. $\mathcal{V}$ will only accept on a challenge J if for all $j \in J$ we have $(e_0 + \gamma e)_j = \tilde{\mathsf{E}}_\mathcal{C}(v_{(\gamma)}, r^*_{(\gamma)})_j$. Since $j, n + j \in \bigcup_{i=1}^{2n} J_i$ we have $(e_0 + \gamma e)_j = \tilde{\mathsf{E}}_\mathcal{C}(v_{(\gamma)}, r^*_{(\gamma)})_j$ and $(e_0 + \gamma e)_{n+j} = \tilde{\mathsf{E}}_\mathcal{C}(v_{(\gamma)}, r^*_{(\gamma)})_{n+j}$ so

$$(e_0)_j - (e_0)_{n+j} + \gamma e_j - \gamma e_{n+j} = \tilde{\mathsf{E}}_\mathcal{C}(v_{(\gamma)}, r^*_{(\gamma)})_j - \tilde{\mathsf{E}}_\mathcal{C}(v_{(\gamma)}, r^*_{(\gamma)})_{n+j}$$
$$= \mathsf{E}_\mathcal{C}(v_{(\gamma)})_j$$
$$= (\mathsf{E}_\mathcal{C}(v_0) + \gamma \mathsf{E}_\mathcal{C}(v))_j$$

that is,

$$\gamma e_j - \gamma e_{n+j} - \gamma \mathsf{E}_\mathcal{C}(v)_j = \mathsf{E}_\mathcal{C}(v_0)_j - (e_0)_j + (e_0)_{n+j}$$

For random γ the left-hand side is uniform and the right-hand side is fixed, hence equality only happens with negligible probability. That proves the claim. $\square$

Since $\mathcal{E}_{\mathsf{ILC}}^{\langle \mathcal{P}^*_{\mathsf{ILC}}(s, pp, u) \overset{\mathsf{ILC}}{\longleftrightarrow} \mathcal{V}_{\mathsf{ILC}}(pp_{\mathsf{ILC}}, u) \rangle}(pp, u)$ is a straight-line extractor, we can simply assume that it gets the transcript as an input, and can be written as $\mathcal{E}_{\mathsf{ILC}}(pp_{\mathsf{ILC}}, u, \mathsf{trans}_{\mathcal{P}_{\mathsf{ILC}}})$. For any PPT $\mathcal{A}$ consider the following experiment.

$$\begin{bmatrix} pp \leftarrow \mathcal{K}(1^\lambda); (pp_{\mathsf{ILC}}, \cdot) = pp; (u, s) \leftarrow \mathcal{A}(pp); \\ \widetilde{\mathsf{trans}_{\mathcal{P}_{\mathsf{ILC}}}} \leftarrow \mathcal{T}^{\langle \mathcal{P}^*(s) \longleftrightarrow \mathcal{V}(pp, u) \rangle}(pp, u); \\ \mathsf{trans}_{\mathcal{P}_{\mathsf{ILC}}} \leftarrow \langle \mathcal{P}^*_{\mathsf{ILC}}(s, pp, u) \overset{\mathsf{ILC}}{\longleftrightarrow} \mathcal{V}_{\mathsf{ILC}}(pp_{\mathsf{ILC}}, u; \rho_{\mathsf{ILC}}) \rangle = b_{\mathsf{ILC}}; \\ w \leftarrow \mathcal{E}_{\mathsf{ILC}}(pp_{\mathsf{ILC}}, u, \mathsf{trans}_{\mathcal{P}_{\mathsf{ILC}}}); \\ \widetilde{w} \leftarrow \mathcal{E}_{\mathsf{ILC}}(pp_{\mathsf{ILC}}, u, \widetilde{\mathsf{trans}_{\mathcal{P}_{\mathsf{ILC}}}}); \end{bmatrix} \quad (3)$$

We have shown that when doing this experiment, the probability that $b = 1 \wedge b_{\mathsf{ILC}} = 0$ and the probability that $b = 1 \wedge \mathsf{trans}_{\mathcal{P}_{\mathsf{ILC}}} \neq \widetilde{\mathsf{trans}_{\mathcal{P}_{\mathsf{ILC}}}}$ are both negligible. By knowledge soundness of $(\mathcal{K}_{\mathsf{ILC}}, \mathcal{P}_{\mathsf{ILC}}, \mathcal{V}_{\mathsf{ILC}})$, the probability that $b_{\mathsf{ILC}} = 1 \wedge (pp, u, w) \notin \mathcal{R}$ is also negligible. Finally, if $\mathsf{trans}_{\mathcal{P}_{\mathsf{ILC}}} = \widetilde{\mathsf{trans}_{\mathcal{P}_{\mathsf{ILC}}}}$ then clearly $w = \widetilde{w}$. Taken together this implies that the probability of $b = 1 \wedge (pp, u, \widetilde{w}) \notin R$ is negligible. We now define $\mathcal{E}^{\langle \mathcal{P}^*(s) \longleftrightarrow \mathcal{V}(pp, u) \rangle}(pp, u)$ to compute $\mathcal{E}_{\mathsf{ILC}}(pp_{\mathsf{ILC}}, u, \mathcal{T}^{\langle \mathcal{P}^*(s) \longleftrightarrow \mathcal{V}(pp, u) \rangle}(pp, u))$. The above experiment shows that $(\mathcal{K}, \mathcal{P}, \mathcal{V})$ is knowledge sound with $\mathcal{E}$ as extractor. $\square$

Theorem 5 (SHVZK). *If $(\mathcal{K}_{\mathsf{ILC}}, \mathcal{P}_{\mathsf{ILC}}, \mathcal{V}_{\mathsf{ILC}})$ is perfect SHVZK and* (Setup, Commit) *is computationally (statistically) hiding then $(\mathcal{K}, \mathcal{P}, \mathcal{V})$ is computationally (statistically) SHVZK.*

Proof To prove we have SHVZK we describe how the simulator $\mathcal{S}(pp, u, \rho)$ should simulate the view of $\mathcal{V}$. Along the way, we will argue why, the variables output by $\mathcal{S}$ have the correct joint distribution. To keep the proof readable, instead of saying that "the joint distribution of [random variable] and all previously defined random variables is identical to the distribution in the real view of $\mathcal{V}$ in $\langle \mathcal{P}(pp, u, w) \longleftrightarrow \mathcal{V}(pp, u) \rangle$" we will simply say that "[random variable] has the correct distribution".

Using the randomness ρ the simulator learns the queries $\rho_{\mathsf{ILC}} = (x_1, \ldots, x_{\mu-1}, Q)$ the internal $\mathcal{V}_{\mathsf{ILC}}$ run by the honest $\mathcal{V}$ will send. $\mathcal{S}$ can therefore run $\mathcal{S}_{\mathsf{ILC}}(pp_{\mathsf{ILC}}, u, \rho_{\mathsf{ILC}})$ to simulate the view of the internal $\mathcal{V}_{\mathsf{ILC}}$. This gives it $(t_1, \ldots, t_\mu, V_{(Q)})$. By the SHVZK property of $(\mathcal{K}_{\mathsf{ILC}}, \mathcal{P}_{\mathsf{ILC}}, \mathcal{V}_{\mathsf{ILC}})$ these random variables will all have the correct joint distribution.

Then $\mathcal{S}$ reads the rest of ρ to learn also the challenges γ and J that $\mathcal{V}$ will send. The simulator picks uniformly at random $\boldsymbol{v}_{(\gamma)} \leftarrow \mathbb{F}^k$. Since in a real proof $\boldsymbol{v}_0$ is chosen at random, we see that the simulated $\boldsymbol{v}_{(\gamma)}$ has the correct distribution. Now $\mathcal{S}$ picks $E_{01}|_J, \ldots, E_\mu|_J$ uniformly at random. Recall that we defined $\tilde{\mathsf{E}}_{\mathcal{C}}(\boldsymbol{v}; \boldsymbol{r}) = (\mathsf{E}_{\mathcal{C}}(\boldsymbol{v}) + \boldsymbol{r}, \boldsymbol{r})$ and by definition of J being allowed, we have for all $j \in J$ that $j + n \notin J$. This means for any choice of $\boldsymbol{v}_0 \in \mathbb{F}^k$ and $V \in \mathbb{F}^{t \times k}$ that when we choose random $\boldsymbol{r}_0 \leftarrow \mathbb{F}^n$ and $R \leftarrow \mathbb{F}^{t \times n}$ we get uniformly random $\tilde{\mathsf{E}}_{\mathcal{C}}(\boldsymbol{v}_0; \boldsymbol{r}_0)|_J$ and $\tilde{\mathsf{E}}_{\mathcal{C}}(V; R)$. Consequently, $E_{01}|_J, \ldots, E_\mu|_J$ have the correct distribution.

Next, the simulator picks $\boldsymbol{r}_{(\gamma)} \in \mathbb{F}^n$ and $R_{(Q)} \in \mathbb{F}^{t \times n}$ one entry and column at a time. For all j such that $j \notin J$ and $j + n \notin J$ the simulator picks random $(\boldsymbol{r}_{(\gamma)})_j \leftarrow \mathbb{F}$ and random $R_j \leftarrow \mathbb{F}^t$. For all j such that $j \in J$ or $j + n \in J$, the simulator then computes the unique $(\boldsymbol{r}_{(\gamma)})_j \in \mathbb{F}$ and $R_j \in \mathbb{F}^t$ such that we get $\tilde{\mathsf{E}}_{\mathcal{C}}(\boldsymbol{v}_{(\gamma)}; \boldsymbol{r}_{(\gamma)}) = \boldsymbol{e}_0|_J + \gamma E|_J$ and $\tilde{\mathsf{E}}_{\mathcal{C}}(V_{(Q)}; R_{(Q)}) = QE|_J$.

Finally, $\mathcal{S}$ defines $E_{01}|_{\bar{J}}, \ldots, E_\mu|_{\bar{J}}$ to be 0 matrices. It then picks $\boldsymbol{s}_1, \ldots, \boldsymbol{s}_\mu$ at random and makes the commitments $\boldsymbol{c}_1, \ldots, \boldsymbol{c}_\mu$ as in the protocol. For $j \in J$ we see that all the $\boldsymbol{c}_i|_j$ commitments are computed as in the real execution from values that have the same distribution as in a real proof. Hence, they will have the correct distribution. The $\boldsymbol{c}_i|_j$s for $j \notin J$ are commitments to different values than in a real proof. However, by the computational (statistical) hiding property of the commitment scheme, they have a distribution that is computationally (statistically) indistinguishable from the correct distribution. $\square$

4.3   Efficiency

We will now estimate the efficiency of a compiled proof of knowledge $(\mathcal{K}, \mathcal{P}, \mathcal{V})$ for $(pp, u, w) \in \mathcal{R}$. Let μ be the number of rounds, $t = \sum_{i=1}^\mu t_i$, k, n given in $\mathsf{E}_{\mathcal{C}}$, and qc the query complexity, i.e., $Q \in \mathbb{F}^{qc \times t}$. Let $T_{\mathcal{P}_{\mathsf{ILC}}}$ be the running time of $\mathcal{P}_{\mathsf{ILC}}(pp_{\mathsf{ILC}}, u, w)$, $T_{\tilde{\mathsf{E}}_{\mathcal{C}}}(k)$ be the encoding time for a vector in $\mathbb{F}^k$, $T_{\mathsf{Commit}}(t_i)$ be the time to commit to t_i field elements, $T_{\mathrm{Mmul}}(qc, t, b)$ be the time it takes to multiply matrices in $\mathbb{F}^{qc \times t}$ and $\mathbb{F}^{t \times b}$, and $T_{\mathcal{V}_{\mathsf{ILC}}}$ is the running time of $\mathcal{V}_{\mathsf{ILC}}(pp_{\mathsf{ILC}}, u)$. Let furthermore C_{ILC} be the communication from the verifier to the prover in $\langle \mathcal{P}_{\mathsf{ILC}} \xleftrightarrow{\mathsf{ILC}} \mathcal{V}_{\mathsf{ILC}} \rangle$, $C_{\mathsf{Commit}}(t_i)$ be the combined size of commitment and randomness for a message consisting of t_i field elements. We give the dominant factors of efficiency of the compiled proof in Fig. 7. The estimates presume $T_{\mathsf{Commit}}(t_1 + 1)$ is not too far from $T_{\mathsf{Commit}}(t_1)$.

Measure	Cost
Prover Computation	$T_{\mathcal{P}_{\mathsf{ILC}}} + t \cdot T_{\tilde{\mathsf{E}}_C}(k) + 2n \cdot \sum_{i=1}^{\mu} T_{\mathsf{Commit}}(t_i) + T_{\mathrm{Mmul}}(\mathrm{qc}+1, t, k+n)$
Verifier Computation	$T_{\mathcal{V}_{\mathsf{ILC}}} + (\mathrm{qc}+1) \cdot T_{\tilde{\mathsf{E}}_C}(k) + 2\lambda \cdot \sum_{i=1}^{\mu} T_{\mathsf{Commit}}(t_i) + T_{\mathrm{Mmul}}(\mathrm{qc}+1, t, 2\lambda)$
Communication	$C_{\mathsf{ILC}} + 2n \cdot \sum_{i=1}^{\mu} C_{\mathsf{Commit}}(t_i) + (\mathrm{qc}+1) \cdot (k+n) + (\mathrm{qc}+1) \cdot t + 2\lambda \cdot t$
Round Complexity	$\mu + 2$

Fig. 7. Efficiency of a compiled proof of knowledge $(\mathcal{K}, \mathcal{P}, \mathcal{V})$ for $(pp, u, w) \in \mathcal{R}$. Communication is measured in field elements and computation in field operations.

5   Instantiations and Conclusion

Putting together the sequence of proofs and sub-proofs in the ILC model, compiling into the standard model using an error-correcting code and a commitment scheme, and finally instantiating the commitment scheme yields special honest-verifier zero-knowledge proofs for arithmetic circuit satisfiability.

Let us now analyze the efficiency of the compilation we get from Fig. 7. If the error-correcting code is linear-time computable, we get $T_{\tilde{\mathsf{E}}_C}(k) = \mathcal{O}(k)$ operations in $\mathbb{F}$, and with the code from Druk and Ishai [DI14] is will actually be $\mathcal{O}(k)$ additions in $\mathbb{F}$.

Let us now plug in the efficiency of our ILC proof given in Fig. 4 into the efficiency formulas in Fig. 7. We use $k \approx \sqrt{N}$, $n = \mathcal{O}(k)$, $t = \mathcal{O}(\sqrt{N})$, $\mu = \mathcal{O}(\log\log N)$, $\mathrm{qc} = 20 = \mathcal{O}(1)$ and assume $k \gg \lambda$. We then get prover computation $T_{\mathcal{P}} = \mathcal{O}(N)$ multiplications $+ 2n \cdot \sum_{i=1}^{\mu} T_{\mathsf{Commit}}(t_i)$, verifier computation $T_{\mathcal{V}} = \mathcal{O}(N)$ additions $+ 2\lambda \cdot \sum_{i=1}^{\mu} T_{\mathsf{Commit}}(t_i)$, communication $C = 2n \cdot \sum_{i=1}^{\mu} C_{\mathsf{Commit}}(t_i) + \mathcal{O}(\lambda\sqrt{N})$ field elements, and round complexity $\mu = \mathcal{O}(\log\log N)$.

Instantiating with the commitment scheme from Applebaum et al. [AHI+17] we get computational knowledge soundness and statistical SHVZK. The commitments are compact, a commitment has size $C_{\mathsf{Commit}}(t_i) = \mathrm{poly}(\lambda)$ regardless of the message size, giving us sub-linear communication. The commitments can be computed in linear time at a cost of $T_{\mathsf{Commit}}(t_i) = \mathrm{poly}(\lambda) + \mathcal{O}(t_i)$ additions., giving us linear time computation for prover and verifier.

Instantiating with the commitment from Ishai et al. [IKOS08] we get statistical knowledge soundness and computational SHVZK. The commitments have linear size $C_{\mathsf{Commit}}(t_i) = \mathrm{poly}\lambda + t_i$ giving us linear communication overall. The commitments can be computed in linear time at a cost of $T_{\mathsf{Commit}}(t_i) = \mathrm{poly}(\lambda) + \mathcal{O}(t_i)$ additions, again giving us linear time computation for prover and verifier.

We summarize the costs in Fig. 8 below and conclude that we now have SHVZK proof systems for arithmetic circuit satisfiability where the prover computation only has constant overhead compared to direct computation of the

Measure\Instantiation	Using [AHI$^+$17]	Using [IKOS08]
Prover Computation	$\mathcal{O}(N)$ multiplications in $\mathbb{F}$	$\mathcal{O}(N)$ multiplications in $\mathbb{F}$
Verifier Computation	$\mathcal{O}(N)$ additions in $\mathbb{F}$	$\mathcal{O}(N)$ additions in $\mathbb{F}$
Communication	poly$(\lambda)\sqrt{N}$ field elements	$\mathcal{O}(N)$ field elements
Round Complexity	$\mathcal{O}(\log\log N)$	$\mathcal{O}(\log\log N)$
Completeness	Perfect	Perfect
Knowledge Soundness	Computational	Statistical
SHVZK	Statistical	Computational

Fig. 8. Efficiency of two instantiations of our SHVZK proofs.

arithmetic circuit given the witness. Moreover, the verifier computation is a linear number of *additions*, which is proportional to the time it takes simply to read the instance.

References

[AHI+17] Applebaum, B., Haramaty, N., Ishai, Y., Kushilevitz, E., Vaikuntanathan, V.: Low-complexity cryptographic hash functions. Cryptology ePrint Archive, Report 2017/036 (2017). http://eprint.iacr.org/2017/036

[BCC+16] Bootle, J., Cerulli, A., Chaidos, P., Groth, J., Petit, C.: Efficient zero-knowledge arguments for arithmetic circuits in the discrete log setting. In: Fischlin, M., Coron, J.-S. (eds.) EUROCRYPT 2016. LNCS, vol. 9666, pp. 327–357. Springer, Heidelberg (2016). doi:10.1007/978-3-662-49896-5_12

[BCCT12] Bitansky, N., Canetti, R., Chiesa, A., Tromer, E.: From extractable collision resistance to succinct non-interactive arguments of knowledge, and back again. In: Innovations in Theoretical Computer Science Conference-ITCS. ACM (2012)

[BCCT13] Bitansky, N., Canetti, R., Chiesa, A., Tromer, E.: Recursive composition and bootstrapping for SNARKS and proof-carrying data. In: ACM Symposium on Theory of Computing-STOC. ACM (2013)

[BCG+13] Ben-Sasson, E., Chiesa, A., Genkin, D., Tromer, E., Virza, M.: SNARKs for C: verifying program executions succinctly and in zero knowledge. In: Canetti, R., Garay, J.A. (eds.) CRYPTO 2013. LNCS, vol. 8043, pp. 90–108. Springer, Heidelberg (2013). doi:10.1007/978-3-642-40084-1_6

[BCI+13] Bitansky, N., Chiesa, A., Ishai, Y., Ostrovsky, R., Paneth, O.: Erratum: succinct non-interactive arguments via linear interactive proofs. In: Sahai, A. (ed.) TCC 2013. LNCS, vol. 7785, p. E1. Springer, Heidelberg (2013). doi:10.1007/978-3-642-36594-2_41

[BFM88] Blum, M., Feldman, P., Micali, S.: Non-interactive zero-knowledge and its applications. In: ACM Symposium on Theory of Computing-STOC. ACM (1988)

[BJY97] Bellare, M., Jakobsson, M., Yung, M.: Round-optimal zero-knowledge arguments based on any one-way function. In: Fumy, W. (ed.) EURO-CRYPT 1997. LNCS, vol. 1233, pp. 280–305. Springer, Heidelberg (1997). doi:10.1007/3-540-69053-0_20

[BR93]   Bellare, M., Rogaway, P.: Random oracles are practical: a paradigm for designing efficient protocols. In: ACM Conference on Computer and Communications Security (ACM CCS). ACM (1993)

[BSCG+16] Ben-Sasson, E., Chiesa, A., Gabizon, A., Riabzev, M., Spooner, N.: Short interactive oracle proofs with constant query complexity, via composition and sumcheck. In: Electronic Colloquium on Computational Complexity (ECCC) (2016)

[BSCGV16] Ben-Sasson, E., Chiesa, A., Gabizon, A., Virza, M.: Quasi-linear size zero knowledge from linear-algebraic PCPs. In: Kushilevitz, E., Malkin, T. (eds.) TCC 2016. LNCS, vol. 9563, pp. 33–64. Springer, Heidelberg (2016). doi:10.1007/978-3-662-49099-0_2

[BSCS16]   Ben-Sasson, E., Chiesa, A., Spooner, N.: Interactive oracle proofs. In: Hirt, M., Smith, A. (eds.) TCC 2016. LNCS, vol. 9986, pp. 31–60. Springer, Heidelberg (2016). doi:10.1007/978-3-662-53644-5_2

[CD98]   Cramer, R., Damgård, I.: Zero-knowledge proofs for finite field arithmetic, or: can zero-knowledge be for free? In: Krawczyk, H. (ed.) CRYPTO 1998. LNCS, vol. 1462, pp. 424–441. Springer, Heidelberg (1998). doi:10.1007/BFb0055745

[CDD+16]   Cascudo, I., Damgård, I., David, B., Döttling, N., Nielsen, J.B.: Rate-1, linear time and additively homomorphic UC commitments. In: Robshaw, M., Katz, J. (eds.) CRYPTO 2016. LNCS, vol. 9816, pp. 179–207. Springer, Heidelberg (2016). doi:10.1007/978-3-662-53015-3_7

[CDP12]   Cramer, R., Damgård, I., Pastro, V.: On the amortized complexity of zero knowledge protocols for multiplicative relations. In: Smith, A. (ed.) ICITS 2012. LNCS, vol. 7412, pp. 62–79. Springer, Heidelberg (2012). doi:10.1007/978-3-642-32284-6_4

[CDS94]   Cramer, R., Damgård, I., Schoenmakers, B.: Proofs of partial knowledge and simplified design of witness hiding protocols. In: Desmedt, Y.G. (ed.) CRYPTO 1994. LNCS, vol. 839, pp. 174–187. Springer, Heidelberg (1994). doi:10.1007/3-540-48658-5_19

[CGM16]   Chase, M., Ganesh, C., Mohassel, P.: Efficient zero-knowledge proof of algebraic and non-algebraic statements with applications to privacy preserving credentials. Cryptology ePrint Archive, Report 2016/583 (2016). http://eprint.iacr.org/2016/583

[Dam00]   Damgård, I.: Efficient concurrent zero-knowledge in the auxiliary string model. In: Preneel, B. (ed.) EUROCRYPT 2000. LNCS, vol. 1807, pp. 418–430. Springer, Heidelberg (2000). doi:10.1007/3-540-45539-6_30

[DI06]   Damgård, I., Ishai, Y.: Scalable secure multiparty computation. In: Dwork, C. (ed.) CRYPTO 2006. LNCS, vol. 4117, pp. 501–520. Springer, Heidelberg (2006). doi:10.1007/11818175_30

[DI14]   Druk, E., Ishai, Y.: Linear-time encodable codes meeting the Gilbert-Varshamov bound and their cryptographic applications. In: Innovations in Theoretical Computer Science Conference-ITCS. ACM (2014)

[DIK10]   Damgård, I., Ishai, Y., Krøigaard, M.: Perfectly secure multiparty computation and the computational overhead of cryptography. In: Gilbert, H. (ed.) EUROCRYPT 2010. LNCS, vol. 6110, pp. 445–465. Springer, Heidelberg (2010). doi:10.1007/978-3-642-13190-5_23

[FNO15]   Frederiksen, T.K., Nielsen, J.B., Orlandi, C.: Privacy-free garbled circuits with applications to efficient zero-knowledge. In: Oswald, E., Fischlin, M. (eds.) EUROCRYPT 2015. LNCS, vol. 9057, pp. 191–219. Springer, Heidelberg (2015). doi:10.1007/978-3-662-46803-6_7

[FS86]   Fiat, A., Shamir, A.: How to prove yourself: practical solutions to identification and signature problems. In: Odlyzko, A.M. (ed.) CRYPTO 1986. LNCS, vol. 263, pp. 186–194. Springer, Heidelberg (1987). doi:10.1007/3-540-47721-7_12

[Gal62] Gallager, R.G.: Low-density parity-check codes. IRE Trans. Inf. Theory 8(1), 21 (1962)

[GGI+14] Gentry, C., Groth, J., Ishai, Y., Peikert, C., Sahai, A., Smith, A.: Using fully homomorphic hybrid encryption to minimize non-interative zero-knowledge proofs. J. Cryptol. (2014)

[GGPR13] Gennaro, R., Gentry, C., Parno, B., Raykova, M.: Quadratic span programs and succinct NIZKs without PCPs. In: Johansson, T., Nguyen, P.Q. (eds.) EUROCRYPT 2013. LNCS, vol. 7881, pp. 626–645. Springer, Heidelberg (2013). doi:10.1007/978-3-642-38348-9_37

[GI01]   Guruswami, V., Indyk, P.: Expander-based constructions of efficiently decodable codes. In: Symposium on Foundations of Computer Science-FOCS. IEEE Computer Society (2001)

[GI02]   Guruswami, V., Indyk, P.: Near-optimal linear-time codes for unique decoding and new list-decodable codes over smaller alphabets. In: ACM Symposium on Theory of Computing-STOC. ACM (2002)

[GI03]   Guruswami, V., Indyk, P.: Linear time encodable and list decodable codes. In: ACM Symposium on Theory of Computing-STOC. ACM (2003)

[GI05]   Guruswami, V., Indyk, P.: Linear-time encodable/decodable codes with near-optimal rate. IEEE Trans. Inf. Theory 51(10), 3393 (2005)

[GKR08] Goldwasser, S., Kalai, Y.T., Rothblum, G.N.: Delegating computation: interactive proofs for muggles. In: ACM Symposium on Theory of Computing-STOC. ACM (2008)

[GMR85] Goldwasser, S., Micali, S., Rackoff, C.: The knowledge complexity of interactive proof-systems (extended abstract). In: ACM Symposium on Theory of Computing-STOC. ACM (1985)

[GQ88]   Guillou, L.C., Quisquater, J.-J.: A practical zero-knowledge protocol fitted to security microprocessor minimizing both transmission and memory. In: Barstow, D., et al. (eds.) EUROCRYPT 1988. LNCS, vol. 330, pp. 123–128. Springer, Heidelberg (1988). doi:10.1007/3-540-45961-8_11

[Gro04] Groth, J.: Honest verifier zero-knowledge arguments applied. BRICS (2004)

[Gro09] Groth, J.: Linear algebra with sub-linear zero-knowledge arguments. In: Halevi, S. (ed.) CRYPTO 2009. LNCS, vol. 5677, pp. 192–208. Springer, Heidelberg (2009). doi:10.1007/978-3-642-03356-8_12

[Gro10] Groth, J.: Short pairing-based non-interactive zero-knowledge arguments. In: Abe, M. (ed.) ASIACRYPT 2010. LNCS, vol. 6477, pp. 321–340. Springer, Heidelberg (2010). doi:10.1007/978-3-642-17373-8_19

[Gro16] Groth, J.: On the size of pairing-based non-interactive arguments. In: Fischlin, M., Coron, J.-S. (eds.) EUROCRYPT 2016. LNCS, vol. 9666, pp. 305–326. Springer, Heidelberg (2016). doi:10.1007/978-3-662-49896-5_11

[GSV98] Goldreich, O., Sahai, A., Vadhan, S.: Honest-verifier statistical zero-knowledge equals general statistical zero-knowledge. In: ACM Symposium on Theory of Computing-STOC. ACM (1998)

[HM96]   Halevi, S., Micali, S.: Practical and provably-secure commitment schemes from collision-free hashing. In: Koblitz, N. (ed.) CRYPTO 1996. LNCS, vol. 1109, pp. 201–215. Springer, Heidelberg (1996). doi:10.1007/3-540-68697-5_16

[HMR15] Hu, Z., Mohassel, P., Rosulek, M.: Efficient zero-knowledge proofs of non-algebraic statements with sublinear amortized cost. In: Gennaro, R., Robshaw, M. (eds.) CRYPTO 2015. LNCS, vol. 9216, pp. 150–169. Springer, Heidelberg (2015). doi:10.1007/978-3-662-48000-7_8

[IKOS08] Ishai, Y., Kushilevitz, E., Ostrovsky, R., Sahai, A.: Cryptography with constant computational overhead. In: ACM Symposium on Theory of Computing-STOC. ACM (2008)

[IKOS09] Ishai, Y., Kushilevitz, E., Ostrovsky, R., Sahai, A.: Zero-knowledge proofs from secure multiparty computation. SIAM J. Comput. **39**(3), 1121 (2009)

[JKO13] Jawurek, M., Kerschbaum, F., Orlandi, C.: Zero-knowledge using garbled circuits: how to prove non-algebraic statements efficiently. In: ACM Conference on Computer and Communications Security (ACM CCS). ACM (2013)

[Kil92] Kilian, J.: A note on efficient zero-knowledge proofs and arguments. In: ACM Symposium on Theory of Computing-STOC. ACM (1992)

[KR08] Kalai, Y.T., Raz, R.: Interactive PCP. In: Aceto, L., Damgård, I., Goldberg, L.A., Halldórsson, M.M., Ingólfsdóttir, A., Walukiewicz, I. (eds.) ICALP 2008. LNCS, vol. 5126, pp. 536–547. Springer, Heidelberg (2008). doi:10.1007/978-3-540-70583-3_44

[MP03] Micciancio, D., Petrank, E.: Simulatable commitments and efficient concurrent zero-knowledge. In: Biham, E. (ed.) EUROCRYPT 2003. LNCS, vol. 2656, pp. 140–159. Springer, Heidelberg (2003). doi:10.1007/3-540-39200-9_9

[MRS17] Mohassel, P., Rosulek, M., Scafuro, A.: Sublinear zero-knowledge arguments for RAM programs. In: Coron, J.-S., Nielsen, J.B. (eds.) EUROCRYPT 2017. LNCS, vol. 10210, pp. 501–531. Springer, Cham (2017). doi:10.1007/978-3-319-56620-7_18

[PHGR13] Parno, B., Howell, J., Gentry, C., Raykova, M.: Pinocchio: nearly practical verifiable computation. In: IEEE Symposium on Security and Privacy. IEEE Computer Society (2013)

[Sch91] Schnorr, C.-P.: Efficient signature generation by smart cards. J. Cryptol. **4**(3), 161 (1991)

[Spi95] Spielman, D.A.: Linear-time encodable and decodable error-correcting codes. In: ACM Symposium on Theory of Computing-STOC. ACM (1995)

Symmetric Key Designs

How to Use Metaheuristics for Design of Symmetric-Key Primitives

Ivica Nikolić[(✉)]

National University of Singapore, Singapore, Singapore
cube444@gmail.com

Abstract. The ultimate goal of designing a symmetric-key cryptographic primitive often can be formulated as an optimization problem. So far, these problems mainly have been solved with trivial algorithms such as brute force or random search. We show that a more advanced and equally versatile class of search algorithms, called metaheuristics, can help to tackle optimization problems related to design of symmetric-key primitives. We use two nature-inspired metaheuristics, simulated annealing and genetic algorithm, to optimize in terms of security the components of two recent cryptographic designs, SKINNY and AES-round based constructions. The positive outputs of the optimization suggest that metaheuristics are non-trivial tools, well suited for automatic design of primitives.

Keywords: Metaheuristic · Simulated annealing · Genetic algorithm · Automatic tool · Cryptographic primitive

1 Introduction

In the past several years we have seen a major development of computer tools for automatic analysis of symmetric-key primitives. The tools cover a wide range of analysis techniques: differential [5–7,16,18,19,21,30–33,38–40], linear [17,30], impossible differential [12,15,26,29,36,43], meet-in-the-middle [8,14,15], etc. Among other applications, the tools can serve as a proof of security of new designs because they can provide resistance of new designs against most (sometimes all) of the known cryptographic attacks.

Advanced computer tools for design of symmetric-key primitives, however, have not been considered. Instead, most of the design problems have been solved either analytically or with trivial computer algorithms such as brute force and random search. Consider, for example, the problem of tweaking AES to make it more resistant against meet-in-the-middle attacks. With automatic tools for analysis against meet-in-the-middle attacks we can check the security margin of each tweaked version of AES. If we tweak only ShiftRows, then we can brute force the space of all tweaks, check each tweak with the above automatic tools, and find the one that provides the highest security. On the other hand, if we decide to tweak both ShiftRows and MixColumns, then the space of tweaks

© International Association for Cryptologic Research 2017
T. Takagi and T. Peyrin (Eds.): ASIACRYPT 2017, Part III, LNCS 10626, pp. 369–391, 2017.
https://doi.org/10.1007/978-3-319-70700-6_13

may be too large for a brute force, and thus we will use a random search. That is, we will check only a subset of randomly chosen tweaks and find the best among them. These two simple algorithms have been basically the only available computer tools to designers.

Assume the goal is to create a tool for automatic design of symmetric-key primitives that is: (1) based on more advanced search methods, and is (2) versatile, can tackle a variety of design problems. Note, brute force and random search do not satisfy the first point because they are trivial, but do satisfy the second point because they can be applied to many design problems. In a nutshell, these two algorithms are the simplest optimization methods. Therefore, to build a better design tool we need to focus on the next class of known optimization algorithms, that is also universal, but more sophisticated. That is the class of metaheuristics.

A metaheuristic is a search algorithm used to find a sufficiently good solution to an optimization problem. It makes almost no assumptions about the optimized (objective) function and it performs equally well when the function is not explicitly defined, but it can be queried. The search strategy implemented in a metaheuristic is often based on some nature-inspired method or technique – metaheuristics are named according to their nature equivalent, for instance, particle swarm optimization, simulated annealing, ant colony optimisation, evolutionary computation, etc. In cryptography, metaheuristics have been used mainly to design Sboxes that satisfy special criteria, such as resistance against cryptographic attacks [1,11,34,42,44].

Our Contributions. Arguably, the design decision behind any part of a symmetric-key cryptographic primitive is driven by the goal of optimization (in terms of security, size, throughput, etc.). Therefore, we regard the problem of design purely as an optimization problem. The computer algorithms that solve the optimization problem we call *tools for automatic design*.

Our tools are based on metaheuristics. These search algorithms are sufficiently universal to solve most of the design optimization problems. We use two nature-inspired metaheuristics: simulated annealing and genetic algorithm. We introduce the metaheuristics in Sect. 2; for each of them we point the main idea, provide a description in pseudo-code, and give a list of the most important parameters. In Sect. 3, we apply the metaheuristics to solve two concrete design optimization problems. To do so, first we identify the optimization problem, then formally defined it (describe the objective function and its input space), and finally we use metaheuristics to find good solutions. Our two problems are related to finding new components in the recently proposed block cipher SKINNY [3] and the AES-round based constructions [23]. We choose these two primitives because they best demonstrate the effectiveness of metaheuristics. Both SKINNY and the AES-round constructions are designed with clear optimization goals and, considering their excellent performance, achieve these goals. Nonetheless, metaheuristics allow even further optimization. We show that simulated annealing and genetic algorithm can be used to find specific components in the two primitives that results in even higher performance according to criteria which was considered important by the designers. More precisely, we use the metaheuristics to find for SKINNY a permutation in the tweakey schedule that leads to a higher resistance against related-tweakey attacks and for the AES-round constructions to find a round transformation that results in a better security against internal collisions.

To summarize, our main objective and contribution is to provide an empirical proof that, due to their simplicity and versatility, metaheuristics are perhaps the most effective tools for automatic design of symmetric-key primitives.

2 Metaheuristics

Consider a simple optimization problem: find optimum (maximum or minimum) of an objective function $f(x) : D \to R$. If $f(x)$ is given as a blackbox, i.e. it can be queried but is not explicitly defined, then mathematical and standard computer science methods for solving optimization problems cannot be applied because they require full definition of $f(x)$. In addition, if the domain D is discrete and has a large size, then the optimization problem cannot be solved by a brute force in practical time.

To cope with these type of problems, we use metaheuristics. They are approximate algorithms – the solution they provide is not guaranteed to be optimal (although some have a theoretical proof of asymptotic convergence). However, metaheuristics output the solution by using only limited computational resources, i.e. they are practical algorithms. Hence, among other applications, metaheuristics are well suited for search of near optimal solutions to optimization problems where the (blackbox) objective function is expensive.

There are various classifications of metaheuristics. According to the search strategy, they are divided into local search (try to find only the local optimum) and global search (global optimum). For instance, one of the most popular metaheuristic is the hill climbing method which tries to find only the local optimum. Another classification is single vs population-based search. A single metaheuristic works with only one candidate solution at a time, while population-based works simultaneously with multiple candidates. Hill climbing, simulated annealing, iterated local search are examples of single search, while genetic algorithm, ant colony optimization are examples of population-based search.

The efficiency of metaheuristics is tested experimentally by comparing the time complexities (measured in calls to $f(x)$) the metaheuristics require to solve some well-known problems. Depending on the problems, the comparative efficiency of two metaheuristics can vary, i.e. for some problems the first may be better, while for others the second. Therefore, the term "best metaheuristic" is meaningless. Testing the efficiency of a metaheuristic is not trivial because each is associated with a set of parameters. A metaheuristic needs a fine tuning of its parameters for each problem – this can be a very long and tedious process but it can have a major impact on its efficiency. For each metaheuristic, there are recommended set of values for its parameters, however, they were deduced empirically from its previous applications and thus provide no guarantee of optimality.

Further we will use two metaheuristics: simulated annealing and genetic algorithm. The choice is not accidental – both of them have been reported as one of the best performing on wide variety of problems in the single-based and population-based categories, respectively. In the sequel we give a minimal description of the two metaheuristics which we believe is sufficient to understand our ideas that follow. An interested reader can find more details about the metaheuristics for instance in [37,41].

2.1   Simulated Annealing

Simulated annealing [9,27] is a single-based, global search metaheuristic. It is a nature-inspired algorithm that mimics a physical process occurring in chemical substances: heating, followed by cooling and crystallizing.

Given an objective function $f(x)$, simulated annealing tries to find its maximum[1] by iteratively improving the potential solution. That is, starting from some random x_0, it builds $x_1, x_2, \ldots$. At iteration i, the value of x_i is produced from the previous value x_{i-1}, with the goal of maximizing further the function $f(x)$, i.e. $f(x_i) \geq f(x_{i-1})$. The main idea of simulated annealing is to allow probabilistic degradation of solutions, i.e. sometimes it accepts x_i even if $f(x_i) < f(x_{i-1})$. However, the probability of acceptance varies: in the early stages (when i is smaller) it accepts more degrading solutions, while later less. Such a strategy allows at the beginning to explore more variety of solutions, including degrading, while later to focus only on local optimization. Note, the degrading solutions allow the algorithm to escape local optima.

A formal description of simulated annealing is given in Algorithm 1. It takes as inputs three parameters: initial temperature T, cooling schedule function $\alpha(T)$, and neighbour function $\epsilon(x)$. In the initialization, it assigns a random value as a best solution x to the maximization problem of $f(x)$. Then it keeps trying to build better solution by iterating the same procedure: from x generate a new candidate x', and if it complies to a certain criteria, accept it as a new solution x. The function $\epsilon(x)$ generates x' from x by slightly changing[2] the value of x. If x' is a better solution than x, then x is updated to x'. However, if x' is worse, then it is not immediately rejected. Rather, it can be accepted, but only with some probability. The probability of acceptance (expressed with $r < e^{\frac{f(x')-f(x)}{T}}$, where $f(x') - f(x)$ is negative) is higher when the temperature T is higher and when the value of the objective function on the new candidate x' is closer to the value of the old candidate x. The iterations are stopped once the termination criteria is met. The criteria can be set differently: through the number of iterations, the value of the temperature, etc.

---

[1] Finding the minimum can be achieved similarly, with minor changes.

[2] For instance, when x is a vector, then $\epsilon(x)$ returns another vector in some predefined ϵ environment of x.

Algorithm 1. Simulated Annealing

Input: temperature T_0, cooling schedule $\alpha(T)$, neighbour function $\epsilon(x)$

$x \leftarrow \$$ ▷ Generate random initial value
$T \leftarrow T_0$
do
    $x' \leftarrow \epsilon(x)$ ▷ Generate random neighbour
    if $f(x') > f(x)$ **then**   ▷ If new maximum then accept it
        $x \leftarrow x'$
    else
        $r \leftarrow U[0,1]$ ▷ Generate uniformly random number
        if $r < e^{\frac{f(x')-f(x)}{T}}$ **then**
            $x \leftarrow x'$ ▷ Accept degrading solution
        end if
    end if
    $T \leftarrow \alpha(T)$ ▷ Reduce the temperature
while (termination criteria not met)

Output: x

Parameters. As mentioned earlier, the main objective when choosing the values of the parameters is to optimize the efficiency of the metaheuristic so that it can produce a solution close to the global maximum in the shortest possible time. Simulated annealing requires the following parameters:

- **Neighbour function:** $\epsilon(x)$ should return x' that is in the neighbourhood of x, i.e. $\|x - \epsilon(x)\|$ should be small. For instance, if x is a vector then we can define $\epsilon(x)$ as a vector that coincides with x on all coordinates except one. Note, if $\|x - \epsilon(x)\|$ is large (or unlimited), then simulated annealing turns into a plain random search. Refer to Appendix B for more discussion on neighbour functions.
- **Cooling schedule:** $\alpha(T)$ should be monotonic (strictly decreasing) function. There are several choices for $\alpha(T)$: linear, exponential, inverse, logarithmic and other cooling schedules. We will use inverse cooling, defined as $\alpha(T) = \frac{T}{1+\beta T}$, where β is small constant, usually of order 0.001. We choose inverse cooling because it outperformed other cooling schedules in our preliminary experiments.
- **Initial temperature:** if T_0 is high then simulated annealing will explore more possibilities, however, it will require more time to converge to a near-optimal solution. Conversely, lower initial temperature leads to faster finding some solution that may not be so optimal. The value of T_0 should be chosen depending on the values of $\epsilon(x)$ and $\alpha(T)$ as well as the allowed time complexity, in order to balance the possibility of exploring more solutions with the maximal allowed time.

2.2   Genetic Algorithm

Genetic algorithm [22] is a population-based, global search metaheuristic. It belongs to the larger family of evolutionary algorithms which simulate natural selection to solve optimization problems.

Algorithm 2. Genetic Algorithm

Input: population size N, selection function $Selection(\{F_i\})$, crossover function $Crossover(P_A, P_B)$, mutation probability $MutationProbability$ and function $Mutate(\{C_i\})$

for $i=1$ to N **do**
    $P_i \leftarrow \$$ ▷ Generate random parents
end for
do
    for $i=1$ to N **do**
        $F_i \leftarrow f(Parent_i)$ ▷ Compute fitness of parents
    end for
    for $i=1$ to $\frac{N}{2}$ **do**
        $(P_A, P_B) = Select(\{F\})$ ▷ Select 2 parents
        $(C_{2i}, C_{2i+1}) \leftarrow Crossover(P_A, P_B)$ ▷ Produce 2 children
    end for
    for $i=1$ to N **do**
        $r \leftarrow U[0,1]$ ▷ Generate uniformly random number
        if $r < MutationProbability$ **then**
            $C_i \leftarrow Mutate(C_i)$ ▷ Mutate child
        end if
    end for
    $\{P_i\} \leftarrow \{C_i\}$ ▷ Update the generation
while (termination criteria not met)

Output: the best parent among $\{P_i\}$

To find maximum of an objective function (called a fitness function), genetic algorithm works in iterations (called generations). At each iteration it tries to improve a set of solutions, rather than a single solution. This set is called a population of individuals. To produce a new population from an old population, i.e. to change the generation, genetic algorithm uses two operations: mutation and crossover. A mutation is applied to one individual and it consists of slightly changing it. On the other hand, crossover is a synonym for reproduction. It takes two individuals (called parents) and produces two new individuals (called children)[3]. The choice of parents is controlled by so-called selection function which decides how to choose the parents. The selection function is biased towards

---

[3] Variations of crossover operators exist where two parents can produce only a single child or more than two children.

individuals with better fitness (higher value of fitness function). This is done to mimic the natural selection of parents – those with better qualities (genes) have higher chance of reproduction. A formal description of genetic algorithm is given in Algorithm 2.

Parameters. Genetic algorithm uses a wide range of parameters:

- **Population size** N: the number of individuals. The recommended value of N is in the range $[\log |D|, 2 \log |D|]$, where $|D|$ is the size of the search space.
- **Selection function**: the most popular types of selections are roulette-wheel (individuals' probabilities of being selected as parents are proportional to their fitness functions), tournament (several individuals are first randomly selected and then in tournament-like fashion the winner is chosen according to his/her fitness value), rank (the individuals are sorted according to their fitness value, and their positions – called ranks– are used to determine their probability of selection), and stochastic (several individuals are simultaneously selected as parents according to their probability distributions). More detailed descriptions of the selection functions are given in Appendix B.
- **Crossover function**: it produces children that share similarities with the parents. For instance, if the two parents are given as vectors (the coordinates of these vectors are called genes), then the corresponding coordinates of the vectors of their children will have values either of the first or of the second parent[4]. Crossover function decides how the children inherit parents' genes. We will use a uniform crossover function, i.e. each gene of the children (each coordinate of the vector) has an equal probability to come from any of the two parents, and this probability is independent of the previous genes.
- **Mutation probability and function**: within one generation, the mutation is applied only to a small number of individuals defined by mutation probability. A recommended value for this probability is in the range $[0.001, 0.01]$, i.e. only around 0.1–1% of the individuals are mutated. The mutation function defines how an individual is changed – it alters slightly the genes of an individual.
- **Elitism**: usually the best individuals within each generation are kept, that is, at the end of a generation, a certain percentage of the best parents progress to the next generation (are copied to children). This is called elitism (from elite). A recommended elitism is in the range $[0.05–0.2]$, i.e. 5–20% of the parents with best fitness progress to the next generation.

3   Applications

Usually the objective of a new cryptographic primitive is to provide at least one better functionality than all known designs. This functionally can vary and may include better throughput, smaller footprint, higher security, etc. Regardless

---

[4] In some cases this is not possible, so some of the genes will be random.

of the chosen functionality, the goal of designers essentially can be seen as an optimization problem.

The optimization of cryptographic designs may or may not be solved with the use of metaheuristics. If the optimization problem is too general or the objective function is not clearly stated, then metaheuristics cannot solve the problem. For instance, trying to tweak somehow the round function of AES to maximize its resistance against impossible differential attacks does not formulate a good objective function. On the other hand, trying to tweak the MixColumns matrix by changing its coefficients, provides a clear objective function: the input to the function is some MixColumns matrix, and the output is the security level against impossible differential attacks[5]. Some optimization problems can be solved better (faster or with higher precision) with methods other than metaheuristics, such as heuristics or even brute force. For instance, trying to tweak the ShiftRows constants in AES to maximize its resistance against impossible differential attack can be solved simply by brute force as the number of all possible variants is small.

From the above discussion it follows that we can *use metaheuristics to design or improve symmetric-key primitives when:*

1. The optimization goal can be quantified (the objective function is clearly stated and can be computed on arbitrary inputs),
2. The search space is relatively large and cannot be covered by a brute force,
3. The solution not necessarily has to be globally/locally optimal (recall, metaheuristics may or may not return optimal solution in feasible time).

Further we give two examples of good optimization goals, that can be tackled with metaheuristics. They are related to improving the security margin[6] of SKINNY [3] and of the AES-round based constructions from [23]. These two primitives are ideal candidates for testing the effectiveness of metaheuristics because they are recent designs, have strong emphasis on optimization of components, and have clear optimization goals. Note, we have considered as well the use of metaheuristics to a few other recent designs, however, for various reasons we omit the details of their applications. For instance, the potential optimization of the functions Simpira v2 [20] and Haraka [28] can be solved with a brute force, therefore the optimization does not satisfy the above second requirement, and hence metaheuristics are not the first choice. On the other hand, optimizing component in the authenticated encryption scheme Deoxys [25] can be done with metaheuristics, however, the problem is too similar to the further analyzed problem of SKINNY, and thus we omit it.

---

[5] Assuming that one can compute the security level against impossible differential attacks with tweaked MixColumns matrix.

[6] However, we remind the reader that this is not necessarily the only use of metaheuristics – they can be applied to optimize designs with respect to throughput, size, etc.

3.1 SKINNY

SKINNY [3] is a family of block ciphers proposed at CRYPTO'16. Its goal is to be on par with NSA cipher SIMON [2] in hardware and software, while providing higher security. The ciphers are tweakable, i.e. besides a key and a plaintext, they have a third input called a tweak. The tweaking is based on a framework [24] that treats the key and the tweak universally, as a single input called tweakey. SKINNY ciphers have state sizes $n = 64$ or $n = 128$ bits, regarded as 4×4 matrices of nibbles. On the other hand, the tweakey sizes t are multiples of the state size n, and have three versions: $t = n, t = 2n, t = 3n$.

SKINNY are iterative substitution-permutation ciphers. In Fig. 1 we give one round of the ciphers when $t = 3n$. A state round consists of five familiar transformations: SubCells is an Sbox layer, AddConstants xors constants, AddRoundTweakey xors the two top rows of each tweakey word to the two top rows of the state, ShiftRows rotates the nibbles of the state rows, and MixColumns multiplies the state columns by some matrix. In the tweakey schedule, the three tweakey words TK_1, TK_2, and TK_3 undergo two transformations: state-wise nibble permutation P_T which is the same for all the tweakeys, and nibble-wise linear transformations l_i.

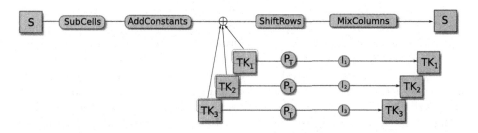

Fig. 1. One round of SKINNY with three tweakeys, $t = 3n$.

To be competitive in hardware and software, SKINNY ciphers have been highly optimized. Most of the transformations used in the ciphers have above average performance according to some design criteria and have been found as a result of some heuristic or a computer search. According to the extended version of the submission document [4], the nibble permutation P_T used in the tweakey schedule "has been chosen to maximize the bounds on the number of active Sboxes ... in the related-tweakey model". The search method used to find P_T is not specified.

With the use of metaheuristics, we will further optimize P_T. Note, the optimization problem has already been well formulated: find P_T to maximize the number of active Sboxes in the best related-tweakey characteristic. To find this number for a particular choice of P_T, as suggested by the designers we use an automatic tool based on integer linear programming (ILP). Therefore, ILP can be seen as the objective function f, which takes as input a permutation P_T and returns the number of active Sboxes. Hence our problem becomes

$$\max_{P_T} f(P_T),$$

where P_T is a permutation of 16 elements with an additional constraint: the first eight elements can be send only to the last eight positions, and vice versa. In fact, besides this constraint, the designers of SKINNY have imposed two additional: (1) P_T must consist of a single cycle, and (2) it sends the first 8 elements to the last 8 positions. In our search, we will relax these two constraints. This increases the search space from slightly under 8! possible choices of P_T to $(8!)^2$. Hence we will operate in a space that cannot be covered by a brute force and that has candidate permutations that may lead to ciphers with higher security margins. However, as we relax constraint (2), our permutations may require higher implementation cost in certain environments. Hence, our search for P_T should be seen in general as tradeoff between possibly higher security and lower speed.

Before we apply the metaheuristics, let us clarify a few points. First, SKINNY has several versions and we will focus on SKINNY-64-192 which is the 64-bit version with three tweakeys ($n = 64, t = 3n = 192$), i.e. on the lightweight version which gives the most freedom to the attacker. Other versions can be processed similarly: moving from 64-bit to 128-bit will require more computational power[7], while reducing the number of tweakey words from three to two or one will require less power[8]. Second, the best characteristics not necessarily have to be found for the full cipher. Rather, once in a round-reduced characteristic the number of active Sboxes reaches some threshold, the cipher is already considered secure. In SKINNY-64-192 this number[9] is 33, and according to [4], for the original choice of P_T it is reached after 18 rounds. We will try to achieve 33 active Sboxes earlier, in 16 rounds[10]. Therefore, our objective function $f(P_T)$ is defined as the number of active Sboxes in the best characteristics on 16 rounds.

Let us clarify the above points. First note that the original permutation P_T^o of SKINNY is defined as

$$P_T^o = \begin{pmatrix} 0 & 1 & 2 & 3 & 4 & 5 & 6 & 7 & 8 & 9 & 10 & 11 & 12 & 13 & 14 & 15 \\ 9 & 15 & 8 & 13 & 10 & 14 & 12 & 11 & 0 & 1 & 2 & 3 & 4 & 5 & 6 & 7 \end{pmatrix} \quad (1)$$

---

[7] The 128-bit version of SKINNY uses 8-bit Sboxes that have the same maximal differential propagation probability of 2^{-2} as the 4-bit Sboxes used in the 64-bit version. Therefore, to achieve 128-bit security (rather than 64-bit security) the number of active Sboxes in the best characteristic has to be much larger, which in return results in higher complexity search.

[8] For results on these versions refer to Appendix A.

[9] The number is defined by the state size and the probability of the best differential transition of the Sbox. The state of SKINNY is 64 bits, and the highest probability of a differential transition in the 4-bit Sbox is 2^{-2}, thus if the number of active Sboxes is $1 + \lfloor \frac{64}{2} \rfloor = 33$, the cipher is resistant against related-tweakey differential attacks.

[10] The number of rounds cannot be predicted a priori. We focus on 16 rounds, but if we do not succeed we can always compare either how many active Sboxes we have reached on 16 rounds, or if we have reached 33 active Sboxes on 17 rounds.

According to the designers, and confirmed with our own ILP tool, $f(P_T^o) = 27$.
We are looking for another permutation P_T

$$P_T = \begin{pmatrix} 0 & 1 & 2 & 3 & 4 & 5 & 6 & 7 & 8 & 9 & 10 & 11 & 12 & 13 & 14 & 15 \\ a_0 & a_1 & a_2 & a_3 & a_4 & a_5 & a_6 & a_7 & a_8 & a_9 & a_{10} & a_{11} & a_{12} & a_{13} & a_{14} & a_{15} \end{pmatrix},$$

such that $f(P_T)$ is as large as possible. Note, there is an *additional* condition,
which requires that $a_i \geq 8$ for $i = 0, 1, \ldots, 7$ and $a_i < 8$ for $i = 8, 9, \ldots, 15$,
which assures that the first 8 elements are sent to the last 8 positions, and vice
versa.

Let us focus on simulated annealing. To solve the optimization problem with
this metaheuristic, we first need to specify the three parameters: $\epsilon(P_T), \alpha(T), T_0$.
As a neighbour function $\epsilon(P_T)$ we use a random transposition. That is, we
randomly choose two indices in P_T and switch the value of the elements with
such indices. Note, however, the choice of indices cannot be completely random
because $\epsilon(P_T)$ must fulfil the additional condition. Hence, to properly implement
$\epsilon(P_T)$, we first choose the half of P_T where the transposition will occur, and only
then the two random elements that belong to the same half. For cooling schedule
$\alpha(T)$, as mentioned before, we use inverse cooling $\alpha(T) = \frac{T}{1+\beta T}$ and experiment
with value of β in the range $[0.001, 0.003]$. Finally, as an initial temperature T_0 we
take values in the range $[1, 2]$. Our termination criteria is time-based, i.e. we stop
the search and output the best found solution after running the metaheuristics
around a day on an 8-core processor.

Further, let us focus on genetic algorithms and the used parameters. In all
of our implementations, the population size is 50. To test the effectiveness and
impact of different selection functions, we used all four of them. In addition, we
use mutation rate of 0.01 and a mutation function that closely resembles $\epsilon(P_T)$
from simulated annealing (i.e. mutation consists of one random transposition).
Finally, we use elitism with 20% rate. The termination criteria is similar to that
of simulated annealing, but we allow more time.

The results of the optimization with the two metaheuristics are as follows.
Both simulated annealing and genetic algorithm were able to find permutations
P_T such that $f(P_T) = 33$. Simulated annealing performed similarly on different
choices of parameters β and T_0, i.e. we did not detect any significant difference.
On average, it required around 1000 calls to the objective function $f(x)$ to find
a permutation P_T such that $f(P_T) = 33$. On the other hand, genetic algorithm
performed better for some choices of selection functions. To find P_T such that
$f(P_T) = 33$ on average over three trials, with stochastic selection it required
950 calls, with rank selection 1380 calls[11], with roulette-wheel 2250 calls, and
with tournament selection 5900 calls. Therefore, we can conclude that simulated
annealing and genetic algorithm with stochastic or rank selection performed
similarly.

---

[11] This number (1380) not necessarily has to be divisible by the population size (50).
The reason is two-fold: (1) we halt the search once a sufficiently good construction is
found, without updating the whole population, and (2) we use elitism, which dictates
that at each generation only $50 \cdot (1 - elitism)$ individuals are updated.

Table 1. Examples of permutations P_T found with simulated annealing (second row) and genetic algorithm (third row) and the resulting security level against related-key tweakey attacks when used in SKINNY-64-192. In the second column are the specifications of the permutations. In the third column is their security, i.e. the number of active Sboxes in the round-reduced characteristics according to the number of rounds. Highlighted numbers correspond to the lowest number of active Sboxes that already match the threshold for related-tweakey security.

Method	P_T	Rounds				
		14	15	16	17	18
Original	9 15 8 13 10 14 12 11 0 1 2 3 4 5 6 7	19	24	27	31	35
Simulated annealing	11 9 14 8 12 10 15 13 2 0 3 6 7 5 1 4	24	28	33	36	39
Genetic algorithm	14 11 8 9 15 13 10 12 1 2 0 7 5 4 3 6	24	28	33	36	39

In Table 1 are shown examples of permutations P_T^{SA}, P_T^{GA} found with simulated annealing and genetic algorithm such that $f(P_T^{SA}) = f(P_T^{GA}) = 33$. For performance measurements, we give in the table as well the number of active Sboxes of the best characteristics reduced to not only 16 rounds, but in the range of 14 to 18 rounds. Evidently, the two new permutations result in higher numbers of active Sboxes in comparison to the original permutation of SKINNY.

We conclude this subsection with a discussion on further use of metaheuristics in SKINNY. One potential direction would be to optimize the resistance against related-tweakey attacks with respect to both P_T and AddRoundTweakey, i.e. by changing the permutation P_T and by identifying which 8 nibbles of the tweakey words should be xored to the state (instead of the 8 nibbles of the first two rows).

3.2 AES-round Based Constructions [23]

Software optimized designs based on the AES round function are presented in [23]. The main objective of the authors of this paper is to provide symmetric-key constructions (as building blocks of MACs and authenticated encryption schemes) that are efficient on the latest Intel processors[12]. The authors show seven constructions that run at only a few tenths of a cycle per byte on Intel processors Ivy Bridge, Haswell and Skylake.

The proposed constructions have a state composed of s words of 128 bits. The state is transformed by a round function given in Fig. 2, where A stands for one AES round. Besides A, the only remaining operation is the xor (of message words M_{i_j} and state words X_{i_j}). Each construction is characterized by a parameter called a rate ρ which is defined as the number of AES rounds required to process a 128-bit message word. That is, ρ is the ratio of the number of calls to A to the number of different message words (in one round). The lower the rate, the

---

[12] These processor have special instructions set called AES-NI, that can execute AES round function as a single instruction.

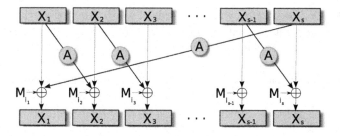

Fig. 2. The round function of the constructions from [23]. The transformations in red and dashed are optional. (Color figure online)

faster the design, hence the goal of the authors has been to reduce the rate as much as possible.

A construction is considered secure if it is free of so-called internal collisions, which are special type of differential characteristics: they start and end in zero differences in the state[13]. A construction should provide 128-bit security, that is, differential characteristics that lead to internal collisions must not have a probability higher than 2^{-128}. To find the best characteristic and its probability, the authors reduce the problem to counting active Sboxes and use the aforementioned integer linear programming tool to get the lower bound on this number. The security level of 128 bits corresponds to at least 22 active Sboxes[14] in the best characteristics.

The seven proposed constructions have different number of state words (7 to 12) and different rates (in the range $[2, 3]$). For a particular choice of state size and rate, the authors use some heuristic (which has not been explained in the paper) to search the space of all constructions as defined in Fig. 2 and consider only those which are resistant against internal collisions, i.e. their best characteristics have at least 22 active Sboxes. Constructions that have the lowest probability of internal collisions (i.e. the highest number of active Sboxes) are considered the best.

Further we use metaheuristics to optimize the constructions according to the design criteria of [23]. The optimization problem is clear: for a particular choice of state size s and rate ρ, *find a round function* as in Fig. 2 that defines a construction whose best differential characteristic that leads to internal collisions *has maximal number of active Sboxes*. Once again, the role of objective function f is played by the integer linear program that returns a lower bound on the number of active Sboxes. To understand what the input to the objective function is, let us focus on Fig. 2. Note, there are three types of red (optional) transformations in the round function. First, each of the s calls to A are optional. Therefore, we can use s-bit vector aes_masks to describe a particular configuration of the calls to A, where i-th bit of aes_masks is set iff in the round function, A is

---

[13] The difference is introduced and later cancelled through the message words.

[14] Because the differential propagation probability of an Sbox in AES is 2^{-6}, thus 128-bit security means $\lfloor \frac{128}{6} \rfloor + 1 = 22$ active Sboxes.

applied to the X_i. Second, all s feedforwards of words (the red vertical lines) can also be described with an s-bit vector $feed\_masks$. Finally, the xors of message words M_{i_j} can be described with vector $messages$ of s coordinates, each an integer value in the range $[0, w]$, where w is the total number of message words in a round. A value of 0 denotes that no message words is xored, while any positive integer value corresponds to the index of the message word being xored. As a result, each potential construction can be described with the three vectors: aes_masks, $feed\_masks$, and $messages$. However, note that not all combinations are possible because the values of aes_masks and $messages$ cannot be arbitrary. Rather, they must agree with the rate ρ. For instance, if $\rho = 2$ and the Hamming weight of aes_masks is 6, then the vector $messages$ can contain the values 1,2 and 3, and it must have each of these values at least once. This assures that the rate of the constructions is indeed 2. Let us assume further that the tuples $(aes\_masks, feed\_masks, messages)$ agree with the predefined rate ρ. Then our optimization problem for fixed state size s and rate ρ can be defined as:

$$\max_{aes\_masks, feed\_masks, messages} f(aes\_masks, feed\_masks, messages)$$

We optimize six of the seven constructions proposed in [23]. We omit one, with rate $\rho = 2$ and size $s = 12$, because it has too expensive objective function – it took us half a day to compute f on an input.

To solve the optimization problems, we run simulated annealing and genetic algorithm with the following parameters. In simulated annealing, the neighbour function $\epsilon(x)$ consists of flipping 1–2 bits in some (or all) of the three vectors aes_masks, $feed\_masks$, $messages$ (with an additional check on the rate ρ). Furthermore, we use inverse cooling $\alpha(T)$ with $\beta = 0.003$, and initial temperature $T_0 = 1.5$. In genetic algorithm, the population size is 30, combined with stochastic selection function, uniform crossover, mutation rate of 0.01, a mutation function based on random flip of bits, and 20% elitism. The termination criteria for both of the metaheuristics is based on number of calls to the objective function, and it is either 500 calls (for smaller search spaces) or 700 calls (for larger).

The outputs of the metaheuristics are given in Table 4. For all six constructions, both of the metaheuristics were able to find better candidates with an increase of 13%–44% to the number of active Sboxes in comparison to the original proposals from [23]. Simulated annealing performed slightly better than genetic algorithm – in limited number of calls to the objective function, it managed to find constructions with higher security margin. We suspect this is due to the termination criteria as genetic algorithm requires more generations to find better solutions (Table 2).

Finally, we note that we have also run metaheuristics to find competing constructions to the published ones [23] not only in terms of higher security, but in terms of efficiency too. We refer the reader to Appendix C for more details.

Table 2. AES-round based constructions. SA and GA stand for simulated annealing and genetic algorithm.

Method	State size	Rate	aes_mask	feed_mask	messages	Active Sboxes
[23]	6	3	111111	100100	011022	22
GA	6	3	111111	100100	122020	26
SA	6	3	111111	011100	102110	27
[23]	7	3	1111110	1101101	1012020	25
GA	7	3	0111111	0100110	0101201	35
SA	7	3	1110111	1111100	0100211	36
[23]	7	2.5	1101110	0100001	1111222	22
GA	7	2.5	1111001	0010110	1201102	25
SA	7	2.5	1011110	0110100	1021102	26
[23]	8	3	11101110	11011101	10102020	34
GA	8	3	11110011	00110000	10102022	42
SA	8	3	00111111	01001000	20221220	45
[23]	8	2.5	11011100	01000011	11112222	23
SA	8	2.5	10011011	10000110	02012021	30
GA	8	2.5	11111000	01100100	11022102	30
[23]	9	3	111111111	100100100	011022033	25
GA	9	3	111111111	111101111	012133031	34
SA	9	3	111111111	100100111	010321121	34

4 Conclusion

Metaheuristics are widely used algorithms for search of solutions to optimization problems. The design of symmetric-key primitives can be seen as one such problem, thus metaheuristics can be used to find better designs. Therefore, metaheuristics can serve as tools for automatic designs of symmetric-key primitives. Unlike brute force and random search, metaheuristics are non-trivial tools which should be scrutinized in absence of better heuristics or of other more advanced search methods.

We used two metaheuristics, simulated annealing and genetic algorithm, to optimize designs with respect to security. Our choice of metaheuristics was guided by their popularity and reported success – both of them are considered among the best performing on well known problems. On the other hand, as an optimization parameter we chose security because that led to well defined and computable objective functions[15]. We wrote the implementations of the two metaheuristics on C – they were straightforward to code. It took us several

---

[15] The objective function is well defined because the security criteria is characterized by a single parameter. On the other hand, it is computable, because there are various tools such as those based on ILP that can produce the output for an arbitrary input.

thousand CPU hours to test for good set of parameters and to find approximate solutions for the design optimization problems in SKINNY and the AES-round constructions. The outputs were positive – the metaheuristics were able to find better components for both of the primitives, sometimes improving the optimized component by more than 40%. Thus we can conclude that metaheuristics can serve as effective tools for automatic design of symmetric-key primitives.

Future research may focus on expanding the area of application and variety of metaheuristics. This includes formulating other design problems as optimization problems and subsequently using the proposed metaheuristics for their solution. We stress out that the optimization problems not necessarily have to be related to an increase in security, but may target better throughput, smaller size, etc. Furthermore, using metaheuristics other than simulated annealing and genetic algorithms may also improve design methods of crypto primitives. Some more advanced metaheuristics, such as the multi-objective genetic algorithm NSGA-II [13], may well excel in solving design problems related to multidimensional optimization, i.e. optimization by several criteria.

Acknowledgments. The author would like to thank the anonymous reviewers of ASIACRYPT'17 for their constructive comments and Yu Sasaki for helping to finalize the paper. This work is supported by the Ministry of Education, Singapore under Grant No. R-252-000-560-112.

A   Applications to **SKINNY-64-64** and **SKINNY-64-128**

In addition to the full search given in Sect. 3.1 on SKINNY-64-192 (i.e. in TK_3), we have also run search for P_T in SKINNY-64-64 and SKINNY-64-128, i.e. in TK_1 and in TK_2. The search criterion for P_T was identical as in Sect. 3.1. With the use of simulated annealing only, we have looked for P_T in three different related-tweakey differential models:

1. **Find P_T for SKINNY-64-64 secure in TK_1.** The search returned several permutations, each resulting in a cipher that has at least 33 active Sboxes in any 11-round related-tweakey differential characteristics.
2. **Find P_T for SKINNY-64-128 secure in TK_2.** Similarly, we found several permutations with at least 34 active Sboxes in any 14-round characteristic.
3. **Find P_T simultaneously for SKINNY-64-64 secure in TK_1 and for for SKINNY-64-128 secure in TK_2.** We found a few permutations P_T that simultaneously provide security in both TK_1 and TK_2. Interestingly, the corresponding characteristics have 33 active Sboxes on 11 rounds in TK_1 and 34 active Sboxes on 14 rounds in TK_2. In other words, any of these permutations can be used as an optimal candidate in scenarios (1) and (2).

Examples of permutations found with the search are given in Table 3.

Table 3. Examples of SKINNY permutations P_T found with simulated annealing that result in TK_1 and TK_2 secure ciphers.

Source	Permutation	Target	Sboxes	Rounds
Original [3]	9 15 8 13 10 14 12 11 0 1 2 3 4 5 6 7	TK_1	32	11
		TK_2	31	14
Model (1)	11 10 15 14 8 9 13 12 6 0 5 1 7 3 2 4	TK_1	33	11
Model (2)	9 8 11 13 14 12 10 15 7 6 4 2 0 5 3 1	TK_2	34	14
Model (3)	12 11 9 8 14 10 13 15 6 2 3 0 7 4 1 5	TK_1	33	11
		TK_2	34	14

B   Specification and Implementation Details of the Metaheuristics

Selection Functions. We work with four types of selection functions:

- **Roulette-wheel selection** is also called furness-proportionate selection. The selection probability of each individual is proportional to its fitness value.
  In Fig. 3, we assume the population is composed of four individuals with fitness measures of 40, 30, 20, 10. Then, the roulette-wheel selection dictates that the first individual has $\frac{40}{100}$ probability of being selected as a parent, the second $\frac{30}{100}$, etc. To select a single parent, we "run the roulette", i.e. uniformly at random choose a number in the range $[0, 100)$, and accordingly choose that individual on which slice the "ball" has landed, e.g. if the number is anywhere in the range $[0, 10)$ then it is the individual 1, if in the range $[10, 30)$ then it is individual 2, etc.
- **Stochastic selection** is similar to roulette-wheel, but it increases the chance of high fitness individuals becoming parents. In stochastic selection the parents are selected in bulk. For example, in Fig. 4 we show how to use the wheel to select four parents at once. We run the roulette once, i.e. select a random number in range $[0, 100]$ and, as in roulette-wheel, choose the corresponding individual as a first parent. Then, the remaining three parents are the ones the correspond to the other three uniformly-spaced numbers. That is, if the ball has landed on 23, then we assume that it has also landed on $23 + \frac{100}{4} = 48$, $23 + 2 \cdot \frac{100}{4} = 73$, and $23 + 3 \cdot \frac{100}{4} = 98$.
- **Rank selection** is as well similar to roulette-wheel. However, instead of using the fitness to determine the portion of the wheel, individuals' rank is used. That is, all individuals within a population are sorted according to their fitness in ascending order, and their position is taken as a fitness measure in roulette-wheel fashion. For instance, individuals with fitnesses of 1,5,20,8, after sorting will be at positions 1, 2, 4, 3, thus have $\frac{1}{10}, \frac{2}{10}, \frac{4}{10}, \frac{3}{10}$ probabilities to be selected as parents.

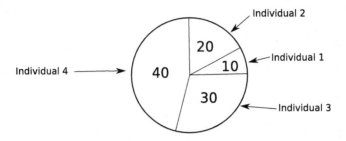

Fig. 3. Roulette-wheel selection with 4 individuals.

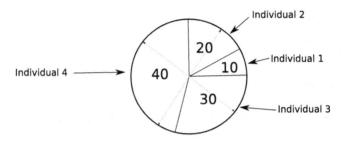

Fig. 4. Stochastic selection with 4 individuals. A spinner (in gray) with for evenly-spaced pointers is spun once to obtain the four parents.

– **Tournament selection** depends on the tournament size and we use the most common size of 2. That is, to select a parent, we uniformly at random choose two individuals, compare their fitness, and choose the one that has higher fitness.

Pseudo code of a full selection procedure (based on roulette-wheel) used by the genetic algorithm for search of P_T in **SKINNY** is given below.

Neighbour Functions $\epsilon(x)$. Intuitively, the task of a neighbour function is to produce a value in the neighbourhood of x. Thus they output values (or vectors) that are very similar to the input.

In SKINNY we define $\epsilon(P_T)$ as a swap of two elements. That is, we randomly choose two positions, and then exchange the elements in these two positions. Since there is an additional requirement on the form of P_T, the swap can occur only between two elements that belong both to the same half. Further we give a simple pseudo-code that accomplishes this.

In the AES-round based construction search, the solution is composed of three vectors, thus $\epsilon(x)$ can be defined as a composition of three separate neighbour functions, one for each of the vectors. A pseudo-code of such function for the vector *aes_masks* is given below.

Algorithm 3. Roulette-wheel selection of parents used in the search for SKINNY

Input: Population of N individuals given as pairs (P_i, F_i), where P_i is an individual (permutation), and F_i is its corresponding fitness

$total\_fitness \leftarrow \sum_1^N F_i$ ▷ Sum up all fitnesses
$left\_range \leftarrow 0$ ▷ Create the roulette-wheel
for i=1 to N **do** ▷ Compute the slices for each ind.
   $range[i]\_left = left\_range$ ▷ Left range
   $range[i]\_right = left\_range + \frac{F_i}{total\_fitness}$ ▷ Right range
   $left\_range = range[i]\_right$
end for

$Parents \leftarrow \emptyset$
for i=1 to $N/2$ **do** ▷ Select $\frac{N}{2}$ pairs of parents
   $C_1 \leftarrow rand()$ ▷ Spin the ball for the first parent
   $C_2 \leftarrow rand()$ ▷ Spin the ball for the second parent
   for j=1 to N **do** ▷ Find the corresponding individuals
      if $range[j]\_left \leq C_1 < range[j]\_right$ **then**
         $parent_1 \leftarrow j$
      end if
      if $range[j]\_left \leq C_2 < range[j]\_right$ **then**
         $parent_2 \leftarrow j$
      end if
   end for
   $Parents = Parents \cup (P_{parent_1}, P_{parent_2})$
end for

Output: $Parents$

Algorithm 4. Neighbour function ϵ for the search in SKINNY

Input: Permutation P_T

$half \leftarrow randInt()\%2$ ▷ Randomly choose a half
$i \leftarrow randInt()\%8$ ▷ Randomly choose the first index
$j \leftarrow randInt()\%8$ ▷ Randomly choose the second index
$P_T' \leftarrow P_T$ ▷ Assign the permutation
$P_T'[8 \cdot half + i] = P_T[8 \cdot half + j]$ ▷ Swap
$P_T'[8 \cdot half + j] = P_T[8 \cdot half + i]$ ▷ Swap

Output: P_T'

C   Efficient AES-Based Constructions

The only goal in the search for AES-based constructions presented in Sect. 3.2 was to improve the security in comparison to the already published constructions in [23], without affecting their efficiency. Further we focus on the latter goal, i.e. our imperative below is to improve the efficiency of the constructions, while still maintaining sufficient security level of at least 22 active Sboxes.

Algorithm 5. Neighbour function for *aes_masks*

Input: Binary array *aes_masks* of s elements

$i \leftarrow randInt() \ \% \ s$ ▷ Randomly choose an index
$aes\_masks' \leftarrow aes\_masks$ ▷ Assign the vector
$aes\_masks'[i] = aes\_masks'[i] \oplus 1$ ▷ Flip the bit

Output: *aes_masks′*

A construction has better efficiency if it has smaller state size, better rate, or both. Constructions that have smaller state size but worse rate, or vice versa, are not considered to be more efficient. To search for efficient constructions, once again we use the two metaheuristics. The formulations of the optimization problems are identical to the formulations given in Sect. 3.2, i.e. we still optimize with respect to the security and with fixed state size and rate. However, once a metaheuristic identifies a construction as optimal, we compare its efficiency to all of the constructions given in [23]. We report in Table 4 the constructions we found to be more efficient than some of the previously known constructions.

Table 4. More efficient AES-round based constructions found with metaheuristics. The numbers in bold denote better efficiency.

Source	State size	Rate	Active Sboxes	*aes_mask*	*feed_mask*	*messages*
[23]	6	3	22	111111	100100	011022
new	**5**	3	30	11001	10100	01101
[23]	7	3	25	1111110	1101101	1012020
new	**6**	3	27	111111	011100	102110
new	**6**	**2.5**	22	101111	110010	102120
[23]	8	3	34	11101110	11011101	10102020
new	8	**2.66**	26	11111111	11000010	10313032
new	**7**	**2.5**	26	1011110	0110100	1021102
[23]	9	3	25	111111111	100100100	011022033
new	9	**2.33**	25	111110011	001010000	201223033
[23]	12	2	28	110110110000	001001001000	111222333123
new	**10**	2	24	1110000100	0100100000	2122102110
new	**8**	2	22	00011011	01000100	12012122
new	**8**	2	26	11000000	01100000	11100010

References

1. Ahmad, M., Bhatia, D., Hassan, Y.: A novel ant colony optimization based scheme for substitution box design. Procedia Comput. Sci. **57**, 572–580 (2015)
2. Beaulieu, R., Shors, D., Smith, J., Treatman-Clark, S., Weeks, B., Wingers, L.: The SIMON and SPECK families of lightweight block ciphers. Cryptology ePrint Archive, Report 2013/404 (2013). http://eprint.iacr.org/2013/404
3. Beierle, C., Jean, J., Kölbl, S., Leander, G., Moradi, A., Peyrin, T., Sasaki, Y., Sasdrich, P., Sim, S.M.: The SKINNY family of block ciphers and its low-latency variant MANTIS. In: Robshaw, M., Katz, J. (eds.) CRYPTO 2016. LNCS, vol. 9815, pp. 123–153. Springer, Heidelberg (2016). doi:10.1007/978-3-662-53008-5_5
4. Beierle, C., Jean, J., Kölbl, S., Leander, G., Moradi, A., Peyrin, T., Sasaki, Y., Sasdrich, P., Sim, S.M.: The SKINNY family of block ciphers and its low-latency variant MANTIS. Cryptology ePrint Archive, Report 2016/660 (2016). http://eprint.iacr.org/2016/660
5. Biryukov, A., Nikolić, I.: Automatic search for related-key differential characteristics in byte-oriented block ciphers: application to AES, Camellia, Khazad and others. In: Gilbert, H. (ed.) EUROCRYPT 2010. LNCS, vol. 6110, pp. 322–344. Springer, Heidelberg (2010). doi:10.1007/978-3-642-13190-5_17
6. Biryukov, A., Nikolić, I.: Search for related-key differential characteristics in DES-Like ciphers. In: Joux, A. (ed.) FSE 2011. LNCS, vol. 6733, pp. 18–34. Springer, Heidelberg (2011). doi:10.1007/978-3-642-21702-9_2
7. Biryukov, A., Velichkov, V.: Automatic search for differential trails in ARX ciphers. In: Benaloh, J. (ed.) CT-RSA 2014. LNCS, vol. 8366, pp. 227–250. Springer, Cham (2014). doi:10.1007/978-3-319-04852-9_12
8. Bouillaguet, C., Derbez, P., Fouque, P.: Automatic search of attacks on round-reduced AES and applications. IACR Cryptol. ePrint Arch. **2012**, 69 (2012)
9. Černý, V.: Thermodynamical approach to the traveling salesman problem: an efficient simulation algorithm. J. Optim. Theory Appl. **45**(1), 41–51 (1985)
10. Cheon, J.H., Takagi, T. (eds.): ASIACRYPT 2016. LNCS, vol. 10031. Springer, Heidelberg (2016). doi:10.1007/978-3-662-53887-6
11. Clark, J.A., Jacob, J.L., Stepney, S.: The design of S-boxes by simulated annealing. In: Congress on Evolutionary Computation, CEC2004, vol. 2, pp. 1533–1537. IEEE (2004)
12. Cui, T., Jia, K., Fu, K., Chen, S., Wang, M.: New automatic search tool for impossible differentials and zero-correlation linear approximations. IACR Cryptol. ePrint Arch. **2016**, 689 (2016)
13. Deb, K., Agrawal, S., Pratap, A., Meyarivan, T.: A fast and elitist multiobjective genetic algorithm: NSGA-II. IEEE Trans. Evol. Comput. **6**(2), 182–197 (2002)
14. Derbez, P., Fouque, P.-A.: Exhausting Demirci-Selçuk meet-in-the-middle attacks against reduced-round AES. In: Moriai, S. (ed.) FSE 2013. LNCS, vol. 8424, pp. 541–560. Springer, Heidelberg (2014). doi:10.1007/978-3-662-43933-3_28
15. Derbez, P., Fouque, P.-A.: Automatic search of meet-in-the-middle and impossible differential attacks. In: Robshaw, M., Katz, J. (eds.) CRYPTO 2016. LNCS, vol. 9815, pp. 157–184. Springer, Heidelberg (2016). doi:10.1007/978-3-662-53008-5_6
16. Dinu, D., Perrin, L., Udovenko, A., Velichkov, V., Großschädl, J., Biryukov, A.: Design strategies for ARX with provable bounds: SPARX and LAX. In: Cheon, J.H., Takagi, T. (eds.) ASIACRYPT 2016. LNCS, vol. 10031, pp. 484–513. Springer, Heidelberg (2016). doi:10.1007/978-3-662-53887-6_18

17. Dobraunig, C., Eichlseder, M., Mendel, F.: Heuristic tool for linear cryptanalysis with applications to CAESAR candidates. In: Iwata, T., Cheon, J.H. (eds.) ASIACRYPT 2015. LNCS, vol. 9453, pp. 490–509. Springer, Heidelberg (2015). doi:10.1007/978-3-662-48800-3_20

18. Emami, S., Ling, S., Nikolić, I., Pieprzyk, J., Wang, H.: The resistance of PRESENT-80 against related-key differential attacks. Crypt. Commun. 6(3), 171–187 (2014)

19. Fouque, P.-A., Jean, J., Peyrin, T.: Structural evaluation of AES and chosen-key distinguisher of 9-round AES-128. In: Canetti, R., Garay, J.A. (eds.) CRYPTO 2013. LNCS, vol. 8042, pp. 183–203. Springer, Heidelberg (2013). doi:10.1007/978-3-642-40041-4_11

20. Gueron, S., Mouha, N.: Simpira v2: a family of efficient permutations using the AES round function. In: Cheon, J.H., Takagi, T. (eds.) ASIACRYPT 2016. LNCS, vol. 10031, pp. 95–125. Springer, Heidelberg (2016). doi:10.1007/978-3-662-53887-6_4

21. Grault, D., Lafourcade, P., Minier, M., Solnon, C.: Revisiting AES related-key differential attacks with constraint programming. Cryptology ePrint Archive, Report 2017/139 (2017). http://eprint.iacr.org/2017/139

22. Holland, J.H.: Adaptation in Natural and Artificial Systems: An Introductory Analysis with Applications to Biology, Control, and Artificial Intelligence. MIT Press, Cambridge (1992)

23. Jean, J., Nikolić, I.: Efficient design strategies based on the AES round function. In: Peyrin, T. (ed.) FSE 2016. LNCS, vol. 9783, pp. 334–353. Springer, Heidelberg (2016). doi:10.1007/978-3-662-52993-5_17

24. Jean, J., Nikolić, I., Peyrin, T.: Tweaks and keys for block ciphers: the TWEAKEY framework. In: Sarkar, P., Iwata, T. (eds.) ASIACRYPT 2014. LNCS, vol. 8874, pp. 274–288. Springer, Heidelberg (2014). doi:10.1007/978-3-662-45608-8_15

25. Jean, J., Nikolić, I., Peyrin, T., Seurin, Y.: Deoxys v1.4. Submitted to CAESAR (2016)

26. Kim, J., Hong, S., Sung, J., Lee, S., Lim, J., Sung, S.: Impossible differential cryptanalysis for block cipher structures. In: Johansson, T., Maitra, S. (eds.) INDOCRYPT 2003. LNCS, vol. 2904, pp. 82–96. Springer, Heidelberg (2003). doi:10.1007/978-3-540-24582-7_6

27. Kirkpatrick, S., Gelatt, C.D., Vecchi, M.P.: Optimization by simulated annealing. Science 220(4598), 671–680 (1983)

28. Kölbl, S., Lauridsen, M.M., Mendel, F., Rechberger, C.: Haraka v2 - efficient short-input hashing for post-quantum applications. IACR Trans. Symmetric Cryptol. 2016(2), 1–29 (2016)

29. Luo, Y., Lai, X., Wu, Z., Gong, G.: A unified method for finding impossible differentials of block cipher structures. Inf. Sci. 263, 211–220 (2014)

30. Matsui, M.: On correlation between the order of S-boxes and the strength of DES. In: De Santis, A. (ed.) EUROCRYPT 1994. LNCS, vol. 950, pp. 366–375. Springer, Heidelberg (1995). doi:10.1007/BFb0053451

31. Moriai, S., Sugita, M., Aoki, K., Kanda, M.: Security of E2 against truncated differential cryptanalysis. In: Heys, H., Adams, C. (eds.) SAC 1999. LNCS, vol. 1758, pp. 106–117. Springer, Heidelberg (2000). doi:10.1007/3-540-46513-8_8

32. Mouha, N., Wang, Q., Gu, D., Preneel, B.: Differential and linear cryptanalysis using mixed-integer linear programming. In: Wu, C.-K., Yung, M., Lin, D. (eds.) Inscrypt 2011. LNCS, vol. 7537, pp. 57–76. Springer, Heidelberg (2012). doi:10.1007/978-3-642-34704-7_5

33. Nikolić, I.: Tweaking AES. In: Biryukov, A., Gong, G., Stinson, D.R. (eds.) SAC 2010. LNCS, vol. 6544, pp. 198–210. Springer, Heidelberg (2011). doi:10.1007/978-3-642-19574-7_14

34. Picek, S., Yang, B., Rozic, V., Mentens, N.: On the construction of hardware-friendly 4x4 and 5x5 S-boxes. Lecture Notes in Computer Science (2016)

35. Robshaw, M., Katz, J. (eds.): CRYPTO 2016. LNCS, vol. 9815. Springer, Heidelberg (2016). doi:10.1007/978-3-662-53008-5

36. Sasaki, Y., Todo, Y.: New impossible differential search tool from design and cryptanalysis aspects. In: Coron, J.-S., Nielsen, J.B. (eds.) EUROCRYPT 2017. LNCS, vol. 10212, pp. 185–215. Springer, Cham (2017). doi:10.1007/978-3-319-56617-7_7

37. Simon, D.: Evolutionary Optimization Algorithms. Wiley, Hoboken (2013)

38. Sun, S., Gerault, D., Lafourcade, P., Yang, Q., Todo, Y., Qiao, K., Hu, L.: Analysis of AES, SKINNY, and others with constraint programming. IACR Trans. Symmetric Cryptol. **2017**(1), 281–306 (2017)

39. Sun, S., Hu, L., Qiao, K., Ma, X., Shan, J., Song, L.: Improvement on the method for automatic differential analysis and its application to two lightweight block ciphers DESL and LBlock-s. In: Tanaka, K., Suga, Y. (eds.) IWSEC 2015. LNCS, vol. 9241, pp. 97–111. Springer, Cham (2015). doi:10.1007/978-3-319-22425-1_7

40. Sun, S., Hu, L., Wang, P., Qiao, K., Ma, X., Song, L.: Automatic security evaluation and (related-key) differential characteristic search: application to SIMON, PRESENT, LBlock, DES(L) and other bit-oriented block ciphers. In: Sarkar, P., Iwata, T. (eds.) ASIACRYPT 2014. LNCS, vol. 8873, pp. 158–178. Springer, Heidelberg (2014). doi:10.1007/978-3-662-45611-8_9

41. Talbi, E.-G.: Metaheuristics: from design to implementation, vol. 74. Wiley, Hoboken (2009)

42. Tesar, P.: A new method for generating high non-linearity S-boxes. Radioengineering (2010)

43. Wu, S., Wang, M.: Automatic search of truncated impossible differentials for word-oriented block ciphers. In: Galbraith, S., Nandi, M. (eds.) INDOCRYPT 2012. LNCS, vol. 7668, pp. 283–302. Springer, Heidelberg (2012). doi:10.1007/978-3-642-34931-7_17

44. Yang, M., Wang, Z., Meng, Q., Han, L.: Evolutionary design of S-box with cryptographic properties. In: 2011 Ninth IEEE International Symposium on Parallel and Distributed Processing with Applications Workshops (ISPAW), pp. 12–15. IEEE (2011)

Cycle Slicer: An Algorithm for Building Permutations on Special Domains

Sarah Miracle[✉] and Scott Yilek

University of St. Thomas, St. Paul, USA
{sarah.miracle,syilek}@stthomas.edu

Abstract. We introduce an algorithm called Cycle Slicer that gives new solutions to two important problems in format-preserving encryption: domain targeting and domain completion. In domain targeting, where we wish to use a cipher on domain $\mathcal{X}$ to construct a cipher on a smaller domain $\mathcal{S} \subseteq \mathcal{X}$, using Cycle Slicer leads to a significantly more efficient solution than Miracle and Yilek's Reverse Cycle Walking (ASIACRYPT 2016) in the common setting where the size of $\mathcal{S}$ is large relative to the size of $\mathcal{X}$. In domain completion, a problem recently studied by Grubbs, Ristenpart, and Yarom (EUROCRYPT 2017) in which we wish to construct a cipher on domain $\mathcal{X}$ while staying consistent with existing mappings in a lazily-sampled table, Cycle Slicer provides an alternative construction with better worst-case running time than the Zig-Zag construction of Grubbs et al. Our analysis of Cycle Slicer uses a refinement of the Markov chain techniques for analyzing matching exchange processes, which were originally developed by Czumaj and Kutylowski (Rand. Struct. & Alg. 2000).

Keywords: Format-preserving encryption · Small-domain block ciphers · Markov chains · Matchings

1 Introduction

Block Cipher designers have traditionally been concerned with constructing ciphers that encrypt bitstrings of a particular, fixed length. The canonical example of such a cipher is AES, which encrypts 128 bit plaintexts into 128 bit ciphertexts. Recently, the practical interest in *format-preserving encryption* (FPE) schemes [2,6,23] for use on data like credit card numbers and US Social Security Numbers (SSNs) has led to the need for ciphers that do not just work on bitstrings of some length, but have more customized domains.

An FPE scheme should have ciphertexts with the exact same format as plaintexts. For example, social security numbers are 9 decimal digit numbers with many additional restrictions, so an FPE scheme for SSNs should result in ciphertexts that are also valid SSNs. Since any 9 digit number can be represented with 30 bits, using a block cipher with block size 30 bits[1] would be an obvious first

---

[1] Even constructing a block cipher with such a small block size and with strong security guarantees is an interesting problem. See [4,12,19,21] for more details.

© International Association for Cryptologic Research 2017
T. Takagi and T. Peyrin (Eds.): ASIACRYPT 2017, Part III, LNCS 10626, pp. 392–416, 2017.
https://doi.org/10.1007/978-3-319-70700-6_14

approach. However, encrypting 30 bits that represent a valid SSN could result in a 30 bit ciphertext that does not correspond to a valid SSN, or even a 9 digit number at all!

FPE is important in practice since practitioners often want to introduce encryption into an existing system while working within constraints like avoiding conflicts with legacy software or existing databases. The first FPE scheme is due to Brightwell and Smith [6], while the term format-preserving encryption was later coined by Spies [23]. Practical interest in this problem has led to a number of recent papers on FPE [2,4,11,15–17] and related papers on constructing new, provably-secure small-domain ciphers [12,18,19,21] that can be used for FPE. The National Institute of Standards and Technology (NIST) has also published a standard for FPE [9] with schemes FF1 [3] and FF3 [5] based on Feistel networks, though recent papers [1,8] have introduced attacks on these standardized schemes.[2]

While many of the recent papers mentioned above focus on the practical issues surrounding FPE, they also raise a number of interesting, more theoretical problems. In this paper, we focus on the theory underlying two important problems in FPE, domain targeting and domain completion, and also describe how our results could be extended to a number of related problems.

DOMAIN TARGETING. Consider again the example above in which we would like an FPE scheme for valid social security numbers. One strategy would be to try to construct a specialized cipher that works on the exact domain we are interested in. The problem with this strategy is that this specialized cipher is specific to one domain and we would likely need to develop a new one if we later needed to encrypt valid credit card numbers, or valid email addresses of some length, or some other uniquely-formatted plaintexts.

If the desired domain has an efficient way to rank and unrank elements, then another solution is to use the *rank-encipher-unrank* approach of [2]. Specifically, to encipher a point in $\mathcal{S}$, this approach first applies a rank algorithm that maps elements in the target domain $\mathcal{S}$ to points in $\{0, \ldots, |\mathcal{S}| - 1\}$, then applies a cipher with domain $\{0, \ldots, |\mathcal{S}| - 1\}$, and finally applies an unrank algorithm to map the result back to $\mathcal{S}$. Efficient ranking and unranking algorithms are known for a number of domains that are important in practice [2,10,15,16], for example, regular languages that can be described by DFAs. Nevertheless, there are domains without efficient rankings [2], and even domains specified by complex regular expressions can prove problematic with this approach. Additionally, the ranking and unranking algorithms cited above can also potentially leak timing information which, looking ahead, is something we would like to avoid with our results. Thus, it is useful to have alternative solutions that only assume the ability to efficiently test membership in $\mathcal{S}$ and do not rely on ranking.

A different strategy that does not rely on efficient ranking is to come up with a general method of transforming a cipher on a larger domain into a cipher on only a subset of the domain points. For example, in the case of social security numbers

---

[2] Bellare, Hoang, and Tessaro [1] give message recovery attacks on both FF1 and FF3, while Durak and Vaudenay [8] detail a more damaging attack on FF3.

such a method could transform a cipher on 30 bit strings into a cipher on 30 bit strings that encode valid SSNs. We call this problem of transforming a cipher on domain $\mathcal{X}$ into a cipher on domain $\mathcal{S} \subseteq \mathcal{X}$ *domain targeting*. This problem is well-known and good solutions exist. A common solution is to use *Cycle Walking*, a folklore technique first formally analyzed by Black and Rogaway [4]. With Cycle Walking, given a permutation P on $\mathcal{X}$, we can map a point $x \in \mathcal{S} \subseteq \mathcal{X}$ by repeatedly applying P until we hit another point in the target set $\mathcal{S}$. In other words, we first check if $P(x) \in \mathcal{S}$, then check $P(P(x)) \in \mathcal{S}$, and so on. Because permutations are made up of cycles, the procedure will eventually either find another point in $\mathcal{S}$ or traverse an entire cycle and return back to x.

Unlike the ranking solutions discussed above, Cycle Walking only requires the ability to test membership in the target domain $\mathcal{S}$. It can also be a good option in practice, since the expected time to encipher a point is small. For example, if the size of the target set $\mathcal{S}$ is at least half the size of the larger set $\mathcal{X}$, then the expected number of applications of the permutation P is at most 2. The worst-case running time of Cycle Walking, however, is much worse, since we might need to walk through a long cycle of points in $\mathcal{X} - \mathcal{S}$ before finding another point in $\mathcal{S}$. The fact that the running time also depends on the specific point being enciphered could also potentially cause headaches for practitioners, due to unpredictable running times or subtle leakage of timing information.

To address these issues, Miracle and Yilek recently introduced an alternative to Cycle Walking called *Reverse Cycle Walking* (RCW) [17]. RCW has lower worst-case running time than traditional Cycle Walking, and the running time does not vary depending on the input. The basic idea underlying RCW is to use permutations on $\mathcal{X}$ to form matchings on $\mathcal{X}$. If, when applying a permutation P on $\mathcal{X}$, two points x and x' from $\mathcal{S}$ appear consecutively in a cycle and the preceding and following points are not in $\mathcal{S}$, then x and x' are matched, and potentially swapped depending on a separate bit flip. Thus, in each round of RCW, many points from $\mathcal{S}$ swap positions. Miracle and Yilek apply a result of Czumaj and Kutylowski [7] to show that repeating this process for enough rounds leads to a random mixing of the points in $\mathcal{S}$.

One disadvantage of RCW is that the number of rounds needed quickly increases as the size of $\mathcal{S}$ gets closer to the size of $\mathcal{X}$. In the same social security number example discussed above, $|\mathcal{S}|/|\mathcal{X}| = 10^9/2^{30} \approx .93$, which is well above $1/2$, so RCW performs poorly, requiring millions of rounds to mix $\mathcal{S}$ sufficiently. The reason for the slowdown is that RCW only swaps points in $\mathcal{S}$ when they appear in a cycle sandwiched between points from outside of $\mathcal{S}$. Specifically, RCW will only potentially swap $x, x' \in \mathcal{S}$ if they are part of a cycle $(\ldots y\ x\ x'\ z \ldots)$ with $y, z \notin \mathcal{S}$. However, if the size of $\mathcal{S}$ is close to the size of $\mathcal{X}$, then we are likely to have cycles that instead have many consecutive points from $\mathcal{S}$ and few points from $\mathcal{X} - \mathcal{S}$, making swaps unlikely with RCW.

Looking forward, our main result will be a new algorithm we call Cycle Slicer that will allow us to take long cycles of points from $\mathcal{S}$ and form many matchings. This will allow us to substantially improve upon RCW for the important cases where $|\mathcal{S}|$ is close to the size of $|\mathcal{X}|$. In particular, in the SSN example, Cycle

Slicer will only need around 12,000 rounds, which is significantly less than needed by RCW.

DOMAIN COMPLETION. The second problem we study is *domain completion*, which has its roots in ad-hoc solutions practitioners used in place of FPE, and which was recently given a formal treatment by Grubbs et al. [11]. Before practical FPE schemes were available, practitioners would solve their ciphertext formatting issues using tokenization systems. Essentially, they would construct a permutation on their desired domain by populating a table with input-output pairs. For example, to encipher an SSN without a good FPE scheme, one can instead add the SSN to a table and randomly sample another social security number to map it to. In other words, in the absence of a good FPE scheme on their desired domain, practitioners would instead construct their own permutation on the domain by using what is typically known as lazy sampling.

If a system contains such a table of input-output pairs, but then at a later point in time a good FPE scheme becomes available for the desired domain, it makes sense to stop adding entries to the table and to start using the new scheme. Yet, to maintain backwards compatibility, the new FPE scheme should only be used on new points, and the table (which can now be made read-only) should still be used on old values. This practical problem leads to an interesting theoretical question: how can we construct a permutation on some domain while staying consistent with a table of existing mappings?

More formally, assuming it is easy to construct permutations on some domain $\mathcal{X}$, we wish to construct a new permutation on $\mathcal{X}$ that preserves an existing set of mappings for points in a *preservation set* $\mathcal{T} \subseteq \mathcal{X}$. Let $\mathcal{U}$ be the set of points that the table maps $\mathcal{T}$ to (i.e., the range of the table mappings). One of the challenges with domain completion is that $\mathcal{T}$ is unlikely to be the same as $\mathcal{U}$, yet the two sets might have some overlap. Thus, the problem is not as easy as just mapping points in $\mathcal{X} - \mathcal{T}$ to other points in $\mathcal{X} - \mathcal{T}$. Some points in $\mathcal{X} - \mathcal{T}$ will need to be mapped to points in $\mathcal{T}$, while some points in $\mathcal{U}$ will need to be mapped to points in $\mathcal{X} - \mathcal{T}$.

Grubbs, Ristenpart, and Yarom (GRY) first describe a solution (attributed to an anonymous reviewer) for this problem when $\mathcal{X}$ can be efficiently ranked. The solution uses the rank-encipher-unrank algorithm discussed earlier in the context of domain targeting, but with the ranking and unranking algorithms additionally performing a binary search through precomputed tables (which increase the space requirements) with the same size as the preservation set $\mathcal{T}$.

As explained above when discussing domain targeting, not all domains can be efficiently ranked, and it may otherwise be desirable to have solutions that avoid ranking. Additionally, GRY point out that the ranking solution could be more susceptible to timing attacks because of both the binary search and the ranking. Their main result is thus an algorithm called Zig-Zag that does not require $\mathcal{X}$ to be efficiently ranked, and only requires the ability to efficiently test membership in the preservation set $\mathcal{T}$. Much like Cycle Walking in the domain targeting setting, Zig-Zag has small expected running time but significantly higher worst-case running time. Thus, Zig-Zag can be a good choice in practice, but has some of the same drawbacks as Cycle Walking. An interesting question, both for theory and for practice, is whether it is possible to do domain completion while avoiding an expected-time process and achieving a lower worst-case running time, while also avoiding ranking. Looking forward, our Cycle Slicer algorithm, when combined with some preprocessing on the table of mappings, will give a new algorithm for domain completion that avoids expected time and has better worst-case running time than Zig-Zag.

OUR MAIN RESULT: THE CYCLE SLICER ALGORITHM. We now introduce our main result, an algorithm we call Cycle Slicer which gives new solutions to the problems introduced above. At a high level, one round of the Cycle Slicer algorithm transforms a permutation P on $\mathcal{X}$ into a matching on some subset of the points of $\mathcal{X}$, where by *matching* we mean a permutation made up of only transpositions (cycles of length 2). By carefully specifying which points should be included in the matching, formalized by an *inclusion function* I, we will be able to use Cycle Slicer to perform both domain targeting and domain completion.

To understand Cycle Slicer, consider the problem of transforming an arbitrary permutation P on $\mathcal{X}$ into another permutation on $\mathcal{X}$ that only consists of length 2 cycles, called transpositions or swaps. If we look at the cycle structure of P, we will likely find many cycles longer than length 2, some substantially longer. The main idea underlying Cycle Slicer is to "slice" up long cycles into a number of smaller, length 2 cycles. To do this, Cycle Slicer uses a *direction function*, Dir, which will simply be a function that gives a random bit for each point in $\mathcal{X}$. If the direction bit for a point x is 1 (i.e., $Dir(x) = 1$), then we say that x is forward-looking. If the direction bit is instead a 0, we call the point backward-looking. If a point x is forward-looking and the next point in the cycle, $P(x)$, is backward-looking, then those points are paired together. Similarly, if x is backward-looking and $P^{-1}(x)$ is forward-looking, then those points are paired up. It is easy to see that the use of the direction function pairs up some of the points in $\mathcal{X}$, while other points whose direction bits were not consistent with the direction bits of the points preceding and following them in their cycles are not paired up. Swapping any number of the paired points now results in a permutation on $\mathcal{X}$ made up of only transpositions.

As an example, suppose permutation P has a cycle $(w\ x\ y\ z)$. This means that $P(w) = x$, $P(x) = y$, $P(y) = z$, and $P(z) = w$. Now suppose our direction function gives bits 0, 1, 0, 0 for these points. This means that x is a forward-looking point, and the three others are all backward-looking. Based on this, x and y will be paired up which, if they are eventually swapped, will lead to a permutation with cycles $(w)(x\ y)(z)$.

The direction function determines which points are paired, but whether a pair is actually swapped also depends on another function we call the inclusion function. An inclusion function I takes two points as input and either outputs 1 or 0, with 0 meaning the points are not swapped. Looking forward, we will be able to apply Cycle Slicer to a number of different problems by specifying different inclusion functions.

One round of Cycle Slicer will result in many points swapping positions, but this alone will not sufficiently mix the points. Thus, for all of our applications we will need many rounds of Cycle Slicer. Since Cycle Slicer, like the Reverse Cycle Walking algorithm described above, mixes points through repeated random matchings, we follow the same approach as Miracle and Yilek for analyzing the number of rounds needed to yield a random permutation on the desired domain. They use techniques introduced by Czumaj and Kutylowski [7] to analyze a matching exchange process. At each step of a matching exchange process, a number k is chosen according to some distribution and then a matching of size is k is chosen uniformly. Both Reverse Cycle Walking and Cycle Slicer can be viewed as a matching exchange process. The analysis given by Czumaj and Kutylowski does not give explicit constants for their bounds, so in order to provide explicit constants for their Reverse Cycle Walking algorithm, Miracle and Yilek reprove several key lemmas from the Czumaj and Kutylowski result. We extend their work to also give explicit constants for general matching exchange processes based on two new parameters which we introduce. The first parameter is the probability that a specific pair of points (x, y) is in the matching and the second parameter is the probability that a second pair (z, w) is also in the matching conditioned on a first pair being in the matching. This analysis allows us to bound the variation distance in each of our applications, and we believe the parameters are general enough that our result is of interest beyond the application to the Cycle Slicer algorithm.

Now that we have described the basic Cycle Slicer algorithm, we will explain how it can be applied to domain targeting and completion.

CYCLE SLICER AND DOMAIN TARGETING. To use Cycle Slicer for domain targeting, constructing a permutation on $S \subseteq X$ out of permutations on X, we simply define an inclusion function I_t that outputs 1 when given two points in the target set S, and outputs 0 otherwise. This means that only points from S will be swapped. Repeating this process for a number of rounds results in a random permutation on the target set.

The resulting process is similar to Reverse Cycle Walking, but the important difference is that when the size of S is close to the size of X, Cycle Slicer is still able to pair many points from S while Reverse Cycle Walking will struggle to pair points. This means substantially fewer rounds of Cycle Slicer are needed. In our running example with social security numbers where $|S| = 10^9$ and $|X| = 2^{30}$, Cycle Slicer will need about $1/2600$ as many rounds as Reverse Cycle Walking. We could improve the performance of RCW by instead letting X be a larger supserset of S; if for RCW we let X be the set of 32-bit values, then its performance significantly increases and we have arguably a fairer comparison. But even with this enhancement for RCW, Cycle Slicer will still need only about $1/4$ as many rounds. We give a more detailed explanation of this in Sect. 5.

CYCLE SLICER AND DOMAIN COMPLETION. Using Cycle Slicer for domain completion is less straightforward, since the table of mappings already imposes constraints on the ultimate cycle structure of any permutation on the entire domain. In particular, the table of mappings can lead to a number of cycles and lines (sequences of points that do not form a cycle). The permutation we wish to build on the entire domain $\mathcal{X}$ needs to have a cycle structure that stays consistent with these. At a high level, our solution is to collapse the lines into a single point and use Cycle Slicer on the resulting induced domain. The "fake" points that represent entire lines from the table will be part of matchings with other points in the domain. After a number of rounds of Cycle Slicer, we can then substitute the entire line back in for the point that represented it. While this is the intuitive idea of what is happening, writing down an algorithm that allows for evaluation of a single point is non-trivial, and we consider this algorithm (found in Sect. 6) one of our main contributions.

This algorithm requires, for each point in a line, the ability to determine the first and last points in the line. This information can be precomputed and also made read-only along with the table. Given this precomputation, our algorithm has lower worst-case running time than the Zig-Zag construction, which could require $|\mathcal{T}|$ loop iterations in the worst case to encipher a single point. Our algorithm is also not expected-time, as enciphering any point outside of $\mathcal{T}$ simply uses the same r-round Cycle Slicer algorithm, and the number of rounds does not vary across different points. This could be of practical significance if reliable execution times are needed, or if there is concern about the potential for timing attacks.

FURTHER USES OF CYCLE SLICER. In addition to the applications of Cycle Slicer discussed above, we can imagine a number of other uses for the new algorithm. In particular, Cycle Slicer would seem to be useful in any situation where we want to use a permutation on some general domain to build a more specialized permutation on a related domain or that is consistent with some additional restrictions. Framed this way, our results are somewhat similar to the work of Naor and Reingold on constructing permutations with particular cycle structure [20]. In Sect. 7 we discuss three settings where we think Cycle Slicer could be useful.

2   Preliminaries

NOTATION. We denote by $x \leftarrow y$ the assignment of the value of y to x. If F is a function, then $y \leftarrow F(x)$ means evaluating the function on input x and assigning the output to y. When talking about a permutation P and its cycle structure, we use the notation $(\ldots\ x\ y\ \ldots)$ to indicate there is a cycle that contains x immediately followed by y, i.e., $P(x) = y$ and $P^{-1}(y) = x$. For any function F (which may or may not be a permutation), we say points $x_1, \ldots, x_j$ form a *line* if for all $1 \leq i \leq j - 1$ it is true that $F(x_i) = x_{i+1}$.

TOTAL VARIATION DISTANCE. In order to generate a random permutation using the Cycle Slicer algorithm it is necessary to have many rounds of Cycle Slicer. To bound the number of rounds needed we will analyze the total variation distance. Let $x, y \in \Omega$, $P^r(x, y)$ be the probability of going from x to y in r steps and μ be a distribution on the state space Ω. Then the *total variation distance* is defined as $||P^r - \mu|| = \max_{x \in \Omega} \frac{1}{2} \sum_{y \in \Omega} |P^r(x, y) - \mu(y)|$. In Sect. 4 we will bound the total variation distance between the distribution after r rounds of the Cycle Slicer algorithm and the uniform distribution.

3   The Cycle Slicer Construction

In this section we introduce our main contribution, the Cycle Slicer algorithm. As described in the introduction, at a high level, the Cycle Slicer algorithm allows us to transform any permutation on a set $\mathcal{X}$ into another permutation on $\mathcal{X}$ that may have additional desirable properties.

FORMAL DESCRIPTION. We now formally describe the Cycle Slicer algorithm. Let $P : \mathcal{X} \to \mathcal{X}$ be a permutation and $P^{-1} : \mathcal{X} \to \mathcal{X}$ be its inverse. Let $Dir : \mathcal{X} \to \{0, 1\}$ be a function called the *direction function*, $I : \mathcal{X} \times \mathcal{X} \to \{0, 1\}$ be a function called the *inclusion function*, and $B : \mathcal{X} \to \{0, 1\}$ be a function called the *swap function*. Given these, we give pseudocode for one round of Cycle Slicer in Algorithm 1. This round of Cycle Slicer results in a permutation on $\mathcal{X}$, and is in fact an involution, serving as its own inverse algorithm. We will

---

Algorithm 1. Cycle Slicer (one round)

1: **procedure** $\mathrm{CS}_{P,P^{-1},I,Dir,B}(x)$
2:     **if** $Dir(x) = 1$ **then**
3:         $x' \leftarrow P(x)$
4:         **if** $Dir(x') = 0$ and $I(x, x') = 1$ and $B(x) = 1$ **then**
5:             $x \leftarrow x'$
6:         **end if**
7:     **else**
8:         $x' \leftarrow P^{-1}(x)$
9:         **if** $Dir(x') = 1$ and $I(x', x) = 1$ and $B(x') = 1$ **then**
10:            $x \leftarrow x'$
11:        **end if**
12:    **end if**
13:    **return** x
14: **end procedure**

In words, to map a point $x \in \mathcal{X}$, Cycle Slicer first applies the direction function Dir to x to see if it should look forward ($Dir(x) = 1$) or backwards. If forward, then it applies the permutation P to x to get a point x' to potentially swap with. If backwards, then it instead applies the inverse permutation P^{-1} to get a point x'. Now, to decide whether or not to swap the positions of x and x', Cycle Slicer applies an inclusion function I and a swap function B. If both output 1, then x and x' are swapped. Note that our algorithm always lets the first input to I be the "forward-looking" point (i.e., the point x with $Dir(x) = 1$) and, similarly, always applies B to the forward-looking point. This ensures we get consistent decision bits for both x and x'.

Algorithm 1 gives code for just one round of Cycle Slicer. In the applications later in the paper, we will need many rounds of Cycle Slicer to ensure points are sufficiently mixed. Towards this, for brevity we let CS^r denote r independent rounds of CS, each with an independent P, P^{-1}, Dir, and B. Each round of CS^r will use the same inclusion function I, however[3]. We then let $CSinv^r$ denote the inverse of CS^r which, since CS is an involution, will just be the r rounds of CS^r applied in reverse order.

Looking forward to the applications of Cycle Slicer discussed in later sections, parameterizing Cycle Slicer with different decision functions I will lead to new solutions to those problems.

DISCUSSION. Intuitively, Cycle Slicer gives us a way to "slice" up the cycles of the underlying permutation P and form a matching. More specifically, consider any point x in our domain $\mathcal{X}$. Suppose the permutation P puts x into a cycle $(\ldots \, w \, x \, y \, \ldots)$. We wish to construct a matching on $\mathcal{X}$, which is a permutation on $\mathcal{X}$ made up of only transpositions (2-cycles). To do this, we apply a direction function Dir to the points in $\mathcal{X}$. If $Dir(x) = 1$, we say x is forward-looking, which means it will potentially pair with $P(x) = y$. If $Dir(x) = 0$, we say it is backward-looking, which means it will potentially pair with $P^{-1}(x) = w$. Of course, the direction function is also being applied to w and y, so they will also be forward-looking or backward-looking. If x is forward-looking and y is backward-looking, they become a pair to potentially swap in a matching; similarly, if x is backward-looking and w is forward-looking, then x and w are a pair to potentially swap in a matching. Whether the points are actually swapped is determined by two other functions, the inclusion function I and the swap function B. The inclusion function allows us to restrict which points will be swapped based on properties like whether or not the two points are part of a particular subset of $\mathcal{X}$. The swap function will simply be a bit flip to determine if the swap should occur or not, and is needed for technical reasons.

Note that at a high-level our analysis of the Cycle Slicer algorithm relies on viewing the algorithm as a Markov chain on the set of permutations of some set $\mathcal{V} \subset \mathcal{X}$ where at each step an independent matching on $\mathcal{V}$ is applied. The independence requirements for P, P^{-1}, B and Dir at each round ensure that an independent matching is applied at each step which is needed to apply the techniques from Czumaj and Kutylowski's analysis which we do in Sect. 4 and Appendix A. The inclusion function I will be used to determine which pairs in the matching are allowed to ensure we generate a matching on the correct set.

PRACTICAL CONSIDERATIONS. There are a number of ways to instantiate the different components of Cycle Slicer in practice, and the specific algorithms used would obviously depend on the application. The most obvious instantiation would use a block cipher $E : \mathcal{K} \times \mathcal{X} \to \mathcal{X}$ for the round permutation P (and the inverse block cipher E^{-1} for P^{-1}), with each round requiring a different

---

[3] It is possible there will be other interesting uses of Cycle Slicer that use different inclusion functions in different rounds. However, our applications do not need this, so to keep things simpler we just use a single inclusion function across all rounds.

block cipher key. Yet, we will soon see that for typical applications we will need thousands of rounds of Cycle Slicer, so a separate key for each round is not ideal. Instead, it would make sense to use a tweakable block cipher [14] with a single key and the round number used as a tweak. Specifically, to evaluate the ith round permutation P_i on a point $x \in \mathcal{X}$, we would compute $E(K, i, x)$, where K is the randomly chosen block cipher key. Since our proposed applications are all in the area of format-preserving encryption, E might be instantiated with one of the standardized Feistel-based modes [3,5,9] (though, based on recent attacks [1,8], care should be taken) or, if stronger provable-security guarantees are desired, perhaps the Swap-or-Not cipher [12] or one of the fully-secure ciphers from [18,21]. All of these options support tweaks.

The direction functions and swap functions could then be instantiated with a pseudorandom function (PRF); a single key could be used for each if the round number is included as an input to the PRFs. Specifically, if we are performing r rounds of Cycle Slicer and have a PRF $F : \mathcal{K}' \times \{1, \ldots, r\} \times \mathcal{X} \rightarrow \{0, 1\}$, we first choose a random key K_1 for the direction functions and a random key K_2 for the swap functions. Then, to evaluate the direction function (resp. swap function) in round j on a point x, we would compute $F(K_1, j, x)$ (resp. $F(K_2, j, x)$).

Looking Forward: Analysis of Cycle Slicer. In the next section, we give an analysis of Cycle Slicer, proving an information theoretic, mixing-time result. More specifically, we give a bound on total variation distance after many rounds of Cycle Slicer, with each round's permutation, direction function, and swap function chosen randomly from the appropriate sets of such functions. We emphasize that this makes our results very modular and applicable to a number of different problems, since typical security proofs in our target application areas first swap out computationally secure components like block ciphers and PRFs with randomly chosen permutations and functions, respectively.

4   Analyzing Cycle Slicer

In order to bound the number of rounds of the Cycle Slicer algorithm that are needed, we will analyze a more general process that generates a permutation on a set of points by applying a matching at each step. Specifically we will analyze a type of matching exchange process first defined and analyzed by Czumaj and Kutylowski [7]. A *matching exchange process* is a Markov chain for generating a permutation on a set of n elements. At every step a number $\kappa \le n/2$ is selected according to some distribution and a matching of size κ is chosen uniformly at random. Independently, for each pair (x, y) of the matching, with probability $1/2$ the elements x and y are exchanged. Czumaj and Kutylowski [7] prove that as long as the expected value of κ is constant then after $\Theta(\log(n))$ steps, the variation distance is $O(1/n)$. We can view the Cycle Slicer algorithm as a matching exchange process and it can be shown that the expected size of a matching is constant. However, the result of [7] does not give explicit bounds on the constants. Miracle and Yilek [17] extend the result of [7] by reproving several key lemmas to give explicit constants for a particular algorithm. Here we extend their work to a more general setting. Given a matching exchange process where the matchings are selected according to the following parameters, we provide explicit bounds, including constants, on the mixing time and variation distance of the matching exchange process. In Sects. 5, 6 and 7 we will show how our analysis can be applied to the Cycle Slicer algorithm in our three different applications. We begin by defining the parameters we will need.

1. For any points x, y the probability that a pair (x, y) is part of a matching is at least p_1.
2. For any points x, y, z, and w conditioned on (x, y) being a pair in the matching, the probability that (z, w) is also in the matching is at least p_2.

Note that p_1 and p_2 refer to the probability these pairs are included in the generated matching (i.e., the matching generated by Cycle Slicer) regardless of the result of the bit flip (or swap function) that determines whether the points are actually flipped. Assuming p_1 and p_2 are $\Omega(1/n)$, we will show the process mixes in time $O(\log n)$. Formally, the *mixing time* of a Markov chain $\mathcal{M}$ with state space Ω is defined as $\tau_{\mathcal{M}}(\epsilon) = \min\{t : ||\mathcal{P}^{t'} - \mu|| \leq \epsilon, \forall t' \geq t\}$ where $||\mathcal{P}^{t'} - \mu|| = \max_{x \in \Omega} \frac{1}{2} \sum_{y \in \Omega} |\mathcal{P}^{t'}(x, y) - \mu(y)|$, $\mathcal{P}^{t'}(x, y)$ is the probability of going from x to y in t steps and μ is the stationary distribution of $\mathcal{M}$. We prove the following result on the mixing time of the matching exchange process with parameters p_1 and p_2 as defined above.

Theorem 1. *For* $T \geq \max\left(40\ln(2n^2), \frac{10\ln(n/9)}{\ln(1 + p_1 p_2 (7/36)((7/9)n^2 - n))}\right) + \frac{72\ln(2n^2)}{p_1 n}$ *the mixing time τ of a matching exchange process with parameters p_1 and p_2 as defined above satisfies*

$$\tau(\epsilon) \leq T \cdot \left\lceil \frac{\ln(n/\epsilon)}{\ln n^2} \right\rceil.$$

When $\epsilon = 1/n$ and $p_1, p_2 = \Omega(1/n)$, the bound simplifies to $\tau(1/n) = \Theta(\ln(n))$.

The proof of Theorem 1 can be found in Appendix A. In order to analyze our Cycle Slicer algorithm a bound on the variation distance will be more useful. This is obtained by a straightforward manipulation of the mixing time bound above. As long as $p_1, p_2 = \Omega(1/n)$ and the number of rounds is at least $T = \Theta(\ln(n))$, the variation distance is less than $1/n$. Let ME^r represent r rounds of the matching exchange (ME) process and ν_{ME^r} be the distribution on permutations of n elements after r rounds of the matching exchange process. More specifically, $\nu_{\mathrm{ME}^r}(x, y)$ is the probability of starting from permutation x and ending in permutation y after r rounds of the matching exchange process. Applying the definition from Sect. 2 gives us $||\nu_{\mathrm{ME}^r} - \mu_s|| = \frac{1}{2} \sum_{y \in \Omega} |\nu_{\mathrm{ME}^r}(x, y) - \mu_s(y)|$ where Ω is the set of all permutations on n elements and μ_s is the uniform distribution Ω. Using this definition and Theorem 1 we have the following.

Corollary 1. *Let* $T = \max\left(40\ln(2n^2), \frac{10\ln(n/9)}{\ln(1 + p_1 p_2 (7/36)((7/9)n^2 - n))}\right) + \frac{72\ln(2n^2)}{p_1 n}$, *then*

$$||\nu_{ME^r} - \mu_s|| \leq n^{1 - 2r/T},$$

where ν_{ME^r} is the distribution after r rounds of the matching exchange process and μ_s is the uniform distribution on permutations of the n elements.

CYCLE SLICER. Next, we use Corollary 1 to bound the total variation distance for the Cycle Slicer algorithm. In the following sections we will look at the Cycle Slicer algorithm in several different applications. In each of these we are effectively generating a permutation on a set $\mathcal{V}$ which is a subset of a larger set $\mathcal{X}$. At each step of Cycle Slicer we use a permutation P on the points in $\mathcal{X}$ to generate a matching on the points in $\mathcal{V}$. Here we prove a general result for this setting. Let $\mathcal{V} \subset \mathcal{X}$ and the inclusion function I be defined as $I(x,y) = 1$ if and only if $x \in \mathcal{V}$ and $y \in \mathcal{V}$. Note that the permutation P chosen at each round is a uniformly random permutation on $\mathcal{X}$ and B and Dir are as defined in Sect. 3 (i.e. independent random bits). In this setting, we prove the following theorem.

Theorem 2. *Let $T = \max \left(40\ln(2|\mathcal{V}|^2), \frac{10\ln(|\mathcal{V}|/9)}{\ln(1+(7/144)((7/9)|\mathcal{V}|^2-|\mathcal{V}|)/|\mathcal{X}|^2)} \right)$*
$+ \frac{144|\mathcal{X}|\ln(2|\mathcal{V}|^2)}{|\mathcal{V}|}$, then

$$||\nu_{CS^r} - \mu_s|| \leq |\mathcal{V}|^{1-2r/T},$$

where ν_{CS^r} is the distribution after r rounds of CS and μ_s is the uniform distribution on permutations on $\mathcal{V}$.

Note that if $|\mathcal{V}|$ is a constant fraction of $|\mathcal{X}|$ then as long as the number of rounds is at least $\Theta(\ln(|\mathcal{V}|))$ then the variation distance will be less than $1/|\mathcal{V}|$.

Proof. In order to apply Corollary 1 to the Cycle Slicer algorithm, we need to bound the parameters p_1 and p_2. Recall that for any pair of points (x,y), p_1 is the probability that the pair (x,y) is part of a potential matching. At any step of Cycle Slicer, for any points $x, y \in \mathcal{V}$, (x,y) is part of the matching if one of two things happen, either $Dir(x) = 1$, $Dir(y) = 0$ and $P(x) = y$ or $Dir(x) = 1$, $Dir(y) = 0$, and $P^{-1}(x) = y$. Each of these events happens with probability $(1/2)(1/2)(1/|\mathcal{X}|)$, implying that $p_1 = (2|\mathcal{X}|)^{-1}$. Note that it is also required that $I(x,y) = 1$. However, this is always true since $x, y \in \mathcal{V}$ and $I(x,y) = 1$ if and only if $x, y \in \mathcal{V}$.

For any points $x, y, z, w \in \mathcal{V}$ conditioned on (x,y) being a pair in the matching, p_2 is the probability that (z,w) is also in the matching. Again, there are two situation where this can happen. Without loss of generality we will assume that $Dir(x) = 1, Dir(y) = 0$ and $P(x) = y$. The other case $(Dir(x) = 0, Dir(y) = 1$ and $P^{-1}(x) = y)$ is almost identical. The pair (z,w) will be in a matching if $Dir(z) = 1$, $Dir(w) = 0$ and $P(z) = w$. There are a total of $|\mathcal{X}| - 1$ points that z could get mapped too (we already know z is not mapped to y since (x,y) is in the matching) so this case happens with probability $(1/2)(1/2)(1/(|\mathcal{X}| - 1))$. Similarly, the pair (z,w) will be in the matching if $G_i(z) = 0$, $G_i(w) = 1$ and $P_i^{-1}(z) = w$ and this occurs with probability $(1/2)(1/2)(1/(|\mathcal{X}|-1))$. Combining these shows $p_2 = (2(|\mathcal{X}| - 1))^{-1} \geq (2|\mathcal{X}|)^{-1}$. Using these bounds on p_1 and p_2 we can now directly apply Corollary 1 to obtain the above theorem. $\square$

5   Domain Targeting

BACKGROUND. Our first application of the Cycle Slicer algorithm is in what we are calling *domain targeting*. In domain targeting, we wish to construct an efficiently-computable permutation on a target set S. Yet, this target set might have "strange" structure, and it might not be clear how to directly build a permutation on such a set.

As described in the introduction, this problem can arise in format-preserving encryption, and a well-known strategy for solving it is to take a permutation P on a larger, "less strange" set $\mathcal{X}$ for which $S \subseteq \mathcal{X}$, and then transform a permutation on $\mathcal{X}$ into a permutation on the target set S. One such transformation is known as *Cycle Walking*, in which the permutation P on the larger set $\mathcal{X}$ is repeatedly applied to a point $x \in S$ until the output of the permutation finally lands back in the target set S. This leads to a small *expected* running time, but the worst-case running time is high and, additionally, enciphering different points may take different amounts of time.

More recently, Miracle and Yilek [17] introduced an alternative to Cycle Walking called *Reverse Cycle Walking* (RCW). RCW uses permutations on the larger set $\mathcal{X}$ to construct matchings on the target set S. In a round of RCW using permutation P on $\mathcal{X}$, if two consecutive points x, x' in a cycle are both in S and are immediately preceded and followed by points *not* in S, then x and x' are paired and, depending on a bit flip, either swapped or not. Said another way, $x, x' \in S$ are only potentially swapped in a round of RCW if they are part of a cycle $(\ldots y \, x \, x' \, z \ldots)$ and $y, z \notin S$.

As a result, in every round of RCW some fraction of the points in the target set S are paired and swapped. Repeating this process for many rounds eventually mixes the points in S sufficiently. RCW has lower worst-case running time than regular Cycle Walking, and the time to encipher a point does not vary depending on the input. As we will see, Cycle Slicer, like RCW, will give us a way to form a matching on the points in S and, with enough rounds, sufficiently mix the points.

The main result in this section is that the Cycle Slicer algorithm can be used to improve upon Reverse Cycle Walking in the important scenario when the size of the target set is a large constant fraction of the size of the larger set $\mathcal{X}$. A simple example of such a situation would be if S is the set of bitstrings that give valid 9 digit decimal numbers and $\mathcal{X}$ is the set of 30 bit strings. In this case, $|S| = 10^9$, $|\mathcal{X}| = 2^{30}$, and the ratio $10^9/2^{30} \approx .93$. In such scenarios, which are important in practice, Cycle Slicer will require significantly less rounds than RCW.

To understand why this will be the case, consider a scenario in which the size of the target set S is a large fraction of the size of $\mathcal{X}$. A random permutation P on $\mathcal{X}$ will likely yield many cycles with points mostly from S and very few points from $\mathcal{X} - S$. For example, we could have a cycle $(x \, x' \, y \, x'' \, x''')$ in which only y is not in the target set and all of the x's are in S. With Reverse Cycle Walking, none of the x's would be swapped, while with Cycle Slicer there will be a reasonable probability some or even all of the x's will be swapped, contributing to the mixing.

DETAILS. More formally, suppose we wish to construct a permutation on a set $\mathcal{S}$ which is a subset of a larger set $\mathcal{X}$ for which we already know how to construct efficient permutations. Then we can use Cycle Slicer with the following inclusion function I_t:

$$I_t(x, y) = \begin{cases} 1 \text{ If } x \in \mathcal{S} \text{ and } y \in \mathcal{S} \\ 0 \text{ Otherwise} \end{cases}$$

In words, we only include two points in a potential swap with the Cycle Slicer algorithm if both points are in the target set $\mathcal{S}$.

Finally, we will use our result from Sect. 4 to bound the number of steps of the Cycle Slicer algorithm needed for domain targeting. We will use Theorem 2 which bounds the variation distance of the Cycle Slicer algorithm in terms of the size of the domain $\mathcal{X}$ and the number of points in the set we are generating a permutation on $\mathcal{V}$. In domain targeting we are generating a permutation on $\mathcal{S}$ with size $|\mathcal{S}|$. This directly gives the following corollary to Theorem 2.

Corollary 2. *Let* $T = \max\left(40\ln(2|\mathcal{S}|^2), \frac{10\ln(|\mathcal{S}|/9)}{\ln(1+(7/144)((7/9)|\mathcal{S}|^2-|\mathcal{S}|)/|\mathcal{X}|^2)}\right)$
$+ \frac{144|\mathcal{X}|\ln(2|\mathcal{S}|^2)}{|\mathcal{S}|}$, *then*

$$\|\nu_{CS^r} - \mu_s\| \leq n^{1-2r/T},$$

where ν_{CS^r} *is the distribution after* r *rounds of* CS *and* μ_s *is the uniform distribution on permutations on* $\mathcal{S}$.

COMPARISON WITH RCW. It is difficult to see from the theorem above but our algorithm gives a significant advantage over RCW in the very typical case where the size of the target set is at least half the size of the domain (i.e. $|\mathcal{S}| \geq |\mathcal{X}|/2$). For example, consider the setting of encrypting social security numbers using an existing cipher for 30 bit strings. Here the size of the target set (i.e., the number of 9 digit numbers) is 10^9 and the domain has size 2^{30}. In this setting our algorithm requires 12,257 rounds until the variation distance is less than $1/10^9$ while RCW requires over 32 million. While these numbers make RCW look completely impractical, it is important to note that since RCW works best when the size of the target set is between $1/4$ and $1/2$ the size of $\mathcal{X}$, in some settings it may be possible to find a larger superset of the target set to use with RCW. If using RCW in our social security number example, it would actually make more sense to choose $\mathcal{X}$ to be the set of 32 bit values instead of 30 bit values. In this case, RCW requires just over 44,000 rounds. Nevertheless, Cycle Slicer is still significantly faster, requiring around $1/4$ the number of rounds.

6 Domain Completion

BACKGROUND. Our second application of Cycle Slicer is in *domain completion*. This problem, recently studied by Grubbs et al. [11], arises when we wish to construct a permutation on some set $\mathcal{X}$, but we need the permutation to maintain some mappings we might already have stored in a table. More formally, let $\mathcal{X}$ be the set we wish to construct a permutation on. Let $\mathcal{T} \subseteq \mathcal{X}$ be called the preservation set, which will be the set of domain points for which we already have a mapping. This mapping is given by a 1-1 and onto function $G : \mathcal{T} \rightarrow \mathcal{U}$ with $\mathcal{T}, \mathcal{U} \subseteq \mathcal{X}$. The domain completion problem is then to construct an efficiently computable permutation $F : \mathcal{X} \rightarrow \mathcal{X}$ (with a corresponding efficiently computable inverse function F^{-1}) such that $F(x) = G(x)$ for all $x \in \mathcal{T}$.

To understand our algorithm, note that the cycle decomposition (or cycle structure) of the partial permutation on $\mathcal{X}$ given by the mapping G consists of a set of cycles, lines and single points (see Fig. 1). The cycles and lines come from the mapping G while the single points are the points in $\mathcal{X} - \mathcal{T} - \mathcal{U}$. At a high-level our algorithm ignores any cycles, collapses any lines to a single point and generates a matching on the remaining set of points using the Cycle Slicer algorithm. When the previously collapsed lines are expanded and the cycles are added back in, this gives a permutation on $\mathcal{X}$ that preserves the mappings given by G. Figure 1 gives an example of this process. Figure 1(b) gives a permutation on the collapsed line $3 \rightarrow 2 \rightarrow 1$ and single points 5 and 6. Figure 1(b) shows how to expand the line and create a permutation on the whole set. To implement this idea we will use procedure CSDC in Algorithm 2 which performs some preprocessing before using Cycle Slicer as a subroutine. We also describe the inverse of this procedure CSDCinv in Algorithm 2.

DETAILS. We will use the Cycle Slicer algorithm as defined in Sect. 3 (Algorithm 1) with the following inclusion function I_c:

$$I_c(x, y) = \begin{cases} 1 \text{ If } x \notin \mathcal{U} \text{ and } y \notin \mathcal{U} \\ 0 \text{ Otherwise} \end{cases}$$

Next, we will describe the procedures $frst$ and lst used by CSDC and CSDCinv respectively. These are again easiest to understand by considering the cycle decomposition of the mapping G. Notice that the function $frst$ is only applied to points x such that $x \in \mathcal{U} - \mathcal{T}$ which implies that in the cycle decomposition x is the last point in a line. We will define $frst(x)$ to be the first point in the line containing x. Note that $frst(x) \in \mathcal{T} - \mathcal{U}$. For example, in Fig. 1 $frst(1) = 3$.

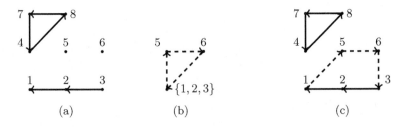

Fig. 1. Cycle Decomposition for $\mathcal{T} = \{2, 3, 4, 7, 8\}$, $\mathcal{U} = \{1, 2, 4, 7, 8\}$ with mappings $3 \rightarrow 2, 2 \rightarrow 1, 4 \rightarrow 8, 8 \rightarrow 7$ and $7 \rightarrow 4$

Algorithm 2. Domain Completion Using Cycle Slicer

1: **procedure** CSDC(x)
2:     **if** $x \in \mathcal{T}$ **then**
3:         $x \leftarrow G(x)$
4:     **else**
5:         **if** $x \in \mathcal{U}$ **then**
6:             $x \leftarrow frst(x)$
7:         **end if**
8:         $x \leftarrow \mathsf{CS}^r(x)$
9:     **end if**
10:     **return** x
11: **end procedure**
12: **procedure** CSDCinv(x)
13:     **if** $x \in \mathcal{U}$ **then**
14:         $x \leftarrow G^{-1}(x)$
15:     **else**
16:         $x \leftarrow \mathsf{CSinv}^r(x)$
17:         **if** $x \in \mathcal{T}$ **then**
18:             $x \leftarrow lst(x)$
19:         **end if**
20:     **end if**
21:     **return** x
22: **end procedure**

Similarly, the function lst is only applied to points at the beginning of a line (i.e., a point $x \in \mathcal{T} - \mathcal{U}$). We will define $lst(x)$ to be the last point in the line containing x (again in Fig. 1 $lst(3) = 1$). If the functions $frst$ and lst are not available they can either be precomputed or computed on the fly. Algorithm 3 gives the detailed procedures for computing $frst$ and lst. Note that in the worst case they each take time $O(|\mathcal{T}|)$. However, $frst$ only needs to be computed for points $x \in \mathcal{U} - \mathcal{T}$ and lst only needs to be computed for points $y \in \mathcal{T} - \mathcal{U}$. This implies that if only the necessary values of $frst$ and lst are precomputed then the overall running time will be $O(|\mathcal{T}|)$ because the algorithm will only traverse each line in the cycle decomposition twice (once for $frst$ of the endpoint and once for lst of the start point).

CORRECTNESS OF CSDC. Next, we will prove that the CSDC algorithm generates a valid permutation F on $\mathcal{X}$ consistent with the mappings given by G. Notice that if we did not use the preprocessing algorithm given by CSDC but instead simply applied the Cycle Slicer algorithm using the inclusion function I_c to the points in $\mathcal{X} - \mathcal{U}$, after enough steps we would generate a random permutation on the set $\mathcal{X} - \mathcal{U}$. This is almost a valid permutation except that it does not give us a mapping for the last point in every line (i.e. points in $\mathcal{U} - \mathcal{T}$) and it does give a mapping for the first point in every line (i.e. points in $\mathcal{T} - \mathcal{U}$) which are already mapped by G. Again, consider the cycle decomposition of G. Each line in this partial cycle decomposition contains exactly one point in $\mathcal{T} - \mathcal{U}$ namely the first point in the line and one point in $\mathcal{U} - \mathcal{T}$ namely the last point in the line. The

---

Algorithm 3. Computing *frst* and *lst*

1: **procedure** $frst(x)$
2:     **while** $x \in \mathcal{U}$ **do**
3:         $x \leftarrow G^{-1}(x)$
4:     **end while**
5:     **return** x
6: **end procedure**
7: **procedure** $lst(x)$
8:     **while** $x \in \mathcal{T}$ **do**
9:         $x \leftarrow G(x)$
10:     **end while**
11:     **return** x
12: **end procedure**

---

preprocessing done by CSDC fixes this by first mapping a point $x \in \mathcal{U} - \mathcal{T}$ to the first point in the line containing x (i.e. $frst(x)$) which is in $\mathcal{T} - \mathcal{U}$ and by using the existing mapping in G for points in $\mathcal{T} - \mathcal{U}$. For the remaining points (i.e., $x \in \mathcal{X} - \mathcal{T} - \mathcal{U}$) no preprocessing is done and the Cycle Slicer algorithm is applied directly. It is straightforward to see from the cycle decomposition view of G that $frst$ and lst give a bijection between the points in $\mathcal{T} - \mathcal{U}$ and the points in $\mathcal{U} - \mathcal{T}$. Thus if Cycle Slicer gives a random permutation on $\mathcal{X} - \mathcal{U}$ than the addition of our preprocessing algorithm CSDC gives a mapping that is uniformly random over all possible ways to extend the mapping G to a permutation F on $\mathcal{X}$.

Finally, we will use our result from Sect. 4 to bound the number of steps of the Cycle Slicer algorithm needed in our domain completion algorithm. We have shown above that our algorithm can be viewed as first using Cycle Slicer to generate a permutation on the set $\mathcal{X} - \mathcal{U}$ which is then mapped bijectively to a permutation on the set $\mathcal{X}$ which is consistent with the mapping G. For simplicity, in the analysis of the Cycle Slicer portion of the algorithm, we view the algorithm in this context as generating a permutation on the set $\mathcal{X} - \mathcal{U}$. Through the bijection the analysis then applies directly to permutations on the set $\mathcal{X}$ consistent with the mapping G. Here however the variation distance is the distance between the distribution after r rounds and the uniform distribution on all permutations consistent with G (instead of uniform on all permutations of $\mathcal{X} - \mathcal{U}$).

Theorem 2 from Sect. 4 bounds the variation distance of the Cycle Slicer algorithm and applies directly to the domain completion setting. In this application, the set $\mathcal{V}$ that we are generating a permutation on has size $|\mathcal{X} - \mathcal{U}|$, which gives the following corollary to Theorem 2.

Corollary 3. *Let* $T = \max\left(40\ln(2|\mathcal{Y}|^2), \frac{10\ln(|\mathcal{Y}|/9)}{\ln(1+(7/144)((7/9)|\mathcal{Y}|^2-|\mathcal{Y}|)/|\mathcal{X}|^2)}\right)$
$+\frac{144|\mathcal{X}|\ln(2|\mathcal{Y}|^2)}{|\mathcal{Y}|}$, *then*

$$\|\nu_{CS^r} - \mu_s\| \leq n^{1-2r/T},$$

where $\mathcal{Y} = \mathcal{X} - \mathcal{U}$, ν_{CS^r} *is the distribution after r rounds of CS and* μ_s *is the uniform distribution on permutations on* $\mathcal{X}$ *consistent with G.*

COMPARISON TO ZIG-ZAG. As we stated in the introduction, one of our goals with Cycle Slicer in domain completion is to improve on the worst-case running time of the Zig-Zag construction from [11], which has a loop that repeats $|\mathcal{T}|$ times in the worst-case. (In each loop iteration, Zig-Zag additionally needs to check membership in $\mathcal{T}$, evaluate the underlying permutation, and do a table look-up.) Additionally, we want to avoid an expected-time procedure to minimize leaked timing information.[4]

At first glance, since CSDC and CSDCinv rely on the *frst* and *lst* functions, it would appear that we also have an expected-time procedure. But, the crucial difference is that we can precompute *frst* and *lst* for each point in the table. This precomputation can either be done (inefficiently) using the procedures in Algorithm 3, or more efficiently as described earlier in this section.

The new table with this precomputed information for each point can still be made read-only, and then evaluating *frst* and *lst* in our CSDC and CSDCinv algorithms will become a simple table look-up. This means the worst-case running time will simply be tied to the number of rounds needed for Cycle Slicer, which will be on the order of $\log |\mathcal{X}|$. Using our running example with social security numbers, if the preservation set has size 1 million (meaning that before we started using a FPE scheme, we manually encrypted 1 million customer's SSNs through lazy sampling and put them in the table), then in the worst-case Zig-Zag needs 1 million loop iterations, while the bounds in our Cycle Slicer results tell us we would need around 11,000 rounds to get a strong level of security. Further, after the precomputation, our algorithms CSDC and CSDCinv will need this many rounds for any point outside of the preservation set, so the running time will not vary widely based on which point is being enciphered.

COMPARISON TO RANKING SOLUTION. The ranking-based construction described by GRY, like our Cycle Slicer construction, requires extra storage in addition to the preservation set table. In particular, it requires computing the rank of each element of $\mathcal{T}$ and $\mathcal{U}$. These two lists of ranks then need to be sorted. Enciphering a point involves applying rank-encipher-unrank, but with the addition of performing a binary search on the sorted lists of ranks. However, this $\log |\mathcal{T}|$ factor is still less than the 11,000 rounds we need with Cycle Slicer in the above example.

Nevertheless, our construction does potentially have some advantages over the ranking-based construction. First, as we have already stated, our results

---

[4] GRY argue that Zig-Zag does not leak damaging timing information, even though enciphering different points may result in very different execution times. We believe there could still be timing attacks in certain applications. For example, if an adversary measures execution time and then learns the corresponding plaintext, and then later observes a different execution time, then the adversary knows the same point was not enciphered this second time, which may or may not be useful and depends on the application.

do not rely on the ability to efficiently rank the desired domain, which can be important for non-regular languages or even regular languages described by a particularly complicated regular expression. Second, our construction could be more resistant to timing attacks. Specifically, GRY explain that implementations of both the binary search and the ranking/unranking[5] could leak important timing information. Our construction, on the other hand, always applies the same number of rounds of Cycle Slicer to points outside of the preservation set, which should be easier to implement without leaking timing information.

7   More Applications of Cycle Slicer

As we mentioned in the introduction, we believe Cycle Slicer will prove useful for any problem requiring the construction of permutations on domains with additional restrictions or constraints. Towards this, in this section we briefly discuss three problems where Cycle Slicer could be applied. We leave further investigation to future work.

DOMAIN EXTENSION. In Sect. 6 we saw that Cycle Slicer is useful for solving the problem of domain completion. Here we argue it would also work well for the closely-related *domain extension* problem, also recently investigated by Grubbs et al. [11].

Formally, let $\mathcal{D}$ be a set for which we already have some mappings. This means there is a preservation set $\mathcal{T} \subseteq \mathcal{D}$ and a 1-1 and onto function $G : \mathcal{T} \to \mathcal{U}$ with $\mathcal{T}, \mathcal{U} \subseteq \mathcal{D}$. The domain extension problem is then to construct an efficiently computable permutation $P : \mathcal{X} \to \mathcal{X}$ on a *larger* set $\mathcal{X} \supseteq \mathcal{D}$ while maintaining the property that $P(x) = G(x)$ for all $x \in \mathcal{T}$. Contrast this with the domain completion problem, in which we would only be interested in constructing a permutation on $\mathcal{D}$. As GRY observe, we cannot hope to construct a random permutation on the larger domain $\mathcal{X}$, since points in the table are all randomly mapped into the smaller domain $\mathcal{D}$, which would be very unlikely in a random permutation on $\mathcal{X}$. Nevertheless, the same algorithm we presented in Sect. 6 would work for this new problem, and multiple rounds would mix the points outside of the table as well as possible.

PARTITIONED DOMAINS. Another application of Cycle Slicer would be in a situation where we need a permutation on a domain $\mathcal{X}$ that can be partitioned into k smaller domains $\mathcal{X}_1, \ldots, \mathcal{X}_k$, and we want the permutation to map points in $\mathcal{X}_i$ to $\mathcal{X}_i$ for all i. For example, we might want a permutation on n bit strings that maps each point to another point with the same first three bits. To apply Cycle Slicer to this problem, we would let the inclusion function allow the swap of two points only if they come from the same domain $\mathcal{X}_i$.

PROGRAMMED IDENTITY MAPPINGS. One last application we will mention is in constructing permutations with some identity mappings that we may want "programmed". For example, we may want points $x, y, z \in \mathcal{X}$ to all be mapped to

---

[5] For example, many ranking algorithms use a procedure similar to cycle walking.

themselves, while all other points should be randomly mixed by the permutation. This is essentially a special case of domain completion, so our techniques in Sect. 6 would apply.

Acknowledgements. We thank the anonymous ASIACRYPT reviewers for providing detailed feedback that really helped improve the paper.

A   Proof of Mixing Time Theorem

Here we provide the details of the proof of Theorem 1 which bounds the mixing time of a matching exchange process with parameters p_1 and p_2. Recall that p_1 is the probability a particular pair (x, y) is in the matching and p_2 is the probability that conditioned on one pair being in the matching, a second pair is also in the matching (this is formally defined in Sect. 4). Our proof relies heavily on the techniques used by Czumaj and Kutylowski [7] and the work by Miracle and Yilek [17] in the context of the Reverse Cycle Walking algorithm. We use the same techniques and thus our main contribution is the definition of the parameters p_1 and p_2 and the determination of explicit constants in terms of p_1 and p_2 which apply to more general matching exchanges processes rather than the specific Reverse Cycle Walking algorithm given in [17]. The proof of Czumaj and Kutylowski [7] gives a result on the mixing time of matching exchange processes that we can directly apply to our settings however their result is an asymptotic result and they do not provide explicit constant. Similar to Miracle and Yilek [17], we provide explicit constants in terms of p_1 and p_2 by reproving several key lemmas. For completeness we give a brief overview of the entire proof here and describe in details the two lemmas we have modified. The complete details of the proof can be found in [17].

As in [17] we will use the delayed path coupling theorem introduced by Czumaj and Kutylowski which builds on the techniques of coupling and path coupling both of which are commonly used techniques in the Markov chain community (see e.g., [13,22]). A *coupling* of a Markov chain with state space Ω is a joint Markov process on $\Omega \times \Omega$ where Ω is the state space. A coupling requires that the marginal probabilities agree with the Markov chain probabilities and that once the processes collide they move together. The expected time until the two processes collide gives a bound on the mixing time. We will let the distance d between two configurations be defined as the minimum number of transpositions (or exchanges of two elements) needed to transition from one configuration to the other. Using delayed path coupling, like with path coupling, we can restrict our attention to pairs of configurations that initially differ by a single transposition (pairs at distance one) and show that with sufficiently high probability after a logarithmic number of steps of a matching exchanges processes, the two processes will have collided.

Theorem 3. *Let d be a distance metric defined on $\Omega \times \Omega$ which takes values in $\{0, \ldots, D\}$, let $U = \{(x, y) \in \Omega \times \Omega : d(x, y) = 1\}$ and let δ be a positive integer.*

Let $(x_t, y_t)_{t \in \mathbb{N}}$ be a coupling for $\mathcal{M}$, such that for every $(x_{t\delta}, y_{t\delta}) \in U$ it holds that $\mathbf{E}\left[d(x_{(t+1)\delta}, y_{(t+1)\delta}) \right] \leq \beta$ for some real $\beta < 1$. Then,

$$\tau_{\mathcal{M}}(\epsilon) \leq \delta \cdot \left\lceil \frac{\ln(D * \epsilon^{-1})}{\ln \beta^{-1}} \right\rceil .$$

As in the result of Miracle and Yilek we will use the same exact coupling introduced by Czumaj and Kutylowski. We provide a very brief description here but a more thorough one can be found in [7,17]. For one process we will matchings $m_1, m_2, \ldots$ that are selected completely randomly according to our matching exchange process probabilities. However for the second process we will use matchings $n_1, n_2, \ldots$ that are very closely related to the first set of matchings $m_1, m_2, \ldots$. Consider a pair of configurations that differ by a single inversion (x, y). If the exact same inversion is selected as a pair in the first matching m_1 then it is possible to couple the processes in one step by using the same matching for both processes but different bit flips for the pair (x, y). However this happens with probability p_1 which in our applications is only $\Theta(1/n)$ and does not give a tight enough bound. Czumaj and Kutylowski observed that it is likely that both x and y will be paired with different elements (x, z) and (y, w). If (x, y) are paired in the next matching m_2 they will choose $n_1 = m_1 - (x, z) - (y, w) + (y, z) + (x, w)$ and then different bit flips for (x, y) in m_2 and n_2. If (z, w) are paired in the next matching m_2 then they will choose $m_1 = n_1$ and then different bit flips for (x, y) in m_2 and n_2. In either case the two processes will have coupled (or collided) after 2 steps. Czumaj and Kutylowski refer to these pairs (x, y) and (y, z) as good pairs and they show that at each step the set of good pairs doubles and after $\Omega(\log(n))$ steps there are a linear number of these and thus it is very likely that a good pair will be an edge in one of the next few matchings. This implies that the matchings $n_1, n_2, \ldots$ can be selected strategically so the processes will collide. More formally a the set of good pairs is defined below.

Definition 1 (Czumaj, Kutylowski). *Without loss of generality, assume X_0 and Y_0 differ by a (x, y) transposition and let $GP_0 = \{(x, y)\}$. For each $(x, y) \in GP_{t-1}$:*

1. *If neither x or y is part of the matching M_t then $(x, y) \in GP_t$.*
2. *If $(x, w) \in M_t$ and y is not part of M_t then $(w, y) \in GP_t$.*
3. *If $(y, w) \in M_t$ and x is not part of M_t then $(w, x) \in GP_t$.*
4. *If (x, w), $(y, z) \in M_t$ then if neither w or z are part of pairs in GP_t then $(w, z) \in GP_t$ and $(x, y) \in GP_t$.*
5. *Otherwise, $(w, z) \in GP_t$.*

In order to prove our theorem we need to show first that the after t_1 steps of the matching exchange process, there are a linear number of good pairs and then that after a t_2 additional steps we will select one of the good pairs as an edge in our matching. These two requirements are formalized in the two lemmas which we prove below. Although we provide more general bounds in terms of p_1 and p_2, the proof of our lemmas below relies heavily on techniques introduced by

Miracle and Yilek [17]. Again for completeness we include the full analysis with our new parameters but the techniques remain the same as in [17]. Combining the lemmas with Czumaj and Kutylowski's coupling [7] and using the delayed path coupling theorem (Theorem 3) proves Theorem 1. More details can be found in [7,17].

Lemma 1. *Let $|GP_t|$ be the number of good pairs at step t, and*
$t_1 = \max(40\ln(2n^2), 10\ln(n/9)/\ln(1 + p_1p_2(7/36)((7/9)n^2 - n)))$ *then*

$$\Pr\left[\,|GP_{t_1}| < n/9\right] \le .5n^{-2}.$$

Proof. Initially (at step $t = 1$) there is one good pair namely, (x, y). At any step t, an existing good pair (x, y) contributes two good pairs to GP_{t+1} if both x and y are mapped to points that are not currently good pairs (see part 4 of Definition 1). We will start by bounding the probability that an existing good pair creates two good pairs at any step in a matching exchange process (i.e., part 4 is selected) in terms of p_1 and p_2. We begin by assuming that there are less than $n/9$ good pairs. If at any step t in the process there are more (i.e., $|GP_t| \ge n/9$) than we are done. Since each good pair contains two points, this implies there are at most $2n/9$ points in good pairs and at least $7n/9$ points not in good pairs. Recall that p_1 is the probability that a particular pair (x, y) is in the matching. Here we require not only that (x, y) is in the matching but that the associated bit-flip is true and the points are exchanged. Since there are least $7n/9$ points not in good pairs, the probability of a point x being match to a point not already in a good pair is $(p_1/2) * (7n/9))$. Note that the Recall that p_2 is the probability that conditioned on a first pair (x, y) being part of a matching, a second pair (w, z) is also in the matching. Given that x is matched to a point that is not a good pair there are at least $7n/9 - 1$ points remaining that are not in good pairs (i.e., not in GP_t) and thus the probability of y also being matched to a point that is not a good pair is at least $(p_2/2)(7n/9-1)$. Let p_4 be the probability that a particular good pair causes part 4 of the good pair definition to be selected (i.e., it results in two good pairs after the next step), then we have:

$$p_4 \ge (p_1 * p_2/4)(7n/9)(7n/9 - 1) = p_1p_2(7/36)((7/9)n^2 - n).$$

As shown in [17] if we let growth rate $G_t = (|GP_{t+1}| - |GP_t|)/|GP_t|$ then by linearity of expectations, $\mathbf{E}\left[G_t\right] = p_4$. Define an indicator random variable Z_t for when G_t exceeds one half it's expectation or $G_t \ge p_4/2$. Each time Z_t is one the number of good pairs increases at least by a factor of $1 + p_4/2$. This implies that if $\sum_{t=0}^{t_1} Z_t \ge \frac{\ln n/9}{\ln(1+p_4/2)}$ then the number of good pairs (i.e., $|GP_{t_1}|$) is at least $(1 + p_4/2)^{(\ln n/9)/\ln(1+p_4/2)} = n/9$, as desired. It is straightforward to show using Markov's inequality that $\Pr\left[Z_t = 0\right] \le 4/5$ (see [17]). It is important to note that the Z_i's are not independent since the growth rate is more likely to be higher when there are fewer good pairs. In our analysis so far we are assuming there are always at most $n/9$ good pairs which holds throughout and thus this process is lower bounded by a process with independent variables $W_1, \ldots W_{t_1}$

where each variable W_i is 1 with probability $1/5$ and 0 with probability $4/5$ which will allow us to apply a Chernoff bound. We will use the following well-known bound $\Pr[W < \mathbf{E}[W]/2] < \exp(-\mathbf{E}[W]/8)$ with $W = \sum_{t=0}^{t_1} W_t$ and $t_1 = \max(40\ln(2n^2), 10\frac{\ln(n/9)}{\ln(1+p_4/2)})$. The selection of t_1 implies that $\mathbf{E}[W] \geq (1/5)40\ln(2n^2) = 8\ln(2n^2)$. Therefore,

$$\Pr[W < \mathbf{E}[W]/2] < \exp(-\mathbf{E}[W]/8) <= \exp(-8\ln(2n^2)/8) = .5n^{-2}.$$

And similarly, $\mathbf{E}[W] \geq (1/5)10\frac{\ln n/9}{\ln(1+p_4/2)} = 2\frac{\ln n/9}{\ln(1+p_4/2)}$. Together these prove Lemma 1,

$$\Pr\left[W < \frac{\ln n/9}{\ln(1+p_4/2)}\right] < \Pr[W < \mathbf{E}[W]/2] < .5n^{-2}.$$

$\square$

The final component of our proof is to show that the matchings over the next t_2 steps will contain a good pair with probability at least $1 - .5n^{-2}$. As in [7,17], we say that a pair (x,y) is part of a *potential* matching if the process maps x to y regardless of the value of the bit-flip. Here, we are interested in the probability that a potential matching contains a good pair.

Lemma 2. *Let $t_2 = 72\ln(2n^2)/(p_1 n)$ then conditioned on $|GP_{t_1}| \geq n/9$, the probability that the next t_2 potential matchings contain no edges from GP_{t_1} is at most $.5n^{-2}$.*

Proof. First, consider a particular good pair (x,y). The probability that x is mapped to y at any step is p_1. There are at last $n/9$ good pairs and thus, by linearity of expectations, the expected number of good pairs in any potential matching is at least $p_1 n/9$. The potential matchings at each step in a matching exchange process are independent and thus we can apply same Chernoff bound above. Define the random variable E_t be the number of edges in the potential matching at time t that correspond to a good pair. Then using Chernoff, we have

$$\Pr\left[\sum_{t=t_1}^{t_1+t_2} E_t < 4\ln(2n^2)\right] < \exp(-8\ln(2n^2)/8) = .5n^{-2}$$

which directly implies that $\Pr\left[\sum_{t=t_1}^{t_1+t_2} E_t < 1\right] < .5n^{-2}$.

$\square$

References

1. Bellare, M., Hoang, V.T., Tessaro, S.: Message-recovery attacks on feistel-based format preserving encryption. In: Weippl, E.R., Katzenbeisser, S., Kruegel, C., Myers, A.C., Halevi, S. (eds.) ACM CCS 2016, pp. 444–455. ACM Press, October 2016
2. Bellare, M., Ristenpart, T., Rogaway, P., Stegers, T.: Format-preserving encryption. In: Jacobson, M.J., Rijmen, V., Safavi-Naini, R. (eds.) SAC 2009. LNCS, vol. 5867, pp. 295–312. Springer, Heidelberg (2009). doi:10.1007/978-3-642-05445-7_19

3. Bellare, M., Rogaway, P., Spies, T.: The FFX mode of operation for format-preserving encryption. http://csrc.nist.gov/groups/ST/toolkit/BCM/documents/proposedmodes/ffx/ffx-spec.pdf
4. Black, J., Rogaway, P.: Ciphers with arbitrary finite domains. In: Preneel, B. (ed.) CT-RSA 2002. LNCS, vol. 2271, pp. 114–130. Springer, Heidelberg (2002). doi:10.1007/3-540-45760-7_9
5. Brier, E., Peyrin, T., Stern, J.: BPS: a format-preserving encryption proposal.http://csrc.nist.gov/groups/ST/toolkit/BCM/documents/proposedmodes/bps/bps-spec.pdf
6. Brightwell, M., Smith, H.: Using datatype-preserving encryption to enhance data warehouse security. In: National Information Systems Security Conference (NISSC) (1997)
7. Czumaj, A., Kutylowski, M.: Delayed path coupling and generating random permutations. Random Struct. Algorithms 17, 238–259 (2000)
8. Durak, F.B., Vaudenay, S.: Breaking the FF3 format-preserving encryption standard over small domains. In: Katz, J., Shacham, H. (eds.) CRYPTO 2017. LNCS, vol. 10402, pp. 679–707. Springer, Cham (2017). doi:10.1007/978-3-319-63715-0_23
9. Dworkin, M.: Recommendation for block cipher modes of operation: methods for format preserving-encryption. NIST Special Publication 800–38G (2016). http://dx.doi.org/10.6028/NIST.Spp.800-38G
10. Dyer, K.P., Coull, S.E., Ristenpart, T., Shrimpton, T.: Protocol misidentification made easy with format-transforming encryption. In: Sadeghi, A.R., Gligor, V.D., Yung, M. (eds.) ACM CCS 2013, pp. 61–72. ACM Press, November 2013
11. Grubbs, P., Ristenpart, T., Yarom, Y.: Modifying an enciphering scheme after deployment. In: Coron, J.-S., Nielsen, J.B. (eds.) EUROCRYPT 2017. LNCS, vol. 10211, pp. 499–527. Springer, Cham (2017). doi:10.1007/978-3-319-56614-6_17
12. Hoang, V.T., Morris, B., Rogaway, P.: An enciphering scheme based on a card shuffle. In: Safavi-Naini, R., Canetti, R. (eds.) CRYPTO 2012. LNCS, vol. 7417, pp. 1–13. Springer, Heidelberg (2012). doi:10.1007/978-3-642-32009-5_1
13. Levin, D.A., Peres, Y., Wilmer, E.L.: Markov Chains and Mixing Times. American Mathematical Society, Providence (2006)
14. Liskov, M., Rivest, R.L., Wagner, D.: Tweakable block ciphers. J. Cryptol. 24(3), 588–613 (2011)
15. Luchaup, D., Dyer, K.P., Jha, S., Ristenpart, T., Shrimpton, T.: LibFTE: a toolkit for constructing practical, format-abiding encryption schemes. In: Proceedings of the 23rd USENIX Security Symposium, pp. 877–891 (2014)
16. Luchaup, D., Shrimpton, T., Ristenpart, T., Jha, S.: Formatted encryption beyond regular languages. In: Ahn, G.J., Yung, M., Li, N. (eds.) ACM CCS 2014, pp. 1292–1303. ACM Press, November 2014
17. Miracle, S., Yilek, S.: Reverse cycle walking and its applications. In: Cheon, J.H., Takagi, T. (eds.) ASIACRYPT 2016. LNCS, vol. 10031, pp. 679–700. Springer, Heidelberg (2016). doi:10.1007/978-3-662-53887-6_25
18. Morris, B., Rogaway, P.: Sometimes-recurse shuffle - almost-random permutations in logarithmic expected time. In: Nguyen, P.Q., Oswald, E. (eds.) EUROCRYPT 2014. LNCS, vol. 8441, pp. 311–326. Springer, Heidelberg (2014). doi:10.1007/978-3-642-55220-5_18
19. Morris, B., Rogaway, P., Stegers, T.: How to encipher messages on a small domain. In: Halevi, S. (ed.) CRYPTO 2009. LNCS, vol. 5677, pp. 286–302. Springer, Heidelberg (2009). doi:10.1007/978-3-642-03356-8_17
20. Naor, M., Reingold, O.: Constructing pseudo-random permutations with a prescribed structure. J. Cryptol. 15(2), 97–102 (2002)

21. Ristenpart, T., Yilek, S.: The mix-and-cut shuffle: small-domain encryption secure against N queries. In: Canetti, R., Garay, J.A. (eds.) CRYPTO 2013. LNCS, vol. 8042, pp. 392–409. Springer, Heidelberg (2013). doi:10.1007/978-3-642-40041-4_22
22. Sinclair, A.: Algorithms for Random Generation and Counting. Progress in Theoretical Computer Science. Birkhäuser (1993)
23. Spies, T.: Format-preserving encryption. Unpublished whitepaper (2008). https://www.voltage.com/wp-content/uploads/Voltage-Security-WhitePaper-Format-Preserving-Encryption.pdf

Symmetrically and Asymmetrically Hard Cryptography

Alex Biryukov[1([⊠])] and Léo Perrin[2,3]

[1] University of Luxembourg, Belval, Luxembourg
alex.biryukov@uni.lu
[2] Inria, Paris, France
leo.perrin@inria.fr
[3] SnT, University of Luxembourg, Belval, Luxembourg

Abstract. The main efficiency metrics for a cryptographic primitive are its speed, its code size and its memory complexity. For a variety of reasons, many algorithms have been proposed that, instead of optimizing, try to increase one of these hardness forms.

We present for the first time a unified framework for describing the hardness of a primitive along any of these three axes: code-hardness, time-hardness and memory-hardness. This unified view allows us to present modular block cipher and sponge constructions which can have any of the three forms of hardness and can be used to build any higher level symmetric primitive: hash function, PRNG, etc.

We also formalize a new concept: *asymmetric hardness*. It creates two classes of users: common users have to compute a function with a certain hardness while users knowing a secret can compute the same function in a far cheaper way. Functions with such an asymmetric hardness can be directly used in both our modular structures, thus constructing any symmetric primitive with an asymmetric hardness. We also propose the first asymmetrically memory-hard function, DIODON.

As illustrations of our framework, we introduce WHALE and SKIPPER. WHALE is a code-hard hash function which could be used as a key derivation function and SKIPPER is the first asymmetrically time-hard block cipher.

Keywords: White-box cryptography · Memory hardness · Big-key encryption · SKIPPER · WHALE · DIODON

1 Introduction

The design of cryptographic algorithms is usually a trade-off between security and efficiency. Broadly speaking, the efficiency of an algorithm is defined along three axes: time, memory and code size. Yet in some scenarios it is desirable to design primitives that are purposefully inefficient for one or several of these

The work of Léo Perrin was supported by the CORE project ACRYPT (ID C12-15-4009992) funded by the Fonds National de la Recherche, Luxembourg.

© International Association for Cryptologic Research 2017
T. Takagi and T. Peyrin (Eds.): ASIACRYPT 2017, Part III, LNCS 10626, pp. 417–445, 2017.
https://doi.org/10.1007/978-3-319-70700-6_15

metrics. This can be done to slow down the attackers, provide different levels of service to privileged and non-privileged users, adjust cost of operation in proof-of-work schemes, etc. Primitives with different forms of computational hardness have been scattered in time and in several seemingly unrelated research/application areas. In this paper we propose a simple unifying framework which allows us to build new provably hard modes of operation and primitives.

The simplest illustration of functions designed to be time consuming to compute is that of key derivation functions (KDF). A KDF is typically built by iterating a one-way function (for example a cryptographic hash function), multiple times. Such functions are intended to prevent an adversary from brute-forcing a small set of keys (corresponding to, say 12 letter strings) by making each attempt very costly.

Time, however, is not the only form of hardness for which an artificial increase can be beneficial. Memory-hardness was one of the design goals of the winner of the Password Hashing Competition, Argon2 [12], the aim being to prevent hardware optimization of the primitive. As another example, one research direction in white-box cryptography is nowadays focusing on designing block ciphers such that the code implementing them is very large in order to prevent duplication and distribution of their functionality [7,11,16,17,22]. In this case, the aim could be to implement some form of Digital Right Management or to prevent the exfiltration of a block cipher key by malware.

Since hardness is an inherently expensive property, there are cases where a trap-door could be welcome. This is the case for the most recent weak white-box block ciphers [11,16,22]: while the white-box implementation requires a significant code size, there exists a functionally equivalent implementation which is much smaller but cannot be obtained unless a secret is known. That way, two classes of users are created: those who know the secret and can evaluate the block cipher efficiently and those who do not and thus are forced to use the code-hard implementation.

The different forms of hardness, their applications and typical instances from the literature are summarized in Table 1.

Table 1. Six types of hardness and their applications.

	Time	Memory	Code size
Applications	KDF, time-lock	Password hashing, egalitarian computing	White-box crypto, big-key encryption
Symmetrically hard functions	PBKDF2 [25]	Argon2 [12], Balloon [18]	XKEY2 [7], WHALE (Sect. 5.2)
Asymmetrically hard functions	RSA-lock [30], SKIPPER (Sect. 5.1)	DIODON (Sect. 2.4)	White-box block ciphers [11,16,17,22]

Our Contribution. Regardless of the form of hardness, the aim of the designer of a hard primitive is to prevent an attacker from by-passing this complexity, even if the attacker is allowed significant precomputation time. Informally, a user cannot decrease the hardness of the computation below a certain threshold. Inspired by the formal definitions of hardness used in white-box cryptography, we provide for the first time a unified framework to study and design cryptographic algorithms with all forms of hardness. Our framework is easily extended to encompass *asymmetric hardness*, a form of complexity which can be bypassed provided that a secret is known.

Our approach consists in combining simple functions called *plugs* having the desired form of hardness with secure cryptographic primitives. Algorithms are then built in such a way as to retain the cryptographic security of the latter while forcing users to pay for the full hardness of the former. In fact, we provide a theoretical framework based on random oracles which reduces the hardness of the algorithms we design to that of their plugs.

Furthermore, we introduce the first asymmetrically memory-hard function, DIODON. It is a function based on `scrypt` [28] modified in such a way as to allow users knowing an RSA private key to evaluate it using a constant amount of memory. It can of course be used as a plug within our framework.

Finally, we used this approach to build a code-hard hash function called WHALE and an asymmetrically time-hard block cipher called SKIPPER. It is impossible to design a functionally equivalent implementation of the WHALE hash function which is much smaller than the basic one. On the other hand, encrypting a block with SKIPPER is time consuming but this time can be significantly decreased if an RSA private key is known.

Outline. First, Sect. 2 provides more details about the different forms of hardness and their current usage for both symmetric and asymmetric hardness. We also introduce the first asymmetrically memory-hard function, DIODON, in Sect. 2.4. Then, Sect. 3 presents our generic approach for dealing with all forms of hardness at once. To this end, plugs achieving all forms of hardness are introduced in the same section. We deduce practical modes of operation for building hard block ciphers and hard sponges which are described in Sect. 4. Our concrete proposals, called SKIPPER and WHALE, are introduced in Sect. 5.

2 Enforcing Hardness

In this section, we argue that many recent ideas in symmetric cryptography can be interpreted as particular cases of a single general concept. The aim of several a priori different research areas can be seen as imposing the use of important resources for performing basic operations or in other words, *bind an operation to a specific form of hardness*. We restrict ourselves to the basic case of a well-defined function mapping each input to a unique output. It means in particular that protocols needing several rounds of communication or randomized algorithms which may return any of the many valid solution to a given problem such as HashCash (see below) are out of our scope. We also tie each function to an algorithm evaluating it: further below, functionally equivalent functions evaluated using different algorithms are considered like different functions. Thus, the "hardness" discussed is the hardness of the algorithm tied to the function.

The three main metrics for assessing the efficiency of an algorithm are its time complexity, its RAM usage and its code size. As we explain below, different lines of research in symmetric cryptography can be interpreted as investigating the design of algorithms such that one of these metrics is abnormally high and cannot be reduced while limiting the impact on the other two as much as possible.

Time-hardness is discussed in Sect. 2.1, *memory-hardness* in Sect. 2.2 and *code-hardness* in Sect. 2.3. Finally, in Sect. 2.4, we present the general notion of *asymmetric hardness*.

It is also worth mentioning that the three forms of hardness are not completely independent from one another. For example, due to the number of memory access needed in order for a function to be memory-hard, a function with this property cannot be arbitrarily fast.

2.1   Time Hardness

While the time efficiency of cryptographic primitives is usually one of the main design criteria, there are cases where the opposite is needed. That is, algorithms which can be made arbitrarily slow in a controlled fashion.

One of the most simple approaches is the one used for instance by the key derivation function PBKDF2 [25]. This function derives a cryptographic key from a salt and a password by iterating a hash function multiple times, the aim being to frustrate brute-force attacks. Indeed, while the password may be from a space small enough to be brute-forced, evaluating the key derivation function for each possible password is made infeasible by its time-hardness.

Somewhat similarly, proofs-of-work such as HashCash (used by the cryptocurrency Bitcoin [26]) consist in finding at least one of many solutions to a given problem. The hardness in this case comes from luck. Miners must find a value such that the hash of this value and the previous block satisfies the difficulty constraint. However, the subset of such valid values is sparse and thus miners have to try many random ones. Two different miners may find two different but equally valid values. Because of this randomness, such puzzles are out of our scope. In this paper, we only consider functions which are equally hard to evaluate on all possible inputs, not puzzles for which finding a solution is hard *on average*.

Furthermore, in order to mitigate the impact of adversaries with vast amount of processors at their disposal, we consider sequential time-hardness. Using a parallel computer should not help an attacker in evaluating the function much quicker. Formalizing parallel time hardness the way we do it for sequential time-hardness is left as a future work.

Overall, the goal of time-hardness is to prevent an adversary from computing a function in a time significantly smaller than the one intended. In other words, it must be impossible to compress the amount of time needed to evaluate the function on a random input.

2.2   Memory Hardness

Informally, a function is memory-hard if even an optimized implementation requires a significant amount of memory. For each evaluation, a large amount of information is written and queried throughout the computation. As a consequence, a memory-hard function cannot be extremely fast.

A function requiring large amounts of RAM for its computation prevents attacker from building ASICs filled with huge number of cores for parallel computations. This implies that memory-hard functions make good password hashing functions and proofs-of-work. One of the first to leverage this insight was the hash function `scrypt` [28] which was recently formally proved to be memory-hard in [3]. More recently several other memory-hard algorithms have been designed, such as the password hashing competition winner Argon2 [12] as well as the more recent Balloon Hashing [18] and Equihash [14]. Those can be used as building blocks to create memory-hard proofs-of-work which can offset the advantage of cryptocurrency miners using dedicated ASICs.

The idea of using memory-hard functions for general purpose computations was further explored in the context of *egalitarian computing* [13]. Similarly, *proofs of space* [5,21] are protocols which cannot be run by users if they are not able to both read and write a large amount of data. However, those are interactive protocols and not functions.

The recent research investigating memory-hardness has lead to several advances in our understanding of this property. For example, the difference between *amortized* and *peak* memory hardness was highlighted in [4].

2.3   Code Hardness

First of all, let us clarify the distinction we make between *memory* and *code*-hardness. With code-hardness, we want to increase the space needed to store information that is needed to evaluate a function on all possible inputs. However, the information itself does not depend on said input. During evaluation of the function, it is only necessary to *read* the memory in which the code is stored. In contrast, memory-hardness deals with the case where we need to store a large amount of information which depends on the function input and which is thus different during each evaluation of the function. In this case, one must be able to both read *and write* to the memory. Furthermore, in a typical code-hard function, only a small fraction of the whole information stored in the implementation is read during each call to the function. On the other hand, if a memory-hard function uses M bytes of memory, then all of those bytes will be written and read at least once.

Code-hardness is very close to what was first defined as *memory-hard white-box implementation (or weak white-box)* in the paper introducing the ASASA crypto-system [11], and later formalized under different names as (M, z)-space hardness [16,17] or as incompressibility in [22] following the more general definition from [20]. In all cases, the aim is the same: the block cipher implementation must be such that it is impossible to write a functionally equivalent implementation with a smaller code. This stands in contrast to *strong* white-box cryptography (as defined in [11]) where inverting a function given its white-box implementation should be impossible. We do not consider this case in this paper.

As was pointed out in [22], what we call code-hardness is also the goal of so-called *big-key encryption*. For instance, the XKEY2 scheme introduced in [7] achieves this goal: it uses a huge table and a nonce to derive a key of regular size (say, 128 bits) to be used in a standard encryption algorithm, e.g. a stream cipher. Bellare et al. show that even if an attacker manages to obtain half of the huge table, i.e. half of the code needed to implement the scheme, then they are still unable to compute the actual encryption key with non-negligible probability. Using our terminology, XKEY2 can be seen as a code-hard key derivation function. A more detailed analysis of the literature on code-hardness is provided in the full version of this paper [15].

The concept of *proof of storage* can also be interpreted as a particular type of code-hard protocol. Indeed, in such algorithms, challengers must prove that they have stored a given file.

2.4 Asymmetric Hardness

In this section, we discuss the concept of *asymmetric hardness* which introduces two classes of users. *Common users* evaluate a hard function but *privileged users*, who know a secret key, can evaluate a functionally equivalent function which is *not* hard. We also introduce DIODON, the first asymmetrically memory-hard function.

Asymmetric Code-Hardness. The most recent white-box block ciphers such as SPACE [16], the PuppyCipher [22] and SPNbox [17] can be seen as providing asymmetric code-hardness. Indeed, while the first aim of these algorithms is to provide regular code-hardness, referred to as "space-hardness" for the former and "incompressibility" for the latter, they both allow the construction of far more code-efficient implementation. For both SPACE and the PuppyCipher, the idea is to compute a large table containing the encryptions of the first 2^t integers with AES-128 for t in $\{8, 16, 24, 32\}$. These tables are then used as the code-hard part of the encryption which cannot be compressed because doing so would require a break of the AES. However, a user knowing the 128-bit AES key can get rid of these tables and merely recompute the entries needed on the fly, thus drastically decreasing the code-hardness of the implementation.

In fact, both constructions can be seen as structures intended to turn an asymmetrically code-hard function into an asymmetrically code-hard block cipher. In both cases, the asymmetrically code-hard function consists in the evaluation of the AES on a small input using either the secret key, in which case the implementation is not code-hard, or using only the public partial codebook which, because of its size, is code-hard.

Time-Hardness. While the asymmetry of its hardness was not insisted upon, there is a known asymmetrically time-hard function, which we call RSA-lock. It was proposed as a time-lock in [30], that is, a function whose output cannot be known before a certain date.

It is based on the RSA cryptosystem [29]. It consists in iterating squarings in an RSA modular ring: a user who does not know the prime factor decomposition of the modulus N must perform t squarings while a user who knows that $N = pq$ can first compute $e = 2^t \mod (p-1)(q-1)$ and then raise the input to the power e. If t is large enough, the second approach is much faster.

Memory-Hardness. We are not aware of any asymmetrically memory-hard function in the existing literature. Thus, we propose the first such function which we call DIODON. It is based on the ROMix function used to build `scrypt` [28] and relies on the RSA crypto-system to provide the asymmetry, much like in RSA-lock.

DIODON maps an element x of $\{0,1\}^t$ to a an element y of $\{0,1\}^u$ using RSA computations and a hash function H. It is parametrized by an RSA modulus N of size n_p, a hash function H and its input and output sizes t and u. Only privileged users know the prime factor decomposition of the RSA modulus. Finally, a parameter M is used to tune its memory-hardness while parameters L and η decide its time complexity. The computation is described in Algorithm 1 and in Fig. 1. In both, T_u denotes the function truncating a bitstring to its u bits of lowest weight.

Algorithm 1. DIODON Asymmetrically memory-hard function
Inputs: t-bit block x; RSA modulus N of n_p bits; M, L;
Output: u-bit output y

$V_0 = x$
for all $i \in \{1, ..., M-1\}$ **do**
$\quad V_i = V_{i-1}^{2^\eta} \mod N$
end for
$S = V_{M-1}$
for all $i \in \{0, ..., L-1\}$ **do**
$\quad j = S \mod M$
$\quad S = H(S, V_j)$
end for
return $T_u(S)$

A user without knowledge of the factorization of N must use Algorithm 1 to evaluate DIODON(x). However, if a user knows the factorization of $N = qq'$, she does not need to store the vector V. Indeed, she can simply evaluate $V_i = x^{2^{i \times \eta}} \mod N$ in constant time by first reducing $2^{i \times \eta}$ modulo $(q-1)(q'-1)$ and then raising x to the corresponding exponent. One may call her ability to access an arbitrary element of V in constant time an *RSA RAM*. Her evaluation strategy

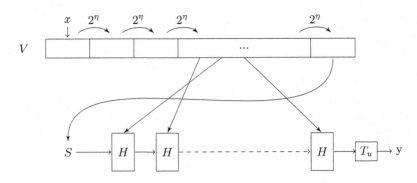

Fig. 1. The evaluation of $y = \text{DIODON}(x)$.

is summarized in Algorithm 2. Basic users need to perform $\eta \times M$ RSA squarings while privileged ones need to perform roughly $(L + 1) \times n_p$ of those as e and each exponent e_j is a priori of length n_p.

Algorithm 2. DIODON for privileged users
Inputs: t-bit block x; RSA factors q, q'; η; M, T;
Output: u-bit output y

$e = 2^{(M-1) \times \eta} \mod (q-1)(q'-1)$
$S = x^e \mod (qq')$
for all $i \in \{0, ..., L-1\}$ **do**
$\quad j = S \mod M$
$\quad e_j = 2^{j \times \eta} \mod (q-1)(q'-1)$
$\quad S = H\big(S, (x^{e_j} \mod (qq'))\big)$
end for
return $T_u(S)$

Let us first consider the simplest parameters, that is $\eta = n_p$ and $L = M$. In this case, the time complexity for both users is comparable as each class of user needs to perform M RSA encryptions.[1] Furthermore, without the knowledge of the secret key, the computation is essentially scrypt ROMix function which was shown to be optimally linearly memory hard [3], keeping time-memory product constant. It means that it is either necessary to store all the values V_i for the computation or any algorithm that saves a factor f in memory will have to pay the same factor in time.

The choice of the parameters η, M, L and n_p has a significant impact on the time efficiency of DIODON. It is investigated in the full version of this paper [15] where we also propose some concrete instances.

---

[1] The privileged user still has a small advantage in this case since knowing the prime factor decomposition of the modulus allows the use of the Chinese Remainder Theorem (DRT) to speed the exponentiation.

3   A Generic Framework

As we have seen, all the techniques presented in the sections above are intended to enforce some form of computational hardness. In this section, we present a unified framework for building any symmetric algorithm with some form of guaranteed hardness. We describe our aim in Sect. 3.1 and our design strategy with the generic hardness definition it is based on in Sect. 3.2. Our constructions need small functions with the intended hardness type to be bootstrapped. We provide examples of those in Sect. 3.3.

3.1   Design Strategy

Our aim is to design a general approach to build any cryptographic primitive with any form of hardness. To achieve this, we will build modes of operations allowing us to combine secure cryptographic primitives, such as the AES block cipher [19] or the Keccak sponge [9], with small functions called *plugs*. These plugs are simple functions with the desired form of hardness.

Our modes of operations, which are presented in Sect. 4, are all based on the same principle: ensuring that enough plug calls with an unpredictable input are performed so as to guarantee that, with overwhelming probability, an adversary cannot bypass the hardness of all plug evaluations. This ensures that the full complexity of a plug evaluation is paid at least once.

Indeed, regardless of the hardness form considered, the strategy of a generic adversary will always be the same. Provided that the plugs are indeed hard to evaluate, the only strategy allowing an adversary to bypass their hardness consists in storing a (feasibly) large number of plug outputs in a database and then querying those. If 2^p outputs of a plug $P : \{0,1\}^t \rightarrow \{0,1\}^v$ have been stored, the plug can be evaluated successfully *without paying for its hardness* with probability 2^{p-t}.

An alternative strategy using the same space consists in storing $2^p(v/d)$ partial outputs of length d. In this case, the success probability becomes $2^{p-t}(v/d) \times 2^{d-v}$: the input is partially known with a higher probability $2^{p-t}(v/d)$ but $(v-d)$ bits of the output remain to be guessed. This method is $(v/d) \times 2^{d-v}$ more efficient than the basic one but, for $1 \leq d < v$, this quantity is always smaller than one. The strategy consisting in storing full outputs is therefore the best.

However, if the output size of the plug is small enough, it might be more efficient for the adversary to directly guess the whole output. The probability that an adversary merely guessing the output of the plug gets it right[2] is 2^{-v}.

Our aim is therefore to guard our constructions from the adversary defined below. Protecting our structure against those is sufficient to reduce their hardness to that of the plug they use. Recall that we tie each function to a specific algorithm evaluating it. Thus, the hardness is actually that of the corresponding algorithm.

---

[2] This is only true under the assumption that the output of the plug is uniformly distributed. As we will see later, the plug can still be used if it is not the case.

Definition 1 (2^p-adversary). *Let f be a time-, memory- or code-hard func-tion. A 2^p-adversary is an adversary trying to generate a function f' which does not have the hardness of f but which does not have access to more memory than needed to store 2^p outputs of f.*

A 2^p-adversary can perform more than 2^p calls to the function it is trying to approximate when generating f', although f' itself cannot have access to more than 2^p of those. Still, f' can perform additional computations using the information stored during its generation.

However, the computational power of this adversary is not unbounded. More precisely we consider only 2^p-adversaries which cannot perform more than 2^{100} operations. This means for example that recovering a 128-bit AES key is out of their reach. We are not interested in guarding against unrealistic, computation-ally unbounded adversaries.

Similarly, the complexity of the plug approximation along the other axes cannot be arbitrarily high. Consider a code-hard function f which uses a table s of M bits and an approximation f' which works as follow:

1. Evaluate and store several values of $f(i)$ for random inputs i using far fewer than M bits of storage.
2. When given a random input x:
   (a) brute-force all possible values of s, i.e. 2^M values using M bits of RAM to store each candidate table s';
   (b) try each candidate s' to see if they yield $f'(i) = f(i)$ using the values of $f(i)$ previously stored;
   (c) once the right s has been found, use it to evaluate $f'(x)$, return it and then erase s'.

Such an approximation f' is not code-hard, it merely needs enough space to store $f(i)$ for several i. However, it is memory-hard since it needs to store the same amount of information as is stored in the implementation of f, although the storage is in RAM rather than code. Much more importantly, its time-hardness is astronomical: M will typically be in the millions, if not the billions, meaning that enumerating all 2^M candidates s' is utterly impossible.

We explicitly do not consider such extreme trade-offs: in our cases, the hard-ness of f remains practical—though expensive—, meaning that a 2^p-adversary trying to bypass its hardness is trying to optimize an already usable imple-mentation. It does not make sense for her to trade code-hardness for a wildly impractical time-hardness.

3.2   Theoretical Framework

Generic Symmetric Hardness. We are now ready to formally define hard-ness. We use a generalization to all forms of hardness of the incompressibility notion from [22].

Definition 2 (R-hardness).   *We say that a function* $f : \mathcal{X} \to \mathcal{Y}$ *is R-hard against* 2^p-*adversaries for some tuple* R $= (\rho, u, \epsilon(p))$ *with* $\rho \in$ *{Time, Code, RAM} if evaluating the function* f *using less than* u *units of the resource* ρ *and at most* 2^p *units of storage is possible only with probability* $\epsilon(p)$. *More formally, the probability for a* 2^p-*adversary to win the efficient approximation game, which is described below, must be upper-bounded by* $\epsilon(p)$.

1. *The challenger chooses a function* f *from a predefined set of functions requiring more than* u *units of* ρ *to be evaluated.*
2. *The challenger sends* f *to the adversary.*
3. *The adversary computes an approximation* f' *of* f *which, unlike* f, *can be computed using less than* u *units of the resource* ρ.
4. *The challenger picks an input* x *of* $\mathcal{X}$ *uniformly at random and sends it to the adversary.*
5. *The adversary wins if* $f'(x) = f(x)$.

This game is also represented in Fig. 2. The approximation f' *computed by the adversary must be evaluated using significantly less than* u *units of the resource* ρ, *although the precomputation may have been more expensive.*

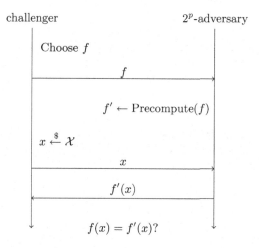

Fig. 2. The game corresponding to the definition of $(\rho, u, \epsilon(p))$-hardness against 2^p-adversaries.

In order for this definition to be relevant, the power of the adversary must be estimated. For example, preventing attacks from 2^{512}-adversaries would most definitely be over engineering and, conversely, preventing attacks from 2^{20}-adversaries would be useless since such precomputation is always feasible.

Our definition is not the strongest in the sense that it does not encompass e.g. "strong space-hardness" [16]. This definition of code-hardness aims at preventing the attacker from encrypting a plaintext *of their choosing*, a far stronger requirement than preventing the encryption of a *random* plaintext.

In the efficient approximation game described above, f' must be less hard than f along the appropriate axis. For example, if f is code-hard then the code implementing f' must be significantly smaller than that implementing f, meaning that the game corresponding to code-hardness when f is an encryption algorithm is essentially the same as the one used in the definition of encryption incompressibility [22]. Indeed, the computation of f' and its use by the adversary corresponds in this case to the computation of the leakage function on the secret large table and its use by the adversary to approximate the original table.

In the case of code-hardness the maximum code size of the implementation of f' must coincide with the power of the 2^p-adversary. Indeed, the implementation of the approximation f' needs at least enough space to store 2^p outputs of the plug.

For time-hardness, the time is measured in number of simple operations. For example, if a function requires evaluating a hash function t times, the unit of time is a hash computation. Memory- and code-hardness are measured in bytes. In our examples, code-hardness is achieved not by using programs with a complex logic but by forcing them to include large and incompressible tables. Much like in most recent white-box block ciphers, the size of the logic of the program is considered to be negligible and the size of the implementation is reduced to that of its incompressible tables.

A function which is easy to compute on a subset of its domain or whose output is not uniformly distributed in its range can still be (ρ, u, ϵ)-hard as such limitations can be taken into account by modifying the ϵ factor. For example, if a function is easy to evaluate on a fraction f of its domain then this limitation is simply captured by the fact that $\epsilon \geq 1/f$.

Generic-Asymmetric Hardness. This generic definition is easily generalized to encompass asymmetric hardness.

Definition 3 (Asymmetric R-hardness). *We say that a function $f : \mathcal{X} \to \mathcal{Y}$ is asymmetrically R-hard against 2^p-adversaries for some tuple $R = (\rho, u, \epsilon(p))$ with $\rho \in \{\,Time, Code, RAM\}$ if it is impossible for a 2^p-adversary to win the approximation game of Definition 2 with probability higher than $\epsilon(p)$, unless a secret K is known.*

If this secret is known then it is possible to evaluate another function f_K which is is functionally equivalent to f but does not have its hardness.

An immediate consequence of this definition is that extracting the secret key K from the description of f must be computationally infeasible. Otherwise, the adversary could simply recover K during the precomputation step, use f_K as their approximation and then win the approximation game with probability 1. This observation is reminiscent of the *unbreakability* notion presented in [20].

White-box block ciphers are simple example of asymmetrically code-hard functions. This concept can also be linked to the *proof of work or knowledge* presented in [6]. It is a proof of work where a solution can be found in a more efficient way if a secret is known.

Asymmetric hardness is a different notion from public key encryption. Indeed, in the latter case, the whole decryption functionality is secret. In our case, the functionality is public. What is secret is a method to evaluate it efficiently.

A Counter-Example. The inversion of a one-way function may seem like a natural example of a time-hard function. However, as described below, it may not satisfy the requirements of our definition of $(\mathsf{Time}, u, \epsilon(p))$-hardness.

Let $h : \{0,1\}^{50} \to \{0,1\}^{50}$ be the function mapping a 50-bit string to the first 50 bits of their SHA-256 digest and let f be the inverse of h. In other words, f returns a preimage of a given digest. This function may seem time-hard at first glance as SHA-256 is preimage resistant. More specifically, it might be expected to be about $(\mathsf{Time}, 2^{50}, 2^{-20})$-hard against a 2^{30}-adversary. However, as is well known, such constructions can be attacked using Hellman's tradeoff [24] in the form of rainbow-tables allowing an attacker to recover a preimage in far less time at the cost of significant but practical pre-computation and storage. If M is the size of this table and T is the time complexity of an inversion using this table then it must hold that $MT^2 = N^2$ where $N = 2^{50}$ in our case. The failure of f to be time-hard in the sense of Definition 2 can be seen in the following strategy to win the approximation game.

1. The challenger chooses a secure hash function (SHA-256) and sends its description to the adversary.
2. The 2^{30}-adversary precomputes Hellman-type rainbow tables with in total 2^{30} entries using about 2^{50} calls to h. This adversary chooses $M = 2^{30}$ and $T = N/\sqrt{M} = 2^{35}$.
3. The challenger chooses a random value $x \in \{0,1\}^{50}$ and sends it to the adversary.
4. With high probability, the adversary computes $f(x)$ using their precomputed table in time $T = 2^{35}$ which is 2^{15} times smaller than the time needed for brute-force.

Thus, such a function is not time-hard in the sense of Definition 2.

3.3 Examples of Plugs

As our modes rely on smaller R-hard function to achieve their goal, we describe an array of such components, one for each hardness goal. A summary of all the plugs we describe, along with what we consider to be their hardness against 2^p-adversaries, is given in Table 2.

While we provide an intuition on why we assume these plugs to have the hardnesses we claim, we do not prove that it is the case.

If the output it is too large to be used in a higher level construction then it is possible to truncate it to v bits. If we denote T_v the function discarding all but the first v bits of its input and if P is a plug with a t-bit input which is $(\rho, u, \epsilon(p))$-hard against 2^p-adversaries, then $x \mapsto T_m (P(x))$ is $(\rho, u, \max(\epsilon(p), 2^{p-t}))$-hard against 2^p-adversaries. Overall, the probability of success of an approximation

Table 2. Possible plugs, i.e. sub-components for our constructions which we assume to be R-hard against 2^p-adversaries.

Hardness	Symmetric	Asymmetric
Time	IterHash$_\eta^t$ (Time, η, 2^{p-t})	RSAlock$_\eta^t$ (Time, η, 2^{p-t})
Memory	Argon2 (RAM, $M/5$, 2^{p-t})	Diodon (RAM, $M/10$, 2^{p-t})
Code	BigLUT$_v^t$ (Code, 2^p, 2^{p-t})	BcCounter$_v^t$ (Code, 2^p, 2^{p-t})

made by a 2^p-adversary of a plug mapping t to v bits is lower-bounded by $\max(2^{-v}, 2^{p-t})$ if its output is uniformly distributed in its range.

Time-Hard Plug. This hardness has been considered in previous works for instance in the context of key stretching and key derivation or for time-lock encryption. In fact, the constructions proposed for each use case can be used to provide time-hardness and asymmetric time-hardness respectively.

Symmetric Hardness. IterHash$_\eta^t$ iterates a t-bit hash function on a t-bit input block η times where η must be much smaller than $2^{t/2}$ to avoid issues related to the presence of cycles in the functional graph of the hash function. If we denote by H the hash function used, then IterHash$_\eta^t(x) = H^\eta(x)$. Evaluating this function requires at least η hash function calls and, provided that the hash function iterated is cryptographically secure, it is impossible for an adversary to guess what the output is after η iterations with probability higher than $2^{-t/2}$.

We consider that this function is (Time, η, 2^{p-t})-hard against 2^p-adversaries, as long as $p \ll t/2$.

Asymmetric Hardness. RSAlock$_\eta^t$ is a function performing η squaring in a RSA modular ring of size $N = qq' \approx 2^t$, where q and q' are secret primes. Using these notations, RSAlock$_\eta^t(x) = x^{2^\eta} \mod N$. The common user therefore needs to perform η squarings in the modular ring.

However, a user who knows the prime decomposition of the RSA modulo can first compute $e = 2^\eta \mod (q-1)(q'-1)$ and thus compute RSAlock$_\eta^t(x) = x^e \mod N$. Furthermore, such a user can also use the Chinese remainder theorem to further speed up the computation which increases their advantage over common users. Thus, as long as $t > n$, the privileged user has an advantage over the common. We consider that RSAlock$_\eta^t$ is asymmetrically (Time, η, 2^{p-t})-hard against 2^p-adversaries.

Code-Hard Plug. As explained in Sect. 2.3, the main goals of code-hardness are white-box and big-key encryption. The structures used for both purposes rely on the same building block, namely a large look-up table where the entries are chosen uniformly at random or as the encryption of small integers. The former, BigLUT$_v^t$, is code-hard. The latter, BcCounter$_v^t$, is asymmetrically code-hard.

Furthermore, an identical heuristic can be applied to both of them to increase the input size of the plug while retaining a practical code size. It is described at the end of this section.

Symmetric Hardness. BigLUT_v^t uses a table K consisting in 2^t entries, each being a v-bit integer picked uniformly at random. Evaluating BigLUT_v^t then consists simply in querying this table: BigLUT_v^t is the function mapping a t-bit integer x to the v-bit integer $K[x]$.

This function is $(\text{Code}, 2^p, 2^{p-t})$-hard against 2^p-adversaries. Indeed, an adversary who has access to 2^p outputs of the function cannot evaluate it efficiently on a random input with probability more than 2^{p-t}. Simply guessing the output succeeds with probability 2^{-v} which is usually much smaller than 2^{p-t}. Thus, we consider that BigLUT_v^t is $(\text{Code}, 2^p, 2^{p-t})$-hard against 2^p-adversaries.

Asymmetric Hardness. BcCounter_v^t is the function mapping a t-bit integer x to the v-bit block $E_k(0^{v-t}||x)$, where E_k is a v-bit block cipher with a secret key k of length at least v. A common user would be given the codebook of this function as a table of 2^t integers while a privileged user would use the secret key k to evaluate this function.

The hardness of BcCounter_v^t is the same as that of BigLUT_v^t for a common user. The contrary would imply the existence of a distinguisher for the block cipher, which we assume does not exist. However, a privileged user with knowledge of the secret key used to build the table can bypass this complexity.

Furthermore, as the key size is at least as big as the block size in modern block ciphers, an adversary guessing the key is not more efficient than one who merely guesses the output of the cipher. Thus, we consider that BigLUT_v^t is asymmetrically $(\text{Code}, 2^p, 2^{p-t})$-hard.

Increasing the Input Size. Both BigLUT_v^t and BcCounter_v^t have a low input size and leave a fairly high success probability for an attacker trying to win the efficient approximation game without using a lot of resource. An easy way to work around this limitation is to use $\ell > 1$ distinct instances of a given function in parallel and XOR their outputs. For example, $x \mapsto f(x)$ where

$$f(x_0||...||x_{\ell-1}) = \oplus_{i=0}^{\ell-1} E_k(\text{byte}(i)||0^{n-t-8}||x_i)$$

and where $\text{byte}(i)$ denotes the 8-bit representation of the integer i combines ℓ different instances of BcCounter_v^t. We consider that, against 2^p-adversaries, it is asymmetrically $(\text{Code}, 2^p, \max(2^{p-v}, (2^{p-t}/\ell)^\ell))$-hard.

Indeed, an attacker could store $2^p/\ell$ entries of each of the ℓ distinct tables, in which case they can evaluate the whole function if and only if all the table entries they need are among those they know. This happens with probability $(2^{p-t}/\ell)^\ell$. Alternatively, they could store the output of the whole function for about 2^p values of the complete input. In that case, they can evaluate the function if and only if the whole input is one that was precomputed, which happens with probability 2^{p-v}. We assume that there is not better attack for a 2^p-adversary than the ones we just described, hence the hardness we claimed.

It is also possible to use a white-box block cipher as an asymmetrically code-hard function as this complexity is precisely the one they are designed to achieve.

Memory-Hard Plug. Definition 2, when applied to the case of memory hardness, corresponds to functions for which a reduction in the memory usage is either impossible or extremely costly in terms of code or time complexity. We are not aware of any function satisfying the first requirement but, on the other hand, there are many functions for which a decrease in memory is bound to cause a quantifiable increase in time complexity. These are the functions we consider here.

Symmetric Hardness. Several recent functions are intended to provide memory-hardness. The main motivation was the Password Hashing Competition (PHC) which favored candidates enforcing memory-hardness to thwart the attacks of adversaries using ASICs to speed up password cracking.

The winner of the PHC competition, Argon2 [12], uses M bytes of memory to hash a password, where M can be chosen by the user. It was designed so that an adversary trying to use less than $M/5$ bytes of memory would have to pay a significant increase in time-hardness. Using our definition, if t is the size of a digest (this quantity can also be chosen by the user) and v is the size of the input, then Argon2 is about (RAM, $M/5, 2^{p-t}$)-hard against 2^p-adversaries as long as enough passes are used to prevent ranking and sandwich attacks [1,2] and as long as $2^{p-t} > 2^{-v}$.

The construction of memory-hard functions is a very recent topic. Only a few such functions are known, which is why Argon2 and DIODON are far more complex than the other plugs proposed in this section. It is an interesting research problem to build a simpler memory hard function with the relaxed constraint that it might be cheap to compute on a part of its domain, a flaw which would easily be factored into the $\epsilon(p)$ probability.

Asymmetric Hardness. The only asymmetrically memory-hard function we are aware of is the one we introduced in Sect. 2.4, DIODON. Because of its proximity with `scrypt`, we claim that it is (RAM, $M/10, 2^{-u}$)-hard for basic users [3]. In other words, we consider that DIODON is asymmetrically (RAM, $M/10, 2^{p-t}$)-hard — under the assumption, as for Argon2, that $2^{p-t} > 2^{-u}$. Note however that [3] guarantees only linear penalties if attacker trades memory for time, while modern memory-hard functions like Argon2 provide superpolynomial penalties. This leads us to the following open problem.

Open Problem 1. *Is it possible to build an asymmetrically memory-hard function for which the time-memory tradeoff is superpolynomial, i.e. such that dividing its memory usage by a factor α must mutiply its execution time by α^d for some $d > 1$?*

4   Modes of Operations for Building Hard Primitives

As said above, our strategy is to combine hard plugs with secure cryptographic primitives in such a way that the input of the plugs are randomized and that enough such calls are performed to ensure that at least one plug evaluation was hard with a high enough probability. The method we use is nicknamed *plug-then-randomize*. It is formalized in Sect. 4.1. Then, the block cipher and the sponge mode of operation based on it are introduced respectively in Sects. 4.2 and 4.3.

Unfortunately, our security arguments are not as formal as those used in the area of provable security. Our main issue is that we are not trying to prevent the adversary from recovering a secret: in fact, there is none in the case of symmetric hardness! Furthermore, since an (inefficient) implementation is public, we cannot try to prevent the attacker from distinguishing the function from an ideal one. It is our hope that researchers from the provable security community will suggest directions for new and more formal arguments.

4.1   Plug-Then-Randomize

Definition 4 (Plugged Function). *Let* $P : \{0,1\}^t \to \{0,1\}^v$ *be a plug and let* $F : \{0,1\}^n \to \{0,1\}^n$ *be a function, where* $t + v \leq n$. *The plugged function* $(F \cdot P) : \{0,1\}^n \to \{0,1\}^n$ *maps* $x = x_t || x_v || x'$ *with* $|x_t| = t$, $|x_v| = v$ *and* $|x'| = m - t - v$ *to* y *defined by:*

$$(F \cdot P)(x_t \, || \, x_v \, || \, x') = y = F\left(x_t \, || \, x_v \oplus P(x_t) \, || \, x'\right).$$

This computation is summarized in Fig. 3.

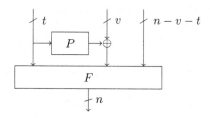

Fig. 3. Evaluating the plugged function $(F \cdot P)$.

Lemma 1 (Plugged Function Hardness). *If* $P : \{0,1\}^t \to \{0,1\}^v$ *is a plug* $(\rho, u, \epsilon(p))$*-hard against* 2^p*-adversaries and if* $F : \{0,1\}^n \to \{0,1\}^n$ *is a public random (permutation) oracle then the plugged function* $(F \cdot P)$ *is* $(\rho, u, \epsilon(p))$*-hard.*

Proof. First, the adversary could try and store 2^p outputs of $(F \cdot P)$. However, such an approximation would work only with probability $2^{p-n} < 2^{p-v} \leq \epsilon$, so that it is less successful than an approximation based on an approximation of the plug.

Without knowledge of the full input of F, it is impossible to predict its output because F is a random (permutation) oracle. Therefore, we simply need to show that the function $\mathcal{F}_P$ mapping (x, y, z) of $\{0,1\}^t \times \{0,1\}^v \times \{0,1\}^{n-t-v}$ to $(x, y \oplus P(x), z)$ is as hard as P itself.

By contradiction, suppose that there is an adversary $\mathcal{A}$ capable of winning the approximation game for $\mathcal{F}_P$. That is, $\mathcal{A}$ can compute an approximation $\mathcal{F}'_P$ of $\mathcal{F}_P$ using less than u units of the resource ρ which works with probability strictly higher than $\epsilon(p)$. Then $\mathcal{A}$ can win the approximation game for P itself as follows. When given P, $\mathcal{A}$ computes the approximation $\mathcal{F}'_P$ of the corresponding function $\mathcal{F}_P$. Then, when given a random input x of P of length t, $\mathcal{A}$ concatenates it with random bitstrings y and z of length v and $n - t - v$ respectively. The output of P is then approximated as the v center bits of $\mathcal{F}'_P(x||y||z) \oplus (x||y||z) = 0^t||P(x)||0^{n-t-v}$. Thus, $\mathcal{A}$ can violate the $(\rho, u, \epsilon(p))$-hardness of P.

We deduce that if P is $(\rho, u, \epsilon(p))$-hard, then so is $\mathcal{F}_P$ and thus $(F \cdot P)$. □

Using this lemma, we can prove the following theorem which will play a key role in justifying the R-hardness of our later constructions.

Theorem 1 (Iterated Plugged Function Hardness). *Let F_i, $i < r$ be a family of r independent random oracles (or random permutation oracles) mapping n bits to n. Let $P : \{0,1\}^t \to \{0,1\}^v$ with $t + v \leq n$ be a plug $(\rho, u, \epsilon(p))$-hard against 2^p-adversaries. Then the function $f : \{0,1\}^n \to \{0,1\}^n$ defined by*

$$f : x \mapsto \big((F_{r-1} \cdot P) \circ ... \circ (F_0 \cdot P)\big)(x)$$

is $(\rho, u, \max(\epsilon(p)^r, 2^{p-n}))$-hard against 2^p-adversaries.

Proof. We denote f_i the function defined by $f : x \mapsto \big((F_{i-1} \cdot P) \circ ... \circ (F_0 \cdot P)\big)(x)$, so that $f = f_r$. We proceed by induction on the number of rounds i, our induction hypothesis being that the theorem holds for $r \leq i$.

Initialization. If $i = 1$ i.e. for $f_1 = (F \cdot P)$, Lemma 1 tells us that this function is (ρ, u, ϵ)-hard. As $\epsilon \geq 2^{p-v} > 2^{p-n}$, the induction holds for $i = 1$.

Inductive Step. Suppose that the theorem holds for i rounds. The attack based on pre-querying 2^p outputs of f_{i+1} and then approximating f_{i+1} using the content of this table would still work. Thus, if $\epsilon^{i+1} \leq 2^{n-p}$ then this strategy is the optimal one. Suppose now that $\epsilon^{i+1} > 2^{n-p}$, which also implies that $\epsilon^i > 2^{n-p}$.

As F_{i+1} is a random (permutation) oracle, the only way to evaluate the output of f_{i+1} is to first evaluate f_i and then to evaluate $(F_{i+1} \cdot P)$. The existence of another efficient computation method would violate the assumption that F_{i+1} is a random oracle.

Thus, the adversary needs first to evaluate f_i and then $(F_{i+1} \cdot P)$. Let f'_j be an approximation of the function f_j computed by a 2^p-adversary and let g_j be an approximation of $(F_j \cdot F)$ computed by the same adversary. The probability of the successful evaluation of f'_{i+1} is:

$$P\left[f'_{i+1}(x) = f_{i+1}(x), \ x \xleftarrow{\$} \{0,1\}^n\right]$$
$$= P\left[g_{i+1}(y) = (F_{i+1} \cdot P)(y) \mid y = f_i(x)\right]$$
$$\times P\left[f'_i(x) = f_i(x), x \xleftarrow{\$} \{0,1\}^n\right].$$

On the other hand, the first term is equal to

$$P\left[g_{i+1}(y) = (F_{i+1} \cdot P)(y) \mid y = f_i(x)\right]$$
$$= P\left[g_{i+1}(y) = (F_{i+1} \cdot P)(y), y \xleftarrow{\$} \{0,1\}^n\right] \quad (1)$$

which, because of Lemma 1, is at most equal to ϵ.

Equation (1) is true. Were it not the case, then F_{i+1} would not be behaving like a random oracle. Indeed, $y = f_i(x)$ is the output of a sequence of random oracle calls sandwiched with simple bijections consisting in the plug calls that are independent from said oracle. Therefore, since x is picked uniformly at random, y must take any value with equal probability. Furthermore, the events $f_i(x) = y$ and $g_{i+1}(y) = (F_{i+1} \cdot P)(y)$ are independent: the latter depends only on the last random (permutation) oracle F_{i+1} while the former depends on all other random (permutation) oracles. As a consequence, the probability that $f'_{i+1}(x) = f_i(x)$ for x picked uniformly at random and for any approximation f'_{i+1} obtained by a 2^p-adversary is upper-bounded by ϵ^{i+1}. □

4.2 Hard Block Cipher Mode (HBC)

Let E_k be a block cipher operating on n-bit blocks using a key of length $\kappa \geq n$. Let P be a plug $(\rho, u, \epsilon(p))$-hard against 2^p-adversaries. The HBC mode of operation iterates these two elements to create an n-bit block cipher with a κ-bit secret key which is $(\rho, u, \max(\epsilon(p)^r, 2^{p-n}))$-hard against 2^p-adversaries. This construction, when keyed by the κ-bit key k, is the permutation $\text{HBC}[E_k, P, r]$ which transforms an n-bit input x as described in Algorithm 3. This process is also summarized in Fig. 4. Below, we describe the hardness (Theorem 2) such an HBC instance achieves. We also reiterate that if an asymmetrically hard plug is used then the block cipher thus built is also asymmetrically hard.

Our proof is in the ideal cipher model, a rather heavy handed assumption. We leave as future work to prove the hardness of this mode of operation in simpler settings.

The role of the round counter XOR in the key is merely to make the block cipher calls independent from one another. If the block cipher had a tweak, these counter additions could be replaced by the use of the counter as a tweak with a fixed key. It is possible to use block ciphers which are not secure in the related-key setting and still retain the properties of HBC by replacing the keys $k \oplus i$ by the outputs of a key derivation function.

Theorem 2 (Hardness of HBC). *If the block cipher E_k used as a component of $\text{HBC}[E_k, P, r]$ is an ideal block cipher and if the plug P is $(\rho, u, \epsilon(p))$-hard against 2^p-adversaries, then the block cipher $\text{HBC}[E_k, P, r]$ is*

Algorithm 3. HBC$[E_k, P, r]$ encryption

Inputs: n-bit plaintext x; κ-bit key k

Output: n-bit ciphertext y

$y \leftarrow E_k(x)$

for all $i \in \{1, ..., r\}$ **do**

$\quad y_t \parallel y_{n-t} \leftarrow y$, where $|y_t| = t$ and $|y_{n-t}| = n - t$

$\quad y_{n-t} \leftarrow y_{n-t} \oplus P(y_t)$

$\quad y \leftarrow E_{k \oplus i}(y_t \| y_{n-t})$

end for

return y

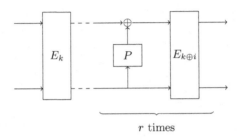

Fig. 4. The HBC block cipher mode.

$$\left(\rho, u, \max(\epsilon(p)^r, 2^{p-n})\right)\text{-}hard\,against\,2^p\text{-}adversaries.$$

Proof. As E_k is an ideal cipher, E_k and $E_{k \oplus i}$ act like two independent random permutation oracles. As a consequence, Theorem 1 immediately gives us the theorem. □

We used the HBC structure to build an asymmetrically time-hard block cipher, SKIPPER, which we describe in Sect. 5.1.

4.3   Hard Sponge Mode (HSp)

The sponge construction was introduced by Bertoni et al. as a possible method to build a hash function [8]. They used it to design Keccak [9] which later won the SHA-3 competition. It is a versatile structure which can be used to implement hash functions, stream ciphers, message authentication codes (MAC), authenticated ciphers as described in [10], pseudo-random number generators (PRNG) and key derivation functions (KDF) as explained for example in [23]. In this section, we first provide a brief reminder on the sponge construction Then, we show how plugs can be combined with secure sponge transformation to build R-hard sponges, thus providing R-hard hash function, MAC, etc.

The Sponge Construction. A sponge construction uses an n-bit public permutation g and is parametrized by its capacity c and its rate r which are such

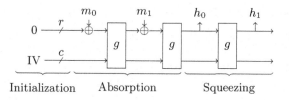

Fig. 5. A sponge-based hash function.

that $r + c = n$. This information is sufficient to build a hash function, as illustrated in Fig. 5. The two higher level operations provided by a sponge object parametrized by the function g, the rate r and the capacity c are listed below.

- **Absorption.** The r-bit block m_i of the padded message m is xored into the first r bits of the internal state of the sponge and the function g is applied.
- **Squeezing.** The first r bits of the internal state are output and the function g is applied on the internal state.

The internal state of the sponge obviously needs to be initialized. It can be set to a fixed string to create a hash function. However, if the initial value is a secret key, we obtain a MAC. Similarly, if the initial value is a secret key/initialization pair, we can generate a pseudo-random keystream by iterating the squeezing operation.

As explained in [10], this structure can be modified to allow single-pass authenticated encryption. This is achieved by using the sponge object to generate a stream with the added modification that, between the generation of the r-bit keystream block and the application of g to the internal state, the r-bit block of padded message is xor-ed into the internal state, just like during the absorption phase. In this case, there is no distinction between the absorption and squeezing phase. Once the whole padded message has been absorbed and encrypted using the keystream, the sponge object is squeezed to obtain the tag.

Finally, a sponge object can also be used to build a KDF or a PRNG using a similar strategy in both cases, as proposed in [23]. The general principle is to absorb the output of the low-entropy source and follow the absorption of each block by many iterations of $x \mapsto 0^r || T_c(g(x))$ on the internal state, where $T_c(x)$ is equal to the last c bits of x. Setting the first r bits of the internal state to zero prevents the inversion of the update function.

Sponge construction are known to be secure as long as the public update function g has no *structural distinguishers* such as high probability differentials or linear approximation with a high bias.

The main advantages of the sponge structure are its simplicity and its versatility. It is simple because it only needs a public permutation and it is versatile because all symmetric primitives except block ciphers can be built from it with very little overhead. As we will show below, the fact that its internal state is larger than that of a usual block cipher also means that attacks based on pre-computations are far weaker.

4.4   The HSp Mode and Its Hardness

Given that a sponge object is fully defined by its rate r, capacity c and public update function g, we intuitively expect that building a R-hard sponge object can be reduced to building a R-hard update function. As stated in the theorem below, this intuition is correct provided that the family of functions $g_k : \{0,1\}^c \to \{0,1\}^c$ indexed by $k \in \{0,1\}^r$ and defined as the capacity bits of $g(x\|k)$ is assumed to be a family of independent random oracles.

We call HSp the mode of operation described in this section. It is superficially similar to a round of the HBC block cipher mode.

An update function g can be made R-hard using the R-hardness of a plug $P : \{0,1\}^t \to \{0,1\}^v$ to obtain a new update function $(g \cdot P)$ as described in Algorithm 4.

Algorithm 4. $(g \cdot P)$ sponge transformation

Inputs: n-bit block x;
Output: n-bit block y

$x_t \| x_v \| x' \leftarrow x$, where $|x_t| = t, |x_v| = v, |x'| = n - t - v$
$x_v \leftarrow x_v \oplus P(x_t)$
$y \leftarrow g(x_t \| x_v \| x')$
return y

This process is summarized in Fig. 6. In order to prevent the adversary from reaching either the input or the output of P, which could make some attacks possible, we impose that $t + v \leq c$ so that the whole plug input and output are located in the capacity.

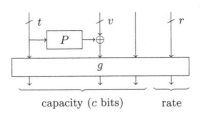

capacity (c bits)   rate

Fig. 6. The hard sponge transformation $(g \cdot P)$.

Theorem 3 (HSp absorption hardness). *Consider a sponge defined by the n-bit transformation $(g \cdot P)$, a rate r and a capacity c so that $r + c = n$ and $r > c$. Let $(g \cdot P)$ be defined as in Algorithm 4, where $P : \{0,1\}^t \to \{0,1\}^v$ is a plug $(\rho, u, \epsilon(p))$-hard against 2^p-adversaries.*

Let Absorb $: \{0,1\}^{\ell \times r} \to \{0,1\}^c$ *be the function mapping an un-padded message m of ℓ r-bit blocks to the capacity bits of the internal state of the sponge after it absorbed m.*

Furthermore, suppose that the n-bit transformation g is such that the family of functions $g_k : \{0,1\}^c \to \{0,1\}^c$ indexed by $k \in \{0,1\}^r$ and defined as $g_k(x) = T_c\left((g(x||k)\right)$ can be modeled as a family of random oracles.

Then Absorb *is $\left(\rho, u, \max(\epsilon(p)^{\ell-1}, 2^{p-c})\right)$-hard against 2^p-adversaries.*

This theorem deals with un-padded messages. The padding of such a message imposes the creation of a new block with a particular shape which cannot be considered to be random.

Proof. Let $g_k : \{0,1\}^c \to \{0,1\}^c$ be as defined in the theorem. Let the message m be picked uniformly at random.

The first call to $(g \cdot P)$ is not $(\rho, u, \epsilon(p))$-hard. Indeed, the content of the message has not affected the content of the capacity yet. However, the capacity bits of the internal state after this first call to $(g \cdot P)$ are uniformly distributed as they are the image of a constant by the function indexed by m_0 from a family of 2^r different random oracles.

Let $m'_i = m_i \oplus z_i$, where m_i is the message block with index $i > 1$ and where z_i is the first r bits of the content of the sponge after the absorption of $m_0, ..., m_{i-1}$. That is, z_i is the content of the rate juste before the call to $(g \cdot P)$ following the addition of the message block m_i. We can therefore represent the absorption function as described in Fig. 7.

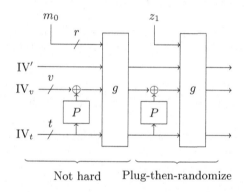

Not hard   Plug-then-randomize

Fig. 7. An alternative representation of the absorption procedure.

Since the message blocks m_i have been picked uniformly at random, so are the values z_i. We can therefore apply Theorem 1, where the independent random oracles are g_{z_i}, the plug is P, the random message is $(g_{m_0} \cdot P)(0||IV)$, the block size is c and the number of rounds is $\ell - 1$.   □

As c is typically much larger than a usual block cipher size of 128 bits, the probability of success of a 2^p adversary can be made much smaller when a sponge is built rather than a block cipher.

Note that if a sponge is used to provide e.g. authenticated encryption, the same bound should be used as the message is absorbed into the state in the same fashion in this case.

The following claim describes the hardness of the squeezing operation.

Claim 1 (HSp squeezing hardness). *Consider a sponge defined by the n-bit transformation $(g \cdot P)$, a rate r and a capacity c so that $r + c = n$ and $r > c$. Let $(g \cdot P)$ be defined as in Algorithm 4, where $P : \{0,1\}^t \to \{0,1\}^v$ is a plug $(\rho, u, \epsilon(p))$-hard against 2^p-adversaries.*

Let $\mathsf{Squeeze}^\ell : \{0,1\}^n \to \{0,1\}^{\ell \times r}$ be the function mapping an internal state of n bits to a stream of ℓ r-bit blocks obtained by iterating ℓ times the $\mathsf{Squeeze}$ operation.

Then $\mathsf{Squeeze}^\ell$ is $\left(\rho, u, \max(\epsilon(p)^\ell, 2^{p-(c+r)})\right)$-hard against 2^p-adversaries.

We cannot prove this hardness using Theorem 1 because the transformations called in each round are all identical. In particular, they cannot be independent. This situation can however be interpreted as a variant of the one in the proof of Theorem 3 where z_i is not formally picked uniformly at random as there is no message absorption but can be interpreted as such because it is the output of the sponge function.

The claimed probability of success bound comes from the hardness of approximating all ℓ calls to the plug composed with the sponge transformation $(\epsilon(p)^\ell)$ and the hardness of using a precomputation of the image of 2^p possible internal states $(2^{p-(c+r)})$.

If the sponge is used to provide a simple stream cipher, the bound of Claim 1 should be used. Indeed, since there is no message absorption in this case, Theorem 3 cannot be used.

5   Practical Instances: Skipper and Whale

We illustrate the versatility and simplicity of the modes of operation described in the previous section by presenting an instance of each. The first is an asymmetrically time-hard block cipher called SKIPPER and the second is a code-hard sponge called WHALE.

5.1   The Skipper Block Cipher

One possible application for *egalitarian computing* which is mentioned but not explored in [13] is *obfuscation*. The idea is to modify a program in such a way that a memory-hard function must be computed in parallel to the execution of the program. Using this very general approach, any program or function could be made memory-hard, not just cryptographic ones. However, a shortcoming in this case is the fact that the compiler returning the obfuscated code must also pay the full price of running this parallel memory-hard function.

Solving this issue requires a primitive with asymmetric memory hardness such as DIODON. However, this primitive is rather slow even for privileged users.

Therefore, we build instead an asymmetrically time-hard block cipher using the HBC mode to combine the AES and the RSA-lock plug. The result is the asymmetrically time-hard block cipher SKIPPER. It could be used to create an efficient obfuscator. The obfuscator would use the fast implementation of the plug to create an obfuscated program which forces common users to evaluate its slow version to run the program. That way, the computational hardness is only paid by the users of the program and not by the compiler. While this cost might be offset through the use of dedicated hardware for the computation of RSA-lock, we note that this function cannot be parallelized.

Our proposal SKIPPER is $HBC[AES - 128, RSAlock_{\eta}^{n_p}, 2]$, that is, a 128-bit block cipher using a 128-bit secret key k, an $RSAlock_{\eta}^{n_p}$ instance truncated to 40 bits as a plug and 3 calls to AES-128 sandwiching 2 calls to the plug. The plug operates modulo $N \geq 2^{n_p}$. The SKIPPER encryption procedure is described in Algorithm 5.

Algorithm 5. SKIPPER encryption

Inputs: n-bit plaintext x; k-bit key k; RSA modulus N

Output: n-bit ciphertext y

$y \leftarrow AES_k(x)$
for all $i \in \{1, 2\}$ **do**
$\quad y_1 \parallel y_2 \leftarrow y$, where $|y_1| = 88$ and $|y_2| = 40$
$\quad y_2 \leftarrow y_2 \oplus T_{40}(y_1^{2^{\eta}} \bmod N)$
$\quad y \leftarrow AES_{k \oplus i}(y_1 \parallel y_2)$
end for
return y

The RSA-based plug we use is asymmetrically $\left(\text{Time}, \eta, \max(2^{p-88}, 2^{-40})\right)$-hard. As said before in Sect. 3.3, we assume that no adversary can evaluate $x^{2^{\eta}} \bmod N$ without performing η squarings in the modular ring. However, a 2^p-adversary can either guess all 40 bits of the output, which succeeds with probability 2^{-40}, or store 2^p out of the 2^{88} possible outputs, in which case a successful evaluation is possible with probability 2^{p-88}.

Merely guessing is the best strategy unless the adversary has access to at least $40 \times 2^{88-40} \approx 2^{53.3}$ bits of storage, i.e. more than a thousand terabytes. Furthermore, the cost of such a pre-computation could only be amortized if more than $2^{48}/2 = 2^{47}$ blocks are encrypted using the same plug, i.e. 2^{54} bits (more than a thousand Tb). Otherwise, the time taken by the precomputation would be superior to the time needed to evaluate the slow function. We therefore consider 2^{48}-adversaries, that is, adversaries capable of pre-computing 2^{48} values of $RSAlock_{\eta}^{n_p}(x)$. Such an adversary is already quite powerful as it has significant computing power and storage in addition to knowing the secret key k. Providing maximum security against more powerful adversaries would probably be over-engineering. Thus, in our setting, the plug is asymmetrically $(\text{Time}, \eta, 2^{-40})$-hard.

Claim 2 (Properties of Skipper). *The block cipher* SKIPPER *is asymmetrically* (*Time*, η, 2^{-80})*-hard and cannot be distinguished from a pseudo-random permutation using less than* 2^{128} *operations.*

SKIPPER is HBC[AES $-$ 128, RSAlock$_\eta^{n_p}$, 2] and its plug is asymmetrically (Time, η, 2^{-40})-hard. Thus, by applying Theorem 2 we obtain that SKIPPER is asymmetrically (Time, η, max $\left(2^{48-128}, (2^{-40})^2\right)$)-hard.

As there is to the best of our knowledge no related-key attack against full-round AES-128 in the case where the related keys are linked by a simple XOR, we claim that SKIPPER cannot be distinguished from a random permutation using much less than 2^{128} operations. Should such distinguishers be found, an alternative key schedule such as the one from [27] could be used.

We implemented SKIPPER on a regular desktop PC. The corresponding benchmarks are provided in the full version of this paper [15].

5.2 The Whale Hash Function

Preventing the leakage of encryption keys is a necessity in order for a system to be secure. A possible method for preventing this was informally proposed by Shamir in a talk at RSA'2013 and then formalized by Bellare et al. in their CRYPTO'16 paper. As the throughput of the exfiltration method used by the attacker is limited, using a huge key would make their task all the harder. To use our terminology, an encryption algorithm with significant code-hardness would effectively be bound to the physical device storing it: since the code cannot be compressed, an attacker would have to duplicate the whole implementation to be able to decrypt the communications. Even a partial leakage would be of little use.

The proposal of Bellare et al., XKEY2, is effectively a code-hard key derivation algorithm which takes as input a random initialization vector and outputs a secret key. Since it is code-hard, an attacker cannot evaluate this function without full knowledge of the source code of the function and cannot extract a smaller (and thus easier to leak) implementation.

We propose the code-hard hash function WHALE as an alternative to XKEY2. It can indeed be used to derive a key by hashing a nonce, a process which cannot be approximated by an attacker unless they duplicate the entirety of the implementation of WHALE. WHALE is a simple sponge-based hash function which uses the XOR of $\lceil 128/t \rceil$ instances of BigLUT$_t^{128}$ as a plug. Different choices of t lead to different levels of code-hardness. It is only parametrized by the input length of the tables t.

It is based on SHA-3-256: it uses the Keccak $- f[1600]$ permutation, the same padding scheme, the same rate $r = 1088$, the same capacity $c = 512$ and the same digest size of 256 bits. There are only two differences:

- the permutation Keccak $- f[1600]$ is augmented with the code-hard plug consisting in the XOR of $\ell = \lceil 128/t \rceil$ distinct instances of BigLUT$_t^{128}$, and
- t blank calls to the transformation are performed between absorption and squeezing.

These parameters were chosen so as to prevent an adversary with access to at most half of the implementation of WHALE to compute the digest of a message with probability higher than 2^{-128}.

Claim 3 (Code-hardness of Whale). *The* WHALE *hash function using tables with t-bit inputs is ($Code, 2^{t+13}/t, 2^{-128}$)-hard against an adversary trying to use only half of the code-space used to implement* WHALE.

WHALE uses $\lceil 128/t \rceil$ tables of 2^t 128-bit entries. Thus, about $2^t \times 128 \times \lceil 128/t \rceil \approx 2^{t+14}/t$ bits are needed to store the implementation of its plug. An adversary trying to compress it and divide its size by 2 therefore has access to $2^{t+13}/t$ bits. Note however that, since the entries in each instance of $BigLUT_t^{128}$ have been picked uniformly at random, it is impossible to actually compress them. The best an attacker can do is therefore to store as many entries as they can.

When hashing a message, at least t calls to the plug are performed during the blank calls to the transformation between the absorption and the squeezing. Therefore, the adversary needs to successfully compute $t \times \lceil 128/t \rceil \geq 128$ entries of the big tables. If they only have half of them stored, then they succeed in computing the digest of a message with probability at most 2^{-128}.

6   Conclusion

We have presented for the first time a unified framework to study all three forms of hardness (time, memory and code) as well as their asymmetric variants. We have proposed DIODON, the first asymmetrically memory-hard function. We have also presented the first general approach for building a cryptographic primitive with *any* type of hardness and illustrated it with two fully specified proposals. The first is the asymmetrically time-hard block cipher SKIPPER which can be made arbitrarily slow for some users while retaining its efficiency for those knowing a secret key. The second is the code-hard hash function WHALE whose implementation cannot be compressed.

Acknowledgements. We thank anonymous reviewers from S&P, USENIX and ASIACRYPT'17 for their helpful comments. The work of Léo Perrin was supported by the CORE project ACRYPT (ID C12-15-4009992) funded by the Fonds National de la Recherche, Luxembourg.

References

1. Alwen, J., Blocki, J.: Efficiently computing data-independent memory-hard functions. In: Robshaw, M., Katz, J. (eds.) CRYPTO 2016. LNCS, vol. 9815, pp. 241–271. Springer, Heidelberg (2016). doi:10.1007/978-3-662-53008-5_9
2. Alwen, J., Blocki, J.: Towards practical attacks on Argon2i and balloon hashing. Cryptology ePrint Archive, Report 2016/759 (2016). http://eprint.iacr.org/2016/759

3. Alwen, J., Chen, B., Pietrzak, K., Reyzin, L., Tessaro, S.: Scrypt is maximally memory-hard. In: Coron, J.-S., Nielsen, J.B. (eds.) EUROCRYPT 2017. LNCS, vol. 10212, pp. 33–62. Springer, Cham (2017). doi:10.1007/978-3-319-56617-7_2

4. Alwen, J., Serbinenko, V.: High parallel complexity graphs and memory-hard functions. In: Proceedings of the Forty-Seventh Annual ACM on Symposium on Theory of Computing, pp. 595–603. ACM (2015)

5. Ateniese, G., Bonacina, I., Faonio, A., Galesi, N.: Proofs of space: when space is of the essence. In: Abdalla, M., De Prisco, R. (eds.) SCN 2014. LNCS, vol. 8642, pp. 538–557. Springer, Cham (2014). doi:10.1007/978-3-319-10879-7_31

6. Baldimtsi, F., Kiayias, A., Zacharias, T., Zhang, B.: Indistinguishable proofs of work or knowledge. In: Cheon, J.H., Takagi, T. (eds.) ASIACRYPT 2016. LNCS, vol. 10032, pp. 902–933. Springer, Heidelberg (2016). doi:10.1007/978-3-662-53890-6_30

7. Bellare, M., Kane, D., Rogaway, P.: Big-key symmetric encryption: resisting key exfiltration. In: Robshaw, M., Katz, J. (eds.) CRYPTO 2016. LNCS, vol. 9814, pp. 373–402. Springer, Heidelberg (2016). doi:10.1007/978-3-662-53018-4_14

8. Bertoni, G., Daemen, J., Peeters, M., Van Assche, G.: Sponge functions. In: ECRYPT hash workshop, vol. 2007. Citeseer (2007)

9. Bertoni, G., Daemen, J., Peeters, M., Van Assche, G.: Keccak specifications. Submission to NIST (Round 2) (2009)

10. Bertoni, G., Daemen, J., Peeters, M., Van Assche, G.: Duplexing the sponge: single-pass authenticated encryption and other applications. In: Miri, A., Vaudenay, S. (eds.) SAC 2011. LNCS, vol. 7118, pp. 320–337. Springer, Heidelberg (2012). doi:10.1007/978-3-642-28496-0_19

11. Biryukov, A., Bouillaguet, C., Khovratovich, D.: Cryptographic schemes based on the ASASA structure: black-box, white-box, and public-key (extended abstract). In: Sarkar, P., Iwata, T. (eds.) ASIACRYPT 2014. LNCS, vol. 8873, pp. 63–84. Springer, Heidelberg (2014). doi:10.1007/978-3-662-45611-8_4

12. Biryukov, A., Dinu, D., Khovratovich, D.: Argon2: new generation of memory-hard functions for password hashing and other applications. In: 2016 IEEE European Symposium on Security and Privacy (EuroS&P), pp. 292–302. IEEE (2016)

13. Biryukov, A., Khovratovich, D.: Egalitarian computing. In: Holz, T., Savage, S. (eds.) 25th USENIX Security Symposium, USENIX Security 16, Austin, TX, USA, 10–12 August 2016, pp. 315–326. USENIX Association (2016). https://www.usenix.org/conference/usenixsecurity16/technical-sessions/presentation/biryukov

14. Biryukov, A., Khovratovich, D.: Equihash: asymmetric proof-of-work based on the generalized birthday problem. In: Proceedings of NDSS 2016, 21–24 February 2016, San Diego, CA, USA. ISBN 1-891562-41-X (2016)

15. Biryukov, A., Perrin, L.: Symmetrically and asymmetrically hard cryptography (full version). Cryptology ePrint Archive, Report 2017/414 (2017). http://eprint.iacr.org/2017/414

16. Bogdanov, A., Isobe, T.: White-box cryptography revisited: Space-hard ciphers. In: Proceedings of the 22nd ACM SIGSAC Conference on Computer and Communications Security, pp. 1058–1069. ACM (2015)

17. Bogdanov, A., Isobe, T., Tischhauser, E.: Towards practical whitebox cryptography: optimizing efficiency and space hardness. In: Cheon, J.H., Takagi, T. (eds.) ASIACRYPT 2016. LNCS, vol. 10031, pp. 126–158. Springer, Heidelberg (2016). doi:10.1007/978-3-662-53887-6_5

18. Boneh, D., Corrigan-Gibbs, H., Schechter, S.: Balloon hashing: a memory-hard function providing provable protection against sequential attacks. In: Cheon, J.H., Takagi, T. (eds.) ASIACRYPT 2016. LNCS, vol. 10031, pp. 220–248. Springer, Heidelberg (2016). doi:10.1007/978-3-662-53887-6_8

19. Daemen, J., Rijmen, V.: The Design of Rijndael: AES-the Advanced Encryption Standard. Springer, Heidelberg (2002). doi:10.1007/978-3-662-04722-4

20. Delerablée, C., Lepoint, T., Paillier, P., Rivain, M.: White-box security notions for symmetric encryption schemes. In: Lange, T., Lauter, K., Lisoněk, P. (eds.) SAC 2013. LNCS, vol. 8282, pp. 247–264. Springer, Heidelberg (2014). doi:10.1007/978-3-662-43414-7_13

21. Dziembowski, S., Faust, S., Kolmogorov, V., Pietrzak, K.: Proofs of space. In: Gennaro, R., Robshaw, M. (eds.) CRYPTO 2015. LNCS, vol. 9216, pp. 585–605. Springer, Heidelberg (2015). doi:10.1007/978-3-662-48000-7_29

22. Fouque, P.-A., Karpman, P., Kirchner, P., Minaud, B.: Efficient and provable white-box primitives. In: Cheon, J.H., Takagi, T. (eds.) ASIACRYPT 2016. LNCS, vol. 10031, pp. 159–188. Springer, Heidelberg (2016). doi:10.1007/978-3-662-53887-6_6

23. Gaži, P., Tessaro, S.: Provably robust sponge-based PRNGs and KDFs. In: Fischlin, M., Coron, J.-S. (eds.) EUROCRYPT 2016. LNCS, vol. 9665, pp. 87–116. Springer, Heidelberg (2016). doi:10.1007/978-3-662-49890-3_4

24. Hellman, M.: A cryptanalytic time-memory trade-off. IEEE Trans. Inf. Theory 26(4), 401–406 (1980)

25. Kaliski, B.: PKCS #5: Password-Based Cryptography Specification Version 2.0. RFC 2898 (Informational) (Sep 2000). http://www.ietf.org/rfc/rfc2898.txt

26. Nakamoto, S.: Bitcoin: A peer-to-peer electronic cash system (2008)

27. Nikolić, I.: Tweaking AES. In: Biryukov, A., Gong, G., Stinson, D.R. (eds.) SAC 2010. LNCS, vol. 6544, pp. 198–210. Springer, Heidelberg (2011). doi:10.1007/978-3-642-19574-7_14

28. Percival, C.: Stronger key derivation via sequential memory-hard functions. Self-published, pp. 1–16 (2009)

29. Rivest, R.L., Shamir, A., Adleman, L.: A method for obtaining digital signatures and public-key cryptosystems. Commun. ACM 21(2), 120–126 (1978). http://doi.acm.org/10.1145/359340.359342

30. Rivest, R.L., Shamir, A., Wagner, D.A.: Time-lock puzzles and timed-release crypto. Technical report, Massachusetts Institute of Technology, Cambridge, MA, USA (1996)

Blockcipher-Based MACs: Beyond the Birthday Bound Without Message Length

Yusuke Naito[✉]

Mitsubishi Electric Corporation, Kanagawa, Japan
Naito.Yusuke@ce.MitsubishiElectric.co.jp

Abstract. We present blockcipher-based MACs (Message Authentication Codes) that have beyond the birthday bound security without message length in the sense of PRF (Pseudo-Random Function) security. Achieving such security is important in constructing MACs using blockciphers with short block sizes (e.g., 64 bit).

Luykx *et al.* (FSE 2016) proposed LightMAC, the first blockcipher-based MAC with such security and a variant of PMAC, where for each n-bit blockcipher call, an m-bit counter and an $(n-m)$-bit message block are input. By the presence of counters, LightMAC becomes a secure PRF up to $O(2^{n/2})$ tagging queries. Iwata and Minematsu (TOSC 2016, Issue 1) proposed F_t, a keyed hash function-based MAC, where a message is input to t keyed hash functions (the hash function is performed t times) and the t outputs are input to the xor of t keyed blockciphers. Using the LightMAC's hash function, F_t becomes a secure PRF up to $O(2^{tn/(t+1)})$ tagging queries. However, for each message block of $(n-m)$ bits, it requires t blockcipher calls.

In this paper, we improve F_t so that a blockcipher is performed only once for each message block of $(n-m)$ bits. We prove that our MACs with $t \leq 7$ are secure PRFs up to $O(2^{tn/(t+1)})$ tagging queries. Hence, our MACs with $t \leq 7$ are more efficient than F_t while keeping the same level of PRF-security.

Keywords: MAC · Blockcipher · PRF · PRP · Beyond the birthday bound · Message length · Counter

1 Introduction

A MAC (Message Authentication Code) is a fundamental symmetric-key primitive that produces a tag to authenticate a message. MACs are often realized by using a blockcipher so that these become secure PRFs (Pseudo-Random Functions) under the standard assumption that the underlying keyed blockciphers are pseudo-random permutations. Hence, in security proofs, these are replaced with random permutations. The advantage of PRF-security is commonly measured by using the parameters: n the block length, q the total number of tagging queries, ℓ the maximum message length (in blocks) of each query and σ the total message length (in blocks) of all queries. Many blockcipher-based MACs are provided with the so-called birthday security. The basic birthday bound looks like $O(\ell^2 q^2/2^n)$ or $O(\sigma^2/2^n)$.

© International Association for Cryptologic Research 2017
T. Takagi and T. Peyrin (Eds.): ASIACRYPT 2017, Part III, LNCS 10626, pp. 446–470, 2017.
https://doi.org/10.1007/978-3-319-70700-6_16

Blockcipher-based MACs are mainly categorized into CBC-type MACs and PMAC-type ones. These MACs are constructed from two functions: hash and finalization functions, where a hash function produces a fixed length hash value from an arbitrary length message; a finalization function produces a tag from a hash value. CBC-type MACs [2,8,15,20,30,31] use hash functions that iterate a keyed blockcipher. The PRF-security bound becomes the birthday one due to the collision in the chaining values. PMAC-type MACs [9,33] use hash functions using a keyed blockcipher parallelly. The following figure shows the structure of PMAC1, where E_K is a keyed blockcipher (K is a secret key), M_1, M_2, M_3 and M_4 are n-bit message blocks and multiplications are performed over the multiplication subgroup of $GF(2^n)$. For collision inputs to the keyed blockcipher, the outputs are canceled out before the finalization function. Hence, the collision might trigger a distinguishing attack. By the birthday analysis for the input collision, the PRF-security bound becomes the birthday one.

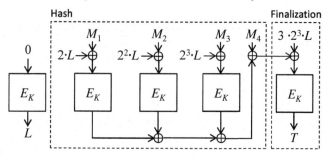

MACs with Beyond the Birthday Bound Security. The birthday bound security may not be enough for blockciphers with short block sizes such as Triple-DES and lightweight blockciphers, as mentioned in [7]. Hence, designing a MAC with *beyond the birthday bound* (BBB) security is an important research of MAC design. Such MACs contribute not only to blockciphers with short block sizes but also to the longevity of 128-bit blockciphers.

Yasuda proposed a CBC-type MAC, called SUM-ECBC [36], and a PMAC-type one, called PMAC_Plus [37]. He proved that the PRF-security bounds become $O(\ell^3 q^3/2^{2n})$. Later, Zhang *et al.* proposed a CBC-type MAC, called 3kf9 [40] that is more efficient than SUM-ECBC. These hash functions have a double length ($2n$ bit) internal state and produce a $2n$-bit value. These finalization functions have the xor of two keyed blockciphers that generates a tag from a $2n$-bit hash value. By the double length internal state, the influences of ℓ and q on the bounds are weakened.

Yasuda designed a PMAC-type MAC, called PMAC with Parity [38], with the aim of weakening the influence of ℓ. He proved that the PRF-security bound becomes $O(q^2/2^n + \ell q\sigma/2^{2n})$. Later, Zhang proposed a PMAC-type MAC with better efficiency, called PMACX [41]. Luykx *et al.* proposed a PMAC-type MAC, called LightMAC [25]. LightMAC is the counter-based construction that is used in the XOR MAC [1] and the protected counter sum [6]. LightMAC can be seen as a counter-based PMAC in which $(i)_m\|M_i$ is input to the i-th keyed blockcipher call, where $(i)_m$ is the m-bit binary representation of i and M_i is the i-th message block of $n - m$ bits. By the presence of counters, the input collision can be avoided, thereby the influence ℓ can completely be removed. They proved that the PRF-security bound becomes $O(q^2/2^n)$, namely, LightMAC is a secure PRF up to $O(2^{n/2})$ tagging queries.

Recently, Iwata and Minematsu proposed MACs with beyond the $O(2^{n/2})$-security, called F_t [16]. F_t is based on t keyed hash functions $H_{L_1}, \ldots, H_{L_t}$ and t keyed blockciphers $E_{K_1}, \ldots, E_{K_t}$, where $L_1, \ldots, L_t$ are hash keys. For a message M, the tag is defined as $F_t(M) = \bigoplus_{i=1}^{t} E_{K_i}(S_i)$ where $S_i = H_{L_i}(M)$. They proved that the PRF-security bound becomes $O(q^{t+1} \cdot \epsilon^t)$ as long as the keyed hash functions are ϵ-almost universal. They pointed out that the hash function of LightMAC is a $O(1/2^n)$-almost universal hash function, and adopting it as these hash functions, the PRF-security bound becomes $O(q^{t+1}/2^{tn})$. Namely, it is a secure PRF up to $O(2^{tn/(t+1)})$ tagging queries.

Why BBB-Security Without Message Length? We explain the importance of achieving BBB-security without message length. Here we consider the following example: the block length $n = 64$, the message length 2^{15} bits (4 Kbytes), and the threshold $1/2^{20}$ (a key is changed when the security bound equals the threshold). The message length is the case of HTTPS connection given in [7] and the threshold is given in [25]. We define the counter size as $m = n/3$ (rounded to the nearest multiple of 8) (in this case, $n = 64$ and $m = 24$). Putting these parameters into security bounds of PMAC_Plus ($O(\ell^3 q^3/2^{2n})$), LightMAC ($O(q^2/2^n)$), and F_t using LightMAC ($O(q^{t+1}/2^{tn})$), a key is changed after the tagging queries given in Table 1 (Line with "Queries"). Then, we consider the case that 2900 tagging queries of message length 4 Kbytes per second can be made. This example is the case of HTTPS connection given in [7]. In this case, a key is changed after the times given in Table 1 (Line with "Times"). Note that the security bound of PMAC_Plus depends on the message length, thereby increasing the length decreases the time. As shown Table 1, PMAC_Plus and Light-MAC require a rekeying within a day, whereas F_t does not require such frequent rekeyings.

Table 1. The numbers of tagging queries of changing a key and the times.

	PMAC_Plus	LightMAC	F_2 $(t = 2)$	F_3 $(t = 3)$	F_4 $(t = 4)$	$\cdots$
Queries	2^{29}	2^{22}	2^{36}	2^{43}	2^{47}	$\cdots$
Times	13 hrs	12 min	274 days	96 years	1539 years	$\cdots$

Question. As mentioned above, achieving BBB-security without message length is important for blockciphers with short block sizes, and F_t using Light-MAC achieves such security. However, it is inefficient because for each input block $(i)_m \| M_i$ it requires t blockcipher calls. It is roughly t times slower than LightMAC. Therefore, the main question of this paper is: *can we design more efficient MACs than F_t while keeping $O(2^{tn/(t+1)})$-security?*

Our Results. Firstly, we focus to design a MAC that is more efficient than F_2 and achieves the $O(2^{2n/3})$-security. As the research direction from PMAC to LightMAC, it is natural to consider a counter-based PMAC_Plus. We call the resultant scheme "LightMAC_Plus". Regarding the efficiency, LightMAC_Plus requires roughly one blockcipher call for each input block $(i)_m \| M_i$, while F_2 requires two blockcipher calls. Hence, LightMAC_Plus is more efficient than F_2. Regarding the PRF-security, by the presence of counters, the influence of ℓ can be removed. We prove that the PRF-security bound becomes $O(q^3/2^{2n})$, namely, LightMAC_Plus is a secure PRF up to $O(2^{2n/3})$ queries.

Next, we focus to design a MAC that is more efficient than F_t and achieves $O(2^{tn/(t+1)})$-security, where $t \geq 3$. Regarding the hash function, we also use that of LightMAC_Plus. Hence, this hash function is roughly t times faster than that of F_t. In order to ensure randomnesses of tags, we use the xor of t keyed blockciphers. However, there is a gap between the output length of the hash function ($2n$ bit) and the input length of the xor function (tn bit). Therefore, we propose a new construction that links between a $2n$-bit output and a tn-bit input. We call the resultant scheme "LightMAC_Plus2", and prove that if $t \leq 7$, then the PRF-security bound becomes $O(q^{t+1}/2^{tn} + q^2/2^{2n})$, namely, it is a secure PRF up to $O(2^{tn/(t+1)})$ tagging queries. In the proof of LightMAC_Plus2, we generalize the hash function by an ϵ-almost universal one, and prove that if $t \leq 7$, then the PRF-security bound is $O(q^{t+1}/2^{tn} + \epsilon)$. We prove that the counter-based hash function is $O(q^2/2^{2n})$-almost universal, which offers the PRF-security bound: $O(q^{t+1}/2^{tn} + q^2/2^{2n})$.

Table 2. Comparison of our MACs and other BBB-secure MACs. Column "# bits/BCs" refers to the number of bits of input message processed per blockcipher call. Column "# BCs in FF" refers to the number of blockcipher calls in a finalization function. F_t uses the hash function of LightMAC. LightMAC_Plus2 has the condition $t \leq 7$.

Scheme	# keys	# bits/BC	# BCs in FF	Security	Ref.
PMAC_Plus	3	n	2	$O(\ell^3 q^3/2^{2n})$	[37]
LightMAC	2	$n - m$	1	$O(q^2/2^n)$	[25]
F_t	$2t$	$(n-m)/t$	t	$O(q^{t+1}/2^{tn})$	[16]
LightMAC_Plus	3	$n - m$	2	$O(q^3/2^{2n})$	This paper
LightMAC_Plus2	$t + 3$	$n - m$	$t + 2$	$O(q^{t+1}/2^{tn} + \epsilon)$	This paper

Finally, in Table 2, we compare our MACs with BBB-secure MACs PMAC_Plus, LightMAC, and F_t. These MACs are PMAC-type ones, and thus parallelizable. We note that the PRF-security bound of LightMAC_Plus2 is satisfied when $t \leq 7$. Proving the PRF-security with $t > 7$ is left as an open problem.

Related Works. The PRF-security bounds of CBC-type MACs and PMAC-type MACs were improved to $O(\ell q^2/2^n)$ [3,27] and $O(\sigma q/2^n)$ [29]. Luykx et al.

studied the influence of ℓ in the PMAC's bound [24]. They showed that PMAC with Gray code [9] may not achieve the PRF-security bound of $O(q^2/2^n)$. Gaži et al. [14] showed that there exists an attack to PMAC with Gray code with the probability of $\Omega(\ell q^2/2^n)$, and instead proved that PMAC with 4-wise independent masks achieves the PRF-security bound of $O(q^2/2^n)$, where the input masks are defined by using 4 random values. Dodis and Steinberger [12] proposed a secure MAC from unpredicable keyed blockciphers with beyond the birthday bound security. Note that the security bound of their MAC includes the message length. Several randomized MACs achieve beyond the birthday bound security [18,19,26]. These require a random value for each query, while our MACs are deterministic, namely, a random value is not required.

Several compression function-based MACs achieve BBB security e.g., [13,21, 35,39]. Naito [28], List and Nandi [22], and Iwata et al. [17] proposed tweakable blockcipher-based MACs with BBB security. These MACs also employ the counter-based PMAC_Plus-style construction, where a counter is input as tweak. Namely, in the security proofs, the power of a tweakable blockcipher is used (distinct tweaks offer distinct random permutations). On the other hand, our MACs do not change the permutation in the hash function for each message block and the permutations in the finalization function. Peyrin and Seurin proposed a nonce-based and tweakable blockcipher-based MAC with BBB security [32]. Several Wegman-Carter-type MACs with BBB security were proposed e.g., [10,11,34]. These MACs use a random value or a nonce, whereas our MACs do not require either of them.

Organization. In Sect. 2, we give notations and the definition of PRF-security. In Sect. 3, we give the description of LightMAC_Plus and the PRF-security bound. In Sect. 4, we give the proof of the PRF-security. In Sect. 5, we give the description of LightMAC_Plus2 and the PRF-security bound. In Sect. 6, we give the proof of the PRF-security. Finally, in Sect. 7, we improve the efficiency of the hash function of LightMAC_Plus2.

2   Preliminaries

Notation. Let $\{0,1\}^*$ be the set of all bit strings. For a non-negative integer n, let $\{0,1\}^n$ be the set of all n-bit strings, and 0^n the bit string of n-bit zeroes. For a positive integer i, $[i] := \{1, 2, \ldots, i\}$. For non-negative integers i, m with $i < 2^m$, $(i)_m$ denotes the m-bit binary representation of i. For a finite set X, $x \xleftarrow{\$} X$ means that an element is randomly drawn from X and is assigned to x. For a positive integer n, $\mathsf{Perm}(n)$ denotes the set of all permutations: $\{0,1\}^n \to \{0,1\}^n$ and $\mathsf{Func}(n)$ denotes the set of all functions: $\{0,1\}^* \to \{0,1\}^n$. For sets X and Y, $X \leftarrow Y$ means that Y is assigned to X. For a bit string x and a set X, $|x|$ and $|X|$ denote the bit length of x and the number of elements in X, respectively. X^s denotes the s-array cartesian power of X for a set X and a positive integer s.

Let $GF(2^n)$ be the field with 2^n points and $GF(2^n)^*$ the multiplication subgroup of $GF(2^n)$ which contains $2^n - 1$ points. We interchangeably think of a point a in $GF(2^n)$ in any of the following ways: as an n-bit string $a_{n-1} \cdots a_1 a_0 \in \{0,1\}^n$ and as a formal polynomial $a_{n-1}\mathbf{x}^{n-1} + \cdots + a_1\mathbf{x} + a_0 \in GF(2^n)$. Hence we need to fix an irreducible polynomial $a(\mathbf{x}) = \mathbf{x}^n + a_{n-1}\mathbf{x}^{n-1} + \cdots + a_1\mathbf{x} + a_0$. This paper uses an irreducible polynomial with the property that the element $2 = \mathbf{x}$ generates the entire multiplication group $GF(2^n)^*$ of order $2^n - 1$. Examples of irreducible polynomial for $n = 64$ and $n = 128$ are given in [33]: $a(\mathbf{x}) = \mathbf{x}^{64} + \mathbf{x}^4 + \mathbf{x}^3 + \mathbf{x} + 1$ and $a(\mathbf{x}) = \mathbf{x}^{128} + \mathbf{x}^7 + \mathbf{x}^2 + \mathbf{x} + 1$, respectively.

PRF-Security. We focus on the information-theoretic model, namely, all keyed blockciphers are assumed to be random permutations, where a random permutation is defined as $P \xleftarrow{\$} \mathsf{Perm}(n)$. Through this paper, a distinguisher $\mathcal{D}$ is a computationally unbounded algorithm. It is given query access to an oracle $\mathcal{O}$, denoted by $\mathcal{D}^{\mathcal{O}}$. Its complexity is solely measured by the number of queries made to its oracles. Let $F[\mathbf{P}]$ be a function using s permutations $\mathbf{P} = (P^{(1)}, \ldots, P^{(s)})$.

The PRF-security of $F[\mathbf{P}]$ is defined in terms of indistinguishability between the real and ideal worlds. In the real world, $\mathcal{D}$ has query access to $F[\mathbf{P}]$ for $\mathbf{P} \xleftarrow{\$} \mathsf{Perm}(n)^s$. In the ideal world, it has query access to a random function $\mathcal{R}$, where a random function is defined as $\mathcal{R} \xleftarrow{\$} \mathsf{Func}(n)$. After interacting with an oracle $\mathcal{O}$, $\mathcal{D}$ outputs $y \in \{0,1\}$. This event is denoted by $\mathcal{D}^{\mathcal{O}} \Rightarrow y$. The advantage function is defined as

$$\mathbf{Adv}^{\mathsf{prf}}_{F[\mathbf{P}]}(\mathcal{D}) = \Pr\left[\mathbf{P} \xleftarrow{\$} \mathsf{Perm}(n)^s; \mathcal{D}^{F[\mathbf{P}]} \Rightarrow 1\right] - \Pr\left[\mathcal{R} \xleftarrow{\$} \mathsf{Func}(n); \mathcal{D}^{\mathcal{R}} \Rightarrow 1\right] .$$

Note that the probabilities are taken over $\mathbf{P}, \mathcal{R}$ and $\mathcal{D}$.

3   LightMAC_Plus

3.1   Construction

Let $\{E_K\}_{K \in \mathcal{K}} \subseteq \mathsf{Perm}(n)$ be a family of n-bit permutations (or a blockcipher) indexed by the key space $\mathcal{K}$, where $k > 0$ is the key length. Let m be the counter size with $m < n$. Let $K, K_1, K_2 \in \mathcal{K}$ be three keys for E. For a message M, the response of LightMAC_Plus$[E_K, E_{K_1}, E_{K_2}]$ is defined by Algorithm 1. Figure 1 illustrates the subroutine Hash$[E_K]$. Here, $M\|10^*$ means that first 1 is appended to M, and if the bit length of $M\|1$ is not a multiple of $n-m$ bits, then a sequence of the minimum number of zeros is appended to $M\|1$ so that the bit length becomes a multiple of $n - m$ bits. Note that $M\|10^* = M_1\|M_2\| \cdots \|M_l$ and $\forall i \in [l] : |M_i| = n - m$. By the counter size m and the padding value 10^*, the maximum message length in bits is at most $(2^m - 1) \times (n - m) - 1$ bit.

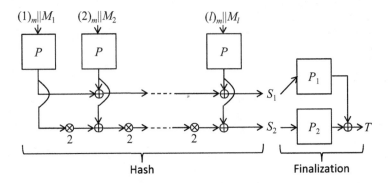

Fig. 1. LightMAC_Plus where $P := E_K$, $P_1 := E_{K_1}$ and $P_2 := E_{K_2}$.

Algorithm 1. LightMAC_Plus

▶ Main Procedure LightMAC_Plus$[E_K, E_{K_1}, E_{K_2}](M)$
1: $(S_1, S_2) \leftarrow \mathsf{Hash}[E_K](M)$
2: $T_1 \leftarrow E_{K_1}(S_1); T_2 \leftarrow E_{K_2}(S_2); T \leftarrow T_1 \oplus T_2$
3: **return** T

▶ Subroutine $\mathsf{Hash}[E_K](M)$
1: Partition $M\|10^*$ into $n - m$-bit blocks $M_1, \ldots, M_l$; $S_1 \leftarrow 0^n$; $S_2 \leftarrow 0^n$
2: **for** $i = 1, \ldots, l$ **do**
3:    $B_i \leftarrow (i)_m\|M_i$; $C_i \leftarrow E_K(B_i)$; $S_1 \leftarrow S_1 \oplus C_i$; $S_2 \leftarrow S_2 \oplus 2^{l-i} \cdot C_i$
4: **end for**
5: **return** (S_1, S_2)

3.2 Security

We prove the PRF-security of LightMAC_Plus in the information-theoretic model, namely, E_K, E_{K_1} and E_{K_2} are replaced with random permutations P, P_1 and P_2, respectively. The upper-bound of the PRF-security advantage is given below, and the security proof is given in Sect. 4.

Theorem 1. *Let $\mathcal{D}$ be a distinguisher making q tagging queries. Then we have*

$$\mathbf{Adv}^{\mathrm{prf}}_{\texttt{LightMAC\_Plus}[P,P_1,P_2]}(\mathcal{D}) \leq \frac{2q^2}{2^{2n}} + \frac{4q^3}{2^{2n}} \ .$$

4 Proof of Theorem 1

Let $F = \texttt{LightMAC\_Plus}$. In this section, we upper-bound the PRF-advantage

$$\mathbf{Adv}^{\mathrm{prf}}_{F[P,P_1,P_2]}(\mathcal{D}) = \Pr[(P, P_1, P_2) \xleftarrow{\$} \mathsf{Perm}(n)^3; \mathcal{D}^{F[P,P_1,P_2]} \Rightarrow 1]$$
$$- \Pr[\mathcal{R} \xleftarrow{\$} \mathsf{Func}(n); \mathcal{D}^{\mathcal{R}} \Rightarrow 1] \ .$$

Without loss of generality, we assume that $\mathcal{D}$ is deterministic and makes no repeated query.

In this proof, we use the following notations. For $\alpha \in [q]$, values defined at the α-th query are denoted by using the superscript character of α such as $B_i^\alpha, C_i^\alpha, S_i^\alpha$, etc., and the message length l at the α-th query is denoted by l_α. For $\alpha \in [q]$ and $j \in [2]$, $\mathsf{Dom}P_j^\alpha := \bigcup_{\delta=1}^\alpha \{S_j^\delta\}$, $\mathsf{Rng}P_j^\alpha := \bigcup_{\delta=1}^\alpha \{T_j^\delta\}$ and $\overline{\mathsf{Rng}P_j^\alpha} := \{0,1\}^n \backslash \mathsf{Rng}P_j^\alpha$.

4.1   Proof Strategy

This proof largely depends on the so-called game-playing technique [4,5]. In this proof, a random permutation P used in Hash is defined before starting the game, whereas other random permutations P_1 and P_2 are realized by lazy sampling. Before starting the game, for $i \in [2]$, all responses of P_i are not defined, that is, $\forall S_i \in \{0,1\}^n : P_i(S_i) = \perp$. When $P_i(S_i^\alpha)$ becomes necessary, if $P_i(S_i^\alpha) = \perp$ (or $S_i^\alpha \notin \mathsf{Dom}P_i^{\alpha-1}$), then it is defined as $P_i(S_i^\alpha) \xleftarrow{\$} \overline{\mathsf{Rng}P_i^{\alpha-1}}$, and otherwise, $P_i(S_i^\alpha)$ is not updated.

The main game is given in Fig. 2, where there are three sub-cases (See lines 2–4 in Fig. 2) and these procedures are defined in Fig. 3. The analysis of Case C is based on the proofs of SUM^2 construction by Lucks [23] and SUM-ECBC by Yasuda [36]. We say a set $\mathsf{Fair}^\alpha \subseteq (\{0,1\}^n)^2$ is fair if for each $T \in \{0,1\}^n$,

$$|\{(T_1, T_2) \in \mathsf{Fair}^\alpha \mid T_1 \oplus T_2 = T\}| = \frac{|\mathsf{Fair}^\alpha|}{2^n}.$$

Let $L^\alpha = \overline{\mathsf{Rng}P_1^{\alpha-1}} \times \overline{\mathsf{Rng}P_2^{\alpha-1}}$. Lucks pointed out that at the α-th query, there exists a set $W \subset L^\alpha$ of size at most $(\alpha-1)^2$ such that $L^\alpha \backslash W$ is fair. In Case C, the fair set is defined as $\mathsf{Fair}^\alpha := L^\alpha \backslash W$. Hence, the α-th output ($T^\alpha = T_1^\alpha \oplus T_2^\alpha$) is uniformly random over $\{0,1\}^n$ as long as $(T_1^\alpha, T_2^\alpha) \in \mathsf{Fair}^\alpha$. See Lemma 2 of [23] or [36] for explicit constructions of fair sets.

Initialization

1: $P \xleftarrow{\$} \mathsf{Perm}(n)$
2: $\forall i \in [2], S_i \in \{0,1\}^n : P_i(S_i) \leftarrow \perp$

Main Game: Upon the α-th query M^α **do**

1: $(S_1^\alpha, S_2^\alpha) \leftarrow \mathsf{Hash}[P](M^\alpha)$
2: **If** $S_1^\alpha \in \mathsf{Dom}P_1^{\alpha-1}$ and $S_2^\alpha \in \mathsf{Dom}P_2^{\alpha-1}$ then goto **Case A**
3: **If** ($S_1^\alpha \in \mathsf{Dom}P_1^{\alpha-1}$ and $S_2^\alpha \notin \mathsf{Dom}P_2^{\alpha-1}$) or ($S_1^\alpha \notin \mathsf{Dom}P_1^{\alpha-1}$ and $S_2^\alpha \in \mathsf{Dom}P_2^{\alpha-1}$)
     then goto Case B
4: **If** $S_1^\alpha \notin \mathsf{Dom}P_1^{\alpha-1}$ and $S_2^\alpha \notin \mathsf{Dom}P_2^{\alpha-1}$ then goto **Case C**
5: **return** T^α

Fig. 2. Main game.

Case A:

1: **If** ¬bad **then** $\mathsf{bad_A} \leftarrow$ true
2: $T^\alpha \xleftarrow{\$} \{0,1\}^n$
3: $\boxed{T_1^\alpha \leftarrow P_1(S_1^\alpha); \; T_2^\alpha \leftarrow P_2(S_2^\alpha); \; T^\alpha \leftarrow T_1^\alpha \oplus T_2^\alpha}$ ▷ Removed in the ideal world

---

Case B: In the following procedure, $S_j^\alpha \in \mathsf{Dom}P_j^{\alpha-1}$ and $S_{j+1}^\alpha \notin \mathsf{Dom}P_{j+1}^{\alpha-1}$, where $j \in [2]$ and if $j = 2$ then $j + 1$ is regarded as 1.

1: $T_{j+1}^\alpha \xleftarrow{\$} \{0,1\}^n$
2: **if** $T_{j+1}^\alpha \in \mathsf{Rng}P_{j+1}^{\alpha-1}$ **then**
3:    **if** ¬bad **then** $\mathsf{bad_B} \leftarrow$ true
4:    $\boxed{T_{j+1}^\alpha \xleftarrow{\$} \overline{\mathsf{Rng}P_{j+1}^{\alpha-1}}}$ ▷ Removed in the ideal world
5: **end if**
6: $P_{j+1}(S_{j+1}^\alpha) \leftarrow T_{j+1}^\alpha; \; T_j^\alpha \leftarrow P_j(S_j^\alpha); \; T^\alpha \leftarrow T_1^\alpha \oplus T_2^\alpha$

Case C:

1: Choose a fair set $\mathsf{Fair}^\alpha \subseteq \overline{\mathsf{Rng}P_1^{\alpha-1}} \times \overline{\mathsf{Rng}P_2^{\alpha-1}}$
2: $(T_1^\alpha, T_2^\alpha) \xleftarrow{\$} \overline{\mathsf{Rng}P_1^{\alpha-1}} \times \overline{\mathsf{Rng}P_2^{\alpha-1}}; \; T^\alpha \leftarrow T_1^\alpha \oplus T_2^\alpha$
3: **if** $(T_1^\alpha, T_2^\alpha) \notin \mathsf{Fair}^\alpha$ **then**
4:    **if** ¬bad **then** $\mathsf{bad_C} \leftarrow$ true
5:    $\boxed{(T_1^\alpha, T_2^\alpha) \xleftarrow{\$} \mathsf{Fair}^\alpha; \; T^\alpha \leftarrow T_1^\alpha \oplus T_2^\alpha}$ ▷ Removed in the real world
6: **end if**
7: $P_1(S_1^\alpha) \leftarrow T_1^\alpha; \; P_2(S_2^\alpha) \leftarrow T_2^\alpha$

Fig. 3. Case A, Case B and Case C.

Let $\mathsf{bad} = \mathsf{bad_A} \vee \mathsf{bad_B} \vee \mathsf{bad_C}$. By the fundamental lemma of game-playing [4,5], we have

$$\mathbf{Adv}^{\mathrm{prf}}_{F[P,P_1,P_2]}(\mathcal{D}) \leq \Pr[\mathsf{bad}] \leq \Pr[\mathsf{bad_A}] + \Pr[\mathsf{bad_B}] + \Pr[\mathsf{bad_C}]. \tag{1}$$

Hereafter, we upper-bound $\Pr[\mathsf{bad_A}]$, $\Pr[\mathsf{bad_B}]$ and $\Pr[\mathsf{bad_C}]$.

4.2   Upper-Bound of $\Pr[\mathsf{bad_A}]$

First we define the following event:

$$\mathsf{coll} \Leftrightarrow \exists \alpha, \beta \in [q] \text{ with } \alpha \neq \beta \text{ s.t. } (S_1^\alpha, S_2^\alpha) = (S_1^\beta, S_2^\beta).$$

Then we have

$$\Pr[\mathsf{bad_A}] \leq \Pr[\mathsf{coll}] + \Pr[\mathsf{bad_A} | \neg \mathsf{coll}] .$$

By Propositions 1 and 2, we have

$$\Pr[\mathsf{bad_A}] \leq \frac{2q^2}{2^{2n}} + \frac{\frac{4}{3}q^3}{2^{2n}} . \tag{2}$$

Proposition 1. $\Pr[\mathsf{coll}] \leq \frac{2q^2}{2^{2n}}$.

Proof. Lemma 1 shows the upper-bound of the probability that for distinct two messages $M^\alpha, M^\beta \in \{0,1\}^*$, $\mathsf{Hash}[P](M^\alpha) = \mathsf{Hash}[P](M^\beta)$, which is at most $4/2^{2n}$. The sum of the upper-bounds for all combinations of message pairs gives

$$\Pr[\mathsf{coll}] \leq \binom{q}{2} \cdot \frac{4}{2^{2n}} \leq \frac{2q^2}{2^{2n}} \ .$$

$\square$

Lemma 1. *For distinct two messages $M^\alpha, M^\beta \in \{0,1\}^*$, the probability that $\mathsf{Hash}[P](M^\alpha) = \mathsf{Hash}[P](M^\beta)$ is at most $4/2^{2n}$.*

Proof. Without loss of generality, we assume that $l_\alpha \leq l_\beta$. $\mathsf{Hash}[P](M^\alpha) = \mathsf{Hash}[P](M^\beta)$ implies that

$$S_1^\alpha = S_1^\beta \text{ and } S_2^\alpha = S_2^\beta \Leftrightarrow$$

$$\underbrace{\bigoplus_{i=1}^{l_\alpha} C_i^\alpha \oplus \bigoplus_{i=1}^{l_\beta} C_i^\beta = 0^n}_{A_{3,1}} \text{ and } \underbrace{\bigoplus_{i=1}^{l_\alpha} 2^{l_\alpha - i} \cdot C_i^\alpha \oplus \bigoplus_{i=1}^{l_\beta} 2^{l_\beta - i} \cdot C_i^\beta = 0^n}_{A_{3,2}}. \quad (3)$$

We consider the following three cases.

1. $\left(l_\alpha = l_\beta\right) \wedge \left(\exists a \in [l_\alpha] \text{ s.t. } B_a^\alpha \neq B_a^\beta\right) \wedge \left(\forall i \in [l_\alpha] \backslash \{a\} : B_i^\alpha = B_i^\beta\right)$.
2. $\left(l_\alpha = l_\beta\right) \wedge \left(\exists a_1, a_2 \in [l_\alpha] \text{ s.t. } B_{a_1}^\alpha \neq B_{a_1}^\beta \wedge B_{a_2}^\alpha \neq B_{a_2}^\beta\right)$
3. $\left(l_\alpha \neq l_\beta\right)$

The first case is that there is just one position a where the inputs are distinct, whereas the second case is that there are at least two positions a_1, a_2 where the inputs are distinct. For each case, we upper-bound the probability that (3) is satisfied.

– Consider the first case: $\exists a \in [l_\alpha]$ s.t. $B_a^\alpha \neq B_a^\beta$ and $\forall i \in [l_\alpha] \backslash \{a\} : B_i^\alpha = B_i^\beta$. Since $B_a^\alpha \neq B_a^\beta \Rightarrow C_a^\alpha \neq C_a^\beta$ and $B_i^\alpha = B_i^\beta \Rightarrow C_i^\alpha = C_i^\beta$, $A_{3,1} \neq 0^n$ and $A_{3,2} \neq 0^n$. Hence, the probability that (3) is satisfied is 0.
– Consider the second case: $\exists a_1, a_2, \ldots, a_j \in [l_\alpha]$ with $j \geq 2$ s.t. $\forall i \in [j] : B_{a_i}^\alpha \neq B_{a_i}^\beta$. Note that $B_{a_i}^\alpha \neq B_{a_i}^\beta \Rightarrow C_{a_i}^\alpha \neq C_{a_i}^\beta$. Eliminating the same outputs between $\{C_i^\alpha : 1 \leq i \leq l_\alpha\}$ and $\{C_i^\beta : 1 \leq i \leq l_\beta\}$, we have

$$A_{3,1} = \bigoplus_{i=1}^{j} \left(C_{a_i}^\alpha \oplus C_{a_i}^\beta\right) \text{ and } A_{3,2} = \bigoplus_{i=1}^{j} 2^{l_\alpha - a_i} \cdot \left(C_{a_i}^\alpha \oplus C_{a_i}^\beta\right) \ .$$

Since in $A_{3,1}$ and $A_{3,2}$ there are at most $l_\alpha + l_\beta$ outputs, the numbers of possibilities for $C_{a_1}^\alpha$ and $C_{a_2}^\alpha$ are at least $2^n - (l_\alpha + l_\beta - 2)$ and $2^n - (l_\alpha + l_\beta - 1)$, respectively. Fixing other outputs, the equations in (3) provide a unique solution for $C_{a_1}^\alpha$ and $C_{a_2}^\alpha$. As a result, the probability that (3) is satisfied is at most $1/(2^n - (l_\alpha + l_\beta - 2))(2^n - (l_\alpha + l_\beta - 1))$.

– Consider the third case. Without loss of generality, assume that $l_\alpha < l_\beta$. Eliminating the same outputs between $\{C_i^\alpha : 1 \leq i \leq l_\alpha\}$ and $\{C_i^\beta : 1 \leq i \leq l_\beta\}$, we have

$$A_{3,1} = \bigoplus_{i=1}^{u} C_{a_i}^\alpha \oplus \bigoplus_{i=1}^{v} C_{b_i}^\beta \ ,$$

where $a_1, \ldots, a_u \in [l_\alpha]$ and $b_1, \ldots, b_v \in [l_\beta]$. By $l_\alpha < l_\beta$, $l_\beta \in \{b_1, \ldots, b_v\}$ and $l_\beta \neq 1$. Since in $A_{3,1}$ and $A_{3,2}$ there are at most $l_\alpha + l_\beta$ outputs, the numbers of possibilities for C_1^β and $C_{l_\beta}^\beta$ are at least $2^n - (l_\alpha + l_\beta - 2)$ and $2^n - (l_\alpha + l_\beta - 1)$, respectively. Fixing other outputs, the equations in (3) provide a unique solution for C_1^β and $C_{l_\beta}^\beta$. As a result, the probability that (3) is satisfied is at most $1/(2^n - (l_\alpha + l_\beta - 2))(2^n - (l_\alpha + l_\beta - 1))$.

The above upper-bounds give

$$\Pr\left[\mathsf{Hash}[P](M^\alpha) = \mathsf{Hash}[P](M^\beta)\right] \leq \frac{1}{(2^n - (l_\alpha + l_\beta))^2} \leq \frac{4}{2^{2n}} \ ,$$

assuming $l_\alpha + l_\beta \leq 2^{n-1}$.

□

Proposition 2. $\Pr[\mathsf{bad_A} | \neg\mathsf{coll}] \leq \frac{\frac{4}{3}q^3}{2^{2n}}$.

Proof. First, fix $\alpha \in [q]$ and $\beta, \gamma \in [\alpha - 1]$ with $\beta \neq \gamma$ (from the condition $\neg\mathsf{coll}$), and upper-bound the probability that $S_1^\alpha = S_1^\beta \wedge S_2^\alpha = S_2^\gamma$, which implies

$$\underbrace{\bigoplus_{i=1}^{l_\alpha - 1} C_i^\alpha \oplus \bigoplus_{i=1}^{l_\beta - 1} C_i^\beta = 0^n}_{A_{4,1}} \text{ and } \underbrace{\bigoplus_{i=1}^{l_\alpha - 1} 2^{l_\alpha - i} \cdot C_i^\alpha \oplus \bigoplus_{i=1}^{l_\gamma - 1} 2^{l_\gamma - i} \cdot C_i^\gamma = 0^n}_{A_{4,2}}. \quad (4)$$

Since M^α, M^β and M^γ are distinct, there are at least two distinct outputs $C^{\alpha,\beta}$ and $C^{\alpha,\gamma}$ where $C^{\alpha,\beta}$ appears in $A_{4,1}$ and $C^{\alpha,\gamma}$ appears in $A_{4,2}$. Fixing other outputs in $A_{4,1}$ and $A_{4,2}$, the equations in (4) provide a unique solution for $C^{\alpha,\beta}$ and $C^{\alpha,\gamma}$. Since there are at most $l_\alpha + l_\beta$ outputs in $A_{4,1}$, the number of possibilities for $C^{\alpha,\beta}$ is at least $2^n - (l_\alpha + l_\beta - 1)$. Since there are at most $l_\alpha + l_\gamma$ outputs in $A_{4,2}$, the number of possibilities for $C^{\alpha,\gamma}$ is at least $2^n - (l_\alpha + l_\gamma - 1)$. Hence, the probability that (4) is satisfied is at most

$$\frac{1}{(2^n - (l_\alpha + l_\beta - 1))(2^n - (l_\alpha + l_\gamma - 1))} \leq \frac{4}{2^{2n}} \ ,$$

assuming $l_\alpha + l_\beta - 1 \leq 2^{n-1}$ and $l_\alpha + l_\gamma - 1 \leq 2^{n-1}$.

Finally, we just run induces α, β, and γ to get

$$\Pr[\mathsf{bad_A} | \neg\mathsf{coll}] \leq \sum_{\alpha=1}^{q} \left(\sum_{\beta, \gamma \in [1, \alpha-1] \text{ s.t. } \beta \neq \gamma} \frac{4}{2^{2n}} \right) \leq \sum_{\alpha=1}^{q} \frac{4(\alpha-1)^2}{2^{2n}} = \sum_{\alpha=1}^{q-1} \frac{4\alpha^2}{2^{2n}}$$

$$\leq \frac{4}{2^{2n}} \times \frac{q(q-1)(2q-1)}{6} \leq \frac{\frac{4}{3}q^3}{2^{2n}} \ .$$

□

4.3   Upper-Bound of $\Pr[\mathsf{bad_B}]$

First, fix $\alpha \in [q]$ and $j \in [2]$, and upper-bound the probability that $\mathcal{D}$ sets $\mathsf{bad_B}$ at the α-th query, namely, $S_j^\alpha \in \mathsf{Dom}P_j^{\alpha-1}$, $S_{j+1}^\alpha \notin \mathsf{Dom}P_{j+1}^{\alpha-1}$, and $T_{j+1}^\alpha \in \mathsf{Rng}P_{j+1}^{\alpha-1}$. Note that if $j = 2$ then $j+1$ is regarded as 1.

- Regarding $S_j^\alpha \in \mathsf{Rng}P_j^{\alpha-1}$, fix $\beta \in [\alpha-1]$ and consider the case that $S_j^\alpha = S_j^\beta$. Since $M^\alpha \neq M^\beta$, there is an output $C^{\alpha,\beta}$ in $\{C_1^\alpha, \ldots, C_{l_\alpha}^\alpha, C_1^\beta, \ldots, C_{l_\beta}^\beta\}$ that is distinct from other outputs. Fixing other outputs, $S_j^\alpha = S_j^\beta$ provides a unique solution for $C^{\alpha,\beta}$. There are at most $2^n - (l_\alpha + l_\beta - 1)$ possibilities for $C^{\alpha,\beta}$. Hence, the probability that $S_j^\alpha \in \mathsf{Dom}P_j^{\alpha-1}$ is at most $|\mathsf{Dom}P_j^{\alpha-1}| \times 1/(2^n - (l_\alpha + l_\beta - 1)) \leq 2(\alpha-1)/2^n$, assuming $l_\alpha + l_\beta - 1 \leq 2^{n-1}$.
- Regarding $T_{j+1}^\alpha \in \mathsf{Rng}P_{j+1}^{\alpha-1}$, T_{j+1}^α is randomly drawn from $\{0,1\}^n$ after $S_j^\alpha \in \mathsf{Rng}P_j^{\alpha-1}$ and $S_{j+1}^\alpha \notin \mathsf{Dom}P_{j+1}^{\alpha-1}$ are satisfied. In this case, T_{j+1}^α is defined independently from S_j^α and S_{j+1}^α. Since $|\mathsf{Rng}P_{j+1}^{\alpha-1}| \leq \alpha - 1$, this probability that $T_{j+1}^\alpha \in \mathsf{Rng}P_{j+1}^{\alpha-1}$ is at most $(\alpha-1)/2^n$.

Hence, the probability that $\mathcal{D}$ sets $\mathsf{bad_B}$ at the α-th query is upper-bounded by the multiplication of the above probabilities, which is $\frac{2(\alpha-1)^2}{2^{2n}}$.

Finally, we just run induces α and j to get

$$\Pr[\mathsf{nosol}] \leq \sum_{\alpha=1}^{q} \sum_{j=1}^{2} \frac{2(\alpha-1)^2}{2^{2n}} \leq \frac{\frac{4}{3}q^3}{2^{2n}} \ . \tag{5}$$

4.4   Upper-Bound of $\Pr[\mathsf{bad_C}]$

For each $\alpha \in [q]$, since $\left|\overline{\mathsf{Rng}P_1^{\alpha-1}} \times \overline{\mathsf{Rng}P_2^{\alpha-1}} \backslash \mathsf{Fair}^\alpha\right| \leq (\alpha-1)^2$, the probability that $(T_1^\alpha, T_2^\alpha) \notin \mathsf{Fair}^\alpha$ is at most

$$\frac{(\alpha-1)^2}{(2^n - (\alpha-1))^2} \leq \frac{4(\alpha-1)^2}{2^{2n}} \ ,$$

assuming $\alpha - 1 \leq 2^{n-1}$. Hence, we have

$$\Pr[\mathsf{bad_C}] \leq \sum_{\alpha=1}^{q} \frac{4(\alpha-1)^2}{2^{2n}} = \sum_{\alpha=1}^{q-1} \frac{4(\alpha-2)^2}{2^{2n}} \leq \frac{\frac{4}{3}q^3}{2^{2n}} \ . \tag{6}$$

4.5   Conclusion of Proof

Putting (2), (5) and (6) into (1) gives

$$\mathbf{Adv}_{F[P,P_1,P_2]}^{\mathsf{prf}}(\mathcal{D}) \leq \frac{2q^2}{2^{2n}} + \frac{\frac{4}{3} \cdot q^3}{2^{2n}} + \frac{\frac{4}{3}q^3}{2^{2n}} + \frac{\frac{4}{3}q^3}{2^{2n}} \leq \frac{2q^2}{2^{2n}} + \frac{4q^3}{2^{2n}} \ .$$

Algorithm 2. LightMAC_Plus2$[H_{K_H}, E_{K_{0,1}}, E_{K_{0,2}}, E_{K_1}, \ldots, E_{K_t}]$

▶ Main Procedure LightMAC_Plus2$[H_{K_H}, E_{K_{0,1}}, E_{K_{0,2}}, E_{K_1}, \ldots, E_{K_t}](M)$
1: $(S_1, S_2) \leftarrow H_{K_H}(M)$
2: $R_1 \leftarrow E_{K_{0,1}}(S_1); R_2 \leftarrow E_{K_{0,2}}(S_2); T \leftarrow 0^n$
3: **for** $i = 1, \ldots, t$ **do**
4:     $X_i \leftarrow R_1 \oplus 2^{i-1} \cdot R_2; Y_i \leftarrow E_{K_i}(X_i); T \leftarrow T \oplus Y_i$
5: **end for**
6: **return** T

---

5 LightMAC_Plus2

5.1 Construction

Let $\mathcal{K}$, $\mathcal{K}_H$ and $\mathsf{Dom}H$ be three non-empty sets. Let $\{E_K\}_{K \in \mathcal{K}} \subset \mathsf{Perm}(n)$ be a family of n-bit permutations (or a blockcipher) indexed by key space $\mathcal{K}$. Let $\{H_{K_H}\}_{K_H \in \mathcal{K}_H}$ be a family of hash functions: $\mathsf{Dom}H \to \{0,1\}^{2n}$ indexed by key space $\mathcal{K}_H$. Let m be the counter size with $m < n$. Let $K_{0,1}, K_{0,2}, K_1, \ldots, K_t \in \mathcal{K}$ be the E's keys and $K_H \in \mathcal{K}_H$ the hash key. For a message M, the response of LightMAC_Plus2$[H_{K_H}, E_{K_{0,1}}, E_{K_{0,2}}, E_{K_1}, \ldots, E_{K_t}]$ is defined by Algorithm 2, where $|S_1| = n$ and $|S_2| = n$. The finalization function is illustrated in Fig. 4.

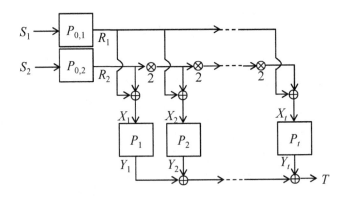

Fig. 4. Finalization function of LightMAC_Plus2, where $P_{0,1} := E_{K_{0,1}}, P_{0,2} := E_{K_{0,2}}, P_1 := E_{K_1}, \ldots, P_t := E_{K_t}$.

5.2 Almost Universal Hash Function

In the security proof, we assume that the hash function H is an almost universal (AU) hash function. The definition is given below.

Definition 1. *Let $\epsilon > 0$. H is an ϵ-AU hash function if for any two distinct messages $M, M' \in \mathsf{Dom}H$, $\Pr[K_H \xleftarrow{\$} \mathcal{K}_H; H_{K_H}(M) = H_{K_H}(M')] \leq \epsilon$.*

5.3   Security

We prove the PRF-security of LightMAC_Plus2 in the information-theoretic model, where permutations $E_{K_{0,1}}, E_{K_{0,2}}, E_{K_1}, \ldots, E_{K_{t-1}}$ and E_{K_t} are replaced with random permutations $P_{0,1}, P_{0,2}, P_1, \ldots, P_{t-1}$ and P_t, respectively, and H is assumed to be an ϵ-AU hash function, where a key is drawn as $K_H \xleftarrow{\$} \mathcal{K}_H$. The upper-bound of the PRF-security advantage is given below, and the security proof is given in Sect. 6.

Theorem 2. *Assume that $t \leq 7$. Let H is an ϵ-AU hash function. Let $\mathcal{D}$ be a distinguisher making q tagging queries. Then we have*

$$\mathbf{Adv}^{\mathsf{prf}}_{\mathtt{LightMAC\_Plus2}[H_{K_H}, P_{0,1}, P_{0,2}, P_1, \ldots, P_{t-1}, P_t]}(\mathcal{D}) \leq 0.5q^2\epsilon + \frac{2^t q^{t+1}}{(2^n - q)^t} .$$

Define the hash function as $H_{K_H} := \mathsf{Hash}[P]$ (given in Algorithm 1). By Lemma 1, Hash is a $4/2^{2n}$-AU hash function, where $\mathcal{K}_H = \mathsf{Perm}(n)$ and $K_H = P$. Hence, combining Lemma 1 and Theorem 2, the following corollary is obtained.

Corollary 1. *Let $H_{K_H} := \mathsf{Hash}[P]$. Then we have*

$$\mathbf{Adv}^{\mathsf{prf}}_{\mathtt{LightMAC\_Plus2}[H_{K_H}, P_{0,1}, P_{0,2}, P_1, \ldots, P_{t-1}, P_t]}(\mathcal{D}) \leq \frac{2q^2}{2^{2n}} + \frac{2^t q^{t+1}}{(2^n - q)^t} .$$

6   Proof of Theorem 2

Assume that $t \leq 7$. Let $F = \mathtt{LightMAC\_Plus2}$ and $\mathbf{P} = (P_{0,1}, P_{0,2}, P_1, \ldots, P_t)$. In this section, we upper-bound the PRF-advantage

$$\mathbf{Adv}^{\mathsf{prf}}_{F[H_{K_H}, \mathbf{P}]}(\mathcal{D}) = \Pr[\mathbf{P} \xleftarrow{\$} \mathsf{Perm}(n)^{t+2}; K_H \xleftarrow{\$} \mathcal{K}_H; \mathcal{D}^{F[H_{K_H}, \mathbf{P}]} \Rightarrow 1]$$

$$- \Pr[\mathcal{R} \xleftarrow{\$} \mathsf{Func}(n); \mathcal{D}^{\mathcal{R}} \Rightarrow 1] .$$

Without loss of generality, we assume that $\mathcal{D}$ is deterministic and makes no repeated query.

In this proof, we use the following notations. For $\alpha \in [q]$, values defined at the α-th query are denoted by using the superscript of α such as $B_i^\alpha, C_i^\alpha, S_i^\alpha$, etc., and the message length l at the α-th query is denoted by l_α. For $\alpha \in [q]$ and $j \in [t]$, $\mathsf{Dom}P_j^\alpha := \bigcup_{\delta=1}^\alpha \{X_j^\delta\}$, $\mathsf{Rng}P_j^\alpha := \bigcup_{\delta=1}^\alpha \{Y_j^\delta\}$ and $\overline{\mathsf{Rng}P_j^\alpha} := \{0,1\}^n \backslash \mathsf{Rng}P_j^\alpha$.

6.1   Proof Strategy

This proof uses the same strategy as in the proof of Theorem 1 (given in Subsect. 4.1). In this proof, random permutations $P_{0,1}$ and $P_{0,2}$ are defined before starting the game, whereas other random permutations are realized by lazy sampling. The main game is given in Fig. 5, where there are three sub-cases defined by inputs to random permutations $X_1^\alpha, \ldots, X_t^\alpha$ (See lines 4–6 in Fig. 5). The sub-cases are given in Fig. 6. Note that for $i \in [t]$, "X_i^α is new" means that $X_i^\alpha \notin \mathrm{Dom}P_i^{\alpha-1}$, and "$X_i^\alpha$ is not new" means that $X_i^\alpha \in \mathrm{Dom}P_i^{\alpha-1}$.

As is the case with the proof of Theorem 1, Case C uses a fair set for the xor of s random permutations with $s \geq 2$. For s random permutations $P_{a_1}, \ldots, P_{a_s}$ at the α-th query, we say a set $\mathsf{Fair}^\alpha \subseteq (\{0,1\}^n)^s$ is fair if for each $T \in \{0,1\}^n$,

$$\left| \left\{ (Y_{a_1}, Y_{a_2}, \ldots, Y_{a_s}) \in \mathsf{Fair}^\alpha \,\middle|\, \bigoplus_{i \in [s]} Y_{a_i} = T \right\} \right| = \frac{|\mathsf{Fair}^\alpha|}{2^n}.$$

Let $L^\alpha := \overline{\mathrm{Rng}P_{a_1}^{\alpha-1}} \times \overline{\mathrm{Rng}P_{a_2}^{\alpha-1}} \times \cdots \times \overline{\mathrm{Rng}P_{a_s}^{\alpha-1}}$. Lucks [23] pointed out that when s is even, there exists a set $W \subset L^\alpha$ of size at most $(\alpha - 1)^s$ such that $L^\alpha \backslash W$ is fair, and when s is odd, there exists a set $W' \subset (\{0,1\}^n)^s$ of size at most $(\alpha - 1)^s$ with $W' \cap L^\alpha = \emptyset$ such that $W' \cup L^\alpha$ is fair. See Lemma 2 of [23] or [36] for explicit constructions of fair sets. In Case C, the fair set is defined as $\mathsf{Fair}^\alpha := L^\alpha \backslash W$ when s is even; $\mathsf{Fair}^\alpha := L^\alpha \cup W'$ when s is odd.

Let $\mathsf{bad} = \mathsf{bad_A} \vee \mathsf{bad_B} \vee \mathsf{bad_C}$. Then by the fundamental lemma of game-playing [4,5], we have

$$\mathbf{Adv}_{F[\mathbf{P}]}^{\mathrm{prf}}(\mathcal{D}) \leq \Pr[\mathsf{bad}] \leq \Pr[\mathsf{bad_A}] + \Pr[\mathsf{bad_B}] + \Pr[\mathsf{bad_C}]. \tag{7}$$

Hereafter, we upper-bound $\Pr[\mathsf{bad_A}]$, $\Pr[\mathsf{bad_B}]$ and $\Pr[\mathsf{bad_C}]$.

Initialization
1: $K_H \overset{\$}{\leftarrow} \mathcal{K}_H$; $(P_{0,1}, P_{0,2}) \overset{\$}{\leftarrow} \mathrm{Perm}(n)^2$
2: $\forall i \in [t], X_i \in \{0,1\}^n : P_i(X_i) \leftarrow\perp$

Main Game: Upon the α-th query M^α **do**
1: $(S_1^\alpha, S_2^\alpha) \leftarrow H_{K_H}(M^\alpha)$
2: $R_1^\alpha \leftarrow P_{0,1}(S_1^\alpha)$; $R_2 \leftarrow P_{0,2}(S_2^\alpha)$;
3: **for** $i \in [t]$ **do** $X_i^\alpha = R_1^\alpha \oplus \left(2^{i-1} \cdot R_2^\alpha\right)$
4: **if** all of $X_1^\alpha, \ldots, X_t^\alpha$ are not new **then** goto **Case A**
5: **if** one of $X_1^\alpha, \ldots, X_t^\alpha$ is new **then** goto **Case B**
6: **if** two ore more of $X_1^\alpha, \ldots, X_t^\alpha$ are new **then** goto **Case C**
7: **return** T^α

Fig. 5. Main Game.

Case A:

1: **if** ¬bad **then** bad$_\mathsf{A}$ ← true
2: $T^\alpha \xleftarrow{\$} \{0,1\}^n$
3: **for** $i \in [t]$ **do** $Y_i^\alpha \leftarrow P_i(X_i^\alpha)$
4: $\boxed{T^\alpha \leftarrow \bigoplus_{i=1}^{t} Y_i^\alpha}$ ▷ Removed in the ideal world

---

Case B: In the following procedure, X_a^α is new, and for all $i \in [t] \backslash \{a\}$ X_i^α is not new.

1: $Y_a^\alpha \xleftarrow{\$} \{0,1\}^n$
2: **if** $Y_a^\alpha \in \mathsf{Rng}P_a^{\alpha-1}$ **then**
3:    **if** ¬bad **then** bad$_\mathsf{B}$ ← true
4:    $\boxed{Y_a^\alpha \xleftarrow{\$} \overline{\mathsf{Rng}P_a^{\alpha-1}}}$ ▷ Removed in the ideal world
5: **end if**
6: $P_i(X_a^\alpha) \leftarrow Y_a^\alpha$
7: **for** $i \in [t] \backslash \{a\}$ **do** $Y_i^\alpha \leftarrow P_i(X_i^\alpha)$
8: $T^\alpha \leftarrow \bigoplus_{i=1}^{t} Y_i^\alpha$

Case C: In the following procedure, $X_{a_1}^\alpha, \ldots, X_{a_s}^\alpha$ are new with $a_1, \ldots a_s \in [t]$ and other inputs are not new where $s \geq 2$.

1: $L^\alpha \leftarrow \overline{\mathsf{Rng}P_{a_1}^{\alpha-1}} \times \overline{\mathsf{Rng}P_{a_2}^{\alpha-1}} \times \cdots \times \overline{\mathsf{Rng}P_{a_s}^{\alpha-1}}$
2: **if** s is even **then**
3:    Choose a fair set $\mathsf{Fair}^\alpha \subseteq L^\alpha$; $(Y_{a_1}^\alpha, Y_{a_2}^\alpha, \ldots, Y_{a_s}^\alpha) \xleftarrow{\$} L^\alpha$
4:    **if** $(Y_{a_1}^\alpha, Y_{a_2}^\alpha, \ldots, Y_{a_s}^\alpha) \notin \mathsf{Fair}^\alpha$ **then**
5:       **if** ¬bad **then** bad$_\mathsf{C}$ ← true
6:       $\boxed{(Y_{a_1}^\alpha, Y_{a_2}^\alpha, \ldots, Y_{a_s}^\alpha) \xleftarrow{\$} \mathsf{Fair}^\alpha}$ ▷ Removed in the real world
7:    **end if**
8: **end if**
9: **if** s is odd **then**
10:    Choose a fair set $\mathsf{Fair}^\alpha \supseteq L^\alpha$; $(Y_{a_1}^\alpha, Y_{a_2}^\alpha, \ldots, Y_{a_s}^\alpha) \xleftarrow{\$} \mathsf{Fair}^\alpha$
11:    **if** $(Y_{a_1}^\alpha, Y_{a_2}^\alpha, \ldots, Y_{a_s}^\alpha) \notin L^\alpha$ **then**
12:       **if** ¬bad **then** bad$_\mathsf{C}$ ← true
13:       $\boxed{(Y_{a_1}^\alpha, Y_{a_2}^\alpha, \ldots, Y_{a_s}^\alpha) \xleftarrow{\$} L^\alpha}$ ▷ Removed in the ideal world
14:    **end if**
15: **end if**
16: **for** $i \in [s]$ **do** $P_i(X_{a_i}^\alpha) \leftarrow Y_{a_i}^\alpha$
17: **for** $i \in [t] \backslash \{a_1, \ldots, a_s\}$ **do** $Y_i^\alpha \leftarrow P_i(X_i^\alpha)$
18: $T^\alpha \leftarrow \bigoplus_{i=1}^{t} Y_i^\alpha$

Fig. 6. Case A, Case B and Case C.

6.2 Upper-Bound of $\Pr[\mathsf{bad_A}]$

First we define the following event:

$$\mathsf{coll} \Leftrightarrow \exists \alpha, \beta \in [q] \text{ with } \alpha \neq \beta \text{ s.t. } (S_1^\alpha, S_2^\alpha) = (S_1^\beta, S_2^\beta).$$

Then we have

$$\Pr[\mathsf{bad_A}] \le \Pr[\mathsf{coll}] + \Pr[\mathsf{bad_A}|\neg\mathsf{coll}] \ ,$$

Regarding $\Pr[\mathsf{coll}]$, since H is an ϵ-AU hash function, the sum of ϵ for all combinations of message pairs gives

$$\Pr[\mathsf{coll}] \le \binom{q}{2} \cdot \epsilon \le 0.5 q^2 \epsilon \ .$$

Regarding $\Pr[\mathsf{bad_A}|\neg\mathsf{coll}]$, for $\alpha \in [q]$, Lemma 2 gives the upper-bound of the probability that all of $X_1^\alpha, \ldots, X_t^\alpha$ are not new, which is $\left(\frac{\alpha-1}{2^n-q}\right)^t$. Then, we run the index α to get

$$\Pr[\mathsf{bad_A}|\neg\mathsf{coll}] \le \sum_{\alpha=1}^{q} \left(\frac{\alpha-1}{2^n-q}\right)^t = \sum_{\alpha=1}^{q-1} \left(\frac{\alpha}{2^n-q}\right)^t \ .$$

Finally we have

$$\Pr[\mathsf{bad_A}] \le 0.5 q^2 \epsilon + \sum_{\alpha=1}^{q-1} \left(\frac{\alpha}{2^n-q}\right)^t \ . \tag{8}$$

Lemma 2. *Assume that* coll *does not occur. Fix* $\alpha \in [q]$, $s \le t$ *and* $a_1, a_2, \ldots, a_s \in [t]$ *such that* $a_1, a_2, \ldots, a_s$ *are distinct. Then the probability that* $\forall i \in [s]: X_{a_i}^\alpha$ *is not new, that is,* $\exists \beta_i \in [\alpha-1]$ *s.t.* $X_{a_i}^\alpha = X_{a_i}^{\beta_i}$ *is at most* $\left(\frac{\alpha-1}{2^n-q}\right)^s$.

Proof. First, fix $\beta_1, \ldots, \beta_s \in [\alpha-1]$, and upper-bound the probability that

$$\underbrace{\forall i \in [s] : X_{a_i}^\alpha \oplus X_{a_i}^{\beta_i} = 0^n.}_{A_9} \tag{9}$$

By Lemma 3, we have only to consider the case where $\beta_1, \ldots, \beta_s$ are distinct. Thus if $\alpha \le s$, then this probability is 0. In the following, we consider the case: $\alpha > s$. Note that A_9 is defined as

$$X_{a_i}^\alpha \oplus X_{a_i}^{\beta_i} = \left(R_1^\alpha \oplus 2^{a_i-1} \cdot R_2^\alpha\right) \oplus \left(R_1^{\beta_i} \oplus 2^{a_i-1} \cdot R_2^{\beta_i}\right)$$

$$= \left(R_1^\alpha \oplus R_1^{\beta_i}\right) \oplus 2^{a_i-1} \cdot \left(R_2^\alpha \oplus R_2^{\beta_i}\right) \ ,$$

where $R_1^\alpha = P_{0,1}(S_1^\alpha)$, $R_2^\alpha = P_{0,2}(S_2^\alpha)$, $R_1^{\beta_i} = P_{0,1}(S_1^{\beta_i})$ and $R_2^{\beta_i} = P_{0,2}(S_2^{\beta_i})$. Then, the number of independent random variables in $\{R_1^\alpha, R_1^{\beta_1}, \ldots, R_1^{\beta_s}, R_2^\alpha, R_2^{\beta_1}, \ldots, R_2^{\beta_s}\}$ that appear in A_9 is counted. Note that $\{R_1^\alpha, R_1^{\beta_1}, \ldots, R_1^{\beta_s}\}$ are independently defined from $\{R_2^\alpha, R_2^{\beta_1}, \ldots, R_2^{\beta_s}\}$.

First, the number of independent random variables in $\{R_1^{\beta_1}, \ldots, R_1^{\beta_s}\}$ and $\{R_2^{\beta_1}, \ldots, R_2^{\beta_s}\}$ is counted. By $\neg\mathsf{coll}$, for all $i, j \in [s]$ with $i \ne j$,

$(S_1^{\beta_i}, S_2^{\beta_i}) \neq (S_1^{\beta_j}, S_2^{\beta_j})$, that is, $(R_1^{\beta_i}, R_2^{\beta_i}) \neq (R_1^{\beta_j}, R_2^{\beta_j})$. Note that if there are z_1 (resp., z_2) independent random variables in $\{R_1^{\beta_1}, \ldots, R_1^{\beta_s}\}$ (resp., $\{R_2^{\beta_1}, \ldots, R_2^{\beta_s}\}$), then the number of distinct pairs for (R_1, R_2) is $z_1 \cdot z_2$ and the number of distinct random variables is $z_1 + z_2$. If $(z_1 \leq 2 \wedge z_2 \leq 2)$ or $(z_1 = 1 \wedge z_2 \leq 4)$, then $z_1 \cdot z_2 \leq z_1 + z_2$, and if $z_1 = 2$ and $z_2 = 3$, then $z_1 + z_2 = 5 < z_1 \cdot z_2 = 6$. Since $s \leq z_1 \cdot z_2$, the sum of the numbers of independent random variables in $\{R_1^{\beta_1}, \ldots, R_1^{\beta_s}\}$ and in $\{R_2^{\beta_1}, \ldots, R_2^{\beta_s}\}$ is at least $\min\{5, s\}$.

By Lemma 4, we have only to consider the case that $\forall i \in [s] : R_1^\alpha \neq R_1^{\beta_i}$ and $R_2^\alpha \neq R_2^{\beta_i}$. Hence, the number of independent random variables in $\{R_1^{\beta_1}, \ldots, R_1^{\beta_s}\}$ and $\{R_2^{\beta_1}, \ldots, R_2^{\beta_s}\}$ is at least $s \leq \min\{5, s\} + 2$. By $s \leq t \leq 7$, there are at least s independent random variables in A_9.

Fixing other outputs in A_9 except for the s outputs, the equations in (9) provide a unique solution for the s outputs. The number of possibilities for the s outputs are at least $2^n - s$. Hence, the probability that (9) is satisfied is at most $(1/(2^n - s))^s$.

Finally, the probability that $\forall i \in [s] : \exists \beta_i \in [\alpha - 1]$ s.t. $X_{a_i}^\alpha = X_{a_i}^{\beta_i}$ is at most

$$(\alpha - 1)^s \cdot \left(\frac{1}{2^n - s} \right)^s \leq \left(\frac{\alpha - 1}{2^n - q} \right)^s .$$

$\square$

Lemma 3. *Assume that* coll *does not occur. For* $\alpha, \beta \in [q]$ *with* $\alpha \neq \beta$, *if there exists* $j \in [t]$ *such that* $X_j^\alpha = X_j^\beta$, *then for all* $i \in [t] \backslash \{j\}$, $X_i^\alpha \neq X_i^\beta$.

Proof. Assume that $X_j^\alpha = X_j^\beta$, which implies

$$X_j^\alpha = X_j^\beta \Leftrightarrow R_1^\alpha \oplus R_1^\beta = 2^{j-1} \cdot \left(R_2^\alpha \oplus R_2^\beta \right) .$$

By $\neg$coll, $R_1^\alpha \oplus R_1^\beta \neq 0^n$ and $R_2^\alpha \oplus R_2^\beta \neq 0^n$. Then, for any $i \in [t] \backslash \{j\}$

$$X_i^\alpha \oplus X_i^\beta = \left(R_1^\alpha \oplus R_1^\beta \right) \oplus 2^{i-1} \cdot \left(R_2^\alpha \oplus R_2^\beta \right)$$
$$= \left(2^{j-1} \oplus 2^{i-1} \right) \cdot \left(R_2^\alpha \oplus R_2^\beta \right) \neq 0^n ,$$

namely, $X_i^\alpha \neq X_i^\beta$.

$\square$

Lemma 4. *For* $\alpha, \beta \in [q]$ *with* $\alpha \neq \beta$, *if* $(R_1^\alpha \neq R_1^\beta \wedge R_2^\beta = R_2^\beta)$ *or* $(R_1^\alpha = R_1^\beta \wedge R_2^\alpha \neq R_2^\beta)$, *then for all* $i \in [t]$ $X_i^\alpha \neq X_i^\beta$.

Proof. Let $\alpha, \beta \in [q]$ with $\alpha \neq \beta$. If $R_1^\alpha \neq R_1^\beta \wedge R_2^\alpha = R_2^\beta$, then for any $i \in [t]$,

$$X_i^\alpha \oplus X_i^\beta = \left(R_1^\alpha \oplus 2^{i-1} \cdot R_2^\alpha \right) \oplus \left(R_1^\beta \oplus 2^{i-1} \cdot R_2^\beta \right) = R_1^\alpha \oplus R_1^\beta \neq 0^n.$$

If $R_1^\alpha = R_1^\beta \wedge R_2^\alpha \neq R_2^\beta$, then for any $i \in [t]$,

$$X_i^\alpha \oplus X_i^\beta = \left(R_1^\alpha \oplus 2^{i-1} \cdot R_2^\alpha \right) \oplus \left(R_1^\beta \oplus 2^{i-1} \cdot R_2^\beta \right) = 2^{i-1} \cdot \left(R_2^\alpha \oplus \cdot R_2^\beta \right) \neq 0^n.$$

$\square$

6.3   Upper-Bound of $\Pr[\mathsf{bad_B}]$

First, fix $\alpha \in [q]$ and $a \in [t]$, and upper-bound the probability that

$$\underbrace{X_a^\alpha \text{ is new, } \forall i \in [t]\backslash\{a\} : X_i^\alpha \text{ is not new,}}_{A_{10,2}} \text{ and } \underbrace{Y_a^\alpha \in \mathsf{Rng}P_a^{\alpha-1}}_{A_{10,3}}. \tag{10}$$

Regarding $A_{10,2}$, by Lemma 2, the probability that $A_{10,2}$ is satisfied is at most $\left(\frac{\alpha-1}{2^n-q}\right)^{t-1}$. Regarding $A_{10,3}$, since Y_a^α is randomly drawn and $|\mathsf{Rng}P_a^{\alpha-1}| \leq \alpha-1$, the probability that $A_{10,3}$ is satisfied is at most $\frac{\alpha-1}{2^n}$. Hence the probability that (10) is satisfied is at most

$$\left(\frac{\alpha-1}{2^n-q}\right)^{t-1} \cdot \frac{\alpha-1}{2^n} \leq \left(\frac{\alpha-1}{2^n-q}\right)^t .$$

Finally, we run induces α and a to get

$$\Pr[\mathsf{bad_B}] \leq \sum_{\alpha=1}^q \sum_{a=1}^t \left(\frac{\alpha-1}{2^n-q}\right)^t \leq \sum_{\alpha=1}^{q-1} t \cdot \left(\frac{\alpha}{2^n-q}\right)^t . \tag{11}$$

6.4   Upper-Bound of $\Pr[\mathsf{bad_C}]$

First, fix $\alpha \in [q]$, $s \in \{2,\ldots,t\}$ and $a_1,\ldots,a_s \in [t]$ such that $a_1,\ldots,a_s$ are distinct, and consider the case that

$$X_{a_1}^\alpha,\ldots,X_{a_{s-1}}^\alpha \text{ and } X_{a_s}^\alpha \text{ are new, } \underbrace{\forall i \in [t]\backslash\{a_1,\ldots,a_s\} : X_i^\alpha \text{ is not new,}}_{A_{12,2}} \text{ and}$$

$$\underbrace{(Y_{a_1}^\alpha,\ldots,Y_{a_{s-1}}^\alpha,Y_{a_s}^\alpha) \notin \mathsf{Fair}^\alpha \text{ if } s \text{ is even}; (Y_{a_1}^\alpha,\ldots,Y_{a_{s-1}}^\alpha,Y_{a_s}^\alpha) \notin L^\alpha \text{ if } s \text{ is odd}}_{A_{12,3}}.$$

$$\tag{12}$$

Regarding $A_{12,2}$, by Lemma 2, the probability that $A_{12,2}$ is satisfied is at most $\left(\frac{\alpha-1}{2^n-q}\right)^{t-s}$. Regarding $A_{12,3}$, if s is even, then since $|L^\alpha\backslash\mathsf{Fair}^\alpha| \leq (\alpha-1)^s$, the probability that $A_{12,3}$ is satisfied is at most $\left(\frac{\alpha-1}{2^n-q}\right)^s$; if s is odd, then since $|\mathsf{Fair}^\alpha\backslash L^\alpha| \leq (\alpha-1)^s$, the probability that $A_{12,3}$ is satisfied is at most $\left(\frac{\alpha-1}{2^n-q}\right)^s$. Hence, the probability that the conditions in (12) are satisfied is at most

$$\left(\frac{\alpha-1}{2^n-q}\right)^{t-s} \cdot \left(\frac{\alpha-1}{2^n-q}\right)^s = \left(\frac{\alpha-1}{2^n-q}\right)^t .$$

Finally, we run induces α and s to get

$$\Pr[\mathsf{bad_C}] \leq \sum_{\alpha=1}^q \sum_{s=2}^t \left(\binom{t}{s} \cdot \left(\frac{\alpha-1}{2^n-q}\right)^t\right) = \sum_{s=2}^t \binom{t}{s} \cdot \left(\sum_{\alpha=1}^{q-1}\left(\frac{\alpha}{2^n-q}\right)^t\right) . \tag{13}$$

6.5   Conclusion of Proof

Putting (8), (11) and (13) into (7) gives

$$\mathbf{Adv}^{\mathsf{prf}}_{F[H_{K_H},\mathbf{P}]}(\mathcal{D})$$

$$\leq 0.5q^2\epsilon + \sum_{\alpha=1}^{q-1}\left(\frac{\alpha}{2^n-q}\right)^t + t\cdot\sum_{\alpha=1}^{q-1}\left(\frac{\alpha}{2^n-q}\right)^t + \sum_{s=2}^{t}\binom{t}{s}\left(\sum_{\alpha=1}^{q-1}\left(\frac{\alpha}{2^n-q}\right)^t\right)$$

$$\leq 0.5q^2\epsilon + \sum_{s=0}^{t}\binom{t}{s}\cdot\left(\sum_{\alpha=1}^{q-1}\left(\frac{\alpha}{2^n-q}\right)^t\right) = 0.5q^2\epsilon + 2^t\cdot\left(\sum_{\alpha=1}^{q-1}\left(\frac{\alpha}{2^n-q}\right)^t\right)$$

$$= 0.5q^2\epsilon + \sum_{\alpha=1}^{q-1}\left(\frac{2\alpha}{2^n-q}\right)^t \leq 0.5q^2\epsilon + \frac{2^t q^{t+1}}{(2^n-q)^t} \quad ,$$

where the last term uses the fact that $\sum_{\alpha=1}^{x}\alpha^t \leq x^{t+1}$ for $x \geq 1$ and $t \geq 1$.

7   Improving the Efficiency of Hash

In this section, we consider a hash function Hash^* with better efficiency than Hash. Hash^* is defined in Algorithm 3 and is illustrated in Fig. 7. Here, $M\|10^*$ means that first 1 is appended to M, and if $|M\|1| \leq n$, then a sequence of the minimum number of zeros is appended to $M\|1$ so that the length in bits becomes n bit; if $|M\|1| > n$, then a sequence of the minimum number of zeros is appended to $M\|1$ so that the total length minus n becomes a multiple of $n - m$.

The difference between Hash and Hash^* is that in Hash the last block message M_l is input to E_K, while in Hash^* it is not input. Therefore, replacing Hash with Hash^*, the efficiency of $\mathtt{LightMAC\_Plus2}$ is improved.

In Lemma 5, the collision probability of Hash^* is given, where E_K is replaced with a random permutation P. Combining Theorem 2 and Lemma 5 offers the following corollary.

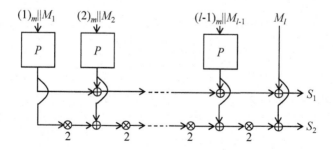

Fig. 7. Hash*.

Algorithm 3. $\mathsf{Hash}^*[E_K](M) = (S_1, S_2)$

1: Partition $M \| 10^*$ into $n - m$-bit blocks $M_1, \ldots, M_{l-1}$ and n-bit block M_l
2: $S_1 \leftarrow 0^n$; $S_2 \leftarrow 0^n$
3: **for** $i = 1, \ldots, l - 1$ **do**
4:     $B_i \leftarrow (1)_m \| M_i$; $C_i \leftarrow E_K(B_i)$; $S_1 \leftarrow S_1 \oplus C_i$; $S_2 \leftarrow S_2 \oplus 2^{l-i} \cdot C_i$
5: **end for**
6: $S_1 \leftarrow S_1 \oplus M_l$; $S_2 \leftarrow S_2 \oplus M_l$
7: **return** (S_1, S_2)

Corollary 2. *Assume that $t \leq 7$. Then we have*

$$\mathbf{Adv}^{\mathrm{prf}}_{\mathtt{LightMAC\_Plus2}[\mathsf{Hash}^*[P], P_{0,1}, P_{0,2}, P_1, \ldots, P_{t-1}, P_t]}(\mathcal{D}) \leq \frac{2q^2}{2^{2n}} + \frac{2^t q^{t+1}}{(2^n - q)^t} \ .$$

Lemma 5. *Let $P \xleftarrow{\$} \mathsf{Perm}(n)$ be a random permutation. For distinct two messages $M^\alpha, M^\beta \in \{0,1\}^*$, the probability that $\mathsf{Hash}^*[P](M^\alpha) = \mathsf{Hash}^*[P](M^\beta)$ is at most $4/2^{2n}$.*

Proof. In this proof, values defined from M^α (resp., M^β) are denoted by using the superscript of α (resp., β), length l of M^α (resp., M^β) is denoted by l_α (resp., l_β). Without loss of generality, we assume that $l_\alpha \leq l_\beta$. $H[P](M^\alpha) = H[P](M^\beta)$ implies that

$$S_1^\alpha = S_1^\beta \text{ and } S_2^\alpha = S_2^\beta \Leftrightarrow$$

$$\underbrace{\bigoplus_{i=1}^{l_\alpha - 1} C_i^\alpha \oplus \bigoplus_{i=1}^{l_\beta - 1} C_i^\beta = Z^{\alpha,\beta}}_{A_{14,1}} \text{ and } \underbrace{\bigoplus_{i=1}^{l_\alpha - 1} 2^{l_\alpha - i} \cdot C_i^\alpha \oplus \bigoplus_{i=1}^{l_\beta - 1} 2^{l_\beta - i} \cdot C_i^\beta = Z^{\alpha,\beta}}_{A_{14,2}} \quad (14)$$

where $Z^{\alpha,\beta} = M_{l_\alpha}^\alpha \oplus M_{l_\beta}^\beta$. We consider the following six cases.

1. $\left(l_\alpha = l_\beta = 1 \right)$
2. $\left(l_\alpha = l_\beta \neq 1 \right) \wedge \left(\forall a \in [l_\alpha - 1] \text{ s.t. } B_a^\alpha = B_a^\beta \right) \wedge \left(M_{l_\alpha} \neq M_{l_\beta} \right)$
3. $\left(l_\alpha = l_\beta \neq 1 \right) \wedge \left(\exists a \in [l_\alpha - 1] \text{ s.t. } B_a^\alpha \neq B_a^\beta \right) \wedge$
   $\left(\forall i \in [l_\alpha - 1] \backslash \{a\} : B_i^\alpha = B_i^\beta \right).$
4. $\left(l_\alpha = l_\beta \neq 1 \right) \wedge \left(\exists a_1, a_2 \in [l_\alpha - 1] \text{ s.t. } B_{a_1}^\alpha \neq B_{a_1}^\beta \wedge B_{a_2}^\alpha \neq B_{a_2}^\beta \right)$
5. $\left(l_\alpha \neq l_\beta \right) \wedge \left(l_\beta = 2 \right)$
6. $\left(l_\alpha \neq l_\beta \right) \wedge \left(l_\beta \geq 3 \right)$

Note that by $l_\alpha \leq l_\beta$, when $l_\alpha \neq l_\beta$, $l_\beta \neq 1$, thereby we do not have to consider the case of $\left(l_\alpha \neq l_\beta \right) \wedge \left(l_\beta = 1 \right)$. The third case is that there is just one position

a where the inputs are distinct, whereas the fourth case is that there are at least two positions a_1, a_2 where the inputs are distinct. For each case we evaluate the probability that the equalities in (14) hold.

- Consider the first and second cases. In these cases, $A_{14,1} = A_{14,2} = 0^n$ and $Z^{\alpha,\beta} \neq 0^n$. Hence (14) is not satisfied.
- Consider the third case. In this case, $A_{14,1} = (C_a^\alpha \oplus C_a^\beta) \neq 2^{l_\alpha - a} \cdot (C_a^\alpha \oplus C_a^\beta) = A_{14,2}$. Hence, in (14) is not satisfied.
- Consider the fourth case. First we eliminate the same outputs between $\{C_i^\alpha, 1 \leq i \leq l_\alpha - 1\}$ and $\{C_i^\beta, 1 \leq i \leq l_\beta - 1\}$ from $A_{14,1}$ and $A_{14,2}$, and then we have

$$A_{14,1} = \bigoplus_{i=1}^{j} \left(C_{a_i}^\alpha \oplus C_{a_i}^\beta \right) \text{ and } A_{14,2} = \bigoplus_{i=1}^{j} 2^{l_\alpha - a_i} \cdot \left(C_{a_i}^\alpha \oplus C_{a_i}^\beta \right) ,$$

where $a_1, \ldots, a_j \in [l_\alpha - 1]$ with $j \geq 2$. Since in $A_{14,1}$ and $A_{14,2}$ there are at most $l_\alpha + l_\beta - 2$ outputs, the numbers of possibilities for $C_{a_1}^\alpha$ and $C_{a_2}^\alpha$ are at least $2^n - (l_\alpha + l_\beta - 3)$ and $2^n - (l_\alpha + l_\beta - 4)$, respectively. Fixing other outputs, the equations in (14) provide a unique solution for $C_{a_1}^\alpha$ and $C_{a_2}^\alpha$. Thus, the probability that (14) is satisfied is at most $1/(2^n - (l_\alpha + l_\beta - 2))(2^n - (l_\alpha + l_\beta - 3))$.
- Consider the fifth case. In this case, $l_\alpha = 1$ and $A_{14,1} = C_1^\beta \neq 2 \cdot C_1^\beta = A_{14,2}$. Hence (14) is not satisfied.
- Consider the sixth case. We eliminate the same outputs between $\{C_i^\alpha : 1 \leq i \leq l_\alpha - 1\}$ and $\{C_i^\beta : 1 \leq i \leq l_\beta - 1\}$ from $A_{14,1}$. By $l_\alpha < l_\beta$, $C_{l_\beta}^\beta$ remains in $A_{14,1}$. Since in $A_{14,1}$ and $A_{14,2}$ there are at most $l_\alpha + l_\beta - 2$ outputs, the numbers of possibilities for $C_{l_\beta}^\beta$ and C_1^β are at least $2^n - (l_\alpha + l_\beta - 3)$ and $2^n - (l_\alpha + l_\beta - 4)$, respectively. Fixing other outputs, the equations in (14) provide a unique solution for $C_{l_\beta}^\beta$ and C_1^β. As a result, the probability of (14) is at most $1/(2^n - (l_\alpha + l_\beta - 3))(2^n - (l_\alpha + l_\beta - 4))$.

Thus, we have

$$\Pr\left[\mathsf{Hash}^*[P](M^\alpha) = \mathsf{Hash}^*[P](M^\beta)\right] \leq \frac{1}{(2^n - (l_\alpha + l_\beta))^2} \leq \frac{4}{2^{2n}} ,$$

assuming $l_\alpha + l_\beta \leq 2^{n-1}$.

$\square$

References

1. Bellare, M., Guérin, R., Rogaway, P.: XOR MACs: new methods for message authentication using finite pseudorandom functions. In: Coppersmith, D. (ed.) CRYPTO 1995. LNCS, vol. 963, pp. 15–28. Springer, Heidelberg (1995). https://doi.org/10.1007/3-540-44750-4_2

2. Bellare, M., Kilian, J., Rogaway, P.: The security of cipher block chaining. In: Desmedt, Y.G. (ed.) CRYPTO 1994. LNCS, vol. 839, pp. 341–358. Springer, Heidelberg (1994). https://doi.org/10.1007/3-540-48658-5_32

3. Bellare, M., Pietrzak, K., Rogaway, P.: Improved security analyses for CBC MACs. In: Shoup, V. (ed.) CRYPTO 2005. LNCS, vol. 3621, pp. 527–545. Springer, Heidelberg (2005). https://doi.org/10.1007/11535218_32

4. Bellare, M., Rogaway, P.: Code-based game-playing proofs and the security of triple encryption. Cryptology ePrint Archive, Report 2004/331 (2004). http://eprint.iacr.org/2004/331

5. Bellare, M., Rogaway, P.: The security of triple encryption and a framework for code-based game-playing proofs. In: Vaudenay, S. (ed.) EUROCRYPT 2006. LNCS, vol. 4004, pp. 409–426. Springer, Heidelberg (2006). https://doi.org/10.1007/11761679_25

6. Bernstein, D.J.: How to stretch random functions: the security of protected counter sums. J. Cryptol. **12**(3), 185–192 (1999)

7. Bhargavan, K., Leurent, G.: On the practical (in-)security of 64-bit block ciphers: collision attacks on HTTP over TLS and OpenVPN. In: Weippl, E.R., Katzenbeisser, S., Kruegel, C., Myers, A.C., Halevi, S. (eds.) Proceedings of the 2016 ACM SIGSAC Conference on Computer and Communications Security, Vienna, Austria, 24–28 October 2016, pp. 456–467. ACM (2016)

8. Black, J., Rogaway, P.: CBC MACs for arbitrary-length messages: the three-key constructions. In: Bellare, M. (ed.) CRYPTO 2000. LNCS, vol. 1880, pp. 197–215. Springer, Heidelberg (2000). https://doi.org/10.1007/3-540-44598-6_12

9. Black, J., Rogaway, P.: A block-cipher mode of operation for parallelizable message authentication. In: Knudsen, L.R. (ed.) EUROCRYPT 2002. LNCS, vol. 2332, pp. 384–397. Springer, Heidelberg (2002). https://doi.org/10.1007/3-540-46035-7_25

10. Brassard, G.: On computationally secure authentication tags requiring short secret shared keys. In: Chaum, D., Rivest, R.L., Sherman, A.T. (eds.) Advances in Cryptology, pp. 79–86. Plenum Press, New York (1982)

11. Cogliati, B., Seurin, Y.: EWCDM: an efficient, beyond-birthday secure, noncemisuse resistant MAC. In: Robshaw, M., Katz, J. (eds.) CRYPTO 2016. LNCS, vol. 9814, pp. 121–149. Springer, Heidelberg (2016). https://doi.org/10.1007/978-3-662-53018-4_5

12. Dodis, Y., Steinberger, J.: Domain extension for MACs beyond the birthday barrier. In: Paterson, K.G. (ed.) EUROCRYPT 2011. LNCS, vol. 6632, pp. 323–342. Springer, Heidelberg (2011). https://doi.org/10.1007/978-3-642-20465-4_19

13. Dutta, A., Nandi, M., Paul, G.: One-key compression function based MAC with security beyond birthday bound. In: Liu, J.K.K., Steinfeld, R. (eds.) ACISP 2016. LNCS, vol. 9722, pp. 343–358. Springer, Cham (2016). https://doi.org/10.1007/978-3-319-40253-6_21

14. Gaži, P., Pietrzak, K., Rybar, M.: The exact security of PMAC. Cryptology ePrint Archive, Report 2017/069 (2017). http://eprint.iacr.org/2017/069

15. Iwata, T., Kurosawa, K.: OMAC: one-key CBC MAC. In: Johansson, T. (ed.) FSE 2003. LNCS, vol. 2887, pp. 129–153. Springer, Heidelberg (2003). https://doi.org/10.1007/978-3-540-39887-5_11

16. Iwata, T., Minematsu, K.: Stronger security variants of GCM-SIV. Cryptology ePrint Archive, Report 2016/853, to appear at IACR Transactions on Symmetric Cryptology. http://eprint.iacr.org/2016/853

17. Iwata, T., Minematsu, K., Peyrin, T., Seurin, Y.: ZMAC: a fast tweakable block cipher mode for highly secure message authentication. In: Katz, J., Shacham,

H. (eds.) CRYPTO 2017. LNCS, vol. 10403, pp. 34–65. Springer, Cham (2017). https://doi.org/10.1007/978-3-319-63697-9_2

18. Jaulmes, É., Joux, A., Valette, F.: On the security of randomized CBC-MAC beyond the birthday paradox limit a new construction. In: Daemen, J., Rijmen, V. (eds.) FSE 2002. LNCS, vol. 2365, pp. 237–251. Springer, Heidelberg (2002). https://doi.org/10.1007/3-540-45661-9_19

19. Jaulmes, E., Lercier, R.: FRMAC, a fast randomized message authentication code. Cryptology ePrint Archive, Report 2004/166 (2004). http://eprint.iacr.org/2004/166

20. Kurosawa, K., Iwata, T.: TMAC: two-key CBC MAC. In: Joye, M. (ed.) CT-RSA 2003. LNCS, vol. 2612, pp. 33–49. Springer, Heidelberg (2003). https://doi.org/10.1007/3-540-36563-X_3

21. Lee, J., Steinberger, J.: Multi-property-preserving domain extension using polynomial-based modes of operation. In: Gilbert, H. (ed.) EUROCRYPT 2010. LNCS, vol. 6110, pp. 573–596. Springer, Heidelberg (2010). https://doi.org/10.1007/978-3-642-13190-5_29

22. List, E., Nandi, M.: Revisiting Full-PRF-secure PMAC and using it for beyond-birthday authenticated encryption. IACR Cryptology ePrint Archive 2016, 1174 (2016)

23. Lucks, S.: The sum of PRPs is a secure PRF. In: Preneel, B. (ed.) EUROCRYPT 2000. LNCS, vol. 1807, pp. 470–484. Springer, Heidelberg (2000). https://doi.org/10.1007/3-540-45539-6_34

24. Luykx, A., Preneel, B., Szepieniec, A., Yasuda, K.: On the influence of message length in PMAC's security bounds. In: Fischlin, M., Coron, J.-S. (eds.) EUROCRYPT 2016. LNCS, vol. 9665, pp. 596–621. Springer, Heidelberg (2016). https://doi.org/10.1007/978-3-662-49890-3_23

25. Luykx, A., Preneel, B., Tischhauser, E., Yasuda, K.: A MAC mode for lightweight block ciphers. In: Peyrin, T. (ed.) FSE 2016. LNCS, vol. 9783, pp. 43–59. Springer, Heidelberg (2016). https://doi.org/10.1007/978-3-662-52993-5_3

26. Minematsu, K.: How to thwart birthday attacks against MACs via small randomness. In: Hong, S., Iwata, T. (eds.) FSE 2010. LNCS, vol. 6147, pp. 230–249. Springer, Heidelberg (2010). https://doi.org/10.1007/978-3-642-13858-4_13

27. Minematsu, K., Matsushima, T.: New bounds for PMAC, TMAC, and XCBC. In: Biryukov, A. (ed.) FSE 2007. LNCS, vol. 4593, pp. 434–451. Springer, Heidelberg (2007). https://doi.org/10.1007/978-3-540-74619-5_27

28. Naito, Y.: Full PRF-secure message authentication code based on tweakable block cipher. In: Au, M.-H., Miyaji, A. (eds.) ProvSec 2015. LNCS, vol. 9451, pp. 167–182. Springer, Cham (2015). https://doi.org/10.1007/978-3-319-26059-4_9

29. Nandi, M.: A unified method for improving PRF bounds for a class of blockcipher based MACs. In: Hong, S., Iwata, T. (eds.) FSE 2010. LNCS, vol. 6147, pp. 212–229. Springer, Heidelberg (2010). https://doi.org/10.1007/978-3-642-13858-4_12

30. NIST: Recommendation for Block Cipher Modes of Operation: the CMAC Mode for Authentication. Spp. 800–38B (2005)

31. Petrank, E., Rackoff, C.: CBC MAC for real-time data sources. J. Cryptol. **13**(3), 315–338 (2000)

32. Peyrin, T., Seurin, Y.: Counter-in-tweak: authenticated encryption modes for tweakable block ciphers. In: Robshaw, M., Katz, J. (eds.) CRYPTO 2016. LNCS, vol. 9814, pp. 33–63. Springer, Heidelberg (2016). https://doi.org/10.1007/978-3-662-53018-4_2

33. Rogaway, P.: Efficient instantiations of tweakable blockciphers and refinements to modes OCB and PMAC. In: Lee, P.J. (ed.) ASIACRYPT 2004. LNCS, vol. 3329, pp. 16–31. Springer, Heidelberg (2004). https://doi.org/10.1007/978-3-540-30539-2_2

34. Wegman, M.N., Carter, L.: New hash functions and their use in authentication and set equality. J. Comput. Syst. Sci. **22**(3), 265–279 (1981)

35. Yasuda, K.: A double-piped mode of operation for MACs, PRFs and PROs: security beyond the birthday barrier. In: Joux, A. (ed.) EUROCRYPT 2009. LNCS, vol. 5479, pp. 242–259. Springer, Heidelberg (2009). https://doi.org/10.1007/978-3-642-01001-9_14

36. Yasuda, K.: The sum of CBC MACs is a secure PRF. In: Pieprzyk, J. (ed.) CT-RSA 2010. LNCS, vol. 5985, pp. 366–381. Springer, Heidelberg (2010). https://doi.org/10.1007/978-3-642-11925-5_25

37. Yasuda, K.: A new variant of PMAC: beyond the birthday bound. In: Rogaway, P. (ed.) CRYPTO 2011. LNCS, vol. 6841, pp. 596–609. Springer, Heidelberg (2011). https://doi.org/10.1007/978-3-642-22792-9_34

38. Yasuda, K.: PMAC with parity: minimizing the query-length influence. In: Dunkelman, O. (ed.) CT-RSA 2012. LNCS, vol. 7178, pp. 203–214. Springer, Heidelberg (2012). https://doi.org/10.1007/978-3-642-27954-6_13

39. Yasuda, K.: A parallelizable PRF-based MAC algorithm: well beyond the birthday bound. IEICE Trans. **96**-**A**(1), 237–241 (2013)

40. Zhang, L., Wu, W., Sui, H., Wang, P.: 3kf9: enhancing 3GPP-MAC beyond the birthday bound. In: Wang, X., Sako, K. (eds.) ASIACRYPT 2012. LNCS, vol. 7658, pp. 296–312. Springer, Heidelberg (2012). https://doi.org/10.1007/978-3-642-34961-4_19

41. Zhang, Y.: Using an error-correction code for fast, beyond-birthday-bound authentication. In: Nyberg, K. (ed.) CT-RSA 2015. LNCS, vol. 9048, pp. 291–307. Springer, Cham (2015). https://doi.org/10.1007/978-3-319-16715-2_16

Author Index

Printed in the United States
By Bookmasters